Developments in Aquaculture and Fisheries Science – 37

EDIBLE SEA URCHINS: BIOLOGY AND ECOLOGY

SECOND EDITION

DEVELOPMENTS IN AQUACULTURE AND FISHERIES SCIENCE

The following volumes are still available:

17. DISEASE DIAGNOSIS AND CONTROL IN NORTH AMERICAN MARINE AQUACULTURE
Edited by C.J. Sindermann and D.V. Lightner 1988 xv + 412 pages

22. FRONTIERS OF SHRIMP RESEARCH
Edited by P.F. DeLoach, W.J. Dougherty and M.A. Davidson 1991 viii + 294 pages

23. MARINE SHRIMP CULTURE: PRINCIPLES AND PRACTICES
By A.W. Fast and L.J. Lester 1992 xvi + 862 pages

24. MODERN METHODS OF AQUACULTURE IN JAPAN (2ND REVISED EDITION)
Edited by H. Ikenoue and T. Kafuku 1992 xiv + 274 pages

25. THE MUSSEL MYTILUS: ECOLOGY, PHYSIOLOGY, GENETICS AND CULTURE
By E. Cosling 1992 xiv + 589 pages

26. PROTOZOAN PARASITES OF FISHES
By J. Lom and I. Dyková 1992 xii + 316 pages

27. AQUACULTURE WATER REUSE SYSTEMS: ENGINEERING DESIGN AND MANAGEMENT
Edited by M.B. Timmons and T. Losordo

28. FRESHWATER FISH CULTURE IN CHINA: PRINCIPLES AND PRACTICE
Edited by J. Mathias and S. Li 1994 xvi + 446 pages

29. PRINCIPLES OF SALMONID CULTURE
Edited by W. Pennell and B.A. Barton 1996 xxx + 1040 pages

30. STRIPED BASS AND OTHER MORONE CULTURE
Edited by R.M. Harrell 1997 xx + 366 pages

31. BIOLOGY OF THE HARD CLAM
Edited by J.N. Kraeuter and M. Castagna 2001 xix + 751 pages

32. EDIBLE SEA URCHINS: BIOLOGY AND ECOLOGY
Edited by J.M. Lawrence 2001 xv + 419 pages

33. DESIGN AND OPERATING GUIDE FOR AQUACULTURE SEAWATER SYSTEM – SECOND EDITION
Edited by J.E. Huguenin and J. Colt 2002 viii + 328 pages

34. BIOLOGY AND CULTURE OF CHANNEL CATFISH
Edited by C.S. Tucker and J.A. Hargreaves 2004 x + 676 pages

35. SCALLOPS: BIOLOGY, ECOLOGY AND AQUACULTURE – SECOND EDITION
Edited by S.E. Shumway and G.J. Parsons 2006 xxxix + 1460 pages

36. THE KNOWLEDGE BASE FOR FISHERIES MANAGEMENT
Edited by L. Motos and D.C. Wilson 2006 xxii + 454 pages

COVER IMAGE

Psammechinus miliaris (Photo courtesy of Maeve Kelly, Scottish Association for Marine Science).

Developments in Aquaculture and Fisheries Science – 37

EDIBLE SEA URCHINS: BIOLOGY AND ECOLOGY

SECOND EDITION

Edited by

JOHN M. LAWRENCE
Department of Biology
University of South Florida
Tampa, Florida
U.S.A.

ELSEVIER

Amsterdam – Boston – Heidelberg – London – New York – Oxford
Paris – San Diego – San Francisco – Singapore – Sydney – Tokyo

Elsevier
Radarweg 29, PO Box 211, 1000 AE Amsterdam, The Netherlands
The Boulevard, Langford Lane, Kidlington, Oxford OX5 1GB, UK

First edition 2007

Library of Congress Cataloging-in-Publication Data
A catalog record for this book is available from the Library of Congress

British Library Cataloguing in Publication Data
A catalogue record for this book is available from the British Library

ISBN-13: 978-0-444-52940-4
ISBN-10: 0-444-52940-3
ISSN: 0167-9309

Printed and bound in The Netherlands
07 08 09 10 11 10 9 8 7 6 5 4 3 2

Contents

Preface to the first edition

Sea urchins have been known to the scientific community since the days of Aristotle. Until about fifty years ago studies on them were primarily anatomical, developmental, and paleontological. Since then, research on their biology and ecology has increased greatly. This was stimulated first by a recognition of their ecological importance and then by a realization of the economic importance.

This book was designed to provide a broad understanding of the biology and ecology of sea urchins. Synthetic chapters consider sea urchins as a whole to give a broad view. Chapters that consider the ecology of individual species integrate the various life-history phases. The goal was to provide both individual and general perspectives. Although the impetus of this book was the interest in fisheries and aquaculture, not all the species considered are important in fisheries or candidates for aquaculture. I believe that an understanding of the biology and ecology of these species will aid in understanding the biology and ecology of those species that are. My only regret is that a wider range of species could not be considered.

Although the authors of the ecology chapters exchanged outlines, they were not required to adhere to a specific one. This recognized the difference in the information available and in the interests of the authors. I found this resulted in a diversity of emphasis and approach that is extremely interesting and enlightening. The treatment of species in both the synthetic and ecology chapters is complementary.

I am grateful to my friends and colleagues, experts in their fields, who have contributed to this book. I thank Wenger Manufacturing, Inc., Sabetha, KS, for funds for the publication of the frontispiece.

John Lawrence
Tampa, Florida

Preface to the second edition

I am gratified the reception to the first edition of Edible Sea Urchins led to the invitation by Elsevier to prepare a revised, updated edition. There have been substantial advances since the first edition.

The basis for the book as described in the preface to the first edition has not changed. Sea urchins are important as a major phylum and for their ecological role in the world's oceans. They are of economic importance in fisheries and have great potential for aquaculture. Their importance as model systems for investigations in developmental biology continues to increase. I still firmly believe it is essential to understand both biology and ecology of species for fisheries and aquaculture.

Again, I am grateful to my friends and colleagues, experts in their fields, who have contributed to this book. I note with sorrow the death of two contributors to the first edition, Larry McEdward and Mia Tegner. I thank Mara Vos-Sarmiento of Elsevier for her supportive efforts in preparing both editions. I am particularly grateful to Kathleen Hotchkiss for her essential assistance in preparing the manuscripts for publication.

John Lawrence
Tampa, Florida

List of contributors

Yukio Agatsuma	Graduate School of Agricultural Science, Tohoku University, Sendai, Miyagi 981-8555, Japan
Neil Andrew	The WorldFish Center, P. O. Box 500 GPO, 10670 Penang, Malaysia
Michael Barker	Department of Marine Science, University of Otago, PO Box 56, Dunedin, New Zealand
Charles François Boudouresque	UMR 6540, Centre d'Océanologie de Marseille, Campus Universitaire de Luminy, 13288 Marseille cedex 9, France
Maria Byrne	Department of Anatomy and Histology, F13, University of Sydney, Sydney, NSW 2006, Australia
Elizabeth Cook	Scottish Association for Marine Science, Oban, Argyll PA37 1QA, Scotland, United Kingdom
Thomas A. Ebert	525 E. Fall Creek Rd., Alsea, Oregon 97324, USA
Bruce Hatcher	Bras d'Or Institute for Ecosystem Research, Cape Breton University, 1250 Grand Lake Road, P. O. Box 5300, Sydney, Nova Scotia B1P 6L2, Canada
Adam D. Hughes	Scottish Association for Marine Science, Oban, Argyll PA37 1QA, Scotland, United Kingdom
John Keesing	Commonwealth Scientific and Industrial Research Organization, Private Bag 5, Wembley, WA 6913, Australia
Maeve Kelly	Scottish Association for Marine Science, Oban, Argyll PA37 1QA, Scotland, United Kingdom
Addison L. Lawrence	Texas Agricultural Experiment Station, Department of Wildlife and Fisheries Sciences, Texas A&M University System, Port Aransas, Texas 78373, USA
John M. Lawrence	Department of Biology, University of South Florida, Tampa, Florida 33620, USA

Michael P. Lesser	Department of Zoology and, Center for Marine Biology University of New Hampshire, Durham, New Hampshire 03824, USA
José Roberto Machado Cunha da Silva	Laboratório de Histofisiologia Evolutiva, Departamento de Biologia Celular e do Desenvolvimento, Instituto de Ciências Biomédicas, Universidade de São Paulo, Av. Prof. Lineu Prestes, 1524, sala 409, CEP 05508-900, São Paulo, SP, Brazil
Adam Marsh	College of Marine Studies, University of Delaware, Lewes, Delaware 19958, USA
T. R. McClanahan	Wildlife Conservation Society, Kibaki Flats no. 12, Bamburi, Kenyatta Beach, P.O. Box 99470, Mombasa 80107, Kenya
James B. McClintock	Department of Biology, University of Alabama at Birmingham, Birmingham, AL 35294-1170, USA
Larry R. McEdward†	Department of Zoology, University of Florida, Gainesville, Florida, USA
Benjamin G. Miner	Department of Biology, Western Washington University, 516 High St., Bellinham, WA 98225, USA
Nyawira Muthiga	Wildlife Conservation Society, Kibaki Flats no. 12, Bamburi, Kenyatta Beach, P.O. Box 99470, Mombasa 80107, Kenya
Laura Rogers-Bennett	California Department of Fish and Game, Bodega Marine Laboratory, 2099 Westside Rd., Bodega Bay, California 94923, USA
Robert Scheibling	Department of Biology, Dalhousie University, Halifax, Nova Scotia B3H 4J1, Canada
Kenichi Tajima	Laboratory of Marine Biotechnology and Microbiology, Division of Marine Life Science, Research Faculty of Fisheries Sciences, Hokkaido University, Hakodate, Hokkaido 041-88611, Japan
Miyuki Tsushima	Educational and Research Center for Clinical Pharmacy, Kyoto Pharmaceutical University, Misasagi, Yamashina-ku, Kyoto, 607-8414, Japan
Tatsuya Unuma	Japan Sea National Fisheries Research Institute, Fisheries Research Agency, Suido-cho, Niigata 951-8121, Japan
Julio Vásquez	Departamento de Biologia Marina, Universidad Católica del Norte, Centros Estudios Avanzados en Zonas Aridas Larronado 1281, Casilla 117, Coquimbo, Chile

† Deceased.

Marc Verlaque UMR 6540, Centre d'Océanologie de Marseille, Campus Universitaire de Luminy, 13288 Marseille cedex 9, France

Charles W. Walker Department of Zoology, Center for Marine Biology and Marine Biomedical Research Group, University of New Hampshire, Durham, NH 03824, USA

Kristina Wasson Department of Biology, University of Alabama at Birmingham, 1300 University Blvd., Birmingham, Alabama 35294-1170, USA

Stephen A. Watts Department of Biology, University of Alabama at Birmingham, 1300 University Blvd., Birmingham, Alabama 35294-1170, USA

Edible Sea Urchins: Biology and Ecology
Editor: John Miller Lawrence

Chapter 1

Edible Sea Urchins: Use and Life-History Strategies

John M Lawrence

Department of Biology, University of South Florida, Tampa, FL (USA).

1. PREHISTORIC FISHING OF SEA URCHINS

Indications of prehistoric human consumption of sea urchins have been discovered in various locations. Best (1929) reported one of the most numerous species of seafood found in middens of Maori on rocky coasts of New Zealand was *Evechinus chloroticus* (*kina*). Almost every excavation at Norfolk Island in the South Pacific contained fragments of one or two species of sea urchins (Campbell and Schmidt 2001). Sand et al. (2002) concluded a strong interest in sea urchins by the first Austronesians settling in the Loyalty Islands by the large amount of sea urchin remains in the lower layers of different first millennium BC sites, in contrast to the near total absence in later horizons. Sea urchin tests appear in Pleistocene middens at New Ireland (Allen et al. 1989). Interestingly, sea urchins are not included in the list of marine fauna found in prehistoric middens in Japan (Imamura 1996), although Yokota (2002) states many sea urchin tests have been found in them. Historic consumption of sea urchins in Japan dates to the ninth century AD (Oshima 1962).

On North American coasts, Black (pers. comm., 2002) found evidence of harvesting *Strongylocentrotus droebachiensis* by native people in southern New Brunswick from about 2400 BP to European contact (about 500 BP). Remains of sea urchins usually ranked third in abundance after shellfish, but predominated occasionally.

Denniston (1974) calculated sea urchins were the primary shellfish eaten by Aleuts at Umnak Island but were a minor part of the diet. Simenstad et al. (1978) concluded Aleut midden remains at Amchitka indicate aboriginal Aleuts overexploited sea otters locally, which led to an abundance of *S. polyacanthus* for consumption. Barsh (2004) suggested heavy, localized seasonal harvesting of sea urchins by aboriginal populations of the Puget Sound-Georgia Strait area. He concluded occasional substantial lens of sea urchin remains in middens indicate the Coast Salish had "sea urchin feasts". Thin, blunt shafts that are relatively common in Aleut archaeological sites may be prongs to obtain sea urchins too deep for hand gathering (D Corbett, pers. comm.).

Sea urchins have been found in middens on the Channel Islands off southern California. Salls (1991) reported a "sea urchin" layer in an Early Holocene midden on San Clemente Island. *Strongylocentrotus purpuratus* is a common species found in middens (primarily

4200–3300 BC) on Santa Cruz Island (Glassow 2002). *Strongylocentrotus* spp. remains, ranging from 0.1% to 39.5% total weight of invertebrate remains, occur in the strata of some Chumash middens on San Miguel Island from 9800 to 220 calculated BP (Vellanoweth et al. 2002; Erlandson et al. 2004).

Tripneustes ventricosus is the primary sea urchin remains in archaeological sites on islands off Venezuela, but *Echinometra lucunter* and *E. viridis* also occur (Antczak 1999). Engel (in Lumbreras 1974: 42) noted sea urchins in Upper Archaic sites on the Peruvian coast. Engel (1963) did not include sea urchins in a list of marine fauna found in middens at a site in Peru although an amazing record of "starfish" is given. Bird and Hyslop (1985) noted sea urchin spines are very common in the preceramic excavations at Huaca Prieta, Peru, and also recorded a "fragment of starfish." *Loxechinus albus* was found in shell middens near Valparaiso, Chile, from the Early Ceramic around 2270 BP (Jeradino et al. 1992). It has been estimated to represent between 0.1% and 2.8% of the diet of the Huentelauquen culture (IV Región, Chile) (Vásquez et al. 1996). Only one of six strata showed evidence of sea urchins on the coast of Arica, Chile (Arriaza et al. 2001). *Paracentrotus lividus* occurred with other invertebrates exploited by Cantabrian Mesolithic groups on the Iberian Peninsula (Arias 1999).

2. CONTEMPORARY COMMERCIAL FISHING OF SEA URCHINS

In recent years, increased demand has led to overfishing and consequently sea urchin stocks have diminished. This was already noted in the first review of sea urchin fisheries (Sloan 1985). Re-analysis has documented the continuing demand for sea urchins that has resulted in continued expansion of fishing grounds and harvest (Keesing and Hall 1998; Andrew et al. 2002). Much interest has focused on the ecological effect of fisheries of predators on sea urchin populations (e.g. Pinnegar et al. 2000; Sala et al. 1998; Vadas and Steneck 1995). This may include increased incidence of disease as populations increase in density (Behrens and Lafferty 2004; Lafferty 2004). An ecological analysis of the direct effect of sea urchin fisheries is pertinent (Guidetti et al. 2004; Pfister and Bradbury 1996). Andrew et al. (2004) pointed out very large indirect effects of sea urchin fisheries on ecology are possible and that such effects will need to be managed. Fisheries management has been the primary response in many parts of the world where overfishing has occurred: Chile (Stotz 2004), Spain (Catoira 2004), Japan (Agatsuma et al. 2004), California (Dewees 2004), and North America (Botsford et al. 2004). Restocking and stock enhancement to replenish depleted populations is another strategy (Bell et al. in press).

3. AQUACULTURE OF SEA URCHINS

As fishing stocks become depleted, interest in aquaculture of sea urchins has considerably increased (Hagen 1996; Lesser and Walker 1998; Keesing and Hall 1998; Andrew et al. 2002; Robinson 2004). Mass production of seed (small sea urchins for outplanting in the field) has been done for *S. intermedius*, *S. nudus* (Sakai et al. 2004), *T. gratilla*,

(Shimabukuro 1991), and *L. albus* (Cárcamo 2004). Reseeding into the field has been done in Japan (Agatsuma et al. 2004; Shimabukuro 1991) and raft culture has been done in China (Chang and Wang 1997). Artificial feeds that support somatic and gonadal growth have been developed (Lawrence and Lawrence 2004). Aquaculture has the potential for production of sea urchins for human consumption and for use of sea urchins as models in research in developmental biology (Lawrence et al. 2001). The entire life cycle of *Lytechinus variegatus* (adults to metamorphosed larvae) has been carried out in artificial seawater with prepared feeds (George et al. 2004).

4. EDIBLE SEA URCHINS

Edible sea urchins are distributed among a number of orders of regular echinoids (Table 1; Matsui 1966; Mottet 1976; Sloan 1985; Saito 1992a,b; Keesing and Hall 1998; Lawrence and Bazhin 1998). I know of no a priori reason why any sea urchin species should not be edible, but the fact remains that relatively few species are eaten. Some major taxa, such as cidaroids, seem not to be eaten at all. Three possibilities could explain why so few species are eaten. The first is accessibility: all the species that are eaten are found in shallow water. A second possibility is palatability. Although *Tetrapygus niger* is abundant along the Chilean coast and the tradition of consumption of sea urchins exists, the species is not consumed as it is not palatable (Lawrence and Bazhin 1998). *Arbacia lixula* is eaten only along the Albanian coast of the Mediterranean (Le Direac'h 1987), presumably because they are not very palatable. Immature gonads of *Hemicentrotus pulcherrimus* are very popular in Japan but the mature ovaries are bitter and not eaten (Murata et al. 2002, 2004). *Diadema setosum* is edible and eaten in some districts of Kyushu Island, but is not very palatable (T Unuma, pers. comm.). A third possibility is historical (cultural). Sea urchin consumption in the Mediterranean countries varies (Le Direac'h 1987). It is high in France but is limited on the North African and Aegean coasts. Historical Greek influence is apparent by the use of the Greek *ritza* for sea urchins collected for eating at Alexandria, Egypt (Roden 1968). Extensive consumption in the Caribbean has been limited to Barbados (Lewis 1962; Scheibling and Mladenov 1987) and, in South America, to Chile (Bustos et al. 1991; Vásquez and Guisado 1992).

5. LIFE-HISTORY STRATEGIES OF SEA URCHINS

Whereas all sea urchin species may be edible, they differ greatly in their biology and ecology. The biological differences include growth, survival and reproduction. These basic components of fitness differ according to the species' life-history strategy and how it functions in its particular environment (Lawrence 1990, 1991; Lawrence and Bazhin 1998). Grime (1974, 1977) considered energy the basis for life-history strategies in his triangular model. Ebert (1982) noted that models for r- and K-strategies and for bet-hedging are based on bioenergetics. Grime (1974, 1977) pointed out stress and disturbance limit energy in biological systems and related life-history strategies to this. Stress limits acquisition and production. This can result from low availability of food or low ability to

Table 1 Abridged classification of echinoids with representative genera (not limited to species that are eaten).

Class	Subclass	Infraclass	Cohort	Superorder	Order	Family	Genus
Echinoidea							
	Perischoechinoidea						
	Cidaroidea						
	Euechinoidea						
		Echinothurioidea	Echinothuriacea		Echinothurioida	Echinothuriidae	*Echinothuria*
		Acroechinoidea	Diadematacea		Diadematoida	Diadematidae	*Centrostephanus*
							Diadema
			Echinacea	Stirodonta	Phymosomatoida	Arbaciidae	*Arbacia*
				Camarodonta	Echinoida	Echinidae	*Echinus*
							Loxechinus
							Paracentrotus
							Psammechinus
						Echinometridae	*Anthocidaris*
							Echinometra
							Evechinus
							Heliocidaris
						Strongylocentrotidae	*Hemicentrotus*
							Strongylocentrotus
						Toxopneustidae	*Lytechinus*
							Pseudoboletia
							Pseudocentrotus
							Toxopneustes
							Tripneustes
			Irregularia				

Modified from Smith (1984).

obtain food. Disturbance removes energy from an organism. In sea urchins, disturbance is usually lethal and results from either predation or abiotic factors (e.g. hydrodynamics). Sea urchin species that experience high disturbance and low stress are *ruderal* species; those that experience low disturbance and low stress are *competitive* or *capitalistic* species; those that experience low disturbance and high stress are *stress-tolerant* species. There is a suite of characteristics associated with these strategies (Table 2). Intermediate strategies would be expected at less than extreme stress and disturbance.

Ebert (1975) showed a positive correlation with sea urchin species between the Brody–Bertalanffy growth constant K and the instantaneous natural mortality rate M as predicted by the model. Similarly, Ebert (1982) also concluded that a positive relationship exists between relative size of the body wall and probability of survival. Individuals of species in the family Toxopneustidae grow the fastest, followed by those in the family Diadematidae and then those in the families Echinidae, Echinometridae and Strongylocentrotidae (Lawrence and Bazhin 1998). *Tripneustes gratilla*, *D. setosum*, and *E. mathaei* maintained together in aquaria and fed the same algal food for 6 months grew at much different rates (Table 3). *Lytechinus variegatus*, *E. mathaei*, and *Eucidaris tribuloides* maintained together in aquaria without food survived for different lengths of time. LT_{50} (lethal time) was ca. 100 days for *L. variegatus*, 175 days for *E. mathaei*, and >200 days for *E. tribuloides* (Lawrence and Bazhin 1998). These are innate differences in the species. *Lytechinus variegatus* and *Tripneustes* spp. seem to be ruderal species while *Echinometra* spp. seem to be stress-tolerant species.

High growth rate, short time to maturity and high reproductive effort and output appear to be desirable traits for fisheries and aquaculture. However, ruderal species with these characteristics may be more vulnerable to predation, high hydrodynamics and disease. Stress-tolerant species would seem less suitable for fisheries and aquaculture according to these criteria. The competitive strategy may exemplify the best combination of characteristics.

Table 2 Characteristics predicted for extreme life-history strategies of sea urchins.

Characteristic	Stress-tolerant	Competitive	Ruderal
Life-history			
Growth rate	Low	High	Very high
Longevity	Very long	Long	Short
Time to maturity	Long	Short	Very short
Reproductive effort	Low	High	Very high
Structural strength	Very strong	Relatively strong	Weak
Physiology			
Effect of low food availability on growth	Small	Great	Very great
Resistance to starvation	High	Moderate	Low
Respiration	Low	High	Very high
Resistance to environmental stress	Very high	High	Low

Modified from Lawrence 1990; Lawrence and Bazhin (1998).

Table 3 Growth (mean mg wet body weight ± SD) of sea urchins maintained together in aquaria and fed algal turf for 6 months at Eilat, Israel (Lawrence, unpublished).

Species	*Echinometra mathaei*	*Diadema setosum*	*Tripneustes gratilla*
23 October	118 ± 12	142 ± 23	246 ± 92
22 April	1175 ± 92	4235 ± 364	25 454 ± 3207
Percentage increase	9.96×10^2	29.82×10^2	103.47×10^2

6. CONCLUSION

Understanding the biology and ecology of the sea urchins is not merely of basic scientific interest. It is impossible to adequately manage fisheries without understanding the biology and ecology of the species involved. Knowledge of the biology and ecology of the species is also important as they indicate the potential requirements of a species for aquaculture.

ACKNOWLEDGMENTS

I thank Y Agatsuma, N Andrew, D Black, G Campbell, D Corbett, J Erlandson, M Robinson, T Unuma, J Vásquez, and K Wasson for assistance with references and helpful comments.

REFERENCES

Agatsuma Y, Sakai Y, Andrew NL (2004) Enhancement of Japan's sea urchin fisheries. In: Lawrence JM, Guzmán O (eds). Sea urchins: fisheries and ecology. DEStech Publications, Lancaster. pp 18–36

Allen J, Gosden C, White JP (1989) Human Pleistocene adaptations in the tropical island Pacific: recent evidence from New Ireland, a Greater Australian outlier. Antiquity 63: 548–561

Andrew NL, Agatsuma Y, Ballesteros E, Bazhin AG, Creaser EP, Barnes DKA, Botsford LW, Bradbury A, Campbell A, Dixon JD, Einarsson S, Gerring P, Hebert K, Hunter M, Hur SB, Johnson CR, Junio-Menez MA, Kalvass P, Miller RJ, Moreno CA, Palleiro JS, Rivas D, Robinson SML, Schroeter SC, Steneck RS, Vadas RI, Woodby DA, Xiaoqi Z (2002) Status and management of world sea urchin fisheries. Oceanogr Mar Biol Annu Rev 40: 343–425

Andrew NL, Agatsuma Y, Dewees CM, Stotz WB (2004) State of sea-urchin fisheries 2003. In: Lawrence JM, Guzmán O (eds). Sea urchins: fisheries and ecology. DEStech Publications, Lancaster. pp 96–98

Antczak, A. (1999) Late prehistoric economy and society of the islands off the coast of Venezuela: a contectual interpretation of the non-ceramic evidence. PhD dissertation. University College, London

Arias P (1999) The origins of the Neolithic along the Atlantic coast of continental Europe. J World Prehistory 13: 403–464

Arriaza BT, Standen VG, Belmonte E, Rosello E, Nials F (2001) The peopling of the Arica coast during the Preceramic: a preliminary view. Chungará (Arica) 33: 31–36

Barsh RL (2004) The importance of human intervention in the evolution of Puget Sound ecosystems. In: Droscher TW, Fraser DA (eds). Proceedings of the 2003 Georgia Basin/Puget sound research conference. http://www.psat.wa.gov/Publications/03_proceedings/start.htm

Behrens MD, Lafferty KD (2004) Effects of marine reserves and urchin disease on southern California rocky reef communities. Mar Ecol Prog Ser 270: 129–139

Bell J, Rothlisberg P, Nash W, Lonergran N, Andrew N (2005) Restocking and stock enhancement of marine invertebrates. Adv Mar Biol 49

Best E (1929) Fishing Methods and Devices of the Maori, Dominion Museum Bull No 12. Government Printer, Wellington

Bird KB, Hyslop J (1985) The preceramic excavations at the Huaca Prieta Chicama Valley, Peru, Anthropological Papers of the American Museum of Natural History. American Museum of Natural History, New York. Vol. 62(1), pp 1–294

Black DW (2002) Out of the blue and into the black: the Middle-Late Maritime Woodland Transition in the Quoddy Region, New Brunswick, Canada. In: Hart JP, Rieth CB (eds). Northeast subsistence-settlement change A.D. 700–1300, New York State Museum Bulletin 496. The University of the state of New York, Albany. pp 301–320

Botsford LW, Campbell A, Miller R (2004) Biological reference points in the management of North American sea urchin fisheries. Can J Fish Aquat Sci 61: 1325–1337

Bustos RE, Godoy AC, Olave MS, Troncoso TR (1991) Desarrollo de técnicas de produción de semillas y repoblación de recursos bentónicos. Instituto de Fomento Pesquero, Santiago

Campbell CR, Schmidt L (2001) Molluscs and echinoderms from the Emily Bay settlement site, Norfolk Island. Rec Australian Mus, Suppl 27: 109–114

Cárcamo PF (2004) Massive production of larvae and seeds of the sea urchin *Loxechinus albus*. In: Lawrence JM, Guzmán O (eds). Sea urchins: fisheries and ecology. DEStech Publications, Lancaster. pp 299–306

Catoira JL (2004) History and current state of sea urchin *Paracentrotus lividus* Lamarck, 1816, fisheries in Galicia, NW Spain. In: Lawrence JM, Guzmán O (eds). Sea urchins: fisheries and ecology. DEStech Publications, Lancaster. pp 64–73

Chang Y-Q, Wang Z (1997) The raft culture of the sea urchin *Strongylocentrotus intermedius*. J Dalian Fish Univ China 23: 69–76

Denniston G (1974) The diet of the ancient inhabitants of Aishishki Point, an Aleut community. Arctic Anthropol XI Suppl. pp 143–152

Dewees CM (2004) Sea urchin fisheries: a California perspective. In: Lawrence JM, Guzmán O (eds). Sea urchins: fisheries and ecology. DEStech Publications, Lancaster. pp 37–55

Ebert TA (1975) Growth and mortality of post-larval echinoids. Amer Zool 15: 755–775

Ebert TA (1982) Longevity, life history, and relative body wall size in sea urchins. Ecol Monogr 52: 353–394

Engel F (1963) A preceramic settlement on the central coast of Peru: Asia, unit 1. Trans Amer Phil Soc, NS 53(3): 1–139

Erlandson, JM, Rick TC, Vellanoweth R (2004) Human impacts on ancient environments: a case study from California's northern Channel Islands. In: Fitzpatrick SM (ed). Voyages of discovery: the archeology of islands. Praeger, Westport, CT. pp 51–83

George SB, Lawrence JM, Lawrence AL (2004) Complete larval development of the sea urchin *Lytechinus variegatus* fed an artificial feed. Aquaculture 242: 217–228

Glassow MA (2002) Prehistoric chronology and environmental change at the Punta Arena site, Santa Cruz Island, California. In: Browne DR, Mitchell KL, Chaney HW (eds). Proceedings of the fifth California islands symposium. Santa Barbara Museum of Natural History, Santa Barbara. p 36

Grime JP (1974) Vegetation classification by reference to strategies. Nature 250: 26–31

Grime JP (1977) Evidence for the existence of three primary strategies in plants and its relevance to ecological and evolutionary theory. Amer Nat 111: 1169–1194

Guidetti P, Terlilzzi A, Boero F (2004) Effects of the edible sea urchin, *Paracentrotus lividus*, fishery along the Apulian rocky coast (SE Italy, Mediterranean Sea). Fish Res 66: 287–297

Hagen NT (1996) Echinoculture: from fishery enhancement to closed cycle cultivation. World Aquacult December 27(4): 6–19

Imamura K (1996) The prehistory of Japan and its position in East Asia. University of Hawai'i Press, Honolulu

Jeradino A, Castilla JC, Ramírez JM, Hermosilla N (1992) Early coastal subsistence patterns in central Chile: a systematic study of the marine-invertebrate fauna from the site of Curaumilla-1. Latin Amer Antiquity 3: 43–62

Keesing JK, Hall KC (1998) Review of harvests and status of world sea urchin fisheries point to opportunities for aquaculture. J Shellfish Res 17: 1505–1506

Lafferty KD (2004) Fishing for lobsters indirectly increases epidemics in sea urchins. Ecol Appl 14: 1566–1573

Lawrence AL, Lawrence JM (2004) Importance, status and future research needs for formulated feeds for sea urchin aquaculture. In: Lawrence JM, Guzmán O (eds). Sea urchins: fisheries and ecology. DEStech Publications, Lancaster. pp 275–283

Lawrence J (1991) Analysis of characteristics of echinoderms associated with stress and disturbance. In: Yanagisawa T, Yasumasu I, Oguro C, Suzuki N, Motokawa T (eds). Biology of Echinodermata. AA Balkema, Rotterdam. pp 11–26

Lawrence JM (1990) The effect of stress and disturbance on echinoderms. Zool Sci 7: 17–28

Lawrence JM, Bazhin A (1998) Life-history strategies and the potential of sea urchins for aquaculture. J Shellfish Res 17: 1515–1522

Lawrence JM, Lawrence AL, McBride SC, George SB, Watts SA, Plank LR (2001) Development in the use of prepared feeds in sea-urchin aquaculture. World Aquacult 32(3): 34–39

Le Direac'h J-P (1987) La pêche des oursins en Médditerranée: historique, techniques, législation, production. In: Boudouresque CF (ed). Colloque international sur *Paracentrotus lividus* et les oursins comestibles. GIS Posidonie, Marseille. pp 335–362

Lesser MP, Walker CW (1998) Introduction to the special section on sea urchin aquaculture. J Shellfish Res 17: 1505–1506

Lewis J (1962) Notes on the Barbados 'sea egg'. J Barbados Mus Hist Soc 29: 79–81

Lumbreras LG (1974) The peoples and cultures of ancient Peru. Smithsonian Institution Press, City of Washington (Translated by Meggers BJ)

Matsui I (1966) Uni no zoshoku. Nihon Suisan Shigen Hogo Kyokai. (The propagation of sea urchins.) Fish Res Bd Canada Trans Ser No 1063

Mottet MG (1976) The fishery biology of sea urchins in the family Strongylocentrotidae. Washington, Dept Fisheries. Tech Rept No. 20

Murata Y, Yokoyama M, Unuma T, Sata NU, Kuwahara R, Kaneniwa M (2002) Seasonal changes of bitterness and pulcherrimine content in gonads of green sea *Hemicentrotus pulcherrimus* at Iwaki in Fukushima Prefecture. Fish Sci 68: 181–189

Murata Y, Kaneniwa M, Oohara I, Kura Y, Yamada H, Sugimoto K, Unuma T (2004) Relationship between the reproductive cycle and the content of pulcherrimine, a novel bitter amino acid, in green sea urchin *Hemicentrotus pulcherrimus* ovaries. In: Heinzeller T, Nebelsick N (eds). Echinoderms: München. AA Balkema, Leiden. p 598

Oshima H (1962) Sea cucumber and sea urchin. Uchida Rokakuho Pub Co Ltd, Tokyo (in Japanese)

Pinnegar JK, Polunin NVC, Francour P, Badalamenti F, Chemello R, Harmelin-Vivien M-L, Hereu B, Milazzo M, Zabala M, D'Anna G, Pipitone C (2000) Trophic cascades in benthic marine ecosystems: lessons for fisheries and protected-area management. Environ Conserv 27: 179–200

Pfister CA, Bradbury A (1996) Harvesting red sea urchins: recent effects and future predictions. Ecol Appl 6: 298–310

Robinson SM (2004) The evolving role of aquaculture in the global production of sea urchins. In: Lawrence JM, Guzmán O (eds). Sea urchins: fisheries and ecology. DEStech Publications, Lancaster. pp 343–357

Roden C (1968) A book of Middle Eastern foods. Thomas Nelson and Sons Ltd, London

Saito K (1992a) Sea urchin fishery of Japan. In: Anonymous. The management and enhancement of sea urchins and other kelp bed resources: a Pacific rim perspective. California Sea Grant College, La Jolla. Rep. No. T-CSGCP-028

Saito K (1992b) Japan's sea urchin enhancement experience. In: Anonymous. The management and enhancement of sea urchins and other kelp bed resources: a Pacific rim perspective. California Sea Grant College, La Jolla. Rep. No. T-CSGCP-028

Sakai Y, Tajima I-I, Agatasuma Y (2004) Mass production of seed of the Japanese edible sea urchins *Strongylocentrotus intermedius* and *Strongylocentrotus nudus*. In: Lawrence JM, Guzmán O (eds). Sea urchins: fisheries and ecology. DEStech Publications, Lancaster. pp 287–298

Sala E, Boudouresque CF, Harmelin-Vivien M (1998) Fishing, trophic cascades, and the structure of algal assemblages: evaluation of an old but untested paradigm. Oikos 82: 425–439

Salls RA (1991) Early Holocene maritime adaptation at Eel Point, San Clemente Island. In: Erlandson JM, Colten RH (eds). Hunter-gatherers of Early Holocene coastal California, Perspectives in California Archaeology. Institute of Archaeology, University of California, Los Angeles. Vol. 1, pp 63–80

Sand C, Bolé J, Ouetcho A (2002) Site LPO023 of Kurin: characteristics of a Lapita settlement in the Loyalty Islands (New Caledonia). Asian Perspect 41: 129–147

Scheibling RE, Mladenov PV (1987) The decline of the sea urchin, *Tripneustes ventricosus,* fishery of Barbados: a survey of fishermen and consumers. Mar Fish Rev 49: 62–69

Shimabukuro, S (1991) *Tripneustes gratilla* (sea urchin). In: Shokita S, Kakazu K, Tomori A, Toma T (eds). Aquaculture in Tropical Areas (English edition: Yamaguchi M). Midori Shobo Co., Ltd., Tokyo

Simenstad CA, Estes JA, Kenyon KW (1978) Aleuts, sea otters, and alternate stable-state communities. Science 200: 403–411

Sloan NA (1985) Echinoderm fisheries of the world: a review. In: Keegan BE, O'Connor BDS (eds). Echinodermata. AA Balkema, Rotterdam. pp 109–124

Smith A (1984) Echinoid palaeobiology. George Allen and Unwin, London

Stotz WB (2004) Sea-urchin fisheries: a Chilean perspective. In: Lawrence JM, Guzmán O (eds). Sea urchins: fisheries and ecology. DEStech Publications, Lancaster. pp 3–17

Vadas RL, Steneck RS (1995) Overfishing and inferences in kelp–sea urchin interactions. In: Skjoldal HR, Hopkins C, Erikstad KE, Leinaas HP (eds). Ecology of fjords and coastal waters. Elsevier Science B.V., Amsterdam. pp 509–524

Vásquez JA, Guisado C (1992) Fishery of sea urchin *(Loxechinus albus)* in Chile. In: Anonymous. The management and enhancement of sea urchins and other kelp bed resources: a Pacific Rim perspective. California Sea Grant College, La Jolla. Rep. No. T-CSGCP-028 (no pagination)

Vásquez, JA, Veliz D, Weisner R (1996) Malacological analysis of an archeological site from Huentelauquen culture, IV Región (Chile). Gayana Oceanol 4: 109–116

Vellanoweth RL, Rick TC, Erlandson JM (2002) Middle and Late Holocene maritime adaptations on northeastern San Miguel Island, California. In: Browne DR, Mitchell KL, Chaney HW (eds). Proceedings of the fifth California islands symposium. Santa Barbara Museum of Natural History, Santa Barbara. pp 607–614

Yokota Y (2002) Fishery and consumption of the sea urchin in Japan. In: Yokota Y, Matranga V, Smolenicka Z (eds). The sea urchin: from basic biology to aquaculture. AA Balkema, Lisse. pp 129–139

Edible Sea Urchins: Biology and Ecology
Editor: John Miller Lawrence

Chapter 2

Gametogenesis and Reproduction of Sea Urchins

Charles W Walker[a], Tatsuya Unuma[b], and Michael P Lesser[a]

[a] *Department of Zoology, University of New Hampshire, Durham, NH (USA).*
[b] *Japan Sea National Fisheries Research Institute Niigata (Japan).*

1. INTRODUCTION

Gametogenesis and intra-gonadal nutrient storage and utilisation are linked processes in sea urchin reproduction. Dynamically interacting germinal and somatic cellular populations make up the germinal epithelium of the sea urchin gonad. Uniquely, sea urchin gonads grow in size not only because gametogenesis increases the size and/or the numbers of germinal cells present, but also because somatic cells within the germinal epithelium, the nutritive phagocytes, store extensive nutrient reserves before gametogenesis begins. Knowledge of these phenomena has led to successful manipulation of sea urchin reproduction and will provide increased opportunities for their aquaculture (Böttger et al. 2004). Our understanding of sea urchin gametogenesis and reproduction should develop rapidly now that sequencing of the genome of the purple sea urchin, *Strongylocentrotus purpuratus*, is completed (www.sugp.caltech.edu).

2. STRUCTURE OF THE GONADS OF THE SEA URCHIN

Branches of the genital coelomic and hemal sinuses (Hamann 1887; Campbell 1966; Strenger 1973; Walker 1982; Pearse and Cameron 1991) interconnect all five gonads of sea urchins. These sinuses project from similar components of the axial complex underneath the madreporite (Fig. 1a). Five branches penetrate the aboral connective tissues and one enters each of the five gonads. A single gonoduct emerges from each gonad. Each gonoduct extends for a substantial distance within the branches of the aboral coelomic sinus before exiting the test through a pentagonal array of gonopores in genital plates surrounding the anus.

In both sexes, the structure of the gonadal wall of sea urchins is similar, but not identical to that seen in sea stars and brittle stars (Walker 1979, 1982; Shirai and Walker 1988; Pearse and Cameron 1991). Two sacs of tissues (outer and inner) compose the gonadal wall. Each consists of several characteristic layers (Fig. 1b). Throughout the gonad, the genital coelomic sinus (GCS) separates the outer sac from the inner sac. The outer sac

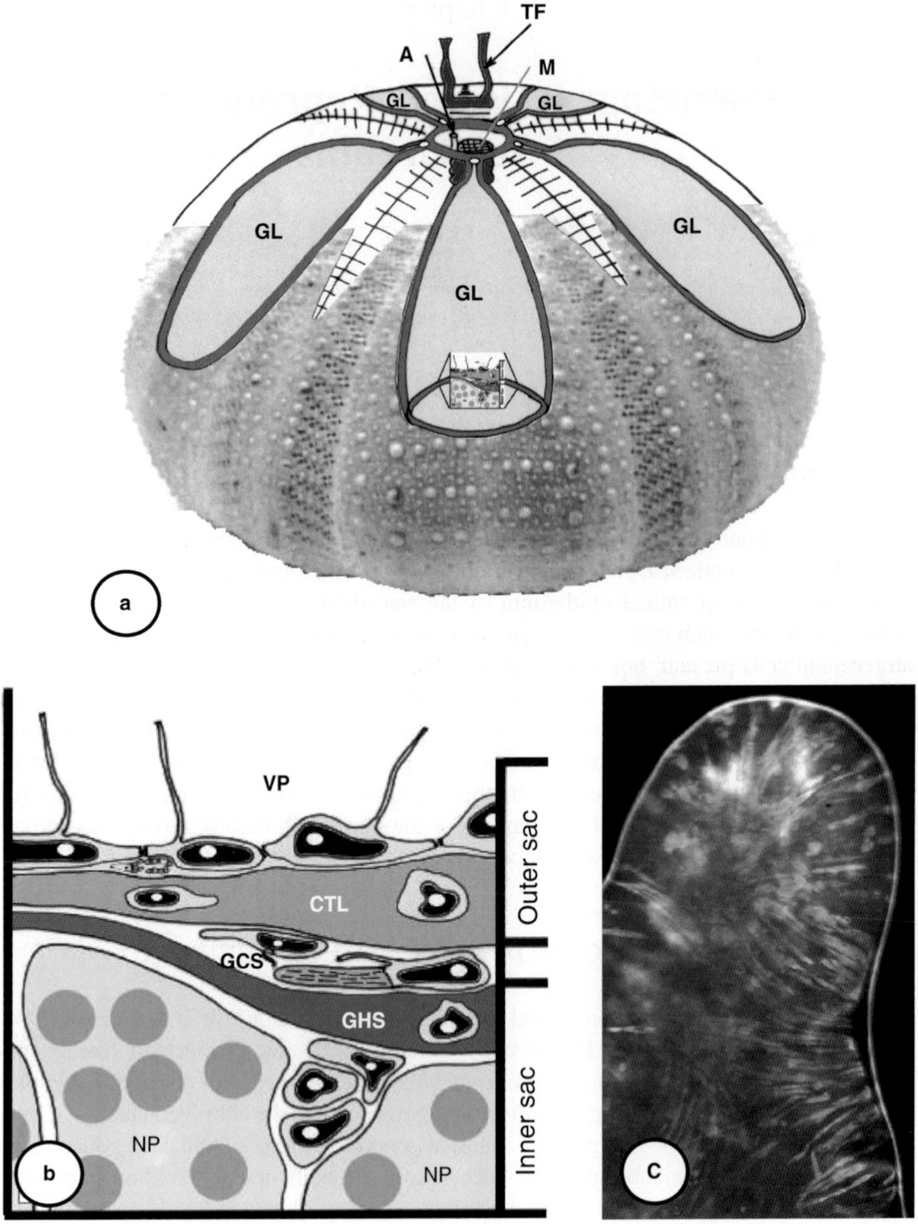

Fig. 1. (a) Diagrammatic representation of the sea urchin reproductive system in aboral side view; ; (b) Diagrammatic representation of the tissues in the sea urchin gonad wall, after Strenger (1973); (c) Lobe of a sea urchin gonad stained with phalloidin to show muscles on the exterior surface of the GHS of the inner sac (white strips). Abbreviations: A – anus; CTL – connective tissue layer; GL – gonad lumen; M – madreporite; NP – nutritive phagocytes; GCS – genital coelomic sinus; GHS – genital hemal sinus; TF – tube feet; VP – visceral peritoneum.

includes a visceral peritoneum (VP) that faces the perivisceral coelom and that is attached to a connective tissue layer (CTL). Nonmuscular epithelial cells also line the CTL on its opposite surface toward the GCS. The inner sac is a genital hemal sinus (GHS) that bears ciliated myoepithelial cells on its outer face like the gonads of sea stars (Figs 1b,c Walker 1979, 1982). These muscles contract rhythmically during gamete release (Okada et al. 1984; Okada and Iwata 1985) in response to the carbohydrate portion of a glycoprotein produced in the intestine and stored in the aboral hemal tissues (Takahashi et al. 1990, 1991). Interconnecting nerves synchronize the activities of the gonads during spawning. On its luminal face, the GHS supports the germinal epithelium. The principal functions of the inner sac are gametogenesis, very limited nutrient storage in the GHS, and extensive nutrient storage in the nutritive phagocytes. Contents of the GHS vary slightly in amount during gametogenesis. Prior to and during gametogenesis, the GHS does contain PAS-positive glycoproteins, although it never becomes engorged like the GHS of sea stars at a similar time during gametogenesis (Walker 1979, 1982; Beijnink et al. 1984; Byrne et al. 1998).

Severing the radial connections between the gonad and the aboral coelomic and hemal branches prevents gonadal development and gametogenesis (Okada 1979). In very small *Lytechinus pictus,* the aboral hemal sinus has five radial extensions (Houk and Hinegardner 1980). These later become the inner sac and force surrounding connective tissue into the form of the outer sac of each gonad as they grow. Expansions of the hemal sinus contain precursors of somatic and gonial cells of the developing germinal epithelia. Both of these precursor cell types are filled with nutrients within membrane-bound vesicles and are found in the echinopluteus before metamorphosis. Very small females may have recognizable oogonia, but oogenesis does not occur. Very small males may contain a few differentiated spermatozoa. The testes of *S. purpuratus* produce spermatozoa within the first 12 months of life (Cameron et al. 1990).

3. INTERACTING GAMETOGENIC AND NUTRITIVE PHAGOCYTE CYCLES IN THE SEA URCHIN GONAD: STAGES, PHYSIOLOGY, AND MOLECULAR BIOLOGY

3.1. Stages in Gametogenesis

During the annual gametogenic cycle, gonads of both sexes of the sea urchin pass through a predictable series of structural changes (Walker 1982; Pearse and Cameron 1991; Walker et al. 1998, 2005). These changes can be classified according to the activities of the two major populations of cells that compose the germinal epithelium. These cellular populations are either: (a) germinal cells (oogonia > ova in the ovary or spermatogonia > fully differentiated spermatozoa in the testis) or (b) somatic cells called nutritive phagocytes (NPs) that are present in both sexes (Caullery 1925; Holland and Giese 1965; Holland and Holland 1969; Kobayashi and Konaka 1971; Walker et al. 2005).

Structural changes in the germinal epithelium of the sea urchin gonad have been described using the staging systems of Fuji (1960 a,b) based on germinal cells and of Nicotra and Serafino (1988) based on NPs (Unuma et al. 1998, 1999; Byrne 1990; Meidel and Scheibling 1998; Walker and Lesser 1998; Walker et al. 1998, 2005).

A valid staging system must simultaneously consider both populations of cells to provide the basis for an understanding of the cell biology of gametogenesis. It is important to recognize that the size of sea urchin gonads does not necessarily relate to the progress of gametogenesis alone. One must very carefully consider which cellular population (germinal or somatic) actually predominates in size and/or numbers within its germinal epithelium in order to determine the stage of gametogenesis that characterizes a particular individual (Walker et al. 1998, 2005; Böttger et al. 2004). In this review, we employ the following stages: (a) inter-gametogenesis and NP phagocytosis, (b) pre-gametogenesis and NP renewal, (c) gametogenesis and NP utilisation, and (d) end of gametogenesis, NP exhaustion and spawning. For additional information, see the Web page: Normal and Manipulated Green Sea Urchin Gametogenesis – (http://zoology.unh.edu/faculty/walker/urchin/gametogenesis.html).

3.2. Nutritive Phagocytes in Ovaries and Testes

Nutritive phagocytes are versatile somatic cells that provide a structural and nutritional microenvironment for germinal cells throughout sea urchin gametogenesis (Walker et al. 2005). They have multiple functions. Nutritive phagocytes may contain amitotic oogonia or spermatogonia and growing primary oocytes or spermatogenic cells basally and simultaneously phagocytize residual ova and spermatozoa luminally (Anderson 1968; Walker et al. 1998, 2005). At full maturity, each ovarian NP encloses a single, growing vitellogenic oocyte in a basal incubation chamber where it may be enveloped by dissolved or particulate nutrients (Walker et al. 2005; Fig. 2a). The same NP may also simultaneously contain additional germ cells at several earlier stages of oogenesis within smaller, discrete basal incubation chambers. Testes also contain numerous NPs that cooperate to provide large basal incubation chambers that ultimately become continuous and together supply nutrients to enormous numbers of spermatogenic cells at diverse stages (Walker et al. 2005; Fig. 2b). One must assume that NPs can recognize gametogenic cells at different stages and react appropriately to them. Errors in this recognition system, especially regarding gonial cells, would lead to sterile gonads. The basis for NP recognition of different stages of gametogenesis is unknown but undoubtedly involves both structural features of the NP (e.g. which regions of the NP and particular gametogenic stages interact) and protein-protein interactions between NP and specific gametogenic stages.

3.3. Inter-gametogenesis and NP Phagocytosis

In *S. droebachiensis*, the inter-gametogenesis and NP phagocytosis stage occurs in the spring and lasts for approximately 3 months (Figs 3a, 4a; Walker 1982; Walker et al. 2005). Ovaries contain residual primary oocytes of various sizes within residual NP incubation chambers, clusters of amitotic oogonia near the GHS, and a few ova within the ovarian lumen (Fig. 3a). In the testes, residual spermatozoa may or may not remain in the testicular lumen and amitotic spermatogonia occur as individual or small clusters of cells beneath NPs (Fig. 4a, arrow).

Toward the end of this stage, NPs in both sexes have increased in length by 20–55 μm as they resume nutrient storage (Walker et al. 2005). NPs may also phagocytize residual

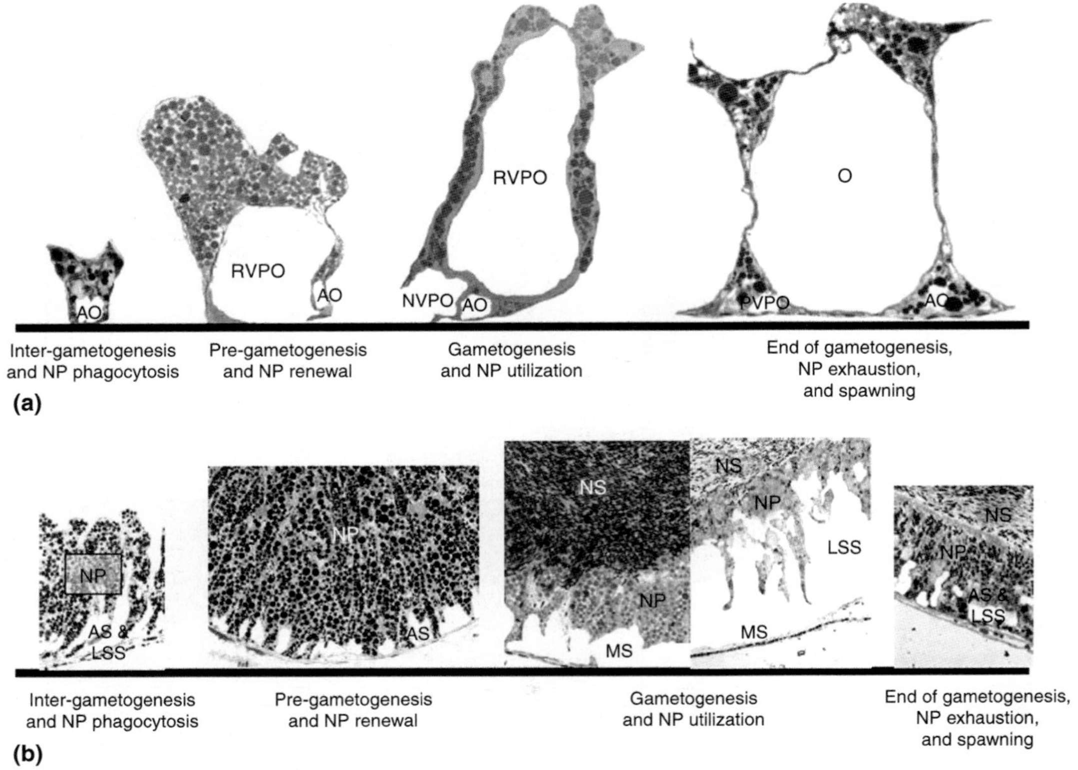

Fig. 2. Nutritive phagocytes in (a) ovaries and (b) testes at each gametogenic stage with germ cells removed from NP incubation chambers (not to scale). For ovaries, the shapes and dimensions of representative, individual NPs are shown, as well as the positions of various germ cell types within discrete NP incubation chambers. For testes, the positions of germ cells relative to groups of NPs are revealed, especially the continuous incubation chamber that forms eventually. Abbreviations: AO – amitotic oogonia; AS – amitotic spermatogonia; NVPO – new vitellogenic primary oocyte; O – ovum; LSS – later spermatogenic stages; RVPO – residual vitellogenic primary oocyte; NP – nutritive phagocyte; NS – new spermatozoa; MS – mitotic spermatogonia. From figure 11, Walker et al. (2005) with permission.

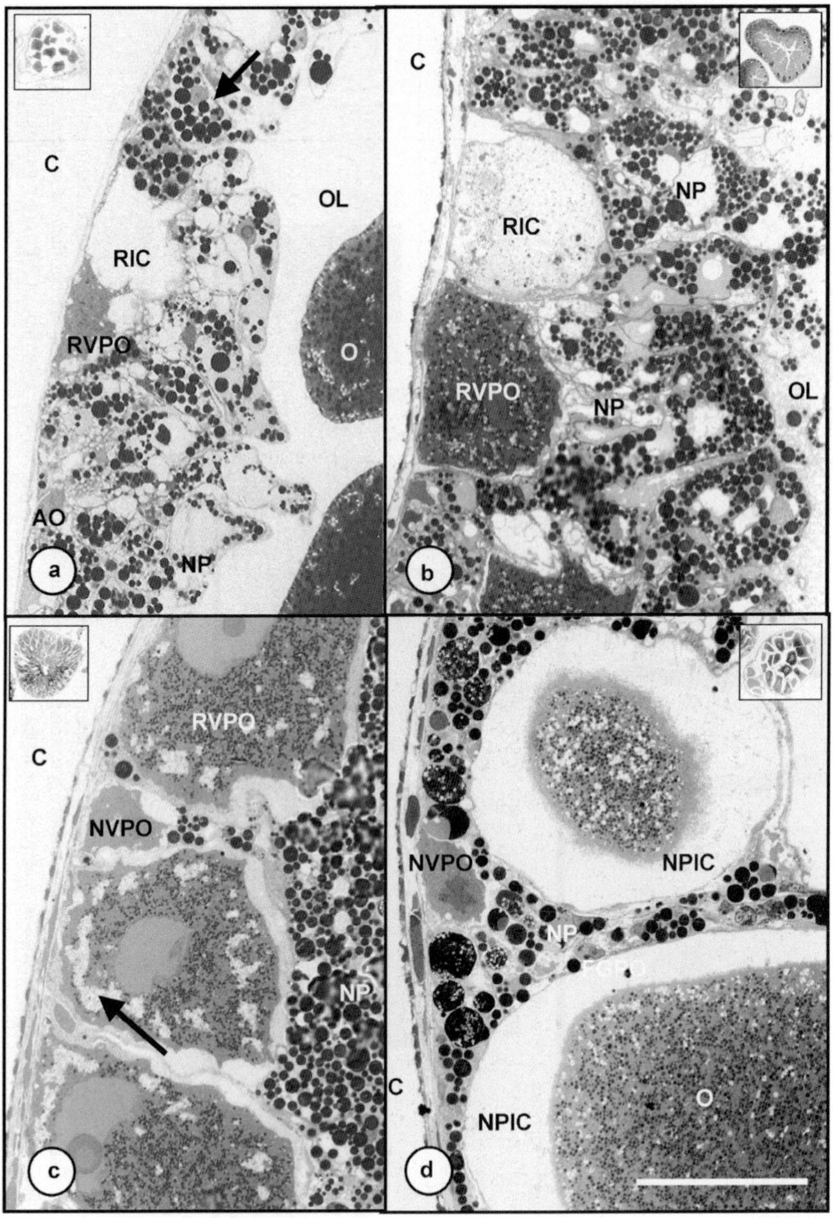

Fig. 3. Stages in the annual gametogenic cycle of female sea urchins. (a) inter-gametogenesis and NP phagocytosis; (b) pre-gametogenesis and NP renewal; (c) gametogenesis and NP utilization; (d) end of gametogenesis, NP exhaustion, and spawning. Abbreviations: AO – amitotic oogonia; C – coelom; O – ovum; OL – ovarian lumen; NP – nutritive phagocyte; NPIC – nutritive phagocyte incubation chamber; NVPO – new vitellogenic primary oocyte; RIC – residual NP incubation chamber; RVPO – residual vitellogenic primary oocyte. Arrow in (a) points out NP nucleus. Arrow in (c) points out vacuoles unique to this stage. Insets show low magnification views of each stage. Scale bar = 50 μm. Insets, a, and b from Walker et al. (2005) with permission.

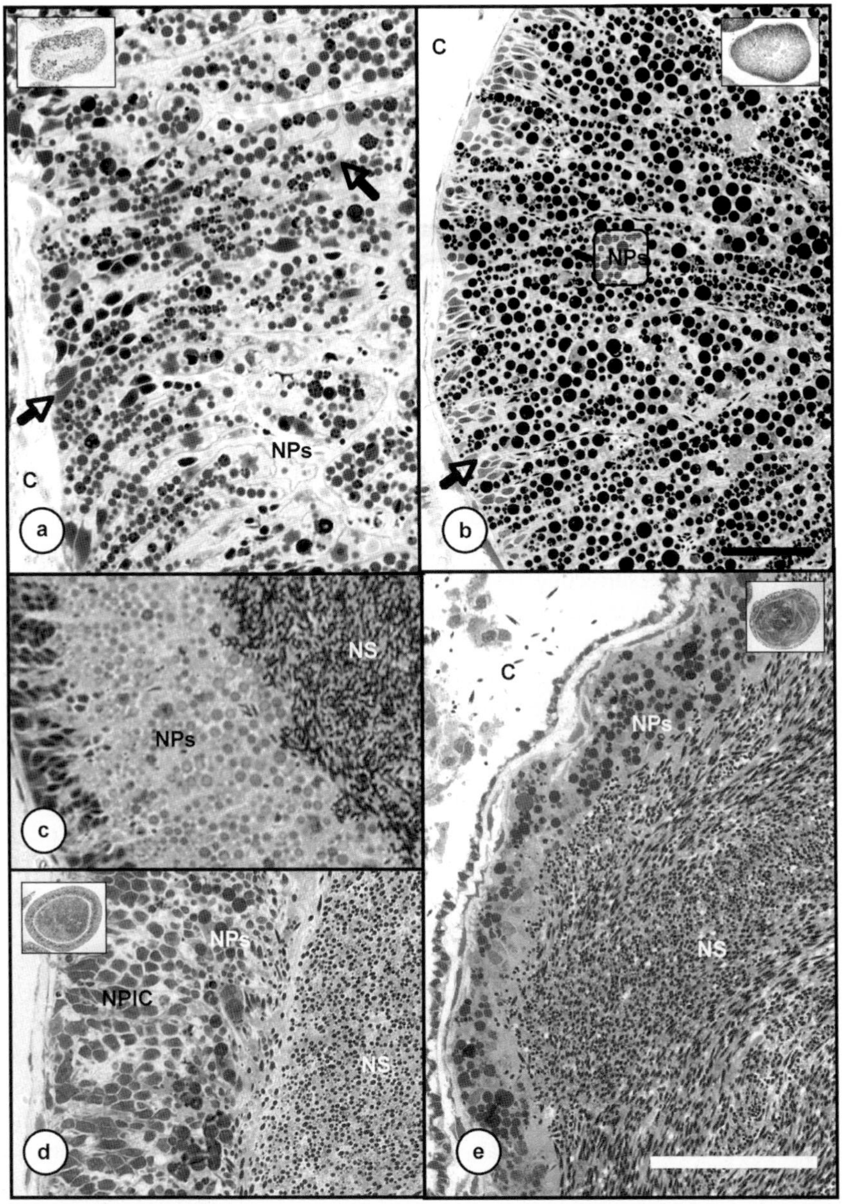

Fig. 4. Stages in the annual gametogenic cycle of male sea urchins. (a) Inter-gametogenesis and NP phagocytosis; (b) pre-gametogenesis and NP renewal; (c) early gametogenesis and NP utilization; (d) late gametogenesis and NP utilization; (e) end of gametogenesis, NP exhaustion, and spawning. Abbreviations: C – coelom; NP – nutritive phagocyte; NPIC – nutritive phagocyte incubation chamber; NS – new spermatozoa. Arrow in (a) within NP points out NP nucleus; that near the coelom points out amitotic spermatogonia. Arrow in (b) points out amitotic spermatogonia. Insets show low magnification view of each stage. Scale bar = 50 μm. Insets and a–c from Walker et al. (2005) with permission.

ova or spermatozoa and may recycle nutrients derived in this way (Pearse 1969a,b, 1970; Reunov et al. 2004b; Walker et al. 2005). Masuda and Dan (1977) suggest that large residual ova cannot be directly phagocytized by smaller NPs. Based on increasing acid phosphatase activity, these authors propose that ova are autophagic, thus aiding in their own digestion. Phagocytosis of spermatozoa and of residual bodies discarded by maturing spermatids has been described in detail for NPs in *Anthocidaris crassispina* (Reunov et al. 2004a,b). Initially, single spermatozoa are endocytosed by NPs into structures called heterophagosomes. Heterophagosomes then aggregate into multi-spermatozoan-containing heterophagosomes and finally become remnant bodies.

3.4. Pre-gametogenesis and NP Renewal

In *S. droebachiensis,* the pre-gametogenesis and NP renewal stage occurs for 3–4 months during the summer (Figs 3b, 4b). In ovaries, the previous generation of germinal cells is retained as a small group of residual immature, primary oocytes. By this time residual ova or spermatozoa are absent from the gonad lumen. Whereas most of these oocytes are growing, some are atretic (Holland and Giese 1965; Gonor 1973; McClary and Barker, 1998; Walker et al. 1998, 2005). Both are found within recently emptied NP incubation chambers (Fig. 3b). Resurrecting residual, immature oocytes from the previous year's gametogenesis may be a mechanism for maximizing oocyte production. In the gonads of both sexes, the next generation of germ cells exists as small clusters of amitotic gonial cells (oogonia or spermatogonia $\sim$ 10–12 μm in diameter). These gonial cell clusters are scattered on the luminal face of the GHS of the gonad underneath the NPs.

Preceding the gametogenesis and NP utilization stage, a variety of proteins, carbohydrates, and lipids are stored in membrane-bound vesicles or otherwise in the cytoplasm of NPs in both sexes. The principal protein accumulated is a glycoprotein with a molecular weight of about 170 kDa (Unuma et al. 1998, 2003). This protein was originally identified as the predominant component of yolk granules in sea urchin eggs (10–15% of total egg protein) and termed the major yolk protein or major yolk glycoprotein (MYP; Harrington and Easton 1982; Kari and Rottmann 1985; Scott and Lennarz 1989; Yokota and Sappington 2002). Unlike other oviparous animals where the yolk protein is female-specific, both male and female sea urchins produce MYP (Shyu et al. 1986; Unuma et al. 2001). The MYP is synthesized in the intestine, coelomocytes, and NPs (Harrington and Ozaki 1986; Shyu et al. 1986; Cervello et al. 1994; Unuma et al. 2001), and then accumulated in NPs (Ozaki et al. 1986; Unuma et al. 1998, 2003). About 80% of the total protein in the gonads at the pre-gametogenesis and NP renewal stage is MYP in both sexes (Unuma et al. 2003), which means that MYP is the protein that makes the greatest contribution to the growth of the gonad at this stage (Unuma 2002).

Another form of MYP with slightly higher molecular weight (about 180 kDa) is contained in the coelomic fluid and considered to be a precursor of MYP in eggs (Harrington and Easton 1982; Giga and Ikai 1985). This higher molecular weight MYP in the coelomic fluid was formerly called sea urchin vitellogenin (Shyu et al. 1986), after the yolk protein precursor, vitellogenin, contained in the blood of oviparous vertebrates. However, the overall sequence of MYP cDNA determined in *Pseudocentrotus depressus* (Unuma et al. 2001), *S. purpuratus* (Brooks and Wessel 2002), and *Hemicentrotus pulcherrimus*

(Yokota et al. 2003) revealed that MYP has no homology to vertebrate vitellogenin. Instead, MYP is slightly homologous to transferrin (an iron transporter) of vertebrates. A new term "echinoferrin" instead of "vitellogenin" has been proposed for the MYP precursor in the coelomic fluid based on the sequence similarlity to transferrin (Yokota et al. 2004). We will use the term "MYP" for both types (170 and 180 kDa) to avoid confusion. The MYP purified from the coelomic fluid potentially binds iron *in vitro* (Brooks and Wessel 2002), but MYP in the coelomic fluid actually binds very small amounts of iron *in vivo* (Unuma et al. 2004). Major yolk protein also binds large amounts of zinc and is considered to have a role as a zinc transporter (Unuma et al. 2004).

Shyu et al. (1987) found a palindromic sequence similar to an estrogen responsive element of vertebrates upstream of the MYP gene in *S. purpuratus* and postulated that MYP synthesis is under the control of estrogens. Experimental administration of estrogen precursors or estrogens themselves resulted in increased gonad size in males (Unuma et al. 1999) and females (Wasson et al. 2000). The significant increase in gonad size might be attributable to an accelerated deposition of MYP as well as other nutrients within NPs of both sexes and not to an increase in the numbers (or size) of gametes. The relationship between estrogens and MYP synthesis, however, remains obscure since the effects of estrogens on MYP gene expression were not examined in these studies. Also, the biosynthesis of estrogen from its precursor has not been demonstrated in sea urchins (Wasson and Watts 2000; Chapter 4).

The PAS-positive contents accumulating within the NP are glycogen, whereas the contents of the membrane-bound vesicles are mucopolysaccharides (Chatlynne 1969; Walker 1982; Laegdsgaard et al. 1991). Glycogen in particular is released from the NP and is apparently taken up early in oogenesis by residual oocytes through pinocytosis (Bal 1970; Tsukahara 1970).

Unidentified lipid droplets occur within the cytoplasm of NP in both ovaries and testes of sea urchins. Their role during gametogenesis is unknown. β-Echinenone and (6′R)-β-carotene-4-one are the major carotenoids in both the ovaries and the testes. The concentration of carotenoids is lower in testes than in ovaries (Griffiths and Perrott 1976; Matsuno and Tsushima 2001). Evidence suggests that echinenone is synthesized from β-carotene in the ovary and not in the gut wall. The origin of echinenone in testes is not known. Carotenoids are also presumably important constituents of the NP, although their intracellular location has not been determined. If carotenoids are only present in oocytes, it is difficult to account for their presence in testes. Consequently, it is likely that carotenoids are first stored in NP and later transferred to the ova in females. The function of carotenoids in males is unknown. In both sexes, carotenoids may be involved in antioxidant protection of the gametes (Shapiro 1991). Increasing carotenoid concentration can substantially improve gonadal color in both sexes. However, carotenoids are not necessary for gonadal growth in either sex (Plank 2000).

3.5. Gametogenesis and NP Utilization

In *S. droebachiensis*, the gametogenesis and NP utilization stage (Figs 3c, 4c) occurs in the fall and early winter and lasts for about 5 months. It differs from the pre-gametogenesis and NP renewal stage since mobilization of nutrients from the NP and initiation of gonial cell mitosis have simultaneously begun. For some time, the pre-gametogenesis and NP renewal stage overlaps with the gametogenesis and NP utilization stage. This overlap can be recognized because the NPs continue to accumulate additional nutrients while they are mobilizing those previously stored (Unuma et al. 2003).

Quantitative changes in MYP and other components in the gonads were examined in the course of gametogenesis in *P. depressus*. These revealed that, as gametogenesis proceeds, MYP in the gonads decreases in both sexes as opposed to an increase of other substances such as proteins other than MYP and nucleic acids (Unuma et al. 2003). At the end of gametogenesis, the amount of MYP was very low in mature testes and less than half of the initial amount in mature ovaries. Based on these results and those from other studies, a model was proposed on the nutritional role of MYP in sea urchin reproduction (Fig. 5; Unuma et al. 2003). Both male and female sea urchins ingest proteins from food, convert some of them to MYP, and store it in NPs. After gametogenesis begins, the MYP stored in both ovarian and testicular NPs degrades to amino acids that are utilized as material for synthesizing new proteins, nucleic acids, and other nitrogen-containing substances that constitute eggs and sperm. Predominant proteins that are newly synthesized and

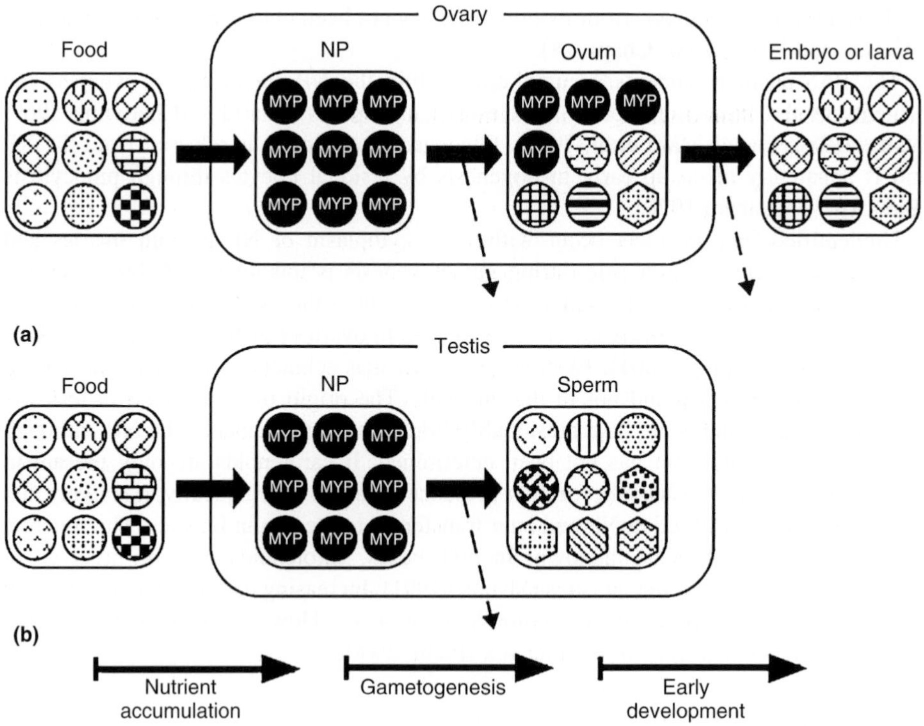

Fig. 5. Nutritional role of the major yolk protein (MYP) in (a) female and (b) male sea urchins (modified from Unuma et al. 2003). MYP has two different functions as a nutrient source: for gametogenesis before spawning and for early development after fertilization. NP – nutritive phagocyte; circles – proteins; hexagons – other molecules. Broken lines with arrows indicate the loss caused by metabolism as an energy source.

accumulated in the gonads in this period are histones (Puigdomenech et al. 1987; Poccia et al. 1989) in males and YP 30 (Wessel et al. 2000) in females. Major yolk protein synthesis probably continues even in the course of gametogenesis, since MYP mRNA expression is detected in the mature males and females (Shyu et al. 1986). Nevertheless, MYP synthesis would be considerably less than its consumption during gametogenesis. In males, most of the stored MYP is consumed by the end of spermatogenesis. In females, however, a portion of the stored MYP is transported to the ova through endocytosis by a dynamin-dependent mechanism (Brooks and Wessel 2003, 2004), and finally forms yolk granules (Ozaki et al. 1986). In terms of its role as a protein reserve, unlike yolk proteins in other animals, MYP has two different functions: for gametogenesis before spawning and for early development after fertilization, the latter being a classical role of yolk protein (Unuma et al. 2003).

MYP in the yolk granules degrades after fertilization and serves as a nutrient source possibly for the larval stage (Scott et al. 1990) and as a cell adhesion molecule (Noll et al. 1985; Cervello and Matranga 1989). MYP is digested by the lysosomal protease, cathepsin B, and possibly other proteases in such early embryos (Yokota and Kato 1988; Mallya et al. 1992; Vafa and Nishioka 1995), whereas proteases involved in degradation of MYP during gametogenesis are unknown.

Oogonial mitoses are identifiable only at the beginning of this stage, whereas spermatogonial mitoses were observed at low levels throughout the year and peak in abundance at the same time the oogonia divide. Perhaps the individualized treatment of germ cells that can occur within discrete NP incubation chambers in females is important in establishing the narrow window of time when oogonia divide. Nutritive phagocyte incubation chambers are maximally distended laterally as oocytes change in shape from conical to nearly round and as they secrete the jelly coat (Fig. 3d). Each NP may simultaneously contain either a fully grown primary oocyte or an ovum as well as previtellogenic primary oocytes and amitotic oogonia in discrete incubation chambers (Fig. 2a). The latter two germ cell types always occur basally.

Numerous ovary-specific proteins are manufactured in the oocyte and are eventually found in the cortical granules of fully mature ova. Among others, these include: hyalin ($\sim$330 kDa; Wessel et al. 1998), glycosaminoglycans (Schuel et al. 1974), serine (25 kDa), proteases (Haley and Wessel 1999), ovoperoxidase and proteoliaisin (Sommers et al. 1989; Nomura et al. 1999), and YP39, a structural protein component of sea urchin yolk platelets (Berg and Wessel 2003). Cortical granules are transferred to the cortical cytoplasm early during oocyte maturation divisions and after breakdown of the germinal vesicle (Wessel et al. 2002; Hinchcliffe et al. 1999). Primary oocytes undergo meiosis in NP incubation chambers to become ova (Fig. 3d). Wessel et al. (2002) document translocation of cortical granules from the general to the cortical cytoplasm early during oocyte maturation and this process must also occur in NP incubation chambers (Walker et al. 2005). Before meiosis, the large vitellogenic oocyte within each NP incubation chamber secretes the jelly coat leaving a large space between the oocyte surface and the NP that surrounds it (Jondeung and Czihak 1982).

As this stage progresses in testes, an increasingly thick, continuous layer of mitotic spermatogonia and all subsequent spermatogenic stages fill a common incubation chamber jointly provided by numerous NPs (Fig. 4c). Spermatogenesis includes the minimal growth in size of primary spermatocytes as they travel along the length of the NP toward the lumen of the gonad. This is followed by meiotic divisions. The result is four spermatids that are interconnected by intercellular bridges. The elaborate interconnections of cohorts of spermatids all derived from a single spermatogonium and characteristic of mammals do not occur in sea urchins (Ward and Nishioka 1993). This may result because the massive NPs stabilize columns of spermatogenic cells during their journey from the luminal face of the GHS toward the testicular lumen (Nicotra and Serafino 1988). Details of spermatogenesis are discussed by Chia and Bickell (1983), Ward and Nishioka (1993), and Simpson and Poccia (1987). Spermiogenesis occurs in the lumen of the testis where immotile, fully differentiated spermatozoa are stored (Longo and Anderson 1969; Cruz-Landin and Beig 1976; Nicotra et al. 1984). In an autoadiographic study of *S. purpuratus*, the duration of spermatogenesis from spermatogonial mitosis to fully differentiated spermatozoan was approximately 12 days (Holland and Giese 1965). Proteins unique to the testis include histone H1, H2B-1, and H2B-2 that are produced in mitotically active spermatogonial cells of the testis (Poccia et al. 1989) and bindin, which is produced by late spermatocytes and early spermatids (Cameron et al. 1990).

3.6. End of Gametogenesis, NP Exhaustion, and Spawning

In *S. droebachiensis,* this stage lasts for 2–3 months (Figs 3d, 4d). NPs have exhausted their supplies of nutrients and are at the smallest size for the year. The gonadal lumen is filled with fully differentiated, stored gametes. In ovaries, stored gametes are ova. In testes, stored gametes are fully differentiated, immotile spermatozoa. In this stage, gamete storage and ultimately spawning are the major activities of gonads of both sexes of the sea urchin. Reunov et al. (2004a,b) suggest that NPs control their own reduction in size during this stage by becoming autophagic through the action of "cell-size-reducing autolysosomes." It will be interesting to determine whether the same nutrient-sensing molecular mechanisms used by other eukaryotes, like that used by the fat bodies of the fruit fly and involving TOR and related genes (Hay and Sonnenberg 2004, Klionsky 2005), are also active during autophagy of the NPs in sea urchins.

Spawning usually occurs during an extended period of several months. During each spawning event, males spawn first. Spermatozoa are activated and begin to swim upon contact with seawater. Fully mature ova are then released into a cloud of actively swimming spermatozoa. The gonads of *Temnopleurus toreumaticus* contract rhythmically following spontaneous or artificially induced spawning. Factors that naturally initiate spawning are relatively unknown (Palmer 1937; Kanatani 1974; Cochran and Engelman 1975, 1976; Starr et al. 1990, 1992; Takahashi et al. 1990, 1991). In *S. droebachiensis*, both sexes spawn in response to a small molecular weight protein resulting from phytoplankton blooms (Himmelman 1975; Starr et al. 1992). Lunar influence over spawning also has been suggested for some species of sea urchins. While some of these latter studies have been discounted (Fox 1924; Yoshida 1952; Kêckês 1966), others seem valid (Kobayashi and Nakamura 1967; Pearse 1975; Kobayashi et al. 1984; Iliffe and Pearse 1982; Lessios 1991).

To understand the biochemical mechanism of spawning in sea urchins, it is instructive to consider the more thoroughly investigated system of sea stars. In the sea star, primary oocytes are arrested at first meiotic prophase and are stored in ovarian follicles within the ovarian lumen until spawning (Kishimoto 1998, 1999). Three interrelated chemical messengers are involved in spawning in sea stars. These are gonad stimulating substance (GSS), maturation inducing substance (MIS; 1-methyl adenine), and maturation-promoting factor (MPF). Just before spawning in sea stars, these messengers lead to germinal vesicle and follicular envelope breakdown (GVBD and FEBD) and to meiosis of primary oocytes (Shirai and Walker 1988).

GSS is a small molecule produced by neuroepithelial cells of the radial nerves on the oral surface of the sea star. Follicle cells surrounding primary oocytes are the targets of GSS and they produce 1-methyl adenine (1-MeAde) as a result of exposure to GSS (Kubota and Kanatani 1975). 1-MeAde is released from follicle cells possibly at the tips of elongated extensions that contact the membrane of the primary oocyte where receptors are located (Shirai and Walker 1988). A signal transduction cascade within oocyte cytoplasm leads to the activation of intracellular cyclin B/Cdc2 complexes (=MPF). When active, these heterodimeric complexes phosphorylate proteins that promote GVBD, FEBD, and advancement of the meiotic divisions of the primary oocyte. The fully mature ova that result are spawned immediately.

Because of structural and functional differences, much of what happens in sea stars during spawning seems impossible in sea urchins. Maturation divisions of urchin primary oocytes occur as soon as these cells are available. This means that the maturation of oocytes and spermatocytes into mature gametes is continuous throughout the gametogenesis and NP utilization stage in sea urchins, and does not occur just prior to spawning of oocytes as in sea stars.

In sea urchins, follicle cells do not surround each primary oocyte. 1-MeAde may be produced in both males and females by NPs (Kubota et al. 1977). The radial nerves of sea urchins do contain a low molecular weight polypeptide similar to GSS (Cochran and Engelman 1976). It is also significant that the NPs of sea urchins are very much like the somatic accessory cells within the spermatogenic columns of male sea stars (Walker 1980, 1982). In male sea stars, the somatic accessory cells within spermatogenic columns are probably the biochemical equivalent of the NP in male sea urchins as both produce 1-MeAde (Kubota et al. 1977).

3.7. Environmental Control of Gametogenesis

The cellular mechanisms that simultaneously stimulate nutrient mobilization from NP and gonial cell mitosis are largely unknown. The same environmental cue may be important for both processes in these different populations of cells or the cues may differ. Shorter photoperiod (Pearse et al. 1986a,b; Walker and Lesser 1998) and decreasing temperature (Masaki and Kawahara 1995; Noguchi et al. 1995; McBride et al. 1997) have both been implicated and co-vary with one another seasonally. For the temperate sea urchin, *S. droebachiensis*, shortening day length in fall is accompanied by a decrease in seawater temperatures.

Although exceptions occur (Cochran and Engelman 1975; Bay-Schmith and Pearse 1987; Yamamoto et al. 1988; Sakairi et al. 1989; Ito et al. 1989), changing photoperiod can be correlated with the initiation of the gametogenesis and NP utilization stage in both sexes of a number of species of sea urchins (Pearse et al. 1986a,b; Walker and Lesser 1998). Little is known about the specific nature of the photoperiod cue. In the Gulf of Maine, the onset of shortening day length in the fall results in the initiation of gametogenesis in *S. droebachiensis* (Walker and Lesser 1998). The underwater light field in the Gulf of Maine is attenuated at both ends of the ultraviolet (UVR: 290–400 nm) and visible spectra (400–750 nm) following absorption and scattering by phytoplankton, suspended particles, and dissolved organic carbon. This results in a light environment that is dominated by wavelengths in the green portion of the spectrum (Fig. 6). The intact sea urchin test exhibits extremely low transparency to both UVR and visible portions of the spectrum, and even at the highest irradiances, the internal organs of *S. droebachiensis* will be exposed to very few photons of these wavelengths. This begs the question of how the shortening day length during the fall is communicated to the somatic and germinal cellular populations within the germinal epithelium of the gonad.

Millott (1968) identified a general surface sensitivity to light (the "dermal light sense"). Millott and Yoshida (in Millott 1968) described putative photoreceptors in the external epithelium of *Diadema antillarum* and noted the photosensitivity of the radial nerve. Their results suggest that a relationship exists between the abundant superficial nerves and dermal photosensitivity, thus implicating these nerves in photoreception (Millott 1968). Microspectrophotometric studies of the spectral sensitivity of these nerves in situ show maxima between 530 and 550 nm (Millott 1968). The pigments associated with echinoderm photoreceptors might be carotenoids (Millott 1968). The isolated "melanophores" of *Centrostephanus longispinus* show maximal sensitivity between 430 and 450 nm (Gras and Weber 1983). These chromatophores are not linked to the nervous system and are

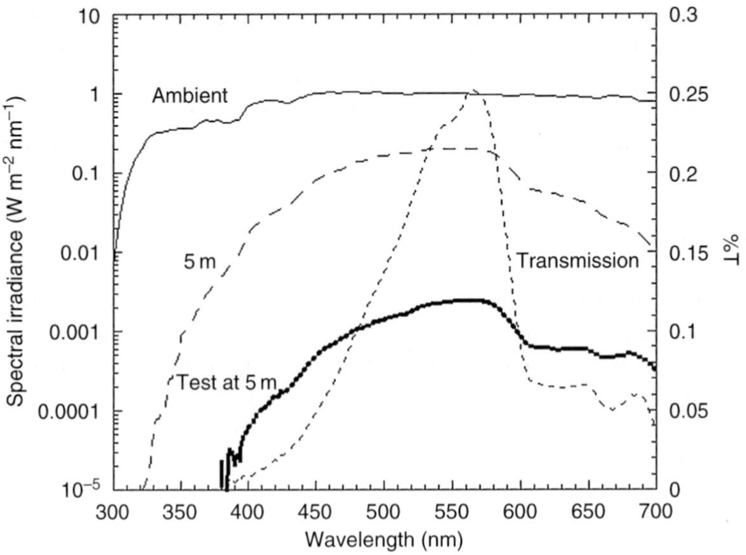

Fig. 6. Ambient and underwater irradiances for the Gulf of Maine. Irradiance was measured through the test of a freshly sacrificed sea urchin held at 5 m.

not directly tied to photoreception involving signal transduction (Gras and Weber 1983). A putative rhodopsin was identified and localized in the sea star, *Asterias forbesi*, and brittlestar, *Ophioderma brevispinum*, utilizing immunocytochemical techniques and an antibody against bovine rhodopsin (Johnsen 1997). The results of this work showed "opsin" expression in brittlestar arms as well as sea star eye spots, but did not characterize the protein or gene sequence of the molecule.

Recently, there has been considerable interest in the evolution of image-forming eyes from non-ocular photoreceptors. A monophyletic origin of eyes has been proposed in which the homeotic master control gene for eye morphogenesis, Pax 6, is present in a wide number of phylogenetically diverse organisms (Gehring and Ikeo 1999). Despite the fact that sea urchins do not have image-forming eyes or even eye spots, like sea stars, Pax-6 is expressed at high concentrations in the tube feet and has substantial amino acid conservation with mouse and human Pax-6 genes (Czerny and Busslinger 1995). It is possible that some ancestral components of a visual system exist in the tube feet and provide a mechanism for sensing changes in ambient irradiances and initiating downstream molecular and biochemical events. Since all metazoans examined share versions of the visual pigment opsin, it is reasonable to predict that an opsin should be present in the tube feet and elsewhere in the sea urchin. Recently, proteins similar to ancient, ciliary and rhabdomeric opsins have been discovered through the purple sea urchin genome project (www.sugp.caltech.edu) and a unique opsin has been isolated from green sea urchin tube feet (Barry, 2005; GenBank accession #DQ285097).

Studies of the effects of photoperiod on gametogenesis in other organisms may help in understanding its effect in sea urchins. Vertebrates with seasonal patterns of gametogenesis are affected by photoperiod in differing ways. Short day lengths typical in the fall promote spermatogenesis in the Atlantic salmon (Björnsson et al. 1994). Fall/winter photoperiod arrests spermatogenesis in amphibians (Paniagua et al. 1990), the weaver bird (Chandola et al. 1976), the bat (Heideman et al. 1992), the bank vole (van Haaster et al. 1993), and the Djungarian hamster (Tähkä et al. 1997). Despite the often conflicting and incomplete explanations of the effects of photoperiod on seasonal gametogenesis, one pattern is apparent. Among those species studied that have seasonal reproduction, animals either respond positively (like echinoderms – Pearse and Walker 1986; Pearse et al. 1986a,b and fish – Nagahama 1994; Miura et al. 1991) or negatively (other vertebrates) to shortening fall photoperiod.

Changing temperature may be the environmental cue that initiates gametogenesis in some species of sea urchins (*P. depressus*, *H. pulcherrimus*, and *A. crassispina*) (Cochran and Engelman 1975; Yamamoto et al. 1988; Sakairi et al. 1989; Ito et al. 1989). In *S. franciscanus*, constant temperature did not affect gametogenesis (McBride et al. 1997). Under ambient temperature, *P. depressus*, *H. pulcherrimus*, and *A. crassispina* reached the gametogenesis and NP utilization stage in October, December, and July, respectively (Yamamoto et al.1988; Sakairi et al.1989). In constant temperature, the gonads of *P. depressus* and *A. crassispina* reached this stage 2 months earlier than at ambient temperature. The gonads of *H. pulcherrimus* remained immature for the entire year, but matured 3 months after a temperature decrease. Various photic conditions (continuous light, continuous darkness, out-of-phase photoperiod) had no effect on the maturation of gonads in these three species at any temperature. Ito et al. (1989) also suggested that a decrease in water temperature initiated gametogenesis in *H. pulcherrimus*. Masaki and Kawahara (1995) reported that a decrease in temperature advanced gonadal maturation of *P. depressus* by 1 month. Noguchi et al. (1995) reported that *P. depressus* matured 2 months earlier than expected when water temperature was kept constant from February to June and then decreased in July. For both *S. droebachiensis* and *Psammechinus miliaris*, the lowering of seawater temperatures is believed to be an important cue for the completion of vitellogenesis (Walker and Lesser 1998; Kelly 2001).

Regardless of the environmental cue that initiates gametogenesis, current evidence suggests that the proximate cue for gonial cell mitosis is the availability of nutrients (Walker and Lesser 1998). This view is supported by the occurrence of spermatozoa in the testicular lumen in the summer, prior to the initiation of the major burst of spermatogenesis in the fall (Walker et al. 1998, 2005), and by the existence of spermatozoa in the earliest embryonic gonads (Houk and Hinegardner 1980). If sufficient nutrients are available and released from NPs, limited numbers of spermatozoa may be made and stored in the testicular lumen. In females, this may account for out-of-season initiation of oogonia mitosis and the growth of immature residual oocytes within NP incubation chambers before gametogenesis has begun for the year (Holland and Giese 1965; Gonor 1973; McClary and Barker 1998).

4. CONCLUSIONS

Gametogenesis and intra-gonadal nutrient storage and utilization are linked processes in sea urchins. Reproduction can be divided into distinct stages based on the prevailing interactions between gametogenic cells and NPs. A more thorough understanding of sea urchin gametogenesis should permit manipulation of the process at several stages in order to accomplish particular goals. For example, it would be advantageous for aquaculturists to know: (a) how to suppress gametogenesis in order to produce high-quality gonads for consumption and (b) how to promote gametogenesis for the increased production of seed stock. Already, manipulation of gametogenesis can be accomplished at various stages, but much more can be accomplished.

Inter-gametogenesis and NP phagocytosis stage: Walker and Lesser (1998) found that prepared food delivered in abundance following spawning and before the out-of-season photoperiod cue resulted in substantial growth of the NP in both males and females. These large NPs were responsible for an increase in size of the gonads to a gonad index of >20%. The NP in gonads will grow if individuals are well fed during the pre-gametogenesis and NP renewal stage. Consequently, gonads can be "bulked" by feeding urchins either natural or prepared diets. Feeding results in increased growth of NP and therefore larger gonads (de Jong-Westman et al. 1995; Lawrence et al. 1997; Walker and Lesser 1998). Ovaries and testes containing fewer gametes relative to somatic cells (= NPs) are preferred in most, but not all cultures. It may be possible to produce such gonads by curtailing gametogenesis before it begins. If the recognition system used by NPs to identify gonial cells could be disrupted, the gonial cells could be completely removed from the gonad and sterile gonads containing only NPs would result. The generation of triploid sea urchins might yield sterile adults as is the case in fish aquaculture. To produce triploid sea urchins, retention of the second polar body, a popular method for generating triploid fish and shellfish, cannot be applied because sea urchin eggs complete both meiotic divisions prior to spawning. Triploid production might be achieved by fusing two fully mature ova followed by successful fertilization through normal sperm. Production of triploid sea urchins might also be accomplished by crossing tetraploids with normal individuals (Unuma 2002). Alternatively, in land-based aquaculture facilities, it is possible to simply maintain sea urchins under an invariant, long-day photoperiod like they normally experience during the inter-gametogenesis and NP phagocytosis stage (Walker and Lesser 1998). In species that respond to photoperiod by initiating the gametogenesis and NP utilization stage, NP should be the predominant constituent of the germinal epithelium.

Pre-gametogenesis and NP renewal stage: Just after spawning and before the pre-gametogenesis and NP renewal stage normally occurs, it is possible to stimulate NP growth by feeding sea urchins a natural or prepared diet. Nutritive phagocyte growth can be initiated during the normal time of the year or much earlier than normal. In the former case, it is possible to produce unusually large gonads in the normal season. In the latter case, it is possible to produce a second, annual crop of individuals that contain gonads with large NPs out-of-season (Walker and Lesser 1998). Adjusting the carotenoid content of prepared diets can be used to modify the color of gonads (Robinson et al. 2002, 2004). Steroids or their agonists (e.g. the phytoestrogen, genistein) have the potential to promote maximum growth of the NP during this stage, but such engineered commercial produce could face public challenges.

Gametogenesis and NP utilization stage: The difference in flavor of ovaries and testes reported during the gametogenesis and NP utilization stage might result from differential metabolism of nutrients mobilized from NPs by females and males. It may be possible to prevent differences in taste between ovaries and testes by suppressing gametogenesis in both sexes using manipulated photoperiod or ploidy. Alternatively, feeding sea urchins *ad libitum* will result in continued accumulation of nutrients in the NP, even though the NPs are mobilizing nutrients for gametogenesis. Knowledge of the nutrients necessary for production of MYP or other NP components would allow preparation of specific diets for this purpose. Existing commercial diet formulations vary widely and consequently have yielded variation in the resulting product. Most diets have been developed without much attention to the actual nutrient requirements of sea urchins at a particular stage of gametogenesis, and instead emphasize maximizing size and color of the commercial product. A thorough knowledge of the biochemistry of NP contents and gamete requirements during this stage should lead to the production of specialized sea urchin diets that support either NP or gamete growth.

End of gametogenesis, NP exhaustion, and spawning: Gametogenesis usually has a negative influence on the quality of sea urchin gonads. This is particularly true near the conclusion of gametogenesis, when gonads are very fragile and contain mainly gametes. Rough handling of sea urchins during harvesting results in disintegration of the gonads as fully mature ova or spermatozoa are released. Regulation of water temperature is used to control the spawning season of *P. depressus* and *H. pulcherrimus* in Japan for the production of seed for enhancement of natural stocks and for aquaculture of sea urchins. Other means for controlling spawning while urchins are harvested and their gonads prepared for sale would be highly useful.

ACKNOWLEDGMENTS

Preparation of this chapter was supported by a National Research Initiative Competitive Grant no. 2002-35206-11631 from the USDA Cooperative State Research, Education, and Extension Service and by a grant from Northeast Regional Aquaculture Center Grant (04–15) to CWW.

REFERENCES

Anderson E (1968) Oocyte differentiation in the sea urchin, *Arbacia punctulata*, with particular reference to the origin of cortical granules and their participation in the cortical reaction. J Cell Biol 37: 514–539

Bal AK (1970) Ultrastructural changes in the accessory-cells and the oocyte surface of the sea urchin *Strongylocentrotus droebachiensis* during vitellogenesis. Z Zellforsch 111: 1–14

Barry TM (2005) The first complete opsin sequence is identified from light sensitive sea urchin tube feet. M Sc thesis, University of New Hampshire, Durham

Bay-Schmith E, Pearse JS (1987) Effect of fixed daylengths on the photoperiodic regulation of gametogenesis in the sea urchin *Strongylocentrotus purpuratus*. Int J Invert Repro Develop 11: 287–294

Beijnink FB, Broertjes FB, Voogt PA (1984) Immunochemical demonstration of vitellogenic substances in the haemal system of the sea star, *Asterias rubens*. Mar Biol Let 5: 303–313.

Berg LK, Wessel GM (2003) Selective transport and packaging of the major yolk protein in the sea urchin. Dev Biol 261: 353–370

Björnsson BT, Taranger GL, Hansen T, Stefansson SO, Haux C (1994) The interrelation between photoperiod, growth hormone, and sexual maturation of adult Atlantic salmon (*Salmo salar*). Gen Comp Endocrinol 93: 70–81

Böttger SA, Walker CW, Unuma T (2004) Care maintenance of adult echinoderms. In: Ettensohn CA, Wessel GM, Wray GA (eds) Development of Sea Urchins, Ascidians, and Other Invertebrate Deuterostomes: Experimental Approaches. Methods in Cell Biology. Academic Press, Boston, Vol. 74, pp 17–38

Brooks JM, Wessel GM (2002) The major yolk protein in sea urchins is a transferrin-like, iron binding protein. Dev Biol 245: 1–12

Brooks JM, Wessel GM (2003) Selective transport and packaging of the major yolk protein in the sea urchin. Dev Biol 261: 353–370

Brooks JM, Wessel GM (2004) The major yolk protein of sea urchins is endocytosed by a dynamin-dependent mechanism. Biol Reprod 71: 705–713

Byrne M (1990) Annual reproductive cycles of the commercial sea urchin *Paracentrotus lividus* from an exposed intertidal and a sheltered habitat on the west coast of Ireland. Mar Biol 104: 275–289

Byrne M, Andrew NL, Worthington DG, Brett PA (1998) Reproduction in the diadematoid sea urchin *Centrostephanus rodgersii* in contrasting habitats along the coast of New South Wales, Australia. Mar Biol 132: 305–318

Cameron RA, Minor JE, Nishioka D, Britten RJ, Davidson EH (1990) Locale and level of bindin mRNA in maturing testis of the sea urchin, *Strongylocentrotus purpuratus*. Dev Biol 142: 44–49

Campbell JL (1966) The haemal and digestive systems of the purple sea urchin, *Strongylocentrotus purpuratus* (Stimpson). PhD dissertation, University of California, Los Angeles

Caullery M (1925) Sur la structure et fonctionnement des gonades chez échinoides. Trav Station Zool Wimereux 9: 21–35

Cervello M, Matranga V (1989) Evidence of a precursor-product relationship between vitellogenin and toposome, a glycoprotein complex mediating cell adhesion. Cell Differ Dev 26: 67–76

Cervello M, Arizza V, Lattuca G, Parrinello N, Matranga V (1994) Detection of vitellogenin in a subpopulation of sea urchin coelomocytes. Eur J Cell Biol 64L: 314–319

Chandola A, Singh R, Thapliyal JP (1976) Evidence for a circadian oscillation in the gonadal response of the tropical weaver bird (*Ploceus phillippinus*) to programmed photoperiod. Chronobiologia 3: 219–227

Chatlynne LG (1969) A histochemical study of oogenesis in the sea urchin, *Strongylocentrotus droebachiensis*. Biol Bull 136: 167–184

Chia FS, Bickell LR (1983) Echinodermata. In: Adiyodi KG, Adiyodi RD (eds) Reproductive Biology of Invertebrates. Vol. II. Spermatogenesis and Sperm Function. Wiley & Sons, Ltd, London, pp 545–619

Cochran RC, Engelman F (1975) Environmental regulation of the annual reproductive season *Strongylocentrotus purpuratus* (Stimpson). Biol Bull 148: 393–401

Cochran RC, Engelman F (1976) Characterization of spawn-inducing factors in the sea urchin, *Strongylocentrotus purpuratus*. Gen Comp Endocrinol 30: 189–197

Cruz-Landin C, Beig D 1976. Spermiogenesis in the sea urchins *Arbacia lixula* and *Echinometra lucunter* (Echinodermata). Cytologia 41: 331–344

Czerny T, Busslinger M (1995) DNA-binding and transactivation properties of Pax-6: three amino acids in the paired domain are responsible for the different sequence recognition of Pax-6 and BSAP (pax-5). Mol Cell Biol 15: 2858–2871

Fox HM (1924) The spawning of echinoids. Proc Camb Phil Soc Biol Sci 1: 71–74

Fuji A (1960a) Studies on the biology of the sea urchin I. Superficial and histological changes in gametogenic process of two sea urchins, *Strongylocentrotus nudus* and *S. intermedius*. Bull Fac Fish Hokkaido Univ 11: 1–14

Fuji A (1960b) Studies on the biology of the sea urchin III. Reproductive cycle of two sea urchins, *Strongylocentrotus nudus* and *S. intermedius*. Bull Fac Fish Hokkaido Univ 11: 49–57

Gehring WJ, Ikeo K (1999) Pax 6 Mastering eye morphogenesis and eye evolution. TIG 15: 371–377

Giga Y, Ikai A (1985) Purification of the most abundant protein in the coelomic of a sea urchin cross reacts with 23S gylcoprotein in the sea urchin eggs. J Biochem 98: 19–26

Gonor JJ (1973) Reproductive cycles in Oregon populations of the echinoid, *Strongylocentrotus purpuratus* (Stimpson). I. Annual gonad growth and ovarian gametogenic cycle. J Exp Mar Biol Ecol 12: 45–64

Gras H, Weber W (1983) Spectral light sensitivity of isolated chromatophores of the sea urchin, *Centrostephanus longispinus*. Comp Biochem Physiol 76A: 279–281

Griffiths M, Perrott P (1976) Seasonal changes in the carotenoids of the sea urchin *Strongylocentrotus droebachiensis*. Comp Biochem Physiol 55B: 435–441

Haley SA, Wessel GM (1999) The cortical granule serine protease CGSP1 of the sea urchin, *Strongylocentrotus purpuratus*, is autocatalytic and contains a low-density lipoprotein receptor-like protein. Dev Biol 211: 1–10

Hamann O (1887) Anatomie and Histologie der Echiniden and Spantangiden. Jen Zeitschr Naturwiss 21: 1–176

Harrington FE, Easton DP (1982) A putative precursor to the major yolk protein of the sea urchin. Dev Biol 94: 505–508

Harrington FE, Ozaki H (1986) The effect of estrogen on protein synthesis in echinoid coelomocytes. Comp Biochem Physiol 84B: 417–421

Hay N, Sonenberg N (2004) Upstream and downstream of mTOR. Genes Develop 18: 1926–1945

Heideman PD, Deoraj P, Bronso FH (1992) Seasonal reproduction of a tropical bat, *Anoura geoffroyi*, in relation photoperiod. J Reprod Fertil 96: 765–773

Himmelman JH (1975) Phytoplankton as a stimulus for spawning in three marine invertebrates. J Exp Mar Biol Ecol 20: 199–214

Hinchcliffe EH, Thompson EA, Miller FJ, Yang J, Sluder G. (1999) Nucleo-cytoplasmic interactions that control nuclear envelope breakdown and entry into mitosis in the sea urchin zygote. J Cell Sci 112: 1139–1148

Holland ND, Giese AC (1965) An autoradiographic investigation of the gonads of the purple sea urchin (*Strongylocentrotus purpuratus*). Biol Bull 128: 241–258

Holland ND, Holland LZ (1969) Annual cycles in germinal and non-germinal cell populations in the gonads of the sea urchin *Psammechinus microtuberculatus*. Pubbl Staz Zool Napoli 37: 394–404

Houk MS, Hinegardner RT (1980) The formation and the early differentiation of sea urchin gonads. Biol Bull 159: 280–294

Iliffe TM, Pearse JS (1982) Annual and lunar reproductive rhythms of the sea urchin *Diadema antillarum* (Philippi) in Bermuda. Internat J Invert Reprod 5: 139–148

Ito S, Shibayama M, Kobayakawa A, Tani Y (1989) Promotion of maturation and spawning of sea urchin *Hemicentrotus pulcherrimus* by regulating water temperature. Nippon Suisan Gakkaishi 55: 757–763 (in Japanese with English abstract)

Johnsen S (1997) Identificaiton and localization of a possible rhodopsin in the echinoderms *Asterias forbesi* (Asteroidea) and *Ophioderma brevispinum* (Ophiuroidea). Biol Bull 193: 97–105

Jondeung A, Czihak G (1982) Histochemical studies of jelly coat of sea-urchin eggs during oogenesis. Histochemistry 76(1): 123–146

de Jong-Westman M, March BE, Carefoot TH (1995) The effect of different nutrient formulations in artificial diets on gonad growth in the sea urchin *Strongylocentrotus droebachiensis*. Can J Zool 73:1495–1502

Kanatani H (1974) Presence of 1-methyladenine in sea urchin gonads and its relation to oocyte maturation. Dev Growth Differ 16: 159–170

Kari BE, Rottmann WL (1985) Analysis of changes in a yolk glycoprotein complex in developing sea urchin embryo. Develop Biol 108: 18–25

Kêckês S (1966) Lunar periodicity in sea urchins. Z Naturforsch 21: 1100–1101

Kelly, MS (2001) Environmental parameters controlling gametogenesis in the echinoid *Psammechinus miliaris*. J Exp Mar Biol Ecol 266: 67–80

Kishimoto T (1998) Cell cycle arrest and release in starfish oocytes and eggs. Seminars Cell Develop 9: 549–557

Kishimoto T (1999) Activation of MPF at meiosis reinitiation in starfish oocytes. Develop Biol 214: 1–8

Klionsky DJ (2005) The molecular machinery of autophagy: unanswered questions. J Cell Sci 118: 7–18

Kobayashi N, Nakamura K (1967) Spawning periodicity of sea urchins at Seto. II. *Diadema setosum*. Publ Seto Mar Biol Lab 15: 173–184

Kobayashi N, Konaka K (1971) Studies on periodicity of oogenesis of sea urchin I. Relation between the oogenesis and the appearance and disappearance of nutritive phagocytes detected by some histological means. Sci Eng Rev Doshisha Univ 12: 131–149

Kobayashi N, Sasao S, Manabe Y (1984) Studies on periodicity in oogenesis of sea urchin. II. Effects of breeding condition upon the oogenesis by some histological methods. Sci Eng Rev Doshisha Univ 25: 199–218

Kubota H, Kanatani H (1975) Production of 1-methyladenine induced by concanavalin A in starfish follicle cells. Dev Growth Diff 17: 177–185

Kubota J, Nakao K, Shirai H, Kanatani H (1977) 1-methyl-adenine producing cell in the starfish testis. Exp Cell Res 106: 63–70

Laegdsgaard P, Byrne M, Anderson DT (1991) Reproduction of sympatric populations of *Heliocidaris erythrogramma* and *H. tuberculata* (Echinoidea) in New South Wales. Mar Biol 110: 359–374

Lawrence JM, Olave S, Otaiza R, Lawrence AL, Bustos E (1997) Enhancement of gonad production in the sea urchin *Loxechinus albus* in Chile fed extruded feeds. J World Aqua Soc 28: 91–96

Lessios HA (1991) Presence and absence of monthly reproductive rhythms among eight Caribbean echinoids off the coast of Panama. J Exp Mar Biol Ecol 153: 27–47

Longo FJ, Anderson EE (1969) Sperm differentiation in the sea urchins *Arbacia punctulata* and *Strongylocentrotus purpuratus*. J Ultrastruct Res 27: 486–509

Mallya SK, Partin JS, Valdizan MC, Lennarz WJ (1992) Proteolysis of the major yolk glycoproteins is regulated by acidification of the yolk platelets in sea urchin embryos. J Cell Biol 117: 1211–1221

Masaki K, Kawahara I (1995) Promotion of gonadal maturation by regulating water temperature in the sea urchin *Pseudocentrotus depressus* – I. Bull Saga Prefect Sea Farm Cen 4: 93–100 (in Japanese)

Masuda R, Dan JC (1977) Studies on the annual reproductive cycle of the sea urchin and the acid phosphatase activity of relict ova. Biol Bull 153: 577–590

Matsuno T, Tsushima M (2001) Carotenoids. In: Lawrence JM (ed.) Edible Sea Urchins: Biology and Ecology. Elsevier Press, Amsterdam, pp 115–138

Mcbride SC, Pinnix WD, Lawrence JM, Lawrence A, Mulligan TM (1997) The effect of temperature on the production of gonads by the sea urchin *Strongylocentrotus franciscanus* fed natural and prepared diets. J World Aqua Soc 28: 357–365

McClary D, Barker M (1998) Reproductive isolation? Interannual variability in the timing of reproduction in sympatric sea urchins, genus *Pseudechinus*. Invert Biol 117: 75–93

Meidel SK, Scheibling RE (1998) Annual reproductive cycle of the green sea urchin, *Strongylocentrotus droebachinesis*, in differing habitats in Nova Scotia, Canada. Mar Biol 131: 461–478

Millott N (1968) The dermal light sense. Symp Zool Soc Lond 23: 1–36

Miura T, Yamaguchi K, Takahashi H, Nagahama Y (1991) Hormonal induction of all stages of spermatogenesis *in vitro* in the male Japanese eel (*Anguilla japonica*). Proc Natl Acad Sci 88: 5774–5778

Nagahama Y (1994) Endocrine regulation of gametogenesis in fish. Int J Dev Biol 38: 217–229

Nicotra A, Arizzi M, Gallo PV (1984) Protein synthetic activities during spermiogenesis in the sea urchin: An high resolution autoradiographic study of ^{3}H-leucine incorporation. Develop Growth Differ 26: 273–280

Nicotra A, Serafino A (1988) Ultrastructural observations on the interstitial cells of the testis of *Paracentrotus lividus*. Int J Invert Repro Develop 13: 239–250

Noguchi H, Kawahara I, Goto M, Masaki K (1995) Promotion of gonadal maturation by regulating water temperature in the sea urchin *Pseudocentrotus depressus* – II. Bull Saga Pref Sea Farm Cen 4: 101–107 (in Japanese)

Noll H, Matranga V, Cervello M, Humphreys T, Kuwasaki B, Adelson D (1985) Characterization of toposomes from sea urchin blastula cells: a cell organelle mediating cell adhesion and expressing positional information. Proc Natl Acad Sci 82: 8062–8066

Nomura K, Hoshino K, Suzuki N (1999) The primary and higher order structures of sea urchin ovoperoxidase as determined by cDNA cloning and predicted by homology modeling. Arch Biochem Biophys 367: 173–184

Okada Y (1979) The central role of the genital duct in the development and regeneration of the genital organs in the sea urchin. Dev Growth Diff 21: 567–576

Okada Y, Iwata KS, Yanagihara M (1984) Synchronized rhythmic contractions among five gonadal lobes in the shedding sea urchin: coordinative function of the aboral nerve ring. Biol Bull 166: 228–236

Okada Y, Iwata KS (1985) A substance inhibiting rhythmic contraction of the gonad in the shedding sea urchin. Zool Sci 2: 805–808

Ozaki H, Moriya O, Harrington FE (1986) A glycoprotein in the accessory cell of the echinoid ovary and its role in vitellogenesis. Roux's Arch Dev Biol 195: 74–79

Palmer L (1937) The shedding reaction in *Arbacia punctulata*. Physiol Zöol 10: 352–367

Paniagua R, Fraile B, Sáez FJ (1990) Effects of photoperiod and temperature on testicular function in amphibians. Histol Histopathol 5: 365–378

Pearse JS (1969a) Reproductive periodicities of Indo-Pacific invertebrates of the Gulf of Suez. I. The echinoids *Prionocidaris baculosa* (Lamark) and *Lovenia elongata* (Gray). Bull Mar Sci 19: 323–350

Pearse JS (1969b) Reproductive periodicities of Indo-Pacific invertebrates of the Gulf of Suez. II. The echinoid *Echinometra mathaei* (DeBlainville). Bull Mar Sci 19: 581–613

Pearse JS (1970) Reproductive periodicities of Indo-Pacific invertebrates of the Gulf of Suez. III. The echinoid *Diadema setosum* (Leske). Bull Mar Sci 20: 679–720

Pearse JS (1975) Lunar reproductive rhythms in sea urchins. A review. J Interdiscipl Cycl Res 6: 47–52

Pearse JS, Eernisse DJ, Pearse VB, Beauchamp KA (1986a) Photoperiodic regulation of gametogenesis in sea stars, with evidence for an annual calendar independent of fixed daylength. Amer Zool 26: 417–431

Pearse JS, Pearse, VB, Davis KK (1986b) Photoperiodic regulation of gametogenesis and growth in the sea urchin, *Strongylocentrotus purpuratus*. J Exp Zool 237: 107–118

Pearse JS, Walker CW (1986) Photoperiodic control of gametogenesis in the North Atlantic sea star, *Asterias vulgaris*. Int J Invert Repro Dev 9: 71–77

Pearse JS, Cameron RA (1991) Echinodermata: echinoidea. In: Giese AC, Pearse JS, Pearse VB (eds) Reproduction of Marine Invertebrates. Vol. VI. Echinoderms and Lophophorates. The Boxwood Press, Pacific Grove, pp 514–662

Plank LR (2000) The effect of dietary carotenoids on gonad growth, development and color in *Lytechinus variegatus* (Lamarck) (Echinodermata: Echinoidea). M Sc thesis, University of South Florida, Tampa

Poccia DL, Lieber T, Childs G (1989) Histone gene expression during sea urchin gametogenesis: an *in situ* hybridization study. Mol Repro Dev 1: 219–229

Puigdomenech P, Romero MC, Allan J, Sautiere P, Giancotti V, Crane-Roginson C (1987) The chromatin of sea urchin sperm. Biochim Biophys Acta 908: 70–80

Reunov AA, Kalachev AV, Yurchenko OV, Au DWT (2004a) Selective resorption in nutritive phagocytes of the sea urchin *Anthocidaris crassispina*. Zygote 12: 71–73

Reunov AA, Yurchenko AV, Kalachev AV, Au DWT (2004b) An ultrastructural study of phagocytosis and shrinkage in nutritive phagocytes of the sea urchin *Anthodicaris crassispina.* Cell Tissue Res 318: 419–428

Robinson SMC, Castell JD, Kennedy EJ (2002) Developing suitable colour in green sea urchins (*Strongylocentrotus droebachiensis*). Aquacultre 206, 289–303

Robinson SMC, Lawrence JM, Burridge L, Haya K, Castell J, Lawrence AL, (2004) Effectiveness of different pigment sources in colouring the gonads of the green sea urchin *Strongylocentrotus droebachiensis.*

In: Lawrence JM, Guzman O. (Eds.), Sea urchins – Fisheries and Ecology, DEStech Publication Inc., Lancaster PA, pp 215–221

Sakairi K, Yamamoto M, Ohtsu K, Yoshida M (1989) Environmental control of gonadal maturation in laboratory-reared sea urchins, *Anthocidaris crassispina* and *Hemicentrotus pulcherrimus*. Zool Sci 6: 721–730

Schuel H, Kelly JW, Berger ER, Wilson WL (1974) Sulfated acid mucopolysaccharides in the cortical granules of eggs. Effects of quarternary ammonium salts on fertilization. Exp Cell Res 88: 24–30

Scott LB, Lennarz WJ (1989) Structure of a major yolk glycoprotein and its processing pathway by limited proteolysis are conserved in echinoids. Dev Biol 132: 91–102

Scott LB, Leahy PS, Decker GL, Lennarz WJ (1990) Loss of yolk platelets and yolk glycoproteins during larval development of the sea urchin embryo. Dev Biol 137: 368–377

Shapiro BM (1991) The control of oxidant stress at fertilization. Science 252: 533–536

Shirai H, Walker C (1988) Chemical control of asexual and sexual reproduction in echinoderms. In: Downer RH and Laufer H (eds) Endocrinology of Selected Invertebrate Types. Alan R. Liss, Inc., New York, pp 453–476

Shyu A, Raff RA, Blumenthal (1986) Expression of the vitellogenin gene in female and male sea urchins. Proc Natl Acad Sci 83: 3865–3869

Shyu A, Blumenthal T, Raff RA (1987) A single gene encoding vitellogenin in the sea urchin *Strongylocentrotus purpuratus*: sequence of the 5' end. Nucleic Acids Res 15: 10405–10417

Simpson MV, Poccia D (1987) Sea urchin testicular cells evaluated by fluorescence microscopy of unfixed tissue. Gamete Res 17: 131–144

Sommers CE, Battaglia DE, Shipiro BM (1989) Localization and developmental fate of ovoperoxidase and proteoliaisin, two proteins involved in fertilization envelope assembly. Dev Biol 131: 226–235

Starr M, Himmelman JH, Therriault J-C (1990) Direct coupling of marine invertebrate spawning with phytoplankton blooms. Science 247: 1070–1074

Starr M, Himmelman JH, Therriault J-C (1992) Isolation and properties of a substance from the diatom *Phaeodactylum tricornutum* which induces spawning in the sea urchin *Strongylocentrotus droebachiensis*. Mar Ecol Prog Ser 79: 275–287

Strenger A (1973) *Sphaerechinus granularis* Violetter Seeigel. Grosses Zool Prak Gustav 18: 1–68

Tähkä KM, Zhuang Y, Tähkä S, Touhimaa P (1997) Photoperiod-induced changes in androgen receptor expression in testes and accessory sex glands of the bank vole, *Clethrionomys glareolus*. Biol Reprod 56: 898–908

Takahashi N, Sato N, Ohtomo N, Kondo A, Takahashi M, Kikuchi K (1990) Analysis of the contraction inducing factor for gonadal smooth muscle contraction in sea urchin. Zool Sci 7: 861–869

Takahashi N, Sato N, Hayakawa Y, Takahashi M, Miyake H, Kikuchi K (1991) Active component of the contraction factor on smooth muscle contraction of the gonad wall in sea urchin. Zool Sci 8: 207–210

Tsukahara J (1970) Formation and behavior of pinosomes in the sea urchin oocyte during oogenesis. Dev Growth Differ 12: 53–64

Unuma T (2002) Gonadal growth and its relationship to aquaculture in sea urchins. In: Yokota Y, Matranga V, Smolenicka Z (eds) The Sea Urchin: From Basic Biology to Aquaculture. Swets & Zeitlinger, Lisse, pp 115–127

Unuma T, Suzuki T, Kurokawa T, Yamamoto T, Akiyama T (1998) A protein identical to the yolk protein is stored in the testis in male red sea urchin, *Pseudocentrotus depressus*. Biol Bull 194: 92–97

Unuma T, Yamamoto T, Akiyama T (1999) Effect of steroids on gonadal growth and gametogenesis in the juvenile red sea urchin *Pseudocentrotus depressus*. Biol Bull 196: 199–204

Unuma T, Okamoto H, Konishi K, Ohta H, Mori K (2001) Cloning of cDNA encoding vitellogenin and its expression in red sea urchin, *Pseudocentrotus depressus*. Zool Sci 18: 559–565

Unuma T, Yamamoto T, Akiyama T, Shiraishi M, Ohta H (2003) Quantitative changes in yolk protein and other components in the ovary and testis of the sea urchin *Pseudocentrotus depressus*. J Exp Biol 206: 365–372

Unuma T, Ohta H, Yamano K, Ikeda K (2004) The transferrin-like protein in the sea urchin is a zinc binding protein. In: Heinzeller T, Nebelsick JH (eds) Echinoderms: München. Balkema, Leiden, p 613

Vafa O, Nishioka D (1995) Developmentally regulated protease expression during sea urchin embryogenesis. Mol Repro Develop 40: 36–47

van Haaster LH, van Eedenburg FJ, de Rooij DG (1993) Effect of prenatal and postnatal photoperiod on spermatogenic development in the Djungarian hamster (*Phodopus sungorus sungorus*). J Reprod Fert 97: 223–232

Walker CW (1979) Ultrastructure of the somatic portion of the gonads in asteroids, with emphasis on flagellated-collar cells and nutrient transport. J Morph 166: 81–107

Walker CW (1980) Spermatogenic columns, somatic cells and the microenvironment of germinal cells in the testes of asteroids. J Morph 166: 81–107

Walker CW (1982) Nutrition of gametes. In: Jangoux M, Lawrence J (eds) Nutrition of Echinoderms. Balkema, Rotterdam, pp 449–468

Walker CW, Lesser MP (1998) Manipulation of food and photoperiod promotes out-of-season gametogenesis in the green sea urchin, *Strongylocentrotus droebachiensis*: implications for aquaculture. Mar Biol 132: 663–676

Walker CW, McGinn NA, Harrington LM, Lesser MP (1998) New perspectives on sea urchin gametogenesis and their relevance to aquaculture. J Shellfish Res 17: 1507–1514

Walker CW, Harrington LM, Lesser MP, Fagerberg WR (2005) Nutritive phagocyte incubation chambers provide a structural and nutritive microenvironment for germ cells of *Strongylocentrotus droebachiensis*, the green sea urchin. Biol Bull 209: 31–48

Ward RD, Nishioka D (1993) Seasonal changes in testicular structure and localization of sperm surface glycoprotein during spermatogenesis in sea urchins. J Histochem Cytochem 41: 423–431

Wasson KM, Gower BA, Watts SA (2000) Responses of the ovaries and testes of *Lytechinus variegatus* Lamarck (Echinodermata: Echinoidea) to dietary administration of progesterone, testosterone and estradiol. Mar Biol 137: 245–255

Wasson KM, Watts SA (2000) Progesterone metabolism in the ovaries and testes of the echinoid *Lytechinus variegatus* (Echinodermata). Comp Biochem Physiology C 127: 263–272

Wessel GM, Berg L, Adelson DL, Cannon G, McClay DR (1998) A molecular analysis on hyalin – A substrate for cell adhesion in the hyaline layer of the sea urchin embryo. Develop Biol 193: 115–126

Wessel GM, Zaydfudim V, Hsu YJ, Laidlaw M, Brooks JM (2000) Direct molecular interaction of a conserved yolk granule protein in sea urchins. Dev Growth Differ 42: 507–517

Wessel GM, Conner SD, Berg L (2002) Cortical granule translocation is microfilament mediated and linked to meiotic maturation in the sea urchin oocyte. Development 129: 4315–4325

Yamamoto M, Ishine M, Yoshida M (1988) Gonadal maturation independent of photic conditions in laboratory-reared sea urchins, *Pseudocentrotus depressus* and *Hemicentrotus pulcherrimus*. Zool Sci 5: 979–988

Yokota Y, Kato KH (1988) Degradation of yolk proteins in sea urchin eggs and embryos. Cell Differ 23: 191–200

Yokota Y, Sappington TW (2002) Vitellogen and vitellogenin in echinoderms. In: Raikhel AS, Sappington TW (eds) Reproductive Biology of Invertebrates, Progress in Vitellogenesis. Vol. XII, Part A. Science Publishers, Enfield, pp 201–221

Yokota Y, Unuma T, Moriyama A, Yamano K (2003) Cleavage site of a major yolk protein (MYP) determined by cDNA isolation and amino acid sequencing in sea urchin, *Hemicentrotus pulcherrimus*. Comp Biochem Physiol 135B: 71–81

Yokota Y, Unuma T, Moriyama A, (2004) Echinoferrin: a newly proposed name for a precursor to yolk protein in sea urchin. In: Heinzeller T, Nebelsick JH (eds) Echinoderms: München. Balkema, Leiden, pp 79–81

Yoshida M (1952) Some observations on the maturation of the sea urchin *Diadema setosum*. Annot Zool Japon 25: 265–271

Edible Sea Urchins: Biology and Ecology
Editor: John Miller Lawrence

Chapter 3

Biochemical and Energy Requirements of Gonad Development

Adam G Marsh[a] and Stephen A Watts[b]

[a]*College of Marine Studies, University of Delaware, Lewes, DE (USA).*
[b]*Department of Biology, University of Alabama at Birmingham, Birmingham, AL (USA).*

1. INTRODUCTION

The importance of understanding energy allocation to developing gonads in cultured sea urchins can be summarized in a simple question: "what is the greatest yield of gonad tissue mass one can obtain with the lowest ration of food?" In considering the growth and development of gonadal tissue as a harvestable resource, the balance between *yield* and *ration* can best be described in terms of the chemical constituents that limit gonad growth, i.e. the biochemical components in the shortest supply (*ration*) that restrict development of the gonad tissues (*yield*). In this chapter we consider the metabolic activities in the sea urchin gonad in relation to an individual's diet to identify the tissue constituents that have the highest anabolic costs and are thus the most likely to restrict or limit tissue production. Although whole-animal energy metabolism (respiration and biochemical pathways) has been well studied in adult urchins (Ellington 1982; Lawrence and Lane 1982), we focus specifically on what little is known about the metabolic activities in sea urchin gonads.

2. CELLULAR ENERGY UTILIZATION

The major metabolic processes that consume large amounts of cellular energy are the same across wide phylogenetic distances. Assessing these general energy requirements is of great interest for understanding growth and development in different organisms, and there has been substantial consideration of how ingested food resources are partitioned among competing metabolic processes, particularly between growth and reproduction (Lawrence and Lane 1982; Thompson 1982; Wieser 1985; Stearns 1992; Lawrence and McClintock 1994; Russell 1998; Guillou et al. 2000; Vadas et al. 2000; Otero-Villanueva et al. 2004).

Generally, the efficiency with which an organism can convert food energy into its own structural energy has been assessed by distinguishing between assimilation (increase in energy content during growth – biomass) and oxidation (energy dissipated during growth – respiration). In a review of data obtained for 17 species of fish and invertebrates, a close correlation was revealed between size-specific growth rates and size-specific respiration rates with a constant ratio of 15 μmol O_2 g DW^{-1} h^{-1} (g dry weight) respired for a deposition of 1 mg DW g $DW^{-1}h^{-1}$ ($r^2 = 0.95$; Wieser 1994). Converting both quantities into energy equivalents (using standard averages of 484 kJ mol^{-1} O_2 respired and 22 J mg DW^{-1} biomass; Gnaiger 1983) reveals a metabolic energy cost of growth of 33% (i.e. for every 100 J of biomass equivalents deposited, an additional 33 J will be consumed by respiration). This translates into a growth efficiency of 75% for these invertebrates and fish, where for every 100 J of biochemical energy that is internally mobilized, 75 J will eventually be deposited into new tissue structures with the other 25 J being lost as heat (Wieser 1994). This general growth efficiency is robust and is equivalent in a variety of other organisms: 75–90% in bacteria (Beauchop and Elsden 1960; Wieser 1994), 68% in heterotrophic protozoans (Fenchel and Findlay 1983), and 70–75% in nonruminate mammals (Blaxter 1989).

The equivalence in growth efficiency across diverse taxa results from the similarity of the metabolic processes that are used for the production and utilization of cellular energy. Cellular metabolism itself comprises a wide range of biochemical activities in any cell; however, in metazoans, the total energy consumption of a cell is dominated by only a few activities (Fig. 1). In mammalian cell lines, energy consumption is dominated by two processes: protein turnover and the activity of the sodium ion pump (Schmidt et al. 1989, 1991; Schneider et al. 1990; Siems et al. 1992). The major energy-consuming process in most eukaryotic cells is protein turnover, accounting for 35% of total cellular metabolism. For all cells to maintain physiological function there is a constant need to replace proteins as they become damaged or nonfunctional.

Sodium ion pump activity (Na^+, K^+-ATPase) is the second largest consumer of cellular energy. The ion gradients established by the sodium pump are critical for maintaining a cell's osmotic balance and resting membrane potential, as well as providing the electrochemical gradients necessary for the uptake of other ions, sugars, amino acids, and neurotransmitters via Na^+ coupled co-transporters (see review by Blanco and Mercer

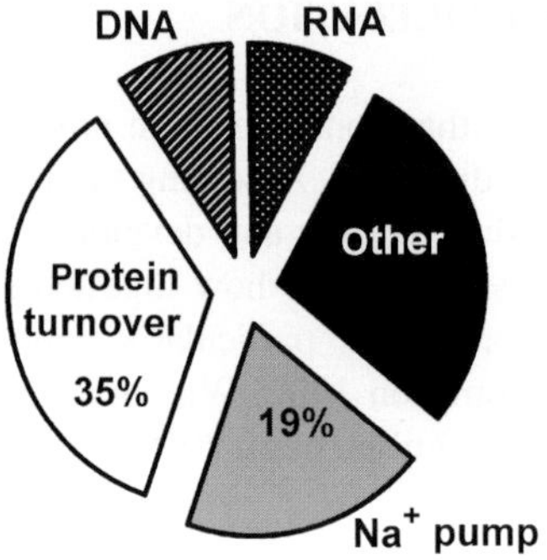

Fig. 1. Metabolic energy utilization in a mammalian cell line. Both protein turnover and sodium pump activity alone can account for 54% of total energy consumption in a cell. DNA and RNA synthesis each represent 8–9% of total metabolism. Siems et al. 1992, with permission.

1998). DNA and RNA synthesis are the other dominant energy-consuming processes and they each account for 8–9% of cellular metabolism (Siems et al. 1992).

3. ENERGY METABOLISM DURING DEVELOPMENT

Echinoid embryos have been used as a model system for biochemical studies since the beginning of this century (see early review by Needham 1931). Embryos present a relatively simple system for assessing cellular biochemistry because food consumption and reproductive growth do not confound their metabolic activities. Studies of protein synthesis in embryos have used different radiolabeled amino acids as tracers for incorporation into protein (Table 1; Berg and Mertes 1970; Fry and Gross 1970; Hogan and Gross 1971; Regier and Kafatos 1977; Goustin and Wilt 1981). During early embryogenesis, the protein content of these echinoid embryos does not change (from fertilization to gastrulation) such that all of the protein degradation activity (utilization ofthe major yolk protein (MYP) (Unuma et al. 1998; Walker et al. 2001) is balanced by an equivalent amount of protein synthetic activity. Thus, measured rates of synthesis are equivalent to the total rates of protein turnover (catabolism + anabolism).

Table 1 The metabolic cost of protein metabolism during early development in echinoids. Rates of protein synthesis have been routinely measured in echinoid embryos. These rates have been converted into respiratory energy equivalents (pmol O_2) to assess the percentage of total aerobic metabolism consumed by protein synthesis and turnover.

Amino acid	Stage	Protein synthesis (ng ind^{-1} h^{-1})	Protein elongation (pmol AA ind^{-1} h^{-1})	% of metabolism	Protein turnover (% protein hr^{-1})
Arbacia punctulata[a]					
Leucine	EC	0.72	5.07	55.5	1.9
Valine	EC	0.70	4.93	54.0	1.9
Strongylocentrotus purpuratus[b]					
Leucine	EC	0.52	3.66	71.0	1.4
Valine	EC	0.47	3.31	64.2	0.76
Lytechinus variegatus[c]					
Valine	GST	0.71	5.00	46.7	0.82
Proline	GST	0.80	5.63	52.6	0.92
Alanine	GST	1.15	8.10	75.7	1.31
Histidine	GST	0.77	5.42	50.7	0.90
Phenylalanine	GST	0.74	5.21	48.7	0.85
Threonine	GST	0.87	6.13	57.3	1.00

EC – early cleavage stage; GST – mid-gastrula stage.
[a]74 μm egg; respiration = 14 pmol O_2 h^{-1} ind^{-1} at 25 °C; protein data from Fry and Gross (1970).
[b]80 μm egg; respiration = 7.9 pmol O_2 h^{-1} ind^{-1} at 16 °C; protein data from Goustin and Wilt (1981).
[c]128 μm egg; respiration = 16.4 pmol O_2 h^{-1} ind^{-1} at 19 °C; protein data from Berg and Mertes (1971).

When these rates of protein synthesis are converted into respiratory equivalents (i.e. using the theoretical energy cost of two GTP for the formation of one peptide bond) and then compared to the total respiration rate of an embryo, an overall average of 58% of metabolism can be accounted for by protein turnover during early development (Table 1). This number is similar to the mammalian number provided in Section 2 (Fig. 1) and we would expect a developmental system to have a high rate of protein synthesis as yolk protein reserves are degraded to provide the amino acid building blocks for new proteins.

An elevation in protein synthesis activity is evident in the total protein turnover when expressed as a fraction of an embryo's total protein mass (Table 1). Here the fractional rate of protein turnover ranges from 0.76 to 1.9% h^{-1} of an embryo's protein mass. For example, embryos of *Arbacia punctulata* have almost 2% of their total protein mass degraded and resynthesized each hour. Similar rates of fractional turnover (1–2% h^{-1}) have also been reported during early development in abalone embryos (Vavra and Manahan 1999). In developmental systems the level of metabolite flux moving from protein reserves to the deposition of new cellular structures is high and can account for the large fraction of cellular energy that is consumed by protein turnover.

The maintenance of Na^+ and K^+ ion gradients consumes a constant fraction of cellular energy as these ions are continually exchanged across the membranes (Blanco and Mercer 1998). Developmental systems (i.e. rapid cell proliferation) impose some unique constraints on the requirements for ion pumping activity. The increase in cell number during early embryogenesis and the consequent increase in cellular membrane surface area necessitate the production of more sodium pumps to regulate intracellular ion flux (Mitsunaga-Nakatsubo et al. 1992; Marsh et al. in press). The *in vivo* physiological activity of the sodium pump (Na^+, Na^+-ATPase) has been characterized during early development in the sea urchins *Strongylocentrotus purpuratus* and *Lytechinus pictus* (Leong and Manahan 1997). Using $^{86}Rb^+$ as a radioactive tracer for K^+ ion transport, Leong and Manahan (1997) have described the ontogenetic changes in activity of Na^+, K^+-ATPase in living embryos. They report physiological activities of the sodium pump that would use 23% and 32% of total metabolic energy consumption at the hatching blastula stage in *S. purpuratus* and *L. pictus*, respectively.

The pattern of total metabolic energy utilization has been assessed in embryos of the Antarctic sea urchin *Sterechinus neumayeri* (at −1.5 °C). Measurements of total metabolic rates (Marsh et al. 1999; Marsh and Manahan 1999) were compiled with measurements of protein turnover (Marsh and Manahan, unpubl. data) and *in vivo* sodium pump activity (Leong and Manahan 1999) to apportion cellular energy consumption. The pattern of energy utilization in embryos of *S. neumayeri* is similar to the pattern described for the temperate urchin embryos. Protein turnover accounts for 53% of cellular metabolism, whereas sodium pump activity consumes 15%. Together, these two processes can account for 68% of the total metabolic rate in *S. neumayeri* embryos at the hatching blastula stage. Despite an environmental temperature difference of over Δ25 °C, the fraction of protein turnover in *S. neumayeri* (2.1% h^{-1}) was equivalent to the levels calculated for other urchin embryos (Table 1), indicative of the significant demand for protein turnover in developmental systems.

Overall, it appears that in systems undergoing rapid development (nutrient translocation and cell proliferation), protein turnover can account for >50% of metabolic energy utilization while sodium pump activities can range from 15% to 30%. Despite the level of complexity in gene expression events and metabolic processes involved with a pattern of active cellular development, 65–80% of a cell's total energy expenditure can be accounted for by these two processes. Although these numbers are for embryonic systems, it is likely that an equivalent pattern of metabolic activity would be active during the initial phases of gonad production when the nutritive phagocytes are growing rapidly. The translocation of nutrients to the gonads, the rapid synthesis of protein (MYP), and the proliferation of cells within those tissues may place a similar degree of physiological requirements on cellular activities. Thus, from a developmental perspective, the general scope of the metabolic pattern depicted in Fig. 1 and Table 1 is likely to be applicable to the production of nutritive phagocytes and gametes. It is in this broad context that we will focus on the energy cost of protein metabolism in relation to the energy produced by carbohydrate catabolism during early production of the nutritive phagocytes.

4. BIOCHEMICAL COMPONENTS OF CELLULAR METABOLISM

In considering the general biochemical composition of developing tissues, there are large differences in energy expenditure associated with the metabolism of proteins, carbohydrates, and lipids. Differences in metabolic costs arise from the ways these biochemical constituents are handled by cells and the specific roles that these components play in cellular metabolism. In general, proteins play a dominant role in cellular metabolism because of their dual roles as structural and functional elements. This dominance is reflected in the metabolic energy expended for protein metabolism (Fig. 1 and Table 1). Carbohydrates serve as the primary fuels used for energy production. Lipids are the major structural elements of cell membranes, but are generally very stable and do not experience the high degree of turnover that is evident in the protein and carbohydrate pools.

The energy costs associated with protein turnover have received considerable attention in ectotherms (Hawkins et al. 1989; Hawkins 1991; Houlihan 1991; McCarthy et al. 1994; Wieser 1994; Fauconneau et al. 1995; Smith and Houlihan 1995; Conceicao et al. 1997; Lyndon and Houlihan 1998). Using whole-animal correlations between oxygen consumption and rates of protein synthesis, it has been found that fish and invertebrates have a metabolic cost of protein turnover in the range of ~13 J mg protein^{-1} (Wieser 1994). The high energy cost associated with protein metabolism primarily results from the degree of recycling that occurs between the synthesis and degradation of peptides. Amino acids are important in terms of nitrogen storage and generally they are not catabolized to component carbon skeletons. In addition, generating energy from amino acid oxidation is inefficient in comparison to the energy yields of carbohydrates (Krebs 1960). Thus, a significant portion of the protein in an animal's diet is directly assimilated into its amino acid and protein pools (85%; Wieser 1994) rather than being directly catabolized.

Most cellular energy is derived from the oxidation of carbohydrates via glycolysis and the Krebs cycle. As such, carbohydrates are usually maintained in readily accessible pools for energy metabolism. The synthesis of glycogen from glucose is the most common pathway metazoans use to store carbohydrates and the anabolic cost of glycogen synthesis is low ($0.42\,J\,mg\,glycogen^{-1}$) when compared to the energy costs of protein synthesis ($13\,J\,mg\,protein^{-1}$). One of the primary advantages of using glycogen for storing energy is that its breakdown is exergonic under physiological conditions ($\Delta G = -8\,kJ\,mol^{-1}$) and thus does not require the input of additional energy to release the stored glucose monomers. In general, most of the carbohydrates assimilated from an animal's diet are not deposited into structural elements or long-term reserves, but are rather maintained in a form in which they are readily available for catabolism to generate cellular energy.

The energy requirements of lipid metabolism are difficult to quantify. Metazoans are particularly efficient at the assimilation and translocation of lipid materials. The conversion efficiency of lipids from dietary sources to tissue deposition can range as high as 99% (Millward and Garlick 1976; Wieser 1994). This is the primary result of the high efficiency at which intact free fatty acids and sterols can be translocated across membranes. In many cases, lipids can be effectively used as biomarkers of dietary sources because of their direct incorporation into an organism's tissues (Harvey et al. 1987), and in particular, ovaries and developing oocytes of invertebrates exhibit lipid profiles that directly reflect their dietary sources (Marsh et al. 1990a,b). Although organisms do modify fatty acid and sterol structures absorbed directly from their diets, the fraction of *de novo* lipid synthesis is generally lower than for carbohydrates and proteins. Consequently, the fraction of cellular energy utilized for lipid metabolism is likely to be low relative to the overall costs of protein, and even carbohydrate, metabolism.

5. GONAD GROWTH

The seasonal gametogenic cycle of echinoids has been a well-studied phenomenon since the 1930s. (Moore 1934; Boolootian 1966; Jangoux and Lawrence 1982; Pearse and Cameron 1991). The convenience of using a gonad index (%GI – the ratio of gonad mass to total body mass) to assess seasonal changes in the gametogenic cycle has produced numerous studies in the literature reporting this value to document relative changes in the size of gonads during gametogenesis. Unfortunately, a %GI value has little further utility because of its dimensionless nature. The biomass information of gonad and body sizes cannot be retrieved from the %GI ratio.

In an attempt to extract as much gonad growth data from previous studies of gametogenesis in urchins, we have calculated a Δ%GI growth index (Fig. 2) from a variety of urchins under different experimental conditions (food, light, and temperature regimes): *S. droebachiensis* (Minor and Scheibling 1997; Walker and Lesser 1998; Havardsson and Imsland 1999; Meidel and Scheibling 1999; Vadas et al. 2000); *Parechinus angulosus* (Greenwood 1980); *Echinometra mathaei* and *Diadema savignyi* (Drummond 1995); *Evechinus chloroticus* (Barker et al. 1998); *S. franciscanus* (McBride et al. 1997); *Paracentrotus lividus* (Spirlet et al. 1998, 2000). From these studies, the %GI data for the period of initial, rapid gonad growth of nutritive phagocytes (either in field populations or during the course of an experiment) were plotted as a function of time, and the slope of a linear regression was used to estimate the daily rate of size-specific growth in gonads (r^2 ranged from 0.846 to 0.999). These $\Delta\%GI\,d^{-1}$ rates were then plotted together against

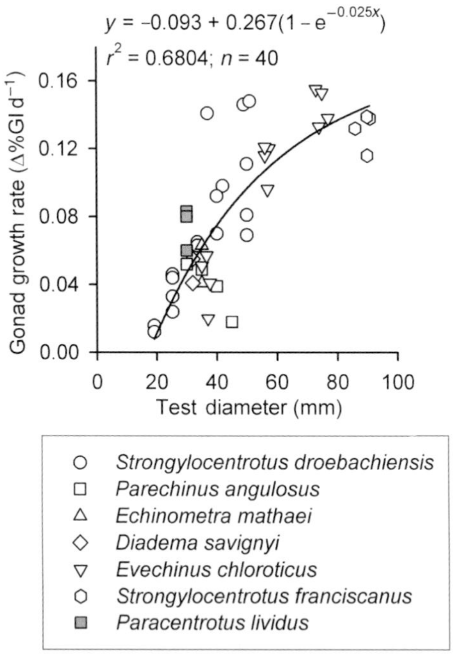

Fig. 2. Gonad growth rates for several urchin species under different experimental conditions are plotted as the daily increase in the gonad index (%GI as wet weight). Data sources for the calculations are described in the text.

test diameter (Fig. 2). Essentially, the *y*-axis in Fig. 2 represents the mass increase in gonad size normalized to both body size and time. If the initial rapid growth phase of the gonad is considered to be about a 3 month period (ca. 100 days), then the *y*-axis can be interpreted to mean that a 50 mm urchin will undergo about an 8% increase in relative gonad mass (wet weight) during this time.

Despite the differences in species and the conditions under which the data were collected, there is a substantial degree of coherence in the relationship between urchin test size and gonad growth rates. The variance in the *y*-axis values for a particular species and size results from the effects of different experimental treatments on rates of gonad growth. However, the growth potential of gonads does not appear to be as variable as one would initially consider from an examination of the diverse array of published studies. In general, *S. droebachiensis* appears to exhibit the largest degree of response in gonad growth to experimental manipulations; however, this could be an artifact of the greater number of data points that were available for this species.

The relevance of using the $\Delta\%\mathrm{GI\,d^{-1}}$ rates was assessed by comparing these values to absolute rates of the mass increase in gonad dry weight (Fig. 3). The work of Spirlet et al. (2000) with *P. lividus* presents sufficient biomass data to assess tissue growth rates. The linear correlation ($r^2 = 0.9654$) between $\Delta\%\mathrm{GI\,d^{-1}}$ (wet weight) and the observed daily increase in gonad biomass ($\mathrm{g\,DW\,d^{-1}}$) further suggests the applicability of this index for a proxy of gonad growth rates.

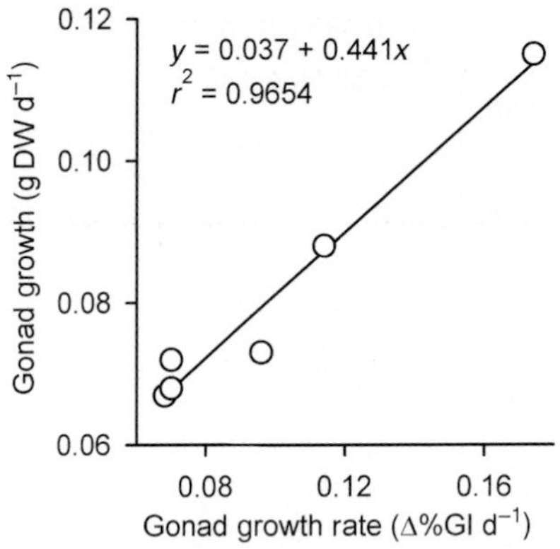

Fig. 3. Direct correlation between the daily index of gonad growth and the mass increase in gonad dry weight for *Paracentrotus lividus*. Calculated from data in Spirlet et al. (2000).

The difficulty of compiling a picture of metabolic activities in sea urchin gonads from all the reproductive studies in the literature is that many different species have been used to measure different variables. There is no single species used as a model system that has received a full metabolic profiling at both physiological and biochemical levels. However, Fig. 2 suggests there may be a common physiological process among these disparate data sets regarding rates of gonad growth. Consequently, cross-comparisons between different studies are likely to yield a broader picture of the general pattern of metabolic processes involved in gonad development than has been previously considered.

6. GONAD ENERGY METABOLISM

The seasonal gametogenic cycle in echinoids has been well described (Walker 1982; Pearse and Cameron 1991; Walker et al. 1998, 2001; Chapter 2). In terms of a marketable product, the best gonad yields are harvested when the nutritive phagocytes of the gonad have attained their greatest degree of mass increase before gametogenic differentiation is initiated. At this point, the nutritive phagocytes have accumulated the biochemical components that will be partitioned to the developing gametes, and the gonad tissues thus have correspondingly high levels of protein, carbohydrate, and lipid reserves. The development of the nutritive phagocytes is dependent on the assimilation of nutrients from the diet, and thus an understanding of the biochemical requirements for growth and development of nutritive phagocytes would facilitate efforts to optimize the feeds of urchins in aquaculture.

It is important to note that the rates of gonad growth utilized in Fig. 2 were obtained for the rapid growth phase of the nutritive phagocytes, during which dietary nutrients are translocated from the gut and deposited in modified stores for later use in the production of gametes (Walker et al. 2001; Chapter 2). Using the linear regression in Fig. 3, a generalized expression (Eq. 1) for the biomass increase in gonads during the development phase of nutritive phagocytes can be derived from the relationship to test diameter shown in Fig. 2:

$$\text{Gonad growth}(\text{g DW d}^{-1}) = -0.0038 + 0.117(1 - e^{(-0.025x)}) \quad (1)$$

From this equation, we can generalize that an average urchin with a test diameter of 50 mm would experience (under *ad libitum* food conditions) a net change in gonad mass of 0.80 mg DW d^{-1} (per g DW-body mass) during the period of rapid growth of the nutritive phagocytes. We will now consider the energy requirements and components of this increase in biomass (0.80 mg DW d^{-1}) for a standard urchin size (50 mm).

6.1. Protein Metabolism

The protein composition of a diet exerts a large influence on the protein content of the developing nutritive phagocytes (Hammer et al. in press) and the energy costs associated with gamete development (Otero-Villanueva et al. 2004). In a unique series of feeding studies maintaining caloric content relatively constant while varying protein content of a diet (Hammer et al. in press), a remarkable degree of interdependence of protein and carbohydrate composition of the gonads was evident (Fig. 4). In particular, the protein pool of the nutritive phagocytes can be dominated by a single protein, MYP (Harrington and Easton 1982; Unuma et al. 1998). This protein functions as a vitellogenin-like storage protein that will be allocated to developing oocytes, serving as a resource of amino acids for an embryo during early development (Walker et al. 2001; Chapter 2). Consequently, MYP synthesis takes a high precedence in terms of apportioning cellular energy.

The feeds used by Hammer et al. (2004) ranged in protein content from 9% to 33%, resulting in a net difference in protein ingestion rates (grams per individual) among the treatments of 450% (0.76–4.2 g protein). The response in gonad protein mass composition was less, spanning 25–40% protein, a difference of only 60%, suggesting that the higher ingestion rates on lower protein diets did compensate for the reduced availability (Hammer et al. 2004).

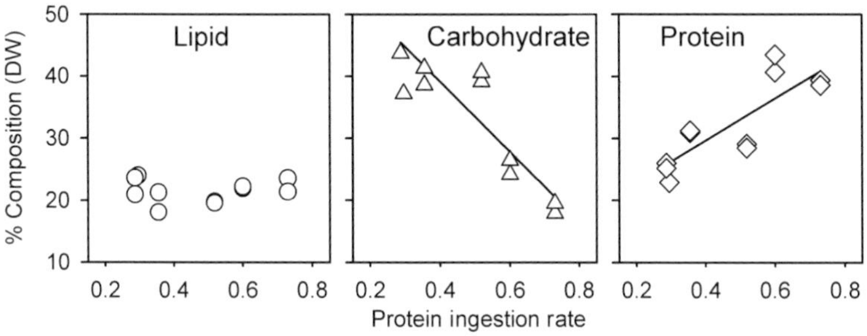

Fig. 4. Biochemical composition of gonad tissue of *Lytechinus variegatus* as a function of dietary protein ration. The negative correlation between protein ration and the carbohydrate composition of the gonad can be described by the regression: $y = 61.56 - 56.23x$; $r^2 = 0.791$. The positive correlation between protein ration and the protein composition of the gonad can be described by the regression: $y = 16.13 + 33.87x$; $r^2 = 0.692$. Data from Hammer et al. (in press).

For the purposes of estimating a cost of protein biosynthesis during development, we will assume that a standard 50 mm urchin has a gonad protein mass composition of 25%. For the 0.8 mg DW d^{-1} of gonad tissue produced (per g DW-body mass) while the nutritive phagocytes were developing ($\Delta\%GI\, d^{-1} = 0.08\%$; Fig. 2), this would equal a net synthesis of 0.2 mg of gonad protein and utilize 2.6 J d^{-1} of cellular energy (assuming 13 J mg $protein^{-1}$, Section 3). It is difficult to assess the total cellular energy budget of a developing gonad. Generally, several early studies of respiration were unable to document a correlation between respiration rates and gametogenesis (Giese et al. 1966; Giese 1967; McPherson 1968; Bellman and Giese 1974). In general, the metabolic rates of echinoderms have received considerable attention (see review Lawrence and Lane 1982), but little work has been published to elucidate the underlying metabolic costs of gametogenesis.

The only approach to assess the energy utilization of gonad development from existing data is to combine the literature studies of different urchins and approach the problem by trying to identify generalizations rather than specific applications. A robust data set relating reproductive stage to respiration rates is available for *P. angulosus* (Stuart and Field 1981). Converting their %GI data into values for gonad growth using the regression equation in Fig. 3 (for *P. lividus*), we find a significant correlation to mass-specific respiration rates (mmol $O_2\, d^{-1}$ g $DW\text{-}body^{-1}$; Fig. 5). The slope of the regression line in Fig. 5 can be interpreted to represent the energy cost (in aerobic equivalents) of mass-specific gonad growth, i.e. the increase in oxygen consumption associated with a unit increase in the gonad mass. Balancing the dimensions in the numerator and denominator, this value equates to 6.96 mmol $O_2\, d^{-1}$ g $DW\text{-}gonad^{-1}$ g $DW\text{-}body^{-1}$. At this point the ratio is normalized to both gonad mass and body mass.

This value can be related to cellular energy utilization by converting moles of oxygen to joules (484 kJ $mol^{-1}\, O_2$; Gnaiger 1983), and then normalizing the rate of energy utilization to the 0.8 mg DW-gonad mass per g DW-body mass estimated earlier for a standard urchin size of 50 mm (Section 4). The result is that the total metabolic energy consumption of that 0.8 mg DW portion of gonad is estimated to be 2.7 J d^{-1} (with

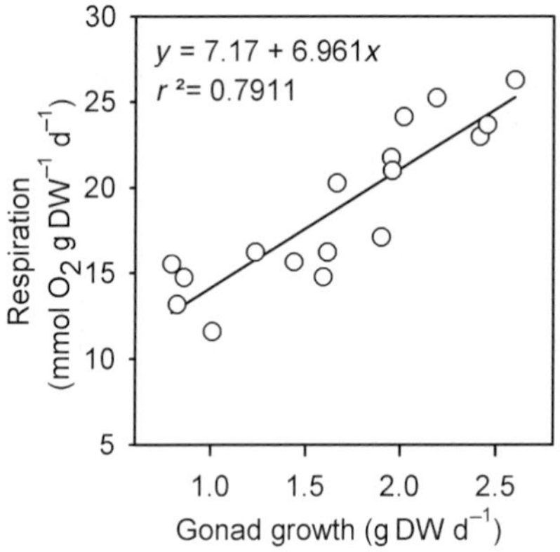

Fig. 5. Correlation of reproductive state with organismal respiration rates for *Parechinus angulosus*. Calculated from Stuart and Field (1981), table 3.

the gonad and body mass units now canceled out of the quotient). This is close to the estimate of cellular energy calculated to be required for protein synthesis in that same 0.8 mg DW-gonad mass unit ($2.6\,J\,d^{-1}$). The general conclusion we can derive from this comparison is that despite the large assumptions drawn in generating these energy utilization values, it is likely that protein metabolism accounts for a large portion of total metabolic energy expenditure in the developing gonads. The high synthetic demands for the production of MYP could potentially result in protein metabolism accounting for near 50% of total cellular energy utilization in the developing nutritive phagocytes. For adult *Psammechinus miliaris* fed a high-protein fish-meal diet, the energy cost of gamete production was estimated to be 45% of the total organismal metabolic energy budget (respiration; Otero-Villanueva et al. 2004).

In looking for a rate-limiting process that sets the maximal production size of a gonad, the synthesis of MYP (and its associated energy costs) could be the biochemical trigger that establishes an upper limit on the growth of the nutritive phagocytes. Overall, we view the correspondence between the estimated respiratory consumption and the estimated demand of protein synthesis only from the perspective that both processes are likely to be of similar magnitude, and that the experimental assessment of this relationship appears to be fruitful ground for further research.

6.2. Carbohydrate Metabolism

Using the most extensive data set to date that relates gonad protein and carbohydrate content relative to different diets (Hammer et al. in press), a significant negative correlation is observed between protein and carbohydrate contents of developing gonads observed in *L. variegatus* (Fig. 6). This relationship suggests a biochemical tradeoff between the synthesis of proteins and carbohydrates in developing gonads. Glycogen is the dominant carbohydrate stored by the nutritive phagocytes as they develop. During this early gonad growth phase before gametogenesis is initiated, glycogen content ranges from 18% to 25% of gonad DW in females and 13%–19% in males for *S. intermedius* (Zalutskaya et al. 1986; Zalutskaya 1988). For tissues that may have a total of 20–40% carbohydrate per g DW, this composition suggests that glycogen alone could account for a substantial portion of the gonad carbohydrates (ca. 50–75% by mass). Once gametogenic differentiation is initiated, glycogen content of the gonads declines. The carbohydrate content of echinoid eggs is low (~5%; Jaeckle 1995; McEdward and Miner 2001; Chapter 5; McEdward and Morgan 2001), indicating that the carbohydrate content of the nutritive phagocytes is not allocated to developing gametes, but rather is used to provide the cellular energy necessary to sustain gametogenesis (Zalutskaya et al. 1986). The synthetic cost of glycogen is relatively low ($0.42\,J\,mg^{-1}$, Section 3), but the cellular potential to generate ATP from 1 mg of glycogen is large ($15.7\,J\,mg\,glycogen^{-1}$ for the complete aerobic catabolism of glucose).

However, oxidizing 1 mg of glucose would require ~32.5 mmol O_2, and would be virtually impossible for any internal tissue of an echinoid given their poor perfusion in the coelomic cavity and hence their exposure to hypoxic conditions (Webster and Giese 1975). Anaerobiosis (the production of cellular ATP without oxygen) is likely to be a dominant metabolic process in the internal tissues of echinoids (Ellington 1982; Shick 1983). In the developing gonads of *S. droebachiensis*, 76–92% of cellular energy can be produced from

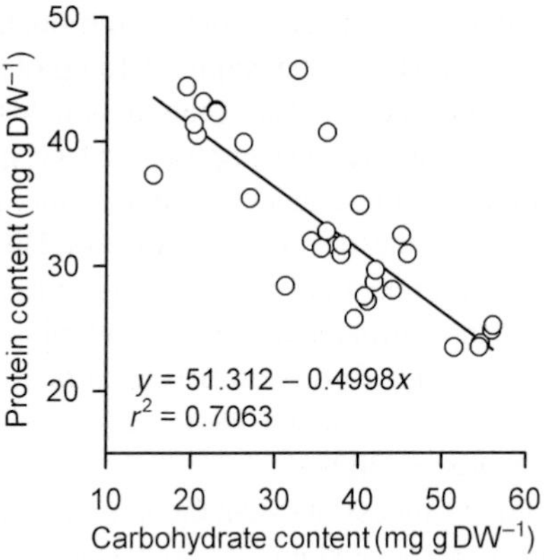

Fig. 6. Negative correlation between protein content and carbohydrate content of developing gonads in *Lytechinus variegatus*. Data from Hammer et al. (in press).

anaerobic metabolism (Bookbinder and Shick 1986). This is an important point because although echinoid gonads store glycogen and thus have a large, potential energy reserve, most of that energy may not be available due to the limited availability of oxygen.

We can calculate the approximate "physiological" energy yield of 1 mg of glycogen (5.38 μmol of glucose) in a developing gonad by assuming that 84% (average of Bookbinder and Shick 1986) would be anaerobically catabolized (4.52 μmol of glucose at 2 ATP per glucose molecule) and the remaining 16% would be aerobically catabolized (0.86μ mol of glucose at 38 ATP per glucose molecule). These processes would generate 0.70 and 2.52 J, respectively, for a total energy yield of 3.22 J mg glycogen^{-1}. This energy yield is equivalent to the cost of protein synthesis for our standard 0.80 mg DW increase in gonad mass (2.6 J), suggesting that the 2.7 J of aerobic oxidation (respiration, determined from the analysis in Fig. 5) could be met by ca. 16% aerobic catabolism of 1 mg glycogen (yielding 2.52 J) to meet the synthesis requirements of 0.2 mg of protein. There is likely to be a close relationship between protein and carbohydrate metabolism in developing gonads because of the energetic interdependence of protein anabolism and carbohydrate catabolism.

6.3. Anaerobic Metabolism

The growth and development of the nutritive phagocytes of the gonad requires a large input of cellular energy. Most considerations of these energy requirements and expenditures utilize estimates based on the full aerobic catabolism of carbohydrates. However, it is likely that the biosynthetic activity of the nutritive phagocytes utilize cellular energy derived from anaerobic glycolysis (Ellington 1982; Shick 1983).

Our derived equivalence between oxygen consumption, catabolism of glycogen, and energy cost of protein synthesis is dependent on the respiration rate extrapolations in Fig. 4. Although these estimates are only approximate, they suggest a significant contribution of anaerobic metabolism supporting the growth of the nutritive phagocytes, as has been previously documented for *S. droebachiensis* (Bookbinder and Shick 1986). On a mass basis, a ratio of 1 mg of glycogen catabolism (84% anaerobic and 16% aerobic) per 0.20 mg protein synthesis balances energy production and consumption. This estimated ratio equates to 3.22 J per 0.2 mg protein synthesis, or 16.1 J mg $protein^{-1}$, which is equivalent to the average cost of protein synthesis presented in Section 3 (13 J mg $protein^{-1}$; Wieser 1994). In fact, our assumption of an average anaerobic partitioning of 84% appears too low in this analysis. If the energy cost of protein synthesis is 13 J mg $protein^{-1}$ in nutritive phagocytes, then this energy cost can be met by the catabolism of 1 mg of glycogen if 88% involves anaerobic pathways and only 12% utilizes aerobic pathways (2.65 J mg $glycogen^{-1}$).

In marine invertebrates, a common alternative for metabolizing pyruvate under anaerobic conditions is its reductive condensation with an amino acid to form an imino acid or opine derivative (Livingstone et al. 1990). These alternate pathways of anaerobic glycolysis often involve the accumulation of alanine, succinate, propionate, and octopine (Hochachka 1975; Ellington 1982). Recent considerations of the phylogenetic occurrence of different anaerobic pathways in invertebrates have shown that deuterostome phyla offer little evidence for opine dehydrogenase pathways found in protostome taxa? (Livingstone et al. 1990; Sato et al. 1993). These results suggest that in echinoderms the dominant pathway for pyruvate metabolism under hypoxic conditions produces lactate via lactate dehydrogenase. Under anoxic conditions, the ovaries of *S. droebachiensis* have been shown to produce large amounts of lactate to maintain cellular ATP levels (Bookbinder and Shick 1986). It is likely that the evolutionary constraints that have limited the scope or diversity of anaerobic pathways in echinoids (and echinoderms in general) have involved the limited body musculature that appears to require high anaerobic capacities in other invertebrates (Shick 1983), which would also account for the lower, organismal respiration rates generally found in echinoderms (Lawrence and Lane 1982).

Although the energy equivalents presented in this chapter have required some extended extrapolations, carbohydrate and protein metabolism appear to be balanced by a significant fraction of anaerobic metabolism. In general, glycogen metabolism is likely to be negatively correlated to protein metabolism in developing gonads where the energy requirements of protein synthesis would be fueled by carbohydrate catabolism. These quantities remain to be measured empirically during nutritive phagocyte growth and development in an urchin, but our estimates of a large anaerobic capacity indicate that such measurements are critical for understanding the potential for gonads to store both carbohydrate and protein nutrient reserves.

7. FEEDING AND METABOLISM

From a survey of recent feeding experiments with artificial diets, a gonad growth index has been calculated ($\Delta\%GI\,d^{-1}$) and correlated to ingestion rates (Fig. 7) for *Loxechinus albus* (Lawrence et al. 1997), *S. droebachiensis* (Klinger 1997), *Evechinus chloroticus* (Barker et al. 1998), and *S. franciscanus* (McBride et al. 1997). It is not surprising that feeding urchins a prepared diet with a high protein content (20–25% dry mass) results in greater yields of gonad mass. However, the pooled data in Fig. 7 do indicate that this

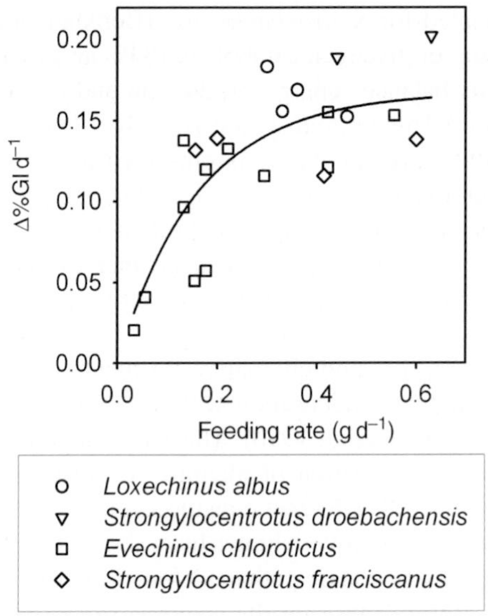

Fig. 7. Gonad growth rates as a function of ingestion rates of formulated diets for several urchin species (source references in text). The regression plotted is $y = 0.1674(1 - e^{(-6.184x)})$, $r^2 = 0.6310$. Experimental data were selected for the time periods of greatest gonad growth (as in Fig. 2).

relationship appears to have a rapid saturation point with a maximal gonad growth rate of $\sim 0.17\%\mathrm{GI\,d}^{-1}$. This relationship indicates that although the absorption efficiency of protein from a diet can be high, there is an absolute limit to the amount of protein that can be synthesized in the nutritive phagocytes during their early growth phase.

An important aspect of considering the efficacy of a diet formulation to enhance gonad yield is identifying the biochemical controls that determine the metabolic relationship between dietary constituents and gonad growth. A potential paradox exists whereby an increase in food quality (protein and amino acids) also increases the energy required to assimilate and incorporate those nutrients (McBride et al. 1997). Under natural food conditions, protein metabolism is likely to already consume 50% of total cellular energy metabolism. At this level of energy allocation it is unlikely that an urchin could produce significant reductions in its other cellular activities in order to provide even more energy for protein metabolism. Feeding an urchin a diet rich in amino acids to enhance protein biosynthesis in the gonads will also require that the diet contains the necessary energy source to fuel that increase in biosynthesis. The level of energy required for protein sparing has not been determined for any echinoid.

How can such a metabolic energy source be provided to a sea urchin? The problem is that most of the calculations of dietary caloric contents are based on the complete aerobic catabolism of the constituent proteins, carbohydrates, and lipids. To identify an appropriate dietary mixture of anabolic substrates and catabolic energy constituents, several areas remain to be functionally assessed:

1. Assimilated proteins are generally not catabolized and the constituent amino acids are recycled for protein biosynthesis. What is the protein contribution to the metabolizable energy pool of a diet? This is important in deciding the optimum protein ration for biosynthesis in a diet.
2. Under hypoxic conditions, assimilated carbohydrates are catabolized with a much lower efficiency for generating cellular energy. What is the efficiency of anaerobic carbohydrate metabolism in developing gonads? This is important in determining how much additional biosynthetic activity a prepared diet could support.
3. Although lipids have higher energy content (per unit mass) than carbohydrates, the oxidation pathways for generating cellular energy from fatty acids are dependent on mitochondrial activity under normoxic conditions. What is the efficiency for metabolizing different lipid classes in developing gonads? Some lipids, carotenoids or other hydrophobic moieties could potentially supply additional (high energy) metabolic fuels.

8. SUMMARY

The macro- and micronutrient composition of a diet has a pronounced effect on the growth and biochemical composition of the nutritive phagocytes during their seasonal growth. There is a direct relationship between the protein content of a diet and the protein content of the developing nutritive phagocytes. We have shown that protein metabolism is likely to dominate the utilization of cellular energy, suggesting that the ability of the nutritive phagocytes to synthesize the MYP protein could be a primary determinant (or limitation) of gonad size. As the protein content of an urchin's diet is increased, protein biosynthesis activity in the gonad increases, but this increase in anabolic activity most likely requires the input of more cellular energy. Consequently, less available energy assimilated from a diet can be stored in the nutritive phagocytes as glycogen. It would appear that the production of gonads with both a high protein and high glycogen content may not be possible because of the mutually exclusive energy tradeoffs they represent.

Optimizing a feed for the maximal production of gonads in a sea urchin requires a consideration of how to balance the energy demands of increased protein metabolism with the energy availability in the diet. The difficulty in supplying metabolic energy in a feed is that the caloric content of a diet should be considered in terms of its energy yield under hypoxic conditions. It is possible that a diet supplemented with substrates that can be metabolized via anaerobiosis may provide more efficient energy substrates under physiological conditions than monosaccharides that require aerobic oxidation. Major yolk protein itself serves as a storage vehicle for the amino acid building blocks that will be needed to form an embryo. Supplying a sea urchin with a diet that has an amino acid composition approximating that of the MYP protein could reduce the requirements of *de novo* synthesis of some amino acids as well as providing sufficient amounts of essential amino acids that they may not be readily capable of synthesizing. Again, the hypoxic conditions of the gonads may alter the standard metabolic activities in the gonads and prevent or limit the ready synthesis of some metabolites (sugars, amino acids). Despite the attention that urchin aquaculture has received in recent years, there is still much to be learned in regard to the metabolic activity of the nutritive phagocytes and the efficient production of a high gonad yield.

ACKNOWLEDGMENTS

We thank M Anguelova for translating portions of the Russian works of EA Zalutskaya. This research was supported in part by the Mississippi Alabama Sea Grant Consortium.

REFERENCES

Barker MF, Keogh JA, Lawrence JM, Lawrence AL (1998) Feeding rate, absorption efficiencies, growth, and enhancement of gonad production in the New Zealand sea urchin *Evechinus chloroticus valenciennes* (Echinoidea: Echinometridae) fed prepared and natural diets. J Shellfish Res 17: 1583–1590

Beauchop T, Elsden SR (1960) The growth of microorganisms in relation to their energy supply. J Gen Microbiol 23: 457–469

Bellman BW, Giese AC (1974) Oxygen consumption of an asteroid and echinoid from the antarctic. Biol Bull 146: 157–164

Berg WE, Mertes DH (1970) Rates of synthesis and degradation of protein in the sea urchin embryo. Exp Cell Res 60: 218–224

Blanco G, Mercer RW (1998) Isozymes of the α-Na^+, K^+-ATPase: Heterogeneity in structure, diversity in function. Amer J Physiol 275: F633–F650

Blaxter K (1989) Energy Metabolism in Animals and Man, Cambridge University Press, Cambridge

Bookbinder LH, Shick JM (1986) Anaerobic and aerobic energy-metabolism in ovaries of the sea urchin *Strongylocentrotus droebachiensis*. Mar Biol 93: 103–110

Boolootian RA (ed.) (1966) Reproductive physiology. In: Physiology of Echinodermata, Interscience, New York, pp 561–613

Conceicao LEC, Houlihan DF, Verreth JAJ (1997) Fast growth, protein turnover and costs of protein metabolism in yolk-sac larvae of the African catfish (*Clarias gariepinus*). Fish Physiol Biochem 16: 291–302

Drummond AE (1995) Reproduction of the sea urchins *Echinometra mathaei* and *Diadema savignyi* on the South African eastern coast. Mar Freshw Res 46: 751–757

Ellington WR (1982) Intermediary metabolism. In: Jangoux M and Lawrence JM (eds) Echinoderm Nutrition, Balkema, Rotterdam, pp 395–415

Fauconneau B, Gray C, Houlihan DF (1995) Assessment of individual protein-turnover in 3 muscle types of rainbow-trout. Comp Biochem Physiol B-Biochem Molec Biol 111: 45–51

Fenchel T, Findlay BJ (1983) Respiration rates in heterotrophic, free-living protozoa. Microb Ecol 9: 99–122

Fry BJ, Gross PR (1970) Patterns of protein synthesis in sea urchin embryos. II. The calculation of absolute rates. Dev Biol 21: 125–146

Giese AC (1967) Changes in body component indexes and respiration with size in the purple sea urchin *Strongylocentrotus purpuratus*. Physiol Zool 40: 194–200

Giese AC, Farmanfarmaian A, Hilden S, Doezema P (1966) Respiration during the reproductive cycle in the sea urchin *Strongylocentrotus purpuratus*. Biol Bull 130: 192–201

Gnaiger E (1983) Calculation of energetic and biochemical equivalents of respiratory oxygen consumption. In: Gnaiger E and Forstner H (eds) Polarographic Oxygen Sensors, Springer-Verlag, Berlin, pp 337–345

Goustin AS, Wilt FH (1981) Protein-synthesis, polyribosomes, and peptide elongation in early development of *Strongylocentrotus purpuratus*. Dev Biol 82: 32–40

Greenwood PJ (1980) Growth, respiration and tentative energy budgets for 2 populations of the sea urchin *Parechinus angulosus* (Leske). Estuar Coast Mar Sci 10: 347–367

Guillou M, Lumingas LJL, Michel C (2000) The effect of feeding or starvation on resource allocation to body components during the reproductive cycle of the sea urchin *Sphaerechinus granularis* (Lamarck). J Exp Mar Biol Ecol 245: 183–196

Hammer BW, Hammer HS, Watts SA, Desmond RA, Lawrence JM, Lawrence AL (2004) The effects of dietary protein concentration on feeding and growth of small *Lytechinus variegatus* (Echinodermata: Echinoidea). Mar Biol 145: 1143–1157

Hammer HS, Hammer BW, Watts SA, Lawrence AL, Lawrence JM (2006) The effect of dietary protein and carbohydrate concentration on the biochemical composition and gametic condition of the sea urchin *Lytechinus variegatus*. J Exp Mar Biol Ecol 334: 109–121

Harrington FE, Easton DP (1982) A putative precursor to the major yolk protein of the sea urchin. Dev Biol 94: 505–508

Harvey HR, Eglinton G, O'Hara SCM, Corner EDS (1987) Biotransformation and assimilation of dietary lipids by *Calanus* feeding on a dinoflagellate. Geochimica et Cosmochimica Acta 51: 3031–3040

Havardsson B, Imsland AK (1999) The effect of astaxanthin in feed and environmental temperature on carotenoid concentration in the gonads of the green sea urchin *Strongylocentrotus droebachiensis* Müller. J World Aquacult Soc 30: 208–218

Hawkins AJS (1991) Protein turnover – a functional appraisal. Funct Ecol 5: 222–233

Hawkins AJS, Widdows J, Bayne BL (1989) The relevance of whole-body protein metabolism to measured costs of maintenance and growth in *Mytilus edulis*. Physiol Zool 62: 745–763

Hochachka PW (1975) Design of metabolic machinery to fit lifestyle and environment. Symp Biochem Soc 41: 3–31

Hogan G, Gross PR (1971) The effect of protein synthesis inhibition on the entry of messenger RNA into the cytoplasm of sea urchin embryos. J Cell Biol 49: 692–701

Houlihan DF (1991) Protein turnover in ectotherms and its relationships to energetics. In: Gilles R (ed.) Advances in Comparative and Environmental Physiology. Vol. 7, Springer-Verlag, Berlin, pp 1–43

Jaeckle WB (1995) Variation in the size, energy content and biochemical composition of invertebrate eggs: Correlates to the mode of larval development. In: McEdward LR (ed.) Marine Invertebrate Larvae, CRC Press, New York, pp 49–78

Jangoux M, Lawrence JM (1982) Echinoderm Nutrition, Balkema, Rotterdam

Klinger TS (1997) Gonad and somatic production of *Strongylocentrotus droebachiensis* fed manufactured feeds. Bull Aquacul Assoc Canada 1: 35–37

Krebs HA (1960) The cause of the specific dynamic action of foodstuffs. Arzneimittel-Forschung 10: 369–373

Lawrence JM, Lane JM (1982) The utilization of nutrients by post-metamorphic echinoderms. In: Jangoux M and Lawrence JM (eds) Echinoderm Nutrition, Balkema, Rotterdam, pp 331–371

Lawrence JM, McClintock JB (1994) Energy acquisition and allocation by echinoderms (Echinodermata) in polar seas: Adaptations for success? In: David B and Féral J-P (eds) Echinoderms Through Time, Balkema, Rotterdam, pp 39–52

Lawrence JM, Olave S, Otaiza R, Lawrence AL, Bustos E (1997) Enhancement of gonad production in the sea urchin *Loxechinus albus* in Chile fed extruded feeds. J World Aquacult Soc 28: 91–96

Leong PKK, Manahan DT (1997) Metabolic importance of Na^+/K^+-ATPase activity during sea urchin development. J Exp Biol 200: 2881–2892

Leong PKK, Manahan DT (1999) Na^+/K^+-ATPase activity during early development and growth of an Antarctic sea urchin. J Exp Biol 202: 2051–2058

Livingstone DR, Stickle WB, Kapper MA, Wang S, Zurburg W (1990) Further studies on the phylogenetic distribution of pyruvate oxidoreductase activities. Comp Biochem Physiol 97B: 661–666

Lyndon AR, Houlihan DF (1998) Gill protein turnover: Costs of adaptation. Comp Biochem Physiol A-Mol Integr Physiol 119: 27–34

Marsh AG, Gremare A, Dawson R, Tenore KR (1990a) Translocation of algal pigments to oocytes in *Capitella* sp. I (Annelida: Polychaeta). Mar Ecol Prog Ser 67: 301–304

Marsh AG, Harvey HR, Gremare A, Tenore KR (1990b) Dietary effects on oocyte yolk-composition in *Capitella* sp. I (Annelida: Polycheata): Fatty acids and sterols. Mar Biol 106: 369–374

Marsh AG, Leong PKK, Manahan DT (1999) Energy metabolism during embryonic development and larval growth of an Antarctic sea urchin. J Exp Biol 202: 2041–2050

Marsh AG, Leong PKK, Manahan DT (2000) Gene expression and enzyme activities of the sodium pump during sea urchin development: Implications for indices of physiological state. Biol Bull 199: 100–107.

Marsh AG, Manahan DT (1999) A method for accurate measurements of the respiration rates of marine invertebrate embryos and larvae. Mar Ecol-Prog Ser 184: 1–10

McBride SC, Pinnix WD, Lawrence JM, Lawrence AL, Mulligan TM (1997) The effect of temperature on production of gonads by the sea urchin *Strongylocentrotus franciscanus* fed natural and prepared diets. J World Aquacult Soc 28: 357–365

McCarthy ID, Houlihan DF, Carter CG (1994) Individual variation in protein-turnover and growth efficiency in rainbow trout, *Oncorhynchus mykiss* (Walbaum). Proc R Soc Lond Ser B-Biol Sci 257: 141–147

McEdward LR, Miner BG (2001) Echinoid larval ecology. In: Lawrence JM (ed.) The Edible Sea Urchin, Elsevier, Amsterdam, pp 59–78

McEdward LR, Morgan KH (2001) Interspecific relationships between egg size and level of parental investment per offspring in echinoderms. Biol Bull 200: 33–50

McPherson BF (1968) Feeding and digestion in the tropical sea urchin *Eucidaris tribuloides*. Biol Bull 135: 308–320

Meidel SK, Scheibling RE (1999) Effects of food type and ration on reproductive maturation and growth of the sea urchin *Strongylocentrotus droebachiensis*. Mar Biol 134: 155–166

Millward DJ, Garlick PJ (1976) The energy cost of growth. Proc Nutrit Soc 35: 339–349

Minor MA, Scheibling RE (1997) Effects of food ration and feeding regime on growth and reproduction of the sea urchin *Strongylocentrotus droebachiensis*. Mar Biol 129: 159–167

Mitsunaga-Nakatsubo K, Kanda M, Yamazaki K, Kawashita H, Fujiwara A, Yamada K, Alasaka K, Shimada J, Yasumasu I (1992) Expression of Na^+, K^+-ATPase α-subunit in animalized and vegetalized embryos of the sea urchin, *Hemicentrotus pulcherrimus*. Development Growth & Differentiation 34(6): 677–684.

Moore HB (1934) A comparison of the biology of *Echinus esculentus* in different habitats. Part I. J Mar Biol Ass UK 19: 869–881

Needham J (1931) Chemical Embryology, Cambridge University Press, London

Otero-Villanueva MDM, Kelly MS, Burnell G (2004) How diet influences energy Partitioning in the regular echinoid *Psammechinus miliaris*; constructing an energy budget. J Exp Mar Biol Ecol 304: 159–181.

Pearse JS, Cameron A (1991) Echinodermata: Echinoidea. In: Giese AC, Pearse JS and Pearse VB (eds) Reproduction of Marine Invertebrates, Echinoderms and Lophophorates. Vol. 7, Boxwood Press, Pacific Grove, pp 513–662

Regier JC, Kafatos FC (1977) Absolute rates of protein-synthesis in sea-urchins with specific activity measurements of radioactive leucine and leucyl-transfer-rna. Dev Biol 57: 270–283

Russell MP (1998) Resource allocation plasticity in sea urchins: Rapid, diet induced, phenotypic changes in the green sea urchin, *Strongylocentrotus droebachiensis* (Müller). J Exp Mar Biol Ecol 220: 1–14

Sato M, Takeuchi M, Kanno N, Nagahisa E, Sato Y (1993) Distribution of opine dehydrogenases and lactate-dehydrogenase activities in marine animals. Comp Biochem Physiol B-Biochem Mol Biol 106: 955–960

Schmidt H, Siems W, Muller M, Dumdey R, Jakstadt M, Rapoport SM (1989) Balancing of mitochondrial and glycolytic ATP production and of the ATP-consuming processes of Ehrlich mouse Ascites tumor cells in a high phosphate medium. Biochem Internat 19: 985–992

Schmidt H, Siems W, Muller M, Dumdey R, Rapoport SM (1991) ATP-producing and consuming processes of Ehrlich mouse ascites tumor cells in proliferating and resting phases. Exp Cell Res 194: 122–127

Schneider W, Siems W, Grune T (1990) Balancing of energy-consuming processes of rat hepatocytes. Cell Biochem Funct 8: 227–232

Shick JM (1983) Respiratory gas exchange in the echinoderms. In: Jangoux M and Lawrence JM (eds) Echinoderm Studies. Vol. 1, Balkema, Rotterdam, pp 67–110

Siems WG, Schmidt H, Gruner S, Jakstadt M (1992) Balancing of energy-consuming processes of K 562 cells. Cell Biochem Funct 10: 61–66

Smith RW, Houlihan DF (1995) Protein-synthesis and oxygen-consumption in fish cells. J Comp Physiol B-Biochem Syst Environ Physiol 165: 93–101

Spirlet C, Grosjean P, Jangoux M (1998) Reproductive cycle of the echinoid *Paracentrotus lividus*: Analysis by means of the maturity index. J Invert Reprod Develop 34: 69–81

Spirlet C, Grosjean P, Janagoux M (2000) Optimization of gonad growth by manipulation of temperature and photoperiod in cultivated sea urchins, *Paracentrotus lividus* (Lamarck) (Echinodermata). Aquaculture 185: 85–99

Stearns SC (1992) The Evolution of Life Histories, Oxford University Press, Oxford

Stuart V, Field JG (1981) Respiration and ecological energetics of the sea urchin *Parechinus angulosus*. South Afr J Zool 16: 90–95

Thompson RJ (1982) The relationship between food ration and reproductive effort in the green sea urchin, *Strongylocentrotus droebachiensis*. Oecologia 56: 50–57

Unuma T, Suzuki T, Kurokawa T, Yamamoto T, Akiyama T (1998) A protein identical to the yolk protein is stored in the testis in male red sea urchin, *Pseudocentrotus depressus*. Biol Bull 194: 92–97

Vadas RL, Beal B, Dowling T, Fegley JC (2000) Experimental field tests of natural algal diets on gonad index and quality in the green sea urchin, *Strongylocentrotus droebachiensis*: A case for rapid summer production in post-spawned animals. Aquaculture 182: 115–135

Vavra J, Manahan DT (1999) Protein metabolism in lecithotrophic larvae (Gastropoda: *Haliotis rufescens*). Biol Bull 196: 177–186

Walker CW (1982) Nutrition of gametes. In: Jangoux M and Lawrence JM (eds) Echinoderm Nutrition, Balkema, Rotterdam, pp 449–468

Walker CW, Lesser MP (1998) Manipulation of food and photoperiod promotes out-of-season gametogenesis in the green sea urchin, *Strongylocentrotus droebachiensis*: Implications for aquaculture. Mar Biol 132: 663–676

Walker CW, McGinn NA, Harrington LM, Lesser MP (1998) New perspectives on sea urchin gametogenesis and their relevance to aquaculture. J Shellfish Res 17: 1507–1514

Walker CW, Unuma T, McGinn NA, Harrington FE, Lesser MP (2001) Reproduction of sea urchins. In: Lawrence JM (ed.) The Edible Sea Urchin, Elsevier, Amsterdam, pp 5–26

Webster SK, Giese AC (1975) Oxygen consumption of the purple sea urchin with special reference to the reproductive cycle. Biol Bull 148: 165–180

Wieser W (1985) A new look at energy conservation in ectothermic organisms and endothermic animals. Oecologia 66: 506–510

Wieser W (1994) Cost of growth in cells and organisms – general rules and comparative aspects. Biol Rev Cambridge Philosophic Soc 69: 1–33

Zalutskaya EA (1988) Seasonal dynamics of glycogen in the testes of the sea-urchin *Strongylocentrotus intermedius*. Biol Morya 14: 65–67

Zalutskaya EA, Varaksina GS, Khotimchenko YS (1986) Glycogen-content in the ovaries of the sea urchin *Strongylocentrotus intermedius*. Biol Morya 12: 38–44

Edible Sea Urchins: Biology and Ecology
Editor: John Miller Lawrence

Chapter 4

Endocrine Regulation of Sea Urchin Reproduction

Kristina M Wasson and Stephen A Watts

Department of Biology, University of Alabama at Birmingham, Birmingham, AL (USA).

1. SEA URCHIN GONAD

As the demands for sea urchin gonads increase and as natural populations decline, a need for sea urchin aquaculture increases (Keesing and Hall 1998; Robinson 2004). To maximize gonad production, it is essential to understand those processes that regulate gonad production. Sea urchin gonads consist of two major cell types: (1) germinal cells, which undergo mitotic proliferation and meiotic reductions to produce large quantities of gametes, and (2) nutritive phagocytes, which undergo cyclic depletion and renewal of macromolecules to provide the nutrients and energy required by the developing gametes (Chapter 2). In this chapter, we consider the putative chemical messengers and their potential regulatory mechanisms in the control of reproduction.

2. EXOGENOUS REGULATION OF REPRODUCTION

2.1. Environmental Factors

Sea urchin reproduction is characterized by specific cycles of gametogenesis and nutritive phagocyte growth and depletion (Chapter 2). Correlative data indicate that seasonal changes in environmental factors, such as photoperiod, temperature, and salinity, influence reproduction within a population (reviewed by Pearse and Cameron 1991). Nutritional condition also affects gametogenesis (Bishop and Watts 1994). The environmental and nutritional effects on the physiology and morphology of the gonads suggest that these exogenous cues are transduced to endogenous factors that regulate reproduction in the adult sea urchin.

2.2. Endocrine Disruptors

Endocrine disruptors are a diverse group of naturally occurring and synthetic compounds that mimic or antagonize the effects of endogenous chemical messengers. Naturally occurring endocrine disruptors include phytoestrogens as well as sex steroids, such as estrogens and androgens. Synthetic compounds include industrial chemicals, herbicides, and pesticides. Endocrine disruptors typically enter marine environments via effluents from sewage treatment plants, and from industry and agricultural runoff (Langston et al. 2005). Few studies have investigated the effects of potential endocrine disruptors on sea urchin reproduction. Lipina et al. (1987) reported an increase in the number of oogonia within the ovaries of *Strongylocentrotus intermedius* after a short exposure (15 days) to cadmium chloride; however, longer exposure (for up to 130 days) appeared to have toxic effects, including decreased number of germ cells, decelerated germ cell growth, and destroyed gametes and nutritive phagocytes. Khristoforova et al. (1984) reported similar effects on the gonads of *S. intermedius* due to cadmium sulfate. Böttger and McClintock (2002) found that chronic exposure of *Lytechinus variegatus* to either inorganic phosphates or organic phosphates inhibited gonadal growth and spawning activities as well as altered the biochemical composition of the gonads. Whether these reported toxicities directly or indirectly affected endocrine function is not known.

3. ENDOGENOUS REGULATION OF REPRODUCTION

Although the endocrine and/or paracrine mechanisms that regulate physiological processes in sea urchins are poorly understood, several classes of putative chemical messengers have been identified. These include sex steroids (Fig. 1), proteins, peptidergic factors, and catecholaminergic and cholinergic factors (Fig. 2). The sex steroids, which are most likely synthesized from cholesterol, consist of progesterone (P4), testosterone (T), and estradiol-17β (E2). Peptidergic factors are small peptides that contain no secondary structure. Catecholaminergic or monaminergic factors that include noradrenaline and dopamine are amino acid derived trophic factors. The putative cholinergic factor in sea urchins is acetylcholine. Examination of these chemical messengers in the context of gonad production, gametogenesis, and nutrient translocation allows the generation of new hypotheses for research into the manipulation of gonad production and regulation of gametogenesis.

3.1. Steroids

3.1.1. Sex steroids in the gonads

Mammalian models of reproduction demonstrate that both sex steroids and peptide hormones regulate gametogenesis. Early studies looked for the presence of sex steroids in sea urchin ovaries. Donahue and Jennings (1937) found that an oil-soluble substance extracted from the ovaries of *L. variegatus* induced vaginal growth in ovariectomized rats. They concluded that this substance was slightly estrogenic in nature, requiring over 4 g of sea urchin ovaries to produce the equivalent of 1 rat unit of estrogenic activity. Donahue (1940) also discovered that both *L. variegatus* and *Echinometra* sp. contained oil-soluble substances that induced estrogenic responses in rat vagina. Twenty years later, Botticelli et al. (1961) demonstrated that ovary extracts from *S. franciscanus* had E2 activity and estimated the E2 concentration to be approximately $100\,pg\,g^{-1}$ of ovaries. Botticelli et al.

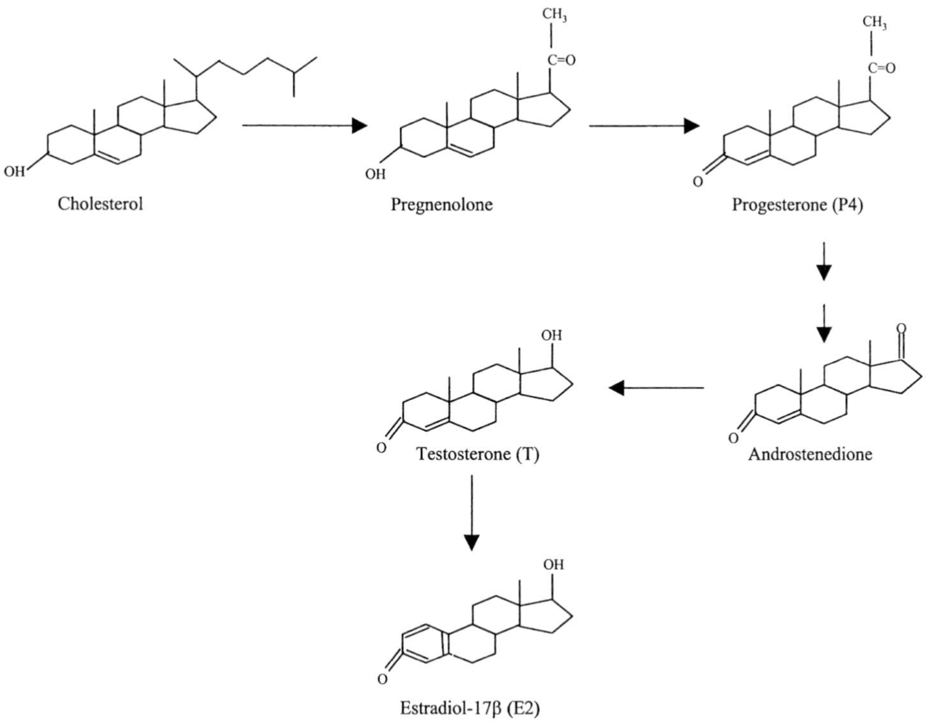

Fig. 1. Pathway of sex steroid synthesis from cholesterol.

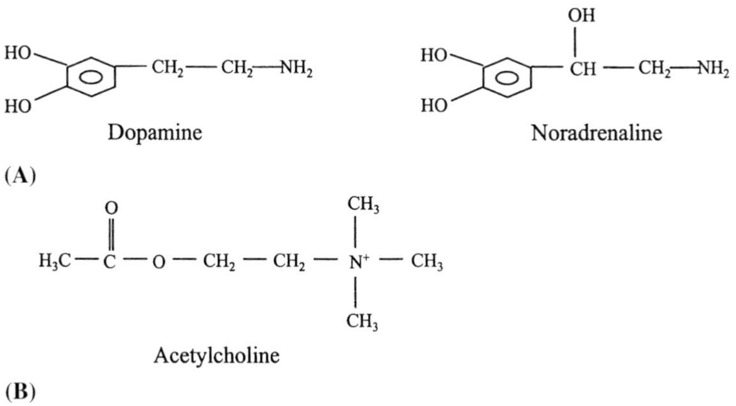

Fig. 2. Putative (A) catecholaminergic factors and (B) cholinergic factors in echinoids.

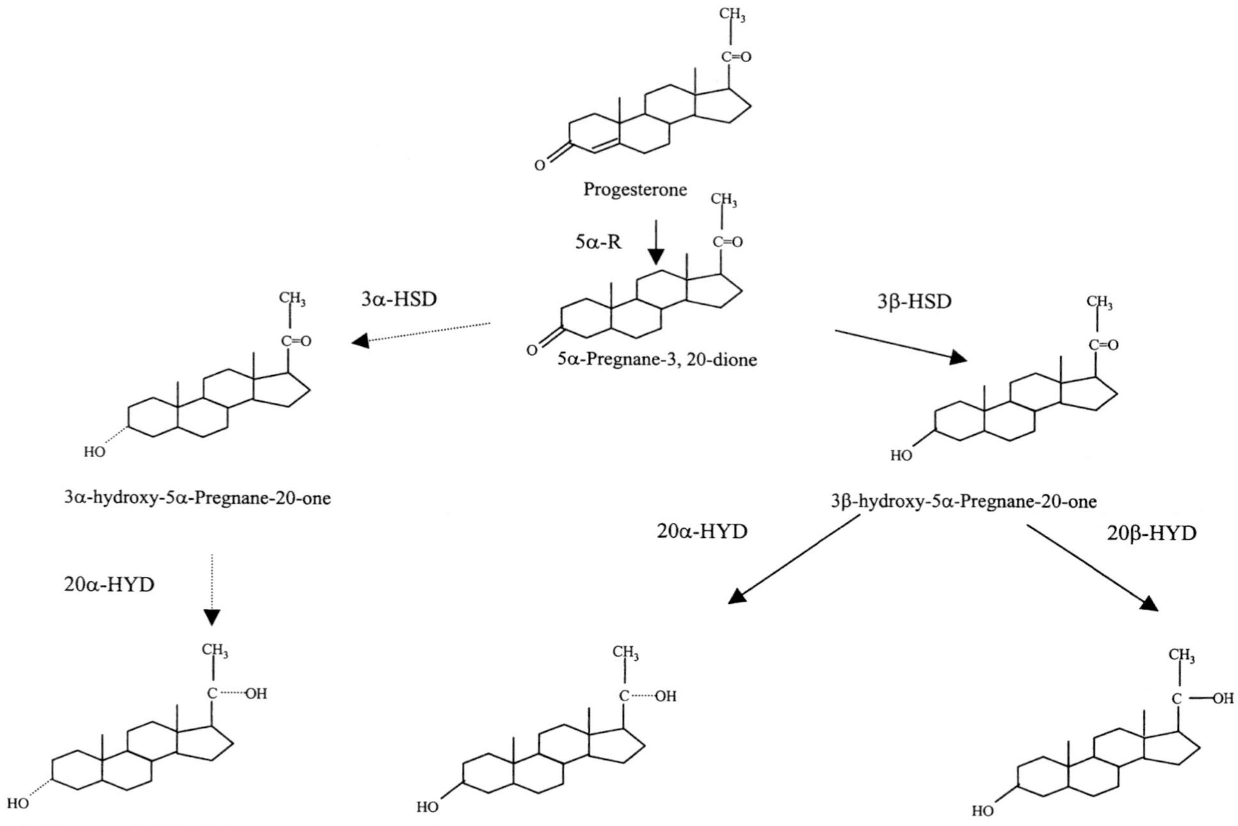

Fig. 4. Pathway of progesterone metabolism in *Lytechninus variegatus*. Solid arrows represent tentatively identified pathways. Dashed arrows represent suggestively identified pathways. 5α-R = 5α-reductase, 3α-HSD = 3α-hydroxysteroid dehydrogenase, 3β-HSD = 3β-hydroxysteroid dehydrogenase, 20α-HYD = 20β-hydroxylase, 20β-HYD = 20β-hydroxylase.

5α-pregnane-3, 20-dione, 3β-hydroxy-5α-pregnan-20-one, 5α-pregnane-3β, 20α-diol, 5α-pregane-3β, 20β-diol, and 5α-pregnane-3α, 20α-diol. Interestingly, the synthesis of androgens from progesterone and sex-specific differences in progesterone metabolism were not observed in the gonads (Wasson and Watts 2000).

Since the majority of steroid metabolites synthesized from the gonads of *L. variegatus* are 5α-reduced steroids, Wasson and Watts (1998) suggested that 5α-reductase is an important enzyme in the steroidogenic pathway of sea urchin gonads. The specific 5α-reductase inhibitor, finasteride (F; Proscar©), inhibited 5α-reductase activity *in vitro* in the gonads. Additionally, the gonads of sea urchins fed for 16 days on a formulated diet supplemented with F demonstrated *in vivo* inhibition of 5α-reductase activity. Specific chemical ablation by F of 5α-reductase could be used as tool to investigate the role of 5α-reduced steroids in the gonads of sea urchins (Wasson and Watts 1998).

It is interesting that the synthesis of estrogens was not observed in any of the studies examining androgen metabolism in sea urchins (Colombo and Belvedere 1976; Watts et al. 1994; Wasson et al. 1998, 2000a). However, Hines et al. (1994) showed that testes of *L. variegatus* converted E2 into an E2 ester, estrone, and aqueous soluble metabolites. Additionally, estrone was converted primarily into E2, and to a lesser extent E2 ester and aqueous soluble metabolites. These data indicate that 17β-HSD is active in the interconversion of E2 and estrone and those estrogens are converted rapidly into aqueous soluble metabolites. Hines et al. (1994) suggested that radiotracer studies might not provide the sensitivity necessary to detect the rapidly metabolized estrogens.

3.1.3. Sex steroids in the gonads

Wasson et al. (2000a) detected immunoreactive E2 in the gonads of *L. variegatus* using a commercial double antibody E2 radioimmunoassay (RIA). Measurement of T and E2 concentrations by RIA in the gonads during one reproductive season demonstrated no significant correlation between reproductive state (as determined by gonad index) and season, reportedly due to high levels of individual variations in steroid concentrations. These variations in steroid concentrations were presumably related to individual variations in specific stages of the gonads (Wasson et al. 2000a). Interestingly, comparison of steroid concentrations indicated significant differences with gonad size, suggesting a relation between gonad production and steroid levels. In both the ovaries and the testes, individuals with low gonad indices (0–3%) had significantly higher T and E2 concentrations than those with higher gonad indices ($>6.1\%$). Whether or not these steroids are involved in the stimulation or maintenance of mitotic and meiotic events in gametes has not been established. The levels of 5α-adiols, one of the major steroid metabolites in sea urchin gonads, are not known. Since the synthesis of the 5α-adiols varied with gonad size (Wasson et al. 2000a), perhaps the 5α-adiols may be more appropriate steroids to measure during the reproductive cycle.

3.1.4. Response to exogenous administration of sex steroids

Varaksina and Varaksin (1987) injected E2 diproprionate into the coelomic fluid of *S. nudus* in November when the ovaries were beginning to grow. Individuals injected with E2 diproprionate produced a higher ovary index and a significantly larger number of mature oocytes than the control treatment. Similar results were obtained when individuals were injected with E2 diproprionate in April (when gonads were maturing). They suggested that E2 diproprionate induced ovarian growth via stimulating RNA and protein synthesis, enhancing vitellogenesis and leading to oocyte maturation.

Wasson et al. (2000b) examined the effects of dietary administration of steroids during the gonadal growth stage. Sex- and steroid-specific effects occurred in the gonads of *L. variegatus* that were fed either E2, E2/ P4, P4, P4/F, T, or T/F for 36 days (Wasson et al. 2000b). Individuals fed E2 had ovary indices 54% larger than individuals fed control diets. Sea urchins fed E2, E2/P4, P4, or P4/F had significantly smaller oocytes than echinoids fed control diets. In addition, individuals fed E2/P4 or P4 alone had a significantly larger volume of the ascini occupied by nutritive phagocytes than control individuals. Since no ovarian growth occurred in individuals fed E2/P4, P4, or P4/F, they concluded that the daily dose of P4 administered was sufficient to inhibit any observable E2 stimulus of ovarian growth. These data suggest further that 5α-reduced progestins may stimulate nutrient accumulation in nutritive phagocytes. Although individuals fed T had significantly smaller oocytes than individuals fed T/F, both treatments had significantly larger oocytes than control individuals. Finally, individuals fed either E2 or T had greater amounts of protein in the ovaries than individuals fed T/F, indicating that E2 and 5α-reduced androgens influence protein accumulation in ovaries. Wasson et al. (2000b) concluded that E2 stimulates ovarian growth but does not stimulate oocyte growth and that E2 and 5α-reduced progestin metabolites may promote nutrient allocation to nutritive phagocytes. Furthermore, T does not affect ovary growth but, upon removal of 5α-reduced androgens (by feeding F), T does promote significant increases in oocyte diameters (Wasson et al. 2000b).

The testes of *L. variegatus* fed P4 exhibited the most pronounced effects (Wasson et al. 2000b). Testis indices were 56% larger in individuals fed P4 than controls; however, no effects on testis growth were detected in individuals fed P4/F, T, or T/F, suggesting that 5α-reduced progestins stimulated testicular growth. Individuals fed P4 also had a significantly larger volume of the ascini occupied by nutritive phagocytes and significantly smaller volumes occupied by spermatogenic columns. Furthermore, individuals fed E2/P4, P4/F, or T/F had greater amounts of carbohydrates in their gonads than control individuals, and individuals fed E2/P4, T, or T/F had significantly greater amounts of lipids in their testes than control individuals. Wasson et al. (2000b) concluded that 5α-reduced progestins may regulate nutrient allocation into nutritive phagocytes and may also influence spermatogenic column formation. Corroboratively, low concentrations of P4 were measured during the early stages of spermatogenesis (as indicated by low testis indices) (Wasson et al. 2000a).

Two studies have examined the effect of sex steroids in juvenile sea urchins. Unuma et al. (1996) demonstrated that juvenile *Pseudocentrotus depressus* fed estrone for 29 days had significantly higher body weights, food intake, and feed efficiencies than control individuals. Feeding of T or E2 had no effect on growth or food utilization of juveniles (Unuma et al. 1996), but juveniles fed AD or estrone for 30 days had significantly higher testis indices than control individuals (Unuma et al. 1999). In addition, estrone promoted spermatogenesis, suggesting that estrone may be important for the initiation of spermatogenesis (Unuma et al. 1999). The ovaries were not affected in the juveniles and

Unuma et al. (1999) concluded the lack of an ovarian response to steroids was due to the lack of sexual maturation of the females.

The role of sterols or steroids obtained from natural dietary sources has not been examined. It is unlikely that predictable levels of steroids could be extracted from the diet to support long-term control over reproductive processes. Anecdotal information suggests that soy, a natural phytoestrogen, inhibits gametogenesis in *S. purpuratus* (J Pearse, pers. comm.), suggesting further that sea urchins can respond to dietary endrocrine disruption, which may provide a potential, noninvasive mechanism to regulate gonad physiology and affect gonad production.

3.2. Protein and Peptidergic Factors

Sea urchin coelomocytes synthesize the precursor to a major yolk glycoprotein (MYP; Chapter 2). Recognizing that estrogens regulate yolk production in vertebrates, Harrington and Ozaki (1986a) examined the response of sea urchin coelomocytes to E2 stimulation. Although they did not observe an effect on the synthesis of the MYP precursor, they did demonstrate the synthesis of a novel protein. This protein varied in size from 78 to 82 kDa depending on species. No other effects on the synthesis of total proteins were observed. This novel protein produced by the coelomocytes may act as a transcription factor to induce transcription of the MYP precursor gene (Harrington and Ozaki 1986b). No other studies have examined the role of other proteins or peptidergic factors on gametogenesis in sea urchins.

Cochran and Engelmann (1972) isolated a spawn-inducing factor from radial nerve extracts of *S. purpuratus*. The supernatant of radial nerves boiled in seawater contained a heat-stable factor that induced spawning by testis fragments. The concentration of radial nerve factor paralleled the reproductive season of *S. purpuratus* (Cochran and Engelmann 1972). Digestion of radial nerve factor with pronase and pepsin destroyed all spawn-inducing activity, indicating that this factor is a peptide. The estimated molecular weight of the radial nerve factor is 5600 Da. This peptide was not glycosylated, but had a persistent yellowish-brown color that was associated with the spawn-inducing activity (Cochran and Engelmann 1976). The radial nerve factor induced the production of a gonad factor, which was not digestible with pronase. This gonad factor also induced spawning in testis fragments of *S. purpuratus*. Although this gonad factor co-migrated with 1-methyladenine in thin layer chromatography (Cochran and Engelmann 1976), this substance was not identified definitively. In starfish, 1-methyladenine was identified as the oocyte maturation inducing substance (reviewed by Kanatani 1979). Since oocyte production is continuous and asynchronous throughout the breeding season in *S. purpuratus* (Chatlynne 1969), Cochran and Engelmann (1976) suggested that the significant levels of this gonad factor must be chronically maintained to induce spontaneous oocyte maturation; however, no seasonal measurement of this gonad factor has been reported.

This was the first demonstration of an "endocrine system" in sea urchins. A neuropeptide produced in radial nerves stimulated the release of a factor by the gonads that induce spawning. The mechanism of action and mode of transfer of this radial nerve factor is unknown. In starfish, radial nerve factor is released into the coelomic cavity and transported via coelomic fluid to the gonads (Cochran and Engelmann 1976); however, such mode of transfer in sea urchins would not account for the rapid (within 1 min) spawning response to radial nerve factor. In addition, the cell type that releases the gonad factor is not known. It is interesting to speculate that nutritive phagocytes would synthesize and store this gonad factor and, upon stimulation by the radial nerve factor, release it. The gonad factor would then stimulate the release of neighboring gametes or stimulate the contraction of muscular epithelium, forcibly releasing gametes from the gonad lumen. Most likely, the gonad factor is synthesized and stored within the neurons or epithelial cells of the peritoneum and, upon stimulation by the radial nerve factor, released to stimulate contraction of the muscular epithelium.

Although Elphick et al. (1992) identified a class of putative neuropeptides, they suggested no physiological function for them. Using antisera raised against the asteroid SALMFamide, they isolated five immunoreactive peptides and putatively identified an amidated nonapeptide, FPVGRVHRFamide. SALMFamide immunoreactivity had widespread distribution throughout the gut of *Echinus esculentus*. Specifically, the pharynx–esophagus portion of the gut displayed extensive immunoreactivity associated with the basi-epithelial nerve plexus, which runs between secretory epithelium and the underlying layers of connective tissue and muscle. Although SALMFamide immunoreactivity was also found in association with the outer sub-coelomic epithelial nerve plexus (which is found between the outer connective tissue layer and coelomic epithelium), no immunoreactive epithelial endocrine or neuroendocrine cells were found (Elphick et al. 1992).

3.3. Catecholaminergic and cholinergic factors

Sea urchin gonads are innervated by monaminergic, cholinergic, and peptidergic nerve fibers but do not contain adrenergic fibers (Cobb 1969). Both the ectoneural and endoneural systems contain the catecholamines noradrenalin and dopamine (Khotimchenko 1983). Thin nerve fibers interlace the ascini of the gonads and may contain biogenic amines (Khotimchenko and Deridovich 1991). Both noradrenalin and dopamine inhibited temperature-stimulated oogenesis in *S. nudus* by suppressing growth and maturation of oocytes (Khotimchenko 1983). Interestingly, the levels of these catecholamines varied with reproductive state. Concentrations were low during periods of minimal gametogenic activity and also before spawning. The initiation of spawning was accompanied by increases in catecholamine concentrations (reviewed by Khotimchenko and Deridovich 1991). Since noradrenaline and dopamine decreased the intensity of uptake of labeled uridine and leucine in large oocytes, Khotimchenko (1983) concluded that the suppression of oocyte growth was due to inhibition of RNA and protein synthesis. They hypothesized that these catecholamines influenced oogenesis via transport through the coelomic fluid. In contrast, acetylcholine was secreted in the regions of neuromuscular contacts of the gonads and was effective at inducing the reduction of muscle elements of ascini, suggesting that the act of gamete release may be regulated by cholinergic mechanisms (Khotimchenko 1983).

4. MECHANISMS OF REGULATION

4.1. Paracrine

Analysis of paracrine mechanisms, in which a chemical messenger induces a response in a cell located in close proximity to the cell that synthesizes it, requires an understanding of gonad morphology. Although nutritive phagocytes lack direct physical contact with developing gametes, thin extensions of nutritive phagocytes surround developing gametes. In male sea urchins, the role of these cells has been equated to that of Sertoli cells in mammalian testes (Pearse and Cameron 1991), providing the necessary local environment to stimulate gamete proliferation and growth. The demonstration of endogenous steroid converting enzymes in nutritive phagocytes (Varaksina and Varaksin 1991) and the E2 and P4 effects on mature gonad growth and gametogenesis (Wasson et al., 2000b) suggest that steroids may be produced by the nutritive phagocytes and influence gametogenesis or nutrient translocation to developing gametes. Histochemical studies also indicate that developing oocytes and spermatids possessed 17β-HSD activity (Varaksina and Varaksin 1991), suggesting that developing gametes also synthesize steroids that may act in paracrine or autocrine mechanisms. The secretion of steroids from either nutritive phagocytes or developing gametes would most likely occur by mechanisms similar to those in vertebrates.

The promotion of gonad growth, gametogenic activity, and nutrient storage by sex steroids and/or their derivatives (Wasson et al. 2000b) suggests that sex steroid receptors are present in echinoid gonads. The presence of steroid receptors in either nutritive phagocytes or germinal cells has not been investigated in any echinoid. Shyu et al. (1987), however, identified an estrogen-responsive element upstream from the vitellogenin gene in *S. purpuratus*, suggesting that the vitellogenin synthesis may be regulated by estrogen.

Direct contacts (tight junctions) are found between nutritive phagocytes (Pearse and Cameron 1991). Similar associations have been reported for mammalian follicle cells, which use these junctions to coordinate cellular communication throughout large numbers of follicle cells. Via these tight junctions, interconnected nutritive phagocytes may communicate the necessary factors to stimulate nutrient translocation to the developing gametes. In vertebrates, the peptides insulin and glucagon regulate nutrient deposition in storage cells, mobilization of nutrients, and translocation to appropriate sites for utilization. It would be of interest to determine whether similar proteins exist in sea urchins and have a role in paracrine regulation of the nutrient translocation activities of nutritive phagocytes.

It is likely that protein or peptidergic factors act by paracrine mechanisms to regulate gonad growth. Mammalian models indicate that a variety of proteins, such as TGF-α (Teerds and Dorrington 1995), inhibin (Hillier and Miro 1993), kit ligand (Joyce et al. 1999), and GDF-9 (Dong et al. 1996), produced by oocytes or follicle cells induce specific cellular responses in follicle cells or oocytes, respectively. Future research on similar proteins and peptidergic factors in sea urchin gonads will be of considerable interest.

4.2. Endocrine

Cochran and Engelmann (1972, 1976) identified a radial nerve factor and a gonad factor that potentially represent a true endocrine system. If this gonad factor induces oocyte maturation (Kanatani 1974), then this system would be chronically stimulated since mature ova are present in the ovarian lumen prior to spawning (Pearse and Cameron 1991). It may be more likely that this gonad factor simply induces a rapid muscular response in the gonads to release gametes. No other studies have examined endocrine mechanisms of regulation in sea urchin reproduction.

Several levels of endocrine regulation become apparent when examining sea urchin reproduction. First, the general synchrony of spawning among individuals within a population indicates that a factor (pheromone?) released into the water by an individual induces spawning in other individuals. The nature and identity of this factor is still unknown. Second, the dramatic influence of the environment on gametogenesis and nutrient storage indicates the existence of a communication system between the environment and the organs involved in reproduction. This should not be surprising since environmental factors have been shown to influence the hypothalamo-hypophyseal-gonadal axis in mammals and other vertebrates. Finally, coordination of gametogenic and nutrient translocation activities among and within each gonad indicates the existence of a physiologically controlled mechanism that coordinates these activities. Chemical messengers, whether of neuronal or non-neuronal origins, would most likely accomplish such communication. Movement of these chemical messengers could be accomplished by using the perivisceral coelom and direct diffusion across the gonad wall or by using the hemal system for direct transport to the gonads. Since the hemal system interconnects with each gonad via the aboral ring complex, this system may provide a more efficient and specific means of control.

5. GENE REGULATION IN REPRODUCTION

Identification of the genes involved in reproduction of adult sea urchins is a vastly untapped area of research. Most of the research involved in sea urchin genomics has dealt primarily with developmental biology and embryogenesis. Recently, genome sequencing of *S. purpuratus* was initiated (http://www.hgsc.bcm.tmc.edu/projects/seaurchin/). A search of the NBCI Entrez gene database (http://www.ncbi.nlm.nih.gov/mapview/map_search.cgi?taxid=7668) for "steroid" and "receptor" revealed only six annotated genes in *S. purpuratus* that resemble steroid hormone receptors. As of September 2005 none of these genes have been ascribed functions. Interestingly, one gene resembles the estrogen-related receptor-β and another resembles the progesterone receptor membrane component 1. Molecular endocrinology will be a fruitful area of sea urchin research in the future, from both a functional and evolutionary standpoint.

6. CONCLUSIONS

Environmental cues influence seasonal changes in gonad morphology and physiology. Knowledge of the physiological mechanisms that stimulate growth and maintenance of nutritive phagocytes and gametes is essential to understand the reproductive biology and ecology of sea urchins in the field. This information is also essential for the development of technologies that will optimize the texture, quality, and quantity of gonads from sea urchins in aquaculture. Diet plays a vital role in stimulation and maintenance of gonad production and is one parameter that is easily manipulated in aquaculture. We have found that formulated feeds supplemented with steroids affect gametogenesis and nutrient allocation to nutritive phagocytes. In the future, phytosteroids may be used to influence gonad growth. For example, Pearse (pers. comm.) attributed the increase in nutritive cells relative to gametes in the gonad to the presence of soy meal in individuals fed a formulated feed. Soy-based diets promote estrogen-dependent as well as estrogen-independent processes (Barnes 1998), and may provide a natural method to adjust a physiological state. Further research is required to fully understand how the dietary administration of these and other specific micronutrients may stimulate or inhibit gonad growth.

Sex steroids, proteins, catecholaminergic, and cholingeric factors all play roles in regulating various aspects of gonad function. No data, however, exist on the chemical messengers required to promote nutrient translocation from the gut to the gonads, the mobilization of nutrients within the nutritive phagocytes, or translocation of nutrients to developing gametes. Understanding the physiological control of these mechanisms will assist in the development of formulated feeds to specifically regulate gonad quality and quantity.

Most of these studies have examined singular reproductive events (i.e. spawning, gonad growth, etc.) in only a few species of sea urchins. Large gaps are still present in our general understanding of the physiological mechanisms that control reproduction and provide direct links between environmental stimuli and gonad function. Studies using molecular biology, genomics, and bioinformatic tools are essential for elucidating the cellular pathways that regulate specific reproductive processes. Such tools will be extremely beneficial to our understanding of the role of chemical messengers and the influence of endocrine disruptors in sea urchin reproduction.

ACKNOWLEDGMENTS

We thank IJ Blader and JM Lawrence for comments on this chapter. Preparation of this review was supported in part by the Mississippi-Alabama Sea Grant Consortium.

REFERENCES

Barnes S (1998) Evolution of the health benefits of soy isoflavones. Proc Soc Exp Biol Med 217: 386–392

Bishop CD, Watts SA (1994) Two-stage recovery of gametogenic activity following starvation in *Lytechinus variegatus* Lamarck (Echinodermata: Echinoidea). J Exp Mar Biol Ecol 177: 27–36

Böttger SA, McClintock JB (2002) Effects of inorganic and organic phosphate exposure on aspects of reproduction in the common sea urchin *Lytechinus variegatus* (Echinodermata: Echnoidea). J Exp Zool 292: 660–671
Botticelli CR, Hisaw FL, Wotiz HH (1961) Estrogens and progesterone in the sea urchin (*Strongylocentrotus franciscanus*) and Pecten (*Pecten hericius*). Proc Soc Exp Biol 106: 887–889
Chatlynne LG (1969) A histochemical study of oogenesis in the sea urchin *Strongylocentrotus purpuratus*. Biol Bull 136: 167–184
Cobb JLS (1969) The distribution of monamines in the nervous system of echinoderms. Comp Biochem Physiol 28: 967–971
Cochran RC, Engelmann F (1972) Echinoid spawning induced by a radial nerve factor. Science 178: 423–424
Cochran RC, Engelmann F (1976) Characterization of spawn-inducing factors in the sea urchin, *Strongylocentrotus purpuratus*. Gen Comp Endocrin 30: 189–197
Colombo L, Belvedere P (1976) Gonadal steroidogenesis in echinoderms. Gen Comp Endocrinol 29: 255–256
Donahue JK (1940) Occurrence of estrogens in the ovaries of certain marine invertebrates. Endocrinology 27: 149–152
Donahue JK, Jennings DE (1937) The occurrence of estrogenic substances in the ovaries of echinoderm. Endocrinology 21: 587–877
Dong J, Albertini DF, Nishimori K, Kuman TR, Lu N, Matzuk MM (1996) Growth differentiation factor-9 is required during early folliculogenesis. Nature 383: 531–535
Elphick MR, Parker K, Thorndyke MC (1992) Neuropeptides in sea urchins. Regul Peptides 39: 265
Harrington FE, Ozaki H (1986a) The effect of estrogen on protein synthesis in echinoid coelomocytes. Comp Biochem Physiol 84B: 417–421
Harrington FE, Ozaki H (1986b) The major yolk glycoprotein precursor in echinoids is secreted by coelomocytes into the coelomic plasma. Cell Different 19: 51–57
Hillier SG, Miro F (1993) Inhibin, activin, and follistatin. Potential roles in ovarian physiology. Ann NY Acad Sci 687: 29–38
Hines GA, Watts SA, McClintock JB (1994) Biosynthesis of estrogen derivatives in the echinoid *Lytechinus variegatus* Lamarck. In: David B, Guille A, Féral J-P, Roux M (eds) Echinoderms through time. Balkema, Rotterdam, pp 711–716
Joyce IM, Pendola FL, Wigglesworth K, Eppig JJ (1999) Oocyte regulation of kit ligand expression in mouse ovarian follicles. Dev Biol 214: 342–353
Kanatani H (1974) Presence of 1-methyladenine in the sea urchin gonad and its relation to oocyte maturation. Dev Growth Diff 16: 157–170
Kanatani H (1979) Hormones in echinoderms. In: Barrington EJW (ed.) Hormones and evolution. Vol. 1. Academic Press, New York, pp 273–307
Keesing JK, Hall KC (1998) Review of harvests and status of world sea urchin fisheries points to opportunities for aquaculture. J Shellfish Res 17: 1597–1604
Khotimchenko YS (1983) Effect of adrenotropic and cholinotropic preparations on growth, maturation, and spawning of the sea urchin *Strongylocentrotus nudus*. Biol Morya 2: 57–63
Khotimchenko YS, Deridovich II (1991) Monoaminergic and cholinergic mechanisms of reproduction control in marine bivalve molluscs and echinoderms: a review. Comp Biochem Physiol 100C: 311–317
Khristoforava NK, Gnezdilova SM, Vlasova GA (1984) Effect of cadmium on gametogensis and offspring of the sea urchin *Strongylocentrotus intermedius*. Mar Ecol Prog Ser 17: 9–14
Langston WJ, Burt GR, Chesman BS, Vane CH (2005) Partitioning, bioavailability and effects of oestrogens and xeno-oestrogens in the aquatic environment. J Mar Biol Ass UK 85: 1–31
Lipina IG, Evtushenko ZS, Gnezdilova SM (1987) Morpho-functional changes in the ovaries of the sea urchin *Strongylocentrotus intermedius* after exposure to cadmium. Ontogenez 18: 269–276
Pearse JS, Cameron RA (1991) Echinodermata: Echinoidea. In: Giese AC, Pearse JS, Pearse VB (eds) Reproduction of marine invertebrates. Vol. VI. Boxwood Press, Pacific Grove, pp 514–664
Robinson SM (2004) The evolving role of aquaculture in the global production of sea urchins. In: Lawrence J, Guzmán O (eds) Sea urchins: fisheries and ecology. DEStech Publications Inc., Lancaster, pp 343–357
Shyu AB, Blumenthal T, Raff RA (1987) A single gene encoding vitellogenin in the sea urchin *Strongylocentrotus purpuratus*: sequence at the 5′ end. Nucleic Acid Res 15: 10405–10417

Teerds KJ, Dorrington JH (1995) Immunolocalization of transformation growth factor-alpha and luteinizing hormone receptor in healthy and atretic follicles of the adult rat ovary. Biol Reprod 52: 500–508

Unuma T, Yamamoto T, Akiyama T (1996) Effects of oral administration of steroids on the growth of juvenile sea urchin *Pseudocentrotus depressus*. Suisanzoshoku 44: 79–83

Unuma T, Yamamoto T, Akiyama T (1999) Effects of steroids on gonadal growth and gametogenesis in the juvenile red sea urchins *Pseudocentrotus depressus*. Biol Bull 196: 199–204

Varaksina GS, Varaksin AA (1987) Effects of estradiol diproprionate on oogenesis of the sea urchin *Strongylocentrotus nudus*. Biol Morya 4: 47–52

Varaksina GS, Varaksin AA (1991) Localization of steroid dehydrogenase in testes and ovaries of sea urchins. Biol Morya 2: 77–82

Wasson KM, Gower BA, Hines GA, Watts SA (2000a) Levels of progesterone, testosterone, and estradiol and the androstenedione metabolism in the gonads of *Lytechinus variegaus* (Echinodermata: Echinoidea). Comp Biochem Physiol 126C: 153–165

Wasson KM, Gower BA, Watts SA (2000b) Responses of ovaries and testes of *Lytechinus variegatus* (Echinodermata: Echinoidea) to dietary administration of estradiol, progesterone and testosterone. Mar Biol 137: 245–255

Wasson KM, Hines GA, Watts SA (1998) Synthesis of testosterone and 5α-androstanediols during nutritionally stimulated gonadal growth in *Lytechinus variegatus* Lamarck (Echinodermata: Echinoidea). Gen Comp Endocrinol 111: 197–206

Wasson KM, Watts SA (1998) Proscar® inhibits 5α-reductase activity in the ovaries and testes of *Lytechinus variegatus* Lamarck (Echinodermata: Echinoidea). Comp Biochem Physiol 120C: 425–431

Wasson KM, Watts SA (2000) Progesterone metabolism in the ovaries and testes of the echinoid *Lytechinus variegatus* Lamarck (Echinodermata). Comp Biochem Physiol 127C: 263–272

Watts SA, Hines GA, Byrum CA, McClintock JB, Marion KR (1994) Tissue- and species-specific variations in androgen metabolism. In: David B, Guille A, Féral J-P, Roux M (eds) Echinoderms through time. Balkema, Rotterdam, pp 155–164

Edible Sea Urchins: Biology and Ecology
Editor: John Miller Lawrence

Chapter 5

Echinoid Larval Ecology

Larry R McEdward[a,†] and Benjamin G Miner[a,b,c]

[a] *Department of Zoology, University of Florida, Gainesville, FL (USA).*
[b] *Bodega Marine Laboratory, University of California, Davis, (USA).*
[c] *Department of Biology, Western Washington University, WA (USA).*

1. INTRODUCTION

Most echinoids, including all but one of the edible species covered in this volume, are free-spawners that produce vast numbers of small, yolk-poor eggs that develop into planktonic, feeding larvae known as echinoplutei. In this review, we describe the diversity of larval stages and the patterns of development in echinoids. We also review the ecological challenges that echinoplutei face, and the physiological, morphological, developmental, and behavioral solutions they utilize to respond to those problems.

2. THE ECHINOID LIFE CYCLE

Echinoids have simple life cycles. Reproduction is confined to the adult stages (see Chapter 2) and generally involves the spawning of eggs and sperm freely into the seawater, followed by external fertilization. In most species, development occurs in the plankton. The exceptions occur in species that brood in which the young are retained in or on the adult body throughout development to the juvenile stage (see review by Poulin and Féral 1996). Developmental biologists have intensively investigated the basic biology of early development through the cleavage stages, blastula formation, gastrulation, and larval morphogenesis. Echinoid development becomes interesting once the characteristic larval form, the echinopluteus (Fig. 1), is attained. Larval development involves growth and elaboration of the larval body, as well as formation of the rudiment of the juvenile echinoid. Development is arrested when metamorphic competence is achieved. Induction of settlement and metamorphosis occur in response to environmental factors that signal the availability of suitable benthic habitat. Post-metamorphic development involves growth and sexual maturation to yield a reproductive adult.

† Deceased.

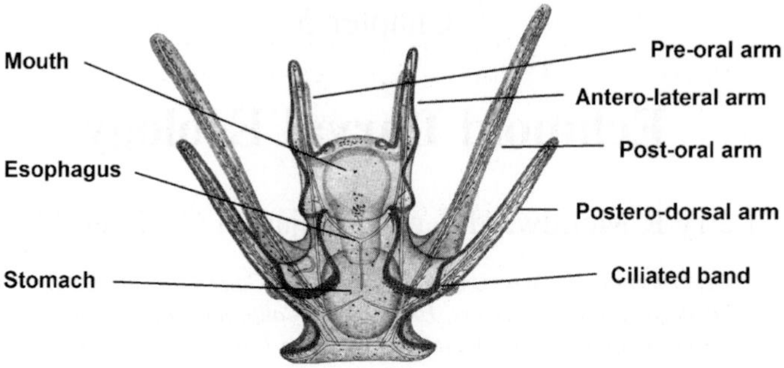

Fig. 1. Dorsal view of the echinopluteus larva of *Echinometra lucunter*, showing the larval arms, ciliated band, and gut. Anterior is oriented towards the top of the figure. Modified from Mortensen (1921).

3. LIFE CYCLE DIVERSITY

The developmental patterns in echinoderms can be described using three characters (McEdward and Janies 1997): mode of nutrition (planktotrophic or lecithotrophic), habitat (pelagic or benthic), and type of morphogenesis (complex larval, simple larval, or nonlarval). Five life cycle patterns have been documented in the echinoids (Table 1). Obligate planktotrophy with a pluteus larva is the most common pattern, occurring in 114 out of 175 species for which data are available (McEdward and Miner 2001). This pattern and larval type are ancestral for echinoids (Strathmann 1978a; Wray 1995a), and also are the most widespread within the class, occurring in 9 of the 11 orders and 26 of the 31 families for which there is published information. This pattern is characterized by very small eggs and juveniles. The larval duration is commonly on the order of 1 month, but this varies widely among species (Emlet et al. 1987). Facultative planktotrophy has been documented in only two species: the clypeasteroid *Clypeaster rosaceus* (Emlet 1986a) and the spatangoid *Brisaster latifrons* (Hart 1996). Although more cases are likely to be discovered, it is unlikely that this pattern is common among echinoids. Egg sizes are intermediate, juveniles are small, and the larval duration is on the order of 1–2 weeks. Species with feeding larvae are common in shallow temperate and tropical waters of both hemispheres (Emlet 2002).

Nonfeeding lecithotrophy involves development via a simplified yolky larva and occurs in 9 out of 175 species, and in 4 orders and 5 families. Nonfeeding lecithotrophs have very large eggs, large juveniles, and larval periods on the order of 1–2 weeks (Emlet et al. 1987). Brooding occurs in 49 out of 175 species and in 7 orders and 8 families (McEdward and Miner 2001). Brooding species retain offspring among the spines (e.g. *Cassidulus caribbearum*, Gladfelter 1978), under the body near the mouth (e.g. *Goniocidaris umbraculum*, Barker 1985), or in depressions of the test (marsupia) (e.g. *Abatus cordatus*, Schatt and Féral 1996). Brooders produce large eggs but there is little information on the sizes of juveniles and duration of development (Emlet et al. 1987). Direct development has been documented in only a single species of echinoid, the Antarctic spatangoid, *A. cordatus* (Schatt and Féral 1996). The eggs are large, juveniles are very large, morphogenesis is highly modified, and development is extremely slow. In some

Table 1 Life cycle patterns and developmental character states in echinoids.

	Obligate Planktotrophy	Facultative Planktotrophy	Nonfeeding Lecithotrophy	Brooding	Direct Development
Example	*Lytechinus variegatus*	*Clypeaster rosaceus*	*Heliocidaris erythrogramma*	*Goniocidaris umbraculum*	*Abatus cordatus*
Mode of nutrition	Planktotrophy	Lecithotrophy	Lecithotrophy	Lecithotrophy	Lecithotrophy
Developmental habitat	Pelagic	Pelagic	Pelagic	Benthic	Benthic
Type of morphogenesis	Complex larval	Complex larval	Simple larval	Simple larval	Nonlarval
Feeding?	Yes	Yes	No	No	No
% Species	65%	1%	5%	28%	<1%
Egg size	70–250	280–350	275–1200	1000–2000	1300
Developmental duration	5–100	5–35	4–30	10–?	250
Juvenile size	250–700	290–410	500–3000	400–?	1800

Egg and juvenile sizes are diameters (μm) and larval development times are given in days between fertilization and settlement/metamorphosis or emergence of the juvenile. Percentages are calculated for 175 species with documented patterns of development (McEdward and Miner 2001).

orders developmental patterns have been particularly well studied (e.g. Clypeasteroida and Echinoida) but other taxa are very poorly known (e.g. Holasteroida and Holectypoida) (McEdward and Miner 2001). Species with nonfeeding, lecithotrophic larvae and either pelagic or benthic development occur in shallow water near southern Australia and Japan, and in deep water in all oceans, except the Arctic Ocean (Emlet 2002).

4. ECHINOID LARVAL DIVERSITY

The echinopluteus (Fig. 1) is a complex, pelagic larva that typically possesses eight anterior-directed arms that bear the ciliated feeding structure and are supported by calcareous skeletal rods (Okazaki 1975). Echinoplutei feed on suspended particulate food and possess a complete, functional gut (Burke 1981). Nerves run along the ciliated band and the esophagus, probably coordinating the production of swimming and feeding currents (Mackie et al. 1969; Strathmann 1971; Burke 1978, 1983a; Hart 1991). A neuropile located at the anterior end of the oral hood probably controls metamorphosis (Burke 1978, 1983a,b). Most of the interior of the larva is gel-filled blastocoelic space (Strathmann 1989). There are three sets of paired coelomic sacs: anterior axocoels, hydrocoels, and posterior somatocoels.

The body forms of plutei are very diverse because of variation in the number (2–13, typically 8) and relative sizes of the larval arms (Onoda 1936, 1938; Mortensen 1921, 1931, 1937, 1938; Wray 1992; Pearse and Cameron 1991; Fig. 2). Larvae of *Diadema setosum* have only two very long, well-developed arms (Fig. 2b). Spatangoids have a unique retrodorsal rod projecting posteriorly, and often as many as 12 anterior arms

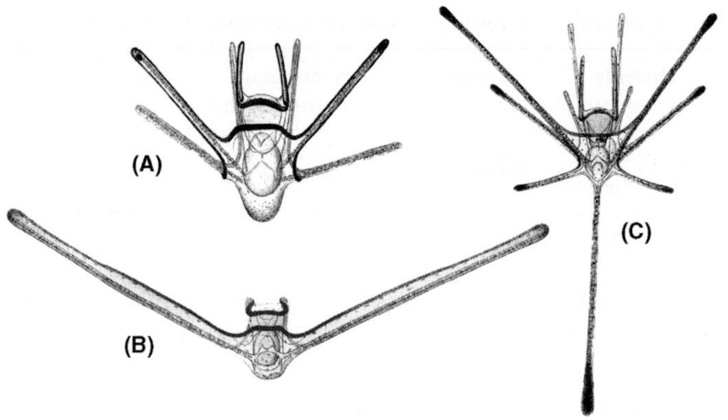

Fig. 2. Diversity of planktotrophic echinopluteus larvae. Mid-stage larvae of (A) *Echinodiscus auritus*, (B) *Diadema setosum* with elongated post-oral arms, and (C) *Lovenia elongata* with an unpaired retrodorsal arm. Modified from Mortensen (1931, 1937).

(Fig. 2c). The skeletal rods supporting the arms can be solid or fenestrated. The elaboration of the posterior skeleton in the main body region is also highly variable.

Yolky, nonfeeding echinoid larvae have much simpler external morphology than echinoplutei (Fig. 3). Derived larvae include reduced pluteus-like forms that have 2–4 larval arms, but lack ciliated bands and a functional gut (e.g. *Phyllacanthus imperialis*, Olson et al. 1993; *Peronella japonica*, Okazaki and Dan 1954) and "schmoos": (*sensu* Wray 1995a, 1996; Fig. 3) that lack arms, ciliated bands, gut, and have only a vestigial skeleton (e.g. *Asthenosoma ijimai*, Amemiya and Tsuchiya 1979; Amemiya and Emlet 1992; *Heliocidaris erythrogramma*, Williams and Anderson 1975; Emlet 1995; *P. parvispinus*, Parks et al. 1989). However, all echinoid larvae share a larval body plan

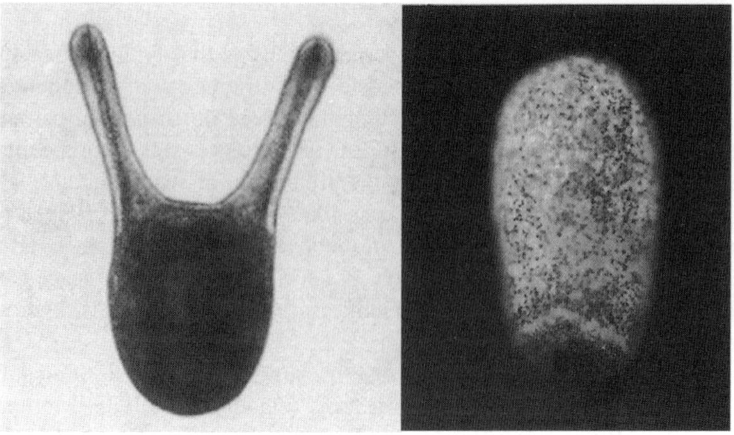

Fig. 3. Nonfeeding echinoid larvae. Left: reduced pluteus of *Peronella japonica*. Modified from Okazaki (1975). Right: "schmoo" larva of *Heliocidaris erythrogramma*. Modified from Emlet (1995).

defined by body symmetry, morphogenetic axes, coelomic organization, nervous system, vestibule, and metamorphic fates of larval structures (McEdward and Miner 2001).

5. REPRODUCTIVE ECOLOGY

5.1. Egg Provisioning

Echinoids produce eggs that range in size by four orders of magnitude (Table 2). Eggs are provisioned largely with protein and lipid; carbohydrate is negligible (Table 2). Of the lipid component, eggs of planktotrophic species are predominantly triglycerides, whereas eggs of lecithotrophic species are predominantly wax esters (Villinski et al. 2002). Egg content represents the level of energy reserves that are available to fuel development of the prefeeding stages, and can supplement feeding stages of larval development. Total energy in echinoid eggs also ranges over four orders of magnitude and is correlated with egg volume (joules $egg^{-1} = 0.094 + 11.494$ egg volume 1.076, $r^2 = 0.997$) and larval nutritional mode (McEdward and Morgan 2001; Table 2).

5.2. Fertilization Ecology

Echinoids have provided a good model for laboratory and field studies of fertilization in free-spawning organisms (see reviews by Levitan 1995; Levitan and Petersen 1995; Yund 2000). Spawning occurs in response to pulses of phytoplankton and elevated temperature (Starr et al. 1990, 1992, 1993). Males typically spawn before females (Levitan 2005), but spawning generally does not involve aggregation of adults (e.g. Levitan 1988). Because of the limited duration of gamete viability after spawning, sperm dilution can severely limit fertilization success, especially in rapid currents or turbulent flow (Pennington 1985; Denny and Shibata 1989; Levitan 2000). However, shear stress imposed by waves can also enhance fertilization (Denny et al. 2002) and elaborate blocks to polyspermy in echinoid

Table 2 Egg characteristics of echinoids categorized by mode of larval nutrition.

	Planktotrophic larvae		Lecithotrophic larvae	
	Range	Mean	Range	Mean
Volume	0.2–11.5	1.75	1000–4000	2100
Protein	35–83%	57%	35–60%	47%
Lipid	18–65%	42%	35–60%	47%
Carbohydrate	4–6%	5%	2–4%	3%
Total energy	0.001–0.046	0.007	16–45	27

Units for measurements: volume = nl; protein, lipid, carbohydrate = % of summed biochemical fractions by weight; total energy = joules egg^{-1}.
Data from McEdward and Morgan (2001).

eggs suggest that high sperm concentrations are encountered. Experiments and models indicate that target size influences the probability of fertilization and could therefore be a selective factor in the evolution of egg size in free-spawning organisms (Levitan 1993, 1996; Levitan and Irvine 2001; but see Podolsky and Strathmann 1996; Podolsky 2004, for a contrary view). Additionally, sexual dimorphism in spawning behavior of sea urchins might be driven by fertilization ecology (Levitan 2005).

The potential for hybridization exists among co-occurring, congeneric species that free-spawn. In some cases, (e.g. *Strongylocentrotus* spp.; *Echinometra* spp.) barriers to hybridization are incomplete (Strathmann 1981; Aslan and Uehara 1997). Fertilization ecology of echinoids is a fascinating and complex topic that deserves much more attention from marine ecologists.

6. LARVAL ECOLOGY

6.1. Reproductive Strategies

Echinoids have provided important insights into the evolutionary ecology of reproductive strategies and developmental patterns of larvae. Because early echinoid embryos are highly regulative, blastomere isolation experiments can be used to test hypotheses about the consequences of different egg sizes on larval function and performance (McEdward 1996). Reductions in egg size result in smaller early stage larvae and slower development through the early larval stages, but do not affect the size or development rate of later larvae, or the size of the post-metamorphic juveniles (Sinervo and McEdward 1988; Hart 1995).

Some (possibly most) echinoid larvae have the capacity to feed facultatively, i.e. before egg reserves are exhausted and exogenous food is necessary for continued growth and development. The duration of this facultative feeding period is correlated with egg size (Miner et al. 2005). At one extreme are *C. roseaceus* and *B. latifrons* that can feed on particulate food, but are capable of developing to metamorphosis without food (= facultative planktotrophy) (Emlet 1986a; Hart 1996). However, some species that require food to complete development (obligate planktotrophy) also have substantial facultative feeding capability (e.g. *Encope michelini*, Eckert 1995; *Clypeaster subdepressus, Encope aberrans, Leodia sexiesperforata*, Herrera et al. 1996; Miner et al. 2005). Surprisingly however, the benefits of facultative feeding appear to be greater for species with small, poorly provisioned eggs than species with large, well-provisioned eggs (Reitzel et al. 2005). The discovery of diversity in nutritional requirements has important implications for understanding the ecology and evolution of feeding larvae (McEdward 1997).

Although nonfeeding larvae are less common in echinoids compared to other echinoderm classes, the developmental differences between feeding and nonfeeding larvae have been well studied, particularly in the case of the Australian sea urchins *H. tuberculata* and *H. erythrogramma* (Wray 1995a). The transition from small-egg planktotrophy to large-egg lecithotrophy is a unidirectional evolutionary change that has occurred at least 14–20 independent times within the class (Strathmann 1978a; Wray 1996; McEdward and Miner 2001). This shift can occur rapidly, in as few as 5–8 million years (estimated from mtDNA data) or 10–13 million years (DNA hybridization estimate) in the case of the *Heliocidaris* spp. (Smith et al. 1990; McMillan et al. 1992) and involves substantial changes in egg size, cleavage pattern, cell fates, timing and location of gene expression, gastrulation, and larval morphogenesis (Williams and Anderson 1975; Raff 1987; Wray and Raff 1989, 1991; Henry et al. 1990; Emlet 1995; Wray 1995b, Zigler et al. 2003).

6.2. Feeding

Echinoplutei feed on particles that are suspended in the surrounding seawater. Currents are produced by the action of the cilia on the ciliated band. Water flows across the band and away from the mouth and circumoral field of the larva at a velocity of 1300–1700 $\mu m\ s^{-1}$. In response to a food particle, a transient (0.1 s), local (100 μm of band), reversal of ciliary beat is induced (Strathmann 1971; Strathmann et al. 1972; Hart 1991) and the particle is retained on the upstream side of the ciliated band. Captured particles are transported to the mouth and ingested. Temperature can influence feeding rates of plutei by altering both the physiology of individuals and the viscosity of water (Podolsky 1994). Additionally, individuals capture larger particles as seawater becomes more viscous (Podolsky 1994).

The natural diet of echinoid larvae is not well known, but larvae in the laboratory eat unicellular algae ranging from 5 to 50 μm in diameter or length (e.g. Rassoulzadegan and Fenaux 1979). Particle size, concentration, flavor, and the stage of larval development all influence larval feeding rates and selectivity (Pedrotti 1995). Plutei preferentially capture or ingest 10 μm plastic beads compared to an equal concentration of 5 μm beads; however, flavoring the smaller beads with algal exudates reverses the size preference (Rassoulzadegan et al. 1984). At low particle concentrations, larvae presumably feed at maximum rates, but at very high concentrations, feeding rates decrease or feeding ceases altogether (Strathmann 1971).

Many species of cultured algae support rapid growth and development through metamorphosis (e.g. *Rhodomonas lens*, *Dunaliella tertiolecta*, *Phaeodactylum tricornutum*, *Isochrysis galbana*; *Cricosphaera elongata*; see Lawrence et al. 1977). A mixture of algal species can lead to either increased or decreased growth (Cárcamo et al. 2005). Echinoplutei can achieve maximum development rates at moderate algal concentrations in the laboratory (e.g. 8 cells μl^{-1} in *Lytechinus variegatus*, Herrera 1998). Algal species differ substantially in nutritional quality for plutei. Larval survival, development rate, growth, and metamorphic success of *Paracentrotus lividus* and *Arbacia lixula* all vary with diet, but juvenile size at metamorphosis does not (Pedrotti and Fenaux 1993). The larval growth and development rates of at least some species of echinoids can be strongly limited by natural concentrations of planktonic food, even in productive coastal waters (Paulay et al. 1985; Lamare and Barker 1999). Whether food limitation is common among marine invertebrate larvae is uncertain (Olson and Olson 1989). Indirect tests of larval growth in the field, using relationships between feeding history and morphology (see Section 6.4), indicate that food limitation occurs and affects numerous species within a geographic area (Fenaux et al. 1994). Intraspecific competition for food is unlikely because planktotrophic larvae are generally very scarce (Strathmann 1996).

Echinoid larvae have the capacity to acquire dissolved organic material (DOM), especially neutral amino acids, directly from seawater (Manahan 1990). In the early stages of development, DOM uptake can provide up to 79% of estimated metabolic requirements of early stage larvae of *S. purpuratus* (Manahan et al. 1983). However, it is unclear whether planktotrophic larvae can successfully complete development and metamorphosis solely on DOM. Likewise, the role of marine bacteria in the nutrition of echinoid larvae is not known.

6.3. Larval Growth

Planktotrophic echinoid larvae grow substantially. On average, juvenile diameters are 3.8 times greater than egg diameters and juvenile test volumes are 94.5 times greater than egg volume (calculated from data in Emlet et al. 1987; Herrera et al. 1996). However, the range is considerable: *L. sexiesperforata* has a diameter ratio (juvenile : egg) of only 1.3 and a volume ratio of 2.0, whereas *Arbacia punctulata* has a diameter ratio of 9 and a volume ratio of 745. Initially, growth increases the size and complexity of the body through the addition and elongation of larval arms (McEdward 1984). The feeding capacity of a pluteus larva is proportional to the total length of the ciliated band because of the mechanics of current production and particle capture (Strathmann 1971; Strathmann et al. 1972; Hart 1991). Early larval growth is designed to build an effective feeding machine. Although the geometric growth of the larva is considerable, the increases in biomass (protein content), energy content, and metabolism are modest, on the order of two- to fourfold. Later in larval life, growth is concentrated in the developing juvenile rudiment. Growth of the juvenile rudiment is energetically expensive, involving a further 9- to 14-fold increase in protein and metabolic activity (McEdward 1984; Fenaux et al. 1985).

Nonfeeding larvae, whether planktonic or brooded, should not grow (add biomass) during development. Utilization of metabolic fuel should result in a decrease in biomass until the juvenile stage when feeding begins. However, the few data that exist for brooding species generally show some increase in biomass or energy content (8–40%) between the egg and juvenile stage (Lawrence et al. 1984; McClintock and Pearse 1986). Two explanations seem likely: maternal transfer of nutrients could have occurred during brooding or juvenile feeding might have begun before the analysis of juvenile biomass. In contrast, pelagic nonfeeding larvae undergo no net change in biomass or energy content during development to metamorphosis (Hoegh-Guldberg and Emlet 1997).

A comparison of co-occurring, congeneric sea urchins illustrates the differences in larval energetics between planktotrophic (*H. tuberculata*) and lecithotrophic (*H. erythrogramma*) development (Hoegh-Guldberg and Emlet 1997). The planktotrophic eggs contain 1.3–1.6% of the energy required for development to metamorphosis, while the lecithotrophic eggs provide all of the energy. The metabolic cost of development is approximately 0.036 J μg^{-1} for juveniles, regardless of the mode of development, suggesting that there are no energetic advantages or disadvantages associated with developmental mode with respect to building a juvenile. The differences are quantitative and they scale with the size of the juvenile. The planktotrophic species produce a juvenile 5.3–7.5 μg DOW (dry organic weight) in 21–30 days. The lecithotrophic species produce a juvenile 11.6–19 μg DOW in 3.5–4 days.

6.4. Phenotypic Plasticity

Echinoplutei vary body form and developmental trajectory in response to the availability of particulate food (Boidron-Metairon 1988; Strathmann et al. 1992; Miner 2005). Larvae that encounter low concentrations of food develop longer arms and smaller stomachs, relative to body size, and delay development of the juvenile rudiment compared to larvae that encounter high concentrations of food. Longer arms increase the capacity to acquire particles from the surrounding seawater (Hart and Strathmann 1994), which is presumably beneficial when food is scarce because acquiring more food should shorten development time and reduce the risk of larval mortality from planktonic predators. The benefits of changes in stomach size and developmental timing of juvenile rudiment formation are less clear, but might result from an energetic tradeoff with arm length (Strathmann et al. 1992; Miner 2005). When food is abundant enough, it might be captured with short arms, and thus energy can be reallocated if shorter arms are produced. Reallocation of this energy to produce a larger stomach and assimilate more food or to develop the juvenile rudiment presumably shortens development time and the time larvae are exposed to planktonic predators. Plutei of *L. variegatus* also display a similar response to the amount of variation in food concentrations (Miner and Vonesh 2004). Larvae produce shorter arms when reared with a fluctuating concentration of food compared to larvae reared with the same mean concentration of food but smaller fluctuations in food concentration. Whether or not this response is adaptive is unclear.

6.5. Swimming

Echinoplutei use the ciliated band feeding structure to generate water currents for locomotion. The band loops around the larval body in an arrangement that results in a posteriorly directed net flow. In still water, larvae with a few long, shallow-angled arms, such as *D. setosum* (Fig. 1b), have greater swimming speed and weight carrying capacity compared to larvae with more arms at greater angles. However, in vertical shear flow, larvae with this stable arm arrangement (fewer, shallow-angled arms) move into downwelling water, which results in a net downward transport (Grunbaum and Strathmann 2003). Late-stage larvae often have parts of the band, known as epaulettes or vibratile lobes, specialized for locomotion, in which the band becomes physically separated from the feeding regions of the band. Locomotory regions of the band are located on prominent ridges or lobes and are oriented transversely so that the current is directed posteriorly. They have much longer cilia that produce faster currents than on the feeding parts of the band. Nonfeeding larvae have much simpler body forms and do not have the conflicting demand of generating both swimming and feeding currents. Cilia tend to be arranged either as a uniform field over the entire surface of the body or in transverse rings (Emlet 1994).

Larvae swim at speeds of 0.2–0.7 μm $s^{-1}(= 0.2 - 2.4$ body lengths per second) and later-stage larvae swim faster (Lee, pers. comm.). Many species are positively phototactic (or negatively geotactic) early in larval life, yet swim towards the bottom as they become competent to undergo metamorphosis and begin to explore the substratum (Chia et al. 1984a). However, late-stage plutei are more common in surface waters than earlier stages

(Emlet 1986b; Pennington and Emlet 1986). Larval swimming certainly moves larvae sufficiently to renew the food and oxygen, as well as dilute excretory and digestive wastes, in the immediately surrounding water. Larval swimming also influences vertical position in the water column and could indirectly affect horizontal transport. Larvae can detect and orient to patches of food (Metaxas and Young 1998a) and can cross moderate haloclines (Metaxas and Young 1998b). Water temperature affects larval swimming speeds by affecting rates of ciliary beating and by altering the viscosity of the seawater, which can have substantial effects on rates of movement (Podolsky and Emlet 1993). Some deep-sea species release eggs that can develop in surface waters, either by floating or upward swimming of early stages (Young et al. 1996). Presumably, late-stage larvae either swim or sink through hundreds to thousands of meters of the water column to settle in the adult habitat. However, swimming speeds are much too slow, compared to horizontal advection by currents, for dispersal of larvae in the sea.

6.6. Mortality and Defense

Very little is known about what kills echinoid larvae in the plankton. Most likely predation is the major cause of larval mortality (Thorson 1950). In the laboratory, a suite of planktonic invertebrate and vertebrate predators (Rumrill and Chia 1985; Pennington et al. 1986), and benthic suspension feeders (Tegner and Dayton 1981; Cowden et al. 1984) consume echinoid larvae. Larval mortality rates have been estimated from field studies to be in the range of 6–27% per day (Rumrill 1990; Lamare and Barker 1999). However, Johnson and Shanks (2003) suggest that predators are not the major cause of larval mortality.

Echinoplutei have several defenses that reduce predation (Young and Chia 1987; Rumrill 1990). In the laboratory, embryos and early larval stages are more readily consumed by common predators than are late-stage larvae (Rumrill and Chia 1985; Pennington et al. 1986). However, this difference is explained by body form (i.e. stage of development) and not size alone and could be the result of skeletal rods in the larval arms or behavioral differences. Defensive behaviors include swimming against quick currents, which are similar to those created by predatory fishes (Young and Chia 1987). Defenses against predation in nonfeeding "schmoo" larvae have not been investigated, but it seems likely that chemistry, rather than morphology would provide greater potential for protection.

It is not known if echinoid larvae starve to death, but they certainly develop slower when food is scarce (Herrera 1998). This could increase the risk of predation (Pearse and Cameron 1991; Strathmann 1996) by extending the duration of the period when the larvae are exposed to planktonic predators. Extreme temperature, salinity, and ultraviolet radiation (UVR) levels potentially have lethal or sublethal effects on echinoid larvae (Pennington and Emlet 1986; Metaxas 1998; see review by Pechenik 1987). Echinoid larvae use behaviors and chemicals to protect against UVR. Mycosporine-like amino acids (MAAs) protect eggs of *S. droebachiensis* against UVR (Adams and Shick 1996). However, embryos of *Dendraster excentricus* are more UVR resistant than embryos of *Strongylocentrotus* spp., but do not have MAAs (Hoffman and Lamare, pers. comm.). Although carotenoids do not appear to function as sunscreen in echinoid embryos, they appear to help mitigate the effects of UVR, possibly by reacting with oxygen species

(Lamare and Hoffman 2004). DNA repair rates might also be important for protecting against UVR (Hoffman and Lamare, pers. comm.). Additionally, larvae that migrate downward reduce the risk of UVR damage (Pennington and Emlet 1986).

7. RECRUITMENT ECOLOGY

Three important processes affect planktonic larvae and thereby influence the abundance of adult echinoids. Larvae are transported (advection and dispersion) passively by water motion throughout the planktonic period and must eventually arrive at a suitable juvenile habitat. Larvae must then detect and settle onto the benthos in the suitable microhabitat. After larvae have settled, juveniles must survive until metamorphosis is complete. At this point larvae are considered to have successfully recruited, ending a critical part of the echinoid life cycle.

The success of each of these three processes is influenced by a suite of abiotic and biotic factors. Abiotic factors include the patterns of water motion on a wide range of spatial scales, temperature, salinity, and ultraviolet light, whereas biotic factors include food, disease, and predation (reviewed in Balch and Scheibling 2001). As a result of these many factors, recruitment in echinoids is spatially and temporally variable (Tegner and Dayton 1981; Paine 1986; Scheibling 1986; Pearse and Hines 1987; Raymond and Scheibling 1987; Sloan et al. 1987; Ebert and Russell 1988; Rowley 1989; Watanabe and Harrold 1991; Ebert et al. 1994; Estes and Duggins 1995; Prince 1995; Leinaas and Christie 1996; Balch and Scheibling 2000; Lambert and Harris 2000; Morgan et al. 2000; Botsford 2001; Lamare and Barker 2001; Wing et al. 2003; Hereu et al. 2004; Tomas et al. 2004). In this section, we will discuss the common challenges larvae face during transport, settlement, and metamorphosis, along with some solutions larvae use to overcome these challenges.

7.1. Larval Transport and Dispersal

Dispersal is a result of two phenomena: advection and dispersion. Advection is dispersal away from a point source, whereas dispersion is the separation of larvae in a cohort (Scheltema 1986a). Both advection and dispersion solve and create problems. Large-scale oceanographic features facilitate advection away from parents. The scale of dispersal should be related to the duration of the planktonic period. However, the actual scale of larval transport is influenced by local hydrodynamics and larval behavior (e.g. vertical migration). For example, the larvae of *Evechinus chloroticus* spawned in a nearly closed fjord with estuarine circulation spend 4–6 weeks in the plankton and recruit locally (Lamare 1998). Molecular studies indicate that echinoid species with planktonic larvae are not subdivided into isolated populations, indicating gene flow throughout the species' broad geographic range (Palumbi and Wilson 1990; Watts et al. 1990; Moberg and Burton 2000). However, molecular studies indicate *P. lividus* has two randomly mating populations in the western Mediterranean and in the eastern Atlantic, with panmixis within each range (Duran et al. 2004).

For echinoids that are free-spawning and have little control over the gametes they fertilize, the risk of inbreeding might be high and dispersal important. Likewise, spread from parental sites or from siblings should reduce crowding and density-dependent mortality. If the quality of the benthic habitat varies temporally and spatially, dispersal is favored. Although the most advantageous scales of dispersal are not known, it is likely that larvae are in the plankton for far longer than is necessary to adequately sample the spatial variance in benthic habitat quality (Palmer and Strathmann 1981). Barnacle larvae can disperse beyond the geographic region where locally adapted settlement responses reliably lead to settlement in suitable habitats (Strathmann et al. 1981). It is not known if echinoid larvae also over-disperse. However, larval stages of coastal echinoid species have been collected from the middle of the Pacific, thousands of kilometers away from their juvenile habitats (Scheltema 1986b). For a larva that cannot swim against the currents, this problem is dire (Jackson and Strathmann 1981). So, why do some larvae spend months in the plankton? Some possible explanations include food limitation, slow development at low temperatures, transport/retention in large-scale oceanographic features, or difficulty in locating suitable benthic habitat. Alternatively, larvae that are swept away from coastal sites and transported long distances probably spend a long time in the plankton and might not contribute substantially to successful recruitment.

Work on *S. purpuratus* in Washington, USA (Paine 1986) and *S. droebachiensis* in Nova Scotia, Canada (Hart and Scheibling 1988; Scheibling 1996) documented settlement events that were correlated with elevated temperatures and presumably shorter larval durations. It is not clear whether elevated temperatures led to successful recruitment through direct effects on larval survival, indirect survival effects via accelerated development rate or increased food supply and hence shorter planktonic duration, or changes in advective transport through altered hydrodynamics. Along the Oregon coast, USA, larvae of *S. purpuratus* and *S. franciscanus* recruited during upwelling relaxations when surface currents were onshore (Ebert and Russell 1988; Miller and Emlet 1997; Morgan et al. 2000; Wing et al. 2003). Upwelling events can transport larvae in surface waters offshore and away from juvenile habitats, which might explain why some sites have low settlement (Ebert 1983; Ebert and Russell 1988; Ebert et al. 1994).

Work on microchemistry of invertebrate larvae is an exciting new area of research and is allowing biologists to answer the difficult question "where do larvae come from?" The basic idea is that larvae of some taxa lay down structures that are retained in juveniles or adults, and carry a chemical signature of where larvae have been (Zacherl et al. 2003; Zacherl 2005; Levin in press). Although methods are currently being tested on other taxa, similar methods should also work for echinoids because the juvenile skeleton of echinoids is partially produced while larvae are in the water column.

7.2. Settlement

Settlement also presents a variety of problems for echinoid larvae. Larvae must become developmentally competent to metamorphose, then detect suitable juvenile habitat, contact that habitat, switch from planktonic living to benthic living, and complete metamorphosis. Larvae possess some interesting adaptations to surmount these obstacles. Some larvae are able to delay settlement increasing the window of opportunity to contact juvenile habitats after attaining metamorphic competency. However, delayed settlement results in decreased growth rate and higher mortality (Highsmith and Emlet 1986). To increase the chances of contacting juvenile habitats, very late-stage plutei swim down or stop swimming in the laboratory (Burke 1980; pers. obs.). Echinoid larvae use a variety of cues to detect suitable microhabitats (reviewed in Strathmann 1978b; Chia et al. 1984b; Rodriguez et al. 1993; Pearce 1997). These cues and behaviors probably inform larvae of food, predator, and conspecific abundances. For example, adults of the sand dollar *Dendraster excentricus* release a chemical that induces larvae to settle and metamorphose near conspecifics (Burke 1984). Larvae of *Anthocidaris crassispina*, *A. punctulata*, *L. pictus*, *Pseudocentrotus depressus*, *S. droebachiensis*, and *S. purpuratus* settle in response to algal films, microbial films, or both (Cameron and Hinegardner 1974; Cameron and Schroeter 1980; Rowley 1989; Pearce and Scheibling 1990, 1991; Kitamura et al. 1993; Rahim et al. 2004), but macroalgae also induce metamorphosis in several species (Williamson et al. 2000; Takahashi et al. 2002; Swanson et al. 2004). Larvae can apparently evaluate the substratum with the juvenile podia and retain the option of swimming away to sample other locations (Burke 1980). Benthic substrata are not an absolute necessity for induction of metamorphosis. At least five species of echinoids are known to metamorphose in offshore surface waters far from the seafloor (depths >2000 m) (Fenaux and Pedrotti 1988).

7.3. Metamorphosis and Recruitment

To successfully recruit, juveniles must survive until metamorphosis is complete. Initially, metamorphosis is marked by a rapid transformation from the larval body form to the juvenile form. The juvenile rudiment everts from the vestibule and larval tissues are absorbed (Okazaki 1975; Chia and Burke 1978; Fig. 4). Nitric oxide appears to coordinate metamorphosis in *Lytechinus pictus* (Bishop and Brandhorst 2001), and might represent an ancient pathway shared among taxa with complex life histories (Bishop and Brandhorst 2003). Although metamorphosis appears to take only minutes, the juvenile gut does not function for several days (Chia and Burke 1978). During this time juveniles cannot eat and must live off maternal reserves or food acquired during larval life.

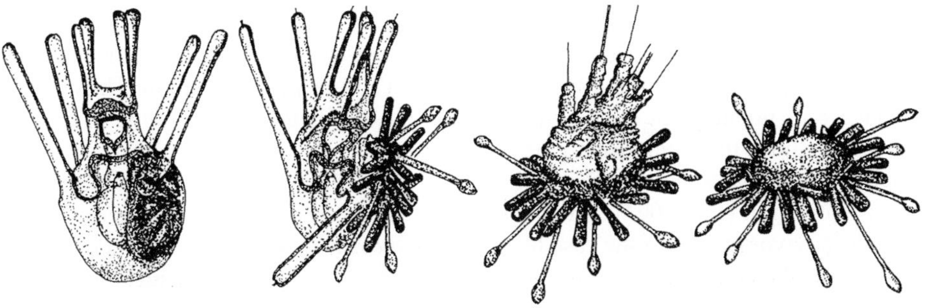

Fig. 4. Eversion of the juvenile rudiment from the vestibule of the pluteus and the degeneration of the larval body during metamorphosis. Modified from Burke (1983c).

Predation, dislodgement, physical conditions, and disease probably influence the success of juveniles after settlement. Several studies have documented predation on juvenile echinoids. On the western Atlantic coast, juvenile *S. droebachiensis* are consumed by rock crabs, lobsters, sculpins, and two species of flounder (Keats et al. 1985; Scheibling and Hamm 1991). Along the eastern Pacific coast, juvenile *S. purpuratus* in kelp forests had higher mortality rates than juveniles in urchin barrens, as a result of predation or disease (Pearse and Hines 1987; Rowley 1990). Densities of new recruits (5–17 days old) averaged more than 1000 individuals m^{-2}, with maximum densities over 2000 individuals m^{-2}. However, very high rates of juvenile mortality eliminated evidence of such recruitment pulses within 10 days (Rowley 1989). In New Zealand, juvenile *E. chloroticus* were more abundant when fish predators were excluded (Andrew and Choat 1982).

Random chance has also emerged as a potentially important factor that influences population of marine invertebrates (Hedgecock 1994). Because fecundity of many marine invertebrates is large ($>10^6$ eggs annually) and larval mortality is thought to be high (see Section 6.6), successful recruitment might just be the result of a few adults getting "lucky" and recruitment can therefore be viewed as a sweepstakes. Flowers et al. (2002) tested the sweepstakes recruitment hypothesis in *S. purpuratus* by comparing the genetic variation in an adult population with the genetic variation of four different cohorts of recruits. They found no evidence to support the sweepstakes hypothesis.

To avoid predators, echinoid juveniles take refuge among conspecific adults or in specific habitats. Juvenile *D. excentricus* that settled among versus away from adult conspecifics had higher survival, resulting from burrowing adults decreasing the abundance of a benthic predator (Highsmith 1982). Juvenile *S. franciscanus* take refuge under conspecific adults, decreasing the risk of predation (Tegner and Dayton 1977, 1981; Tegner and Levin 1983; Breen et al. 1985; Sloan et al. 1987; Rogers-Bennett et al. 1995). Mussel beds and cobbles protect juvenile *S. droebachiensis* from fishes, crabs, and lobsters (Witman 1985; Scheibling and Hamm 1991).

Massive concentrations of sea urchins (urchin fronts) have been documented in kelp beds and seagrass beds around the world (Maciá and Lirman 1999). Fronts often consist of individuals of uniform size which presumably comprise a single year class, possibly a single recruitment event (Camp et al. 1973; Watanabe and Harrold 1991; Hagen 1995; Maciá and Lirman 1999; Rose et al. 1999). Do urchin fronts indicate unusually high larval or juvenile survivorship? Our calculations show that the entire urchin front (often with average densities over 100 individuals per square meter and extending for several kilometers) could be accounted for by the spawn of one or a very few adults, given typical estimated rates of larval mortality (see Section 6.6) and planktonic duration. It is not the abundance of urchins that requires explanation but rather the concentration of animals into aggregations that comprise a grazing front.

Disease and parasites can also kill juvenile echinoids (Scheibling and Stephenson 1984; Vaïtilingon et al. 2004). Although more pronounced in adults, diseases can reduce juvenile survivorship, as in the case of *S. droebachiensis* in Nova Scotia (Scheibling and Stephenson 1984). Lower infection rates in juveniles as compared to adults might result from the location of juveniles in crevices that are less exposed to water-borne pathogens (Scheibling and Stephenson 1984). Juveniles must also remain attached to the substratum. This is a problem for juveniles that live in wave-exposed or on sandy-bottom habitats.

Specific habitats, like mussel beds, might reduce the risk of dislodgement for juveniles settling in the intertidal. Juvenile *D. excentricus* weigh themselves down in sandy habitats by ingesting large sand particles (Chia 1973).

8. CONCLUSIONS

Some aspects of echinoid reproductive and larval ecology have been reasonably well studied, such as fertilization ecology, feeding mechanisms, phenotypic plasticity, and evolution of nonfeeding larvae. On the other hand, many aspects of the basic biology and ecology of larvae, such as nutritional requirements, environmental tolerances, disease resistance, settlement preferences, dispersal ecology, and population dynamics, which are particularly important for aquaculture and commercial harvesting, remain very poorly known.

ACKNOWLEDGMENTS

M Wilson provided comments on the manuscript. Facilities were provided by the Department of Zoology, University of Florida and by the Friday Harbor Laboratories, University of Washington. Support was provided by grant OCE-9819593 from the National Science Foundation. Contribution number 2296 Bodega Marine Laboratory, University of California, Davis.

REFERENCES

Adams NL, Shick JM (1996) Mycosporine-like amino acids provide protection against ultraviolet radiation in eggs of the green sea urchin *Strongylocentrotus droebachiensis*. Photochem Photobiol 64: 149–158

Amemiya S, Emlet RB (1992) The development and larval form of an echinothurioid echinoid, *Asthenosoma ijimai*, revisited. Biol Bull 182: 15–30

Amemiya S, Tsuchiya T (1979) Development of the echinothurid sea urchin *Asthenosoma ijimai*. Mar Biol 52: 93–96

Andrew NL, Choat JH (1982) The influence of predation and conspecific adults on the abundance of juvenile *Evechinus chloroticus* (Echinoidea: Echinometridae). Oecologia 54: 80–87

Aslan LM, Uehara T (1997) Hybridization and F1 backcrosses between two closely related tropical species of sea urchins (genus *Echinometra*) in Okinawa. Invert Reprod Dev 31: 319–324

Balch T, Scheibling RE (2000) Temporal and spatial variability in settlement and recruitment of echinoderms in kelp beds and barrens in Nova Scotia. Mar Ecol Prog Ser 205: 139–154

Balch T, Scheibling RE (2001) Larval supply, settlement and recruitment in echinoderms. Echino Stud 6: 1–83

Barker MF (1985) Reproduction and development in *Goniocidaris umbraculum*, a brooding echinoid. In: Keegan BF, O'Connor BDS (eds) Echinodermata. A.A. Balkema, Rotterdam, pp 207–214

Bishop CD, Brandhorst BP (2001) NO/cGMP signaling and HSP90 activity represses metamorphosis in the sea urchin *Lytechinus pictus*. Biol Bull 201: 294–304

Bishop CD, Brandhorst BP (2003) On nitric oxide signaling, metamorphosis, and the evolution of biphasic life cycles. Evol Dev 5: 542–550

Boidron-Metairon IF (1988) Morphological plasticity in laboratory-reared echinoplutei of *Dendraster excentricus* (Eschscholtz) and *Lytechinus variegatus* (Lamarck) in response to food conditions. J Exp Mar Biol Ecol 119: 31–41

Botsford LW (2001) Physical influences on recruitment to California Current invertebrate populations on multiple scales. ICES J Mar Sci 58: 1081–1091

Breen PA, Carolsfeld W, Yamanaka KL (1985) Social behavior of juvenile red sea urchins, *Strongylocentrotus franciscanus* (Agassiz). J Exp Mar Biol Ecol 92: 45–61

Burke RD (1978) Structure of the nervous system of pluteus larva of *Strongylocentrotus purpuratus*. Cell Tiss Res 191: 233–247

Burke RD (1980) Podial sensory receptors and the induction of metamorphosis in echinoids. J Exp Mar Biol Ecol 47: 223–234

Burke RD (1981) Structure of the digestive tract of the pluteus larva of *Dendraster excentricus* (Echinodermata, Echinoida). Zoomorph 98: 209–225

Burke RD (1983a) Development of the larval nervous system of the sand dollar, *Dendraster excentricus*. Cell Tiss Res 229: 145–154

Burke RD (1983b) Neural control of metamorphosis in *Dendraster excentricus*. Biol Bull 164: 176–188

Burke RD (1983c) The induction of metamorphosis of marine invertebrate larvae: stimulus and response. Can J Zool 61: 1701–1719

Burke RD (1984) Pheromonal control of metamorphosis in the Pacific sand dollar, *Dendraster excentricus*. Science 225: 442–443

Cameron RA, Hinegardner RT (1974) Initiation of metamorphosis in laboratory cultured sea urchins. Biol Bull 146: 335–342

Cameron RA, Schroeter SC (1980) Sea urchin recruitment: effect of substrate selection on juvenile distribution. Mar Ecol Prog Ser 2: 243–247

Camp DK, Cobb SP, Van Breedveld JF (1973) Overgrazing of seagrasses by a regular urchin, *Lytechinus variegatus*. BioSci 23: 37–38

Cárcamo PF, Candia AI, Chaparro OR (2005) Larval development and metamorphosis in the sea urchin *Loxechinus albus* (Echinodermata: Echinoidea): effects of diet type and feeding frequency. Aquaculture 249: 375–386

Chia FS (1973) Sand dollar: a weight belt for the juvenile. Science 181: 73–74

Chia FS, Buckland-Nicks J, Young CM (1984a) Locomotion of marine invertebrate larvae: a review. Can J Zool 62: 1205–1222

Chia FS, Burke RD (1978) Echinoderm metamorphosis: fate of larval structures. In: Chia FS, Rice ME (eds) Settlement and metamorphosis of marine invertebrate larvae. Elsevier and North Holland, New York, NY, pp 219–234

Chia FS, Young CM, McEuen FS (1984b) The role of larval settlement behavior in controlling patterns of abundance in echinoderms. In: Engels W, Clark Jr WH, Fisher A, Olive PJW, Wend DF (eds) Advances in invertebrate reproduction 3. Elsevier Science Publishers, Amsterdam, pp 409–424

Cowden C, Young CM, Chia FS (1984) Differential predation on marine invertebrate larvae by two benthic predators. Mar Ecol Prog Ser 14: 145–149

Denny MW, Nelson EK, Mead KS (2002) Revised estimates of the effects of turbulence on fertilization in the purple sea urchin, *Strongylocentrotus purpuratus*. Biol Bull 203: 275–277

Denny MW, Shibata MF (1989) Consequences of surf-zone turbulence for settlement and external fertilization. Am Nat 134: 859–889

Duran S, Palacin C. Becerro MA, Turon X, Giribet G (2004) Genetic diversity and population structure of the commercially harvested sea urchin *Paracentrotus lividus* (Echinodermata: Echinoidea). Mol Ecol 13: 3317–3328

Ebert TA (1983) Recruitment in echinoderms. Echino Stud 1: 169–203

Ebert TA, Russell MP (1988) Latitudinal variation in size structure of the west coast purple sea urchin: a correlation with headlands. Limnol Oceanogr 33: 286–294

Ebert TA, Schroeter SC, Dixon JD, Kalvass P (1994) Settlement patterns of red and purple sea urchins (*Strongylocentrotus franciscanus* and *Strongylocentrotus purpuratus*) in California, USA. Mar Ecol Prog Ser 111: 41–52

Eckert GL (1995) A novel larval feeding strategy of the tropical sand dollar, *Encope michelini* (Agassiz): adaptation to food limitation and an evolutionary link between planktotrophy and lecithotrophy. J Exp Mar Biol Ecol 187: 103–128

Emlet RB (1986a) Facultative planktotrophy in the tropical echinoid *Clypeaster rosaceus* (Linnaeus) and a comparison with obligate planktotrophy in *Clypeaster subdepressus* (Gray) (Clypeasteroida, Echinoidea). J Exp Mar Biol Ecol 95: 183–202

Emlet RB (1986b) Larval production, dispersal, and growth in a fjord: a case study on larvae of the sand dollar *Dendraster excentricus*. Mar Ecol Prog Ser 31: 245–254

Emlet RB (1994) Body form and patterns of ciliation in nonfeeding larvae of echinoderms: functional solutions to swimming in the plankton. Am Zool 34: 570–585

Emlet RB (1995) Larval spicules, cilia, and symmetry as remnants of indirect development in the direct-developing sea urchin *Heliocidaris erythrogramma*. Dev Biol 167: 405–415

Emlet RB (2002) Sea urchin larval ecology: food rations, development, and size at metamorphosis. In: Yokota Y, Matranga V, Smolenicka Z (eds) The sea urchin: from basic biology to aquaculture. A.A. Balkema, Lisse, pp 105–110

Emlet RB, McEdward LR, Strathmann RR (1987) Echinoderm larval ecology viewed from the egg. Echino Stud 2: 55–136

Estes JA, Duggins DO (1995) Sea otters and kelp forests in Alaska: generality and variation in a community ecological paradigm. Ecol Monogr 65: 75–100

Fenaux L, Cellario C, Etienne M (1985) Croissance de la larve de l'oursin *Paracentrotus lividus*. Mar Biol 86: 151–157

Fenaux L, Pedrotti ML (1988) Metamorphose des larves d'echinides en pleine eau. PSZNI Mar Ecol 9: 93–107

Fenaux L, Strathmann MF, Strathmann RR (1994) Five tests of food-limited growth of larvae in coastal waters by comparisons of rates of development and form of echinoplutei. Limnol Oceanogr 39: 84–98

Flowers JM, Schroeter SC, Burton RS (2002) The recruitment sweepstakes has many winners: genetic evidence form the sea urchin *Strongylocentrotus purpuratus*. Evolution 56: 1445–1453

Gladfelter WB (1978) General ecology of cassiduloid urchin *Cassidulus caribbearum*. Mar Biol 47: 149–160

Grunbaum D, Strathmann RR (2003) Form, performance and trade-offs in swimming and stability or armed larvae. J Mar Res 61: 659–691

Hagen NT (1995) Recurrent destructive grazing of successionally immature kelp forests by green sea urchins in Vestfjorden, Northern Norway. Mar Ecol Prog Ser 123: 95–106

Hart MW (1991) Particle captures and the method of suspension feeding by echinoderm larvae. Biol Bull 180: 12–27

Hart MW (1995) What are the costs of small egg size for a marine invertebrate with feeding planktonic larvae. Am Nat 146: 415–426

Hart MW (1996) Evolutionary loss of larval feeding: development, form and function in a facultatively feeding larva, *Brisaster latifrons*. Evolution 50: 174–187

Hart MW, Scheibling RE (1988) Heat waves, baby booms, and the destruction of kelp beds by sea urchins. Mar Biol 99: 167–176

Hart MW, Strathmann RR (1994) Functional consequences of phenotypic plasticity in echinoid larvae. Biol Bull 186: 291–299

Hedgecock D (1994) Does variance in reproductive success limit effective population size of marine organisms? In: Beaumont A (ed) Genetics and evolution of aquatic organisms. Chapman and Hall, London, pp 122–134

Henry JJ, Wray GA, Raff RA (1990) The dorsoventral axis is specified prior to first cleavage in the direct developing sea urchin *Heliocidaris erythrogramma*. Development 110: 875–884

Hereu B, Zabala M, Linares C, Sala E (2004) Temporal and spatial variability in settlement of the sea urchin *Paracentrotus lividus* in the NW Mediterranean. Mar Biol 144: 1011–1018

Herrera JC (1998) Nutritional strategies of echinoplutei. PhD, University of Florida, Gainesville

Herrera JC, McWeeney SK, McEdward LR (1996) Diversity of energetic strategies among echinoid larvae and the transition from feeding to nonfeeding development. Ocean Acta 19: 313–321

Highsmith RC (1982) Induced settlement and metamorphosis of sand dollar (*Dendraster excentricus*) larvae in predator-free sites: adult sand dollar beds. Ecology 63: 329–337

Highsmith RC, Emlet RB (1986) Delayed metamorphosis: effect on growth and survival of juvenile sand dollars (Echinoidea: Clypeasteroida). Bull Mar Sci 39: 347–361

Hoegh-Guldberg O, Emlet RB (1997) Energy use during the development of a lecithotrophic and a planktotrophic echinoid. Biol Bull 192: 27–40

Jackson GA, Strathmann RR (1981) Larval mortality from offshore mixing as a link between pre-competent and competent periods of development. Am Nat 118: 16–26

Johnson KB, Shanks AL (2003) Low rates of predation on planktonic marine invertebrate larvae. Mar Ecol Prog Ser 248: 125–139

Keats DW, South GR, Steele DH (1985) Ecology of juvenile green sea urchins (*Strongylocentrotus droebachiensis*) at an urchin dominated subtidal site in eastern Newfoundland. In: Keegan BF, O'Connor BDS (eds) Echinodermata. A.A. Balkema, Rotterdam, pp 295–302

Kitamura H, Kitahara S, Koh HB (1993) The induction of larval settlement and metamorphosis of two sea urchins, *Pseudocentrotus depressus* and *Anthocidaris crassispina*, by free fatty-acids extracted from the coralline red alga *Corallina pilulifera*. Mar Biol 115: 387–392

Lamare MD (1998) Origin and transport of larvae of the sea urchin *Evechinus chloroticus* (Echinodermata: Echinoidea) in a New Zealand fiord. Mar Ecol Prog Ser 174: 107–121

Lamare MD, Barker MF (1999) In situ estimates of larval development and mortality in the New Zealand sea urchin *Evechinus chloroticus* (Echinodermata: Echinoidea). Mar Ecol Prog Ser 180: 197–211

Lamare MD, Barker MF (2001) Settlement and recruitment of the New Zealand sea urchin *Evechinus chloroticus*. Mar Ecol Prog Ser 218: 153–166

Lamare MD, Hoffman J (2004) Natural variation in carotenoids in the eggs and gonads of the echinoid genus, *Strongylocentrotus*: implication for their role in ultraviolet radiation photoprotection. J Exp Mar Biol Ecol 312: 215–233

Lambert DM, Harris LG (2000) Larval settlement of the green sea urchin, *Strongylocentrotus droebachiencsis*, in the southern Gulf of Maine. Invert Biol 119: 403–409

Lawrence JM, Fenaux L, Jangoux M (1977) Qualitative nutritional requirements of echinoderm larvae. In: Recheige M (ed) CRC handbook series in nutrition and food, section D. Nutritional requirements. CRC Press, Inc., Cleveland, pp 401–407

Lawrence JM, McClintock JB, Guille A (1984) Organic level and caloric content of eggs of brooding asteroids and an echinoid (Echinodermata) from Kerguelen (South Indian Ocean). Int J Invert Reprod Dev 7: 249–257

Leinaas HP, Christie H (1996) Effects of removing sea urchins (*Strongylocentrotus droebachiensis*): stability of the barren state and succession of kelp forest recovery in the east Atlantic. Oecologia 105: 524–536

Levin LA (2006) Recent progress in understanding larval dispersal: new directions and digressions. Integr Comp Biol 46: 282–297

Levitan DR (1988) Asynchronous spawning and aggregative behavior in the sea urchin *Diadema antillarum* (Philippi). In: Burke RD, Mladenov PV, Lambert P, Parsley RL (eds) Echinoderm biology. A.A. Balkema, Rotterdam, pp 181–186

Levitan DR (1993) The importance of sperm limitation to the evolution of egg size in marine invertebrates. Am Nat 141: 517–536

Levitan DR (1995) The ecology of fertilization in free-spawning invertebrates. In: McEdward LR (ed) Ecology of marine invertebrate larvae. CRC Press, Boca Raton, FL, pp 123–156

Levitan DR (1996) Predicting optimal and unique egg sizes in free-spawning marine invertebrates. Am Nat 148: 174–188

Levitan DR (2000) Sperm velocity and longevity trade off each other and influence fertilization in the sea urchin *Lytechinus variegatus*. Proc Roy Soc Lond B 267: 531–534

Levitan DR (2005) Sex-specific spawning behavior and its consequences in an external fertilizer. Am Nat 165: 685–694

Levitan DR, Irvine SD (2001) Fertilization selection on egg and jelly-coat size in the sand dollar *Dendraster excentricus*. Evolution 55: 2479–2483

Levitan DR, Petersen C (1995) Sperm limitation in the sea. Trends Ecol Evol 10: 228–231

Maciá S, Lirman D (1999) Destruction of Florida Bay seagrasses by a grazing front of sea urchins. Bull Mar Sci 65: 593–601

Mackie GO, Spencer AN, Strathmann RR (1969) Electrical activity associated with ciliary reversal in echinoderm larvae. Nature 223: 1384–1385

Manahan DT (1990) Adaptations by invertebrate larvae for nutrient acquisition from seawater. Am Zool 30: 147–160

Manahan DT, Davis JP, Stephens GC (1983) Bacteria-free sea urchin larvae: selective uptake of neutral amino acids from seawater. Science 220: 204–206

McClintock JB, Pearse JS (1986) Organic and energetic content of eggs and juveniles of Antarctic echinoids and asteroids with lecithotrophic development. Comp Biochem Physiol A 85: 341–345

McEdward LR (1984) Morphometric and metabolic analysis of the growth and form of an echinopluteus. J Exp Mar Biol Ecol 82: 259–287

McEdward LR (1996) Experimental manipulation of parental investment in echinoid echinoderms. Am Zool 36: 169–179

McEdward LR (1997) Reproductive strategies of marine benthic invertebrates revisited: facultative feeding by planktotrophic larvae. Am Nat 150: 48–72

McEdward LR, Janies DA (1997) Relationships among development, ecology, and morphology in the evolution of echinoderm larvae and life cycles. Biol J Linn Soc 60: 381–400

McEdward LR, Miner BG (2001) Larval and life cycle patterns in echinoderms. Can J Zool 78: 1125–1170

McEdward LR, Morgan KH (2001) Interspecific relationships between egg size and the level of parental investment per offspring in echinoderms. Biol Bull 200: 33–50

McMillan WO, Raff RA, Palumbi SR (1992) Population genetic consequences of developmental evolution in sea urchins (genus *Heliocidaris*). Evolution 46: 1299–1312

Metaxas A (1998) The effect of salinity on larval survival and development in the sea urchin *Echinometra lucunter*. Invert Reprod Dev 34: 323–330

Metaxas A, Young CM (1998a) Responses of echinoid larvae to food patches of different algal densities. Mar Biol 130: 433–445

Metaxas A, Young CM (1998b) Behavior of echinoid larvae around sharp haloclines: effects of the salinity gradient and dietary conditioning. Mar Biol 131: 443–459

Miller BA, Emlet RB (1997) Influence of nearshore hydrodynamics on larval abundance and settlement of sea urchins *Strongylocentrotus franciscanus* and *S. purpuratus* in the Oregon upwelling zone. Mar Ecol Prog Ser 148: 83–94

Miner BG (2005) Evolution of feeding structure plasticity in marine invertebrate larvae: a possible trade-off between arm length and stomach size. J Exp Mar Biol Ecol 315: 117–125

Miner BG, Vonesh JR (2004) Effects of fine grain environmental variability on morphological plasticity. Ecol Lett 7: 794–801

Miner BG, McEdward LA, McEdward LR (2005) The relationship between egg size and the duration of the facultative feeding period in marine invertebrate larvae. J Exp Mar Biol Ecol 321: 135–144

Moberg PE, Burton RS (2000) Genetic heterogeneity among adult and recruit red sea urchins, *Strongylocentrotus franciscanus*. Mar Biol 136: 773–784

Morgan LE, Wing SR, Botsford LW, Lundquist CJ, Diehl JM (2000) Spatial variability in red sea urchin (*Strongylocentrotus franciscanus*) cohort strength relative to alongshore upwelling variability in northern California. Fish Oceanogr 9: 83–98

Mortensen T (1921) Studies of the development and larval forms of echinoderms. GEC Gad, Copenhagen

Mortensen T (1931) Contributions to the study of the development and larval forms of echinoderms. I–II. D Kgl Danske Vidensk Selsk Biol Skr, Naturvidensk og Mathem Afd 9, 4

Mortensen T (1937) Contributions to the study of the development and larval forms of echinoderms. III. D Kgl Danske Vidensk Selsk Biol Skr, Naturvidensk og Mathem Afd 9, 7

Mortensen T (1938) Contributions to the study of the development and larval forms of echinoderms. IV. D Kgl Danske Vidensk Selsk Biol Skr, Naturvidensk og Mathem Afd 9, 7

Okazaki K (1975) Normal development to metamorphosis. In: Czihak G (ed) The sea urchin embryo. Biochemistry and morphogenesis. Springer-Verlag, New York, NY, pp 177–232

Okazaki K, Dan K (1954) The metamorphosis of partial larvae of *Peronella japonica* Mortensen, a sand dollar. Biol Bull 106: 83–99

Olson RR, Cameron JL, Young CM (1993) Larval development (with observations on spawning) of the pencil urchin *Phyllacanthus imperialis*: a new intermediate larval form. Biol Bull 185: 77–85

Olson RR, Olson MH (1989) Food limitation of planktotrophic marine invertebrate larvae: does it control recruitment success. Ann Rev Ecol Syst 20: 225–247

Onoda K (1936) Notes on the development of some Japanese echinoids with special reference to the structure of the larval body. Jap J Zool 6: 637–654

Onoda K (1938) Notes on the development of some Japanese echinoids, with special reference to the structure of the larval body. Report II. Jap J Zool 8: 1–13

Paine RT (1986) Benthic community-water column coupling during the 1982–1983 El Nino. Are community changes at high-latitudes attributable to cause or coincidence? Limnol Oceanogr 31: 351–360

Palmer AR, Strathmann RR (1981) Scale of dispersal in varying environments and its implications for life histories of marine invertebrates. Oecologia 48: 308–318

Palumbi SR, Wilson AC (1990) Mitochondrial DNA diversity in the sea urchins *Strongylocentrotus purpuratus* and *S. droebachiensis*. Evolution 44: 403–415

Parks AL, Bisgrove BW, Wray GA, Raff RA (1989) Direct development in the sea urchin *Phyllacanthus parvispinus* (Cidaroidea): phylogenetic history and functional modification. Biol Bull 177: 96–109

Paulay G, Boring L, Strathmann RR (1985) Food limited growth and development of larvae: experiments with natural sea water. J Exp Mar Biol Ecol 93: 1–10

Pearce CM (1997) Induction of settlement and metamorphosis in echinoderms. In: Fingerman M, Nagabhushanam R, Thompson M-F (eds) Recent advances in marine biotechnology. Oxford and IBH Publishing Co. Pvt. Ltd., New Delhi, pp 283–341

Pearce CM, Scheibling RE (1990) Induction of metamorphosis of larvae of the green sea urchin, *Strongylocentrotus droebachiensis*, by coralline red algae. Biol Bull 179: 304–311

Pearce CM, Scheibling RE (1991) Effect of macroalgae, microbial films, and conspecifics on the induction of metamorphosis of the green sea urchin *Strongylocentrotus droebachiensis* (Muller). J Exp Mar Biol Ecol 147: 147–162

Pearse JS, Cameron RA (1991) Echinodermata: Echinoidea. In: Giese AC, Pearse JS, Pearse VB (eds) Reproduction of marine invertebrates. VI. Echinoderms and lophophorates. The Boxwood Press, Pacific Grove, CA, pp 513–662

Pearse JS, Hines AH (1987) Long-term population dynamics of sea urchins in a central California kelp forest: rare recruitment and rapid decline. Mar Ecol Prog Ser 39: 275–283

Pechenik JA (1987) Environmental influences on larval survival and development. In: Giese AC, Pearse JS, Pearse VB (eds) Reproduction of marine invertebrates. IX. General aspects: seeking unity in diversity. Blackwell Scientific Publications and The Boxwood Press, Palo Alto and Pacific Grove, CA, pp 551–608

Pedrotti ML (1995) Food selection (size and flavor) during development of echinoderm larvae. Invert Reprod Dev 27: 29–39

Pedrotti ML, Fenaux F (1993) Effects of food diet on the survival, development and growth rates of two cultured echinoplutei (*Paracentrotus lividus* and *Arbacia lixula*). Invert Reprod Dev 24: 59–70

Pennington JT (1985) The ecology of fertilization of echinoid eggs: the consequences of sperm dilution, adult aggregation, and synchronous spawning. Biol Bull 169: 417–430

Pennington JT, Emlet RB (1986) Ontogenic and diel vertical migration of a planktonic echinoid larva, *Dendraster excentricus* (Eschscholtz): occurrence, causes, and probable consequences. J Exp Mar Biol Ecol 104: 69–95

Pennington JT, Rumrill SS, Chia FS (1986) Stage-specific predation upon embryos and larvae of the Pacific sand dollar, *Dendraster excentricus*, by 11 species of common zooplanktonic predators. Bull Mar Sci 39: 234–240

Podolsky RD (1994) Temperature and water viscosity: physiological versus mechanical effects of suspension feeding. Science 265: 100–103

Podolsky RD (2004) Life-history consequences of investment in free-spawning eggs and their accessory coats. Am Nat 163: 735–753

Podolsky RD, Emlet RB (1993) Separating the effects of temperature and viscosity on swimming and water movement by sand dollar larvae (*Dendraster excentricus*). J Exp Biol 176: 207–221

Podolsky RD, Strathmann RR (1996) Evolution of egg size in free-spawners: consequences of the fertilization fecundity tradeoff. Am Nat 148: 160–173

Poulin E, Féral JP (1996) Diversity of Antarctic echinoids: importance of dispersal strategies. Ocean Acta 19: 464

Prince J (1995) Limited effects of the sea urchin *Echinometra mathaei* (de Blainville) on the recruitment of benthic algae and macroinvertebrates into intertidal rock platforms at Rottnest Island, Western Australia. J Exp Mar Biol Ecol 186: 237–258

Raff RA (1987) Constraint, flexibility, and phylogenetic history in the evolution of direct development in sea urchins. Dev Biol 119: 6–19

Rahim SAKA, Li JY, Kitamura H (2004) Larval metamorphosis of the sea urchins, *Pseudocentrotus depressus* and *Anthocidaris crassispina* in response to microbial films. Mar Biol 144: 71–78

Rassoulzadegan F, Fenaux L (1979) Grazing of echinoderm larvae (*Paracentrotus lividus* and *Arbacia lixula*) on naturally occurring particulate matter. J Plank Res 1: 215–223

Rassoulzadegan F, Fenaux L, Strathmann RR (1984) Effect of flavor and size on selection of food by suspension-feeding plutei. Limnol Oceanogr 29: 357–361

Raymond BG, Scheibling RE (1987) Recruitment and growth of the sea urchin *Strongylocentrotus droebachiensis* (Müller) following mass mortalities off Nova Scotia, Canada. J Exp Mar Biol Ecol 108: 31–54

Reitzel AM, Miles CM, Heyland A, Cowart JD, McEdward LR (2005) The contribution of the facultative feeding period to echinoid larval development and size at metamorphosis: a comparative approach. J Exp Mar Biol Ecol 317: 189–201

Rodriguez SR, Ojeda FP, Inestrosa NC (1993) Settlement of benthic marine invertebrates. Mar Ecol Prog Ser 97: 193–207

Rogers-Bennett L, Bennett WA, Fastenau HC, Dewees CM (1995) Spatial variation in red sea urchin reproduction and morphology: implications for harvest refugia. Ecol Appl 5: 1171–1180

Rose CD, Sharp WC, Kenworthy WJ, Hunt JH, Lyons WG, Prager EJ, Valentine JF, Hall MO, Whitfield PE, Forqurean JW (1999) Overgrazing of a large seagrass bed by the sea urchin *Lytechinus variegatus* in outer Florida Bay. Mar Ecol Prog Ser 190: 211–222

Rowley RJ (1989) Settlement and recruitment of sea urchins (*Strongylocentrotus* spp) in a sea urchin barren ground and a kelp bed: are populations regulated by settlement or post-settlement processes. Mar Biol 100: 485–494

Rowley RJ (1990) Newly settled sea urchins in a kelp bed and urchin barren ground: a comparison of growth and mortality. Mar Ecol Prog Ser 62: 229–240

Rumrill SS (1990) Natural mortality of marine invertebrate larvae. Ophelia 32: 163–198

Rumrill SS, Chia FS (1985) Differential mortality during the embryonic and larval lives of northeast Pacific echinoids. In: Keegan BF, O'Connor BDS (eds) Echinodermata. A.A. Balkema, Rotterdam, pp 333–338

Schatt P, Féral JP (1996) Completely direct development of *Abatus cordatus*, a brooding schizasterid (Echinodermata: Echinoidea) from Kerguelen, with description of perigastrulation, a hypothetical new mode of gastrulation. Biol Bull 190: 24–44

Scheibling RE (1986) Increased macroalgal abundance following mass mortalities of sea urchins (*Strongylocentrotus droebachiensis*) along the Atlantic coast of Nova Scotia. Oecologia 68: 186–198

Scheibling RE (1996) The role of predation in regulating sea urchin populations in eastern Canada. Ocean Acta 19: 421–430

Scheibling RE, Hamm J (1991) Interactions between sea urchins (*Strongylocentrotus droebachiensis*) and their predators in field and laboratory experiments. Mar Biol 110: 105–116

Scheibling RE, Stephenson RL (1984) Mass mortality of *Strongylocentrotus droebachiensis* (Echinodermata: Echinoidea) off Nova Scotia, Canada. Mar Biol 78: 153–164

Scheltema RS (1986a) On dispersal and planktonic larvae of benthic invertebrates: an eclectic overview and summary of problems. Bull Mar Sci 39: 290–322

Scheltema RS (1986b) Long-distance dispersal by planktonic larvae of shoal-water benthic invertebrates among central Pacific islands. Bull Mar Sci 39: 241–256

Sinervo B, McEdward LR (1988) Developmental consequences of an evolutionary change in egg size: an experimental test. Evolution 42: 885–899

Sloan NA, Lauridsen CP, Harbo RM (1987) Recruitment characteristics of the commercially harvested red sea urchin *Strongylocentrotus franciscanus* in southern British Columbia, Canada. Fisheries Res 5: 55–69

Smith MJ, Boom JDG, Raff RA (1990) Single-copy DNA distance between two congeneric sea urchin species exhibiting radically different modes of development. Mol Biol Evol 7: 315–326

Starr M, Himmelman JH, Therriault J-C (1990) Direct coupling of marine invertebrate spawning with phytoplankton blooms. Science 247: 1071–1074

Starr M, Himmelman JH, Therriault J-C (1992) Isolation and properties of a substance from the diatom *Phaeodactylum tricornutum* which induces spawning in the sea urchin *Strongylocentrotus droebachiensis*. Mar Ecol Prog Ser 79: 275–287

Starr M, Himmelman JH, Therriault J-C (1993) Environmental control of green sea urchin, *Strongylocentrotus droebachiensis*, spawning in the St. Lawrence estuary. Can J Fish Aquat Sci 50: 894–901

Strathmann RR (1971) The feeding behavior of planktotrophic echinoderm larvae: mechanisms, regulation, and rates of suspension-feeding. J Exp Mar Biol Ecol 6: 109–160

Strathmann RR (1978a) Evolution and loss of feeding larval stages of marine invertebrates. Evolution 32: 894–906

Strathmann RR (1978b) Larval settlement in echinoderms. In: Chia FS, Rice ME (eds) Settlement and metamorphosis of marine invertebrate larvae. Elsevier and North Holland, Amsterdam, pp 235–246

Strathmann RR (1981) On barriers to hybridization between *Strongylocentrotus droebachiensis* (O.F. Müller) and *Strongylocentrotus pallidus* (G.O. Sars). J Exp Mar Biol Ecol 55: 39–47

Strathmann RR (1989) Existence and functions of a gel filled primary body cavity in development of echinoderms and hemichordates. Biol Bull 176: 25–31

Strathmann RR (1996) Are planktonic larvae of marine benthic invertebrates too scarce to compete within species. Oceanol Acta 19: 399–407

Strathmann RR, Branscomb ES, Vedder K (1981) Fatal errors in set as a cost of dispersal and the influence of intertidal flora on set of barnacles. Oecologia 48: 13–18

Strathmann RR, Fenaux L, Strathmann MF (1992) Heterochronic developmental plasticity in larval sea urchins and its implications for evolution of nonfeeding larvae. Evolution 46: 972–986

Strathmann RR, Jahn TL, Fonseca JRC (1972) Suspension feeding by marine invertebrate larvae: clearance of particles by ciliated bands of a rotifer, pluteus, and trochophore. Biol Bull 142: 505–519

Swanson RL, Williamson JE, De Nys R, Kumar N, Bucknall MP, Steinberg PD (2004) Induction of settlement of larvae of the sea urchin *Holopneustes purpurascens* by histamine from a host alga. Biol Bull 206: 161–172

Takahashi Y, Itoh K, Ishii M, Suzuki M, Itabashi Y (2002) Induction of larval settlement and metamorphosis of the sea urchin *Strongylocentrotus intermedius* by glycoglycerolipids from the green alga *Ulvella lens*. Mar Biol 140: 763–771

Tegner MJ, Dayton PK (1977) Sea urchin recruitment patterns and implications of commercial fishing. Science 196: 324–326

Tegner MJ, Dayton PK (1981) Population structure, recruitment and mortality of two sea urchins (*Strongylocentrotus franciscanus* and *S. purpuratus*) in a kelp forest. Mar Ecol Prog Ser 5: 255–268

Tegner MJ, Levin LA (1983) Spiny lobsters and sea urchins: analysis of a predator prey interaction. J Exp Mar Biol Ecol 73: 125–150

Thorson G (1950) Reproductive and larval ecology of marine bottom invertebrates. Biol Rev 25: 1–45

Tomas F, Romero J, Turon X (2004) Settlement and recruitment of the sea urchin *Paracentrotus lividus* in two contrasting habitats in the Mediterranean. Mar Ecol Prog Ser 282: 173–184

Vaïtilingon D, Eeckhaut I, Fourgon D, Jangoux M (2004) Population dynamics, infestation and host selection of *Vexilla vexillum*, an ectoparastic muricid of echinoids, in Madagascar. Dis Aquat Org 61: 241–255

Villinski JT, Villinski JC, Byrne M, Raff RA (2002) Convergent maternal provisioning and life-history evolution in echinoderms. Evolution 56: 1764–1775

Watanabe JM, Harrold C (1991) Destructive grazing by sea urchins *Strongylocentrotus* spp in a central California kelp forest: potential roles of recruitment, depth, and predation. Mar Ecol Prog Ser 71: 125–141

Watts RJ, Johnson MS, Black R (1990) Effects of recruitment on genetic patchiness in the urchin *Echinometra mathaei* in western Australia. Mar Biol 105: 145–151

Williams DHC, Anderson DT (1975) The reproductive system, embryonic development, larval development and metamorphosis of the sea urchin *Heliocidaris erythrogramma* (Val.) (Echinoidea: Echinometridae). Aust J Zool 23: 371–403

Williamson JE, De Nys R, Kumar N, Carson DG, Steinberg PD (2000) Induction of metamorphosis the sea urchin *Holopneustes purpurascens* by the metabolite complex from the algal host *Delisea pulchra*. Biol Bull 198: 332–345

Wing SR, Botsford LW, Morgan LE, Diehl JM, Lundquist CJ (2003) Inter-annual variability in larval supply to populations of three invertebrate taxa in the northern California Current. Estuar Coast Shelf Sci 57: 859–872

Witman JD (1985) Refuges, biological disturbance, and rocky subtidal community structure in New England. Ecol Monogr 55: 421–445

Wray GA (1992) The evolution of larval morphology during the post-Paleozoic radiation of echinoids. Paleobiol 18: 258–287

Wray GA (1995a) Evolution of larvae and developmental modes. In: McEdward LR (ed) Ecology of marine invertebrate larvae. CRC Press, Boca Raton, FL, pp 413–447

Wray GA (1995b) Punctuated evolution of embryos. Science 267: 1115–1116

Wray GA (1996) Parallel evolution of nonfeeding larvae in echinoids. Syst Biol 45: 308–322

Wray GA, Raff RA (1989) Evolutionary modification of cell lineage in the direct-developing sea urchin *Heliocidaris erythrogramma*. Dev Biol 132: 458–470

Wray GA, Raff RA (1991) Rapid evolution of gastrulation mechanisms in a sea urchin with lecithotrophic larvae. Evolution 45: 1741–1750

Young CM, Chia FS (1987) Abundance and distribution of pelagic larvae as influenced by predation, behavior, and hydrographic factors. In: Giese AC, Pearse JS, Pearse VB (eds) Reproduction of marine invertebrates IX. General aspects: seeking unity in diversity. Blackwell Scientific Publications, Palo Alto, CA, pp 385–463

Young CM, Devin MG, Jaeckle WB, Ekaratne SUK, George SB (1996) The potential for ontogenetic vertical migration by larvae of bathyal echinoderms. Oceanol Acta 19: 263–271

Yund PO (2000) How severe is sperm limitation in natural populations of marine free-spawners? Trends Ecol Evol 15: 10–13

Zacherl DC (2005) Spatial and temporal variation in statolith and protochonch trace elements as natural tags to track larval dispersal. Mar Ecol Prog Ser 290: 145–163

Zacherl DC, Manriquez PH, Paradis G, Day RW, Castilla JC, Warner RR, Lea DW, Gaines SD (2003) Trace elemental fingerprinting of gastropod statoliths to study larval dispersal trajectories. Mar Ecol Prog Ser 248: 297–303

Zigler KS, Raff EC, Popodi E, Raff RA, Lessios HA (2003) Adaptive evolution of bindin in the genus Hekiocidaris is correlated with the shift to direct development. Evolution 57: 2293–2302

Edible Sea Urchins: Biology and Ecology
Editor: John Miller Lawrence

Chapter 6

Growth and Survival of Postsettlement Sea Urchins

Thomas A Ebert

Department of Zoology, Oregon State University, Corvallis, OR (USA)

1. GROWTH

1.1. Introduction

Growth in echinoids means change in mass, diameter, and shape of the test, which requires expansion, calcification, and production of soft tissues. Skeletal growth is based on cellular processes that result in both shape changes during growth and rates of diameter change. Various forces that could determine shape during growth have been explored (e.g. Dafni 1986; Ellers and Telford 1992), but shape seems best described as a response to forces generated by skeletal weight (Ellers 1993). Echinoid skeletons consist of ossicles that include spines, elements of Aristotle's lantern, and plates of the test that are attached to each other by small projections and collagen threads (Telford 1985). Growth is by calcification around individual plates and addition of new plates at the aboral ends of ambulacral and interambulacral rows. Ossicles are calcite that contains various amounts of magnesium (e.g. Clarke and Wheeler 1922; Chave 1954) and, mostly, are constructed as a fenestrated stereom that varies in porosity and construction (e.g. Nissen 1969; Weber et al. 1969; Smith 1980). An organic matrix forms the structure for calcification and ossicles contain a coherent organic phase in the form of concentric laminae and radial threads (e.g. Dubois 1991; Ameye et al. 1998, 1999). These skeletal attributes provide information for interpreting overall growth of the body as well as measures of environmental conditions.

1.2. Skeletal Composition

Using X-ray diffraction, Weber (1969a) analyzed different skeletal elements of echinoids from a wide range of habitats. Ossicles showed statistical differences in magnesium content with spines having the lowest levels (Fig. 1). As magnesium contents in body parts of an individual are different, it is clear that content is not a passive result of environmental conditions and strongly suggests that there is substantial physiological modification of magnesium incorporation.

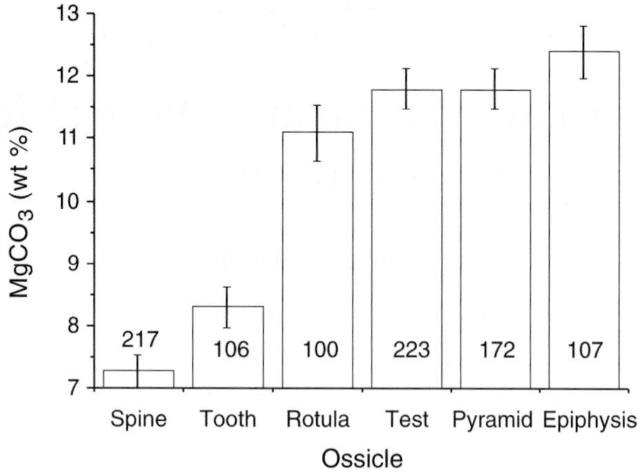

Fig. 1. Magnesium in echinoid ossicles. Number of determinations is given; bars are 95% limits. Data from Weber (1969a).

Davies et al. (1972) showed that magnesium content changed in spine tips of *Strongylocentrotus purpuratus* during regeneration. The youngest regenerating calcite had the lowest percent magnesium and magnesium content increased as the regenerating tip matured. Sumich and McCauley (1972) found a pattern of change in magnesium content of plates from the oral to aboral surface of *Allocentrotus fragilis*. Plates closest to the peristome had the highest magnesium content, which decreased towards the aboral surface. The oldest and slowest growing plates are at the oral side of the test.

The magnesium content of ossicles increases with temperature (Chave 1954; Weber 1969a; Davies et al. 1972), but also shows variation between species at a single location (Raup 1966; Weber 1969a; Table 1). The magnesium content is correlated with K, the Brody–Bertalanffy growth constant, so fast-growing species have less magnesium in their tests. The analysis, however, is crude in the sense that estimates of K are sometimes for related species and are based on studies at locations other than Enewetak Atoll and Fiji. The pattern of magnesium content and K are similar for the two sites but the measured magnesium content for samples from Fiji are higher than those from Enewetak. It is not known whether this is because of differences in the selection of plates to be analyzed, temperature, or some other factor. The correlation between magnesium content and K should be viewed as an interesting hypothesis that, if corrected for temperature, magnesium content may provide an estimate of growth rate.

In addition to magnesium differences, echinoid skeletal elements differ in isotopic composition of carbon and oxygen. Mostly, parts differ from inorganically precipitated $CaCO_3$ in seawater and hence they differ from such groups as mollusks, brachiopods, and foraminiferans (Weber and Raup 1966a, 1968). Spines are similar to inorganic precipitates or mollusk shell but other parts of the skeleton differ in a systematic pattern across most species. In particular, ossicles of Aristotle's lantern show a consistency in which parts are more enriched by ^{13}C and ^{18}O (Weber and Raup 1966a,b).

Table 1 Weight percent $MgCO_3$ in interambulacral plates of regular echinoid species from Enewetak Atoll and Singatota, Fiji; K is the Brody–Bertalanffy growth-rate constant.

Species	**%$MgCO_3$**		**K(yr^{-1})**	**Species for comparison**
	Enewetak	**Fiji**		
Tripneustes gratilla	9.75	11.7	1.24[a,e], 1.39[a,f], 1.80[3]	*T. ventricosus*[a], *T. gratilla*[c]
Mespilia globulus	10.4			
Echinothrix calamaris	10.4			
Echinothrix diadema	10.6		0.82[a], 0.38[b]	*E. diadema*
Diadema savignyi	10.8		0.79[d], 1.18[a,g]	*D. savignyi*[d], *D. antillarum*[a]
Diadema setosum		14.8	0.50[b]	*D. setosum*[b]
Eucidaris metularia	11.6		0.67[a]	*E. tribuloides*
Parasalenia gratiosa	12.2			
Echinometra mathaei	13.5	15.6	0.46[a], 0.51[b], 0.32[d]	*E. mathaei*
Heterocentrotus mammillatus		15.6	0.24[b]	*H. mammillatus*
Heterocentrotus trigonarius	14.0		0.24[b]	*H. mammillatus*
Echinostrephus aciculatus	14.3			

Data for $MgCO_3$: Enewetak Atoll (Raup 1966); Singatota, Fiji (Weber 1969a); estimates of the growth parameter $K\,yr^{-1}$ from [a]Ebert (1975), [b]Ebert (1982), [c]Bacolod and Dy (1986), [d]Drummond (1994), [e]McPherson (1965), [f]Lewis (1958), and [g]Randall et al. (1964) for the same species or species in the same genus; for estimates using data in Ebert (1982), K was recalculated using the Brody–Bertalanffy model; correlation between %$MgCO_3$ and $K\,yr^{-1}$ is significant at $p = 0.01$.

Isotope ratios were reported by Weber and Raup (1966a) in delta notation (Eq. (1)), which shows the change in $^{13}C/^{12}C$ relative to a standard and expressed in parts per thousand (‰),

$$\delta^{13}C = \left(\frac{^{13}C/^{12}C_{\text{sample}}}{^{13}C/^{12}C_{\text{standard}}} - 1 \right) \times 1000. \tag{1}$$

Two examples illustrate isotope fractionation in echinoid ossicles. Data for *Strongylocentrotus droebachiensis* (Fig. 2) are from a single site (Weber and Raup 1966b). Spines were least enriched with ^{12}C (i.e. the most positive $\delta^{13}C$) and so were similar to isotope patterns shown in mollusk calcite. Ossicles that show increased ^{12}C also show increased ^{16}O and the fractionation pattern of $\delta^{13}C$ and $\delta^{18}O$ is similar to the pattern displayed by a wide range of species. The apparent explanation for differences, though not the pattern, is that calcification of different ossicles draws on pools with somewhat different mixes of metabolic CO_2 and bicarbonate from seawater. As spines are directly exposed to seawater it is not surprising that they would be most like mollusk calcite. Calcification of the test would represent a combination of bicarbonate directly from seawater but also bicarbonate from the perivisceral fluid or from the water vascular system, which would include metabolic CO_2 as well as diffusion of bicarbonate across the tube feet. Ossicles of Aristotle's lantern are inside the peripharyngeal coelom and therefore isolated from the large coelom of the body cavity that contains the perivisceral fluid. The peripharyngeal coelom is in direct contact with seawater at the edge of the peristome across the thin-walled gills. There also are tube feet with ampullae on the peristomal membrane and a water ring at the top of Aristotle's lantern just inside the peripharyngeal coelom. Radial water canals pass beneath the rotules and down the lantern to the inside of the peristomal membrane and then along the inside of the test with branches to tube feet. Fluid is moved slowly in the water vascular system and diffusion across its walls can provide bicarbonate of both metabolic and seawater origins to the top of Aristotle's lantern. It is possible that the differences in isotope fractionation that occur during growth of the ossicles are just reflections of the complex set of distances of ossicles from membranes and delivery systems. The order of isotopic fraction in ossicles is similar to the ordering for magnesium content so ossicles with high magnesium content also are high in ^{12}C.

A second example of isotopic fractionation demonstrates an additional complication that does not appear to be just the result of how fluid is moved and transported across membranes of coelomic cavities. The basic relationship between $\delta^{13}C$ and $\delta^{18}O$ in ossicles of *S. purpuratus* at Sunset Bay, Oregon, is similar to that shown by *S. droebachiensis* (Ebert, unpubl.). Individuals from two microhabitats, however, differed in ^{12}C enrichment so the pattern of $\delta^{18}O$ vs. $\delta^{12}C$ showed a clear separation. Owing to the differences in (1) the amounts and types of food available to microhabitats and (2) the growth rates of sea urchins, the slow-growing individuals were most enriched with ^{12}C.

Growth of echinoid ossicles is a result of the presence of organic molecules and an organic matrix in the calcification process (Dubois and Chen 1989; Ameye et al. 1998, 1999). Differences in magnesium content may reflect differences in the organic environment of the calcification sites of the growing skeleton. Organic differences of the matrix or in the vacuolar fluid surrounding calcifying spicules also may be the explanation

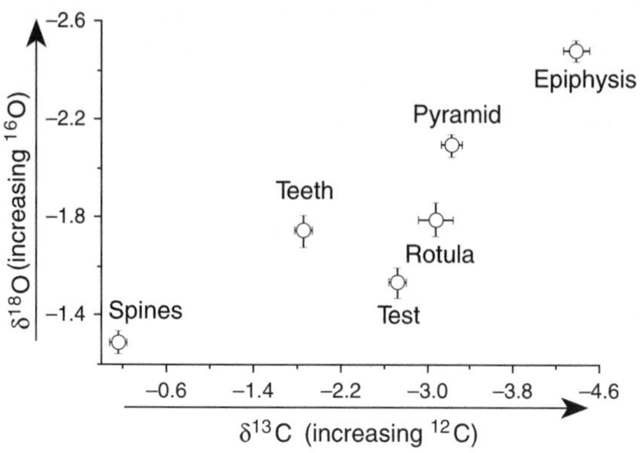

Fig. 2. Isotopic composition of calcareous ossicles relative to a standard for *Strongylocentrotus droebachiensis* from Waldoboro, ME, USA; error bars = 1 se; $N = 12$; Weber and Raup (1966b).

for isotopic differences. The author knows of no studies that have attempted to deduce growth rates from chemical analyses of skeletal parts but this could potentially be a powerful tool for the evaluation of growth transitions from single samples at a field site and particularly where mark–recapture studies are not possible.

1.3. Resorption

Resorption as well as deposition occurs in the endoskeleton. Possibly the earliest demonstration of this was in cidaroid spines (Prouho 1887), which grow and stop when the spine is an appropriate size for the size of the associated tubercle and overall size of the test. When a cidaroid spine stops growing, the dermis dies and usually the surface becomes covered with fouling organisms except for a small collar immediately above the milled ring, which remains covered by an epidermis. The problem in growth is that small individuals of a cidarid species have fouled spines yet the spines must grow because large individuals have large spines. The solution to the puzzle is that spine growth takes place by shedding and regenerating the spine tip. The spine breaks above the milled ring near the junction with the tip that lacks an epidermis. A new tip grows to a larger size before growth stops and the epidermis dies again. Shedding the spine tip involves dissolution of calcite in the collar along what is now called Prouho's membrane. The cellular processes associated with activity within Prouho's membrane have been described by Märkel and Roser (1983a,b).

Resorption of calcite appears to be a general phenomenon in echinoids. It has been reported in ossicles of Aristotle's lantern (Märkel 1979), tubercles on test plates of spatangoids (David and Néraudeau 1989), and rods of pedicellariae (Bureau et al. 1991; Dubois and Ghyoot 1995). Other than work on Prouho's membrane (Märkel and Roser 1983a,b) and pedicellariae (Bureau et al. 1991; Dubois and Ghyoot 1995) that directly link cellular activity with resorption, other accounts of resorption are interpretations of aspects of structure that appear to require resorption.

Märkel's (1979) analysis of growth of ossicles of Aristotle's lantern for *Arbacia lixula* and *Eucidaris tribuloides* showed that the polycrystalline coverings of ossicles as well as the monocrystalline stereom must be resorbed, particularly in the rotulae. Growth and resorption in rotulae also are interesting because the growth center is eventually lost during ontogeny as the youngest plate areas are resorbed.

David and Néraudeau (1989) described curious circular marks on the test surface of spatangoids. The marks are not randomly distributed on tests and are distinctive features of species. The best interpretation of these marks is that they form by autotomy of tubercles and associated spines, which is a process that would require calcite resorption to undercut tubercles. In this respect, the process would be more comparable to the formation of Prouho's membrane than to the reshaping of ossicles in Aristotle's lantern.

Resorption of calcite in response to starvation or low food has been reported for test size both in the field and in the laboratory. A decrease in test diameter has been reported for *S. purpuratus* (Ebert 1967a, 1968; Pearse and Pearse 1975), *Heliocidaris erythrogramma* (Constable 1993), and *Diadema antillarum* (Levitan 1988, 1989). The magnitude of decrease varies from a few millimeters to 7 mm (Levitan 1989). Small changes of less than a millimeter in ambital diameter could result from tightening of sutures that are relatively open during rapid growth as proposed by Pearse and Pearse (1975) and Constable (1993); no resorption of calcite would be required. Large decreases, on the other hand, would require resorption. The maximum shrinkage shown for *S. purpuratus* in Ebert's (1968) study was about 3 mm. In contrast, no statistically significant decrease in test diameter was found in *S. purpuratus* fed only one day every 8 weeks in the laboratory (Fansler 1983; Ebert 1996). As the field individuals had some food, it is not probable they would shrink while those starved in the laboratory did not.

In 1964, tagging of *S. purpuratus* was done with nylon monofilament threaded through the test (Ebert 1965). Three diameter measurements were made on small sea urchins (≤2 cm) and all five ambulacral–interambulacral diameters were measured on larger ones. The ambulacrum with the tag showed reduced growth, was identified on data cards for most sea urchins, and so could be removed from analysis of growth (Ebert 2004). Because of variation in diameter measurements, the smallest diameter was then selected rather than the mean of the measurements. Making these corrections, however, did not remove obvious negative growth of as much as 3.5 mm (Fig. 3A). Negative growth of more than 1 mm, however, did disappear if the initial size measurements in July 1964 were removed from analysis (Fig. 3B).

Tagging in July 1964 involved a crew of several individuals measuring sea urchins. Measuring a sea urchin accurately is not as easy as measuring a billiard ball. It is a bit of an art because it is necessary to get the jaws of calipers positioned between spines while making sure that the measurement is from the center of an ambulacrum to the opposite interambulacrum and the sea urchin is not tipped. Five measurements that fail to get between spines will result in a mean that is biased high even though variance may be small. Subsequent measurements where calipers are positioned more carefully will give a smaller size and can lead to an erroneous conclusion that shrinkage occurred. The conclusion (Ebert 2004) based on reanalysis of Ebert (1967a, 1968) was that measurement

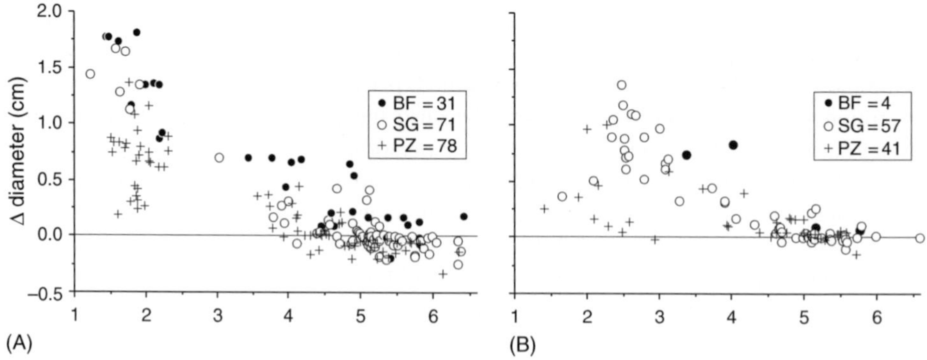

Fig. 3. Change in diameter as a function of minimum diameter of *Strongylocentrotus purpuratus* tagged and measured in July 1994 in three areas of Sunset bay, Oregon (Ebert 1968). BF = Boulder Field, SG = Surfgrass Area, PZ = Postelsia Zone; tag interference corrected by removing the diameter measurement with the monofilament tag in July 1965 and finding a new minimum diameter; (A) starting date July 1964 and end date July 1965; (B) data combined using size pairs with time close to one year but excluding Jul. 1964–1965, Dec. 1964 to Nov. 1965, Mar. 1965 to Apr. 1966, and Jul. 1965 to Jul. 1966.

bias particularly in July 1964 was the basis of concluding that shrinkage had occurred. Neither calcite resorption nor changes in suture size would be required.

Errors in measurement are also probable in studies of *Strongylocentrotus franciscanus*. Smith and Botsford (1998) and Smith et al. (1998) showed growth increment data that indicate negative values as large as 5 mm over a period of 23 days, although most increments were positive. The authors provided an estimate of measurement error equal to about 2%, which for a 6 cm animal would be about 1.2 mm. Kalvass *et al.* (1998) showed eight negative increments out of 38 measurements of tagged *S. franciscanus*. The largest negative increment was about 2.5 mm over a period of 111 days. The same individual showed a negative growth of about 4 mm over 205 days. As the sea urchins were well fed and gonads increased in size, the authors attributed the decrease to be due to a measurement error. It would be better, however, to refer the negative growth estimates as due to bias because with just measurement error, taking a mean of 5 or 10 diameter measurements would provide an unbiased estimate of the true mean.

Measurement errors may seem trivial but changes of just a few millimeters have been reported as evidence of shrinkage (e.g. Ebert 1967a, 1968; Levitan 1988, 1989). Errors also can influence choice of a growth model and variance of parameter estimates. There also are consequences for the selection of a tagging method because growth less than measurement error will not be detected. Methods that can detect real changes in size must be used.

The changes in diameter for *S. purpuratus* (Pearse and Pearse 1975; Ebert 2004), *H. erythrogramma* (Constable 1993), and *D. setosum* (Ebert, unpubl.) all suggest that shrinkage of suture spaces coupled with measurement errors and bias are sufficient to account for negative growth.

1.4. Natural Growth Lines

There has been substantial interest in the use of lines in skeletal ossicles to determine age of individuals and hence growth and survival parameters. There are a number of problems, however, in using natural lines and, accordingly, their use to evaluate age and growth (Ebert 1988b; Russell and Meredith 2000).

In cross section, growth lines in spines look like rings in tree trunks; lines are large elements of the stereom. These lines have been interpreted as age related (Carpenter 1870; Deutler 1926; Swan 1952; Moore 1966; Weber 1969b; Dotan and Fishelson 1985) or as repair lines following breaks (Ebert 1967b). The relationship between test diameter and ring number is linear or close to linear for *H. erythrogramma* (Moore 1966), *S. purpuratus* (Ebert 1968), *Heterocentrotus trigonarius* (Weber 1969b) and *H. mammillatus* (Dotan and Fishelson 1985; Ebert 1985). Growth as a function of age is not linear for sea urchins and so the lines in spines cannot indicate age (Ebert 1985). Lines can indicate breaks but they also will form without breaks (Weber 1969b; Heatfield 1971; Dotan and Fishelson 1985; Ebert 1986, 1988a). The linear relationship may indicate the number of growth episodes when a nongrowing spine no longer fits its ever-growing tubercle (Ebert 1986).

Growth lines in other ossicles continue to be used to estimate age and parameters of growth functions (Table 2). Lines exist in ossicles of the test and in Aristotle's lantern and have been interpreted as annual since first examined in detail by Deutler (1926). They

Table 2 Examples of age estimates for sea urchins based on the maximum observed growth zones in skeletal ossicles.

Species	Years	Reference
Order Temnopleuroida		
Family Toxopneustidae		
Lytechinus variegatus	4	Beddingfield and McClintock (2000)
	10	Hill et al. (2004)
Sphaerechinus granularis	9	Lumingas and Guillou (1994a)
Order Arbacioida		
Family Arbaciidae		
Arbacia punctulata	7	Hill et al. (2004)
Order Echinoida		
Family Echinidae		
Echinus acutus var. *norvegicus*	11	Gage et al. (1986)
Echinus affinis	28	Gage and Tyler (1985)
Echinus elegans	21	Gage et al. (1986)
Echinus esculentus	12	Nichols et al. (1985)
Loxechinus albus	11	Gebauer and Moreno (1995)
Psammechinus miliaris	7	Jensen (1969)
	10	Gage (1991)
Paracentrotus lividus	8	Crapp and Willis (1975)
Sterechinus neumayeri	40	Brey et al. (1995)
Sterechinus antarcticus	75	Brey (1991)
Family Echinometridae		
Evechinus chloroticus	10	Dix (1972)
Family Strongylocentrotidae		
Allocentrotus fragilis	15	Sumich and McCauley (1973)
Strongylocentrotus droebachiensis	24	Robinson and MacIntyre (1997)
Strongylocentrotus intermedius	10	Agatsuma (2001)
Strongylocentrotus nudus	6	Kawamura (1966)

are visible as pigmented bands and, at least in some cases, as density differences revealed by X-rays (Pearse and Pearse 1975; Jones 1970 cited in Dix 1972). The appropriateness of lines as indicators of age has been studied by documenting the seasonal progression of pigmented and clear bands being formed at plate edges (Moore 1935; Kawamura 1966, 1973; Taki 1972a,b; Dix 1972; Turon et al. 1995), holding individuals in the field or in tanks (Dix 1972; Gage 1991), by following the progression of modes in size–frequency distributions (Crapp and Willis 1975; Lumingas and Guillou 1994a), tagging methods such as tetracycline (Ebert 1988b; Gage 1991, 1992a,b; Brey et al. 1995; Russell and Meredith 2000), invasive tags (Walker 1981), and changes in calcium:magnesium across ossicles (Robinson and MacIntyre 1997).

Growth bands exist as alternating dark and light or translucent and opaque zones, depending on observations made using transmitted or reflected light. Preparation of plates varies across studies but most workers use some modification of the method used by Moore (1935) or Jensen (1969). Jensen used preserved specimens or ones freshly killed in alcohol. An interambulacral series was dissected from a preserved specimen and the interior cleaned using fine forceps and a soft brush. The cleaned interambulacrum was heated over an alcohol lamp to char the inside and, after cooling, the series was immersed in xylene or methyl benzoate to reveal the growth zones. Modifications of the method include using a muffle furnace rather than an alcohol lamp and first cleaning the plates using sodium hypochlorite (Pearse and Pearse 1975; Gage 1991), and etching (Smith 1984), or etching and staining (Dix 1972; White et al. 1985). Brey (1991) ground jaws to a thickness of $\leq$1 mm and then cleared with tert-butyl-methyl-ether before using with transmitted light to count growth lines.

Seasonal changes in light and dark bands show a seasonal signal. For example, *Evechinus chloroticus* showed average development of a light zone at the plate margin from narrow to wide from May through August (winter), at which time about 70% of the sample had a wide light zone at the margin (Dix 1972). A dark zone began to develop during September (spring) first as a narrow band but then increasing in width reaching a summer maximum in February when over 90% of the sample had a wide dark zone at the margin. Other examples could be given all showing a similar pattern. The annual signal is clear in many studies, at least in medium-sized individuals, but the physiological mechanisms of formation are much less certain and under laboratory conditions feeding experiments can induce band formation that is not annual (Pearse and Pearse 1975). Also, major bands usually contain smaller bands of uncertain origin (e.g. Smith 1984; Gage 1991).

The validation methods all work, in the sense that an annual signal can be detected for a range of sizes. A problem exists for some species, such as *S. purpuratus* (Ebert 1988b) and *S. droebachiensis* (Russell and Meredith 2000) where individuals may add more than one line per year. A more serious problem is that it is not possible to recognize natural lines in very slow-growing individuals because natural lines may be too thin (Ebert 1988b; Gage 1991, 1992b; Brey et al. 1995; Russell and Meredith 2000). Consequently, true ages of large individuals may be seriously underestimated. Although for medium-sized individuals one line may be formed each year, a priori it is not possible to decide when individuals might be too small or too large for the method to provide unbiased results.

1.5. Tagging

Use of natural growth lines is attractive because just a single sample is required to be gathered at a site. Sometimes, such as for deep-sea species, this is the only way at present that growth estimates can be obtained. In general, however, growth is best studied by tagging and various approaches have been tried (reviewed by Hagen 1996). Five general approaches have been used: (1) plastic tubes slipped over spines (McPherson 1968); (2) a tag inserted into a hole drilled or punched through the test (e.g. Fuji 1962; Ebert 1965; Lees 1968; Dix 1970; Olsson and Newton 1979; Hur et al. 1985; Cuenca 1987; Neill 1987); (3) chemicals such as tetracycline or calcein (e.g. Kobayashi and Taki 1969; Pearse and Pearse 1975; Ebert 1980b, 1982; Russell 1987; Kenner 1992; Russell et al. 1998;

Ebert et al. 1999; Lamare and Mladenov 2000); (3) passive integrated transponder (PIT) tags (Hagen 1996; Kalvass et al. 1998); and (4) coded wire tags (Kalvass et al. 1998). All of these methods have negative as well as positive attributes.

Plastic tubes were successfully used by McPherson (1968) to tag *E. tribuloides*. The technique is simple but a key point is that it was used on a cidarid. As primary spines of cidarids have no epidermis, a plastic tube would act as just another fouling organism on a spine. It would, however, be lost when the spine tip shed during growth. Ebert (unpubl. obs.) found that thin rings cut from vinyl tubing used to insulate electrical wires (spaghetti tubing) were lost from spines of *S. purpuratus* within 6 weeks and so could be used for short-term studies.

Tags inserted into the test with an external number or color-code, as well as PIT tags, permit multiple measurements of individuals. Following tagging, individuals can be measured at a frequency of less than 1 year so reasonable returns can be obtained even if annual loss rates are high. A major problem, as shown above, is that there can be substantial measurement errors of live sea urchins. Also, with sub-annual measurements, growth models must be used that can accommodate seasonal effects (e.g. Cloern and Nichols 1978; Pauly 1981; Sager 1982; Sager and Gosselck 1986). Additional parameters associated with time of year of minimum growth and the strength of the seasonal signal are included in the analysis, which increases the overall problems of parameter estimation. Available growth models with seasonal parameters all use size at age data; no model with seasonal parameters currently exists that uses an initial size and Δsize with initial sizes taken at different times of year and Δsize being less than 1 year.

Invasive tags inserted through the test are prone to loss and also may decrease growth and survival rates. Neill (1987) found that sea urchins (species not given) tagged with Floy anchor tags survived in the laboratory for up to 60 days but a tagged cohort in the field had a half-life of only 11 days. The half-life of a cohort of *Centrostephanus coronatus* tagged with stainless steel wire anchored either in or through tests was only about 15 days (Nelson and Vance 1979).

Lees (1970) tagged *S. purpuratus* with stainless steel wire and reported a loss of 92% after 9 months in the field. Assuming an exponential rate of loss, the half-life of the tagged cohort was about 1.8 months. Loss was primarily due to mortality, but a small loss due to movement also was possible. Hur et al. (1985) tagged *Hemicentrotus pulcherrimus* with nylon line threaded through the test. Over a period of 77 days, 37 of 90 tagged sea urchins died compared with 7 of 45 for non-tagged individuals (41% vs. 16%).

Re-analysis of data for *S. purpuratus* from the Postelsia Zone of Sunset Bay, Oregon, that were tagged with nylon monofilament (Ebert 1968) provides an estimate of a survival rate of $0.395\,yr^{-1}$. The instantaneous mortality rate, M, is $0.928\,yr^{-1}$, i.e. $-\ln(0.395)$. Assuming exponential decay, cohort half-life is about 0.75 years, which includes losses due both to mortality and tag loss. This is a very rapid loss compared with estimates of M for *S. purpuratus* tagged with tetracycline (Russell 1987). Mean estimates of M were 0.12, 0.20, and $0.31\,yr^{-1}$ for sites on Vancouver Island (Canada), San Diego (USA), and Punta Baja (Mexico) respectively. These translate to cohort half-lives of 5.8, 3.5, and 2.2 years.

Sea urchins marked with PIT tags have a high survival rate (Hagen 1996). In a laboratory study, 13 out of 16 tagged *S. droebachiensis* were alive after 1 year, which is a survival rate of 81% and a cohort half-life of 4.8 years. Kalvass et al. (1998) reported a 10% loss of PIT tags from 30 tagged *S. franciscanus*. Loss rate was not different than for coded wire tags, which was 23 out of 131 tagged. Mortality was 8 out of 159 sea urchins. Their study, however, was confounded by having all sea urchins tagged with coded wire and some also tagged with either PIT tags or tetracycline.

Few studies document the effect of invasive tagging methods on growth. Hagen (1996) showed that growth of *S. droebachiensis* did not differ between individuals tagged with PIT tags and controls. After 93 days, Kalvass et al. (1998) found no differences in the growth of *S. franciscanus* tagged with PIT tags plus coded wire tags vs. just coded wire tags. There were, however, differences after 205 days. They concluded that differences were probably not biologically important because the difference between adjusted mean test diameters of final size vs. original size was only 2 mm. The individuals they used had mean diameters of ca. 50 mm and so 2 mm could be biologically significant although it also falls within the errors associated with measurement.

Nylon monofilament line threaded through the test has a negative effect on growth. All five ambulacral to interambulacral diameters were measured for tagged *S. purpuratus* at Sunset Bay, Oregon, (Ebert 1968) and the diameter measurement containing the monofilament line was noted. Small animals, which showed the greatest diameter increase during a year, consistently showed the tagged diameter measurement as the smallest of the five. The relationship broke down with increasing size and decreasing growth until measurement errors equaled or exceeded actual growth (Ebert 2004).

In general, invasive tags such as dart or nylon monofilament appear to modify both survival and growth, and should be avoided. PIT and coded wire tags appear to be much better but all of these methods suffer because of difficulties in measuring diameters of live sea urchins. Errors can be substantial and tend to increase with increasing sea urchin size and lack of worker experience. If measurements of large individuals contain an error of several millimeters, growth will go undetected and become part of variance surrounding a maximum size estimate (cf. Smith and Botsford 1998) or will be interpreted as negative growth.

The antibiotic tetracycline and calcein have been used for estimating growth in sea urchins. Both of these chemicals bind to calcium ions and are incorporated into the skeleton during calcification. Tetracycline has been used by Kobayashi and Taki (1969), Taki (1972a,b), Ebert (1977), Schroeter (1978), Russell (1987, 2001), Rowley (1990), Kenner (1992), Gage (1992a,b), Estes and Duggins (1995), Ebert and Russell (1992, 1993), Russell et al. (1998), and Ebert et al. (1999). Calcein tagging has been used less frequently (Rowley 1990; Rogers-Bennett 1994; Ebert 1998; Lamare and Mladenov 2000; Russell and Urbaniak 2004). Under ultraviolet illumination, tetracycline fluoresces yellow and calcein fluoresces green. There are additional chemicals that have been used for other organisms with calcium-based skeletons that probably could be used on sea urchins. These include alizarin complexone (Tsukamota et al. 1989; Sanchez-Lamadrid 2001), calcein blue (Brooks et al. 1994), alizarin red S (Lagardêre et al. 2000), and xylenol orange (Day et al. 1992); there may be others chemicals as well (cf. Lee et al. 2003).

The advantages of tetracycline or calcein tagging are: (1) large numbers of individuals can be tagged very rapidly in the field, underwater if necessary; (2) very small growth increments (fractions of a millimeter) can be detected and measured including zero growth; and (3) very small individuals can be tagged by soaking them in a bath containing the tagging agent. The disadvantages are: (1) individuals must be killed in order to detect the mark; (2) although measuring growth increments is rapid, sample preparation is time consuming because skeletal plates must be scrupulously clean and if sodium hypochlorite is used to remove soft tissue, precautions are necessary against the chlorine gas that is released; (3) growth in test diameter cannot be measured directly but must be estimated from growth increments of individual ossicles, which can be test plates or, more frequently, demi-pyramids (jaws) of Aristotle's lantern; and (4) negative growth involving skeletal resorption (should it actually occur) cannot be detected.

Treatment of 1–8 mg of tetracycline had no effect on the growth or survival of 3 cm *S. purpuratus* (Ebert 1988b). The volume of solution injected was 0.1–0.2 ml. Similar lack of growth effects for tetracycline have been reported for *S. franciscanus* (Bureau 1996; Kalvass et al. 1998) and *Psammechinus miliaris* (Gage 1991) although Gage reported an apparent transient effect, which may have been due to the tetracycline, the invasive nature of the injection, or could have been an artifact associated with repeated measurements of a cohort that had experienced loss of some individuals (e.g. loss of one large individual in a small sample reduces mean size). Volume of fluid injected is important for survival. For *S. purpuratus* with a mean diameter of ca. 3 cm, injecting 0.5 ml increased mortality whereas injections of 0.1 and 0.2 ml were similar to controls (Ebert, unpubl.). The physical nature of damage is unknown but sea urchins injected with large volumes often show flaccid tube feet and are unable to attach to the substrate.

A recommended protocol is 1 g tetracycline dissolved in 100 ml of seawater and 0.1-0.2 ml injected into individuals that range from 2 to 5 cm with less for smaller and somewhat more for larger individuals. This formulation is not critical and species may differ in volumes and possibly concentrations that are tolerated. Calcein as a bath for small individuals has been used following the directions of Wilson et al. (1987). A stock solution is made by dissolving 6.25 g of calcein in 1 l of tap water and buffered to a pH of 5 to 6 using sodium bicarbonate. The bath is made by adding stock solution to seawater to obtain a final concentration of $125\,\mathrm{mg\,l^{-1}}$; i.e. 20 ml of stock is diluted to make 1 l of bath. Calcein at this concentration does not interfere with growth of juvenile *S. droebachiensis* (Russell and Urbaniak 2004). Concentrations of calcein of 160 and $240\,\mathrm{ml\,l^{-1}}$ have adverse effects on fish larvae (Bumguardner and King 1996) but have not been tested on sea urchins. Calcein also can be injected with good results (Lamare and Mladenov 2000). These authors used 0.5 gm of calcein in 1 l of seawater buffered to pH 8 with NaOH.

Laboratory studies on growth using *S. purpuratus* have spanned all seasons and have shown 100% tagging success with tetracycline (Fansler 1983; Russell 1987; Ebert 1988b; Edwards and Ebert 1991). On the other hand, Gage (1991) had only partial success in tagging *P. miliaris* in the laboratory from November through March and attributed the poor success to a seasonal effect. Taki (1971) found maximum uptake by *S. intermedius* in the winter and little or zero uptake of tetracycline during the summer. Differences in success of tagging may represent differences in the strength of seasonal effects for different species and may indicate that calcification is not just passively controlled by temperature but may be compensated (cf. Ebert et al. 1999).

Growth comparisons based on chemically tagged individuals can be made by using just growth changes of demi-pyramids or other ossicles. For certain purposes, such as estimating survival from growth and size structure (Ebert 1987, 1998; Ebert et al. 1999; Smith and Botsford 1998; Smith et al. 1998), it is necessary to transform growth parameters for jaws into test diameter growth. This can be done by using allometric parameters either after growth parameters have been estimated for jaw growth or the allometric parameters can be used to estimate original test diameter (D) from change in diameter estimated from changes in jaw length (J) based on tetracycline or calcein tags (Eqs (2) and (3)) and then growth parameters estimated from test diameters or original test diameter and ΔD:

$$D_t = D_{t+1} - \Delta D, \tag{2}$$

$$\Delta D_t = \alpha\left(J_{t+1}^{\beta} - J_t^{\beta}\right). \tag{3}$$

Either way makes an assumption that the allometric relationship does not change during the period of time spent in the field. Usually the allometric parameters are determined from the large collection made after tagged individuals have been in the field for 1 year. Allometry parameters are plastic and respond to available food (cf. Ebert 1980a; Fansler 1983; Levitan 1991), and so it is assumed that food availability following tagging was typical of previous years, which may or may not be true. If additional information were known about the sea urchins at the time of tagging, Eq. (3) could be modified to account for changes in allometry. Either a common α with different β's could be used or both α and β could be different.

A test of the assumption of unchanged parameters could be made using a single collection of tagged sea urchins by measuring growth increments in test plates as well as demi-pyramids (jaws). The additional work of measuring growth in test plates is not trivial because of preparation, which may include grinding, and the fact that the largest plate at the ambitus of a collected individual may not have been the largest plate when the sea urchin was tagged. A change in allometric parameters can be a problem in estimating original size if individuals are collected and tagged, and then moved to an area with different food availability, if cultured sea urchins are tagged and released into the field, or in laboratory experiments. Algal production changes with ocean conditions (e.g. Murray and Horn 1989; Dayton et al. 1999) and so year to year changes in jaw allometry are reasonable but have not been explored.

In a test of tagging methods, Kalvass et al. (1998) used tetracycline to tag *S. franciscanus* and reported that the back-calculated growth rate of the test was only 80% of directly measured growth based on original test diameters. Measurement errors certainly were present but the problem could be associated with changes in the allometric relationship between jaw and test due to dramatic changes in food availability. A change in allometry would change the estimate of initial size as shown in Fig. 4. Line OA represents the original allometric relationship and OB is the new relationship after animals were well fed and when jaws became relatively smaller; i.e. the diameter is larger for a given jaw size. The growth and change of allometry is indicated by the line labeled "growth trajectory." If the new line, OB, were used to estimate the original test diameter, D_t, the estimate would be larger than the actual value of D_t, i.e. ΔD would be too small, which is what Kalvass et al. (1998) actually found.

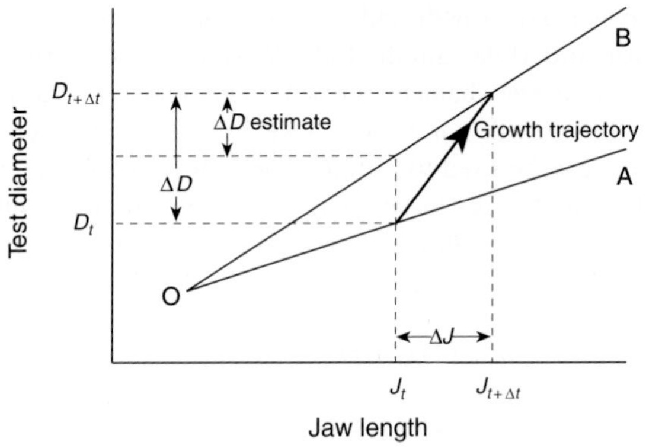

Fig. 4. Consequences of back-calculating original diameter from change in jaw length (ΔJ) when sea urchins have been moved to a habitat with a greater food supply; the allometric relationship changes as jaws become relatively smaller and back-calculation gives an estimate of original size larger than the actual original size.

1.6. Growth Models

There are at least four uses of growth models. First, they provide a way of summarizing growth data either from tagging, natural lines, or changes in size structure in a way that, one hopes, reflects underlying biological mechanisms. Second, models are useful for generating reasonable projections of $\text{size}_{t+\Delta t}$ as a function of size_t or size_t as a function of age (t) yielding a visual display of results of field or laboratory studies. Third, growth parameters provide a means for comparing different species or populations from different areas; and, fourth, growth models are useful when other parameters must be estimated such as survival rates from size structure or estimates of yield from a stock. Different models have been used and the following examples illustrate how increasing the number of observations and spread of size to include small individuals changes the model that best describes the data.

The most widely used growth model for describing size, D_t, as a function of age, t, is the Brody–Bertalanffy model (Eq. (4)) with three parameters that are easy to understand in terms of growth:

$$D_t = D_\infty\left(1 - b\mathrm{e}^{-Kt}\right), \tag{4}$$

or

$$D_t = D_\infty\left(1 - \mathrm{e}^{-K(t-t_0)}\right), \tag{5}$$

where D_∞ is asymptotic size (growth is 0), K is a rate constant with time-inverse units; if time is in years, $K\,\mathrm{yr}^{-1}$,

$$b = \frac{D_\infty - D_0}{D_\infty}, \tag{6}$$

D_0 is the size at $t = 0$, t_0 is the age when size is zero, and

$$b = e^{Kt_0}. \tag{7}$$

The Brody–Bertalanffy model is useful (e.g. Ebert 1975; Charnov 1993) as a means of making broad comparisons across species but it cannot describe growth that includes inflexion points.

The Brody–Bertalanffy model is just one of a family of curves (Richards 1959) now collectively called the Richards function (Ebert 1999) or generalized von Bertalanffy function (Piennar and Thomson 1973):

$$D_t = D_\infty(1 - be^{-Kt})^{-n}. \tag{8}$$

When $n = -1$, Eq. (8) is the Brody–Bertalanffy equation. When n is +1 the function is the logistic and as $|n| \to \infty$ the function approaches the Gompertz equation. The function has a discontinuity at $n = 0$. The parameter b in Eq. (8) is modified for the Richards function,

$$b = \frac{D_\infty^{-1/n} - D_0^{-1/n}}{D_\infty^{-1/n}}. \tag{9}$$

The choice of the appropriate model from this family is based on the shape of a graph of size at $t + \Delta t$ as a function of size at t or Δsize as a function of size$_t$. If the relationship is approximately linear, $n = -1$, the Brody–Bertalanffy model is best. If it is not linear, some other model is more appropriate.

With mark–recapture data of the sort obtained from chemical tagging, parameter estimation for the Brody–Bertalanffy, logistic, and Gompertz equations can be done by transforming data and using linear regression. The disadvantage of this approach is that confidence limits for K and D_∞ are not obtained. A much better approach is to use nonlinear regression. For the general Richards function, the nonlinear formulation is

$$D_{t+1} = \left[e^{-K} D_t^{-1/n} + D_\infty^{-1/n}(1 - e^{-K}\right]^{-n} \tag{10}$$

or, if the Gompertz model is more appropriate,

$$D_{t+1} = \exp\left[e^{-K} \ln D_t^{-1/n} + \ln D_\infty^{-1/n}(1 - e^{-K}\right]. \tag{11}$$

Data for *Sterechinus neumayeri* from Antarctica (Brey et al. 1995) provide an illustration of estimating parameters as well as showing the relationship between tetracycline data and natural growth lines.

Data for diameter (D) and jaw length (J) (Figure 5A) were used to determine the allometric relationship so diameter at time of tagging could be estimated. A GM functional regression was used (Ricker 1973; Barker et al. 1988; Jolicoeur 1990), which includes measurement errors in both D and J:

$$D = 3.9412 J^{1.0985} \tag{12}$$

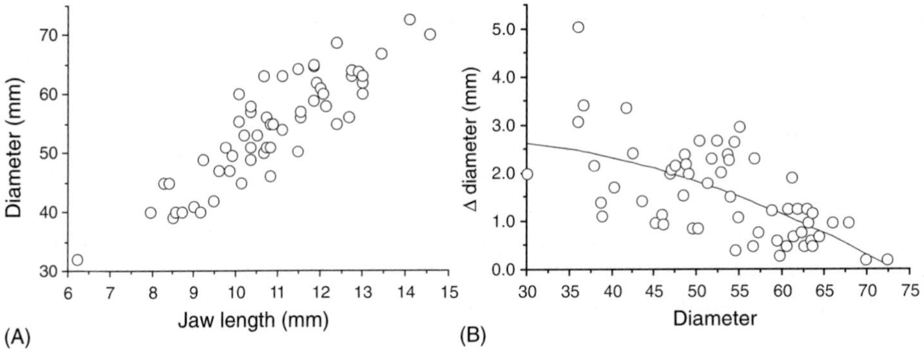

Fig. 5. Growth analysis of *Sterechinus neumayeri* from Antarctica. (A) Relationship between diameter and jaw (demi-pyramid) length; (B) ΔD vs. original diameter based on individuals tagged with tetracycline and remaining in the field for 1 year; fitted curve is the Gompertz function. Data from Brey et al. (1995).

and the parameters used with Eqs (2) and (3) to estimate diameter at tagging, D_t and then ΔD (Fig. 5B).

Initial estimates of parameters showed that in the Richards function n was large, indicating that the Gompertz equation was appropriate; final results are shown in Table 3.

Table 3 Growth analysis of *Sterechinus neumayeri* at McMurdo Station; the row "years to b" is the time required to attain estimated size at t_0 were the true D_0 0.5 mm.

Model	Parameter	Estimate	se.	Lower $<$ 95% $>$ Upper	
(A) *Tetracycline tagging*, $N = 59$, $r^2 = 0.39$					
Gompertz	D_∞ (mm)	73.39	2.752	67.88	78.90
	K (yr^{-1})	0.098	0.012	0.075	0.121
Brody–Bertalanffy	D_∞ (mm)	78.03	4.45	69.11	86.95
	K (yr^{-1})	0.064	0.011	0.042	0.086
(B) *Natural growth bands*, $N = 83$, $r^2 = 0.61$					
Gompertz	D_∞ (mm)	67.72	3.317	61.12	74.33
	K (yr^{-1})	0.117	0.030	0.05	0.176
	b	0.225	0.035	0.155	0.294
	Size at t_0 (mm)	26.25			
	Years to b	16			
Brody–Bertalanffy	D_∞ (mm)	68.92	3.929	61.11	76.74
	K (yr^{-1})	0.092	0.027	0.039	0.146
	b	0.669	0.079	0.512	0.826
	Size at t_0 (mm)	22.81			
	Years to b	4			

Data from Brey et al. (1995).

Brey et al. (1995) also found that the Gompertz model was best for their size increment data but based their calculations on growth increments of the jaws rather than first converting to diameters. Their estimate of J_∞ was 14.02 mm, which using Eq. (12) gives an estimate of $D_\infty = 71.7$ mm and $K = 0.107\,\text{yr}^{-1}$. Both of these parameters are within the 95% confidence limits shown in Table 3.

Analysis of size-at-age based on natural lines (Fig. 5) again was done using the Gompertz equation and the parameters are similar to those obtained from tagged individuals (Table 3). The growth curve for size vs. the number of growth lines is drawn in Fig. 6 together with the curve based on individuals tagged with tetracycline. Because the parameter b cannot be estimated using tagged animals but is one of the estimated parameters of size-at-age data, the same value of b was used for both lines in Fig. 6. The agreement between the two methods is very good.

The estimated value for parameter b is something of a problem and suggests that the estimated ages based on natural lines may be too low or that the Gompertz model is not appropriate. With the Gompertz equation,

$$b = \frac{\ln D_\infty - \ln D_0}{\ln D_\infty}, \tag{13}$$

where D_0 is the size at age 0, which should be some value close to 0.5 mm and is a fairly typical size at settlement. The estimated value of b (Table 3) was 0.2248 and so D_0 is 26 mm. The estimate of D_0 from Brey et al. (1995) is 18 mm. Both values are far too high and suggest that although growth parameters D_∞ and K were reasonably estimated from natural lines based on tetracycline tagging, actual age probably was too low. If the Gompertz function is appropriate and D_0 actually is 0.5 mm, the new estimate of b is 1.161 35 and it would take about 16 years to reach a size of 26 mm; all age estimates in Fig. 6 would have to be incremented by 16 years. This seems like a very large correction and so actually pegging the growth curve to the age axis remains a challenge. If the correct

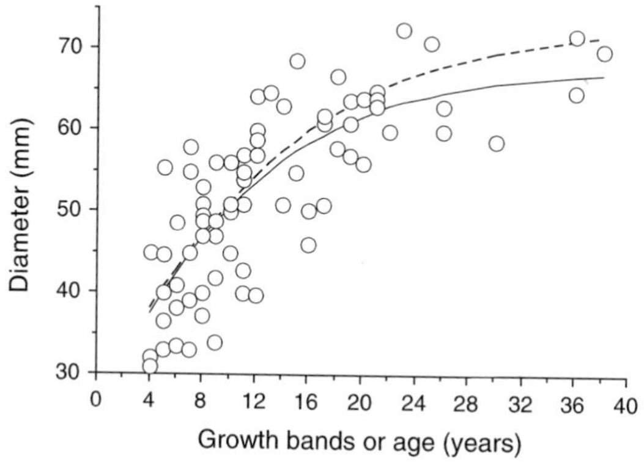

Fig. 6. Size as a function of age for *Sterechinus neumayeri* from Antarctica based on natural growth lines (solid line) compared with estimated growth parameters from individuals tagged with tetracycline (dashed line); it is assumed that both curves have the same size at $t = 0$. Data from Brey et al. (1995).

growth model were the Brody-Bertalanffy (Table 3), with $b = 0.6691$, $D_0 = 22.81$ mm and time to this size if $D_0 = 0.5$ mm, is 4.3 years, this still indicates a problem with determining actual age.

Growth of *Echinus esculentus* has been described using natural growth lines in genital plates or plates of the test. There are substantial differences in the shapes of the growth curves that have been published for size vs. age ranging from sigmoid to nearly linear and comparison of these studies provides insights into general problems of using natural lines as well as growth studies and models in general.

The use of natural lines in genital plates to estimate age was developed by Moore (1935). Preparation methods have varied over time but all workers (White et al. 1985; Nichols et al. 1985; Comely and Ansell 1988; Gage 1992a,b) abrade the surface of a plate to expose lines that can be counted. White et al. (1985) focused their work on etched plates of the test but also examined genital plates and concluded that there was no difference in estimating age. Scanned figures from the above papers were used to reconstruct data sets so comparisons could be made (Table 4).

The largest individuals in samples studied by Moore (1935) were between 110 and 120 mm in diameter. These sizes are typical of those reported in studies conducted since 1985 from other sites in the British Isles (White et al. 1985; Nichols et al. 1985; Comely and Ansell 1988; Gage 1992a,b), but are substantially smaller than sizes reported from other areas The largest test diameter given by Mortensen (1943) is 160 mm from Iceland

Table 4 Summary of growth parameter estimates for *Echinus esculentus*.

	n	D_∞	K (yr^{-1})	D_1	D_7	Comments
Moore (1935)	−1	120.1	0.288	30.1	104.1	Δd vs. d
White et al. (1985)	−1	155.8	0.119	58.6	105.8	$D_0 = 47.14$ if first age class = 3, $D_0 = 0.5$ if first age class = 6
		155.1	0.112	16.4	84.3	Δd vs. mean d (weighted)
Nichols et al. (1985)	+1	98.7	0.797	7.0	88.9	$b = -29.213$, $D_0 = 0.31$ mm
	−1	100.3	0.563	43.5	98.4	For ages ≥ 4 years, $D_0 = 0.5$ mm
Comely and Ansell (1988)[a]	−1	211.0	0.124	24.6	122.4	Cuan d vs. age, $D_0 = 18.56$
		155.2	0.175	24.9	109.6	Duan ML, $D_0 = 18.87$
		249.1	0.081	19.4	107.8	Duan LL, $D_0 = 7.64$
		254.4	0.085	20.7	114.1	Duan DW, $D_0 = 3.81$
		197.0	0.117	21.8	110.1	Duan sites Δd vs. d
Gage (1992a)	−1	77.5	0.827	43.6	77.3	Δd vs. d
		109.9	0.261	25.2	92.2	Initial age = 2, final age = 4
		76.8	0.871	44.7	76.6	Initial age = 1, final age = 3

n is the Richards function parameter: −1 = Brody–Bertalanffy, +1 = logistic; D_1 and D_7 are diameter estimates at ages 1 and 7 years.

[a]Cuan and Duan are sites in the Oban area of Scotland; ML = mid-*Laminaria*, LL = lower *Laminaria*, DW = deep water well below the Laminaria zone.

and Clark (1925) reported a diameter of 176 mm. Grieg (1931) reported individuals with diameters "up to 200 mm" in the Varangerfjord of Norway. Grieg, however, did not actually observe these large *E. esculentus* in Varangerfjord but referred to the 1901 trawl journal of the research vessel "Michael Sars" (N Hagen, pers. comm.).

The size–age relationship given by Moore (1935, figure 10) is close to linear (Fig. 7A) but converting data to Δdiameter vs. diameter (Fig. 7B) provides a better view of change in size. A single regression was estimated for three of Moore's sites, Breakwater, Breast, and Keppel (the Chickens site was excluded because growth appeared to be substantially lower). The regression equation has a slope of −0.2501 and an intercept of 30.036 mm:

$$K = -\ln(1 + \text{slope}), \tag{14}$$

$$D_\infty = -\text{intercept}/\text{slope} \tag{15}$$

so $K = 0.288\,\text{yr}^{-1}$ and $D_\infty = 120.10\,\text{mm}$.

If *E. esculentus* grew according to these parameters of the Brody–Bertalanffy equation starting at settlement size of 0.5 or 0.8 mm,

$$b = \frac{D_\infty - 0.5}{D_\infty} = 1.0,$$

at 1 year,

$$D_1 = 120.10(1 - \exp(-0.288)),$$

and

$$D_1 = 30.0\,\text{mm},$$

which is closer to the size at age 2 suggested by Moore.

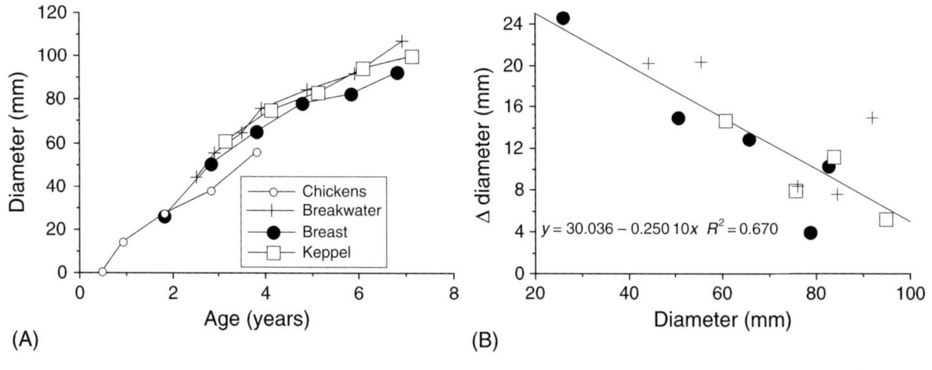

Fig. 7. Growth of *Echinus esculentus* using natural growth lines in genital plates using data from Moore (1935). (A) Copy of Moore's figure 10; (B) conversion of Moore's data to Δ diameter with a time interval of 1 year; Chickens area excluded from regression.

White et al. (1985) used growth lines revealed when plates of the test were etched with 10% HCl and stained with toluidine blue. The smallest *E. esculentus* (≈80 mm) were estimated to be 3 years old. Using the Brody–Bertalanffy model and accepting the size at age estimates, $D_\infty = 155.8$ mm and $K = 0.119\,\mathrm{yr}^{-1}$. The size at $t = 0$, D_0, however, is too large, 47.14 mm (Fig. 8A). If the analysis is changed to adjust age so $D_0 = 0.5$ mm, the adjustment leaves D_∞ and K unchanged, but shifts age by 3 years so what is identified as age 3 should be age 6 years.

Using means for each year, a plot of Δd vs. diameter shows a nonlinear trend (Fig. 8B), which suggests that the Brody–Bertalanffy model may not be particularly good but the size range is restricted to individuals >80 mm and the initial and final points at 80 and 115 mm in diameter have very small values of N. A linear regression with mean values weighted by N in each size category, shows $D_\infty = 155.1$ and $K = 0.112\,\mathrm{yr}^{-1}$. With these parameters and $D_0 = 0.5$ mm, a 1 year old would have a diameter of 16.5 mm, which is about the size indicated by Moore for the Chickens site.

Nichols et al. (1985) used genital plates and reported very different results from other workers primarily with respect to early growth. The cause of this rests on judging the annual nature of lines near the centers of genital plates. The plot of their data is sigmoid and they modeled growth using the logistic equation (Fig. 9A), which is the Richards function with a shape parameter equal to +1. The Richards formula for the logistic is

$$D_t = \frac{D_\infty}{1 - be^{-Kt}} \tag{16}$$

and

$$b = 1 - \frac{D_\infty}{D_0}.$$

By nonlinear regression, $b = -29.213$ and so $D_0 = 0.31$ mm, which is a very reasonable value for the size at $t = 0$ although smaller than the value of 0.81 mm reported by Nichols

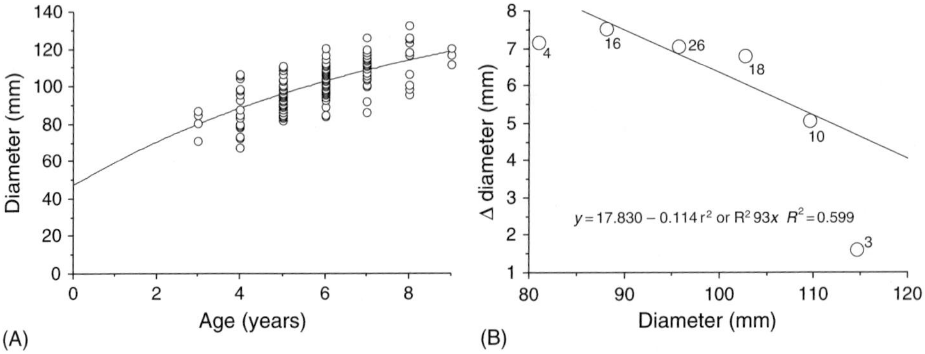

Fig. 8. Growth of *Echinus esculentus* using natural growth lines in test plates using data from White et al. (1985) for Kilkieran Bay. (A) Copy of figure 4 from White et al.; (B) conversion of data from White et al. to Δ diameter for means with a time interval of 1 year; numbers by symbols are the number of observations in each putative age class used to estimate Δ diameter and weight values for regression.

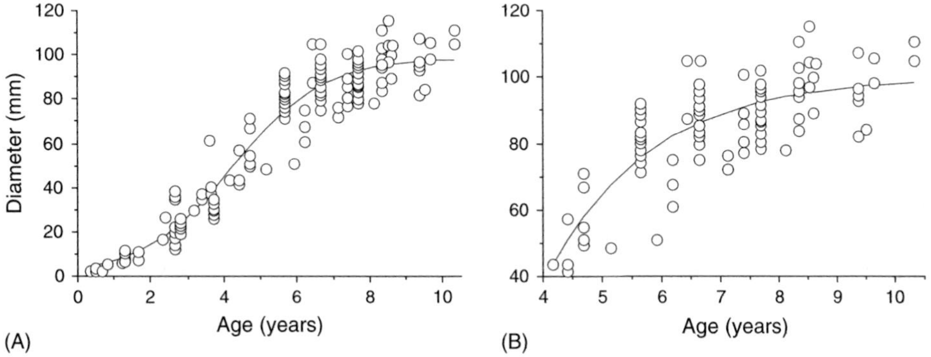

Fig. 9. Growth of *Echinus esculentus* in natural growth lines in genital plates using data from Nichols et al. (1985). (A) copy of figure 3a in Nichols et al.; (B) truncation of data to start at putative age 4, when the data approximately follow a Brody-Bertalanffy curve.

et al. (1985). The central problem is whether growth lines counted by Nichols et al. are actually annual. If individuals older than 4 years are selected (Fig. 9B) and parameters estimated for the Brody–Bertalanffy model, $D_{\infty} = 100.3$ mm and $K = 0.563\,\mathrm{yr}^{-1}$. Size at age 4 was estimated to be 37.9 mm, and the shape is much like that reported by White et al. (1985).

Comely and Ansell (1988) estimated age from genital plates and like Moore (1935) reported a relationship that is close to linear (Fig. 10A). Individual regressions of diameter vs. age provide estimates of maximum diameter that range from 155 mm (ML Duan) to 254 mm (DW Duan). The estimates of $D_{\infty} = 155$ mm with $K = 0.175\,\mathrm{yr}^{-1}$ are similar to the results published by White et al. (1985) but parameters for other samples are well outside estimates published by others. A plot of Δdiameter vs. original diameter (Fig. 10B) shows a negative slope but the Cuan sample is higher than the samples from

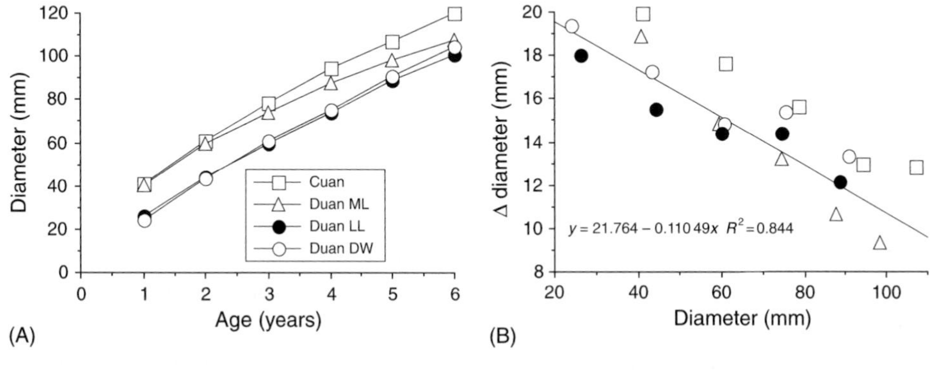

Fig. 10. Growth of *Echinus esculentus* in natural growth lines in genital plates using data from Comely and Ansell (1988). (A) Copy of Comely and Ansell's figure 12; (B) conversion of Comely and Ansell's data to Δdiameter with a time interval of 1 year; regression is for Duan sites only and excludes the Cuan site.

Duan and so was excluded from the regression. Parameter estimates for the combined Duan samples are $K = 0.117\,\text{yr}^{-1}$ and $D_\infty = 197\,\text{mm}$, which is very high. Estimated size at age 1 was 21.8, which is similar to Moore's Chickens site.

Gage (1992a) tagged *E. esculentus* with tetracycline and kept individuals that ranged from 12 to 54 mm in cages on the sea floor. When collected 2 years later, the estimates of original size based on a regression of diameter vs. genital plate width and tetracycline marks, ranged from 28 to 56 mm. Set up was on 23 November 1988 and collections were made on 10 November 1990 so $\Delta t = 2$ years.

There are at least two ways of analyzing the data. One is to use Δ diameter vs. original diameter. The other is to use all individuals collected in 1988 and assign them an age so there is a size at $t = \text{age}$ and a second size at $t = \text{age} + 2$. Gage assigned the original age as 2 years even though he says on p608 that they probably were settlers from 1987. If they started at age 2 the results are very different from the analysis using Δdiameter vs. original diameter. If they started at age 1, the results are very similar.

Using Δ diameter vs. original diameter (Fig. 11A), the slope of the regression is 0.809 and so $K = 1.6544\,\text{yr}^{-2}$; for a 1 year period

$$K\,\text{yr}^{-1} = 1.6544/2 = 0.827\,\text{yr}^{-1}$$

and

$$D_\infty = -\text{Intercept/slope} = 77.46\,\text{mm}.$$

Gage selected age 2 for all of the individuals collected in 1988, which provides a different analysis. The nonlinear model is

$$D_{\text{age}} = D_\infty(1 - e^{-K^*\text{age}})$$

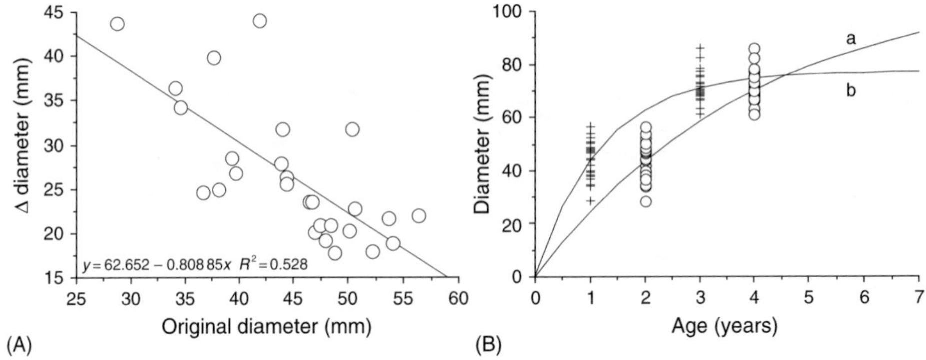

Fig. 11. Growth of *Echinus esculentus* in natural growth lines in genital plates and tetracycline tagging using data from Gage (1992a). (A) Copy of Gage's figure 5 but with linear regression; time interval = 2 years; (B) copy of Gage's figure 6a with solid circles the same as given by Gage so at time of collection sea urchins were 2 years old and fitted with line a; + symbols are the same data shifted to start at age 1 year fitted by line b.

with ages 2 and 4 for initial size and sizes of tagging individuals 2 years later. Note that this analysis did not use size pairs but rather just size at age. The results of a nonlinear regression are:

$$D_\infty = 109.92\,\text{mm}$$
$$K\,\text{yr}^{-1} = 0.261$$

which are similar to the results presented by Gage. If, on the other hand, sea urchins were age 1 when tagged and age 3 when collected and processed (Fig. 11B), results are quite different:

$$D_\infty = 76.82\,\text{mm}$$
$$K\,\text{yr}^{-1} = 0.871,$$

which are similar to estimates obtained from the linear regression of ΔD vs. original D. The estimates hinge on whether the Brody–Bertalanffy equation holds over the entire life of *E. esculentus* and whether a limited size range is sufficient to estimate parameters. No very small individuals or any large ones were included in the analysis.

It is difficult to resolve differences among the four papers estimating growth of *Echinus esculentus* from natural lines (Table 4). Because it is not known how old an individual with one growth line actually is, an important issue is how to fix the growth curve to the age axis. Tag-recapture data, such as that of Gage (1992a), provides the best estimate of growth changes for individuals that started at sizes ranging from 28 to 56 mm in diameter. A linear regression is appropriate for Δ diameter vs. original diameter, which means that a Brody–Bertalanffy model would be appropriate. It is necessary, however, to assume that a linear relationship would hold over the entire size range. This is a problem for all studies that use mark–recapture. The assertion in choosing the Brody–Bertalanffy model is that growth rate is maximum at the time of settlement; there is no lag period, which seems unlikely and so the logistic model of Nichols et al. (1985) seems better.

Judging age based on natural lines is subject to substantial errors (cf. Ebert 1988b; Russell and Meredith 2000). Validation of lines by various methods such as tetracycline as done by Gage (1992a) or by following the seasonal progression of dark and light bands at the edge of plates (Moore 1935 and others) have focused on medium-sized individuals. Small sea urchins may add more than one line per year and lines cannot be resolved in large, slow-growing sea urchins (Ebert 1988b; Gage 1992b; Russell and Meredith 2000). Decisions concerning which lines to count as annual is a problem and so Nichols et al. (1985) counted lines in genital plates that other workers did not. Some of these lines may be annual but without validation there is no way of knowing. Results of Comely and Ansell (1988) provide estimates of maximum diameter that far exceed other studies. This probably is because they did not count some lines that others would have. It is expected that natural lines that crowd together in large sea urchins would not be counted and so ages of large individuals will be underestimated. Some workers for other sea urchin species, however, have obtained results from tetracycline tagging that are consistent with natural lines (Brey et al. 1995).

Comparison of the studies of *Echinus esculentus* over the past 70 years fail to arrive at a convincing size at age relationship. Consequently, it is unknown whether an initial lag exists it probably does as suggested by Nichols et al. (1985), how old large individuals might be, and how variable growth might be across depth and geographic region.

There are other ways of estimating parameters as well as other models. One modification of the Richards function was developed by Schnute (1981). Schnute's model replaces parameters K and D_∞ with parameters that have units of growth associated with them; g_1 is the growth rate of individuals of size y_1 and g_2 is the growth rate of individuals size y_2; y_1 and y_2 are arbitrary sizes that are selected to be near the extremes of the data. Francis (1995) developed a difference equation version of Schnute's model that is appropriate for mark-recapture data:

$$\Delta D = -D_t\left[\mathrm{e}^{-K\Delta t}D_t^b + c(1-\mathrm{e}^{-K\Delta t})\right]^{1/b} \tag{17}$$

where b is $-1/n$ (from Eq. (8)),

$$K = \ln\left[\frac{y_2^b - y_1^b}{\lambda_2^b - \lambda_1^b}\right]. \tag{18}$$

where $\lambda_1 = y_1 + g_1$, $\lambda_2 = y_2 + g_2$ and

$$c = \frac{y_2^b\lambda_1^b - y_1^b\lambda_2^b}{\lambda_1^b - y_1^b + y_2^b - \lambda_2^b}, \tag{19}$$

where c is D_∞^b or $D_\infty^{-1/n}$in Eq. (8).

Assembling all of these pieces into a single model is necessary for statistical packages such as SYSTAT (1992). In the SYSTAT model each line ends with a comma and carriage return , ¶. The comma and carriage return mean that the next line is a continuation and so the model actually is a single line. In the code provided here, the values of y_1 and y_2 are set equal to 0.4 and 1.7 but would be different for different data sets and ΔS is DS.

```
MODEL DS = −S + (S^B*EXP(−LOG((1.7^B − .4^B)/((1.7+,¶
G2)^B − (.4 + G1)^B))) + (1.7^B*(.4 + G1)^B − .4^B*(1.7+,¶
G2)^B)/((.4 + G1)^B − .4^B + 1.7^B − (1.7 + G2)^B)*(1 − EXP,¶
(−LOG((1.7^B − .4^B)/((1.7 + G2) B − (.4 + G1)^B)))))^(1/B)¶
```

If the Gompertz model is appropriate, which would be indicated by estimates of n that are large either positive or negative, and where the 95% confidence limits for n include both positive and negative values:

$$\Delta D = -D + D^{\exp(-K\Delta t)}\exp\left[c(1-\mathrm{e}^{-K\Delta t})\right], \tag{20}$$

$$K = \ln\left[\frac{\ln(y_2/y_1)}{\ln(\lambda_2/\lambda_1)}\right], \tag{21}$$

and

$$c = \frac{\ln(y_2)\ln(\lambda_1) - \ln(y_1)\ln(\lambda_2)}{\ln(\lambda_1 y_2) - \ln(\lambda_2 y_1)}. \tag{22}$$

The following example using *E. chloroticus* tagged with calcein (Lamare and Mladenov 2000) illustrates application of Francis' growth analysis. In order to determine changes in diameters, the relationship between demi-pyramid (jaw) length and test diameter was established using a GM functional regression:

For Doubtful Sound,

$$D = 5.7470J^{1.04683} \tag{23}$$

and for Tory Channel,

$$D = 5.41585J^{0.96110}. \tag{24}$$

Initial jaw size was converted to original test size and the difference between initial and final test size was determined using Eqs (2) and (3). Initial estimates of parameters using the Richards function showed that the 95% confidence limits included both positive and negative values of n and so the Gompertz equation was considered an appropriate alternative to the more general Richards function. The scatter of data and the fitted Gompertz lines are shown in Fig. 12. The estimates for annual growth rate at the selected sizes of 40 and 100 mm (y_1 and y_2) (Table 5) show the same pattern as indicated in the figures presented by Lamare and Mladenov (2000); namely, that growth of small individuals is possibly slightly higher at the Tory Channel site but large sea urchins growth better in Doubtful Sound. The 95% confidence limits for g_1 show that growth may be the same for small animals at both sites; there are few small individuals, however, at both sites and a larger sample would be needed to resolve possible differences.

With larger numbers of individuals for all sizes, the shape of the growth curve is not adequately described by the Richards family of curves because large individuals appear to continue to growth at a very slow and nearly linear rate for many years and small individuals show a lag and accelerating growth phase. *Strongylocentrotus franciscanus* from the west coast of North America shows this pattern. The Tanaka growth model (Eq. (25) Tanaka 1982, 1988) seems more appropriate for these data (Ebert 1998; Ebert et al. 1999) and is a four-parameter equation (f, c, a, and d) that accommodates an early lag and exponential phase followed by a declining growth rate. The function does not always have an asymptotic size and so growth can continue throughout life:

$$D_t = \frac{1}{\sqrt{f}} \ln\left[2f(t-c) + 2\sqrt{f^2(t-c)^2 + fa}\right] + d \tag{25}$$

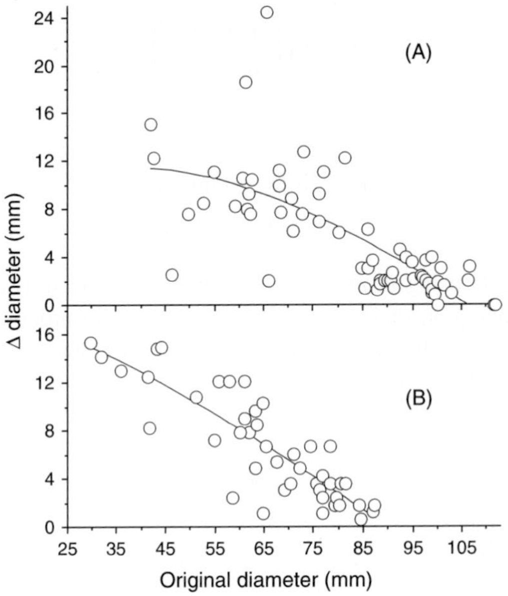

Fig. 12. Change in diameter, ΔD, as a function of original diameter for *Evechinus chloroticus*. (A) Doubtful Sound, $N = 63$; (B) Tory Channel, $N = 45$; fitted lines are the Gompertz growth model. Data from Lamare and Mladenov (2000).

Table 5 Growth analysis of *Evechinus chloroticus* from Doubtful Sound and Troy Channel, South Island, New Zealand, using the Francis analogue of Schnute's growth model.

Site	*N*	Parameter	Estimate	Lower <95%>	Upper
Doubtful Sound	63	g_1	11.946 mm	10.008 mm	13.884 mm
		g_2	1.809	0.589	3.028
Tory Channel	45	g_1	12.755	11.405	14.1059
		g_2	−4.584	−6.387	−2.780

g_1 is the growth rate per year at $y_1 = 40$ mm and g_2 at $y_2 = 100$ mm.
Data from Lamare and Mladenov (2000).

where

$$c = \frac{a}{E} - \frac{E}{4f}$$

and

$$E = \exp\left(\sqrt{f}(D_0 - d)\right).$$

D_0 is size at time 0.

The difference equation for the Tanaka function is

$$D_{t+\Delta t} = \frac{1}{\sqrt{f}} \ln\left[2G + 2\sqrt{G^2 + fa}\right] + d \tag{26}$$

where

$$G = E/4 - fa/E + f\Delta t$$

and

$$E = \exp\left(\sqrt{f}(D_t - d)\right)$$

Equation (26) also can be written with the growth increment as a function of original size:

$$\Delta D = -D_t + \frac{1}{\sqrt{f}} \ln\left[2G + 2\sqrt{G^2 + fa}\right] + d. \tag{27}$$

Further details of the Tanaka equation are given in Tanaka (1988), Ebert (1999) and Ebert et al. (1999).

Tetracycline tagged *S. franciscanus* from Shaw Island, Washington, provide an example (Ebert et al. 1999) for use of the Tanaka model. The growth increments were measured and the allometric relationship between diameter and jaw length (Fig. 13A) was determined:

$$D = 5.5483J^{1.2368} \tag{28}$$

Equations (2) and (3) were used to estimate diameter at the time of tagging and then ΔD was used as a function of size (Fig. 13B) to estimate growth parameters. The long period of very slow growth of large individuals is apparent. Small individuals do not grow as

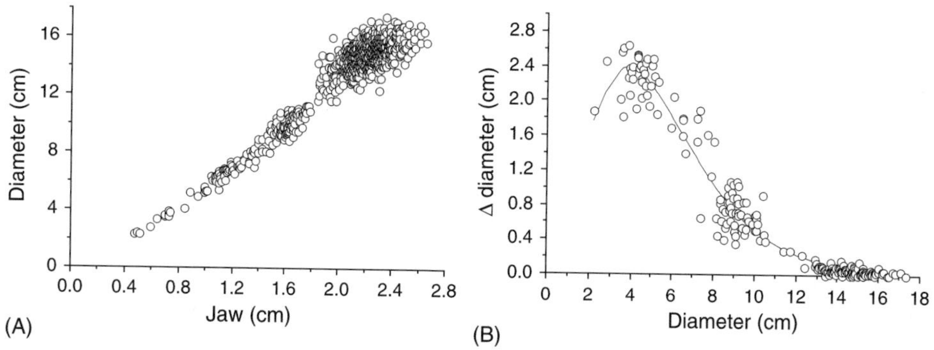

Fig. 13. Growth analysis of *Strongylocentrotus franciscanus* tagged at Shaw Island, WA, USA and recaptured 1 year later. (A) Relationship between test diameter and jaw length (cm); $N = 822$; (B) ΔD vs. original test diameter (cm); fitted line based on the Tanaka function; $N = 248$; $f = 0.189$ (± 0.010 se), $d = 7.485$ (± 0.073 se), and $a = 0.161$ (± 0.005 se). Data from Ebert et al. (1999).

rapidly as would be expected based on the trajectory of individuals at about 6 cm and so it is apparent that there is a maximum growth rate that is attained at a size greater than newly settled individuals, which is an attribute of the Brody–Bertalanffy model. This pattern is typical of a curve that is fit better with the Tanaka function than by the Richards family of curves.

A modification of the Brody–Bertalanffy growth model developed by Grosjean et al. (2003) has some of the features of the Tanaka function but retains an asymptotic size. These authors followed a cohort of *Paracentrotus lividus* starting at $t = 6$ months for 7 years and measured diameters of all survivors every 6 months. As part of this long-term study they found that large sea urchins inhibited the growth of small ones (Grosjean et al. 1996) and so inhibition was included in the model by use of a transfer function from an inhibited state (S) to the state that inhibited others (L). Large individuals in the cohort at 6 months started in the L-state and grew according to the standard Brody–Bertalanffy model (Eq. (4)). Smaller sea urchins started in the S-state but transferred to the L-state based on a logistic function. Transfer was just a matter of time and so at the end of 7 years all sea urchins were in the L-state and had attained maximum sizes. A restriction was that individuals never changed their relative positions in the size hierarchy so the largest individual at 6 months, should it survive, would be the largest individual at 7 years and have an asymptotic size larger than individuals that were small at 6 months. Nonlinear quantile regression (Koenker and Park 1996) was used to estimate growth parameters for three quantiles: the largest 2.5%, the smallest 2.5%, and 95% of individuals of medium size. Different growth models were tried for the three quantiles, but the overall best model was the new one proposed by the authors:

$$D_t = D_0 + (D_\infty - D_0)\,\frac{1 - \mathrm{e}^{-Kt}}{1 + l_\mathrm{i}\mathrm{e}^{-K_\mathrm{i}t}} \tag{29}$$

where l_i and K_i are parameters of inhibition and determine transfer from the S-state to the L-state.

The model proposed by Grosjean et al. (2003) has not been explored for use with field data where single cohorts seldom can be followed for many years. Also, it is unclear whether two assumptions are met by most species; namely, (1) maintenance of relative rank in size; and (2) actual cessation of growth. If there is no asymptotic size, the solution is relatively simple because the Brody-Bertalanffy model in Eq. (29) could be replaced with some other growth model such as the Tanaka. Maintenance of relative rank in size would be more difficult to resolve for field populations.

The pattern that emerges from the analysis of different growth models is that as number of measurements and size range increase, growth models for sea urchins must become more complicated in order to accommodate the pattern of data points. Tagging with tetracycline or calcein provides resolution that is not possible with invasive tags, including PIT tags, and avoids problems of resolution of slow growth in large individuals where diameter increments are less than 0.1 mm. Small increments cannot be determined when measurement errors for large individuals can be several millimeters. In Fig. 13B, 2 mm of measurement error would lump all individuals above 12 cm. It is expected that the Tanaka model, or something similar such as adding additional parameters to the

Brody–Bertalanffy model, will probably be models of choice as studies of other sea urchin species include methods that provide high resolution of growth over a wide range of sizes.

2. SURVIVAL

Survival rates can be based on following cohorts in the field over many years, which does not appear to have been done with sea urchins, or can be based on combining information on growth transitions with size–frequency data (Beverton and Holt 1956; Van Sickle 1977; Ebert 1987). These methods have been applied to sea urchins (e.g. Ebert 1975; Ebert et al. 1999; Russell 1987; Russell et al. 1998; Smith et al. 1998) and in the context of estimating productivity/biomass by Brey (1991; Brey et al. 1995). The survival model that is used in all of these analyses is the simple decaying exponential,

$$N_t = N_0 \mathrm{e}^{-Mt}. \qquad (30)$$

The annual survival rate, p, is e^{-M} and so the annual mortality rate is $1-\mathrm{e}^{-M}$. This model is appropriate for organisms that do not show effects of aging during their lives; the probability that a 5-year-old individual survives for 1 year is the same as the probability that a 20-year-old or 100-year-old individual survives. The simple model is used not because it is known that survival is age invariant but rather because it is very difficult to obtain good age- or size-specific data on survival and so the simplest model is selected.

The Beverton and Holt (1956) model makes use of the mean of a size–frequency distribution and has been modified (Ebert 1987) to be used with the Richards function parameters and with a size at recruitment to the population being studied, D_R:

$$M = \frac{K\left(D_\infty^{-1/n} - \overline{X}\right)}{\overline{X} - D_R^{-1/n}}. \qquad (31)$$

$\overline{X}$ is the mean of transformed sizes in the frequency distribution using $D_t^{-1/n}$ and it is necessary to assume constant and continuous recruitment, which may be reasonable if size at recruitment to the sampled population is large relative to the size at settlement so individuals are several years old when recruited.

If recruitment to the sampled population is seasonal, a different model is needed:

$$\overline{D_t} = D_\infty\left(1-\mathrm{e}^{-M}\right)\sum_{t=0}^{\omega} \mathrm{e}^{-Mt}\left[1 - b\mathrm{e}^{-K(T+t)}\right]^{-n} \qquad (32)$$

for growth modeled with the Richards function. Derivation of Eqs (30) and (32) are given elsewhere (Ebert 1987, 1999).

Application of Eqs (30)–(32) provide estimates of $M\,\mathrm{yr}^{-1}$ and hence estimates of annual survival and mortality (Table 6). Even though the Brody–Bertalanffy model may not be the best one to use for sea urchin growth, it is widely used and probably captures

Table 6 Summary of Brody–Bertalanffy parameters (K and D_∞) and annual survival rate (p); time units are years, # is number in Figure 6.14.

Family	Species	#	K yr^{-1}	D_∞	p	Method	Location	Reference
Cidaridae	*Eucidaris tribuloides*	7	0.47	50.93	0.87	2[b]	Virginia Key, FL USA	McPherson (1968)
Diadematidae	*Diadema antillarum*	6	0.49	95.09	0.70	2	US Virgin Islands	Karlson and Levitan (1990)
			0.49	88.40		2	Barbados, BWI	Lewis (1966)
	Diadema savignyi	4	0.79	84.9	0.52	2	South Africa	Drummond (1994)
	Centrostephanus rodgersi	12	0.24	94.23	0.92	1[a]	Port Jackson, Australia	Ebert (1982)
Toxopneustidae	*Lytechinus pictus*	3	0.99	45.53	0.64	1	Agua Hedionda, San Diego, CA, USA	Detwiler (1996)
		5	0.66	28.16	0.38	2	Point Loma, San Diego, CA, USA	Detwiler (1996)
	Lytechinus variegatus		1.27	64.51		2	Miami, FL, USA, and Bermuda	Moore et al. (1963)
			1.82	44.36		2	Jobos Bay, Puerto Rico	Rivera (1979)
			1.10	58.79		2	Mammee Shallows, Jamaica	Greenway (1977)
	Tripneustes ventricosus	2	1.55	92.29	0.30		Barbados, BWI	Scheibling and Mladenov (1988)
			1.66	102.3		2	Barbados, BWI	Lewis (1958)
	Tripneustes gratilla	1	1.8	108	0.01	2*	Danahon Reef, Central Philippines	Bacolod and Dy 1986
	T. gratilla elatensis		0.87	95.0		2	Eilat, Israel	Dafni (1992)
	Sphaerechinus granularis		0.31	119.8		3[c]	Bay of Brest, France [Plougastel]	Lumingas and Guillou (1994b)

		11	0.26	123.6	0.54	3	Bay of Brest, France	Lumingas and Guillou (1994a)
Echinometridae	*Echinometra mathaei*	9	0.32	86.8	0.63	2	South Africa	Drummond (1994)
	Heterocentrotus mammillatus	13	0.23	53.41	0.93	1	Honaunau Bay, HI, USA	Ebert (1982)
	Evechinus chloroticus	10	0.28	104.89	0.95	1	Doubtful Sound, NZ	Lamare and Mladenov (2000)
		8	0.39	85.07	0.90	1	Tory Channel, NZ	
Stomopneustidae	*Stomopneustes variolaris*	17	0.11	145.4	0.80	2	South Africa	Drummond (1994)
Echinidae	*Sterechinus antarcticus*	19	0.02	82.4	0.93	3	Weddell Sea, Antarctica	Brey (1991)
	Sterechinus neumayeri	18	0.06	78.03	0.94	1	McMurdo Sound, Antarctica	Brey et al. (1995)
			0.09	68.92		3		Brey et al. (1995)
	Echinus esculentus	15	0.15	140.0	0.84	3	Kilkieran Bay, Ireland	White et al. (1985)
Strongylocentrotidae	*Strongylocentrotus pallidus*	20	0.01	102.3	0.92	3	Barents Sea	Bluhm et al. (1998)
	Strongylocentrotus droebachiensis	16	0.15	70.61	0.66	2,3	Tromsø and Bod, Norway	Sivertsen and Hopkins (1995)
		21	0.28	89.4	0.74	2	Womens Bay, Kodiak, Alaska, USA	Munk (1992)
	Strongylocentrotus franciscanus	14	0.22	118.1	0.92	1	Bodega Marine Reserve, CA, USA	Morgan et al. (2000)

[a] Tagging; [b] Size-frequency analysis; 2* is the program ELEFAN; [c] Natural growth lines.

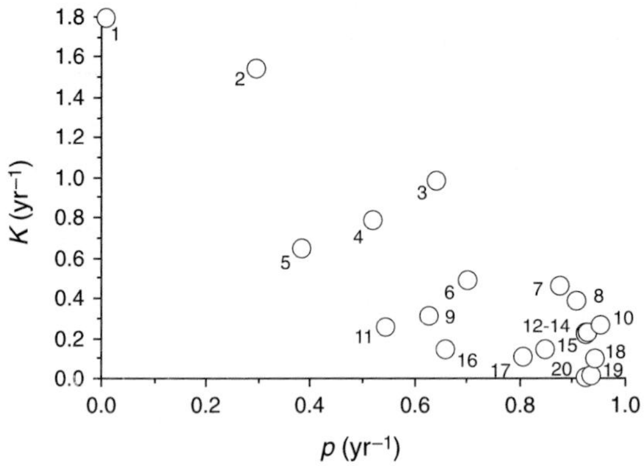

Fig. 14. Relationship between K yr^{-1} of the Brody–Bertalanffy growth model and annual survival probability, p; numbers by circles refer to species in Table 6.

the broad aspects of growth in the sense that fast- and slow-growing species can be identified. The author has sampled the literature and combined growth using the Brody–Bertalanffy model with size structure to estimate M and p (Table 6). Species in the table are just a sample of available published studies but are sufficient to demonstrate the overall relationship between the growth rate constant K and p (Fig. 14), which is similar to the pattern previously published (Ebert 1975). Species that grow rapidly also are short-lived. Moreover, there are some similarities of species within families. The fastest-growing and shortest-lived are toxopneustids followed by diadematids. There may be additional familial rankings but at present the data do not seem adequate for such analysis. It should be emphasized that short life in such species as *Tripneustes gratilla* does not indicate senescence because the body wall is very thin and hence more easily damaged than in species with thicker bodies (Ebert 1982). This is like comparing the survival rates of heavy beer steins with champagne flutes in a tavern; with equal use, more flutes break.

If the Tanaka model is more appropriate for describing growth,

$$\overline{D_t} = \left(1 - e^{-M}\right) \sum_{t=0}^{\omega} e^{-Mt} \alpha \left\{ \frac{1}{\sqrt{f}} \ln \left[2f(t+T-c) + 2\sqrt{f^2(t+T-c)^2 + fa} \right] + d \right\}^{\beta}. \quad (33)$$

In studies with sufficient data so that the Brody–Bertalanffy model is inappropriate, survival estimates in some cases are very high (Table 7), sufficiently so to make one wonder whether they can be true. Data gathered for *S. franciscanus* (Ebert et al. 1999) indicate that the estimates should be considered as probable. Sample sizes were large and showed very low annual transition rates for diameter growth across centimeter size classes (Ebert 1998). Only 3% per year moved from the 13–14 cm size class to the 14–15 cm size class, and 1% moved from 14–15 cm to 15–16 cm and from 15–16 cm to ≥16 cm. The estimates based on tetracycline tagging were validated by predicting the amount of a jaw that would have to be milled off to go from modern ^{14}C to pre A-bomb

Table 7 Estimates of $M\text{yr}^{-1}$ and annual survival rate (p) for selected echinoid species where growth was estimated using a model other than Brody–Bertalanffy.

Species	M (yr^{-1})	p	Source
Temperate			
Centrostephanus rodgersi	0.06	0.94	Ebert (1982)
Evechinus chloroticus	0.05–0.10	0.95–0.90	Lamare (1997)
Heliocidaris erythrogramma	0.10	0.91	Ebert (1982)
Strongylocentrotus droebachiensis	0.04–0.30	0.96–0.74	Russell et al. (1998)
Strongylocentrotus franciscanus			Ebert et al. (1999)
Southern sites	0.09–0.40	0.91–0.67	
Northern sites	0.02–0.20	0.98–0.82	
Strongylocentrotus purpuratus			Russell (1987)
Southern sites	0.12–0.33	0.89–0.72	
Northern site	0.10–0.16	0.90–0.85	
Tropical			
Diadema setosum	0.29	0.75	Ebert (1982)
Echinometra mathaei	0.15	0.86	Ebert (1982)
Echinothrix diadema	0.32	0.73	Ebert (1982)
Heterocentrotus mammillatus	0.11	0.89	Ebert (1982)
Salmacis belli	0.20	0.82	Ebert (1982)
Stomopneustes variolaris	0.19	0.83	Ebert (1982)

^{14}C (Ebert and Southon 2003). The estimates of very slow growth appear to be valid and *S. franciscanus* with a test diameter of 17 cm would be well over 100 years old. The largest individuals in three samples from British Columbia reported by Bureau (1996) were 18.3, 19.0, and 19.8 cm. Given low transfer rates, it must take a very long time to reach these sizes and so 200+ years to 19 cm does not seem unreasonable. Many other echinoids may have similar growth and survival rates. This has profound implications for fishery management.

ACKNOWLEDGMENTS

Portions of work presented here were supported by the US National Science Foundation and by the sea urchin fishermen and state agencies of California, Oregon, Washington, and Alaska. Work with *Strongylocentrotus franciscanus* was done with S Schroeter, J Dixon, and M Russell. N Hagen kindly obtained and translated the 1931 Grieg paper for me. Data were very graciously provided by M Lamare for *Evechinus chloroticus*, T Brey for *Sterechinus neumayeri*, and J Munk for *Strongylocentrotus droebachiensis*. I greatly appreciate their willingness to share data with me. The manuscript benefited from comments from J Lawrence.

REFERENCES

Agatsuma A (2001) Ecology of *Strongylocentrotus nudus*. In: Lawrence JM (ed.) Edible sea urchins: biology and ecology. Elsevier, Amsterdam, pp 347–361

Ameye L, Compère P, Dille J, Dubois P (1998) Ultrastructure and cytochemistry of the early calcification site and of its mineralization organic matrix in *Paracentrotus lividus* (Echinodermata: Echinoidea). Histochem Cell Biol 110:285–294

Ameye L, Hermann R, Killian C, Wilt F, Dubois P (1999) Ultrastructural localization of proteins involved in sea urchin biomineralization. J Histochem Cytochem 47:1189–1200

Bacolod PT, Dy DT (1986) Growth, recruitment pattern and mortality rate of the sea urchin, *Tripneustes gratilla* Linnaeus, in a seaweed farm at Danahon Reef, Central Philippines. Philippine Scient 23:1–14

Barker F, Soh YC, Evans RJ (1988) Properties of the geometric mean functional relationship. Biometrics 44:279–281

Beddingfield SD, McClintock JB (2000) Demographic characteristics of *Lytechinus variegatus* (Echinoidea: Echinodermata) from three habitats in a north Florida bay, Gulf of Mexico. PSZN: Mar Ecol 21:17–40

Beverton RJ, Holt SJ (1956) A review of methods for estimating mortality rates in exploited fish populations, with special reference to sources of bias in catch sampling. Cons Perm Interna Explor Mer 140 (pt I):67–83

Bluhm BA, Piepenburg D, von Juterzenka K (1998) Distribution, standing stock, growth, mortality and production of *Strongylocentrotus pallidus* (Echinodermata: Echinoidea) in the northern Barents Sea. Polar Biol 20:325–334

Brey T (1991) Population dynamics of *Sterechinus antarcticus* (Echinodermata: Echinoidea) on the Weddell Sea shelf and slope, Antarctica. Antarctic Sci 3:251–256

Brey T, Pearse J, Basch L, McClintock J (1995) Growth and production of *Sterechinus neumayeri* (Echinoidea: Echinodermata) in McMurdo Sound, Antarctica. Mar Biol 124:279–292

Brooks RC, Heidinger RC, Kohler CC (1994) Mass-marking otoliths of larval and juvenile walleyes by immersion in oxytetracycline, calcein, or calcein blue. North Amer J Fish Manag 14:143–150

Bumguardner BW, King TL (1996) Toxicity of oxytetracycline and calcein to juvenile striped bass. Trans Am Fish Soc 125:143–145

Bureau D (1996) Relationship between feeding, reproductive condition, jaw size and density in the red sea urchin, *Strongylocentrotus franciscanus*. MS Thesis, Simon Fraser University, Burnaby, BC

Bureau F, Dubois P, Ghyoot M, Jangoux M (1991) Skeleton resorption in echinoderms: regression of pedicellarial stalks in *Sphaerechinus granularis* (Echinoidea). Zoomorphology 110:217–226

Carpenter WB (1870) On the reparation of the spines of echinida. Monthly Micros J 3:225–228

Charnov EL (1993) Life history invariants: some explorations of symmetry in evolutionary ecology. Oxford University Press, New York

Chave KE (1954) Aspects of the biogeochemistry of magnesium 1. Calcareous marine organisms. J Geol 62:266–283

Clark HL (1925) A catalogue of the recent sea-urchins in the collection of the British Museum (Nat. Hist.). London. Trustees of the British Museum.

Clarke FW, Wheeler WC (1922) The inorganic constituents of marine invertebrates. US Geol Surv Prof Pap 124:1–62

Cloern JE, Nichols FH (1978) A von Bertalanffy growth model with a seasonally varying coefficient. J Fish Res Bd Can 35:1479–1482

Comely CA, Ansell AD (1988) Population density and growth of *Echinus esculentus* L. on the Scottish west coast. Estuar, Coast Shelf Sci 27: 311–334

Constable AJ (1993) The role of sutures in shrinking of the test in *Heliocidaris erythrogramma* (Echinoidea: Echinometridae). Mar Biol 117:423–430

Crapp GB, Willis ME (1975) Age determination in the sea urchin *Paracentrotus lividus* (Lamarck), with notes on the reproductive cycle. J Exp Mar Biol Ecol 20:157–178

Cuenca C (1987) Quelques méthodes de marquages des oursins échinidés (Échinodermes). Bull Soc Sc nat Ouest France ns 9:26–37

Dafni J (1986) A biomechanical model for the morphogenesis of regular echinoid tests. Paleobiol 12:143–160

Dafni J (1992) Growth rate of the sea urchin *Tripneustes gratilla elatensis*. Israel J Zool 38:25–33

David B, Néraudeau D (1989) Tubercle loss in spatangoids (Echinodermata, Echinoidea): original skeletal structures and underlying processes. Zoomorphology 109:39–53

Davies TT, Crenshaw MA, Heatfield BH (1972) The effect of temperature on the chemistry and structure of echinoid spine regeneration. J Paleo 46:874–883

Day RW, Williams C, Hawkes GP (1992) A comparison of flurochromes for marking abalone shells. Mar Freshwater Res 46:599–605

Dayton, PK, Tegner MJ, Edwards PB, Riser KL (1999) Temporal and spatial scales of kelp demography: the role of oceanographic climate. Ecol Monog 59:219–250

Detwiler PM (1996) Demography, growth, mortality and resource allocation in the white sea urchin *Lytechinus pictus*. MS Thesis, San Diego State University, San Diego, CA

Deutler F (1926) Über das Wachstum des Seeigelskeletts. Zool Jahrb Abt Anat Ontog 48:119–200

Dix TG (1970) Biology of *Evechinus chloroticus* (Echinoidea: Echinometridae) from different localities 2. Movement. N Z J Mar Freshwater Res 4:267–277

Dix TG (1972) Biology of *Evechinus chloroticus* (Echinoidea: Echinometridae) from different localities. 4. Age, growth, and size. N Z J Mar Freshwater Res 6:48–68

Dotan A, Fishelson L (1985) Morphology of spines of *Heterocentrotus mammillatus* (Echinodermata, Echinoidae) and its ecological significance In: Keegan BF, O'Connor DBS (eds) Echinodermata. Balkema, Rotterdam, pp 253–260

Drummond AE (1994) Aspects of the life history biology of three species of sea urchin on the South African east coast. In: David B, Guille A, Féral J-P, Roux M (eds) Echinoderms through time. Balkema, Rotterdam, pp 637–641

Dubois P (1991) Morphological evidence of coherent organic material within the stereom of postmetamorphic echinoderms. In: Suga S, Nakahara H (eds) Mechanisms and phylogeny of mineralization in biological systems. Springer-Verlag, Tokyo, pp 41–45

Dubois P, Chen CP (1989) Calcification in echinoderms. Echino Stud 3:109–178

Dubois P, Ghyoot M (1995) Integumentary resorption and collagen synthesis during regression of headless pedicellariae in *Sphaerechinus granularis* (Echinodermata: Echinoidea). Cell Tissue Res 282:297–309

Ebert TA (1965) A technique for the individual marking of sea urchins. Ecology 46:193–194

Ebert TA (1967a) Negative growth and longevity in the purple sea urchin *Strongylocentrotus purpuratus* (Stimpson). Science 157:557–558

Ebert TA (1967b) Growth and repair of spines in the sea urchin *Strongylocentrotus purpuratus* (Stimpson). Biol Bull 133:141–149

Ebert TA (1968) Growth rates of the sea urchin *Strongylocentrotus purpuratus* related to food availability and spine abrasion. Ecology 49:1075–1091

Ebert TA (1975) Growth and mortality in postlarval echinoids. Am Zool 15:755–775

Ebert TA (1977) An experimental analysis of sea urchin dynamics and community interactions on a rock jetty. J Exp Mar Biol Ecol 27:1–22

Ebert TA (1980a) Relative growth of sea urchin jaws: an example of plastic resource allocation. Bull Mar Sci 30:467–474

Ebert TA (1980b) Estimating parameters in a flexible growth equation, the Richards function. Can J Fish Aquat Sci 37:687–692

Ebert TA (1982) Longevity, life history, and relative body wall size in sea urchins. Ecol Monogr 52:353–394

Ebert TA (1985) The non-periodic nature of growth rings in echinoid spines. In: Keegan BF, O'Connor DBS (eds) Echinodermata. Balkema, Rotterdam, pp 261–267

Ebert TA (1986) A new theory to explain the origin of growth lines in sea urchin spines. Mar Ecol Prog Ser 34:197–199

Ebert TA (1987) Estimating growth and survival parameters by nonlinear regression using average size in catches. In: Pauly D, Morgan GR (eds) Theory and application of length-based methods in stock assessment, ICLARM Conference Proceedings 13, pp 35–44

Ebert TA (1988a) Growth, regeneration and damage repair in spines of the slate-pencil sea urchin *Heterocentrotus mammillatus* (L.) (Echinodermata: Echinoidea). Pac Sci 42:160–172

Ebert TA (1988b) Calibration of natural growth lines in ossicles of two sea urchins, *Strongylocentrotus purpuratus* and *Echinometra mathaei*, using tetracycline. In: Burke RD, Mladenov PV, Lambert P, Parsley RL (eds) Echinoderm biology. Balkema, Rotterdam, pp 435–443

Ebert TA (1996) Adaptive aspects of phenotypic plasticity in echinoderms. Oceanolo Acta 19:347–355

Ebert TA (1998) An analysis of the importance of Allee effects in management of the red sea urchin *Strongylocentrotus franciscanus*. In: Mooi R, Telford M (eds) Echinoderms: San Francisco. Balkema, Brookfield, VT, pp 619–627

Ebert TA (1999) Plant and animal populations. Methods in demography. Academic Press, San Diego

Ebert TA (2004) Shrinking sea urchins and the problems of measurement In: Heinzeller T, Nebelsick J (eds) Echinoderms: München. Balkema, London, pp 321–325

Ebert TA, Dixon JD, Schroeter SC, Kalvass PE, Richmond NT, Bradbury WA, Woodby DA (1999) Growth and mortality of red sea urchins *Strongylocentrotus franciscanus* across a latitudinal gradient. Mar Ecol Prog Ser 190:189–209

Ebert TA, Russell MP (1992) Growth and mortality estimates for red sea urchin, *Strongylocentrotus franciscanus*, from San Nicolas Island, California. Mar Ecol Prog Ser 81:31–41

Ebert TA, Russell MP (1993) Growth and mortality of subtidal red sea urchins (*Strongylocentrotus franciscanus*) at San Nicolas Island, California, USA: problems with models. Mar Biol 117:79–89

Ebert TA, Southon JR (2003) Red sea urchins (*Strongylocentrotus franciscanus*) can live over 100 years: confirmation with A-bomb ^{14}carbon. Fish Bull 101:915–922

Edwards PB, Ebert TA (1991) Plastic responses to limited food availability and spine damage in the sea urchin *Strongylocentrotus purpuratus*. J Exp Mar Biol Ecol 145:295–220

Ellers O (1993) A mechanical model of growth in regular sea urchins: predictions of shape and a developmental morphospace. Proc R Soc Lond B 254:123–129

Ellers O, Telford M (1992) Causes and consequences of fluctuating coelomic pressure in sea urchins. Biol Bull 182:424–434.

Estes JA, Duggins DO (1995) Sea otters and kelp forests in Alaska: generality and variation in a community ecological paradigm. Ecol Monogr 65:75–100

Fansler SC (1983) Phenotypic plasticity of skeletal elements in the purple sea urchin, *Strongylocentrotus purpuratus*.. MS Thesis, San Diego State University, San Diego, CA

Francis RICC (1995) An alternative mark-recapture analogue of Schnute's growth model. Fish Res 23:95–111

Fuji AR (1962) A new tagging method for the sea urchin, *Hemicentrotus pulcherrimus* (A. Agassiz). The Aquiculture (Sendai). 10:11–14

Gage JD (1991) Skeletal growth zones as age-markers in the sea urchin *Psammechinus miliaris*. Mar Biol 110:217–228

Gage JD (1992a) Natural growth bands and growth variability in the sea urchin *Echinus esculentus*: results from tetracycline tagging. Mar Biol 114:607–616

Gage JD (1992b) Growth bands in the sea urchin *Echinus esculentus*: results from tetracycline-mark/recapture. J Mar Biol Ass UK 72:257–260

Gage JD, Tyler PA (1985) Growth and recruitment of the deep-sea urchin *Echinus affinis*. Mar Biol 90:41–53

Gage JD, Tyler PA, Nichols D (1986) Reproduction and growth of *Echinus acutus* var. *norvegicus* Duben & Kören and *E. elegans* Duben & Kören on the continental slope off Scotland. J Exp Mar Biol Ecol 101:61–83

Gebauer P, Moreno CA (1995) Experimental validation of the growth rings of *Loxechinus albus* (Molina, 1782) in southern Chile (Echinodermata: Echinoidea). Fish Res 21:423–435

Greenway M (1977) The production of Thalassiaa testudinum Konig in Kingston Harbour, Jamaica. PhD Thesis, University of the West Indies

Grieg, JA (1931) Echinodermer fra den norske Kyst. Bergens Museums Årbok 1930:10:4–13

Grosjean Ph, Spirolet Ch, Jangoux M (1996) Experimental study of growth in the echinoid Paracentrotus lividus (Lamarck, 1816) (Echinodermata). J Exp Mar Biol Eco 201: 173–184

Grosjean Ph, Spirolet Ch, Jangoux M (2003) A functional growth model with intraspecific competition applied to a sea urchin, Paracentrotus lividus. Can J Fish Aquat Sci 60:237–246

Hagen NT (1996) Tagging sea urchins: a new technique for individual identification. Aquaculture 139:271–284

Heatfield BM (1971) Growth of the calcareous skeleton during regeneration of spines of the sea urchin, *Strongylocentrotus purpuratus* (Stimpson): a light and scanning electron microscopic study. J Morph 134:57–90

Hill SK, Aragona JB, Lawrence JM (2004) Growth bands in test plates of the sea urchins *Arbacia punctulata* and *Lytechinus variegatus* (Echinodermata) on the central Florida Gulf Coast shelf. Gulf of Mexico Sci 2004: 96–100

Hur S-B, Yoo S-K, Rho S (1985) Laboratory tagging experiment of sea urchin, *Hemicentrotus pulcherrimus* (A. Agassiz). Bull Korean Fish Soc 18:363–368

Jensen M (1969) Age determination of echinoids. Sarsia 37:41–44

Jolicoeur P (1990) Bivariate allometry: interval estimation of the slopes of the ordinary and standardized normal major axes and structural relationship. J Theo Biol 144:275–285

Kalvass PE, Hendrix JM, Law PM (1998) Experimental analysis of 3 internal marking methods for red sea urchins. Calif Fish Game 84:88–99

Karlson RH, Levitan DR (1990) Recruitment-limitation in open populations of *Diadema antillarum*: an evaluation. Oecologia 82:40–44

Kawamura K (1966) On the age determining character and growth of a sea urchin, *Strongylocentrotus nudus*. Sci Rpts Hokkaido Fish Exp Stat 6:56–61

Kawamura K (1973) Fishery biological studies on a sea urchin, *Strongylocentrotus intermedius* (A. Agassiz). Sci Rpts Hokkaido Fish Exp Stat 16:1–54

Kenner MC (1992) Population dynamics of the sea urchin *Strongylocentrotus purpuratus* in a central California kelp forest: recruitment, mortality, growth, and diet. Mar Biol 112:107–118

Kobayashi S, Taki J (1969) Calcification in sea urchins I. A tetracycline investigation of growth of the mature test in *Strongylocentrotus intermedius*. Calcif Tissue Res 4:210–223

Koenker, R, Park BJ (1996) An interior point algorithm for nonlinear quantile regression. J Economet. 71:265–283

Lagardêre F, Thibaudeau K, Bégout-Anras, ML (2000) Feasibility of otolith markings in large juvenile turbot, *Scophthalmus maximus*, using immersion in alizarin red S solutions. ICES J Mar Sci 57:1175–1181

Lamare MD (1997) Population biology, pre-settlement processes and recruitment in the New Zealand sea urchin *Evechinus chloroticus* Valenciennes (Echinoidea: Echinometridae). Ph D Thesis, University of Otago, Dunedin

Lamare MD, Mladenov PV (2000) Modelling somatic growth in the sea urchin *Evechinus chloroticus* (Echinoidea: Echinometridae). J Exp Mar Biol Ecol 243:17–44

Lee TC, Mohsin S, Taylor D, Parkesh R, Gunnlaugsson T, O'Brian FJ, Giehl M, Gowin W (2003) Detecting microdamage in bone. J Anat 203:161–172

Lees DC (1968) Tagging subtidal echinoderms. Underwater Nat 5:16–19

Lees DC (1970) The relationship between movement and available food in the sea urchins *Strongylocentrotus franciscanus* and *Strongylocentrotus purpuratus*. MS Thesis, San Diego State University, San Diego, CA

Levitan DR (1988) Density-dependent size regulation and negative growth in the sea urchin *Diadema antillarum* Philippi. Oecologia 76:627–629

Levitan DR (1989) Density-dependent size regulation in *Diadema antillarum*: effects on fecundity and survivorship. Ecology 70:1414–1424.

Levitan DR (1991) Skeletal changes in the test and jaws of the sea urchin *Diadema antillarum* in response to food limitation. Mar Biol 111:431–435

Lewis JB (1958) The biology of the tropical sea urchin *Tripneustes esculentus* Leske in Barbados, British West Indies. Can J Zool 36:607–621

Lewis JB (1966) Growth and breeding in the tropical echinoid *Diadema antillarum* Phillipi. Bull Mar Sci 16:151–158

Lumingas LJL, Guillou A (1994a) Growth zones and back-calculation for the sea urchin, *Sphaerechinus granularis*, from the Bay of Brest, France. J Mar Biol Ass UK 74:671–686

Lumingas LJL, Guillou A (1994b) Plasticité de l'oursin, *Sphaerechinus granularis* (Lamarck), face aux variations de l'environnement. In: David B, Guille A, Féral J-P, Roux M (eds) Echinoderms through time. Balkema, Rotterdam, pp 757–762

Märkel K (1979) Structure and growth of the cidaroid socket-joint lantern of Aristotle of non-cidaroid regular echinoids (Echinodermata, Echinoidea). Zoomorphologiy 94:1–32

Märkel K, Roser U (1983a) The spine tissues in the echinoid *Eucidaris tribuloides*. Zoomorphology 103:25–41

Märkel K, Roser U (1983b) Calcite-resorption in the spine of the echinoid *Eucidaris tribuloides*. Zoomorphology 103:43–58

McPherson BF (1965) Contributions to the biology of the sea urchin *Tripneustes ventricosus*. Bull Mar Sci 15:228–244

McPherson BF (1968) Contributions to the biology of the sea urchin *Eucidaris tribuloides* (Lamarck). Bull Mar Sci 18:400–443

Moore GP (1966) The use of trabecular bands as growth indicators in spines of the sea urchin, *Heliocidaris erythrogramma*. Aust J Sci 29:52–54

Moore HB (1935) A comparison of the biology of *Echinus esculentus* in different habitats. Part II. J Mar Biol Ass UK 20:109–128

Moore HB., Jutare T, Bauer JC, Jones JA (1963) The biology of *Lytechinus variegatus*. Bull Mar Sci Gulf and Caribbean 13:23–53

Morgan LE, Botsford LW, Wing SR, Smith BD (2000) Spatial variability in growth and mortality of the red sea urchin, *Strongylocentrotus franciscanus*, in northern California. Can J Fish Aquat Sci 57:980–992

Mortensen Th (1943) A monograph of the Echinoidea III.3. Reitzel, Copenhagen

Munk JE (1992) Reproduction and growth of green urchins *Strongylocentrotus droebachiensis* (Müller) near Kodiak, Alaska. J Shellfish Res 11:245–254

Murray SN, Horn MH (1989) Variations in standing stocks of central California macrophytes from a rocky intertidal habitat before and during the 1982–1983 El Niño. Mar Ecol Prog Ser. 58:113–122

Neill JB (1987) A novel technique for tagging sea urchins. Bull Mar Sci 41:92–94

Nelson BV, Vance RR (1979) Diel foraging patterns of the sea urchin *Centrostephanus coronatus* as a predator avoidance strategy. Mar Biol 51:251–258

Nichols D, Sime AAT, Bishop GM (1985) Growth in populations of the sea-urchin *Echinus esculentus* L. (Echinodermata: Echinoidea) from the English Channel and Firth of Clyde. J Exp Mar Biol Ecol 86:219–228

Nissen H-U (1969) Crystal orientation and plate structure in echinoid skeletal units. Science 166:1150–1152

Olsson, M, Newton G (1979) A simple, rapid method for marking individual sea urchins. Calif Fish Game 65:58–62

Pauly D (1981) The relationships between gill surface area and growth performance in fish: a generalization of von Bertalanffy's theory of growth. Meeresforschung 28:251–282

Pearse JS, Pearse VB (1975) Growth zones in the echinoid skeleton. Am Zool 15:731–753

Piennar LV, Thomson JA (1973) Three programs used in population dynamics WVONB - ALOMA - BHYLD (FORTRAN 1130). Fish Res Bd Canada Tech Rpt 367:1–32

Prouho H (1887) Reserches sur le *Dorocidaris papillata* et quelques autres échinides de la Méditerranée. Arch Zool Exp Gén 15:213–380

Randall JE, Schroeder RE, Starck II WE (1964) Notes on the biology of the echinoid *Diadema antillarum*. Caribbean J Sci 4:421–433

Raup DM (1966) The endoskeleton. In: Boolootian RA (ed.) Physiology of Echinodermata. Interscience Publishers, New York, pp 379–395

Richards FJ (1959) A flexible growth function for empirical use. J Exp Bot 10:290–300

Ricker WE (1973) Linear regression in fishery research. J Fish Res Bd Can 30:409–434

Rivera JA (1979) Aspects of the biology of *Lytechinus variegatus* (Lamarck, 1816) at Jobos Bay, Puerto Rico (Echinoidea: Toxopneustidae). MS Thesis, University of Puerto Rico, Mayaguez

Robinson SMC, MacIntyre AD (1997) Aging and growth of the green sea urchin. Bull Aquacul Ass Can 91:56–60

Rogers-Bennett L (1994) Spatial patterns in the life history characteristics of red sea urchins, *Strongylocentrotus franciscanus*: implications for recruitment and the California fishery. PhD Dissert, University of California, Davis

Rowley RJ (1990) Newly settled sea urchins in a kelp bed and urchin barren ground: a comparison of growth and mortality. Mar Ecol Prog Ser 62:229–240

Russell MP (1987) Life history traits and resource allocation in the purple sea urchin *Strongylocentrotus purpuratus* (Stimpson). J Exp Mar Biol Ecol 108:199–216

Russell MP (2001) Spatial and temporal variation in growth of the green sea urchin, *Strongylocentrotus droebachiensis*, in the Gulf of Maine, USA. In: Barker M (ed.) Echinoderms 2000. Balkema, Rotterdam, pp 533–538

Russell MP, Ebert TA, Petraitis PS (1998) Field estimates of growth and mortality of the green sea urchin, *Strongylocentrotus droebachiensis*. Ophelia 48:137–153

Russell MP, Meredith RW (2000) Natural growth lines in echinoid ossicles are not reliable indicators of age: a test using *Strongylocentrotus droebachiensis*. Invert Biol 119:410–420

Russell MP, Urbaniak LM (2004) Does calcein affect estimates of growth rate in sea urchins? In: Heinzeller T, Nebelsick J (eds) Echinoderms: München. Balkema, London, pp 53–57

Sager G (1982) Das Längenwachstum der Nordsee Seezunge (*Solea vulgaris* Quensel) und die Problematik der Jahresschwankungen. Anat Anz 151:160–178

Sager G, Gosselck F (1986) Investigation into seasonal growth of *Branchiostoma lanceolatum* off Heligoland, according to data by Courtney (1975). Internat Rev Hydrobiol 71:701–707

Sanchez-Lamadrid A (2001) The use of alizarin complexone for immersion marking of otoliths of larvae of gilthead sea bream, *Spartus aurata* L. Fish Manag Ecol 8:279–281

Scheibling RE, Mladenov PV (1988) Distribution, abundance and size structure of *Tripneustes ventricosus* on traditional fishing grounds following the collapse of the sea urchin fishery in Barbados. In: Burke RD, Mladenov PV, Lambert P, Parsley RL (eds) Echinoderm biology. Balkema, Rotterdam, pp 449–455

Schnute J (1981) A versatile growth model with statistically stable parameters. Can J Fish Aquat Sci 38:1128–1140

Schroeter SC (1978) Experimental studies of competition as a factor affecting the distribution and abundance of purple sea urchins, *Strongylocentrotus purpuratus* (Stimpson). PhD Thesis, University of California, Santa Barbara, CA

Sivertsen K, Hopkins CCE (1995) Demography of the echinoid *Strongylocentrotus droebachiensis* related to biotope in northern Norway. In: Skjoldal HR, Hopkins C, Erikstad KE, Leinaas HP (eds) Ecology of fjords and coastal waters. Elsevier, Amsterdam, pp 549–571

Smith AB (1980) Stereom microstructure of the echinoid test. Spec Pap Palaeo 25:1–81

Smith AB (1984) Echinoid palaeobiology. George Allen & Unwin, London

Smith BD, Botsford LW (1998) Interpretation of growth, mortality, and recruitment patterns in size-at-age, growth increment, and size frequency data. Can Spec Pub Fish Aquat Sci 125:125–139

Smith BD, Botsford LW, Wing SR (1998) Estimation of growth and mortality parameters from size frequency distributions lacking age patterns: the red sea urchin (*Strongylocentrotus franciscanus*) as an example. Can J Fish Aquat Sci 55:1236–1247

Sumich JL, McCauley JE (1972) Calcium magnesium ratios in the test plates of *Allocentrotus fragilis*. Mar Chem 1:55–59

Swan EF (1952) Regeneration of spines by sea urchins of the genus *Strongylocentrotus*. Growth 16:27–35

SYSTAT (1992) SYSTAT. Statistics, Version 5.2 Edition SYSTAT, Inc., Evanston, IL

Taki J (1971) Tetracycline labelling of test plates in *Strongylocentrotus intermedius*. Sci Rpts Hokkaido Fish Exp Stat 13:19–29

Taki J (1972a) A tetracycline labelling observation on growth zones in the jaw apparatus of *Strongylocentrotus intermedius*. Bull Jap Soc Sci Fish 38:181–188

Taki J (1972b) A tetracycline labelling observation on growth zones in the test plate of *Strongylocentrotus intermedius*. Bull Jap Soc Sci Fish 38:117–125

Tanaka M (1982) A new growth curve which expresses infinite increase. Publ Amakusa Mar Biol Lab 6:167–177

Tanaka M (1988) Eco-physiological meaning of parameters of ALOG growth curve. Publ Amakusa Mar Biol Lab 9:103–106

Telford M (1985) Domes, arches and urchins: the skeletal architecture of echinoids (Echinodermata). Zoomorphology 105:114–124

Tsukamota K, Seki Y, Oba T, Oya M, Iwahashi M (1989) Application of otolith to migration study of salmonids. Physiol Ecol Japan Spec 1:119–140

Turon X, Giribet G, López S, Palacin C (1995) Growth and population structure of *Paracentrotus lividus* (Echinodermata: Echinoidea) in two contrasting habitats. Mar Ecol Prog Ser 122:193–204

Van Sickle J (1977) Mortality rates from size distributions. The application of a conservation law. Oecologia 27:311–318

Walker MM (1981) Influence of season on growth of the sea urchin *Evechinus chloroticus*. N Z J Mar Freshwater Res 15:201–205

Weber J, Greer R, Voight B, White E, Roy R (1969) Unusual strength properties of echinoderm calcite related to structure. J Ultrastruct Res 26:355–366

Weber JN (1969a) The incorporation of magnesium into the skeletal calcites of echinoderms. Am J Sci 267:537–566

Weber JN (1969b) Origin of concentric banding in the spines of the tropical echinoid *Heterocentrotus*. Pac Sci 23:452–466

Weber JN, Raup DM (1966a) Fractionation of the stable isotopes of carbon and oxygen in marine calcareous organisms – the Echinoidea. Part I. Variation of C^{13} and O^{18} content within individuals. Geochim Cosmochim Acta 30:681–703

Weber JN, Raup DM (1966b) Fractionation of the stable isotopes of carbon and oxygen in marine calcareous organisms – the Echinoidea. Part II. Environmental and genetic factors. Geochim Cosmochim Acta 30:705–736

Weber JN, Raup DM (1968) Comparison of C^{13}/C^{12} and O^{18}/O^{16} in the skeletal calcite of recent and fossil echinoids. J Paleo 42:37–50

White M, Keegan BF, Leahy Y (1985) growth retardation in the regular echinoid *Echinus esculentus* Linnaeus. In: Keegan BF, O'Connor BDS (eds) Echinodermata. Balkema, Boston, pp 369–375

Wilson CA, Beckman DW, Dean JM (1987) Calcein as a fluorescent marker of otoliths of larval and juvenile fish. Trans Am Fish Soc 116:668–670

Edible Sea Urchins: Biology and Ecology
Editor: John Miller Lawrence

Chapter 7

Feeding, Digestion, and Digestibility

John M Lawrence[a], Addison L Lawrence[b], and Stephen A Watts[c]

[a] *Department of Biology, University of South Florida, Tampa, FL (USA).*
[b] *Texas A&M Experimental Agricultural Station, Port Aransas, TX (USA).*
[c] *Department of Biology, University of Alabama at Birmingham, Birmingham, AL (USA).*

1. INGESTION

Ingestion is an important aspect of nutrition. The amount and frequency of ingestion of food by sea urchins is affected by physical and chemical characteristics of the food, physiological state of the individual, and abiotic environmental conditions. There are two sources of nutrients, dissolved organic matter and ingestible matter. The utilization of dissolved organic matter, particularly in terms of amino acids, is well documented for several life stages and several species (Stewart 1979; Davis et al. 1985; de Burgh et al. 1977; de Burgh and Burke 1983; de Burgh 1978; Ferguson 1982). This review will be restricted to ingestible matter.

1.1. Food Chemistry

The response of animals to the chemical composition of food that results in consumption is complex and has several stages (Table 1; Lindstedt 1971). An animal may be attracted to food but not stimulated to eat it. An animal may be incited to eat food but deterred from ingestion. These stages are not always explicitly recognized in studies on feeding in sea urchins. Klinger and Lawrence (1984) suggested differences in consumption of foods by *Lytechinus variegatus* are determined primarily by degree of persistence in feeding and not by selection of food. Cronin and Hay (1996) pointed out the feeding response of sea urchins to foods is affected by their nutritional state: recently fed individuals are less responsive.

1.1.1. Attractants

Movement of sea urchins toward food indicates attraction. Experimental demonstration of aggregation of sea urchins on food in the field, perhaps involving attraction, has been shown for *Strongylocentrotus droebachiensis* (Himmelman and Steele 1971; Scheibling and Hamm 1991; Vadas et al. 1986), *Strongylocentrotus franciscanus* (Mattison et al. 1977), *Tetrapygus niger* (Rodríguez and Ojeda 1998; Rodríguez and Fariña 2001) and *L. variegatus* (Beddingfield and McClintock 2000; Vadas and Elner 2003).

Table 1 Types of behavioral responses to chemicals in food (from Lindstedt 1971).

Category	Behavioral response
Attractant	Movement toward food
Repellant	Movement away from food
Arrestant	Stopped movement toward food
Incitant	Initiation of feeding
Suppressant	No initiation of feeding
Stimulant	Ingestion
Deterrent	No ingestion

Duggan and Miller (2001) and Dumont et al. (2004) found no movement by *S. droebachiensis* toward or away from a kelp bed. Similarly, Lauzon-Guay et al. (2006) found *S. droebachiensis* showed no strong directionality in foraging movement. Distance attraction has been documented in the laboratory by studies in which sea urchins moved in direction of currents flowing through algae. Leighton (1971) found *Macrocystis* was attractive to *Strongylocentrotus purpuratus* but not to *Lytechinus anamesus*. *Strongylocentrotus droebachiensis* is attracted to water flowing over algae (Garnick 1978; Larson et al. 1980; Mann et al. 1984; Scheibling and Hamm 1991; Vadas 1977). *Strongylocentrotus droebachiensis* shows discrimination among algae in the laboratory (Larson et al. 1980; Vadas 1977). These differences could be due to either quantitative or qualitative differences in the attractants. These studies did not measure concentration or chemical nature of organic molecules in the seawater. Machiguchi (1987) reported *S. intermedius* were attracted to water passed over *Laminaria angustata* being eaten by sea urchins but not to water passed over the algal or sea urchins alone, indicating the importance of concentration of attractants.

Lytechinus variegatus perceives agar blocks containing dried seagrass and algae at only 3–5 cm distance (Klinger and Lawrence 1985). *Tripneustes ventricosus* detects food, indicated by waving of spines, and extension and waving of tube feet in the direction of food, in still water at a distance of up to 30 cm (Lewis 1958).

1.1.2. Stimulants and deterrents

The persistence of sea urchins in contact with food suggests stimulation of feeding. Association with food in the field has been reported for *S. droebachiensis* (Garnick 1978; Himmelman and Steele 1971; Vadas et al. 1986) and *S. franciscanus* (Mattison et al. 1977). Persistence on food in the laboratory has been shown for *S. droebachiensis* (Scheibling and Hamm 1991; Vadas 1977; Vadas et al. 1986) and *L. variegatus* (McClintock et al. 1982; Klinger and Lawrence 1985). *Lytechinus variegatus* and *Echinometra lucunter* consumed more of a practical food containing bivalve tissue than that containing seagrass tissue (McClintock et al. 1982). Consumption of a practical food by *L. variegatus* was directly dependent on concentration of either seagrass or the alga *Gracilaria verrucosa* (Klinger and Lawrence 1984). Identified stimulants are listed in Table 2.

Table 2 Examples of feeding stimulants and deterrents of sea urchins.

Compound	Species	Reference
Feeding Stimulants		
Monosaccharides		
Galactose	*L. variegatus*	Klinger and Lawrence (1984)
Polysaccharides		
Alginate	*S. purpuratus*	Leighton (1984)
Amino acids		
L-phenylalanine	*L. variegatus*	Klinger and Lawrence (1984)
Lipids		
Glycolipids	*S. intermedius*	Sakata et al. (1989)
Terpenoids		
Sesquiterpene	*D. antillarum*	Hay et al. (1987b)
Diterpenes	*D. antillarum*	Hay et al. (1987b)
Prenylated hydroquinones	*D. antillarum*	Hay et al. (1987b)
Phenols		
Phlorotannins	*T. gratilla*	Steinberg and van Altena (1992)
Feeding deterrents		
Monosaccharides		
Galactopyranosylglycerol	*S. intermedius*	Ishii et al. (2004)
Terpenoids		
Diterpenes	*A. punctulata*	Hay et al. (1987a); Cronin and Hay (1996); Cronin et al. (1997)
	S. nudus, S. intermedius	Kurata et al. (1998)
	D. antillarum	Hay et al. (1987b)
	L. variegatus	Barbosa et al. (2004)
Sesquiterpenes	*D. antillarum*	Hay et al. (1987b)
	S. intermedius	Sakata et al. (1991b)
Prenylated hydroquinones	*D. antillarum*	Hay et al. (1987b)
Phenols		
Phlorotannins	*S. purpuratus*	Steinberg (1988)
	T. gratilla	Steinberg and van Altena (1992)
Phenylpropanoid	*S. intermedius*	Ishii et al. (2004)
Bromodiphenyl ethers	*S. intermedius*	Kurata et al. (1997)
Tannic acid	*S. purpuratus*	Steinberg (1988)
Phloroglucinol	*S. purpuratus*	Steinberg (1988)
Polyketide		
Pyrone	*S. intermedius*	Sakata et al. (1991a)

(*Continued*)

Table 2 Continued

Compound	Species	Reference
Miscellaneous		
Sulfuric acid	*S. droebachiensis*	Pelletreau and Muller-Parker (2002)
Dimethyl sulfide	*S. droebachiensis*	Van Alstyne and Houser (2003)
Calcium carbonate	*D. antillarum*	Hay et al. (1994)
	D. setosum	Pennings and Svedberg (1993)

Most interest has focused on feeding deterrents as chemical defenses of plants to sea urchins (Hay 1988; Hay and Fenical 1992). Identified deterrents are listed in Table 2). Correlations between secondary metabolites in food and feeding by sea urchins suggests that function and isolation of and tests with the secondary metabolites are essential to establish their role (Hay and Fenical 1988). Hay and Fenical (1992) concluded neither chemical structure nor pharmacological activity of a compound could be used to predict its effect on a herbivore. Hay et al. (1994) reported powered calcium carbonate in food with low organic content deterred feeding in *D. antillarum*, but not in food with high organic content. Pennings and Svedberg (1993) reported *Diadema setosum* and *Echinometra* sp. feed preferentially on food containing powdered calcite. They found aragonite deterred feeding in *D. setosum* and had no effect on *Echinometra* sp. Steinberg and van Altena (1992) found phlorotannins from alga did not deter feeding by *Tripneustes gratilla* but total phenols from *Sargassum vestitum* did. Phlorotannis from the alga *Carpophyllum maschalocarpum* actually enhanced feeding by *T. gratilla*.

The response to a food is affected by the alternatives available and nutritional state. Deterrents do not necessarily prevent ingestion. Irvine (1973) reported *S. droebachiensis* and *S. franciscanus* consumed *Nereocystis luetkeana* and *Agarum* spp. but not the sulfuric acid containing alga *Desmarestia viridis* when all algae were provided in the laboratory. When *D. viridis* was the only alga provided, *S. franciscanus* never consumed it but *S. droebachiensis* began consuming it after several days and had eroded teeth after several months.

1.2. Environmental Conditions

1.2.1. Hydrodynamics

Hydrodynamics in the field affects activity of *S. droebachiensis* (Himmelman and Steele 1971; Siddon and Witman 2003), *S. franciscanus* (Cowen et al. 1982), *S. nudus* (Agatsuma et al. 1996), *Echinometra oblonga*, *Echinometra mathaei* (Russo 1977), *Loxechinus albus* (Dayton 1985), and *Centrostephanus coronatus* (Lissner 1980). High hydrodynamics decreases food consumption by sea urchins, either by a direct effect on activity or indirectly through an effect on algae. A direct decrease with increased hydrodynamics has been suggested for *S. droebachiensis* (Himmelman and Steele 1971) and an indirect effect through increased motion of algae for *L. albus* (Vasquez 1992).

Water movement affects consumption as consumption rate increases with water circulation in *L. variegatus* (Moore and McPherson 1965). Consumption rate of small (Horizontal Diameter, HD = 53 mm) *S. nudus* in the laboratory was greatest in still water and decreased ca. 50% in oscillating flow at ca. 0.1 m s^{-1} (Kawamata 1998). In contrast, consumption rate of large (HD = 80 mm) *S. nudus* was highest in oscillating flow at ca. 0.25 m s^{-1}. Water movement increased gonad production of *Evechinus chloroticus* cultured in cages (James in press).

1.2.2. Light

A diel (24 h) rhythm in food consumption, usually nocturnal (nighttime), has been recorded for many species such as *Paracentrotus lividus* (Kemp 1962) *Diadema antillarum* (Lewis 1964), *D. setosum* (Lawrence and Hughes-Games 1972), *T. gratilla* (Lison de Loma et al. 1999; Vaïtilingon et al. 2003) and *E. mathaei* (Mills et al. 2000). Light is the obvious stimulus. This probably is a response to a diurnal (daytime) activity rhythm of predators with light as a learned conditioned stimulus for the sea urchins. The most convincing evidence is that *D. setosum* shows nocturnal activity in a site with diurnal predators but no diel rhythm of activity in a site without predators (Fricke 1974). *P. lividus* is nocturnal in the Mediterranean Sea (Kemp 1962) and diurnal in a population off the Irish coast, apparently due to difference in diel behavior of predators (Ebling et al. 1966; Crook et al. 2000).

An effect of light on activity could be due to level or variation in intensity. The response has been recorded as positive, negative, and variable, possibly resulting from immediate past exposure (Millott 1975). Spines and podia react to a sudden increase or decrease in illumination (Millott 1975). Fuji (1967) observed *S. droebachiensis* in the field fed less during the day. Less variation occurred in the laboratory between daytime consumption (2000 lux) and nighttime. He showed that consumption activity decreased with either an increase or decrease in light intensity. Lewis (1958) reported *T. ventricosus* in aquaria were inactive during the day and responded to both increase and decrease in light intensity. No diel rhythm in feeding was found for *T. gratilla* and *Salmacis sphaeroides* in the Philippines (Klumpp et al. 1993). *T. gratilla* in Papua New Guinea were reported to be active or feeding during the day (Nojima and Mukai 1985). In contrast, Lison de Loma et al. (1999) reported nocturnal feeding by *T. gratilla* at La Réunion (Indian Ocean), with maximum consumption before dawn and after sunset. Food consumption by *P. lividus* is the same under laboratory conditions of long-day or short-day photoperiods (Spirlet et al. 2000). This suggests light intensity level per se (except extremely high intensity) does not affect sea urchin behavior (including consumption) but changes in intensity do.

1.2.3. Temperature

The effect of short-term and long-term changes in temperature on rate of food consumption differs. Immediate changes in response to temperature change have been reported for *L. variegatus* (Moore and McPherson 1965; Klinger et al. 1986; Hofer 2002), *T. ventricosus* (Moore and McPherson 1965), *S. purpuratus* Leighton (1971),

S. franciscanus (Leighton 1971; McBride et al. 1997) and *Eucidaris tribuloides* (Lares and McClintock 1991).

Long-term change in temperature appears to result in acclimation of rate of food consumption. Moore and McPherson (1965) reported complete seasonal acclimatization to temperature consumption rate of *T. ventricosus*. Their data for acclimatization of consumption by *L. variegatus* are less convincing. However, Klinger et al. (1986) and Hofer (2002) showed acclimation of consumption rate of *L. variegatus* occurs within 3–5 weeks. Fuji (1967) found food consumption of large *S. intermedius* was similar in June-July and in December. In contrast, higher food consumption in summer than winter was reported for *S. droebachiensis* (Larson et al. 1980) and *S. nudus* (Machiguchi 1993). The rate of food consumption by *E. tribuloides* at 18 °C was significantly less than at 27 °C for 5 months (Lares and McClintock 1991). Food consumption by *P. lividus* held at 16–24 °C were similar and remained significantly higher than that at 12 °C for 45 days, indicating lack of acclimation (Spirlet et al. 2000).

1.3. Food Shape

Lowe (1974) and Lilly (1975) concluded consumption rates of *T. gratilla* and *L. variegatus* on differing algal types is affected by the ease of algal manipulation. Himmelman and Carefoot (1975) proposed the effectiveness of the feeding apparatus of sea urchins affects feeding rate. *Lytechinus variegatus* feeds faster on large blocks of food than on terete, stalk-like forms and slower on flat, blade-like forms (Klinger 1982). Klinger suggested the Aristotle's lantern is more effective in rasping and that blade- and sheet-like plants are unwieldy food items.

1.4. Physiological State

1.4.1. Nutritional state

Even when food is continuously available and environmental conditions do not change, sea urchins do not necessarily feed continuously. The consumption rate of *P. lividus* in beds of *Posidonia oceanica* fluctuates daily (Nedelec et al. 1983). Consumption rate by *L. variegatus* is high when food is supplied after starvation in the laboratory and decreases to a lower rate when individuals are fed *ad libitum* (Lawrence et al. 2003). This indicates sea urchins can become satiated. Miller and Mann (1973) reported the ingestion rate of *S. droebachiensis* over a 10 day period varied over an order of magnitude. Klinger et al. (1994) found average ingestion and fecal production rates of *L. variegatus* were constant, but individual rates of both ranged from <1 to 7 g wet weight per day, respectively. This variation in consumption rate by individuals under constant conditions indicates an effect of internal, physiological state.

Distal control is also suggested by differences in consumption rate with quality of food. Lower consumption rates of formulated feed than natural food have been reported for *E. tribuloides* (Lares and McClintock 1991), *L. albus* (Lawrence et al. 1997), and *S. franciscanus* (McBride et al. 1997). This difference may be related to greater digestibility of formulated feeds and/or nutritional quality. McBride et al. (1998) found an inverse relation between consumption rate of *S. franciscanus* and concentration of protein in formulated food.

Hammer et al. (in press) found a similar relation with *L. variegatus*. Fernandez and Boudouresque (2000) found *Paracentrotus lividus* consumes more formulated feed prepared with vegetable meal than feed prepared with animal meal. This suggests distal control of food consumption and indicates innate, finite capacity for production. However, Senaratna et al. (2005) reported greater food consumption and lower absorption efficiency of formulated food than an alga by *Heliocidaris erythrogramma*. This could be the result of a nutritionally inadequate formulated food.

1.4.2. Body size

The rate of food consumption is allometric to body size. Faster rates of consumption per unit size have been reported for *S. intermedius* (Kunetzov 1946; Fuji 1967), *L. variegatus* (Moore and McPherson 1965; Klinger 1982), *T. ventricosus* (Moore and McPherson 1965), *T. gratilla* (Mukai and Nojima 1985; Klumpp et al. 1993), *S. sphaeroides* (Klumpp et al. 1993), *S. droebachiensis (*Miller and Mann 1973, *S. purpuratus* (Leighton 1968, 1971), *P. angulosus* (Buxton and Field 1983), and *P. lividus* (Nedelec et al. 1983). The amount of food contained in the gut of *P. lividus* is directly related to body size (Nedelec 1983).

1.4.3. Reproductive state

Food consumption seems related to reproductive state. Kunetzov (1946) and Klinger et al. (1997) reported lowest consumption of food by *S. droebachiensis* during the season of gametogenesis. Fuji (1962) reported a correlation between the lowest rate of food consumption by *S. intermedius* in the summer between July and October and when individuals have large, mature gonads or are in the spawning period. Kawamura and Hayashi (1965) provided experimental evidence to support this by showing both test and gonad growth of *S. intermedius* ceased in summer even when individuals were maintained under constant temperature.

Agatsuma and Sugawara (1988) and Agatsuma et al. (1993, 1996) found a similar inverse relation between gut contents and gonad index in *S. nudus*. Food consumption was high in spring when gonads were growing and low in fall when gonads were large and still premature. It was suggested that low food consumption in winter was due to low temperature and high hydrodynamics. Fernandez and Boudouresque (1997) noted the inverse relation between the repletion index (amount of food in the gut) and the gonad index they found for *P. lividus* contrasted to Régis' (1979) report of a parallel pattern of seasonal variation in gut contents and gonad size. The latter studies were also done in the field where other environmental factors such as temperature co-vary.

Fuji (1962) suggested an internal physiological state associated with the gonads was the cause of the inverse relation between food consumption and gonad size. Leighton (1968) suggested space available for food is inversely related to the size of the gonads.

2. DIGESTION

2.1. Structure of the Gut

De Ridder and Jangoux (1982) concluded that the general anatomy and histology of the gut of sea urchins is remarkably consistent in the group, consisting of a pharynx, esophagus, stomach, intestine, and rectum. The greatest variation among taxa is the degree of development of the siphon associated with the stomach. Holland and Ghiselin (1970) reported only a siphonal groove along the stomachal lumen in cidaroids and diadematoids but a distinct tube bypassing the stomach from the esophageal junction to the intestine in the other regular sea urchins. However, Campos-Creasey (1992) discovered siphons in diadematoids and echinothuroids. Cuénot (1948) suggested the movement of water through the siphon prevented dilution of enzymes in the stomach during digestion. Perrier (1875) had suggested it transports oxygenated water to the intestine. This was confirmed by Claereboudt and Jangoux (1985) who reported oxygen concentration in the intestine is higher than in the stomach of *P. lividus*. It is possible the siphon provides oxygen required for active transport of nutrients in this region (Lawrence 2001).

Mucus cells are found in the pharynx and esophagus, and throughout the entire gut in cidaroids, but not in the stomach and in the intestine of some diadematoids and other regular sea urchins (Holland and Ghiselin 1970). The stomach and intestine are specialized (Fig. 1). The stomach has exocrine cells consistent with presumptive enzyme production and secretion (Anisimov 1981; Fuji 1961; Holland and Lauritis 1968; Powis in De Ridder and Jangoux 1982; Sweijd 1990; Tokin and Filimonova 1977). The intestine has numerous enterocytes and well-developed mitochondria consistent with absorption

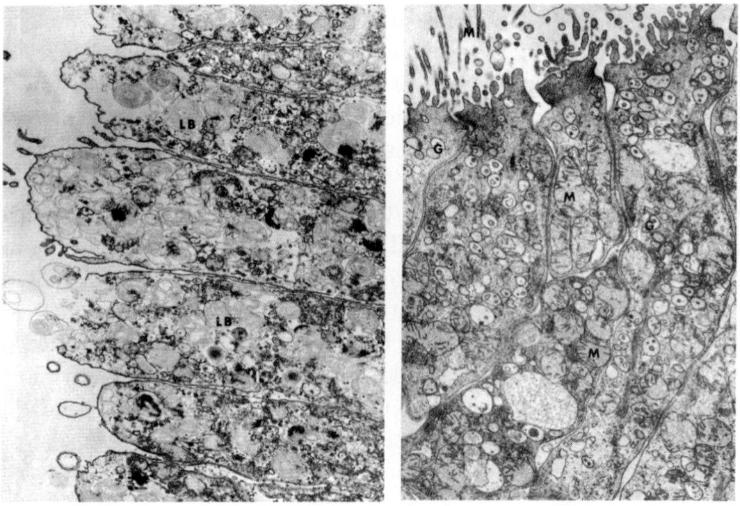

Fig. 1. *Strongylocentrotus droebachiensis*. Left: Stomach with storage and zymogen cells, LB – lamellar bodies. Right: intenstine with absorbing cells, Mi – microvilli, M – mitochondria, G – apical granules. Tokin and Filimonova (1977), with permission.

(Powis in De Ridder and Jangoux 1982; Tokin and Filimonova 1977). Differences in digestive enzyme activity and absorption (James and Bamford 1974) confirm this regional specialization.

De Ridder and Jangoux (1982) concluded all sea urchins except cidaroids surround ingested food with mucus in the pharynx to form a pellet that remains intact through defecation. However, Vaïtilingon et al. (2003) reported food was not contained in pellets in *T. gratilla,* but was in loose agglomerates held together by mucus. Observations on these pellets are poor and their role is not clear. This mucus cover must be porous as digestion of their contents and absorption of the digestive products occur. Buchanan (1969) suggested the pellets are involved in internal volume control or in packaging the feces. De Ridder and Jangoux (1982) implied a role in the digestion of plant material, pointing out that pellets are not formed in the carnivorous cidaroids.

2.2. Digestive Enzymes

Regional differences in digestive enzymatic activity are consistent with regional differences in structure. The stomach has much higher amylase activity than the intestine (Liemans and Dandrifosse 1972). Cornet and Jangoux (1974) reported lower enzyme activity of β-glucuronidase in the pharynx than in the stomach and intestine for five species of sea urchins. Activities were higher in the stomach than in the intestine in *Sphaerechinus granularis* and *P. lividus*, similar in both regions in *Psammechinus miliaris*, and higher in the intestine than in the stomach in *Arbacia lixula* and *Echinus esculentus*. Sweijd (1990) found digestive enzyme activity is greater in the stomach than in the intestine in *P. angulosus*.

Almost all studies on digestive enzymes in sea urchins concern carbohydrates. No study has shown cellulase activity on native cellulose (Lewis 1964; Suzuki et al. 1984) although the linear, soluble carboxymethylcellulose is digested (Claereboudt and Jangoux 1985; Elyakova 1972; Gómez-Pinchetti and García-Reina 1993; Obrietan et al. 1991; Suzuki et al. 1984; Sweijd 1990; Yamaguchi et al. 1989; Yokoe and Yasumasu 1964). The appearance of intact algal and seagrass cells in sea urchin feces (Cabral de Oliveira 1991; Lawrence 1976; Scott 1902) indicates that cellulase activity can be minimal. Klinger (1984) concluded sea urchins are not able to digest structural plant polysaccharides such as β-glucan cellulose. However, β-glucosidases that can hydrolyze cellobiose do occur (Lewis 1964; Molodtsov et al. 1974).

Reserve polysaccharides are digested. Amylase occurs in *P. lividus* (Claereboudt and Jangoux 1985), *S. purpuratus* (Lasker and Giese 1954), *S. droebachiensis* (Obrietan et al. 1991), *D. antillarum* (Lewis 1964), and *P. angulosus* (Sweijd 1990), and glycogenase in *S. purpuratus* (Lasker and Giese 1954), *D. antillarum* (Lewis 1964), and *P. angulosus* (Sweijd 1990).

Various levels of hydrolysis of algal polysaccharides have been found. Agarase has been reported for *S. purpuratus* (Farmanfarmaian and Phillips 1962), *D. antillarum* (Gómez-Pinchetti and García-Reina 1993; Lewis 1964), and *Anthocidaris crassispina* (Yamaguchi et al. 1989). Hydrolase activity for algin, the main structural polysaccharide of brown algae, is variable. Huang and Giese (1958) did not find alginase in *S. purpuratus*, but Eppley and Lasker (1959) did. Franssen and Jeuniaux (1965) reported alginase in *P. miliaris*. Favarov and Vaskovsky (1971) reported very low alginase activity in *S. intermedius* and *S. nudus*, much less than they found in mollusks. Yamaguchi et al. (1989) found low alginase activity in *A. crassispina*. Suzuki et al. (1984) also found no alginase in *S. nudus* and Sweijd (1990) found none in *P. angularis*. In contrast, Gómez-Pinchetti and García-Reina (1993) reported high alginase activity for *D. antillarum*. Laminarin and fucoidan are digested by *S. nudus* (Sova et al. 1970; Suzuki et al. 1984), *S. granularis*, *Psammechinus microtuberculatus*, *P. lividus* (Piavaux 1977) and *P. angularis* (Sweijd 1990). Fucoidan is also digested by *A. crassispina* (Yamaguchi et al. 1989). Carrageenan is digested by *D. antillarum* (Benitez and Macaranas 1978), *A. crassispina* (Yamaguchi et al. 1989), and *P. angularis* (Sweijd 1990). General low activity of hydrolases have been reported for β-1,4-mannan, porphylan, (β-1,3- and β-1,4-xylan in *A. crassispina* (Yamaguchi et al. 1989) and xylan in *P. angulosus* (Sweijd 1990).

Enzymes hydrolyzing oligosaccharides have been reported. Molodtsov and Vafina (1972) and Molodtsov et al. (1974) found α- and β-glucosidase, α- and β-galactosidase, α- and β-xylosidase, and α- and β-mannosidase in *S. intermedius*. Curiously, only β-glucosidase, β-galactosidase, and α-mannosidase were found in *S. nudus*. Johnson et al. (1980) reported amylase, maltase, arylsulphatase and β-glucuronicase from *E. esculentus* and *P. lividus*. Klinger (1984) found α-glucosidase activity was higher than that of β-glucosidase, and β-galactosidase and α-galactosidase activity was very low in *E. tribuloides*, *Arbacia punctulata*, *D. antillarum*, *E. lucunter*, and *L. variegatus*. Activity was highest in *L. variegatus*. Membrane-bound β-glucuronidase, β-fructosidase, and β-galactosidase have been reported for *E. esculentus* (Clifford et al. 1982). α-Glycosidase and β-glycosidase activity of the Antarctic *Sterechinus neumayeri* is several-fold lower than that of temperate species when measured at environmental temperature (Klinger et al. 1997).

Klinger (1984) concluded sea urchins have carbohydrases that are broadly appropriate for their herbivorous diet. This does not include a great capacity for digestion of structural carbohydrates. Certainly they can hydrolyze water-soluble polysaccharides and oligosaccharides, but it is possible these must be made available by mechanical disruption of cell walls or other hydrolytic origin. Sweijd (1990) suggested that high activities of enzymes hydrolyzing structural polysaccharides are not necessary for them to be functional in disrupting the cell wall to allow access to the cell contents. Thus, their role would not be in providing carbon units for anabolism.

Fewer data exist for proteases and lipases in sea urchins. Gelatin is digested by *D. antillarum* (Lewis 1964), and casein is digested by *S. purpuratus* (Lasker and Giese 1954) and *S. droebachiensis* (Obrietan et al. 1991). Chymotrypsin occurs in *S. neumayeri* (Klinger et al. 1997) and dipeptidase in *F. esculentus* (Clifford et al. 1982) Lewis (1964) found a weak lipase in *D. antillarum*, and Suppes et al. (1977) found phospholipase in *S. intermedius* and *S. nudus*. Emulsifiers have not been investigated in sea urchins.

Studies suggest nutritional condition has an effect on enzymatic activity. The thickness of the luminal epithelium of the gut of *S. purpuratus* decreases in starved individuals and shows a seasonal cycle in size, being thinnest in the winter (Lawrence et al. 1965) The cell number and size in the gut of *L. variegatus* decrease with starvation (Klinger et al. 1988). The size of the gut of starved *S. purpuratus* and *L. variegatus* respond quickly with re-feeding (Bishop and Watts 1992; Klinger et al. 1988; Lawrence et al. 1965). The total digestive enzyme activity varies directly with the size of the gut in *L. variegatus* (Klinger 1989).

2.3. Gut Transit Time

The time food remains in the gut is important in considering digestion and absorption. In sea urchins, it is affected primarily by the amount of food consumed and frequency of feeding.

Gut transit time is also affected by the structure of the sea urchin gut. Animal guts have been modeled as batch reactors, plug-flow reactors, and continuous-flow, stirred-tank reactors (Penry and Jumars 1987). The type of gut affects gut transit time and digestion. Guts of sea urchins are certainly not batch reactors that process food in discrete batches. The generous dimensions of the gut do not give it the tubular characteristics associated with the plug-flow reaction in which a continuous, orderly flow of material occurs. The gut seems most similar to the continuous-flow, stirred-tank reactor in which a continuous flow of material and a mixing of material occurs. Consequently, the first appearance of a meal in feces (Lares and McClintock 1991) does not indicate the transit time of the entire meal.

Lewis (1964) reported a gut transit time of 8–12 h in *D. antillarum* fed *ad libitum* in the laboratory. However, the gut transit time can vary within a day if feeding shows a diel rhythm as in *D. setosum* (Lawrence and Hughes-Games 1972). Food is present in the stomach only during the night when feeding is occurring. It is always present in the intestine, being retained during the daytime period of nonfeeding. A similar diel rhythm in the amount of food in the stomach but not in the intestine occurs in *T. gratilla* (Lison de Loma et al. 1999). In contrast, Vaïtilingon et al. (2003) found opposite diel rhythms in the stomach and intestine of *T. gratilla*.

Gut transit time is ca. 12–40 h in *L. variegatus* and is affected by temperature (Klinger et al. 1986, 1994) and 8–50 h in *P. lividus* (Lawrence et al. 1989). Marked feces first began to appear ca. 7.5 h after feeding in *E. tribuloides* regardless of temperature (Lares and McClintock 1991). The gut transit time is usually <1 day but longer transit times indicate considerable mixing of the food occurs in the gut. Similarly, gut transit time is ca. 2 days for *S. droebachiensis* but small amounts of marked food continue to appear in the feces for up to 20 days after ingestion (John et al. 1990). In contrast, Bedford and Moore (1985) found that *P. miliaris* had a longer gut transit time when eating fresh rather than rotting *Laminaria saccharina*. Otero-Villanueva et al. (2004) also found prolonged residence time in *P. miliaris,* slightly longer in small (>42 h) than in large (38–42 h) individuals.

Lasker and Giese (1954) reported defecation is continuous in *S. purpuratus* fed continuously, but that it slowed when food was removed so that the gut still retained some food after 2 weeks In contrast, *L. variegatus* retained little food in the gut only 3 days after food was removed (Klinger et al. 1994). Gut transit time is not affected by quality of the food in *P. lividus* (Lawrence et al. 1989), *E. tribuloides* (Lares and McClintock 1991), *S. droebachiensis* (John et al. 1990), or *L. variegatus* (Klinger et al. 1994).

2.4. Role of Microorganisms in Digestion

Lasker and Giese (1954) raised the possibility that bacteria could be involved in digestion in sea urchins. They estimated the bacteria in the gut contents of *S. purpuratus* in the order of $10^6\,ml^{-1}$. Uncles (1977) made the only thorough analysis of the bacterial flora in the gut of a sea urchin. The gut contents of *E. esculentus* had an average bacterial viable count of 2×10^7 per 3 cm section, with *Vibrio* predominating and *Pseudomonas* and *Aeromonas* next in abundance. Fong and Mann (1980) reported between 2×10^8 and 6×10^9 bacteria per ml gut contents of *S. droebachiensis*. Bauer and Agerter (1994) found Gram-positive bacilli were present more frequently than *Vibrio* in the guts of a number of shallow-water, tropical sea urchins, but did not quantify their abundance.

Gut bacteria from *L. albus* (García-Tello and Baya 1973), *L. variegatus*, *A. punctulata* (Prim and Lawrence 1975), and *P. angulosis* (Sweijd 1990) can digest a variety of algal polysaccharides. Bacteria from the gut of *L. variegatus* and *A. punctulata* also degraded two algae (*Ulva lactuca* and *Eucheuma nudum*) but not *Caulerpa prolifera*, and had only a slight effect on two seagrasses (*Diplanthera wrightii* and *Thalassia testudinum*). Prim and Lawrence (1975) and Sweijd (1990) concluded the bacteria in the gut of sea urchins are fairly nonselective and stochastic in occurrence. Fungi occur in the gut contents of *L. variegatus* (Wagner-Merner et al. 1980; Wagner-Merner and Lawrence 1980) but they have not been investigated.

Lasker and Giese (1954) noted feces are produced slowly when *S. purpuratus* is starved and heavily laden with bacteria. They reported the mucus membrane of almost every algal pellet was lined with coccus and rod-shaped bacteria, and described bacteria around algal cells damaged by the sea urchin's teeth. They suggested that since the digestive enzymes did not appear to act on intact algal cells, bacteria were responsible for their digestion. Bacteria from the gut contents did digest intact algae and agar *in vitro*. Lasker and Giese suggested the sea urchin's enzymes digest the available organic material in the cell wall before the bacteria had multiplied sufficiently to be competitive, and that the residue is then gradually decomposed by the bacteria. This does not seem to be important in *L. variegatus* on a short-term basis, as absorption efficiency did not differ between individuals fed continuously or only once every 4 days (Lawrence et al. 2003).

Lasker and Giese (1954) raised the possibility that sea urchins harbor bacteria or protozoans that digested polysaccharides. Farmanfarmaian and Phillips (1962) stated the feces of *S. purpuratus* feeding continuously showed no bacterial enrichment and that digestion and absorption occurred in the stomach before bacterial enrichment occurred, but apparently they made no measurements. They concluded algal digestion in *S. purpuratus* is not dependent on bacteria and more likely the result of the sea urchin's digestive enzymes. Their suggestion is supported by the observation of Claereboudt and Jangoux (1985) that a good correlation exists between amylase activity in the tissue and gut contents in the stomach but not in the intestine. They suggested amylase activity in food pellets in the intestine could be microbial. Sawabe et al. (1995) reported alginate and its constituents, polymannuronate and polyguluronate, are degraded by bacteria isolated from the gut of *S. intermedius*. The bacteria were primarily *Vibrio* and fermentative.

Prim and Lawrence (1975) hypothesized that bacterial enzymes probably contribute little to digestion when food is abundant as food could pass through the gut too quickly for a bacterial flora to develop. An alternative hypothesis is that of Bjorndal (1980), who noted the great contribution bacteria make to the digestion of seagrass in sea turtles. She proposed the homogeneous diet of sea turtles allowed a bacterial flora specific for seagrass to develop and that this might not occur in sea urchins with a heterogeneous diet. Guerinot and Patriquin (1981a) argued that bacteria would not pass through with the food if they are part of an attached, endemic microflora. Such a flora has not been demonstrated. The bacteria of *S. droebachiensis* that Fong and Mann (1980) reported as responsible for nitrogen fixation were in the gut contents.

Guerinot et al. (1977) reported nitrogen fixation in *S. droebachiensis* feeding in the field on seagrass and several algal species but not in individuals collected from bare rock. Odintsov (1981) expressed doubt that individuals from bare rock would lack nitrogen-fixing bacteria. Guerinot et al. showed this fixation occurred with dissected gut tissue and not with the remainder of the body. Guerinot and Patriquin (1981a,b) subsequently demonstrated nitrogen fixation is widespread, reporting it for *D. antillarum*, *E. lucunter*, and *T. ventricosus*. Guerinot and Patriquin (1981a) reported fixation in *S. droebachiensis* was seasonal, being inversely correlated with the nitrogen concentration in kelp (*Laminaria* spp.) and seagrass (*Zostera marina*). They showed gut microflora are responsible for this fixation. They suggested this could result from suppression of bacterial nitrification in the presence of metabolic products of nitrogen-containing compounds in the sea urchin's gut.

Fixed nitrogen is assimilated by sea urchins (Fong and Mann 1980; Guerinot and Patriquin 1981a). Fong and Mann (1980) found radioactive carbon appeared in all of the protein amino acids of the gonads of *S. droebachiensis* fed labeled glucose or cellulose. Labeled glucose injected into the coelom after suppression of the bacteria resulted in labeling of some amino acids (nonessential amino acids) but not others (essential amino acids). They concluded *S. droebachiensis* can digest cellulose, probably with bacterial aid, and that gut bacteria can synthesize essential amino acids that are available to the sea urchin. They also pointed out it would be necessary to do feeding experiments to confirm the bacterially produced essential amino acids limit sea urchin growth. Guerinot et al. (1977) calculated that 8–15% of the daily nitrogen requirements of *S. droebachiensis* could come from nitrogen fixation. Odintsov (1981) thought this was an overestimation as he found that nitrogen fixation could contribute no more than 0.03% in *S. nudus* and 0.08% in *S. intermedius*. Guerinot and Patriquin (1981a) subsequently estimated the yearly rates of nitrogen fixation by the gut bacteria are more than adequate to account for the nitrogen needs of both growth and daily maintenance in *Strongylocentrotus* (spp.).

Ciliates are commensals in the gut of sea urchins (Berger 1964). Farmanfarmaian and Phillips (1962) stated protozoan in addition to bacterial enrichment occurred in guts of *S. purpuratus* after feeding but reported no data. Lasker and Giese (1954) reported several hundred ciliates per milliliter can be found in guts contents of *S. purpuratus* and were observed to ingest bacteria and disintegrated algae. They concluded it is unlikely protozoans contributed to the digestion of algae because of their small numbers and feeding habits. It is possible protozoans crop the bacteria and affect bacterial population dynamics.

De Ridder and Foret (2001) concluded the bacteria in the gut of regular sea urchins are not host-specific as they are common in the environment and are also found in the guts of other marine invertebrates. Their review indicated the bacteria are usually opportunistic in the utilization of marine plant polysaccharides, and that there is little direct evidence of an effective contribution to digestion.

3. DIGESTIBILITY

Digestibility (absorption efficiency) (Lawrence 1975, 1987) is important in interpreting the nutritional quality of feeds for sea urchins. It represents the difference between the amount of feed ingested and amount of feces. The nature of feed used in nutritional studies, including digestibility, has varied (Table 3). Natural feeds, practical or semi-purified formulated feeds typically have been used in digestibility studies. Only semi-purified and purified feeds can be used to determine specific nutrient requirements. Digestibility is usually referred to as apparent digestibility since the contribution of endogenous material such as digestive enzymes and intestinal tissue is not considered. Apparent feed digestibilities can be expressed in general terms such as apparent dry matter digestibility, apparent crude protein digestibility, apparent energy digestibility, and apparent nitrogen digestibility or for specific nutrients such as apparent methinonine digestibility and apparent magnesium digestibility, depending upon the class of nutrient or specific nutrient quantified in both feed and feces. Total apparent digestibility, apparent organic digestibility, and apparent energy digestibility all have information content that is useful for evaluating nutritional quality of foods for sea urchins.

Many factors influence apparent feed digestibility values. These include species and age of the animal, prior history of the animal (such as starved or fed), environmental conditions (particularly salinity, temperature, oxygen, pH and water quality parameters) and nutrient associations. Nutrient associations refer to associative effects among different levels and kinds of nutrients in the feed presented to the animal. For example, a totally different apparent protein digestibility probably would result from a feed containing 10% vs. 90% kelp meal or a feed containing 10% vs. 50% protein. Other examples are apparent fiber

Table 3 Terminology used in this paper for feed types.

Feed type	Characteristic
Natural feed	Any natural food regardless of source (marine or terrestrial), alive or dead, and not dried or processed.
Formulated feed	
Practical feed	Dried unpurified ingredients. Cannot be prepared to determine specific nutrient requirements.
Semi-purified feed	Dried purified and unpurified ingredients. Can be prepared to determine specific nutrient requirements.
Purified feed	Dried purified ingredients. Can be prepared to determine specific nutrient requirements.

digestibility for a feed containing 2% vs. 10% fiber or apparent methionine digestibility of a feed containing 0.1% vs. 5% methionine. Another factor often overlooked is feed palatability. The amount of nutrient loss by leaching may be important for feed having low palatability with a slow ingestion rate.

There are several methods for determining apparent digestibility. These are: gravimetric or total collection, chromic oxide marker, indigestible fiber, *in vitro* enzyme digestion, nutrient composition correlations (i.e. fiber, soluble/insoluble fiber ratio or nitrogen), and radiolabeled tracers. The gravimetric method for estimating the digestibility of food by sea urchins through direct calculation based on the difference in quantity of food eaten and feces produced is by far the best. Major problems are the large amount of food and feces necessary for weighing, difficulty in complete recovery of feces, variation in individual ingestion rate, and prolonged gut retention time. An additional problem is the continuous-flow, stirred-tank reactor nature of the gut that results in mixing of food ingested over time and prolonged defecation of that food (op. cit.).

Because of these problems, direct measurement of digestibility by quantification usually involves combining sea urchins and increasing the time over which feces are collected. Miller and Mann (1973) combined three or more *S. droebachiensis* in a container over 6 days and found that 95% confidence limits were usually within 30% of the mean of container replicates. Lowe and Lawrence (1976) similarly combined 2–3 *L. variegatus* but measured food ingested and fecal production daily. Hawkins (1981) collected feces from individual *D. antillarum* for up to 8 days. Frantzis and Grémare (1992) and Fernandez and Boudouresque (2000) combined feces collected daily for three days from 10 *P. lividus*. Otero-Villanueva et al. (2004) made daily estimates of the food provided each day and feces collected the following morning by individual *P. miliaris* for each month of their experiment.

To avoid the problems associated with complete recovery of feces and the amount of feces necessary for direct calculation of digestibility, an indirect method based on the difference in concentration of a marker (McGinnis and Kasting 1964) or ash (Conover 1966) in the food and feces has been used. The marker is presumed to be nondigested and nonabsorbed in the gut. Lowe and Lawrence (1976) used ash in food as the marker. They recognized absorption or other loss of ash from the gut contents could occur. Leighton (1968) found minimal or no changes in the proportion of calcium and magnesium carbonate in the ash fraction of algae ingested by *S. purpuratus*. Trace metals are absorbed from algal food by sea urchins (Stevenson and Ufret 1966) but they are probably insufficient to have an important effect. The fate of soluble inorganic compounds in the cytoplasm upon disruption of algal or seagrass cells is unknown. It is probable they are not present in the feces. To avoid this problem, Schlosser et al. (2005) used acid insoluble ash as the marker in their study with *P. lividus*.

Larson et al. (1980) reported digestibility of five algal species by *S. droebachiensis* calculated directly and indirectly (using ash as a marker). Amounts of organic matter that would be absorbed calculated on the basis of gravimetric digestibility are greater than amounts calculated indirectly, even if all organic matter were absorbed (Lawrence 1987). This indicates loss of ash during passage of food through the gut.

Klinger et al. (1994) estimated apparent dry matter digestibility of *L. variegatus* calculated indirectly with ash as a marker was significantly less (12.5%) than with chromic oxide as a marker, interpreted due to loss of ash. Lares (1999) found food pellets in the gut of *L. variegatus* fed prepared food marked with chromic acid differed in the amount of chromic oxide they contained. He suggested chromic oxide may separate from the feed with digestion, as it is not bound physically or chemically to the organic molecules of the feed, and may accumulate differentially in the pellets. Pellets differing in chromic oxide content would differ in density and could pass through the gut differently as found for particles of different densities in sand dollars (Chia 1973). Otero-Villanueva et al. (2004) found digestibility in *P. miliaris* measured indirectly with ash as a marker was much higher than that measured directly.

Frequency of feeding is important in digestibility in sea urchins because it affects the quantity of food present in the gut; gut residence time is greater in the absence or low rates of feeding. Boolootian and Lasker (1964) found no difference in absorption efficiencies of *S. purpuratus* fed brown algae continuously or once in 2 weeks. This is surprising as Lasker and Giese (1954) reported that algal pellets produced by *S. purpuratus* with continual feeding were less completely digested than those produced after initiation of starvation. Fuji (1967) found the absorption efficiency of *S. intermedius* for *Ulva pertusa* remained at ca. 80% whether individuals were fed 83 mg per day or 20 mg per day; and that for *Laminaria japonica* increased only from 66% when individuals were fed 204 mg per day to 82% when fed 31 mg per day. However, Fuji did not measure gut transit time for these individuals, but used his observation that, with starvation, gut contents decreased to an asymptote after 1 day for individuals fed *Laminaria japonica* and after 3 days for those fed *Ulva pertusa*. An inverse relationship between rate of food consumption and digestibility was reported for *P. lividus* (Frantzis and Grémare 1992) and *S. franciscanus* (McBride et al. 1999). Lawrence et al. (2003) reported digestibility in *L. variegatus* increased with increased time interval between feeding.

Estimating nutritional quality of a feed by its organic and energy composition (Paine and Vadas 1969; Himmelman and Carefoot 1975) can be misleading as it does not consider digestibility (McClintock 1986). Fleming (1995) found no strong relationship between digestibility in the abalone *Haliotis rubra* and organic content and energy of algae. The appearance of intact cells of seagrass in feces of *L. variegatus* suggests low digestion of structural carbohydrates in cell walls (Lawrence 1976). Santelices et al. (1983) reported algal spores survive digestion in *T.niger*. Cabral de Oliveira (1991) cultured feces produced from sea urchins collected from the field. She reported that ca. 24% of the algal species in the guts of *L. variegatus* and *Echinometra lucunter*, and ca. 50% of those in the guts of *A. lixula* survived digestion. Lasker and Giese (1954) found cell walls of algae in the gut of *S. purpuratus* are mostly digested only if inside pellets. The amount of structural carbohydrates in prepared feeds would be expected to affect digestibility. Lowe and Lawrence (1976) showed a ranking of algae and seagrass in terms of organic content differed from a ranking in terms of digestibility. However, Chang et al. (2005) found no difference in digestibility by *S. intermedius* fed an alga or a prepared feed despite the lack of cell walls and a much lower concentration of structural polysaccharides in the prepared feed.

4. CONCLUSIONS

It is essential to have knowledge of feeding and digestion in sea urchins in order to understand their biology and ecology. Information on feeding and digestion will also provide guidelines in the development of prepared feeds for aquaculture. This information is inadequate at this time. Much opportunity exists for study of the complex behavior associated with attraction to food and feeding. More information on digestion is needed. Even for carbohydrases, information is restricted to a few obvious polysaccharides and oligosaccharides and to very few species of sea urchins. A major gap in our knowledge is information about proteases and lipases. Even the information on digestive enzymes that does exist usually simply confirms activity. Systematic, comparative studies of digestive enzymes and conditions in the gut during digestion are imperative. The role of microorganisms in the digestive processes of sea urchins must be clarified.

ACKNOWLEDGMENT

We are grateful to BJ Baker for assistance with biochemical classification.

REFERENCES

Agatsuma Y, Matsuyama K, Nakata A (1996) Seasonal changes in feeding activity of the sea urchin *Strongylocentrotus nudus* in Oshoro Bay, southwestern Hokkaido. Nippon Suisan Gakkaishi 62: 592–597 (in Japanese, English abstract)

Agatsuma Y, Nakata A, Matsuyama K (1993) Feeding and assimilation of the sea urchin, *Strongylocentrotus nudus* for *Laminaria religiosa*. Sci Repts Hokkaido Fish Expt Sta 40: 21–29 (in Japanese, English abstract)

Agatsuma Y, Sugawara Y (1988) Reproductive cycle and food ingestion of the sea urchin, *Strongylocentrotus nudus,* (A. Agassiz), in southern Hokkaido. II. Seasonal changes of the gut content and test weight. Sci Repts Hokkaido Fish Expt Sta 30: 43–49 (in Japanese, English abstract)

Anisimov AP (1981) Morphological and cytochemical characteristics of the alimentary canal epithelium of sea urchin *Strongylocentrotus intermedius* and *Strongylocentrotus nudus* (Echinodermata: Echinoidea). Biol Morya 3: 32–42

Barbosa JP, Teixeira VL, Pereira RC (2004) A dolabellane diterpene from the brown algae *Dictyota pfaffii* as chemical defense against herbivores. Bot Mar 47: 147–151

Bauer JC, Agerter CJ (1994) Isolation of potentially pathogenic bacterial flora from tropical sea urchins in selected west Atlantic and east Pacific sites. Bull Mar Sci 55: 142–150

Beddingfield SD, McClintock JB (2000) Demographic characteristics of *Lytechinus variegatus* (Echinoidea: Echinodermata) from three habitats in a North Florida Bay, Gulf of Mexico. Mar Ecol PSZN 21: 17–40

Bedford AP, Moore PG (1985) Macrofaunal involvement in the sublittoral decay of kelp debris: the sea urchin *Psammechinus miliaris* (Gmelin) (Echinodermata: Echinoidea). Est Coastal Shelf Sci 20: 19–40

Benitez LV, Macaranas JM (1978) Purification and characterization of kappa-carrageenase from the tropical sea urchin, *Diadema setosum*. International symposium on marine biogeography and evolution in the Southern Hemisphere, p 13

Berger J (1964) The morphology, systematics, and biology of the entocommensal ciliates of echinoids. PhD thesis. University of Illinois, Urbana-Champaign

Bishop CD, Watts SA (1992) Biochemical and morphometric study of growth in the stomach and intestine of the echinoid *Lytechinus variegatus* (Echinodermata). Mar Biol 114: 459–467

Bjorndal KA (1980) Nutrition and grazing behavior of the green turtle. Mar Biol 56, 147–154

Boolootian RA, Lasker R (1964) Digestion of brown algae and the distribution of nutrients in the purple sea urchin *Strongylocentrotus purpuratus*. Comp Biochem Physiol 11: 273–289
Buchanan JB (1969) Feeding and the control of volume within the tests of regular sea urchins. J Zool Lond 159: 51–64
de Burgh ME (1978) Specificity of l-alanine transport in the spine epithelium of *Paracentrotus lividus* (Echinoidea). J Mar Biol Ass UK 58: 425–440
de Burgh ME, Burke RD (1983) Uptake of dissolved amino acids by embryos and larvae of *Dendraster excentricus* (Eschscholtz) (Echinodermata: Echinoidea). Can J Zool 61: 349–354
de Burgh ME, West AB, Jeal F (1977) Absorption of L-alanine and other dissolved nutrients by the spines of *Paracentrotus lividus* (Echinoidea). J Mar Biol Ass UK 57: 1031–1045
Buxton CD, Field JG (1983). Consumption, defaecation and absorption efficiency in the sea-urchin, *Parechinus angulosus* Leske S Afr J Zool 18: 11–14
Cabral de Oliveira M (1991) Survival of seaweeds ingested by three species of tropical sea urchins from Brazil. Hydrobiologia 222: 13–17
Campos-Creasey ES (1992) A study of the feeding biology of deep-sea echinoids from the North Atlantic. PhD thesis. University of Southampton, Southampton
Chang Y-Q, Lawrence JM, Cao X-B, Lawrence AL (2005) Food consumption, absorption, assimilation and growth of the sea urchin *Strongylocentrotus intermedius* fed a prepared feed and the alga *Laminaria japonica*. J World Aquacult Soc 36: 68–75
Chia FS (1973) Sand dollar: a weight belt for the juvenile. Science 181: 73–74
Claereboudt M, Jangoux M (1985) Conditions de digestion et activité de quelques polysaccharidases dans le tube digestif de l'oursin *Paracentrotus lividus* (Echinodermata). Biochem Syst Ecol 13: 51–54
Clifford C, Walsh J, Reidy N, Johnson DB (1982) Digestive enzymes and subcellular localization of disaccharidases in some echinodoerms. Comp Biochem Physiol 71B: 105–110
Conover RJ (1966) Assimilation of organic matter by zooplankton. Limnol Oceanogr 11: 338–345
Cornet D, Jangoux M (1974) Arylsulphatases and β-glucuronidase in the digestive system of some echinoderms. Comp Biochem Physiol 47B: 45–52
Cowen RK, Agegian CR, Foster MS (1982) The maintenance of community structure in a central California giant kelp forest. J Exp Mar Biol Ecol 64: 189–201
Cronin G, Hay ME (1996) Susceptibility to herbivores depends on recent history of both the plant and animal. Ecology 77: 1531–1543
Cronin G, Paul VJ, Hay ME, Fenical W (1997) Are tropical herbivores more resistant than temperate herbivores to seaweed chemical defenses? Dieterpenoid metabolites from *Dictyota acutiloba* as feeding deterrents for tropical versus temperate fishes and urchins. J Chem Ecol 23: 289–302
Crook AC, Long M, Barnes DKA (2000) Quantifying daily migration in the sea urchin *Paracentrotus lividus*. J Mar Biol Ass UK 80: 177–178
Cuénot L (1948) Anatomie, ëthologie et systématique des echinodermes. In: Grassé P-P (ed.) Traité de zoologie 11. Échinodermes – Stomatocordés – Procordés. Masson et Cie, Paris, pp 1–363
Davis JP, Keenan CL, Stephens GC (1985) Na^+-dependent amino acid transport in bacteria-free sea urchin larvae. J Comp Physiol B 156: 121–127
Dayton PK (1985) The structure and regulation of some South American kelp communities. Ecol Monogr 55: 447–468
De Ridder C, Foret T (2001) Non-parasitic symbioses between echinoderms and bacteria. Echino Stud 6: 111–169
De Ridder C, Jangoux M (1982) Digestive systems: Echinoidea. In: Jangoux M, Lawrence JM (eds) Echinoderm nutrition. Balkema, Rotterdam, pp 213–234
Duggan RE, Miller RJ (2001) External and internal tags for the green sea urchin. J Exp Mar Biol Ecol 258: 115–122
Dumont C, Himmelman JH, Russell MP (2004) Size-specific movement of green sea urchins *Strongylocentrotus droebachiensis* on urchin barrens in eastern Canada. Mar Ecol Prog Ser 273: 93–101
Ebling FJ, Hawkins AD, Kitching JA, Muntz L, Pratt VM (1966). The ecology of Lough Ine. XVI. Predation and diurnal migration in the *Paracentrotus* community. J An Ecol 35: 559–566
Elyakova LA (1972) Distribution of cellulases and chitinases in marine invertebrates. Comp Biochem Physiol 438: 67–70

Eppley RW, Lasker R (1959) Alginase in the sea urchin, *Strongylocentrotus purpuratus*. Science 129: 214–215
Farmanfarmaian A, Phillips JH (1962) Digestion, storage, and translocation of nutrients in the purple sea urchin (*Strongylocentrotus purpuratus*). Biol Bull 123: 105–120
Favarov VV, Vaskovsky VE (1971) Alginases of marine invertebrates. Comp Biochem Physiol 38B: 689–696
Ferguson JC (1982) A comparative study of the net metabolic benefits derived from the uptake and release of free amino acids by marine invertebrates. Biol Bull 162: 1–17.
Fernandez C, Boudouresque, C-F (1997) Phenotypic plasticity of *Paracentrotus lividus* Echinodermata: Echinoidea) in a lagoonal environment. Mar Ecol Prog Ser 152: 245–154
Fernandez C, Boudouresque C-F (2000) Nutrition of the sea urchin *Paracentrotus lividus* (Echinodermata: Echinoidea) fed different artificial food. Mar Ecol Prog Ser 204: 131–241
Fleming AE (1995) Digestive efficiency of the Australian abalone *Haliotis rubra* in relation to growth and feed preference. Aquaculture 134: 279–293
Fong W, Mann KH (1980) Role of gut flora in the transfer of amino acids through a marine food chain. Can J Fish Aquat Sci 37: 88–96
Franssen J, Jeuniaux C (1965) Digestion de l'acide alginique chez les invertébrés. Can Biol Mar 6: 1–21
Frantzis A, Grémare A (1992) Ingestion, absorption, and growth rates of *Paracentrotus lividus* (Echinodermata: Echinoidea) fed different macrophytes. Mar Ecol Prog Ser 95: 169–183
Fricke HW (1974) Möglicher Einfluβ von Feinden auf das Verhalten von *Diadema*-Seeigeln. Mar Biol 27: 59–62
Fuji A (1961) Studies on the biology of the sea urchin IV. Histological observation of the food canal of *Strongylocentrotus intermedius*. Bull Fac Fish Hokkaido Univ 11: 195–202
Fuji A (1962). Studies on the biology of the urchin. V. Food consumption of *Strongylocentrotus intermedius*. Jap J Ecol 12: 181–186
Fuji A (1967) Ecological studies on the growth and food consumption of Japanese common littoral sea urchin, *Strongylocentrotus intermedius* (A. Agassiz). Mem Fac Fish, Hokkaido Univ 15: 83–160
García-Tello P, Baya AM (1973) Acerca de la posible función de bacterias agaroliticas aisladas del erizo blanco (*Loxechinus albus* (Mol.)). Mus Nac Hist Nac Pub Oc 15: 3–8
Garnick E (1978) Behavioral ecology of *Strongylocentrotus droebachiensis* (Muller) (Echinodermata: Echinoidea). Oecologia 37: 77–84
Gómez-Pinchetti JL, García-Reina G (1993) Enzymes from marine phycophages that degrade cell walls of seaweeds. Mar Biol 116: 553–558
Guerinot ML, Fong W, Patriquin DG (1977) Nitrogen fixation (acetylene reduction) associated with sea urchins (*Strongylocentrotus droebachiensis*) feeding on seaweeds and eelgrass. J Fish Res Bd Can 34: 416–420
Guerinot ML, Patriquin DG (1981a) The association of N_2-fixing bacteria with sea urchins. Mar Biol 62: 197–207
Guerinot ML, Patriquin DG (1981b) N_2-fixing vibrios isolated from the gastrointestinal tract of sea urchins. Can J Microbiol 27: 311–317
Hammer HS, Watts SA, Lawrence AL, Lawrence JM, Desmond R (2006) The effect of dietary protein on consumption, survival, growth and production of the sea urchin *Lytechinus variegatus*. Aquaculture 254: 483–495
Hawkins, CM (1981) Efficiency of organic matter absorption by the tropical echinoid *Diadema antillarum* Philippi fed non-macrophytic algae. J Exp Mar Biol Ecol 49: 245–253
Hay M (1988) Marine plant-herbivore interactions: the ecology of chemical defense. Ann Rev Ecol Syst 19: 111–145
Hay ME, Fenical W (1992) Chemical mediation of seaweed-herbivore interactions. In: John DM, Hawkins SJ, Price JH (eds) Plant-animal interactions in the marine benthos. Clarendon Press, Oxford, pp 319–337
Hay ME, Fenical W (1988) Marine plant-herbivore interactions: the ecology of chemical defense. Ann Rev Ecol Syst 19: 111–145
Hay ME, Fenical S, Gustafson K (1987) Chemical defense against diverse coral-reef herbivores. Ecology 68: 1581–1591
Hay ME, Kappel QE, Fenical W (1994) Synergisms in plant defenses against herbivores: interactions of chemistry, calcification, and plant quality. Ecology 75: 1714–1726
Hay ME, Piel J, Boland W, Schnitzler I (1998) Seaweed sex pheromones and their degradation products frequently suppress amphipod feeding but rarely suppress sea urchin feeding. Chemoecology 8: 91–98

Himmelman JH, Carefoot TH (1975) Seasonal changes in calorific values of three Pacific coast seaweeds, and their significance to some marine invertebrate herbivores. J Exp Mar Biol Ecol 18: 139–151

Himmelman JH, Steele DH (1971) Foods and predators of the green sea urchin *Strongylocentrotus droebachiensis* in Newfoundland waters. Mar Biol 9: 315–322

Hofer SC (2002) The effect of temperature on feeding and growth characteristics of the sea urchin *Lytechinus variegatus* (Echinodermata: Echinoidea). MSc thesis. University of Alabama at Birmingham, Birmingham

Holland ND, Ghiselin M (1970) A comparative study of gut mucous cells in thirty-seven species of the class Echinoidea (Echinodermata). Biol Bull 138: 286–305

Holland ND, Lauritis JA (1968) The fine structure of the gastric exocrine cells of the purple sea urchin, *Strongylocentrotus purpuratus*. Trans Amer Microsc Soc 87: 201–209

Huang H, Giese AC (1958) Tests for digestion of algal polysaccharides by some marine herbivores. Science 127: 475

Irvine G (1973) The effect of selective feeding by two species of sea urchins on the structuring of algal communities. MSc thesis. University of Washington, Seattle

Ishii T, Okino T, Suzuki M, Machiguchi Y (2004) Tichocarpols A and B, two novel phenylpropanoids with feeding-deterrent activity from the red alga *Tichocarpus crinitus*. J Nat Prod 67: 1764–1766

James DW, Bamford DR (1974) Regional variation in alanine absorption in the gut of *Echinus esculentus*. Comp Biochem Physiol 49A: 101–113

James PJ (2006) The effects of wave and feeding disturbance on roe enhancement of the sea urchin *Evechinus chloroticus* held in sea-cages. Aquaculture 252: 361–371

John DA, Mencken TJ, Klinger TS (1990) Feeding and digestion of prepared protein and carbohydrate rich diets by *Strongylocentrotus droebachiensis* (O.F. Müller) (Echinodermata: Echinoidea). Northeast Gulf Sci 11: 89

Johnson DB, Rushe S, Glynn B, Canning M, Smyth T (1980) Hydrolases in the digestive tracts of some echinoderms. In: Jangoux M (ed.) Echinoderms: present and past. Balkema, Rotterdam, pp 313–317

Kawamata S (1998) Effect of wave-induced oscillatory flow on grazing by a subtidal sea urchin *Strongylocentrotus nudus* (A. Agassiz). J Exp Mar Biol Ecol 224: 31–48

Kawamura K, Hayashi T (1965). Influence of temperature on feeding, growth and gonad development of *Strongylocentrotus intermedius*. J Hokkaido Fish Sci Inst 22: 2–39 (in Japanese)

Kemp M (1962) Recherches d'écologie compare sur *Paracentrotus lividus* (Lmk) et *Arbacia lixula* (L). Rec Trav St Mar Endoume 25: 47–116

Klinger TS (1982) Feeding rate of *Lytechinus variegatus* (Lamarck) (Echinodermata: Echinoidea) on differing physiognomies of an artificial food of uniform composition. In: Lawrence JM (ed.) Echinoderms: proceedings of the international conference, Tampa Bay. Balkema, Rotterdam, pp 29–32

Klinger TS (1984) Activities and kinetics of digestive α- and β-glucosidase and β-galactosidase of five species of echinoids (Echinodermata). Comp Biochem Physiol 78A: 597–600

Klinger TS (1989) A preliminary investigation of the effects of food quality and quantity on the levels of α- and β-glucosidase and nucleic acids in the gut tissues *of Lytechinus variegatus* (Lamarck) (Echinodermata: Echinoidea). Amer Zool 29: 172A

Klinger TS, Hsieh HL, Pangallo RA, Chen CP, Lawrence JM (1986) The effect of temperature on feeding, digestion, and absorption of *Lytechinus variegatus* (Lamarck) (Echinodermata: Echinoidea). Physiol Zool 59: 332–336

Klinger TS, Lawrence JM (1984) Phagostimulation of *Lytechinus variegatus* (Lamarck) (Echinodermata: Echinoidea) Mar Behav Physiol 11: 49–67

Klinger TS, Lawrence JM (1985) Distance perception of food and the effect of food quantity on feeding behavior of *Lytechinus variegatus* (Lamarck) (Echinodermata: Echinoidea). Mar Behavior Physiol 11: 327–344

Klinger TS, Lawrence JM, Lawrence AL (1994) Digestive characteristics of the sea-urchin *Lytechinus variegatus* (Lamarck) (Echinodermata: Echinoidea) fed prepared feeds. J World Aquacult Soc 25: 489–496

Klinger TS, Lawrence JM, Lawrence AL (1997) Gonad and somatic production of *Strongylocentrotus droebachiensis* fed manufactured feeds. Bull Aquacul Assoc Canada 97-1: 35–37

Klinger TS, McClintock JB, Watts SA (1997) Activities of digestive enzymes of polar and subtropical echinoderms. Polar Biol 18, 154–157

Klinger TS, Watts SA, Forcucci D (1988) Effect of short-term feeding and starvation on storage and synthetic capacities of gut tissues of *Lytechinus variegatus* (Lamarck) (Echinodermata: Echinoidea). J Exp Mar Biol Ecol 117: 187–195

Klumpp DW, Salita-Espinosa JT, Fortes MD (1993) Feeding ecology and trophic role of sea urchins in a tropical seagrass community. Aquat Bot 45: 205–229

Kurata K, Taniguchi K, Agatsuma Y, Suzuki M (1998) Diterpenoid feeding-deterrents from *Laurencia saitoi*. Phytochemistry 47: 363–369

Kurata K, Taniguchii K, Takashima K, Hayashi I, Suzuki M (1997) Feeding-deterrent bromophenols from *Odonthalia corymbifera*. Phytochemistry 45: 485–487

Kunetzov VV (1946) Nutrition and growth of plant feeding marine invertebrates of the eastern Murman. Bull Acad Sci USSR 4: 431–452 (in Russian, English abstract)

Lares MT (1999) Evaluation of direct and indirect techniques for measuring absorption efficiencies of sea urchins (Echinodermata: Echinoidea) using prepared feeds. J World Aquacult Soc 10: 201–207

Lares MT, McClintock JB (1991) The effects of food quality and temperature on the nutrition of the carnivorous sea urchin *Eucidaris tribuloides* (Lamarck). J Exp Mar Biol Ecol 149: 279– 286

Larson BR, Vadas RL, Keser M (1980) Feeding and nutritional ecology of the sea urchin *Strongylocentrotus droebachiensis* in Maine, USA. Mar Biol 59: 49–62

Lasker R, Giese AC (1954) Nutrition of the sea urchin, *Strongylocentrotus purpuratus*. Biol Bull 106: 328–340

Lauzon-Guay J-S, Scheibling RE, Barbeau MA (2006) Movement patterns in the green sea urchin, *Strongylocentrotus droebachiensis*. J Mar Biol Ass UK 886: 167–174

Lawrence JM (1975) On the relationships between marine plants and sea urchins. Oceanogr Mar Biol Ann Rev 13: 213–286

Lawrence JM (1976) Absorption efficiencies of four species of tropical echinoids fed *Thalassia testudinum*. Thal Jugoslavica 12: 201–205

Lawrence JM (1987) Echinodermata. In: Pandian TJ, Vernberg FJ (eds) Animal energetics: v. 2. Bivalvia through Reptilia. Academic Press, Inc., San Diego, pp 229–321

Lawrence JM (2001) Function of eponymous structures in echinoderms: a review. Can J Zool 79: 1251–1264

Lawrence JM, Hughes-Games L (1972) The diurnal rhythm of feeding and passage of food through the gut of *Diadema setosum* (Echinodermata: Echinoidea). Isr J Zool 21: 13–16

Lawrence JM, Lawrence AL, Holland ND (1965) Annual cycle in the size of the gut of the purple sea urchin, *Strongylocentrotus purpuratus* (Stimpson). Nature 205: 1238–1239

Lawrence JM, Olave S, Otaiza R, Lawrence AL, Bustos E (1997) Enhancement of gonad production in the sea urchin *Loxechinus albus* in Chile fed extruded feeds. J World Aquacul Soc 28: 91–96

Lawrence JM, Plank LF, Lawrence AL (2003) The effect of consumption frequency on consumption of food, absorption efficiency, and gonad production in the sea urchin *Lytechinus variegatus*. Comp Biochem Physiol 134A: 69–75

Lawrence J, Régis M-B, Delmas P, Gras G, Klinger T (1989) The effect of quality of food on feeding and digestion in *Paracentrotus lividus* (Lamarck) (Echinodermata: Echinoidea). Mar Behav Physiol 15: 137–144

Leighton DL (1968) A comparative study of food selection and nutrition in the abalone *Haliotis rufescens* (Swainson) and the sea urchin *Strongylocentrotus purpuratus* (Stimpson). PhD thesis. University of California, San Diego

Leighton DL (1971). Grazing activities of benthic invertebrates in southern California kelp beds. Nova Hedwigia 32: 421–453

Lewis JB (1958). The biology of the tropical sea urchin, *Tripneustes esculentus* Leske in Barbados, British West Indies. Can J Zool 36: 607–621

Lewis JB (1964) Feeding and digestion in the tropical sea urchin *Diadema antillarum*. Philippi. Bull Mar Sci 16: 151–158

Liemans M, Dandrifosse GD (1972) Equipement digestif enzymatique de quelques échinodermes. Arch Int Physiol Biochem 80: 847–851

Lilly G (1975) The influence of diet on the growth of the tropical sea urchin *Tripneustes ventricosus* (Lamarck). Doctoral dissertation. University of British Columbia, Vancouver

Lindstedt KJ (1971) Chemical control of feeding behavior. Comp Biochem Physiol 39A: 553–581

Lison de Loma T, Harmelin-Vivien ML, Conand C (1999) Diel feeding rhythm of the sea urchin *Tripneustes gratilla* (L.) on a coral reef at La Reunion, Indian Ocean. In: Candia Carnevali MD, Bonosoro F (eds) Echinoderm research 1998. Balkema, Rotterdam, pp 87–92

Lissner AL (1980) Some effects of turbulence on the activity of the sea urchin *Centrostephanus coronatus* Verrill. J Exp Mar Biol Ecol 48: 185–193

Lowe E (1974) Absorption efficiencies, feeding rates, and food preferences of *Lytechinus variegatus* (Echinodermata: Echinoidea) for selected marine plants. Master's thesis. University of South Florida, Tampa

Lowe EF, Lawrence JM (1976) Absorption efficiencies of *Lytechinus variegatus* (Lamarck) (Echinodermata: Echinoidea) for selected marine plants. J Exp Mar Biol Ecol 21: 223–234

Machiguchi Y (1987) Feeding behavior of sea urchin *Strongylocentrotus intermedius* (A. Agassiz) observed in Y-shaped chamber. Bull Hokkaido Reg Fish Res Lab 51: 33–37 (in Japanese, English abstract)

Machiguchi Y (1993) Growth and feeding behavior of the red sea urchin *Strongylocentrotus nudus* (A. Agassiz). Bull Hokkaido Natl Fish Res Inst 57: 81–86

Mann KH, Wright JLC, Welsford BE, Hatfield E (1984) Responses of the sea urchin *Strongylocentrotus droebachiensis* (O.F. Müller) to water-borne stimuli from potential predators and potential food algae. J Exp Mar Biol Ecol 79: 233–244

Mattison JE, Trent JD, Shanks AL, Akin TB, Pearse JS (1977) Movement and feeding activity of red sea urchins (*Strongylocentrotus franciscanus*) adjacent to a kelp forest. Mar Biol 39: 25–30

McBride SC, Lawrence JM, Lawrence AL, Mulligan TJ (1998) The effect of protein concentration in prepared feeds on growth, feeding rate, total organic absorption, and gross assimilation efficiency of the sea urchin *Strongylocentrotus franciscanus*. J Shellfish Res 17: 1563–1570

McBride SC, Lawrence JM, Lawrence AL, Mulligan TM (1999) Ingestion, absorption and gonad production of adult *Strongylocentrotus franciscanus* fed different rations of a prepared diet. J World Aquacult Soc 30: 364–370

McBride SC, Pinnix WD, Lawrence JM, Lawrence AL, Mulligan TM (1997) The effect of temperature on production of gonads by the sea urchin *Strongylocentrotus franciscanus* fed natural and prepared diets. J World Aquacult Soc 28: 357–365

McClintock JB (1986) On estimating energetic values of prey: implications in optimal diet models. Oecologia 70: 161–162

McClintock JB, Klinger TS, Lawrence JM (1982) Feeding preferences of echinoids for plant and animal food models. Bull Mar Sci 32: 365–369

McGinnis AJ, Kasting R (1964) Colorimetric analysis of chromic oxide used to study food utilization by phytophagous insects. Agricul Food Chem 12: 259–262

Miller RJ, Mann KH (1973) Ecological energetics of the seaweed zone in a marine bay on the Atlantic coast of Canada. III. Energy transformations by sea urchins. Mar Biol 99: 99–114

Millott N (1975) The photosensitivity of echinoids. Adv Mar Biol 13: 1–52

Mills SC, Peyrot-Clausade-Peyrot M, Fontaine MF (2000) Ingestion and transformation of algal turf by *Echinometra mathaei* on Tiahura fringing reef (French Polynesia). J Exp Mar Biol Ecol 254: 71–84

Molodtsov NV, Vafina MG (1972) The distribution of β-N-acetylglucosaminidase in marine invertebrates. Comp Biochem Physiol 41B: 113–120

Molodtsov NV, Vafina MG, Kim A, Sundukova EV, Artyukov AA, Bliknov YG (1974) Glycosidases of marine invertebrates from Posiet Bay, Sea of Japan. Comp Biochem Physiol 48B: 463–470

Moore HB, McPherson BF (1965) A contribution to the study of the productivity of the urchins *Tripneustes esculentus* and *Lytechinus variegatus*. Bull Mar Sci 15: 855–871

Mukai H, Nojima S (1985) A preliminary study on grazing and defecation rates of a seagrass grazer, *Tripneustes gratilla*, (Echinodermata; Echinoidea) in a Papua New Guinean seagrass bed. Spec Publ Mukaishima Mar Biol Sta 1985: 185–192

Nedelec H (1983) Sur un nouvel indice de réplétion pour les oursins régulier. Rapp Comm Int Mer Médit 28(3): 149–151

Nedelec H, Verlaque M, Dallot S (1983) Note preliminaire sur les fluctuations de l'activité trophique de *Paracentrotus lividus* dans l'herbier de posidonies. Rapp Comm Int Mer Médit 28(3): 153–155

Nojima S, Mukai H (1985) A preliminary report on the distribution pattern, daily activity and moving pattern of a seagrass grazer, *Tripneustes gratilla* (L.) (Echinodermata: Echinoidea), in Papua New Guinean seagrass beds. Spec Pub Mukaishima Mar Biol Sta 1985: 173–183

Obrietan K, Drinkwine M, Williams DC (1991) Amylase, cellulase and protease activities in surface and gut tissues of *Dendraster excentricus, Pisaster ochraceus* and *Strongylocentrotus droebachiensis* (Echinodermata). Mar Biol 109: 53–57

Odintsov VS (1981) Nitrogen fixation (acetylene reduction) in the digestive tract of some echinoderms from Vostok Bay in the Sea of Japan. Mar Biol Lett 2: 259–263

Otero-Villanueva MM, Kelly MS, Burnell G (2004) How diet influences energy partitioning in the regular echinoid *Psammechinus miliaris*; constructing an energy budget. J Exp Mar Biol Ecol 304: 159–181

Paine RT, Vadas (1969) Calorific values of benthic marine algae and their postulated relation to invertebrate food preference. Mar Biol 4: 79–86

Pelletreau KN, Muller-Parker G (2002). Sulfuric acid in the phaeophyte alga *Desmarestia munda* deters feeding by the sea urchin *Strongylocentrotus droebachiensis*. Mar Biol 141: 1–9

Pennings SC, Svedberg J (1993) Does $CaCO_3$ in food deter feeding by sea urchins? Mar Ecol Prog Ser 101: 163–167

Penry DL, Jumars PA (1987) Modeling animal guts as chemical reactors. Amer Nat 129: 69–96

Perrier F (1875) Recherches sur l'appareil circulatoire des Oursins. Arch Zool Exp Gen 4: 604–643

Piavaux A (1977) Distribution and localization of the digestive laminarinases in animals. Biochem Syst Ecol 5: 213–239

Prim P, Lawrence JM (1975) Utilization of marine plants and their constituents by bacteria isolated from the gut of echinoids (Echinodermata). Mar Biol 33: 167–173

Régis MB (1979) Analyse des fluctuations des indices physiologiques chez deux echinoids (*Paracentrotus lividus* (Lmk) et *Arbacia lixula* L.) du Golfe de Marseille. Tethys 9: 167–181 (in French, English abstract)

Rodríguez SR, Fariña JM (2001) Effect of drift kelp on the spatial distribution pattern of the sea urchin *Tetrapygus niger*: a geostatistical approach. J Mar Ass UK 81: 179–180

Rodríguez SR, Ojeda FP (1998). Behavioral responses of the sea urchin *Tetrapygus niger* to predators and food. Mar Fresh Behav Physiol 31: 21–37

Russo AS (1977) Water flow and the distribution and abundance of echinoids (genus *Echinometra*) on an Hawaiian reef. Aust J Mar Freshwater Res 28: 693–702

Sakata K, Iwase Y, Ina K, Fijuita D (1991b) Halogenated terpenes isolated from the red alga *Plocamiuim leptophyllum* as feeding inhibitors for marine herbivores. Nippon Suisan Gakkashi 57: 743–746

Sakata K, Iwase Y, Kato K, Ina K, Machiguchi Y (1991a) A simple feeding inhibitor assay for marine herbivorous gastropods and the sea urchin *Strongylocentrotus intermedius* and its application to unpalatable algal extracts. Nippon Suison Gakkaishii 57: 261–265

Sakata K, Kato K, Ina K, Machiguchi Y (1989) Glycerolipids as potent feeding stimulants for the sea urchin, *Strongylocentrotus intermedius*. Agric Biol Chem 53: 1457–1459

Santelices B, Correa J, Avila M (1983) Benthic algal spores surviving digestion by sea urchins. J Exp Mar Biol Ecol 70: 263–269

Sawabe O, Oda Y, Shiomi Y, Ezura Y (1995) Alginate degradation by bacteria from the gut of sea urchins and abalones. Microb Ecol 30: 193–202

Scheibling RE, Hamm J (1991) Interactions between sea urchins (*Strongylocentrotus droebachiensis*) and their predators in field and laboratory experiments. Mar Biol 110: 105–116

Schlosser SC, Lupatsch I, Lawrence JM, Lawrence AL, Shpigel M (2005) Protein and energy digestibility and gonad development of the European sea urchin *Paracentrotus lividus* (Lamarck) fed algal and prepared diets during spring and fall. Aquacult Res 36: 972–982

Scott FH (1902) Food of the sea-urchin (*Strongylocentrotus droebachiensis*). Contrib Can Biol 1901–1902: 49–54

Senaratna M, Evans LH, Southam L, Tsvetnenko E (2005) Effect of different food formulations on feed efficiency, gonad yield and gonad quality in the purple sea urchin *Heliocidaris erythrogramma*. Aquacul Nutr 11: 199–207

Siddon CE, Witman JD (2003) Influence of chronic, low-level hydrodynamic forces on subtidal community structure. Mar Ecol Prog Ser 261: 99–110

Sova VV, Elyakova LA, Vaskovsky VE (1970) The distribution of laminarins in marine invertebrates. Comp Biochem Physiol 32: 459–464

Spirlet C, Grosjean P, Jangoux M (2000) Optimization of gonad growth by manipulation of temperature and photoperiod in cultivated sea urchins, *Paracentrotus lividus*(Lamarck) (Echinodermata). Aquaculture 185: 85–99

Steinberg PD (1988) Effects of quantitative and qualitative variation in phenolic compounds on feeding in three species of marine invertebrate herbivores. J Exp Mar Biol Ecol 120: 221–237

Steinberg PD, van Altena I (1992) Tolerance of marine invertebrate herbivores to brown algal phlorotannins in temperate Australia. Ecol Monogr 62: 189–222

Stevenson RA, Ufret SL (1966) Iron, manganese, and nickel in skeletons and food of the sea urchins *Tripneustes esculentus* and *Echinometra lucunter*. Limnol Oceanogr 1: 11–17

Stewart MG (1979) Absorption of dissolved organic nutrients by marine invertebrates. Oceanogr Mar Biol Ann Rev 17: 163–192

Suppes ZS, Ziashko SV, Vaskovskky VE (1977) Phospholipases of marine invertebrates. III. Distributions of phospholipases A_1 and A_2. Biol Moroyha 6: 68–70

Suzuki M, Kikuchi R, Ohnishi T (1984) The polysaccharide degradation activity in digestive tract of sea urchin *Strongylocentrotus nudus*. Bull Jap Soc Sci Fish 50: 1255–1260

Sweijd NA (1990) The digestive mechanisms of an intertidal grazer, the sea urchin *Parechinus angulosus*. MSc thesis. Rhodes University, Grahamstown

Tokin TB, Filimonova FG (1977) Electron microscope study of the digestive system of *Strongylocentrotus droebachiensis* (Echinodermata: Echinoidea). Mar Biol 44: 143–155

Uncles SE (1977) Bacterial flora of the sea urchin *Echinus esculentus*. App Env Microbiol 34: 347–350

Vadas RL (1977) Preferential feeding: an optimization strategy in sea urchins. Ecol Mongr 47: 337–371

Vadas RL, Sr, Elner RW (2003) Responses to predation cues and food in two species of sympatric, tropical sea urchins. PSZN Mar Ecol 24: 101–121

Vadas RL, Elner RW, Garwood PE, Babb LG (1986) Experimental evaluation of aggregation behavior in the sea urchin *Strongylocentrotus droebachiensis*. Mar Biol 90: 433–448

Vaïtilingon D, Rasolofonirina R, Jangoux M (2003) Consumption preferences, seasonal gut repletion indices, and diel consumption patterns of the sea urchin *Tripneustes gratilla* (Echinodermata: Echinoidea) on a coastal habitat off Toliara (Madagascar). Mar Biol 143: 451–458

Van Alstyne KL, Houser LT (2003) Dimethylsulfide release during macroinvertebrate grazing and its role as an activated chemical defense. Mar Ecol Prog Ser 250: 175–181

Vasquez JA (1992) *Lessonia trabeculata*, a subtidal bottom kelp in northern Chie: a case study for a structural and geographical comparison. In: Seeliger U (ed.) Coastal plant communities of Latin America. Academic Press, Inc., San Diego, pp 77–89

Wagner-Merner DS, Duncan WR, Lawrence JM (1980) Preliminary comparison of Thraustochytriacea in the guts of a regular and irregular echinoid. Bot Mar 23: 95–97

Wagner-Merner DS, Lawrence JM (1980) Occurrence of fungi (Thraustochytriacea) in the gut *of Lytechinus variegatus* (Lamarck) (Echinodermata: Echinoidea). Fla Sci 43: 62–63

Yamaguchi K, Araki I, Aoki T, Tseng C-H, Kitamikado M (1989) Algal cell wall-degrading enzymes from viscera of marine animals. Nip Suisan Gak 55: 105–110

Yokoe Y, Yasumasu I (1964) The distribution of cellulases in the invertebrates. Comp Biochem Physiol 13: 323–338

Edible Sea Urchins: Biology and Ecology
Editor: John Miller Lawrence

Chapter 8

Carotenoids in Sea Urchins

Miyuki Tsushima

Kyoto Pharmaceutical University, Kyoto (Japan).

1. INTRODUCTION

Carotenoids are widely distributed, naturally occurring pigments, usually red, orange, or yellow in color (Goodwin 1980, 1984; Matsuno and Hirao 1989). They occur not only as free forms but also as esters, glycosides, sulfates, and carotenoproteins. Carotenoids consisting of eight isoprenoid units in a molecule are called carotenes, e.g. β-carotene. The oxidized derivatives are called xanthophylls, e.g. β-echinenone, astaxanthin, lutein, zeaxanthin, and fucoxanthin. More than 700 different carotenoids are now known (Britton et al. 2004).

Some of the main roles of carotenoids in organisms are provitamin A activity, photoprotection, radical quenching, pigments, and immunological modulation (Bendich 1994; Krinsky 1994; Matsuno 1991). Carotenoid concentrations are occasionally high in the reproductive organs of fungi, algae, and animals, which suggests they participate in reproduction (Goodwin 1980, 1984).

This chapter deals with carotenoid distribution and metabolism in sea urchins, the effect of dietary carotenoids on gonad color, the role of carotenoids on egg production and development, and biological functions in sea urchins.

2. CAROTENOID DISTRIBUTION IN SEA URCHINS

Most of the pigmentation in sea urchins occurs in the gonads. β-Caroten-4-one was first isolated from the gonads of *Paracentrotus lividus* (Lederer 1935) and named echinenone (Goodwin and Taha 1950). β-Echinenone along with α-echinenone (=phoenicopterone) (Fox and Hopkins 1966) (α-caroten-4-one) has been isolated as a major carotenoid in the gonads of many species of edible sea urchins (Table 1). However, β-carotene is dominant in the gonads of *Prionocidaris baculosa*, *Phyllacanthus dubius*, and *Eucidaris metularia* (order Cidaroida), the most primitive sea urchin group. In *Asthenosoma ijimai* and *Araeosoma owstoni* (order Echinothurioida), β-carotene, β-echinenone, canthaxanthin, and astaxanthin are predominant carotenoids, and in *Peronella japonica* (order Clypeasteroida) astaxanthin is a principal carotenoid. The carotenoid patterns of *P. baculosa*, *P. dubius*, *E. metularia*, *A. ijimai*, *A. owstoni* and *P. japonica,* which are not palatable, are quite different from those of edible sea urchins (Matsuno and Tsushima 2001; Table 1).

Table 1 Carotenoids in the gonads (ovaries and testes) of sea urchins.

Species	Carotenoid concentration (% of the total carotenoid)						Reference
	Car.	Ech.	Can.	Ast.	L+Z	Others	
Cidaroida							
Prionocidaris baculosa	56.7	32.2	+	–	–	11.1	1
Phyllacanthus dubius	60.4	21.5	–	–	–	18.1	1
Eucidaris metularia	76.4	18.2	–	–	–	5.4	1
Echinothurioida							
Asthenosoma ijimai	20.0	35.5	10.1	4.6	–	29.8	1
Araeosoma owstoni	19.6	39.3	5.8	12.0	–	23.3	1
Diadematoida							
Diadema savigny	40.5	46.0	1.2	–	5.5	6.8	1
Clypeasteroida							
Clypeaster japonicus	5.0	77.0	11.8	+	1.2	5.0	1
Scaphechinus mirabilis	5.0	85.0	+	+	2.4	7.6	1
Astriclypeus manni	16.9	70.2	1.1	–	4.0	7.8	1
Peronella japonica	1.0	10.0	2.2	65.7	–	21.1	1
Spatangoida							
Brissus agassizi	4.5	80.0	+	+	0.7	14.8	1
Arbacioida							
Glyptocidaris crenularis	22.4	56.5	1.5	–	–	19.6	1
Echinoida							
*Paracentrotus lividus**	4.5	65.0	–	–	3.4	27.1	2
*Echinostrephus aciculatus**	13.0	70.3	2.0	–	3.7	11.0	1
*Anthocidaris crassispina**	16.0	66.1	2.7	–	6.0	9.2	1
*Echinometra mathaei**	7.2	73.8	–	–	6.5	12.5	1
*Heliocidaris erythrogramma**							
Ovaries	1.3	93.4	0.7	–	–	4.6	1
Testes	4.0	91.8	0.6	–	–	3.6	1
*H. tuberculata**							
Ovaries	11.1	70.6	0.7	–	6.0	11.6	1
Testes	19.1	65.5	1.5	–	3.1	10.8	1
*Temnopleurus toreumaticus**	9.5	72.4	2.3	1.1	2.3	12.4	1
*Mespilia globulus**	31.2	39.9	14.4	0.2	4.7	9.6	1
*Pseudocentrotus depressus**	5.1	79.6	1.2	–	0.6	13.5	1
*Tripneustes gratilla**	21.8	66.0	+	–	1.9	10.3	1
*Lytechinus variegatus**							
Ovaries	13.2	74.7	–	–	2.2	9.9	3
Testes	20.9	65.3	–	–	3.9	9.9	3
Eggs	5.8	83.5	–	–	3.2	7.5	3
*Hemicentrotus pulcherrimus**	9.9	45.6	2.2	–	17.0	25.3	1
*Strongylocentrotus intermedius**	12.1	70.3	1.6	–	1.6	14.4	1
*S. nudus**	33.3	57.0	+	–	1.0	8.7	1

Table 1 Continued

Species	Carotenoid concentration (% of the total carotenoid)						Reference
	Car.	Ech.	Can.	Ast.	L+Z	Others	
*S. droebachiensis**							
Ovaries	21.2	68.7	–	–	5.2	4.9	3
Eggs	8.0	77.5	–	–	9.8	4.7	3
*S. franciscanus**							
Ovaries	15.6	74.3	–	–	0.7	9.4	3
Testes	19.4	75.5	–	–	1.0	4.1	3
Eggs	16.2	77.3	–	–	–	6.5	3
*S. purpuratus**							
Eggs	18.8	74.3	–	–	2.5	4.4	4

Key to abbreviations: Car. – β- and α-Carotene; Ech. – β- and α-Echinenone; Can. – Canthaxanthin; Ast. – Astaxanthin; L + Z – Lutein + Zeaxanthin. Symbols: + trace; – not detected; * edible sea urchin.
References: 1. Matsuno and Tsushima (2001), 2. Rodríguez-Bernaldo de Quirós et al. (2001), 3. Lawrence et al. (2004), 4. Griffiths (1966).

Most studies characterizing carotenoids in sea urchin gonads do not distinguish between testes and ovaries or between gametogenic stages. The concentrations of carotenoids in ovaries of *Heliocidaris erythrogramma* and *H. tuberculatus* are much higher than those in testes although the profiles are similar (Matsuno and Tsushima 2001). Meanwhile, in *Strongylocentrotus franciscanus* and *Lytechinus variegatus*, carotenoid concentrations and profiles of testes and ovaries do not differ (Lawrence et al. 2004; Table 1).

Griffiths and Perrott (1976), however, report preferential deposition of β-echinenone in eggs rather than in the entire ovary. β-Echinenone is about 90% and β-carotene about 8% of egg carotenoids of *S. droebachiensis*. A similar increased storage of β-echinenone at the expense of β-carotene occurs with maturation of eggs in *Tripneustes gratilla* (Shina et al. 1978), *S. droebachiensis*, *S. franciscanus*, and *L. variegatus* (Lawrence et al. 2004; Table 1). It is presumed that carotenoids are transferred with other nutrients to developing eggs. The fate of the carotenoid in nongametogenic cells of testes is not known as sperm do not contain carotenoids.

In the case of the gut, β-carotene, fucoxanthin, fucoxanthinol, and fucoxanthinol ester are the common major carotenoids in most sea urchins except *A. ijimai* (Matsuno and Tsushima 2001). In *A. ijimai*, β-echinenone, α-echinenone, astaxanthin, and alloxanthin are major carotenoids along with 7, 8-didehydroastaxanthin and 7, 8, 7′, 8′-tetradehydroastaxanthin. Metabolic products of fucoxanthin from the natural diet (brown algae and diatoms) of herbivorous sea urchins have not been found in carnivorous *A. ijimai*. The carotenoid patterns of the gut of sea urchins species are influenced by the carotenoid profiles in the diets.

It is generally accepted that the major pigment in the test and spines of sea urchins is naphthoquinone (Fox and Hopkins 1966). But contrary to expectation, astaxanthin is a principal pigment in the test of *P. japonica* (Kawaguti and Yamasu 1954). Carotenoids occur along with naphthoquinones in the test and spines of all species of sea urchins examined. The carotenoid patterns of the test and spines of sea urchins species are very similar to those of their gonads (Matsuno and Tsushima 2001).

β-Echinenone fraction in the gonads, gut, test, and spines of *Pseudocentrotus depressus* consists of all-*E*- and 9′*Z*-β-echinenone (Tsushima and Matsuno 1997). Most of the β-echinenone fraction in the gonads (ovaries and testes) consists of 9′*Z*-isomer. Vershinin and Lukyanova (1993), however, report that only all-*E*-β-echinenone is detected in the embryos of *S. intermedius*. It should be made clear that these studies suggest that the *Z*-carotenoid may have a specific function in the sea urchin, possibly related to reproduction.

Occurrence of carotenoids in sea urchins and their distribution in organs have been described in detail (Matsuno and Tsushima 2001).

3. METABOLISM OF CAROTENOIDS IN SEA URCHINS

In general, animals do not synthesize carotenoids *de novo* and those found in bodies of animals are either a direct accumulation of carotenoids from the food or are partly modified through metabolic reactions. β-Echinenone can be derived from dietary β-carotene via β-isocryptoxanthin in the gonad of sea urchins, based on speculations from the results of carotenoid analysis (Griffiths and Perrott 1976). The presence of this oxidative metabolic pathway is proved by feeding experiments in *P. depressus* (Tsushima et al. 1993). Furthermore, another possible oxidative metabolic pathway of α-carotene to α-echinenone via α-isocryptoxanthin has been found (Tsushima, unpubl. data).

4. THE EFFECT OF DIETARY CAROTENOIDS ON GONAD COLOR IN SEA URCHINS

There is considerable interest in enhancing the diets of sea urchin to produce acceptable commercial color of the gonads. A number of scientific studies have examined carotenoids in sea urchins fed artificial diets (Goebel and Barker 1998; Matsuno and Tsushima 2001; McLaughlin and Kelly 2001; Pearce et al. 2003; Plank and Lawrence 2002; Robinson et al. 2002, 2004; Shpigel et al. 2005).

A diet of the microalgae *Phaeodactylum tricornutum* improves gonad color of *Psammechinus miliaris* in comparison with the macroalgae *Laminaria saccharina* (McLaughlin and Kelly 2001). In *S. droebachiensis*, the microalgae *Dunaliella salina*, containing all-*E*-and 9*Z*-β-carotene, is more effective at providing good coloration than comparable concentrations of synthetic β-carotene (all-*E* forms only) (Robinson et al. 2002). Similarly, the synthetic β-carotene diet does not significantly affect gonad color of *Evechinus chloroticus* (Goebel and Barker 1998). It is not clear why natural β-carotene is more effective than the synthetic all-*E*-β-carotene, or whether gonad coloration is related to bioavailability, bioabsorption, or bioconversion of the added pigments in the gut and in the gonad.

Dietary xanthophylls such as capsanthin, zeaxanthin, lutein, astaxanthin, and fucoxanthin are not very effective in producing color in sea urchin gonads in comparison to the β-carotene diet (Tsushima et al. 1993; Kawakami et al. 1998; Plank and Lawrence 2002; Robinson et al. 2004). In contrast, β-echinenone is more effective in producing good color in *P. depressus* (Tsushima et al. 1997). These studies indicate that β-carotene and β-echinenone are required for carotenoid accumulation in gonads and that xanthophylls are not appropriate except for β-echinenone.

5. THE ROLE OF CAROTENOIDS IN SEA URCHINS

Carotenoids are important in the nutrition of animals, having roles as provitamin A activity (Matsuno 1991), immunological modulation (Bendich 1994), antioxidants and protectants against photosensitive damage (Krinsky 1994). Carotenoids are important in egg production and development, and biological functions of sea urchins.

5.1. Egg Production and Development

The role of dietary β-carotene, β-echinenone, and astaxanthin in the development of *P. depressus* has been investigated and compared with the role of vitamin A and vitamin E in growth, reproduction, and development of sea urchins by feeding experiments (Tsushima et al. 1997). The concentration of carotenoids increases more in individuals fed β-echinenone than in those fed β-carotene. In both cases, β-echinenone has a higher concentration than β-carotene and other carotenoids are minimal. β-Carotene and β-echinenone have significant effects on larval survival. A vitamin E diet, which affects the number of the ova, has no significant effects on larval survival. Therefore, the β-echinenone group is the best of all the groups tested in both number of ova and larval survival. Both vitamin A and astaxanthin have no effect on either. In the case of *S. droebachiensis*, de Jong-Westman et al. (1995a,b) report that β-carotene increases gonad growth, egg-energy content and rate of larval development. These studies indicate that carotenoids, especially β-carotene and β-echinenone, exert a significant influence on egg production and development.

It has, however, been reported that xanthophylls also affect fecundity and the quality of the eggs of *P. depressus* (Kawakami et al. 1998). Fucoxanthin diet does not accumulate in the gonad, but affects the ova. Furthermore, it is very interesting that the number of ova is higher in fucoxanthin group than that of β-carotene group. Xanthophylls (lutein and zeaxanthin) in the adult diet enhance egg and juvenile production of *L. variegatus* in comparison with β-carotene diet (George et al. 2001). Xanthophylls such as fucoxanthin, lutein, and zeaxanthin do not affect gonad color, but this is important for sea urchin aquaculture as a diet containing xanthophylls would enable production of large numbers of juveniles with high survival rates.

5.2. Biological Functions

A disease characterized by spine loss occurs in *P. depressus* fed a carotenoid-free diet. Furthermore, the gonads of most individuals are underdeveloped (Tsushima, unpubl. obs.). These symptoms in sea urchins may be due to a decrease in biological defense caused by carotenoid deficiency. In an investigation of biological defense reactions in *P. depressus*, Kawakami et al. (1998) examined the effects of β-carotene, β-echinenone, astaxanthin, fucoxanthin, vitamin A, and vitamin E on phagocytic activities. They found that carotenoids, especially β-carotene and β-echinenone, facilitate phagocytosis. Furthermore, fucoxanthin also affects phagocytic activities (Kawakami et al. 1998). Phagocytic cells of *S. nudus*, produces reactive oxygen species when the phagocytic cells engulf foreign material (Ito et al. 1992). These results indicate that the carotenoids, β-carotene, β-echinenone, and fucoxanthin, may play an important role in the biological defense of sea urchins by quenching the reactive oxygen species, as well as increasing the production of phagocytic cells.

Lamare and Hoffman (2004) report the role that carotenoids may act as photoprotectants in four species of *Strongylocentrotus* against the damaging effects of UV radiation on biological functions. They suggest that carotenoids are photoprotective in echinoid eggs, probably by mitigating effects of reactive oxygen species.

6. CONCLUSIONS

β-Echinenone, isolated as a major carotenoid in the gonads of the edible sea urchins, is oxidatively formed from βcarotene via β-isocryptoxanthin. Furthermore, it exerts influences on gonad color, reproduction, development, and phagocytosis in the sea urchin. On the other hand, xanthophylls such as fucoxanthin, lutein, and zeaxanthin do not affect gonad color but have significant effects on fecundity and survival of larvae.

Most of the β-echinenone fraction in the gonads (ovaries and testes) consists of 9′*Z*-isomer. It must be made clear that the *Z*-carotenoid may have a specific function in the sea urchin, possibly related to reproduction.

ACKNOWLEDGMENTS

I am grateful to Takao Matsuno and John Lawrence for comments on the manuscript.

REFERENCES

Bendich A (1994) Recent advances in clinical research involving carotenoids. Pure Appl Chem 66: 1017–1024

Britton G, Liaaen-Jensen S, Pfander H (eds) (2004) Carotenoids. Handbook. Birkhäuser, Basel

de Jong-Westman M, March BE, Carefoot TH (1995a) The effect of different nutrient formulations in artificial diets on gonad growth in the sea urchin *Strongylocentrotus droebachiensis*. Can J Zool 73: 1495–1502

de Jong-Westman M, Qian P-Y, March BE, Carefoot TH (1995b) Artificial diets in sea urchin culture: effects of dietary protein level and other additives on egg quality, larval morphometrics, and larval survival in the green sea urchin, *Strongylocentrotus droebachiensis*. Can J Zool 73: 2080–2090

Fox DL, Hopkins TS (1966) The comparative biochemistry of pigments. In: Boolootian RA (ed.) Physiology of Echinodermata. Interscience Publishers, New York, pp 277–300

George SB, Lawrence JM, Lawrence AL, Smiley J, Plank L (2001) Carotenoids in the adult diet enhance egg and juvenile production in the sea urchin *Lytechinus variegatus*. Aquaculture 199: 353–369

Goebel N, Barker MF (1998) Artificial diets supplemented with carotenoid pigments as feeds for sea urchins. In: Mooi R, Telford M (eds) Echinoderms: San Francisco. Balkema, Rotterdam, pp 667–672

Goodwin TW (1980) The biochemistry of the carotenoids. Vol. 1. Plants. Chapman and Hall, London

Goodwin TW (1984) The biochemistry of the carotenoids. Vol. 2. Animals. Chapman and Hall, London

Goodwin TW, Taha MM (1950) The carotenoids of the gonads of the limpets *Patella vulgata* and *Patella depressa*. Biochem J 47: 244–249

Griffiths M (1966) The carotenoids of the eggs and embryos of the sea urchin *Strongylocentrotus purpuratus*. Dev Biol 13: 296–309

Griffiths M, Perrott P (1976) Seasonal changes in the carotenoids of the sea urchin *Strongylocentrotus droebachiensis*. Comp Biochem Physiol 55B: 435–441

Ito T, Matsutani T, Mori K, Nomura T (1992) Phagocytosis and hydrogen peroxide production by phagocytes of the sea urchin *Strongylocentrotus nudus*. Develop Comp Immunol 16: 287–294

Kawaguti S, Yamasu T (1954) Carotenoid pigment in the test of cake-urchin, *Peronella japonica*. Biol J Okayama Univ 3: 150–158

Kawakami T, Tsushima M, Katabami Y, Mine M, Ishida A, Matsuno T (1998) Effect of β, β-carotene, β-echinenone, astaxanthin, fucoxanthin, vitamin A and vitamin E on the biological defense of the sea urchin *Pseudocentrotus depressus*. J Exp Mar Biol Ecol 226: 165–174

Krinsky NI (1994) The biological properties of carotenoids. Pure Appl Chem 66: 1003–1010

Lamare MD, Hoffman J (2004) Natural variation of carotenoids in the eggs and gonads of the echinoid genus, *Strongylocentrotus*: implications for their role in ultraviolet radiation photoprotection. J Exp Mar Biol Ecol 312: 215–233

Lawrence JM, Montoya R, McBride SB, Harris LG (2004) Carotenoid concentrations and profiles in testes, ovaries and eggs of the sea urchins *Strongylocentrotus droebachiensis*, *Strongylocentrotus franciscanus* and *Lytechinus variegatus*. In: Lawrence JM, Guzmán O (eds) Sea urchins: fisheries and ecology. DEStech Publications, Lancaster, pp 173–178

Lederer E (1935) Echinenone and pentaxanthin; two new carotenoids in the sea urchin (*Echinus esculentus*). Compt Rend 201: 300–302

Matsuno T (1991) Xanthophylls as precursors of retinoids. Pure Appl Chem 63: 81–88

Matsuno T, Hirao S (1989) Marine carotenoids. In: Ackman RG (ed.) Marine biogenic lipids, fats, and oils. Vol. 1. CRC Press, Boca Raton, FL, pp 251–388

Matsuno T, Tsushima M (2001) Carotenoids in sea urchins. In: Lawrence JM (ed.) Edible sea urchins: biology and ecology. Elsevier Science B.V., Amsterdam, pp 115–138

McLaughlin G, Kelly MS (2001) Effect of artificial diets containing carotenoid-rich microalgae on gonad growth and color in the sea urchin *Psammechinus miliaris* (Gmelin). J Shellfish Res 20: 377–382.

Pearce CM, Daggett TL, Robinson SMC (2003) Effects of starch type, macroalgal meal source, and β-carotene on gonad yield and quality of the green sea urchin, *Strongylocentrotus droebachiensis* (Müller), fed prepared diets. J Shellfish Res 22: 505–519

Plank LR, Lawrence JM (2002) The effect of dietary carotenoids on gonad production and carotenoid profiles in the sea urchin *Lytechinus variegates*. J World Aquacult Soc 33: 127–137

Robinson SMC, Castell JD, Kennedy EJ (2002) Developing suitable colour in the gonads of cultured green sea urchins (*Strongylocentrotus droebachiensis*). Aquaculture 206: 289–303

Robinson SM, Lawrence JM, Burridge L, Haya K, Martin J, Castell J, Lawrence AL (2004) Effectiveness of different pigment sources in colouring the gonads of the green sea urchin (*Strongylocentrotus droebachiensi)*. In: Lawrence JM, Guzmán O (eds) Sea urchins: fisheries and ecology. DEStech Publications, Lancaster, pp 215–221

Rodríguez-Bernaldo de Quirós A, López-Hernández J, Simal-Lozano J (2001) Determination of carotenoids and liposoluble vitamins in sea urchin (*Paracentrotus lividus*) by high performance liquid chromatography. Eur Food Res Technol 212: 687–690.

Shina A, Gross J, Lifshitz A (1978) Carotenoids of the invertebrates of the red sea (Eilat shore)-II. Carotenoid pigments in the gonads of the sea urchin *Tripneustes gratilla* (Echinodermata). Comp Biochem Physiol 61B: 123–128

Shpigel M, McBride SC, Marciano S, Ron S, Ben-Amotz A (2005) Improving gonad colour and somatic index in the European sea urchin *Paracentrotus lividus*. Aquaculture 245: 101–109

Tsushima M, Kawakami T, Matsuno T (1993) Metabolism of carotenoids in sea-urchin *Pseudocentrotus depressus.* Comp Biochem Physiol 106B: 737–741

Tsushima M, Kawakami T, Mine M, Matsuno T (1997) The role of carotenoids in the development of the sea urchin *Pseudocentrotus depressus.* Invert Reprod Develop 32: 149–153

Tsushima M, Matsuno T (1997) Occurrence of 9′Z-β-echinenone in the sea urchin *Pseudocentrotus depressus.* Comp Biochem Physiol 118B: 921–925

Vershinin A, Lukyanova ON (1993) Carotenoids in the developing embryos of sea urchin *Strongylocentrotus intermedius.* Comp Biochem Physiol 104B: 371–373

Edible Sea Urchins: Biology and Ecology
Editor: John Miller Lawrence

Chapter 9

Disease in Sea Urchins

Kenichi Tajima[a], José Roberto Machado Cunha da Silva[b], and John M Lawrence[c]

[a]*Laboratory of Marine Biotechnology and Microbiology, Hokkaido University, Hokkaido (Japan).*
[b]*Departamento de Biologia Celular e do Desenvolvimento, Universidade de São Paulo, São Paulo (Brazil).*
[c]*Department of Biology, University of South Florida, Tampa, FL (USA).*

1. INTRODUCTION

Jangoux (1987a,b,c,d) documented the occurrence of a variety of disease organisms in sea urchins. Most accounts are simply descriptive without any consideration of the biology or ecology of the relationship. Mass mortalities presumably resulting from microorganisms began to be reported in the 1970s (Maes and Jangoux 1984). However, it was the catastrophic mass mortality of *Diadema antillarum* in the Caribbean Sea in 1983 (Chapter 11) and of *Strongylocentrotus droebachiensis* off the coast of Nova Scotia between 1980 and 1983 (Chapter 18) of large magnitude and major ecological consequences that caught biologists' attention. Occurrence of disease has been noted in *Paracentrotus lividus* (Chapter 13), *Strongylocentrotus franciscanus*, and *Strongylocentrotus purpuratus* (Chapter 19). Fisheries of predators may increase incidence of disease of sea urchins in the field as populations increase in density (Behrens and Lafferty 2004; Lafferty 2004). Advent of aquaculture of sea urchins has increased interest in disease as it is one of the biggest threats to sea urchin producers.

Most reports concern disease in regular sea urchins, but disease also occurs in irregular echinoids. Schwammer (1989) recorded bald sea urchin disease in the heart urchin *Spatangus purpureus*. Specimens of the heart urchin *Meoma ventricosa* with spine loss and discoloration of the test had gram-negative bacteria in the catch connective tissue of the spines (Nagelkerken et al. 1999). Innoculates of bacteria isolated from affected *M. ventricosa* infected the sea urchin *Lytechinus variegatus*. Specimens of the sand dollar *Encope micropora* with discolored epithelium and an eroded test suggesting disease have been reported (Sonnenholzner and Lawrence 1998). Lesions in the fossil holasteroid *Collyrites dorsalis* from the Middle Jurassic were considered identical to those produced by bald sea urchin disease in extant regular echinoids (Radwańska and Radwański 2005).

2. BACTERIAL DISEASE IN JAPANESE SEA URCHINS

2.1. Evidence for Bacterial Disease

To promote production of sea urchins, local government and fishermen's cooperative associations have established sea urchin breeding centers in many parts of Japan. Before being released into the sea, juveniles are grown for 7–8 months to about 5 mm diameter in rearing tanks. Each rearing tank consists of several cages, each with several thousands of juveniles. Disease occurs during the growth period in the tank.

Two types of diseases occur: one at a high seawater temperature (summer disease), the other at a low temperature (spring disease). In northern parts of Kyusyu, *togenukesho*, a disease that infects juvenile *Pseudocentrotus depressus* and *Hemicentrotus pulcherrimus*, has occurred from January to April each year since 1981 (Kanai 1993). A mass mortality of juvenile *P. depressus* occurred in 1992 at the Saga Prefectural Sea Farming Center in Kyusyu (Hamaguchi et al. 1993). A filiform gliding bacterium was isolated as the causative agent.

A mass mortality occurred for the first time in cultured juvenile *Strongylocentrotus intermedius* at Shakotan-Cho fisheries breeding center southeast of Hokkaido in August 1993 (Tajima et al. 1997b). The disease killed almost all the sea urchins (about 800 000 individuals) in the center. The disease was called the "spotting disease." It has occurred every year since then on a smaller scale in several breeding centers in central and southern parts of Hokkaido. The disease also occurred after handling the sea urchins for selection from the end of May to the beginning of June when the seawater temperature ranged from 11 to 13 °C at Date, Shiriuchi, and Shikabe in southern parts of Hokkaido in 1994. In 1995, a short rod psychrophilic bacterium was isolated as the causative agent of the disease.

2.2. Symptoms of the Disease

Green or black spots on the surfaces of diseased animals were described by Kanai (1993) as *togenukesho*. Partial spine loss, which gradually spread, occurs at the early stage. External signs of the disease in *P. depressus* in the summer are exfoliation of epidermis of the oral surface and discoloration of the white portion of the peristome and dark green of the mouth or lantern, and discoloration and extension of the tube feet.

Strongylocentrotus intermedius with the spotting disease develop blackish-red lesions on the test, spine loss, and tube feet that cannot attach (Shimizu et al. 1995). The epidermal lesions of the test and appendages are disorganized and infiltrated by coelomic red spherule cells and brown granules. Muscle fibers at the base of the spines, tube feet and ampullae are fragmented. These pathological conditions are similar to those of the "bald sea-urchin" disease (Jangoux 1987a), but may be a rather generic response to secondary infection of epidermal lesions (Roberts-Regan et al. 1988). Typical external signs of disease in *S. intermedius* at low water temperature are the reddish color of the peristome in the early stage and a blue-black color in the late stage. No gross signs occur on the body surface (Tajima et al. 1998b).

2.3. Isolation and Description of the Disease-Causing Bacterium

Kanai (1993) isolated a gliding bacterium (2–6 × 0.4–0.6 μm) as the causal agent of the disease. The colorless bacterium, slightly raised with a thinly spread margin, grew on Uni (sea urchin) agar medium, which includes seawater extracts of the test of juvenile *P. depressus and H. pulcherrimus*. It grew from 10 to 20 °C, with an optimum at 14 to 16 °C. It did not grow on media containing only sodium chloride or < 80% seawater.

Hamaguchi et al. (1993) reported a gliding bacterium (20–60 μm in length) was the causal agent of disease of *P. depressus* in summer (seawater temperature 23–25 °C) at Saga. It reacted to the anti-*Flexibacter maritimus* strain 2408 rabbit serum. The bacterium was isolated from spines with lesions on *Cytophaga* medium prepared with seawater. The characteristics of the causal bacterium of the disease in Kyusyu were not described.

2.3.1. *Summer disease*

Summer disease usually occurs from late July to the end of August when the coastal seawater temperature becomes >20 °C. Spotting lesions with a blackish-purple color appear on the surface, followed by progressive spine loss. The causative bacterium (strain 12-N) of the disease was isolated from the coelomic fluid and body surface of diseased sea urchins in 1993 (Tajima et al. 1997a). Colonies of strain 12-N on solid medium produce a bright yellow pigment. The bacterium has slightly gliding motility. The bacteria are long, flexible rods, about 4–15 μm in length, in young cultures that become short rods in old cultures. The bacterium is a Gram-negative filamentous rod that requires seawater for growth. It grows weakly at 37 °C and does not grow at 10 °C. The optimum temperature for growth is thought to be between 25 and 30 °C. It does not degrade agar, cellulose, chitin, esculin, or trybutyrin. It does degrade starch, casein, gelatin, tyrosine, DNA, and lysis of dead bacterial (*E. coli*) cells. The change in morphology during the growth cycle and degradative abilities indicate strain 12-N belongs to the genus *Flexibacter* and not *Cytophaga*.

Strain 12-N differs from *F. maritimus* in its growth temperature range, degradation of starch and tween 20. Other characteristics are the same as those of *F. maritimus*. Other reference strains differ from strain 12-N in many characteristics. *Flexibacer maritimus* and reference strains of *Cytophaga* did not react with anti-12-N rabbit serum. Kanai (1993) and Hamaguchi et al. (1993) reported a kind of gliding organism was associated with the disease. As neither of them gave characteristics it is not possible to compare strain12-N to their strains. Hamaguchi et al. (1993) stated that their strain reacted with anti- *F. maritimus* strain 2408 rabbit serum, which contrasts to strain 12-N.

The mol% G + C content of strain 12-N was 40.5. The mol% G + C content in the genus *Flexibacter* range from 37 to 47, and that of *Cytophaga* range from 30 to 45 (around 35 in many strains) (Krieg and Holt 1984). The mol% G + C content indicates strain 12-N belongs to the genus *Flexibacter*. The phenotypic characteristics of strain 12-N most closely resemble those of *F. maritimus* IAM 14317, but they differ in mol% G + C. Hansen et al. (1992) reported that *F. ovolyticus*, which resembles *F. maritimus* in the mol% G + C content and phenotypic characteristics, cause disease in the eggs and larval stages of Atlantic halibut (*Hippoglossus hippoglossus*). However, its G + C content are lower than that of strain 12-N. Moreover, the G + C content of *C. latercula*, *C. aprica*, *F. litoralis*, and *F. polymorphus* (Holt et al. 1994) are different from that of strain 12-N. 16S rDNA sequence of the strain 12-N was also investigated. Strain 12-N was phylogenetically distinct from any other *Flexibacter* sp. and *Cytophaga* sp. (unpubl.). Recently, Suzuki et al. (2001) proposed a new genus *Tenacibaculum*, including *T. mesophilum* and *T. amylolyticum,* which were isolated from sponges and macroalgae by phylogenetic, analysis based on 16S rDNA and GyrB sequences. *Flexibacter maritimus* and *F. ovolyticus* were also transferred to the new genus. In the paper they mentioned that *Flexibacter* sp. strain 12-N should be also included in this genus.

DNA–DNA homology values of *F. maritimus* against the strain 12-N was 4.6%. But DNA–DNA hybridization experiment against *T. mesophilum* and *T. amyplyolyticum* was not yet determined. There is a possibility that the strain 12-N is a new species of the genus *Tenacibaculum*.

2.3.2. *Spring disease*

Spring disease usually occurs from the end of May to the beginning of June when the seawater temperature ranges from 11 to 13 °C. The reddish color of the peristome in affected individuals appears at the early stage of the disease. The causative bacterium isolated from diseased *S. intermedius* sampled at Date, Shiriuchi, and Shikabe breeding center are Gram-negative, oxidase positive, slightly curved, short-rod facultative anaerobes with a single polar flagellum (Tajima et al. 1998b).

Translucent white colonies were formed in almost pure state on SA medium inoculated with coelomic fluids of diseased individuals of the three breeding centers. Similar results were also found on plates inoculated with samples of the body surfaces of diseased individuals. They fermented glucose without gas production. The G + C content of all 15 strains ranged from 43.3 to 46.4 mol%. These three characteristics confirmed that these strains belong to the genus *Vibrio*. The biochemical properties of the strains from Shiriuchi are completely identical with those from Shikabe. They differ from those from Date by the indole production test. However, *V. campbellii*, *V. proteolyticum*, *V. harveyi*, and *Vibrio* spp. isolated from the intestine of sea urchins were relatively closer to the strains, even though they differed in 4–5 properties from the 15 strains.

All strains from Date are genotypically closely related to strain Da-2, with homology values ranging from 80.6% to 100%. But, strain Da-2 shows relatively low relatedness, ranging from 44.1% to 56.4%, with strains from Shiriuchi and Shikabe. The strains from Shiriuchi and Shikabe are also genotypically closely related to each other, with significant homology values from 78% to 100%, but not with those of Date, with low values from 33.1% to 46.9%. These results indicate the strains of Shiriuchi and Shikabe are genotypically identical and different from those of Date at the genospecies level. Three representative strains (Da-2, Sr-3, and Sk-1) differ from the 17 reference strains because of their relatively low homology values ranging from 18.2% to 64.1%.

The slide-agglutination reactions of anti-Da-2, -Sr-3, and -Sk-1 sera, and formalin-killed cells of each strain isolated from the three breeding centers indicated that the strains from Date are different from those from Shiriuchi and Shikabe (Takeuchi et al. 1999). None of the reference strains (formalin-killed cells) reacted with any of the antisera

(Takeuchi et al. 1999). This suggests the causative organisms also differ in serology from the reference strains.

Cross-agglutination reaction between antisera absorbed with homologous and heterologous antigens of strains Da-2, Sr-3, and Sk-1, and their O- or F-antigens were used to determine thermostable and thermolabile antigenic compositions. The thermostable antigenic composition suggests strain Da-2 differs from strains Sr-3 and Sk-1 and strains Sr-3 and Sk-1 are identical to each other. Strains Sr-3 and Sk-1 have common thermolabile antigens that differ from those of strain Da-2.

Analysis of 16S rDNA sequence of Da-2, Sr-3, and Sk-1 show that strain Da-2 is phylogenetically different from the closely related Sr-3 and Sk-1. They are clearly distinct from any other *Vibrio* sp. found in DNA databases (Tajima, unpubl.).

The alginolytic *Vibrio* strains isolated from the intestine of sea urchins (Sawabe et al. 1995) were suspected at an early stage to be the causative organisms as strains isolated from the coelomic fluid of diseased individuals showed a high alginolytic activity (Takeuchi et al. 1999). However, DNA-DNA homology shows the causative agents differs from that of intestinal *Vibrio* sp. (Takeuchi et al. 1999).

2.4. Biological Responses to Bacterial Infection

Acid phosphatase (ACP) is generally divided into tartrate-resistant ACP (TRACP) and tartrate-sensitive ACP (TSACP). Tartrate-sensitive ACP is a major enzyme of the lysosome, which contains many digestive enzymes capable of breaking down autogenous and exogenous material and is involved in the biological defense systems of animals (Brightwell and Tappel 1968). Shimizu and Nagakura (1993) found that *S. intermedius* with the spotting disease show more intense total ACP activity than healthy individuals due to an increase in TSACP. In addition, tissues of diseased individuals have higher TSACP activity than those of healthy ones. Shimizu and Nagakura concluded that diseased tissues have enhanced ACP activity to resist infection from external organisms such as bacteria or protozoans. Individuals with spotting disease show reddish or brownish discoloration on their body surface. This results from infiltration of many red spherule cells and brown granules into the epidermis. The ACP-positive reactions are prominent in the brown granule masses (Shimizu and Nagakura 1993).

Phagocytes and *Vibrio* were mixed *in vitro* to study their interaction (Tajima, unpubl.). The number of phagocytes does not change but the number of *Vibrio* cells decreases gradually when they are nonpathogenic. On the contrary, both phagocytes and *Vibrio* initially decrease in number when the *Vibrio* is pathogenic and then the bacterial cells increase in number and the phagocytes gradually disappear. This suggests *Vibrio* produces an extracellular material that overcomes phagocytes.

2.5. Bacterial Control

Tajima et al. (1998a) investigated the control of *Flexibacter* strain 12-N of *S. intermedius*. The bacterium does not infect sea urchins at 15 °C, a temperature near the lowest growth temperature of the bacterium. The bacterium is highly sensitive to erythromycin, oleandomycin, and lincomycin but resistant to kanamycin and nalidixic acid. Susceptibility of the bacterium to ultraviolet irradiation and oxidants is similar to those of fish pathogenic bacteria. Among the eight disinfectants tested *in vitro*, 300 ppm hydrogen peroxide for 5 min at 15 °C is most effective against the bacterium without detrimental effects on the sea urchin. The bacterium enters into the viable but non-culturable (VBNC) state in 75% artificial seawater at 5 °C; this generates the speculation that this bacterium may resuscitate from VBNC state during summer (Masuda et al. 2004). Yamase et al. (2006) clarified that the bacterium resuscitated from VBNC state by adding iron chloride in final concentration of $0.34\,mg\,l^{-1}$ in microcosm (i.e. suspension) and pathogenicity of the VBNC cells of the bacterium under iron concentration of $3.40\,mg\,l^{-1}$ revealed that VBNC cells were able to regain the culturable state and sea urchins were found dead with the symptoms of the disease. These results suggest a possibility of control of the disease by removing iron in rearing seawater without using antibiotics and disinfectants.

The causative *Vibrio* sp. was found in seawater at Date City fisheries breeding center, as well as the surface of the sea urchins and rearing cages throughout the year in 1996 (Tajima et al. 2000). *Vibrio* strains Da-2 and Sr-3 are highly sensitive to benzyl penicillin and erythromycin at 20 °C, but resistant to cloxacillin, spiramycin, and chemotherapeutics. They are more sensitive to ultraviolet irradiation than the *Flexibacter* strain 12-N. Treatment with 500 ppm chorine dioxide for 30 min at 15 °C killed the bacteria without injury to sea urchins (Tajima et al. 2000).

3. IMMUNOLOGICAL RESPONSE TO BACTERIAL DIESASES IN SEA URCHINS

3.1. General Concept of the Immune Response

Immunology is concerned with the native or acquired resistance to infection with microorganisms or those processes that identify the pathogen and defend against it, namely maintaining organism integrity in normal physiological growth, repair, senility, and disease (Tauber and Chernyak 1991).

In the late nineteenth century, the Russian biologist Elie Metchnikoff (1854–1916) observed that some cells present in the perivisceral coelom and mesenchymal tissue of echinoderms were able to move and engulf inert or even live particles. Metchnikoff observed this phenomenon in various taxa, although not in cephalochordates (Metchnikoff 1891; Silva et al. 1995, 1998) and developed the phagocytic theory: “I used the term phagocytosis to designate the amoeboid cells able to capture and digest microorganisms and other elements. I called it the Phagocytic Theory based on this defensive cell property” (Metchnikoff 1891; 1905). Metchnikoff is the father of the metaphysical idea of the immune system as an active process mediated by the host (the cells, according to Metchnikoff), which was later incorporated by the humoralists (where only the humors have the protective function, later described as complement system and circulating antibodies) (Tauber and Chernyak 1991).

Metchnikoff also reported that during echinoderm gastrulation mesenchyme cells are capable of phagocytosis almost immediately after they reach the coelomic cavity. These cells are also capable of moving in the direction of the invading object, enveloping it completely through their fusion into a plasmodium. The stage at which phagocytosis can first be characterized in the embryos of the sea urchin was demonstrated in *L. variegatus* by microinjection of the yeast *Saccharomyces cerevisiae* into the blastocoele (Silva 2000). Secondary mesenchyme cells were first observed phagocytosing during the mid-gastrula stage. As incubation time increased, the number of yeasts per phagocyte rose. Using vital fluorescence dyes, stained free yeast were seen in the blastocoele during late-gastrula stage, indicating cell death and suggesting specific factors, such as proteases, in the extracellular environment. This study (Silva 2000) indicates that mid-gastrula is the development stage at which phagocytosis first takes place and thus "self-identity" begins to be established in the embryo of *L. variegatus* and probably among the others echinoids. The starting point of phagocytic activity reflects a biological capacity for distinguishing between self and nonself, capacity to distinguish between what belongs to the individual and what does not (Tauber and Chernyak 1991; Silva 2000, 2001).

3.2. Coelomocyte Types

The importance of coelomocytes has been described in relation to immunity (Isaeva and Korembaum 1990), wound repair, coagulation, and rejection of grafts (Coffaro and Hinegardner 1977; Smith 1981; Isaeva and Korembaum 1990; Silva and Peck 2000; Silva et al. 2001). In Echinoidea, the literature describes four coelomocyte types: phagocytic amoebocytes (60–70%), vibratile cells (15–20%), red spherule cells (5–10%), and colorless spherule cells (5–10%) (Johnson 1969a; Bertheussen and Seljelid 1978; Isaeva and Korembaum 1990; Mangiaterra and Silva 2001).

Phagocytic amoebocyte (PA) phagocytosis has been used as a tool to evaluate the immune response of sea urchins against abiotic factors. Some substances can regulate phagocytosis. The sea urchin *P. depressus* kept at 16 °C and fed fucoxantine, β-carotene, and β-echinenone, increased phagocytosis by the PA (Kawakami et al. 1998).

3.3. Phagocytosis

Phagocytosis by echinoderm coelomocytes was first observed in the perivisceral coelom of *Asterias rubens* by Durham in 1891 (in Smith 1981: 532), after injection of India ink or blue aniline. Phagocytic amebocytes were also found in the tube feet, gut wall, and other organs by Cuénot in 1891 (in Smith 1981: 532). Since then, numerous studies have reported many species of echinoderms taking up various substances, including bacteria, inert particles, foreign cells, and senescent cells (Cuénot 1948; Millott 1950; Boolootian and Giese 1958; Johnson and Beeson 1966; Johnson 1969c; Hobaus 1978). Among echinoderms, the uptake of particles seems to take place only by amebocytes (Johnson 1969c; Bertheussen and Seljelid 1978; Smith 1981; Isaeva and Korenbaum 1990; Edds 1993; Plytycz and Seljelid 1993; Chia and Xing 1996; Xing et al. 1998; Mangiaterra and Silva 2001; Silva et al. 2001), and only by the petaloid form of this cell type (Boolootian and Giese 1958; Johnson 1969a,c; Smith 1981). The amebocytes are also the only coelomocytes containing intranuclear iron bodies (Millott 1950; Burton 1966; Vevers 1967; Millott and Vevers 1968; Johnson 1969b; Hobaus 1978), which can be considered a typical feature of these cells. Hydrogen peroxide production by sea urchin phagocytes has been described only in *Strongylocentrotus nudus in vitro* (Ito et al. 1992; Gross et al. 1999a). Phagocytosis of erythrocytes by phagocytes is enhanced by opsonization with the coelomic fluid of *S. nudus*. Phagocytes during the stimulated state produce more hydrogen peroxide than resting phagocytes; however, hydrogen peroxide production by phagocytes is not affected by osponic activity of the coelomic fluid (Gross et al. 1999).

3.4. Inflammatory Process

The definition of an inflammatory process, according to Metchnikoff, is that of "cellular migration followed by phagocytosis" (Metchnikoff 1891). In invertebrates that lack a secondary immune response, the inflammatory process is essential in natural resistance against bacteria and potential pathogenic agents. The inflammatory process is a well-known process described in mammals and is the key mechanism in the resolution of bacterial diseases in tissues like integument and in the perivisceral coelom. Metchnikoff asserted that inflammation was the most important phenomenon in pathology with only one exception (Gross et al. 1999a; Silva et al. 1995, 1998) and proposed that the primary effectors of the immune response were circulating amoeboid-like phagocytes.

After the injection of yeast into the peristomial membrane, cellular migration of amebocytes followed by phagocytosis of yeast was demonstrated in the first report of an induced inflammatory process in the peristomial membrane of *L. variegatus* (Mangiaterra and Silva 2001). Other studies describe the presence of red spherule cells, in addition to amebocytes, during abnormal epithelial growth in *S. franciscanus* (Johnson and Chapman 1970), during development of the spotted gonad disease in *S. intermedius* (Shimuzu 1994), in the inflammatory-like reaction of individuals experimentally infested with the bald-sea-urchin disease (Maes and Jangoux 1984), and in the disease causing mass mortality of sea urchins (*S. droebachiensis*) in Nova Scotia (Jones et al. 1985). These findings, however, all involve infections associated with inflammatory processes in other tissues and for longer experimental periods. The absence of spherule cells in the induced inflammatory process suggests one of two possibilities: either that the amebocyte is the first coelomocyte cell type to migrate to the injury site during the inflammatory process perhaps followed by the red spherule cell in longer experimental time spans, or that in this tissue and with this noninfective stimulus, it is actually the only cell type to migrate. A follow-up to the present study, using longer time spans and different types of stimuli, might illuminate the roles of the various cell types.

In some studies, the introduction of foreign particles into the general coelom induces a considerable increase in the coelomocyte concentration in the perivisceral coelom (Bossche and Jangoux 1976; Mangiaterra and Silva 2001; Borges et al. 2005). In other studies this increase is not significant (Silva et al. 2001).

Beck and Habicht (1991) biochemically purified interleucin 1 (IL1) with a molecular weight similar to vertebrate IL1 suggesting a conservatism of IL1 among echinoderms. Later, Beck and Habicht (1993) isolated molecules of IL1 and demonstrated the phagocytosis inibition through the inactivation of this interleucin in the coelomic fluid with antibodies anti-IL1 from rabbits.

3.5. Origin of the Coelomocytes

The origin of coelomocytes in echinoderms is controversial (Smith 1981). Liebman (1950) suggests the peritoneum epithelial is the source of coelomocytes in *Arbacia punctulata*. Bosshe and Jangoux (1976) suggest a coelom epithelial origin for the coelomocytes of the starfish *A. rubens* but point out two other possible origins, the coelomopoietic organ and the mitotic division of coelomocytes. Nevertheless, there is no evidence for mitotic activity of coelomocytes (Chia and Xing 1996).

Holland et al. (1965) suggest different coelomocyte origin sites in *S. purpuratus*. In *A. rubens*, these cells come from the coelomic epithelium (Bosshe and Jangoux 1976; Maes and Jangoux 1984). Nevertheless, Brillouet et al. (1981), also studying *A. rubens*, suggest that the axial organ is able to produce progenitor cells and coelomocytes when stimulated with mitogens.

A significant increase of free PA was observed after ferritin irritant injection into the perivisceral coelom of *L. variegatus* (Mangiaterra and Silva 2001). There was no multiplication of these cells in the perivisceral coelom, but PA proliferation occurred in the axial organ independent of the irritant stimulus in the perivisceral coelom.

The origin and the storage site of PA are suggested to be the axial organ (Millott and Vevers 1968). The PA of this organ proliferates mainly at its periphery. Nevertheless, the proliferative activity in the axial organ associated with the morphological and histochemical characteristics of immature cells (presence of small intranuclear iron bodies) present in this organ demonstrate that the axial organ is one of, if not the only, organ that produces PA. Proliferative activity even in the control animals indicates a continuous production of these cells. The absence of a strong difference between the axial organ proliferation in the animals previously injected with irritant 6 and 24 h before the sacrifice confirms the recruitment of pre-existing PA instead of a proliferation response in the axial organ to increase the number of PA.

Studies in echinoderms describe the axial organ at the anal region, close to the madreporite. Nevertheless, these data are available almost only for holothurians and starfishes (Bossche and Jangoux 1976; Brillouet et al. 1981; Smith 1981; Leclerc et al., 1993; Chia and Xing 1996). The axial organ location in *L. variegatus* is either close to the Aristotle's lantern or in the oral region. The physiological implications of this anatomical difference needs to be investigated and may be related to the body structure diversity among the groups of echinoderm. The morphology of the axial organ and their channels described by Leake (1975) in different species of echinoids agrees with our data for *L. variegatus*. However, this study suggests only a secretory function for this organ.

The morphology of the different channels in the axial organ suggests an easier pathway of the PA through the central channel and the thinner ones, especially due to the smaller resistance of the squamous epithelium and to the lack of observed junctions. The presence of cilia in the prismatic and cubic epithelium of the main channel suggests a coelomocyte transport function of this channel to the digestive tube.

Despite the suggestion that free coelomocytes proliferate in sea urchins (Holland et al. 1965), some data confirm the lack of cell division in the perivisceral coelom (Bossche and Jangoux 1976; Chia and Xing 1996).

The presence of isolated and intranuclear iron bodies in the digestive tube contents indicates that this is an important excretion pathway for these cells. The connection of the axial organ with the digestive tube indicates two main functions for this organ: first, and mainly, the production of new PA that are released into the perivisceral coelom and second, their excretion.

3.6. Coelomic Fluid and Coelomocyte Concentration

Coelomic fluid fulfills many functions including translocation, excretion, locomotion, protection of the viscera, and humoral immunity (Chia and Xing 1996). Perivisceral coelomic fluid resembles seawater, but contains a higher potassium concentration, low quantities of lipids, proteins, sugars, and the coelomocytes (Smith 1981; Chia and Xing 1996). The structure and function of coelomocytes in different classes of echinoderms remain controversial because of many classifications and names ascribed to similar cells (Boolootian and Giese 1958; Smith 1981; Isaeva and Korembaum 1990; Edds 1993; Chia and Xing 1996; Gross et al. 1999).

The number of coelomocytes varies substantially between species and also among individuals according to the size and physiological conditions (Smith 1981; Laughlin 1989; Isaeva and Korembaum 1990; Edds 1993; Chia and Xing 1996; Gross et al. 1999). These variations between relative proportions of different coelomocyte types are described in only a few studies (Smith 1981). As these works usually do not mention methods or location of collection, season of the year, or animal size, it is quite difficult to make comparative analyses. Despite the importance of comparative studies of coelomocyte populations from different sea urchins species, these studies are insufficient for understanding the biology of coelomocytes of echinoderms.

Echinoderms possess an open circulatory system that includes the perivisceral coelomic system, water vascular system, perihemal system, and hemal system. The first coelom is the main one and can be divided into the oral region, near the peristomial membrane and the aboral region, close to madreporite (Smith 1981). Given the few data available regarding variation in coelomocyte populations of individuals from the same species, Borges et al. (2005) quantitatively and qualitatively analyzed the coelomocyte population of *L. variegatus* in the oral and aboral regions of the perivisceral coelom. The perivisceral coelom constitutes the main coelomic volume of the sea urchin without mechanical, physiological, or physical divisions as demonstrated by the radiological studies and Indian ink dispersion analyses (Borges et al. 2005).

Types of coelomocytes of *L. variegatus* are like those described for other species, but their proportions differ in oral and aboral regions, where PA predominate (Mangiaterra and Silva 2001). The significant variability in total coelomocyte number in different sea urchins has been described for other echinoderms (Johnson 1969a; Bertheussen and Seljelid 1978; Smith 1981; Isaeva and Korenbaum 1990; Edds 1993; Plytycz and Seljelid 1993; Chia and Xing 1996; Gross et al. 1999; Silva and Peck 2000; Silva et al. 2001; Borges et al. 2002), However, these studies have not distinguished coelomocytes in the oral and aboral regions of the perivisceral coelom.

That quantitative differences in the PA of the oral and aboral of sea urchins kept in the normal position (mouth down) did not change in individuals kept inverted (mouth up) for

24 h, clearly showed a different behavior for PA *in vivo* in the perivisceral coelom and free PA described *in vitro* in hanging drops by Johnson (1969a) where the coelomocytes were deposited on the menisci of the drops.

Circulation of the coelomic fluid in sea urchins was believed to be through the ciliated duct in the axial organ (Smith 1981). The circulation in the perivisceral coelom is believed to be achieved by movements of the animal together with contractions of some of the internal structures, like Aristotle's lantern and beating of the endothelial cilia. Rapid movement of the vibratile cells also assists in coelomic fluid circulation (Smith 1981). Study of the principal driving mechanism of perivisceral fluid circulation in *L. variegatus* showed that the ciliated epithelium is an ineffective driving mechanism for mixing of the perivisceral fluid. Instead, the Aristotle's lantern is the major driving force of circulation of the perivisceral fluid (Hanson and Gust 1986). Nevertheless, their relative importance and the actual volume moved have not been described.

The ability of PA to migrate toward a gradient and also their capacity to respond to different substances in different ways is suggested by many authors (Beck and Habichtl 1991; Beck and Habicht 1993; Gross et al. 1999; Mangiaterra and Silva 2001). Directed cell migration toward a specific stimulus (chemotaxis) is one of the central components of the immune and inflammatory process among metazoans. Cell migration in invertebrates can be modulated by bacterial components as lipopolisacarides (LPS) and phormil, methil, leucil phenilanine (fMLP), mammalian cytokines as interleukin-1, interleukin-2, and tumor necrosis factor (TNF), hormones such corticotropin and adrenocorticotrophic, and tumoral neuropeptides (Burke and Watkins 1991; Dureus et al. 1993; Fawcett and Tripp 1994; Ottaviani et al. 1995). This suggested chemotatic capacity of PA might be the explanation for differences found in *L. variegatus* (Borges et al. 2005), underlining the need for more studies.

Harrington and Ozaki (1986) described a 200 kDa glygoprotein secreted in the coelom of echinoid coelomocytes (*Dendraster excentricus* and *S. purpuratus*), important during the vitellogenesis.

The morphological difference of the granule content of coelomocytes were observed under transmission electron microscopy (TEM) (Smith 1981), also suggesting different molecular weight. The total general protein content higher in the aboral region may be due the higher percentage of red spherulocytes and vibratile cells in this region, since these cells had much more electron dense granules than the PA.

Our morphometric results of the coelomocytes present with *L. variegatus* in the perivisceral coelom agree with the literature (Chia and Xing 1996). There was no significant difference between the oral and aboral regions indicating the same morphometric patterns in both regions but with different cell proportions.

3.7. Coagulation and Encapsulation

Coagulation in sea urchins is a stereotyped response after contact between the coelomic fluid and the external environment (Isaeva and Korenbaum 1990). Classification of clots was revised by Endean (1966) and Isaeva and Korenbaum (1990), identifying three main groups (Boolootian and Giese 1958). One type, petaloid phagocytic amoebocyte, forms an aggregate that induces the filopodial form to become very compact, without periplasmatic fusion. Other coelomocytes do not seem to be active in this kind of formation. The second type results from the fusion of coelomocytes forming a syncytium. In the third type, explosive cells form a fibrilar net together with phagocytic amoebocytes resulting in a jelly structure. These explosive cells are similar in behavior to the spherulocytes that undergo morphological change and lysis.

When the foreign particle is larger than phagocytes themselves, the latter can accumulate around them, resulting in encapsulation (Chia and Xing 1996). Encapsulation may precede agglutination of bacteria, when morular cells degranulate and release acid phosphatase in the aggregated mass of bacteria. Possibly phagocytic amoebocytes are able to release lysosome enzymes in the brown bodies forming as the result of encapsulation (Isaeva and Korenbaum 1990).

3.8. The Complement System and Humoral Factors

The complement system in echinoderms functions as an opsonin (Smith and Davidson 1994) in *S. purpuratus*, having a homology with the complement system of vertebrates. The capacity of humoral factors to destroy or damage invading cells has been shown in many echinoids (Gross et al. 1999a), like hemolysin, agglutinins and hemaglutinins. Aglutinins are suggested to be involved in wound repair, opsonization, and encapsulation. Lectins are important to identify foreign cells thought opsonization and play an important role in lytic functions during clot formation and wound repair (Gross et al. 1999a).

The phenoloxidase system is an important system in many invertebrate groups (like arthropods) and has been identified in certain circulating coelomocytes and coelomic fluids of few echinoderms (Jans et al. 1996).

4. CONCLUSIONS

Disease can have important consequences on sea urchins in the field and in culture. The epidemiology of disease in the field will be difficult to study, as will its effects on individuals and populations there. It is likely the experience of disease with seeds in Japan will be repeated with the aquaculture of larger individuals. It is essential that investigations begin in anticipation of this. The immunological response to disease is little known. Its study will be of considerable interest in regard to sea urchins themselves and for evolutionary implications.

REFERENCES

Beck G, Habicht GS (1991) Purification and biochemical characterisation of an invertebrate interleukin 1. Mol Immunol 28: 577–584

Beck G, Habicht, GS (1993) Invertebrate cytoquines III: invertebrate interleukin-1-like molecules stimulate phagocytosis by tunicate and echinoderm cells. Cell Immunol 146: 284–299

Behrens MD, Lafferty KD (2004) Effects of marine reserves and urchin disease on southern California rocky reef communities. Mar Ecol Prog Ser 270: 129–139

Bertheussen K, Seljelid R (1978) Echinoid phagocytes *in vitro*. Exp Cell Res 111: 401–412

Boolootian RA, Giese AC (1958) Coelomic corpuscles of echinoderms. Biol Bull 115: 53–63

Borges JCS, Jensch-Junior BE, Garrido P, Mangiaterra MBBCD, Silva JRMC (2005) Phagocytic amoebocyte sub populations in the perivisceral coelom of the sea urchin *Lytechinus variegatus* (Lamarck, 1816). J Exp Zool 303A: 241–248

Borges JCS, Porto-Neto LR, Mangiaterra MBCD, Jensch-Junior BE, Silva JRMC (2002) Phagocytosis *in vivo* and *in vitro* in the Antarctic sea urchin *Sterechinus neumayeri* (Meissner) at 0 °C. Polar Biol 25: 891–897

Bossche JP, Jangoux M (1976) Epithelial origin of starfish coelomocytes. Nature 261: 227–228

Brightwell R, Tappel L (1968) Lysosomal acid phosphatases. Arch Biochem Biophys 124: 333–343

Brillouet C, Leclerc M, Panijel J, Binaghi R (1981) *In vitro* effect of various mitogens on starfish (*Asterias rubens*) axial organ cells. Cell Immunol 57: 136–144

Burke RD, Watkins FF (1991) Stimulation of starfish coelomocytes by interleukin-1. Biochemical and Biophysical Research Communications, Vol. 180(2): 579–584, 1991.

Burton MPM (1966) Echinoid coelomic cells. Nature 211: 1095–1096

Chia F, Xing J (1996) Echinoderm coelomocytes. Zool Stud 35: 231–254

Coffaro KA, Hinegardner RT (1977) Immune response in the sea urchin *Lytechinus pictus*. Science 197: 1389–1390

Cuénot L (1948) Anatomie éthologie, et systématique, des échinodermes. In: Grassé P (ed.) Traité de Zoologie, Échinodermes, Stomocordes, Procordes, Vol. XI. Masson et Cie, Paris, pp 3–363

Dureus P, Louis D, Grant AV, Bilfinger TV, Stefano GB (1993) Neuropeptide Y inhibits human and invertebrate immunocyte chemotaxis, chemokinesis, and spontaneous activation. Cell Mol Neurobiol 13: 541–546

Edds KT (1993) Cell biology of echinoid coelomocytes. I. Diversity and characterization of cell types. J Invert Pathol 61: 173–178

Endean R (1966) The coelomocytes and coelomic fluids. In: Physiology of Echinodermata. Interscience, New York

Fawcett LB, Tripp MR (1994) Chemotaxis of *Mercenaria mercenaria* hemocytes to bacteria *in vitro*. J Invert Pathol 63: 275–284

Gross PS, Al-Sharif WZ, Clow LA, Smith LC (1999) Echinoderm immunity and evolution of the complement system. Dev Comp Immunol 23: 429–442

Hamaguchi M, Kawahara 1, Usuki H (1993) Mass mortality of *Pseudocentrotus depressus* caused by a bacterial infection in summer. Suisanzoshoku 41: 189–193 (in Japanese)

Hansen GH, Bergh O, Michaelsen J, Knappskog D (1992) *Flexibacter ovolyticus* sp. nov., a pathogen of eggs and larvae of Atlantic halibut, *Hippoglossus hippoglossus* L. Int J Syst Bacteriol 42: 451–458

Hanson JL, Gust G (1986) Circulation of perivisceral fluid in the sea urchin Lytechinus variegatus. Mar Biol 92: 125–134.

Harrington FE, Ozaki H (1986) The major yolk glycoprotein precursor in echinoids is secreted by coelomocytes into the coelomic plasma. Cell Differ 19: 51–57

Hobaus E (1978) Studies on phagocytes of regular sea urchins (Echinoidea, Echinodermata). 1. The occurrence of iron containing bodies within the nuclei of phagocytes. Zool Anz 200: 31–40

Holland ND, Phillips JH, Giese AC (1965) An autoradiographic investigation of coelomocyte production in the purple sea urchin (*Strongylocentrotus purpuratus*). Biol Bull 128: 259–270

Holt JO, Krieg NR, Sneath PHA, Staley JT, Williams ST (1994) Bergey's manual of determinative bacteriology, 9th edn. Williams and Wilkins Co., Baltimore, p 787

Isaeva VV, Korenbaum ES (1990) Defense functions of coelomocytes and immunity of echinoderms. Sov Mar Biol 15: 353–363

Ito T, Matsutani T, Mori K, Normurat T (1992) Phagocytosis and hydrogen peroxide production by phagocytes of the sea urchin *Strongylocentrotus nudus*. Dev Comp Immunol 16: 287–294

Jangoux M (1987a) Diseases of Echinodermata. I. Agents microorganisms and protistans. Dis Aquat Org 2: 147–162

Jangoux M (1987b) Diseases of Echinodermata. II. Agents metazoans (Mesozoa to Bryozoa). Dis Aquat Org 2: 205–234

Jangoux M (1987c) Diseases of Echinodermata. III. Agents metazoans (Annelida to Pisces). Dis Aquat Org 3: 59–83

Jangoux M (1987d) Diseases of Echinodermata. IV. Structural abnormalities and general considerations on biotic diseases. Dis Aquat Org 3: 221–229

Jans, D, Dubois P, Jangoux M (1996) Defensive mechanism of holothuroids (Echinodermata): formation, role, and fate of intracoelomic brown bodies in the sea cucumber *Holothuria tubulosa*. Cell Tissue Res 283: 99–106

Johnson PT (1969a) The coelomic elements of sea urchins (*Strongylocentrotus*). I. The normal coelomocytes, their morphology and dynamics in hanging drop. J Invert Pathol 13: 25–41

Johnson PT (1969b) The coelomic elements of sea urchins (*Strongylocentrotus*). II. Cytochemistry of the coelomocytes. Histochemie 17: 213–231

Johnson PT (1969c) The coelomic elements of sea urchins (*Strongylocentrotus*). III. *In vitro* reaction to bacteria. J Invert Pathol 13: 42–62

Johnson PT, Beeson RJ (1966) *In vitro* studies on *Patiria miniata* (Brandt) coelomocytes, with remarks on revolving cysts. Life Sci 5: 1641–1666

Johnson PT, Chapman FA (1970) Abnormal epithelial growth in sea urchin spines (*Strongylocentrotus franciscanus*). J Invert Pathol 16: 116–122

Jones GM, Hebda AJ, Scheibling RE, Miller RJ (1985) Histopathology of the disease causing mass mortality of sea urchin (*Strongylocentrotus droebachiensis*) in Nova Scotia. J Invert Pathol 45: 260–271

Kanai K (1993) "Togenukesho" of sea urchins. Proceedings: symposium on disease on fish and shellfish culture in Kyusyu and Okinawa. The Japanese Society of Fish Pathology, Tokyo, p 7

Kawakami T, Tsushima M, Katabami Y, Mine M, Ishida A, Matsuno T (1998) Effect of ββ-carotene, β-echinenone, astaxanthin, fucoxanthin, vitamin A and vitamin E on the biological defense of the sea urchin *Pseudocentrotus depressus*. J Exp Mar Biol Ecol 226: 165–174

Krieg NR, Holt JO (1984) Bergey's manual of systematic bacteriology, Vol. 1. Williams and Wilkins Co., Baltimore

Lafferty KD (2004) Fishing for lobsters indirectly increases epidemics in sea urchins. Ecol Appl 14: 1566–1573

Lawrence JM (1996) Mass mortality of echinoderms from abiotic factors. Echino Stud 5: 103–137

Leake LD (1975) Comparative histology. Academic Press, London

Leclerc M, Bajelan M, Barot R, Tlaskalova-Hogenova, H (1993) Effect of silica on the spontaneous cytotoxity of axial organ cells from *Asterias rubens*. Cell Biol Internat 17: 787–789

Liebman E (1950) The leucocytes of *Arbacia punctulata*. Biol Bull 98: 46–59

Maes P, Jangoux M (1984) The bald-sea-urchin disease: a biopathological approach. Helgol Meeresunters 37: 217–224

Mangiaterra MBBCD, Silva JRMC (2001) Induced inflammatory process in the sea urchin (*Lytechinus variegatus*). J Invert Biol 120: 178–184

Masuda Y, Tajima K, Ezura Y (2004) Resuscitation of *Tenacibaculum* sp., the causative bacterium of spotting disease of sea urchin *Strongylocentrotus intermedius*, from the viable but non-culturable state. Fisheries Sci 70: 277–284

Metchnikoff E (1891) Lectures on the comparative pathology of inflammation. Dover, New York (republished in 1968)

Metchnikoff E (1905) Immunity in infective diseases. Dover, New York (republished in 1968)

Millott N (1950) Integumentary pigmentation and the coelomic fluid of *Thyone briareus* (Lesueur). Biol Bull 99: 343–344

Millott N, Vevers HG (1968) The morphology and histochemistry of the echinoid axial organ. Phil Trans R Soc Lond 253B: 201–230

Nagelkerken I, Smith GW, Snelder E, Karel M, James S (1999) Sea urchin *Meoma ventricosa* die-off in Curaçao (Netherlands Antilles) associated with a pathogenic bacterium. Dis Aquat Org 38: 71–74

Ottaviani E, Franchini AC, Cassanelli S, Genedani S (1995) Cytokines and invertebrate immune responses. Biol Cell 85Ç: 87–91

Plytycz B, Seljelid R (1993) Bacterial clearance by the sea urchin *Strongylocentrotus droebachiensis*. Dev Comp Immunol 17: 283–289

Radwańska U, Radwański A (2005) Myzostomid and copepod infestation of Jurassic echinoderms: a general approach, some new occurrences, and/or re-interpretation of previous reports. Acta Geol Polonica 55: 109–130

Roberts-Regan DL, Scheibling RE, Jellett JF (1988) Natural and experimentally induced lesions of the body wall of the sea urchin *Strongylocentrotus droebachiensis*. Dis Aquat Org 5: 51–62

Sawabe T, Oda Y, Shiomi Y, Ezura Y (1995) Alginate degradation by bacteria isolated from the gut of sea urchins and abalones. Microb Ecol 30: 193–202

Schwammer HM (1989) Bald-sea-urchin disease: record of incidence in irregular echinoids – *Spatangus purpureus*, from the SW-coast of Krk (Croatia – Jugoslavia). Zool Anz 223: 100–106

Shimuzu M (1994) Histopathological investigation of the spotted gonad disease in the sea urchin *Strongylocentrotus intermedius*. J Invert Pathol 63: 182–187

Shimizu M, Nagakura K (1993) Acid phosphatase activity in the body wall of the sea urchin, *Strongylocentrotus intermedius*, cultured at varying water temperature. Comp Biochem Physiol 106B: 303–307

Shimizu M, Takaya Y, Ohsaki S, Kawamata K (1995) Gross and histopathological signs of the spotting disease in the sea urchin *Strongylocentrotus intermedius*. Fisheries Sci 61: 608–613

Silva JRMC (2000) The onset of phagocytosis and identity in the embryo of *Lytechinus variegatus*. J Dev Comp Immunol 24: 733–739

Silva JRMC (2001) The role of the phagocytes on embryos some morphological aspects. Microsc Res Tech 57: 498–506

Silva JRMC, Hernandez-Blazquez FJ, Porto-Neto LR, Borges JCS (2001) Comparative study of *in vivo* and *in vitro* phagocytosis including germicide capacity in *Odontaster validus* (Koehler, 1906) at 0 °C. J Invert Path 77: 180–185

Silva JRMC, Mendes EG, Mariano M (1995) Wound repair in the amphioxus (*Branchiostoma platae*), an animal deprived of inflammatory phagocytes. J Invert Pathol 65: 147–151

Silva JRMC, Mendes EG, Mariano M (1998) Regeneration in the amphioxus (*Branchiostoma platae*). Zool Anz 237: 107–111

Silva JRMC, Peck L (2000) Induced *in vitro* phagocytosis of the antarctic starfish *Odontaster validus* (Koehler, 1906) at 0 °C. Polar Biol 23: 225–230

Smith LC, Davidson EH (1994) The echinoderm immune system: characters shared with vertebrate immune system and characters arising later in deuterostome phylogeny, Ann NY Acad Sci 712: 213–226

Smith VJ (1981) The echinoderms. In: Ratcliffe NA, Rowley AF (eds) Invertebrate blood cells, Vol. 2. Academic Press, London, pp 513–562

Sonnenholzner J, Lawrence JM (1998) Disease and predation in *Encope micropora* (Echinoidea: Clypeasteroida) at Playas, Ecuador. In: Mooi R, Telford M (eds) Echinoderms: San Francisco. Balkema, Rotterdam, pp 829–833

Suzuki M, Nakagawa Y, Yamamoto S (2001) Phylogenetic analysis and taxonomic study marine *Cytophaga*-like bacteria: proposal for *Tenacibaculum* gen. nov. with *Tenacibaculum maritimum* comb. nov., and *Tenacibaculum ovolyticum* gen. nov., and description of *Tenacibaculum mesophilum* sp. nov. and *Tenacibaculum amylolyticum* sp. nov. Int J Syst Evol Bacteriol 51: 1639–1652

Tajima K, Hirano T, Nakano K, Ezura Y (1997a) Taxonomical study on the causative bacterium of spotting disease of sea urchin *Strongylocentrotus intermedius*. Fisheries Sci 63: 897–900

Tajima K, Hirano T, Shimizu M, Ezura Y (1997b) Isolation and pathogenicity of the causative bacterium of spotting disease of sea urchin *Strongylocentrotus intermedius*. Fisheries Sci 63: 249–252

Tajima K, Hirano H, Fujimoto S, Itoh S, Ezura Y (1998a) Control methods for spotting disease of sea urchin *Strongylocentrotus intermedius*. Nippon Suisan Gakkaishi 64: 65–68 (in Japanese)

Tajima K, Takeuchi K, Nakano, K, Shimizu M, Ezura Y (1998b) Studies on a bacterial disease of sea urchin *Strongylocentrotus intermedius* occurring at low water temperatures. Fisheries Sci 64: 918–920

Tajima K, Takeuchi K, Takahata M, Hasegawa M, Watanabe S, Eqbal MM, Ezura Y (2000) Seasonal occurrence of the pathogenic *Vibrio* sp. of the disease of sea urchin *Strongylocentrotus intermedius* occurring at low water temperatures and the prevention methods of the disease. Nippon Suisan Gakkaishi 66: 799–804 (in Japanese)

Takeuchi K, Tajima K, Iqbal MM, Sawabe T, Ezura Y (1999) Studies on the taxonomy and serology of the causative bacteria of the disease of sea urchin *Strongylocentrotus intermedius* occurring at low water temperatures. Fisheries Sci 65: 264–268

Tauber AI and Chernyak L (1991) Metchnikoff and the Origins of Immunology, from Metaphor to Theory. New York: Oxford University Press

Vevers HG (1967). The histochemistry of the echinoid axial organ. Symp Zool Soc Lond 20: 65–74

Xing J, Leung MF, Chia FS (1998) Quantitative analysis of phagocytosis by amebocytes of a sea cucumber, *Holothuria leucospilota*. Invert Biol 117: 67–74

Yamase T, Sawabe T, Kuma K, Tajima K (2006) Effect of iron on resuscitation of *Tenacibaculum* sp., the causative bacterium of spotting disease of short-spined sea urchin *Strongylocentrotus intermedius*, from the viable but non-culturable (VBNC) state. Fish Pathol 41: 1–6 (in Japanese)

Colour Plates

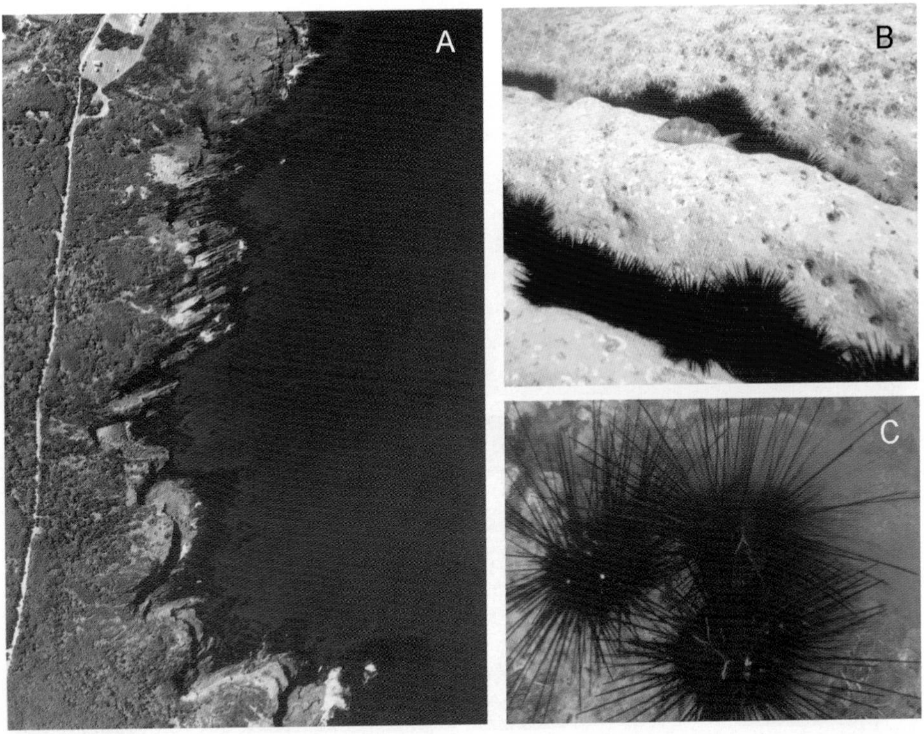

Plate 1. (A) Detail of a 1:8000 photograph of Green Cape, New South Wales. The site shown is approximately Disaster Bay Site 1 from Andrew and O'Neil (2000). The darker margin of the reef is Fringe habitat and expansive areas of pale gray are Barrens habitat (photo: N. Andrew). (B) *Centrostephanus rodgersii* at Jervis Bay Marine Park, NSW, Australia (photo: T Lynch). (C) *Diadema setosum* (left) and two *Diadema savignyi* (spawning) (right) (photo: T McClanahan).

Plate 2. (A) *Loxechinus albus* (photo: J Vásquez). (B) *Hemicentrotus pulcherrimus* (photo: Y Agatsuma). (C) *Pseudocentrotus depressus* (photo: Y Agatsuma). (D) *Strongylocentrotus intermedius* (photo: Y Agatsuma).

Plate 3. (A) *Evechinus chloroticus* (photo: M Barker). (B) *Psammechinus miliaris* (photo: M Kelly). (C) *Strongylocentrotus franciscanus* (adult and juvenile) (photo: L Rogers-Bennett). (D) *Strongylocentrotus nudus* (photo: Y Agatsuma).

Plate 4. (A) *Paracentrotus lividus*, reproductive aggregation in bed of *Posidonia oceanica* at Port Cros, France (photo: C-F Boudouresque). (B) *Strongylocentrotus droebachiensis*, grazing front consuming kelp (*Laminaria digitata* and *Laminaria longicruris*) at 10 m depth, Ketch Harbour, Nova Scotia, Canada (photo: R Scheibling). (C) *Echinometra mathaei* (photo: T McClanahan).

Plate 5. (A) *Tripneustes gratilla* (photo: Y Hiratsuka). (B) Small *Tripneustes gratilla* on *Sargassum* in culture at the Okinawa Prefectural Sea Farming Center (photo: Y Hiratsuka). (C) Aerial view of barren grounds created by moving fronts (from left to right) of *Lytechinus variegatus* in beds of the seagrass *Thalassia testudinum* at Steinhatchee, Florida in the Gulf of Mexico, 1971 (photo: D Camp). (D) *Lytechinus variegatus* in beds of the seagrass *Syringodium fililforme* in Florida Bay (photo: D Lirman).

Edible Sea Urchins: Biology and Ecology
Editor: John Miller Lawrence

Chapter 10

Ecology of *Centrostephanus*

Neil L Andrew[a] and Maria Byrne[b]

[a]*The WorldFish Center, Penang (Malaysia)*
[b]*Department of Anatomy and Histology, University of Sydney, Sydney, NSW (Australia)*

1. BIOGEOGRAPHY

Sea urchins in the family Diadematoida, particularly those in the genus *Diadema*, are among the most ecologically important echinoids in tropical regions. Less familiar are the temperate and subtropical diadematoids in the genus *Centrostephanus*. There are 10 nominal species of *Centrostephanus* (Pawson and Miller 1983). The three best studied are: *C. rodgersii* (A Agassizi) (Plate 1B) in southeastern Australia, *C. coronatus* (Verrill) off the California coast and *C. longispinus* (Philippi) in the European Mediterranean. Of these three species *C. rodgersii* attains the highest local abundance and is the only one fished (Pawson and Miller 1983; Francour 1991; Andrew 1993, 1994; Andrew et al. 2002).

Centrostephanus rodgersii is found on subtidal rocky reefs in southeastern Australia, northern New Zealand (Choat and Schiel 1982) and the Kermadec Islands (Schiel et al. 1986; Cole et al. 1992). In Australia, its distribution is centred on New South Wales where it is most abundant between 2 and 20 m (Underwood et al. 1991). *Centrostephanus rodgersii* is also found in eastern Victoria, on the eastern Bass Strait islands and along the east coast of Tasmania (Johnson et al. 2005). At its northern limit in Australia, *C. rodgersii* co-occurs with hard corals at the Solitary Islands in northern New South Wales. At its southern limit, along the Tasman Peninsula in Tasmania it is found with cold-water algae such as *Macrocystis pyrifera* and *Durvillaea potatorum*. There has been an increase in the populations of *C. rodgersii* in Tasmania since the 1980s. Prior to 1978, the species was recorded only from the islands in eastern Bass Strait and northeastern Tasmania but it now extends down most of the eastern coast (Johnson et al. 2005). The reasons for this range extension are probably due to changes in the strength of the East Australian Current and associated warming of waters (Johnson et al. 2005). This range extension may only be an episode in a history of fluxes in the range of *C. rodgersii*.

Centrostephanus coronatus is known from southern California and offshore islands where it is common on subtidal rocky reefs at 3–10 m depth (Nelson and Vance 1979; Vance 1979; Hartney and Grorud 2002) and is a rare species at the Galapagos Islands (Lawrence and Sonnenholzner 2004). *Centrostephanus longispinus* occurs in rocky habitats and *Posidonia* beds in the Mediterranean, usually at low densities ($<0.5\,m^2$) and most commonly at 20–30 m depth (Francour 1991). This species also occurs in the Eastern Atlantic from Morroco to Gabon (Pawson and Miller 1983). Increases in the local abundance of *C. longispinus* may be associated with warming of Mediterranean waters (Francour et al. 1994).

2. ECOLOGICAL IMPACTS

2.1. Habitat Structure

As is characteristic of many diadematoids, *Centrostephanus* species are highly light sensitive, often foraging at night and remaining hidden in rocky crevices and holes during the day (Kennedy and Pearse 1975; Nelson and Vance 1979; Lissner 1980; Andrew 1993, 1994). Where they exhibit this behaviour, distinct 'halos' of crustose coralline algae form around their shelters (Nelson and Vance 1979; Andrew 1993, 1994). Individuals may show shelter fidelity with individuals returning to the same shelter in 8 months of observations (Nelson and Vance 1979). *Centrostephanus* species exhibit day–night differences in colour, being dark black during the day and changing to a paler colour at night (Pearse 1972; Francour 1991). This is due to contraction of skin pigment cells (chromatophores) at dusk and extension of these cells at dawn (Pearse 1972; Gras and Weber 1983). When they contract at dusk these cells are likely to uncover sensory cells involved in light sensitivity (Gras and Weber 1983).

Centrostephanus rodgersii is locally abundant and may form dense aggregations. On reefs in southern New South Wales, densities of more than 60 individulas per square metre have been recorded. At these extreme densities, the sea urchins occur as aggregations of interwoven spines and may be observed outside shelters during the day. The substratum beneath such aggregations is often grazed to the underlying rock. More usually, *C. rodgersii* is hidden in shelters during the day and mean densities range between 0.5 and 6 individulas per square metre (Andrew and Underwood 1989). No other *Centrostephanus species* are reported to reach the densities observed in Australia for C. rodgersii.

In New South Wales, *C. rodgersii* is the dominant herbivore on shallow rocky reefs. On the margins of these reefs in the shallowest water (ca. 2 m depth), large brown algae are abundant (Plate 1A), particularly *Phyllospora comosa* and *Sargassum* spp. – forming the fringe habitat described by Underwood et al. (1991; see also Andrew 1999). Within the fringe, *C. rodgersii* may be abundant and maintain patches free of large brown algae, but more often occurs in crevices and depressions. *Centrostephanus rodgersii* forms extensive barrens habitat that starts at approximately 5 m depth. The factors that determine the location and depth of the transition between the fringe and barrens habitat types have not been studied. We hypothesise that the depth at which the boundary occurs is related to wave exposure and the degree of topographic complexity on the reef. A sharp boundary in the distribution of *C. coronatus* is influenced by wave exposure (Lissner 1980, 1983). At depths greater than approximately 20 m (depending on wave exposure), *C. rodgersii* becomes less dense and individuals are found in depressions and crevices amongst sponges and other sessile animals that are increasingly abundant at those depths. Extensive barrens formed by *C. rodgersii* in Tasmania occur at 20–30 m depth on boulder and flat rock habitats, much deeper than in New South Wales (Johnson et al. 2005). The reason for this difference in depth distribution is not known, but it may be influenced by the sweeping action of algal fronds in shallow water (Johnson et al. 2005).

Centrostephanus rodgersii barrens habitat varies in size from the small 'halos' surrounding single sea urchins to large areas of many hectares (Fletcher 1987; Andrew and Underwood 1989, 1992, 1993; Underwood et al. 1991; Andrew 1993; Andrew et al. 1998). This habitat-type shares feature in common with barrens in other parts of the world, notably the absence of large brown algae, high though variable densities of sea urchins and a different assemblage of algae, invertebrates and fish (either in the sizes of individuals or species) from those in neighbouring areas of reef with a large biomass of algae (Underwood et al. 1991 and Holbrook et al. 1994; Gillanders and Kingsford 1998). Areas of barrens are found throughout the range of *C. rodgersii* and particularly in central and southern New South Wales where slightly more than 50% of near-shore reef (less than 150 m from shore) is barrens (Fig. 1A) (Andrew and O'Neill 2000). Towards the limits of the species distribution, in the north of New South Wales barrens are less common. At the southern end of its range in Tasmania, the presence of *C. rodgersii* and barrens habitat is influenced by larval supply (Johnson et al. 2005). In northern New Zealand and the Kermadec Islands, *C. rodgersii* occurs as scattered individuals and has not been observed to form barrens (Choat and Schiel 1982; Schiel et al. 1986; Cole et al. 1992).

Although the density of *C. rodgersii* within barrens may vary widely, its ecological impact is similar because of the high vulnerability of small kelp plants to grazing. The density of limpets and smaller algae show more subtle differences across the range of sea urchin densities (Fletcher 1987; Andrew and Underwood 1989, 1993). When the density of *C. rodgersii* is experimentally reduced, the cover of crustose corallines declines as filamentous and turfing algae increase but, even when densities are reduced to a third of original densities, foliose algae do not successfully recruit (Andrew and Underwood 1993; Hill et al. 2003). The impact of grazing by *C. rodgersii* is not proportional to sea urchin density in either the fringe or barrens habitats indicating that these habitat types will remain stable unless there is a dramatic change in urchin density (Andrew and Underwood 1993; Andrew 1994; Hill et al. 2003). The foraging effort of *C. rodgersii* reduced in response to increased population density in manipulative experiments (Hill et al. 2003).

The most abundant algae in areas grazed by *Centrostephanus* species are crustose and turf coralline algae and these algae dominate barrens habitat (Nelson and Vance 1979; Vance 1979; Fletcher 1987; Andrew and Underwood 1989; Francour 1991). In New South Wales several species of limpet graze the films of diatoms and small ephemeral algae growing on the surface of the corallines (Fletcher 1987). Limpets depend on grazing by *C. rodgersii* to keep the substratum free of larger turf-forming and foliose algae. When sea urchins are removed, turf-forming and large brown algae grow in profusion and limpets slowly disappear (Fletcher 1987; Andrew and Underwood 1993). Sessile animals, such as sponges and ascidians, occupy space in barrens areas only where sea urchins do not graze heavily and are often common on rock walls in association with the large barnacle *Austrobalanus imperator*.

Centrostephanus species are also habitat providers. In California, *C. coronatus* significantly effects recruitment, migration and survival of gobies (*Lythrypnus dalli*) through provision of habitat structure and refuge from predation (Hartney and Grorud 2002). Similarly, *C. rodgersii* is associated with a suite of benthic and mobile invertebrates (Worthington and Blount 2003). *Clarkoma pulcra*, an ophiuroid similar in colour to *C. rodgersii*, is found under this urchin during the day and emerges from its urchin cover at night (M Byrne, pers. obs.).

2.2. Food and Feeding Ecology

The diet of *Centrostephanus* species includes algae, seagrass, tunicates and encrusting invertebrates such as bryozoans and sponges (Vance 1979; Francour 1991; Andrew 1993, 1994; Hill et al. 2003). *Centrostephanus coronatus* exploits a wide variety of food with three plant and eight animal taxa most common in gut contents (Vance 1979).

In areas where *Centrostephanus* species shelter in crevices, the urchins become active after dusk and forage through the night before returning to the same crevice several hours before dawn (Fig. 1). Individual *C. rodgersii* can move up to 10 m from their crevices during excursions, whereas *C. coronatus* forages within a meter of its shelter (Nelson and Vance 1979; Jones and Andrew 1990). This foraging behaviour can cause sharp boundaries between barrens and kelp habitat in New South Wales (Andrew 1994). Although in manipulative experiments grazing by different densities of *C. rodgersii* resulted in different algal composition (Hill et al. 2003), this urchin appears to play a relatively small role in determining the distribution and abundance of large brown algae outside the well-defined patches surrounding their shelters (Andrew 1993, 1994). The experimental provision of shelters within kelp forests led to the creation of new

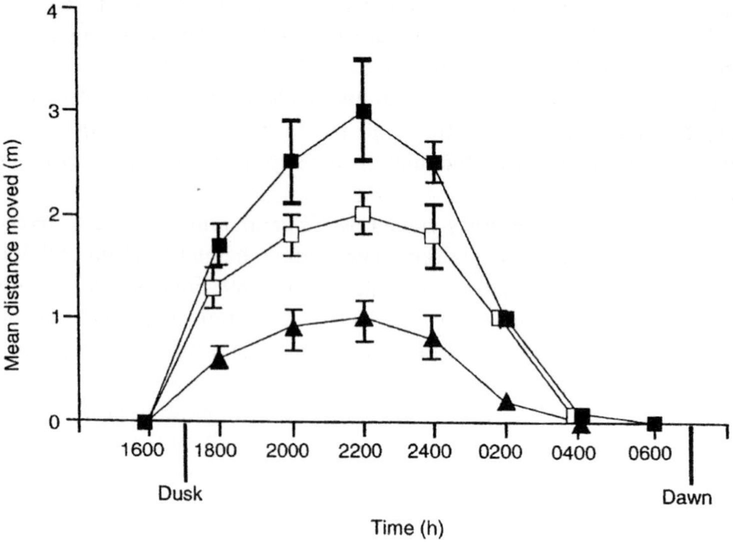

Fig. 1. Mean distance (±SE, $n = 5$) moved by individual *C. rodgersii* from their crevices as a function of the time of night. Curves represent different patches of barrens habitat. Redrawn from Jones and Andrew (1990).

patches of barrens, suggesting that the availability of shelter limits the extent of barrens (Andrew 1993).

Nothing is known of the foraging behaviour of *C. rodgersii* in locations where they are not found in shelters during the day. There is nothing in patterns of aggregation observed during daylight to suggest they form grazing fronts as several strongylocentrotid species do in North America (Chapters 18 and 19) or *Evechinus chloroticus* does in New Zealand (Chapter 16). Large algae have some refuge in size because *C. rodgersii*, unlike strongylocentrotids, does not appear capable of climbing onto large plants.

3. POPULATION REGULATION

3.1. Recruitment

Processes influencing larval settlement of *C. rodgersii* are poorly understood. The supply of larvae is likely to be strongly influenced by the dominant East Australian Current that sweeps southward along the coast. This current is relatively predictable as far south as Seal Rocks (latitude ~32°S), after which its path becomes more erratic and may break away from the coast. As the current moves offshore it fractures into eddies, sometimes returning to the coast and at other times spinning off into the Tasman Sea or continuing down the coast. Sea conditions in southern New South Wales are influenced more by winds, swells generated by distant storms, tidal currents and river outflows. The East Australian Current, its eddies and associated counter currents have a major influence on water temperature and probably the transport of larvae along the coast. The expansion of *C. rodgersii* in Tasmania is thought to be caused by incursion of larvae from New South Wales in the East Australian Current across Bass Strait (Johnson et al. 2005). This is supported by the lack of genetic differentiation among populations of *C. rodgersii* in New South Wales and Tasmania (Johnson et al. 2005).

Recruitment in *C. rodgersii* varies in space and time. For example, at sites in Sydney, sea urchins <40 mm in test diameter were consistently observed during the mid-1980s (Andrew and Underwood 1989) but were absent in 1997. At other sites, such as Port Stephens, north of Sydney few sea urchins in their first year of life have been observed. Most populations appear dominated by larger and older sea urchins (Andrew and Underwood 1989; Andrew and O'Neill 2000). These population structures suggest sporadic recruitment. The need of small individuals to find larger crevices as they increase in size may lead to a demographic bottleneck. Care is required in interpreting population structure, however, as juveniles are cryptic, and may be under-represented in size-frequency distributions.

Small *C. rodgersii* are found almost exclusively under small boulders and in crevices (Andrew and Underwood 1989). Although no evidence indicates an association between juveniles and the spine canopy of conspecific adults, juveniles may be found in the backs of crevices and under the same boulders as larger sea urchins. The greatest recorded recruitment of *C. rodgersii* was in Botany Bay, Sydney and followed a mass mortality of conspecific adults (Andrew 1991). Under experimental conditions, *C. rodgersii* recruits to boulders irrespective of habitat, whether these boulders are covered by kelp or whether conspecific adults are present (Andrew 1993). Recruitment in *C. rodgersii* may increase when adult densities are reduced (Worthington and Blount 2003).

3.2. Predation and Disease

The nocturnal foraging habits of *C. coronatus* are suggested to be a predator avoidance mechanism in response to the sheephead wrasse (*Pimelometopon pulchrum*), a day active fish that preys on urchins and other benthic organisms (Nelson and Vance 1979). The heterodontid shark *Heterodontus francisci* may also prey on *C. coronatus* (Pearse and Timm 1971). Andrew and Underwood (1989) and Andrew (1993) suggested that day sheltering behaviour of *C. rodgersii* may be a predator-avoidance behaviour, although this has not been directly studied.

Few studies directly assess the relative importance of predation in regulating populations of *C. rodgersii.* Although many species, notably the wrasse *Achoerodus viridis* (Gillanders 1995) and *Heterodontus portusjacksoni* (McLaughlin and O'Gower 1971), consume *C. rodgersii*, the impact of their predation is unknown. In addition to these large predators, numerous species of smaller wrasses, octopus and a suite of microcarnivores such as amphipods occur on rocky reefs in New South Wales: nothing is known of their impact on *C. rodgersii*.

The presence of expansive barrens areas in southeastern Australia may be due to a reduction in abundance, or lack, of sea urchin predators (Johnson et al. 2005; Wright et al. 2005). It could be argued that present-day populations of predators, particularly the rock lobster *Jasus verreaux*, *Achoerodus viridis* and *Heterodontus portusjacksoni*, are greatly diminished from fishing and that they historically played a greater role in limiting sea urchin numbers. *Achoerodus viridis* has been partially protected from fishing in New South Wales since 1969 and, since the late 1970s, has been protected from all fishing except recreational line fishing (it is not susceptible to being taken by line). *Heterodontus portusjacksoni* is not protected but, although taken as a by-catch in prawn trawl fisheries, it is not targeted by commercial or recreational fishers on rocky reefs. *Jasus verreauxi* occurs on rocky reefs in southern New South Wales and Tasmania but are heavily fished and usually small. Little is known of the interaction between rock lobsters and sea urchins in New South Wales. In Tasmania legal sized rock lobsters are reported to be important predators of *C. rodgersii* (Johnson et al. 2005). None of these potential predators of *C. rodgersii* are monitored in marine reserves in New South Wales so there remains no basis for inferences about their role in the ecology of near-shore rocky reefs, particularly their role in regulating sea urchin populations. Both rock lobsters and heterodontid sharks are nocturnal foragers and therefore potentially have access to a wider size range of sea urchins than do the diurnally active fishes.

When provided with shelter in the form of large boulders in a kelp forest, adult *C. rodgersii* recruited in sufficient numbers to create small patches of barrens (Andrew 1993). The inference was that *C. rodgersii* foraged out from shelters at night and returned before predators such as *A. viridus* became active. This constraint on the behaviour of large *C. rodgersii* may be important in determining the extent and persistence of patches of barrens.

It is now evident that shelter dependence by large *Centrostephanus* is not obligate. In the south of New South Wales and Tasmania, sea urchins are often found in the open during the day, particularly at sites where they are at high densities (Andrew 1999; Andrew and O'Neill 2000; Johnson et al. 2005). The ecological significance of these differences in daytime shelter dependence remains largely unexplored. Andrew and Underwood (1989) and Andrew (1993) concluded that shelter dependence confers a greater degree of stability and predictability to the location of barrens habitat, and contrasted these patterns with patterns of dispersion and behaviour of strongylocentrotid urchins. It is now apparent that results from studies in Hawkesbury sandstone areas where crevice dwelling is common cannot be generalised across the geographical range of *C. rodgersii*. There is no evidence that the densities of predators differ along the New South Wales coast (e.g. Gillanders and Kingsford 1998). Similarly, it remains an open question whether sites with sea urchins that do not show shelter dependence are less predictable in space and time. Understanding the processes that constrain foraging behaviour of *C. rodgersii* remains an important area of research (Jones and Andrew 1990).

Whereas massive mortalities of strongylocentrotid urchins from disease have led to major community changes, e.g. in Nova Scotia (Chapter 18), there are no reported instances of mortality causing disease in *Centrostephanus* species. Parasitic nematodes are common in the gonads of *C. coronatus* (40–85% infection, $n = 153$) where they interfere with gameteogenesis or partially castrate the host (Pearse and Timm 1971).

3.3. Competition

The abalone *Haliotis rubra* is rare in barrens habitat and, given the predominance of this habitat in central and southern New South Wales, it is reasonable to conclude that *C. rodgersii* has a large negative influence on the abundance of abalone (Shepherd 1973; Andrew 1993; Andrew and Underwood 1989; Andrew et al. 1998; Andrew and O'Neill 2000; Johnson et al. 2005). Andrew et al. (1998) removed sea urchins from areas of reef in southern New South Wales and reported that densities of *H. rubra* rose from near zero to a mean of >1 indivisulas per square metre over the following 3.5 years, in the absence of *C. rodgersii*. The ecological cause of the absence of abalone in barrens remains unknown. Possible hypotheses include competition for food (Shepherd 1973), incidental mortality of juvenile abalone by grazing *C. rodgersii* (McShane 1991), and loss to fishes and other predators caused by the reduced structural complexity of the habitat. No evidence separates these nonexclusive hypotheses. For example, competition for food may cause large abalone to move away from areas of barrens but shelter is more important earlier in life. Andrew's (1993) results are consistent with a size-specific relationship between *C. rodgersii* and *H. rubra*. When shelter was provided in patches of barrens, densities of abalone were initially higher in the barrens than in *Ecklonia* forest but, after 2.5 years, few abalone remained in the barrens. As Andrew (1993) did not separate *H. rubra* from the smaller abalone, *H. coccoradiata,* the results cannot unambiguously separate the above hypotheses.

Centrostephanus rodgersii and *H. rubra* co-occur in the algal fringe, and mean densities of sea urchins may be relatively large (Andrew and Underwood 1989). However, densities of the two species are negatively correlated at the spatial scale of $10\,m^2$ and also on a smaller, nearest-neighbour scale (Andrew and Underwood 1992). Although strong evidence indicates that *C. rodgersii* has a negative impact on the dispersion and local abundance of *H. rubra*, none indicates that the reverse is true.

3.4. Physical Factors

Centrostephanus rodgersii is most abundant on subtidal rocky reefs not subjected to influxes of freshwater. Two mass mortality events have been observed on shallow reefs inside Botany Bay immediately after floods (Andrew 1991). Commercial abalone divers also report die-offs in the south of New South Wales following heavy rainfalls over short periods. Severe storms on the open coast can move large boulders around the reef, and crushed *C. rodgersii* are found following such storms (NL Andrew, pers. obs.). Such extreme storms are not frequent and their impact on populations of *C. rodgersii* is unknown.

4. REPRODUCTION

4.1. Reproductive Cycle

The sexes are separate in *Centrostephanus* species although hermaphrodites are occasionally encountered (Pearse 1972; Kennedy and Pearse 1975; King et al. 1994). Sexual dimorphism is evident in these urchins (Pawson and Miller 1983). Male *C. rodgersii*, *C. coronatus* and *C. longispinus* have conspicuous tubular genital papillae while females have short conical papillae.

Centrostephanus rodgersii becomes sexually mature at 40–60 mm test diameter (King et al. 1994). Small *C. rodgersii* (30–50 mm males, 50 mm females) can be induced to spawn but their gametes are not reliably fertile. Males may reproduce earlier than females. Mature females spawn an abundance of small (110–120 μm diameter) eggs. Reproduction in *C. rodgersii* was examined at fringe and barrens habitats along the New South Wales coast to examine the influence of latitude (an indirect factor) and food availability on gonad growth (Byrne et al. 1998). This study spanned the distribution of *C. rodgersii* from the northern subtropical region to the southern temperate region (over 7° of latitude). Gametogenesis was synchronous in all populations: gamete development increased in May followed by maturation in early June and the onset of spawning in late June. This synchrony indicates that gametogenesis and spawning are cued by exogenous factors present at all the populations studied. The most consistent and likely factor to control gametogenesis is photoperiod. The increased tempo of gametogenesis in May coincided with decreasing day length. Mature gametes are not readily obtained from *C. rodgersii* until the second week in June. Short days and lunar conditions coinciding with the winter solstice appear likely proximate factors that might cue the onset of spawning. Control of reproductive activity by photoperiod and/or lunar cues is similar to that documented for *C. coronatus* where the gonad develops with a monthly pattern (Pearse 1972; Kennedy and Pearse 1975). Vitellogenesis in *C. rodgersii* takes approximately 1 month (Byrne et al. 1998). In *C. coronatus* gametogenesis is entrained by lunar cues, as is seen in tropical *Diadema* species (Pearse 1970; Pearse and Cameron 1991).

Centrostephanus rodgersii has a major winter spawning period the length of which depends on location. Northern populations have a short 1-month breeding period with complete spawn-out by July (O'Connor et al. 1978; Byrne et al. 1998). Spawning in the middle and southern parts of the range extends for several months. The longest spawning period (5–6 months) occurs in the southern temperate region where gamete development through August replenishes the supply of mature gametes. Spawning in the south continues through November due to prolonged gamete storage.

4.2. Habitat Related Patterns

Comparison of reproduction of *C. rodgersii* in fringe and barrens areas indicates that, in addition to latitude, gonad growth is strongly influenced by food quality and quality. Sea urchins from fringe habitats have larger gonads than those from barrens (Fig. 2). At some locations gonad yield from fringe urchins is approximately twice that obtained from the barrens urchins. Lower reproductive output in barrens areas is likely due to the

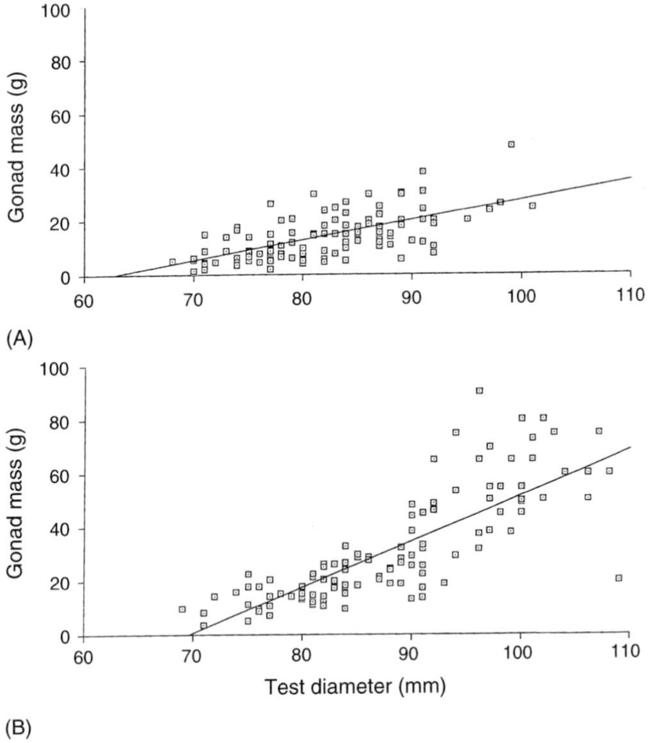

Fig. 2. Relationship between maximum gonad weight and test diameter in *C. rodgersii* from (A) Barrens and (B) Fringe habitats. (A) $R^2 = 0.39$; (B) $R^2 = 0.63$. Redrawn from Byrne et al. (1998).

low abundance of macroalgal food compared with the fringe habitat. The digestive tract of *C. rodgersii* from barrens typically contains an assortment of food pellets indicative of grazing in a coralline habitat, whereas the digestive tract of urchins from fringe areas is filled with pellets derived from a macroalgal food source.

The relationship between size and reproductive output by *C. rodgersii* at fringe and barrens, assessed at gonad peak, shows that gonad weight increases with increasing diameter (Fig. 2). This relationship however is highly variable. Size and maximum gonad yield showed no significant relationship (Byrne et al. 1998).

Maximum gonad yield from the barrens habitat were on average $< 10\%$, while that of urchins from the fringe habitat was $> 10\%$. Clearly focusing on *C. rodgersii* from fringe areas would be most profitable for fishers. Reductions in the density of individuals in barrens habitat may help overcome the poor roe quality of urchins from these areas (Byrne et al. 1998; Blount and Worthington 2002). An improvement in roe colour can be obtained by reducing density in barrens areas and by transplanting urchins to fringe habitat (Worthington and Blount 2003).

On the central coast of New South Wales roe recovery is greatest in autumn. During this season most *C. rodgersii* from both barrens and fringe areas have yellow to gold coloured gonads which are suitable for the market (colour #106/7, 110, 119; Sea Urchin Color Card, University of Maine). Mustard or tan coloured gonads (colour #108,118) are less frequent. Because gonad colour is the result of assimilation and conversion of carotenoid pigments from the diet (Griffiths and Perrott 1976; Tsushima and Matsuno 1990), these colours are likely to reflect different diets.

4.3. Development and Larval Ecology

Centrostephanus species have planktotrophic development involving a feeding echinopluteus larva with the long postoral arms characteristic of diadematoids (Emlet et al. 2002). Early larvae have a four-armed appearance (Figs 3A,B). As the postoral arms grow the larva develops a 2-armed profile (Fig. 3D). It was some decades before the identity of this larval type, called *Echinopluteus transverses* by Mortensen (1921), was determined. The anterolateral arms remain small and the posterodorsal and preoral arms do not develop (Figs 3C,D). The distinctive features of *Centrostephanus* larvae facilitate their identification in plankton samples.

In *C. rodgersii* normal development occurs within 31–37% salinity and 12–25 °C temperature (King 1992, 1999). At 20 °C the blastula stage is reached at 12 h post fertilisation and the blastulae hatch at 20 h. The pluteus stage is reached at approximately 65 h. Total development time to the competent larval stage in *C. rodgersii* takes 3–5 months (M Byrne, unpubl.), indicating the potential for long-distance dispersal in this urchin. In contrast to other sea urchins, the larvae of *Centrostephanus* do not exhibit phenotypic plasticity in arm length in response to variation in feeding regime (Shilling 1995). Larval nutrition does however influence the appearance of the larvae of *C. rodgersii*. Larvae fed a low ration or starved larvae have thin arms, whereas those fed a high ration have a thickened arm epithelium (Figs 3A,B).

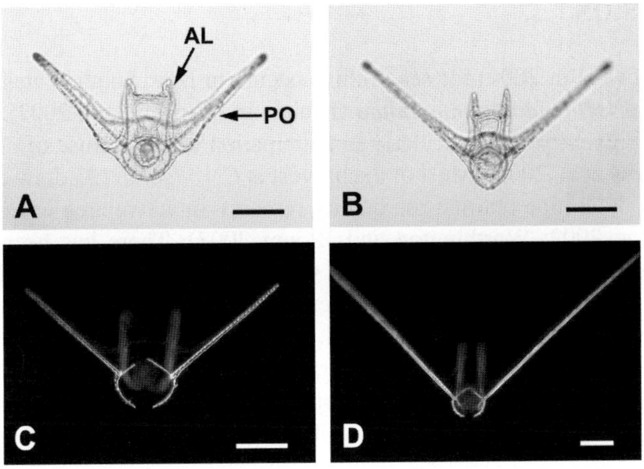

Fig. 3. *Echinopluteus transverses* (A,B) 6-day-old plutei of *C. rodgersii.* (A) Fed larvae have wider arm epithelia than (B) starved larvae. In these larvae the disproportionate growth of the postoral (PO) relative to the anterolateral (AL) arms is evident. (C) 5-week-old and (D) 6-week-old larvae viewed with polarized light showing that the posterodorsal and preoral arms do not develop. The postoral arms continue to grow giving the larva a two-armed profile. Scale bars = 100 μm.

5. GROWTH AND AGE

Centrostephanus rodgersii is a large (maximum test diameter = 120 mm), fast growing sea urchin (Ebert 1982). Patterns in size structure suggest that they may grow as much as 40 mm in their first year (Andrew 1993). Growth in larger individuals has been estimated using mark recapture techniques and direct ageing using validated growth checks visible on the demipyramids (Ebert 1982; C Blount and N Andrew, unpubl.). The largest sea urchins found (>110 mm test diameter), may be up to 20 years old. Size-frequency distribution of *C. rodgersii* is often dominated by a single mode between 70 and 90 mm test diameter (Andrew and Underwood 1989; Andrew and O'Neill 2000). Sea urchins in this size range probably are 4–10 years old (C Blount and N Andrew, unpubl.). Age frequency analysis of *C. rodgersii* populations in Tasmania indicates an age span of 8–16 years with a mean of 12 years old (Johnson et al. 2005). Relatively little information exists on variability in growth with location and habitat. Individuals in the fringe appear to have shorter spines and thicker tests than those in the barrens. Growth seems slower in the barrens than in near-shore areas with an abundance of large brown algae (C Blount and N Andrew, unpubl.). Although growth is faster at Sydney, near the centre of the species distribution than at Eden (approximately 350 km to the south), maximum age shows little difference between these localities (C Blount and N Andrew, unpubl.). The growth and mortality rates reported by Ebert (1982) are relatively low and may not be representative of the more exposed localities where *C. rodgersii* is most abundant. Similarly, the allometric relationships Ebert reported may vary with wave exposure but there are no comparative data.

6. CONCLUSION

A small fishery (50 t in 2000) for sea urchins occurs in New South Wales, split between *C. rodgersii* and *Heliocidaris tuberculata* (Blount and Worthington 2002). The fishery is likely to expand eventually, particularly in anticipation of the demise of other sea urchin fisheries (Andrew et al. 2002). Most of the harvest is *C. rodgersii* (Andrew et al. 1998) and the factors that determine optimal roe quality remains an active area of research (Blount and Worthington 2002; Worthington and Blount 2003). There has been an enormous increase in the commercial exploitation of sea urchins in recent years. The history of several major urchin fisheries has, however, been one of explosive growth followed by over-fishing and decline (Andrew et al. 2002). To our knowledge the fishery for *C. rodgersii* is the only fishery for a temperate diadematoid. *Centrostephanus rodgersii* exerts a profound influence on the ecology of the shallow subtidal in southeast Australia; it will be critical to integrate this ecological knowledge into management of the fishery.

ACKNOWLEDGEMENTS

We are grateful to John Himmelman for detailed comments on this chapter. Thanks also to John Lawrence and Craig Blount for comments on the manuscript. Raphael Morgan and Thomas Prowse provided assistance.

REFERENCES

Andrew NL (1991) Changes in subtidal habitat following mass mortality of sea urchins in Botany Bay, New South Wales. Aust J Ecol 16: 353–362

Andrew NL (1993) Spatial heterogeneity, sea urchin grazing and habitat structure on reefs in temperate Australia. Ecology 74: 292–302

Andrew NL (1994) Survival of kelp adjacent to areas grazed by sea urchins in New South Wales, Australia. Aust J Ecol 16: 353–362

Andrew NL (1999) New South Wales. In: Andrew NL (ed.) Under Southern Seas: The Ecology of Australia's Rocky Reefs. University of New South Wales Press, Sydney, pp 8–19

Andrew NL, Underwood AJ (1989) Patterns of abundance in the sea urchin *Centrostephanus rodgersii* on the central coast of New South Wales, Australia. J Exp Mar Biol Ecol 131: 61–80

Andrew NL, Underwood AJ (1992) Associations and abundance of sea urchins and abalone on shallow subtidal reefs in southern New South Wales. Mar Freshwater Res 43: 1547–1559

Andrew NL, Underwood AJ (1993) Density-dependent foraging in the sea urchin *Centrostephanus rodgersii* on shallow subtidal reefs in New South Wales, Australia. Mar Ecol Prog Ser 99: 89–98

Andrew NL, O'Neill T (2000) Large-scale patterns in habitat structure on subtidal rocky reefs in New South Wales. Mar Freshwater Res 51: 255–263

Andrew NL, Worthington, DG, Brett PA, Bentley N (1998) Interactions between the abalone fishery and sea urchins in new South Wales. Final Report to Fisheries Research and Development Corporation (Australia). No. 93/102. 64 pp

Andrew NL, Agatsuma Y, Ballesteros E, Bazhin AG, Creaser EP, Barnes DKA, Botsford LW, Bradbury A, Campbell A, Dixon JD, Einarsson S, Gerring P, Hebert K, Hunter M, Hur SB, Johnson CR, Juinio-Meñez MA, Kalvass P, Miller RJ, Moreno CA, Palleiro JS, Rivas D, Robinson SML, Schroeter SC, Steneck RS, Vadas RL, Woodby DA, Xiaoqi Z (2002) Status and management of world sea urchin fisheries. Oceanogr Mar Biol Annu Rev 40: 343–425

Blount C, Worthington DG (2002) Identifying individuals of the sea urchin *Centrostephanus rodgersii* with high-quality roe in New South Wales, Australia. Fisheries Res 58: 341–348

Byrne M, Andrew NL, Worthington DG, Brett PA (1998) Reproduction in the diadematoid sea urchin *Centrostephanus rodgersii* in contrasting habitats along the coast of New South Wales, Australia. Mar Biol 132: 305–318

Choat JH, Schiel DR (1982) Patterns of distributions and abundance of large brown algae and invertebrate herbivores in subtidal regions of northern New Zealand. J Exp Mar Biol Ecol 60: 129–162

Cole RG, Creese RG, Grace RV, Irving P, Jackson BR (1992) Abundance patterns of subtidal benthic invertebrates and fishes at the Kermadec Islands. Mar Ecol Prog Ser 82: 207–218

Ebert TA (1982) Longevity, life-history, and relative body wall size in sea urchins. Ecol Monogr 52: 353–394

Emlet RB, Young CM, George SB (2002) Phylum Echinodermata: Echinoidea. In: Young CM (ed) Atlas of Marine Invertebrate Larvae. Academic Press, London, pp 531–552.

Fletcher WJ (1987) Interactions among subtidal Australian sea urchins, gastropods and algae: effects of experimental removals. Ecol Monogr 57: 89–109

Francour P (1991) Statut de *Centrostephanus longispinus* en Mediterranee. In: Boudouresque CF, Avon M, Gravez V (eds) Les Espéces Marines á Protéger en Méditerranée. GIS Posidonie publ, Marseille, France, pp 187–202

Francour P, Boudouresque CF, Harmelin JG, Harmelin-Vivien ML, Quignard JP (1994) Are the Mediterranean waters becoming warmer? Information from biological indicators. Mar Poll Bull 28: 523–526

Gillanders BM (1995) Feeding ecology of the temperate marine fish *Achoerodus viridis* (Labridae): size, seasonal and site-specific differences. Mar Freshwater Res 46: 1009–1020

Gillanders BM, Kingsford MJ (1998) Influence of habitat on abundance and size structure of a large temperate-reef fish, *Achoerodus viridis* (Pisces: Labridae). Mar Biol 132: 503–514

Gras H, Weber W (1983) Light-induced alterations in cell shape and pigment displacement in chromatophores of the sea urchin *Centrostephanus longispinus*. Cell Tiss Res 182: 165–176

Griffiths M, Perrott P (1976) Seasonal changes in the carotenoids of the sea urchin *Strongylocentrtotus droebachiensis*. Comp Biochem Physiol 55B: 435–441

Hartney KB, Grorud KA (2002) The effect of sea urchins as biogenic structures on the local abundance of temperate fish. Oecologia 131: 506–513

Hill A, Blount C, Poore AGB, Worthington DG, Steinberg PD (2003) Grazing effects of the sea urchin *Centrostephanus rodgersii* in two contrasting rocky reef habitats: effect of urchin density and its implications for the fishery. Mar Freshwater Res 54: 691–700

Holbrook SJ, Kingsford MJ, Schmitt RJ, Stephens JS (1994) Spatial and temporal patterns in assemblages of temperate reef fish. Amer Zool 34: 463–475

Jones GP, Andrew NL (1990) Herbivory and patch dynamics on rocky reefs in temperate Australasia: the roles of fish and sea urchins. Aust J Ecol 15: 505–520

Johnson C, Ling S, Ross J, Shepherd S, Miller K (2005) Establishment of the long-spined sea urchin (*Centrostephanus rodgersii*) in Tasmania: first assessment of potential threats to fisheries. Final Report to Fisheries Research and Development Corporation (Australia). No. 2001/044. 79 pp

Kennedy B, Pearse JS (1975) Lunar synchronisation of the monthly reproductive rhythm in the sea urchin *Centrostephanus coronatus* Verrill. J Exp Mar Biol Ecol 17: 323–331

King CK (1992) Reproduction and development of *Centrostephanus rodgersii* – a potential bioassay organism. BSc Thesis, University of Sydney

King CK (1999) The impact of metals and organic contaminants on the development of marine invertebrates from the Cl. Echinoidea and Cl. Bivalvia. PhD Thesis, University of Sydney

King CK, Hoegh-Guldberg O, Byrne M (1994) Reproductive cycle of *Centrostephanus rodgersii* (Echinoidea), with recommendations for the establishment of a sea urchin fishery in New South Wales. Mar Biol 120: 95–106

Lawrence JM, Sonnenholzner J (2004) Distribution and Abundance of Asteroids, Echinoids, and Holothuroids in Galápagos. In: Heinzeller T, Nebelsick JH (eds) Echinoderms: München. AA Balkema, Leiden, pp 239–244

Lissner AL (1980) Some effects of turbulance on the activity of the sea urchin *Centrostephanus coronatus* Verrill. J Exp Mar Biol Ecol 48: 185–193

Lissner AL (1983) Relationship of water motion to the shallow water distribution and morphology of two species of sea urchins. J Mar Res 41: 691–709

McLaughlin RH, O'Gower AK (1971) Life-history and underwater studies of a heterodontid shark. Ecol Monogr 41: 271–289

McShane PE (1991) Density-dependent mortality of recruits of the abalone *Haliotis rubra* (Mollusca: Gastropoda). Mar Biol 110: 385–389

Mortensen T (1921) Studies of the Development and Larval Forms of Echinoderms. G E C Gad, Copenhagen

Nelson BV, Vance RR (1979) Diel foraging of the sea urchin *Centrostephanus coronatus* as a predator avoidance strategy. Mar Biol 51: 251–258

O'Connor C, Riley G, Lefebvre S, Bloom D (1978) Environmental influences on histological changes in the reproductive cycle of four NSW sea urchins. Aquaculture 15: 1–17

Pawson DL, Miller JE (1983) Systematics and ecology of the sea urchin genus of *Centrostephanus* (Echinodermata: Echinoidea) from the Atlantic and Eastern Pacific Oceans. Smith Contr Mar Sci 20: 1–15

Pearse JS (1970) Reproductive periodicities of Indo-Pacific invertebrates in the Gulf of Suez. III. The echinoid *Diadema setosum* (Leske). Bull Mar Sci 20: 697–720

Pearse JS (1972) Monthly reproductive rhythm in the diademed sea urchin (*Centrostephanus coronatus*). Bull Mar Sci 20: 697–720

Pearse JS, Timm RW (1971) Juvenile nematodes (*Echinocephalus pseudouncinatus*) in the gonads of the sea urchin *Centrostephanus coronatus* Verrill. J Exp Mar Biol Ecol 8: 167–186

Pearse JS, Cameron RA (1991) Echinodermata: Echinoidea. In: Giese AC, Pearse JS, Pearse VB (eds) Reproduction of Marine Invertebrates, Volume VI, Echinoderms and Lophophorates. Boxwood Press, Pacific Grove, CA, pp 514–662

Schiel DR, Kingsford MJ, Choat JH (1986) Depth distribution and abundance of benthic organisms and fishers at the subtropical Kermadec islands. NZ J Mar Freshwater Res 20: 521–535

Shepherd SA (1973) Competition between sea urchins and abalone. Aust Fish 32: 4–7

Shilling FM (1995) Morphological and physiological responses of echinoderm larvae to nutritive signals. Amer Zool 35: 399–414

Tsushima M, Matsuno T (1990) Comparative biochemical studies of carotenoids in sea urchins – I. Comp Biochem Physiol 96B: 801–810

Underwood AJ, Kingsford MJ, Andrew NL (1991) Patterns in shallow subtidal marine assemblages along the coast of New South Wales. Aust J Ecol 16: 231–249

Vance RR (1979) Effects of grazing by the sea urchin, *Centrostephanus coronatus*, on prey community composition. Ecology 60: 537–546

Worthington DG, Blount C (2003) Research to develop and manage the sea urchin fisheries of NSW and eastern Victoria. Final Report to Fisheries Research and Development Corporation (Australia). No. 1999/128. 182 pp

Wright JT, Dworjanyn SA, Rogers CN, Steinberg PD, Williamson JE, Poore AGB (2005) Density-dependent sea urchin grazing: differential removal of species, changes in community composition and alternative community states. Mar Ecol Prog Ser 298: 143–156

Edible Sea Urchins: Biology and Ecology
Editor: John Miller Lawrence

Chapter 11

Ecology of *Diadema*

NA Muthiga and TR McClanahan

Wildlife Conservation Society, Marine Programs, Bronx, NY (USA).

1. INTRODUCTION

1.1. Species of *Diadema*

Six species of *Diadema* (Humphreys 1797, Gray 1825) are recognised: *D. antillarum* (Phillipi 1845); *D. mexicanum* (A Agassiz 1863); *D. savignyi* (Michelin 1845; *D. setosum* (Leske 1778); *D. palmeri* Baker 1967 and *D. paucispinum* A Agassiz 1863 (Pearse 1998). Long black spines characterise the genus and individual species are sometimes difficult to identify in the field except for *D. palmeri* that has distinct red colouration. The sympatric species *D. savignyi* and *D. setosum* (Plate 1C) may occur in single species groups or in mixed groups and may be distinguished by test colour pattern, spine morphology and pedicellarie shape. Most populations of *D. setosum* in Kenya have an orange ring around the anus and five bright (rarely pale) white spots on the aboral ends of the interambulacrals (Muthiga 2003). *Diadema savignyi*, on the other hand, has iridescent blue or green lines along the interambulacrals and around the periproct and sometimes five very pale white spots on the aboral ends of the interambulacrals (Pearse and Arch 1969).

These characteristics are apparently common for other populations of these species throughout the Indo-Pacific (Clark 1921, 1946; Mortensen 1940; Pearse and Arch 1969; Lessios and Pearse 1996). However, individuals with intermediate characteristics are sometimes encountered and populations of *D. setosum* in the Red Sea show more variability in the colour pattern (Clark 1966; Pearse 1970). Mortensen (1940) suggested that the intermediate characteristics are due to hybridisation. Lessios and Pearse (1996) confirmed that hybridisation and introgression occurs between *D. savignyi* and *D. setosum* that could result in intermediate characteristics. According to Lessios and Pearse (1996), individuals of *D. paucispinum* are characterised by having no orange ring, no white spots and no iridescent blue dots. Coppard and Campbell (2004) described spine morphology of *Diadema* species and showed that the spines of *D. paucispinum* are visibly more robust than the other species and this is reflected in the internal structure of the spine with large closely packed solid wedges, small axial cavity and strengthening structures such as spokes and trabeculae. The spines of *D. setosum* are also distinct, especially in relation to the length relative to test size. The other members of *Diadema* have very similar spines that are difficult to differentiate at the macroscopic level.

1.2. Biogeography and Large-Scale Distribution

Sea urchins of the genera *Diadema* are some of the most widespread, abundant and ecologically important sea urchins in tropical regions (reviews in Lawrence and Sammarco 1982; Lessios 1988; Birkeland 1989; Carpenter 1997; Tuya et al. 2005). *Diadema* populations occur in all tropical seas to a depth of up to 70 m (Mortensen 1940). Lessios et al. (2001a) summarise the geographical distributions for the different species of *Diadema* as follows: *D. antillarum* primarily occurs in the tropical Atlantic, from Florida and Bermuda to Brazil and from Madeira (but contrary to Mortensen 1940, not in Azores; Wirtz and Martins 1993) to the Gulf of Guinea; *D. mexicanum* occurs in the tropical eastern Pacific from the Sea of Cortez to Ecuador, including the Islands of Revillagigedos, Clipperton, Isla del Coco, and Galapagos (Maluf 1988); *D. palmeri* is only known from the north coast of New Zealand (Baker 1967) and from the south coast of Australia (Rowe and Gates 1995); *D. paucispinum* is thought to be primarily a Hawaiian species occurring in the South Pacific and Hawaiian islands (Clark and Rowe 1971) but Pearse (1998) has indicated that it may occur in other parts of the tropical Indo-Pacific. Clark and Rowe (1971) reported that both sympatric species *D. savignyi* and *D. setosum* occur in the Mascarene Islands, Eastz Africa and Madagascar, southeast Arabia, Ceylon, East Indies, North Australia, Philippine Islands, China and south Japan, South Pacific Islands. In addition, only *D. setosum* was reported in the Red Sea, Persian Gulf and Bay of Bengal, whereas *D. savignyi* was reported in the islands of the western Indian Ocean.

Because some confusion exists as to the true distributions of *D. savignyi* and *D. setosum*, Pearse (1998) examined populations across the tropical Pacific from Honshu Island Japan to Moorea, French Polynesia and found the distribution of these species occurred along an inshore/landmass to offshore/oceanic gradient suggesting that *D. setosum* is more tightly associated with land masses in the tropical Pacific whereas *D. savignyi* is more widely spread throughout. In East Africa, *D. setosum* populations are more tightly associated with islands, whereas both *D. savignyi* and *D. setosum* are found along the mainland (McClanahan, unpubl.). Genetic studies on the mDNA of *Diadema* (Lessios et al. 1996, 2001b) indicate that *D. setosum* is the outgroup among species of *Diadema*, with the exception of *D. palmeri* whose genetic position remains unknown. *Diadema antillarum* and *D. savignyi* are nearly indistinguishable genetically. Hence, the *D. antillarum*/*D. savignyi* clade is pan-tropical in distribution, with the exception of the Red Sea, whereas *D. setosum* is more restricted to continental shelves and adjacent islands of the Indian and west Pacific oceans.

1.3. Local Distribution and Abundance Patterns

Diadema species are ubiquitous and common throughout their range and are found primarily on shallow reef habitats on sand, rubble or coral (Bauer 1980), although *D. savignyi* and *D. setosum* also occur in the channels of mangroves and in estuaries and creeks along the Kenyan coast (Muthiga, pers. obs.). The upper limit of *D. setosum* is restricted by desiccation. While the Indo-Pacific species are reported to depths of 10 m (Pearse 1998), *D. antillarum* has been reported to depths of 70 m (The Nature Conservancy 2004). *Diadema* distribution is patchy although some populations across the

Indo-Pacific occur in large aggregations (Fox 1926; Stephenson et al. 1931; Onada 1936; Pearse and Arch 1969; McClanahan 1988), whereas in the Red Sea the animals are cryptic and hide in crevices and under ledges during the day (Magnus 1967; Pearse 1970). Aggregations have also been observed for *D. antillarum* in the Caribbean (Randall et al. 1964; Kier and Grant 1965) and *D. mexicanum* in the Eastern Tropical Pacific (Pearse 1968a).

Prior to 1983, *D. antillarum* was the most dominant herbivore in tropical western Atlantic reef habitats (Ogden 1977; Sammarco 1980; Hawkins and Lewis 1982). Bauer (1980) surveyed *D. antillarum* at 30 sites across the western North Atlantic and reported densities ranging from 17 to 593 individulas per square metre. In general highest densities were found in shallow intertidal communities with relatively high water movement. Other factors controlling this variability included seasonal changes in salinity and turbidity, sewage outfalls (that generally increased densities) and silty conditions (that decreased densities). The Indo-Pacific species of *Diadema* along the coasts of Kenya and Tanzania occur at densities ranging from 0.003 to 4.3 individulas per square metre for *D. savignyi* and 0.006 to 0.3 individulas per square metre for *D. setosum* (Table 1). *Diadema savignyi* has been reported to be more abundant than *D. setosum* in Kenya and Tanzania (McClanahan and Shafir 1990; McClanahan 1998).

Table 1 Population density (mean number $10\,m^{-2}$ (±sem)) and biomass estimates (mean kg wet weight ha^{-1}) of *Diadema savignyi* and *D. setosum* on Kenyan and Tanzanian reefs.

Reef sites	*D. savignyi*		*D. setosum*	
	Density	Biomass	Density	Biomass
Kenyan reefs				
Malindi	0.03 ± 0.02 (16)	3.30		
Watamu	0.05 ± 0.03 (8)	6.25		
Vipingo	9.17 ± 0.86 (19)	1131.00	2.13 ± 0.26 (19)	319.50
Kanamai	5.36 ± 0.63 (20)	670.00	2.78 ± 0.39 (20)	416.70
Cannon Pt.	5.1 ± 0.97 (10)	637.50		
Malaika	0.5 ± 0.2 (16)	62.50	0.13 ± 0.08 (16)	19.50
Shanzu	1.13 ± 0.59 (15)	141.30	0.2 ± 0.11 (15)	30.00
Bamburi	6.86 ± 1.51 (20)	857.50	0.39 ± 0.1 (20)	58.50
Ras Iwatine	4.63 ± 1.10 (7)	578.80	0.32 ± 0.15 (7)	48.00
Diani	0.44 ± 0.22 (17)	55.00	0.06 ± 0.02 (17)	9.00
Kisite	0.54 ± 0.19 (54)	67.50		
Tanzanian reefs				
Funguni	21.9 ± 4.6 (9)	273.60	0.11 ± 0.1 (9)	16.50
Makome	0.11 ± 0.02 (9)	13.80		
Taa	11.89 ± 4.25 (9)	1486.30	0.33 ± 0.1 (9)	49.50
Chumbe	1.1 ± 0.7 (18)	137.50		
Mbudya	8.53 ± 2.86 (12)	1066.30	0.13 ± 0.09 (12)	19.50
Bongoyo	43.4 ± 10.45 (30)	5425.00	2.87 ± 1.1 (30)	430.50

Density estimates were compiled from McClanahan et al. (1998) except for Shanzu, Malaika and Cannon Pt. which were sampled by Muthiga.

The numbers in parentheses indicate the number of quadrates sampled per reef.

2. POPULATION BIOLOGY AND ECOLOGY

2.1. Reproductive Biology and Ecology

Diadema are gonochoric with separate sexes and hermaphroditism is rare (Pearse and Cameron 1991). Bak et al. (1984) reported blastulae in the gonads of two individuals of *D. antillarum* and suggested that this was due to parthenogenesis; however, Pearse and Cameron (1991) suggested that the blastulae more likely resulted from self-fertilisation within hermaphroditic gonads as reported for other species of echinoids. Although some sea urchins show some external sexual differences including differences in the size of genital papillae, no external sexual characteristics have been reported in *Diadema*. It is expected that *Diadema* populations mainly have 1:1 sex ratios but little data are available to verify this supposition. *D. savignyi* and *D. setosum* populations at Kanamai Kenya have a 1:1 sex ratio (Muthiga, unpubl.) whereas Hori et al. (1987) found more males than female *D. setosum* (1:0.7) out of a total of 487 individuals in Singapore.

The developmental stage of the gametogenic cycle in *Diadema* is similar to that of most echinoids; itconsists of a regular sequence of changes of the gametogenic cells and nutritive phagocytes, and differs only in the timing of the different stages amongst species (Pearse and Cameron 1991). The morphology of nutritive phagocytes also varies with stage, position in the gonad and species. Nutritive phagocytes appear particularly distinct among diadematoids including *Centrostephanus* and *Diadema* (Pearse 1970; Pearse and Cameron 1991). The cells are flat or squat in shape and line the gamete filled lumen of gonadal tubules after gamete production has terminated (Pearse 1970). The sequence of events during the gametogenic cycle is described in detail for *D. setosum* in the Gulf of Suez (Pearse 1970), *D. savignyi* in South Africa (Drummond 1995), and for both species in Kenya (Muthiga 2003).

The process of spermatogenesis from production of spermatocytes to accumulation of numerous sperm is completed in 1–2 months in *D. setosum* (Pearse 1970) and 1 month in *D. setosum* and *D. savignyi* (Muthiga 2003). Oogenesis also follows a regular sequence of events and is completed in 1 month in both *D. savignyi* and *D. setosum* (Muthiga 2003). The length of the vitellogenetic stage varies among different species and has been estimated to be less than a month in *D. setosum* (Iliffe and Pearse 1982) and in *D. setosum* and *D. savignyi* (Muthiga 2003). Ova range in size from 67 to 73 μm in *D. antillarum*, 67 μm in *D. mexicanum* and 92 μm in *D. setosum* (Emlet et al. 1987). Lessios (1987) showed that there was great variability in egg size and found differences between individuals of the same species, within a sample, between days, months, years and habitats and with no apparent correlation with time, body or gonad size or fecundity in 13 echinoid species in Panama. *Diadema antillarum* showed significant daily variation in egg size, whereas *D. mexicanum* showed no differences in egg volume between days.

2.2. Reproductive Cycles

The reproductive cycles of *Diadema* species can be continuous or restricted and can vary from one geographic location to another (Table 2), indicating that several factors influence reproduction. For example, *D. antillarum* reproduces continuously throughout

Table 2 Summary table of the seasonal and lunar spawning cycles for *Diadema*.

Species	Location	Spawning season	Lunar period	References
D. antillarum	Bermuda	Apr–Oct	New moon	Iliffe and Pearse (1982)
	South Florida	Oct–Dec	New moon	Bauer (1976)
	Virgin Islands	Continuous	New moon	Randall et al. (1964)
	Barbados	Jan–Apr		Lewis (1966)
	Curaçao	Continuous (peak Jan–Mar)		Randall et al. (1964)
	Panama (Caribbean)	Continuous		Lessios (1981)
	Panama		New moon	Lessios (1984)
D. mexicanum	Panama (Pacific)	Sep–Nov		Lessios (1981)
	Panama		Full moon	Lessios (1984)
D. savignyi	Kenya	Continuous (peak Feb–May)	After full moon (days 17–18)	Muthiga (2003)
	South Africa	Dec–Apr	After full moon	Drummond (1995)
D. setosum	Philippines	Continuous	Full moon	Tuason and Gomez (1979)
	Singapore	Continuous		Hori et al. (1987)
	NW Red Sea	Jun–Sep		Pearse (1970)
	Japan		Full moon	Yoshida (1952)
	Guam, Madang, Moorea and Samoa		After full moon	Pearse, pers. comm.
	Kenya	Continuous (peak Feb–May)	Lunar day 8–10	Muthiga (2003)

the year in the Virgin Islands and Curacao (Randall et al. 1964) and on the Caribbean side of Panama (Lessios 1981), whereas spawning is seasonal in Bermuda (Iliffe and Pearse 1982), Barbados (Lewis 1966), and south Florida (Bauer 1976). In addition, *D. savignyi* and *D. setosum*, at the same localesalong the Kenyan coast, reproduce throughout the year but with peaks at different times of the year (Muthiga 2003).

Although gametogenesis is well described for *Diadema* and echinoids in general, the factors that regulate the timing of gametogenesis and spawning are poorly understood. Studies that have compared conspecific populations of *Diadema* from different latitudes (Pearse 1968b, 1970) show that reproductive cycles tend to become more synchronised with increased distance from the equator and seasonality presumably becomes more pronounced. Pearse (1974) compiled information on *D. setosum* and suggested that reproduction is controlled by temperature, since populations in northern Japan and Suez spawned during the boreal summer while populations in Australia spawned in the austral summer. Pearse (1974) concluded that gametogenesis proceeds when temperatures are above ~25 °C in this species.

Correlative evidence of the influence of temperature has also been shown at localities with similar latitudes but different degrees of seasonality. For example, *D. mexicanum* from the seasonal environment of the Bay of Panama displayed a synchronous well-defined reproductive cycle, whereas *D. antillarum* from the less seasonal Caribbean shore of Panama showed less synchrony (Lessios 1981). Moreover, information from populations of *D. setosum* that occur near the equator (where presumably less seasonality occurs) supports this suggestion since populations in the Philippines (Tuason and Gomez 1979), Singapore (Hori et al. 1987) and Kenya (Muthiga 2003) reproduce throughout the year. Many of the studies that report the influence of temperature are correlative; the exact mechanism that temperature influences gametogenesis remains unclear. For example, Pearse (1974) failed to induce gametogenesis in *D. setosum* individuals that were moved from the temperate Gulf of Suez and maintained at warmer summer temperatures.

Most species of *Diadema,* like most diadematoids, have monthly reproductive rhythms with spawning mainly occurring around the full moon in *D. mexicanum* and *D. setosum,* whereas *D. antillarum* and *D. savignyi* spawn around the new moon (Table 2). Interestingly, comparisons of coexisting populations of *D. savignyi* and *D. setosum* along the Kenyan coast showed a contrasting cycle, with *D. savignyi* spawning after the full moon (between lunar days 17 and 18) and *D. setosum* spawning before the full moon (lunar days 8-10). Temporal separation of reproductive cycles has also been reported in *D. mexicanum* (Lessios 1984). The factors affecting the spawning behaviour of *Diadema* are not always very clear; however, *D. setosum,* has been reported to spawn at different phases of the moon (including full moon (Yoshida 1952; Pearse 1972) and both full and new moon (Fox 1924; Kobayashi and Nakumura 1967)) and in some cases failed to show lunar periodicity (Stephenson 1934; Mortensen 1937; Pearse 1968b, 1970). Monthly cycles may be controlled by moonlight. However, the lack of synchrony among different populations reported in *D. setosum* in the Red Sea (Pearse 1972) suggests that synchrony within populations is either not under worldwide lunar control or it can be adjusted by local environmental conditions. Pearse (1972) suggested that spawning in *D. setosum* and *Centrostephanus coronatus* could be explained by tidal rhythms.

Other factors influencing gametogenesis include age and size of individuals. These influence the age and size at sexual maturity as well as the quantity of gametes produced. Lawrence (1987) reported that gonads of extremely large *D. setosum* show little reproductive activity. Lack of food can also affect gonad growth, and poorly fed individuals produce small or no gonads (Levitan 1988a). However, there is little evidence that nutrition plays a role in regulating the onset of gametogenesis (Pearse and Cameron 1991).

2.3. Feeding Ecology

Species of *Diadema* are mainly omnivorous grazers and detritus feeders, ingesting substrate and scraping algal films off hard substratum (Mortensen 1940; Randall et al. 1964;

Lewis 1964; Pearse 1970). *D. antillarum* preys on live coral as well as coral spat (Bak and van Eys 1975; Carpenter 1981; Sammarco 1980, 1982b). According to Lewis (1964) the feeding activity of *D. antillarum* is greatest in the afternoon and early evening. However, Tuya et al. (2004a) found nocturnal activity in *D. antillarum* in the eastern Atlantic. Lawrence and Hughes-Games (1972) reported a nocturnal activity rhythm in *D. setosum* at Elat, Red Sea, with feeding occurring only at night. The foraging behaviour of *D. antillarum* maintains 'halos' a common feature associated with patch reefs in the Caribbean that consist of a band of bare sand between the base of the reef and the adjacent *Thalassia* and *Syringodium* beds (Ogden et al. 1973).

Intestinal contents of *Diadema* species reflect the type of food available. For example, *D. antillarum* collected from *Thalassia* beds have gut contents that are largely *Thalassia* with smaller amounts of fine silt, whereas guts of individuals collected from reef areas contain algae and detritus (Randall et al. 1964). On Kenyan reefs, *D. savignyi* and *D. setosum* have gut contents composed of coral sediment (48–52%), algae (~28%), seagrass (20%), and invertebrates (2%) (McClanahan 1988). Tuya et al. (2001) divided algae eaten by *D. antillarum* in the Canary Islands into three groups: *Halopteris, Lobophora* and *Dictyota* were preferred; *Padina* was intermediate, and *Cystoseira* was least preferred. Solandt and Campbell (2001) found that *D. antillarum* in Jamaica was attracted towards water passing over heavily calcified *Halimeda opuntia* and less-calcified algae such as *Lobophora variegata*. They were not attracted to *Sargassum* sp. or *Galaxaura* sp. Despite this difference in attraction to algae, there was no selection when a variety of algae were provided for consumption.

Defence chemicals produced by marine algae mediate the rate of hebivory by *Diadema*. *Laurencia obstusa, Stypopodium zonale* and various species of *Dictyota* are consumed at relatively low rates by herbivores including *Diadema* in Caribbean reefs (Littler et al. 1983; Hay 1984; Hay and Paul 1986), which suggests that secondary metabolites in these seaweeds are defences against herbivory. Hay et al. (1987) extracted secondary metabolites from tropical marine algae and found that stypotriol from *S. zonale,* pachydictyol-A from *Dictyota* and other algae, elatol from *L. obstusa* and isolaurinterol from species of *Laurencia* reduced the amount of *Thalassia* eaten by *D. antillarum*. Cymopol, from the green alga *Cymopolia barbata,* stimulated feeding by *Diadema*.

Food passes through the guts of *D. antillarum* in 8–12 h, and enzymes that digest carbohydrates, proteins and lipids are present in the gut (Lewis 1964). In *D. setosum* from Elat, food is present in the stomach only at night when feeding occurs and is retained in the intestine throughout indicating that digestion mainly occurs in the intestine (Lawrence and Hughes-Games 1972). Lewis (1964) describes the morphology of the gut of *D. antillarum* and suggests that digestive enzymes are mainly produced in the caecum and foregut while absorption occurs in the hindgut.

Although seasonal changes have been reported in the size of the gut of some sea urchins related to changes in food availability (Lawrence et al. 1965), no seasonal changes have been reported in *D. setosum* in the Red Sea or on Kenyan reefs (Pearse 1970; Muthiga 2003). On the Kenyan coast, the relative size of the gut of *D. savignyi* increased when gonad size decreased suggesting that a seasonal feeding pattern may occur in this species (Muthiga 2003). Some species of *Diadema* have also been shown to regulate body size when food is limited increasing the size of the jaw and decreasing the size of the test in *D. setosum* (Ebert 1980) and *D. antillarum* (Levitan 1989). The increase in jaw size is primarily due to a decrease in size of the test (Levitan 1991). A larger jaw under conditions of food limitation is an adaptive behavior as it increases efficiency in food collection.

2.4. Growth and Longevity

Estimates of growth in *Diadema* (Table 3) have been reported mainly from laboratory-reared animals, size frequency measurements in natural populations, and observations of individuals held in aquaria or cages (Randall et al. 1964; Lewis 1966; Bauer 1976, 1982; Ebert 1982; Drummond 1994). Tagging using monofilament line was not successful on *D. antillarum* (Randall et al. 1964). Ebert (1982) found that tetracycline was unreliable with *D. setosum*, possibly due to the low levels of incorporation of tetracycline or reworking of calcite in the skeletal elements. Ebert (1982) reported a growth parameter K of 0.008 for a few individuals of *D. setosum* in Zanzibar and Elat, Israel. These estimates are lower than growth estimates reported for the conspecific *D. savignyi* in South Africa ($K = 0.79$; Drummond 1994). Ebert (1982) reported asymptotic diameters of 91.9 and 83.6 mm for *D. setosum* at Zanzibar and Elat, respectively. These estimates are close to the largest *D. setosum* (88.5 mm) and *D. savignyi* (90 mm) measured at Kanamai, Kenya (Muthiga, unpubl.). These estimates indicate that *D. setosum* has a very slow initial growth rate in contrast to its conspecific *D. savignyi* that shows rapid growth and a short life span of 3–5 years (Drummond 1994).

Growth studies of Caribbean *Diadema* have mainly been on caged or laboratory reared individuals. Randall et al. (1964) recorded growth rates of 3.5–6.7 mm test diameter per month in caged *D. antillarum* of 4–25 mm from the Virgin Islands. Lewis (1966) recorded similar rates of growth in laboratory-reared individuals in Barbados. Urchins between 10 and 30 mm grew at a rate of 3.2 mm per month, a rate that was faster than for larger individuals (30–50 mm) that grew at a rate of 1.8 mm per month. Bauer (1976, 1982) recorded similar growth rates for *D. antillarum*. These growth rates are

Table 3 Summary table of the growth parameter (K), maximum length (L_{inf}), and mortality estimates (Z) for species of *Diadema*.

Species	Location	N	K	L_{inf} (mm)	Z (yr^{-1})	Reference
D. antillarum	Virgin Islands	44	1.18	75.6	1.3	Randall et al. (1964)
	Florida Keys				1.4	Lewis (1966)
D. setosum	Zanzibar°, Tanzania	55	0.008	91.92	0.29	Ebert (1982)
	Eilat, Israel	7	0.019	83.57	0.65	Ebert (1982)
D. savignyi	Isipingo Beach, South Africa		0.79			Drummond (1994)
D. paucispinum	Kealakekua Bay, Hawai				1.9	Ebert (1982)

considerably lower than reported for other Caribbean urchins (Lewis 1958; Moore et al. 1963).

2.5. Pelagic Larval Dynamics

Mortensen (1931, 1937) has described the structure of the planktotrophic larvae of *D. antillarum* and *D. setosum*. The members of the order Diadematoida have larval forms that are characterised by long fenestrated postoral arms that are 7 times the length of the body, or more than 1.5 mm long (Pearse and Cameron 1991). The significance of these long arms in *Diadema* has not been explored although shape and size of feeding larvae may have significance for larval development. Longer arms, for example, may provide a more effective defence against some predators, or increase efficiency of feeding since longer arms will have a larger ciliated surface area that may result in increased clearing rates for feeding larvae (Strathmann et al. 1992).

The factors that influence the rate of larval development in echinoids are mainly abiotic and include temperature, salinity and food availability (Pearse and Cameron 1991); however, few studies have been carried out on *Diadema*. The larval period has been estimated for *D. setosum* at 6 weeks, which is about average for many echinoids (Emlet et al. 1987). In species that have feeding larvae, egg size may be an important factor controlling larval growth. The eggs of *Diadema* are some of the smallest reported for species with planktotrophic larvae (Emlet et al. 1987). *Diadema* reared at 26 °C develop into blastulae within 6 h and early plutei within 35 h (Amy 1983), which is a relatively fast rate compared to the cidaroid *Eucidaris tribuloides* (egg diameter 95 μm). However, as noted by Lessios (1987), egg size is very variable and is not a good predictor of larval success.

2.6. Benthic Population Dynamics

The single most significant event that affected *D. antillarum* populations was the 1983–1984 mass mortality, the most extensive die-off of a marine animal ever reported. Lessios (1988) described the advance of the disease that was first reported in Panama; it spread 2000 km east to Tobago and 4000 km west along the coastlines of central America, through the Gulf of Mexico to Bermuda generally following the surface currents, indicating that the agent that caused the mass mortality was waterborne. The disease, however, also spread from Florida to the Lesser Antilles against the major offshore currents. The disease caused the death of more than 93% of *D. antillarum* individuals over a 13-month period. Interestingly, despite the first records of the die-off occurring at the mouth of the Panama Canal, the disease did not infect the Panamanian populations of *D. mexicanum* (Lessios et al. 1984).

In general, populations that initially had high densities of *D. antillarum*, such as Barbados, experienced less mortality (Hunte and Younglao 1988). The causal agent of the mass mortality remains unknown. Bauer and Ageter (1987) cultured two species of bacteria *Clostridium* from infected individuals that caused mortality in healthy individuals but cautioned that the evidence is not sufficient to conclude that these bacteria caused the mass mortality. In addition, although the die-off coincided with the 1983 El Niño that caused some mortality of corals in the Caribbean, there was no direct link between the El Niño and the die-off (Lessios 1988). A similar mortality event occurred in the Florida Keys in 1991 (Forcucci 1994) that reduced densities of *D. antillarum* by 97% in the offshore reefs off Key West and by 83% at a patch reef offshore of Long Key in the Florida Keys. The cause of this decline is unknown (Forcucci 1994).

Although many reefs experienced recruitment a few months after the mass die-off, *D. antillarum* populations continued to decline in Panama (Lessios 1988), Jamaica (Hughes et al. 1985) and Curaçao (Bak 1985) a few years after the die-off as a result of recruitment failure (Karlson and Levitan 1990). Recovery of *D. antillarum* populations has not occurred throughout most of its range (Lessios 1988; Carpenter 1990a; Karlson and Levitan 1990) except in Barbados where most reefs showed up to 57% recovery by 1985 (Hunte and Younglao 1988) and some sites along the northern coast of Jamaica that are showing signs of recovery after a recruitment failure of over a decade with abundance being close to densities recorded in the early 1970s and 1980s (Edmunds and Carpenter 2001). In Barbados where the density of *D. antillarum* had initially been high, the reefs with the highest population density had the highest recruitment, an indication that larvae preferentially settle on reefs with high adult densities. Localised recruitment of *D. antillarum* to St. Croix in 2000–2001 (Miller et al. 2003) and Jamaica (Edmunds and Carpenter 2001) has been reported. Knowlton (2004) pointed out that the failure of widespread recovery of *D. antillarum* over two decades was not expected. She suggested that there might be a threshold density below which population growth is negative. Low recovery rates despite some recruitment could also be due to high post-settlement mortality (Chiappone et al. 2002). Recent surveys at six locations across the Caribbean from Jamaica to Bonaire indicate recovery of *D. antillarum* populations to post-die off densities in shallow waters on outer reef communities (Carpenter and Edmunds 2006).

The Pacific species of *Diadema* have not received as much attention as *D. antillarum* and few long-term population studies exist. McClanahan (1998) compared the abundance, distribution and diversity of nine species of sea urchins including *D. savignyi* and *D. setosum* at several locations over a 7-year period in Kenya. The abundance of these species showed variability in space and time. For example, *D. savignyi* populations showed an average overall increase in density of 37% across all sites except at Kanamai where a decrease of 7% occurred over the 7-year period. *D. setosum* also showed a similar overall average increase of 36%; however, two sites showed decreases, a newly established marine protected area (MPA), the Mombasa Marine Park (4%) and Kanamai (3%). At Diani, a heavily fished reef, an increase of ~136% is the reported for several sites.

The major factor controlling abundance of sea urchins in this study was the level of predation, with protected sites that had the highest abundance of predators and the highest predation index showing the lowest densities of urchins (McClanahan and Shafir 1990; McClanahan and Mutere 1994; McClanahan 1998, 2000). This relationship between sea urchin abundance and predation remained robust across geographical regions, as comparisons between reefs in Kenya and Tanzania showed similar responses (McClanahan et al. 1999). Although fringing reefs had more *Echinometra*, *Diadema* were common on the patch and island reefs of Tanzania where *Echinometra* were rare. Tuya et al. (2004b) reported fisheries in the Canary Islands resulted in increased density of *D. antillarum* dominated by small to intermediate sized individuals.

3. COMMUNITY ECOLOGY AND COEXISTENCE

3.1. Ecosystem Effects

Diadema and its effects on coral reef ecosystems have been well studied. Field studies and experimental manipulations of density have shown that *D. antillarum* has a large effect on algal biomass, diversity and productivity (Sammarco et al. 1974; Ogden and Lobel 1978; Carpenter 1981; Sammarco 1982a; Sammarco et al. 1986). Tuya et al. (2004c) reported the increase in density of *D. antillarum* in the Canarian Archipelago due to fishing that resulted in a decrease in fleshy macroalgae. Brown algal assemblages reached deeper waters at unfished sites (Tuya et al. 2005). In the Caribbean, the intensity of the effects of the die-off of *D. antillarum* was also greater in those areas subjected to intense fishing (Hay 1984; Carpenter 1990b; Robertson 1991). At intermediate densities, *D. antillarum* has a positive effect on coral recruitment, reducing competition for space with algae (Sammarco 1980). Reefs that are showing recovery in Jamaica, for example, had an 11-fold increase in juvenile corals (Edmunds and Carpenter 2001) and similar patterns have been reported on other Caribbean reefs (Carpenter and Edmunds 2006). However, at very high densities, above 7 adult individuals per square metre, coral spat are preyed upon by *D. antillarum* with negative effects on coral recruitment (Sammarco 1980).

Although *D. savignyi* and *D. setosum* are common and often the most abundant sea urchins on reefs in the Indo-Pacific (Pearse 1998; McClanahan et al. 1999), few detailed studies have been carried out on their effect, except in East Africa. McClanahan et al. (1996), in a sea urchin reduction experiment that included reduction of *D. savignyi* and *D. setosum*, reported large effects on the fish and algal biomass. The biomass of fish nearly tripled after reduction with population densities increasing by 65% and species richness increasing by 30% compared to adjacent control plots. Algae and seagrass cover increased more in the reefs that were fished than in the unfished reefs following the reduction. Dart (1972) suggested grazing by echinoids, including *D. setosum*, facilitated coral colonisation in the Sudanese Red Sea.

3.2. Competitive Interactions with Other Sea Urchins

A comparison of differences in body morphology, distribution, diet, intra- and inter-competition between *D. savignyi*, *D. setosum* and *E. mathaei* showed that *E. mathaei* is the competitive dominant at reefs in Kanamai and Vipingo, Kenya (McClanahan 1988). *D. savignyi* tended to occur in small crevices, whereas *D. setosum* occurred in the open spaces between coral, usually in groups. Experimental manipulations showed that *E. mathaei*, though smaller, outcompeted both *Diadema* species for crevice space. Although interactions between *D. setosum* and *D. savignyi* always resulted in the larger individual winning, *D savignyi* was the competitive dominant for crevice space due to the reduced spine length to test size ratio. This gives it a larger test size for the same crevice space requirement (McClanahan 1988). Competitive interactions have also been reported between the Caribbean species of *Diadema* and *Echinometra* with *D. antillarum* seemingly the superior competitor (Williams 1977, 1980; Shulman 1990), but as pointed out in the *Echinometra* chapter (Chapter 15), this is unlikely as *Echinometra* always has refuge in its smaller burrows.

D. savignyi and *D. setosum* sometimes form large aggregations throughout their range (Fox 1926; Stephenson et al. 1931; Onada 1936; Dakin and Bennett 1963). Similar formations have been reported for *D. antillarum* and *D. mexicanum* (Randall et al. 1964; Kier and Grant 1965; Pearse 1968a). Pearse and Arch (1969) suggested these aggregations function as protective social units and form when population densities are so high that crevices are limited. Levitan (1988b) reported that the aggregative behaviour of *D. antillarum* was not related to reproduction and Pearse (1968a) and McClanahan (1988) suggested that this might be a predator–avoidance behavior.

D. savignyi and *D. setosum* also occur in large groups and coexist on the Kenyan coast because of a lunar spawning rhythm and seasonal reproductive cycle that separate these sympatric species reproductively (Muthiga 2003). *D. savignyi*, which is more abundant in East Africa (McClanahan et al. 1999), has a reproductive pattern that is highly synchronous during the lunar period and a peak reproductive period that coincides with the time most favourable for larval development. These two factors together could serve to make *D. savignyi* more reproductively successful than *D. setosum*.

3.3. Competitive Interactions with Fish

Diadema directly compete with several species of fish, including damselfish and other herbivorous fish (Kaufman 1977; Williams 1977, 1978, 1980; Sammarco and Williams 1982). The three-spot damselfish *Eupomacentrus planifrons* actively excludes *D. antillarum* from its algal lawns during the day at Discovery Bay Jamaica and influences the local distribution pattern and abundances of this urchin (Sammarco and Williams 1982). Algal lawns encourage the growth of some species of coral such as *Favia fragum* but are also sites of high spat mortality of some coral species, such as *Agaricia* and *Porites* spp. Algal lawns may, however, have an overall net effect of increasing coral diversity within the reef community (Sammarco and Williams 1982). Damselfishes also exclude *Diadema* in algal lawns in the Gulf of Thailand (Kamura and Choonhabandit 1986), Tahiti (Glynn and Colgan 1988), Panama (Eakin 1987; Glynn and Colgan 1988) and the Galapagos (Glynn and Wellington 1983).

The three-spot damselfish also plays a mediating role by altering the competitive interactions between *D. antillarum* and *E. viridis* in a back-reef environment in Jamaica (Williams 1977, 1980). The more mobile but less abundant *D. antillarum* elicited a stronger territorial response from the damselfish than the more abundant *E. viridis* that resulted in reduced competitive interactions between the urchin species. Competitive interactions also occur between *D. mexicanum* and the Acapulco damselfish *Stegastes acapulcoensis* at Uva island Panama (Eakin 1988). The urchin was excluded from algal lawns through aggression and actively avoided algal lawns even when lawns were undefended at night. This suggests that apart from the aggressive behaviour of damselfish, algal lawns are not preferred either due to unpalatability or accessibility (Eakin 1988). Since the algal lawns and underlying substratum remain protected from grazing by *D. mexicanum*, this interaction potentially influences the reef framework by reducing bioerosion of the substratum.

3.4. Predation and Predators

Remains of *Diadema* have been identified in 15 species of finfish, including balistids, carangids, diodontids, labrids, ostraciids, pomadasyids, sparids and tetraodontids (mostly species with hard palettes), as well as two species of gastropods (*Cassis madagascariensis* and *C. tuberosa*) and the spiny lobster *Panulirus argus* (Randall et al. 1964). The toadfishes *Amphichthys cryptocentrus* and *Sanopus barbatus* as well as the queen triggerfish, *Balistes vetula*, were reported to feed almost exclusively on *D. antillarum* before the die-off but switched to small fishes and mobile benthic invertebrates (Robertson 1987; Reinthal et al. 1984). Remains of *D. antillarum* have also been reported in small wrasses including *Halichoeres bivittatus*, *H. poeyi* and *Thalassoma bifasciatum*; however, this is thought to be from the scavenging activities of these wrasses. The West Indian butterflyfish *Prognathodes aculeatus* and the Indo-Pacific butterflyfish (*Forcipiger*) that have long snouts feed on the tube feet of *D. antillarum.* The clingfish *Diademichthys lineatus* has also been reported to feed on *Diadema* tube feet and pedicellariae (Sakashita 1992).

No gut analysis or direct observation of predation on *D. savignyi* or *D. setosum* has been reported. McClanahan (1988) tethered both these species and found that predation rates were low compared to rates on *E. mathaei*, the competitive dominant, at Kanamai and Vipingo reef lagoons in Kenya and suggested that this led to species coexistence. In addition, apart for reports on predation on coral and coral spat by *D. antillarum* (Bak and van Eys 1975; Carpenter 1981; Sammarco 1980, 1982b), few reports on predation by *Diadema* are available.

4. HERBOVORY AND GRAZING EFFECTS

4.1. Herbivory

Grazing by *D. antillarum* has an important ecological function on reefs and benthic algal communities throughout the Caribbean and western Atlantic (Ogden et al. 1973; Sammarco et al. 1974; Carpenter 1981, 1986; Sammarco 1982a; Lessios et al. 1984). It is estimated that prior to the die-off, approximately 20% of the monthly net benthic primary production of the fringing reef in Barbados was consumed by *D. antillarum* (Hawkins and Lewis 1982). This is considerably higher than the estimate of 7% for the sea urchin *Strongylocentrotus droebachiensis* in kelp beds but lower than estimates of 47% for *L. variegatus* in seagrass habitats (Hawkins and Lewis 1982). Tuya et al. (2004c) reported a similar important role for *D. antillarum* on algal assemblages in the eastern Atlantic.

In an experiment that eliminated an entire population of *D. antillarum* on a reef at St. Croix, Virgin Islands, algal biomass increased by an order of magnitude, species composition was altered, shifts in dominance occurred (*Halimeda opuntia*, the dominant algae, was replaced in dominance by other species of algae, especially *Padina sanctae-crucis*), an increase in species numbers (at least five more algal species not found in grazed controls), and a decrease in equitability due mainly to the high dominance of *P. sanctae-crucis*.

The well-documented mass mortality of *D. antillarum* of 1983–1984 and the subsequent studies on the effects of the die-off emphasised the importance of grazing by this species. During this period, the population density of *D. antillarum* was reduced by 90–100% with dramatic increases of algal biomass (42–93%) and changes in community structure (Bak et al. 1984; Lessios et al. 1984; Carpenter 1985, 1990b; de Ruyter van Steveninck and Bak 1986; Liddell and Ohlhorst 1986; Hughes et al. 1985, 1987). In addition to the large increase in algal biomass, coral and crustose coralline algae and coral recruitment decreased and trapping of sediments by algal turfs increased. This supports the hypothesis that grazing by *D. antillarum* is the primary determinant of algal distribution and biomass on shallow reefs throughout the Caribbean (Sammarco et al. 1974; Carpenter 1986; Foster 1987).

Herbivory by *D. antillarum* enhances the quality of algal turfs, increasing productivity 2- to 10-fold more per unit chlorophyll a than turfs that are not grazed (Carpenter 1981, 1985b, 1986). Ammonium excretion by *D. antillarum* supplies up to 19% of the total nitrogen requirement of these algal turfs (Williams and Carpenter 1988). Because algal turfs account for the bulk of primary productivity in coral reefs, this enhanced productivity has a large effect on the community of reefs in the Caribbean.

Diadema compete for food with herbivorous fishes (Ogden 1976; Wanders 1977; Williams 1981; Hay and Taylor 1985). Therefore, the die-off of *D. antillarum* led to substantial increases of algal biomass and had the potential to increase abundances of herbivorous fishes. Robertson (1991) showed that the numbers of surgeonfishes *Acanthurus coeruleus*, and *A. chirurugus* that feed almost exclusively in reefs where *D. antillarum* occur increased by 250% and 160%, respectively, between 1979 and 1989. No increase occurred in *A. bahianus*, a species that feeds in offshore habitats free from *D. antillarum*. These results are consistent with the suggestion that these two species competed for food with *D. antillarum* (Hay and Taylor 1985) and that competition limited their population size (Carpenter 1990b). Increases in herbivorous fishes including surgeonfishes and parrotfishes were also reported in St. Croix after the die-off (Carpenter 1990b). These findings are consistent with sea urchin reduction experiments reported from Kenya (McClanahan et al. 1996).

4.2. Bioerosion

Prior to the die-off, the grazing activity of *D. antillarum* had a substantial effect on calcium carbonate budgets of reefs in the Caribbean (Ogden 1977; Stearn and Scoffin 1977; Stearn et al. 1977; Scoffin et al. 1980). It is estimated, for example, that at pre-mortality densities of 3–12 individulas per square metre, *D. antillarum* eroded the reef at rates of 3.6–9.1 $kg\,m^{-2}\,yr^{-1}$ (Ogden 1977; Bak 1994). Given that the average coral reef accretes at ~4 $kg\,m^{-2}\,yr^{-1}$ (Bak 1994), the populations of *D. antillarum* had a significant influence on calcium carbonate budgets of Caribbean reefs. In addition, *D. antillarum* occasionally preys on coral spat and influences coral recruitment. Some grazing is beneficial for coral recruitment through reduction of algal growth but coral recruitment is greatly depressed at high densities of urchins (Sammarco 1980).

In the Indian Ocean, a study of bioerosion and herbivory on Kenyan reefs showed bioerosion to be greater than herbivory by sea urchins and was proportional to body size of the urchin species (Carreiro-Silva and McClanahan 2001). The larger bodied *D. setosum* exhibited bioerosion rates of 1.8 ± 0.3 g $CaCO_3$ ind^{-1} d^{-1} and herbivory rates of 1.1 ± 0.2 g algae ind^{-1} d^{-1}, whereas the smaller *D. savignyi* exhibited bioerosion rates of 0.7 ± 0.2 $CaCO_3$ ind^{-1} d^{-1} and herbivory rates of 0.4 ± 0.1 g algae ind^{-1} d^{-1}. As the highest densities of urchins were recorded in fished reefs, rates of bioerosion (1180 ± 230 g $CaCO_3$ m^{-2} yr^{-1}) and herbivory (450 ± 77 g algae m^{-2} yr^{-1}) were highest on fished reefs (Carreiro-Silva and McClanahan 2001). Protected reefs showed 20 times lower rates of bioerosion and herbivory (50.3 ± 25.8 g $CaCO_3$ m^{-2} yr^{-1} and 20.7 ± 10.4 g algae m^{-2} yr^{-1}). The findings from this study suggest that echinoids are important in the carbon cycle and reef development of Indian Ocean reefs and that fishing can influence these ecological processes. Uy et al. (2001) reported that bioerosion by *D. setosum* in the Philippines was greater in areas without than with seagrass.

5. CONCLUSIONS

Sea urchins in the genus *Diadema* are among the dominant grazers and forces of erosion of the substratum in shallow tropical reef environments. They are likely to compete with other herbivores and also influence the interactions between algae and coral (Carpenter and Edmunds 2006). Consequently, they can play an important role in the ecology of tropical reefs and the factors that influence their population dynamics and coexistence with others species are of significance to reef ecology and fisheries management (McClanahan et al. 1996, 1999). In some places and at some population densities, they are considered pests as they dominate benthic primary productivity and exclude other herbivores (Tuya et al. 2005), while they can also be seen as important in maintaining and restoring coral populations (Carpenter and Edmunds in press) and, therefore, have been considered candidates for aquaculture and reef restoration. Consequently, the historical levels of their populations are of interest as it may indicate the degree to which they are either required for the maintenance of coral reef ecology (Jackson et al. 2001; Lessios et al. 2001b) or pests that have been released from predation by overfishing (Hay 1984; Levitan 1992). This is likely to differ with biogeographic regions, evolutionary history, and habitats such that it is difficult to conclude what are the appropriate or historical levels of abundance. Modelling studies have shown, however, that at biomass of around 400 g m^{-2} (about 4 adults per square meter) *Diadema* dominate benthic primary productivity and exclude other herbivores (McClanahan 1992, 1995). Consequently, for management purposes, at half these levels they may keep algal biomass sufficiently low to allow for coral recruitment and create surplus benthic productivity to allow for coexistence with other herbivores.

REFERENCES

Amy RL (1983) Gamete sizes and developmental time tables of five tropical sea urchins. Bull Mar Sci 33: 173–176

Bak RPM (1985) Recruitment patterns and mass mortalities in the sea urchin *Diadema antillarum*. Proc 5th Int Coral Reef Congr 5: 267–272

Bak RPM (1994) Sea urchin bioerosion on coral reefs; place in carbonate budget and relevant variables. Coral reefs 13: 99–103

Bak RPM, Van Eys G (1975) Predation of a sea urchin *Diadema antillarum* on living coral. Oecologia 20: 111–115

Bak RPM, Carpay MJE, de Ruyter van Steveninck ED (1984) Densities of the sea urchins *Diadema antillarum* before and after mass mortalities on the coral reefs of Curacao. Mar Ecol Prog Ser 17: 105–108

Bauer JC (1976) Growth, aggregation and maturation in the echinoid *Diadema antillarum* Phillipi. Bull Mar Sci 26: 273–277

Bauer JC (1980) Observations on geographical variations in population density of the echinoid *Diadema antillarum* within the western north Atlantic. Bull Mar Sci 30: 509–515

Bauer JC (1982) On the growth of a laboratory-reared sea urchin, *Diadema antillarum* (Echinodermata: Echinoidea). Bull Mar Sci 32: 643–645

Bauer, JC, Agerter CJ (1987) Isolation of bacteria pathogenic for the sea urchin *Diadema antillarum* (Echinodermata: Echinoidea). Bull Mar Sci 40: 161–165

Birkeland C (1989) The influence of echinoderms on coral reef communities. In: Jangoux M, Lawrence JM (eds) Echinoderm studies, Vol. 3. AA Balkema, Rotterdam, pp 1–79

Carpenter RC (1981) Grazing by *Diadema antillarum* (Philippi) and its effects on the benthic algal community. J Mar Res 39: 749–765

Carpenter RC (1985) Sea urchin mass mortality: effect of reef algal abundance, species composition and metabolism and other coral reef herbivores. Proc 5th Int Coral Reef Congr 4: 53–60

Carpenter RC (1986) Partitioning herbivory and its effects on coral reef algal communities. Ecol Mono 56: 345–363

Carpenter RC (1990a) Mass mortality of *Diadema antillarum* I. Long-term effects on sea urchin population-dynamics and coral reef algal communities. Mar Biol 104: 67–77

Carpenter RC (1990b) Mass mortality of *Diadema antillarum*: II. Effects on population densities and grazing intensities of parrotfishes and surgeonfishes. Mar Biol 104: 79–86

Carpenter RC (1997) Invertebrate predators and grazers. In: Birkeland C (ed.). Life and Death of Coral Reefs. Chapman & Hall, New York, pp 198–229

Carpenter RC, Edmunds PJ (2006) Local and regional scale recovery of *Diadema* promotes recruitment of scleractinian corals. Ecol Letts 9(3): 271–280

Carreiro-Silva M, McClanahan TR (2001) Echinoid bioerosion and herbivory on Kenyan coral reefs: the role of protection from fishing. J Exp Mar Biol Ecol 262: 133–153

Chiappone M, Swanson DW, Miller SL, Smith SG (2002) Large-scale surveys on the Florida Reef Tract indicate poor recovery of the long-spined sea urchin *Diadema antillarum.* Coral Reefs 21: 155–159

Clark H (1921) The echinoderm fauna of Torres Strait; its composition and its origin, Carnegie Institution Publisher, Washington, DC, pp 214–224

Clark, H (1946) The echinoderm fauna of Australia; its composition and its origin. Carnegie Institution Publisher, Washington, DC, pp 566–567

Clark A (1966) Echinoderms from the Red Sea. Part 2. Bull sea fish. Res Stn Israel 41: 26–58

Clark A, Rowe FWE (1971) Monograph of the shallow water Indo-West Pacific Echinoderms. British Museum (Natural History), London

Coppard SE, Campbell AC (2004) Taxonomic significance of the spine morphology in the echinoid genera *Diadema* and *Echinothrix*. Invert Biol 123(4): 357–371

Dakin W, Bennett I (1963) The Great Barrier Reef and some mention of other Australian Coral reefs. 2nd ed. Une Smith, Sydney, p. 176

Dart JKG (1972) Echinoids, algal lawn and coral recolonization. Nature 239: 50–51

de Ruyter van Steveninck ED, Bak RPM (1986) Changes in abundance of coral reef bottom components related to mass mortality of the sea urchin *Diadema antillarum.* Mar Ecol Prog Ser 34: 87–94

Drummond AE (1994) Aspects of the life history biology of three species of sea urchins on the South African east coast. In: David B, Guille A, Feral JP, Roux M (eds). Echinoderms through Time. AA Balkema, Rotterdam, pp 637–641

Drummond AE (1995) Reproduction of the sea urchins *Echinometra mathaei* and *Diadema savignyi* on the South African Eastern Coast. Mar Freshwater Res 46: 751–757

Eakin CM (1987) Damselfish and their algal lawns, a case of plural mutualism. Symbiosis 4: 275–288

Eakin CM (1988) Avoidance of damselfish lawns by the sea urchin *Diadema mexicanum* at Uva Island, Panama. 6th Int Coral Reef Symp 2: 21–26
Ebert TA (1980) Relative growth of sea urchin jaws: an example of plastic resource allocation. Bull Mar Sci 30: 467–474
Ebert TA (1982) Longevity, life history, and relative body wall size in sea urchins. Ecol Monogr 52: 353–394
Edmunds PJ, Carpenter RC (2001) Recovery of *Diadema antillarum* reduces macroalgal cover and increases abundance of juvenile corals on a Caribbean reef. Proc Natl Acad Sci 98: 5067–5071
Emlet RB, McEdward LR, Strathmann RR (1987) Echinoderm larval ecology viewed from the egg. In: Jangoux M, Lawrence JM (eds) Echinoderm studies, Vol. 2. AA Balkema, Rotterdam, pp 55–136
Forcucci D (1994) Population density, recruitment and 1991 mortality event of*Diadema antillarum* in the Florida Keys. Bull Mar Sci 54: 917–928
Foster SA (1987) The relative impacts of grazing by Caribbean coral reef fishes and *Diadema*: effects of habitat and surge. J Exp Mar Bio Ecol 105: 1–20
Fox HM (1924) Lunar periodicity in reproduction. Proc R Soc Lond B 95: 523–550
Fox HM (1926) Cambridge expedition to the Suez Canal, 1924. Appendix to the report on the echinoderms. Trans Zool Soc 22: 129
Glynn P, Wellington GM (1983) Corals and coral reefs of the Galapagos Islands. University of California Press, Berkeley
Glynn PW, Colgan, MW (1988) Defense of corals and enhancement of coral diversity by territorial damselfishes. Proc 6th Int Coral Reef Sym 2: 157–163
Hawkins CM, Lewis JB (1982) Ecological energetics of the tropical sea urchin *Diadema antillarum* Philippi in Barbados, West Indies. Estuar Coast Shelf Sci 15: 645–669
Hay ME (1984) Patterns of fish and urchin grazing on Caribbean coral reefs: are previous results typical? Ecology 65: 446–454
Hay ME, Paul VJ (1986) Seaweed susceptibility to herbivory: chemical and morphological correlates. Mar Ecol Prog Ser 33: 255–264
Hay ME, Taylor PR (1985) Competition between herbivorous fishes and urchins on Caribbean reefs. Oecologia 65: 591–598
Hay ME, Fenical W, Gustafson K (1987) Chemical defense against diverse coral-reef herbivores. Ecology 68: 1581–1591
Hori R, Phang VPE, Lam TJ (1987) Preliminary study on the pattern of gonadal development of the sea urchin *Diadema setosum* off the coast of Singapore. Zool Sci 4: 665–673
Hughes TP, Keller BD, Jackson JBC, Boyle MJ (1985) Mass mortality of the echinoid *Diadema antillarum* Philippi in Jamaica. Bull Mar Sci 36: 377–384
Hughes TP, Reed DC, Boyle, MJ (1987) Herbivory on coral reefs: community structure following mass mortalities of sea urchins. J Exp Mar Biol Ecol 133: 39–59
Hunte W, Younglao D (1988) Recruitment and population recovery of *Diadema antillarum* (Echinodermata: Echinoidea) in Barbados. Mar Ecol Prog Ser 45: 109–119
Iliffe TM, Pearse JS (1982) Annual and lunar reproductive rhythms of the sea urchin, *Diadema antillarum* (Philippi) in Bermuda. Int J Invert Reprod 5: 139–148
Jackson JBC, Kirby MX, Berger WH, Bjorndal KA, Botsford LW, Bourque BJ, Bradbury RH, Cooke R, Erlandson J, Estes JA, Hughes TP, Kidwell SM, Lange CB, Lenihan HS, Pandolfi JM, Peterson CH, Steneck RS, Tegner MJ, Warner RR (2001) Historical overfishing and the recent collapse of coastal ecosystems. Science 293: 629–638
Kamura S, Choonhabandi S (1986) Algal communities within the territories of the damselfish, *Stegastes apicalis* and the effects of grazing by the sea urchin *Diadema* species in the Gulf of Thailand. Galaxea 5: 175–193
Karlson RH, Levitan DR (1990) Recruitment limitation in open populations of *Diadema antillarum*: an evaluation. Oecologia 82: 40–44
Kauffman L (1977) The threespot damselfish: effects on benthic biota of Caribbean Coral reefs. Third int. Coral Reef Symposium, Miami, Florida: 559–564
Kier P, Grant RE (1965) Echinoid distribution and habits, Key Largo Coral Reef Preserve, Florida. Smith Misc Coll 149: 1–68
Knowlton N (2004) Multiple "stable" states and the conservation of marine ecosystems. Prog Oceanogr 60: 387–396

Kobayashi N, Nakamura K (1967) Spawning periodicity of the sea urchins at Seto. II. *Diadema Setosum*. Publs Seto mar. biol. lab 15: 173–184

Lawrence JM (1987) A functional biology of echinoderms. The John Hopkins University Press, Baltimore

Lawrence JM, Hughes-Games (1972) The diurnal rhythm of feeding and passage of food through the gut of *Diadema setosum* (Echinodermata: Echinoidea). Israel J Zool 21: 13–16

Lawrence JM, Sammarco PW (1982) Effects of feeding on the environment: echinoidea In: Jangoux M, Lawrence JM (eds) Echinoderm nutrition. AA Balkema, Rotterdam, pp 499–519

Lawrence JM, Lawrence AL, Holland ND (1965) Annual cycle in the size of the gut of the purple sea urchin, *Strongylocentrotus purpuratus* (Stimpson). Nature 205: 1238–1239

Lessios HA (1981) Reproductive periodicity of the echinoids *Diadema* and *Echinometra* on the two coasts of Panama. J Exp Mar Biol Ecol 50: 47–61

Lessios HA (1984) Possible prezygotic reproduction isolation in sea urchins separated by the Isthmus of Panama. Evolution 38: 1144–1148

Lessios HA (1987) Temporal and spatial variation in egg size of 13 Panamanian echinoids. J Exp Mar Biol Ecol 114: 217–239

Lessios HA (1988) Mass mortality of *Diadema antillarum* in the Caribbean: what have we learned? Ann Rev Ecol Syst 19: 371–393

Lessios HA, Pearse JS (1996) Hybridization and introgression between Indo-pacific species of the sea urchin *Diadema*. Mar Biol 126: 715–723

Lessios HA, Robertson DR, Cubit JD (1984) Spread of *Diadema* mass mortality through the Caribbean. Science 226: 335–337

Lessios HA, Kessing BD, Wellington GM, Greybeal A (1996) Indo-Pacific echinoids in the tropical eastern Pacific. Coral Reefs 15: 133–142

Lessios HA, Kessing BD, Pearse JS (2001a) Population structure and speciation in tropical seas: global phylogeography of the sea urchin Diadema. Evolution 55: 955–975

Lessios HA, Garrido MJ, Kessing BD (2001b) Demographic history of *Diadema antillarum*, a keystone herbivore on Caribbean reefs. Proc R Soc Lond B 268: 1–7

Levitan DR (1988a) Density-dependent size regulation and negative growth in the sea urchin *Diadema antillarum* Philippi. Oecologia 76: 627–629

Levitan DR (1988b) Asynchronous spawning and aggregative behavior in the sea urchin *Diadema antillarum* Philippi. Proc 6th Int Echinoderm Conf 181–186

Levitan DR (1989) Density-dependent size regulation in *Diadema antillarum*: effects on fecundity and survivorship. Ecology 70: 1414–1424

Levitan DR (1991) Skeletal changes in the test and jaws of the sea urchin *Diadema antillarum* in response to food limitation. Mar Biol 111: 431–435

Levitan DR (1992) Community structure in times past: influence of human fishing pressure on algal–urchin interactions. Ecology 73: 1597–1605

Lewis JB (1958) The biology of the tropical sea urchin *Tripneustes esculentus* Leske in Barbados, British West Indies. Can J Zool 36: 607–621

Lewis JB (1964) Feeding and digestion in the tropical sea urchin *Diadema antillarum* Philippi Can J Zool 42: 550–557

Lewis JB (1966) Growth and breeding in the tropical echinoid *Diadema antillarum* Philippi. Bull Mar Sci 16: 151–157

Liddell WD, Ohlhorst SL (1986) Changes in benthic community composition following the mass mortality of *Diadema antillarum*. J Exp Mar Biol Ecol 95: 271–278

Littler MM, Taylor PR, Littler DS (1983) Algal resistance to herbiuory on a Caribbean barrier reef. Coral Reefs 2: 111–118

Magnus DBE (1967) Ecological and ethological studies and experiments on the echinoderms of the Red Sea. In: Studies in Tropical Oceanography, 5. Miami, pp 635–664

Maluf LV (1988) Composition and distribution of the central eastern Pacific echinoderms. Technical Reports, Number 2, Natural History Museum of Los Angeles County, Los Angeles

McClanahan TR (1988) Coexistence in a sea urchin guild and its implications to coral reef diversity and degradation. Oecologia 77: 210–218

McClanahan TR (1992) Resource utilization, competition and predation: a model and example from coral reef grazers. Ecol Mod 61: 195–215

McClanahan TR (1995) A coral reef ecosystem-fisheries model: impacts of fishing intensity and catch selection on reef structure and processes. Ecol Mod 80: 1–19

McClanahan TR (1998) Predation and the distribution and abundance of tropical sea urchin populations. J Exp Mar Biol Ecol 221: 231–255

McClanahan TR (2000) Recovery of the coral reef keystone predator, *Balistapus undulatus*, in East African marine parks. Biol Cons 94: 191–198

McClanahan TR, Shafir SH (1990) Causes and consequences of sea urchin abundance and diversity in Kenyan coral reef lagoons. Oecologia 83: 362–370

McClanahan TR, Mutere JC (1994) Coral and sea urchin assemblage structure and interrelationships in Kenyan reef lagoons. Hydrobiologia 286: 109–124

McClanahan TR, Kamukuru AT, Muthiga NA, Yebio MG, Obura D (1996) Effect of sea urchin reductions on algae, coral, and fish populations. Conserv Biol 10: 136–154

McClanahan TR, Muthiga NA, Kamukuru AT, Machano H, Kiambo RW (1999) The effects of marine parks and fishing on coral reefs of northern Tanzania. Biol Conserv 89: 161–182

Miller RJ, Adams AJ, Ogden NB, Ogden JC, Ebersole JP (2003) *Diadema antillarum* 17 years after mass mortality: is recovery beginning on St. Croix? Coral Reef 22: 181–187

Moore HB, Jutare T, Bauer JC, Jones JA (1963) The biology of *Lytechinus variegatus*. Bull Mar Sci 13: 23–53

Mortensen T (1931) Contributions to the study of the development of the larval forms of the echinoderms I–II. Mem de l'Academie Royale des Sciences et des Lettres de Danemark, Copenhague, Section des Sciences, 9me serie t. IV. No.1 1–39

Mortensen T (1937) Contributions to the study of the development of the larval forms of the echinoderms III. Mem de l'Academie Royale des Sciences et des Lettres de Danemark, Copenhague, Section des Sciences, 9me serie t. VII. No. 1 1–65

Mortensen T (1940) A monograph of the Echinoidea III. Aulodonta with additions to Vol. II (Lepidocentroida and Stirodonta). Reitzel, Copenhagen, pp 392

Muthiga NA (2003) Coexistence and reproductive isolation of the sympatric echinoids *Diadema savignyi* Michelin and *Diadema setosum* (Leske) on Kenyan coral reefs. Mar Biol 143: 669–677

Ogden JC (1976) Some aspects of herbivore-plant relationship on Caribbean reefs and seagrass beds. Aquat Bot 2: 103–116

Ogden JC (1977) Carbonate sediment production by parrotfish and sea urchins on Caribbean reefs. Tulsa 4: 281–288

Ogden JC, Lobel PS (1978) The role of herbivorous fishes and urchins in coral reef communities. Environ Biol Fish 3: 49–63

Ogden JC, Brown RA, Salesky N (1973) Grazing by the Echinoid *Diadema antillarum* Philippi: formation of halos around West Indian Patch Reefs. Science 182: 715–717

Onada, K (1936) Notes on the development of some Japanese echinoids with special reference to the structure of the larval body. Jap J Zool 6: 635–654

Pearse JS (1968a) Gametogenesis and reproduction in several abyssal and shallow water echinoderms of the Eastern Tropical Pacific. Stanford Oceanogr Exped Rpt Te Vega Cruise, pp 225–234

Pearse JS (1968b) Patterns of reproductive periodicities in four species of Indo-Pacific echinoderms Ind Acad Sci B 67: 247–279

Pearse JS (1970) Reproductive periodicities of Indo-Pacific invertebrates in the Gulf of Suez. III. The Echinoid *Diadema Setosum* (Leske). Bull Mar Sci 20: 697–720

Pearse JS (1972) A monthly reproductive rhythm in the Diadematid sea urchin *Centrostephanus coronatus* verrill. J Exp Mar Biol Ecol 8: 167–186

Pearse JS (1974) Reproductive patterns of tropical reef animals: three species of sea urchins. Proc 2nd Int Coral Reef Symp 1: 235–240

Pearse JS (1998) Distribution of *Diadema savignyi* and *D. setosum* in the tropical Pacific. In: Mooi R, Telford M (eds) Echinoderms. AA Balkema, Rotterdam, pp 777–782

Pearse JS, Arch SW (1969) The aggregation behavior of *Diadema* (Echinodermata: Echinoidea). Micronesia 5: 165–171

Pearse JS, Cameron RA (1991) Echinodermata, Echinoidea. In: Giese AC, Pearse JS, Pearse VB (eds) Reproduction of marine invertebrates, Echinoderms and Lophophorates, Vol. 6. The Boxwood Press, Pacific Grove, CA, pp 514–662

Randall JE, Schroeder RE, Starck WA (1964) Notes on the biology of the echinoid *Diadema antillarum*. Caribb J Sci 4: 421–433

Reinthal PN, Kensley B, Lewis SM (1984) Dietary shifts in the queen triggerfish *Balistes vetula* in the absence of its primary food item, *Diadema antillarum*. Mar Ecol 5: 191–195

Robertson DR (1987) Responses of two coral reef toadfishes (Batrachoididae) to the demise of their primary prey, the sea urchin *Diadema antillarum*. Copeia 3: 637–642

Robertson DR (1991) Increases in surgeonfish populations after mass mortality of the sea urchin *Diadema antillarum* in Panama indicate food limitation. Mar Biol 111: 437–444

Rowe F, Gates J (1995) Echinodermata. In: Wells A (ed.) Zoological Catalogue of Australia, Vol. 33. CSIRO, Melbourne, pp 510

Sakashita H (1992) Sexual dimorphism and food habits of the clingfish, Diademichthys lineatus and its dependence on host sea urchin. Environ. Biol. Fish. 34: 95–101

Sammarco PW (1980) Diadema and its relationship to coral spat mortality: grazing, competition, and biological disturbance. J Exp Mar Biol Ecol 45: 245–272

Sammarco PW (1982a) Effects of grazing by *Diadema antillarum* Philippi (Echinodermata: Echinoidea) on algal diversity and community structure. J Exp Mar Biol Ecol 65: 83–105

Sammarco PW (1982b) Echinoid grazing as a structuring force in coral communities: whole reef manipulations. J Exp Mar Biol Ecol 61: 31–55

Sammarco PW, Williams AH (1982) Damselfish territoriality: influence on *Diadema* distribution and implications for coral community structure. Mar Ecol Prog Ser 8: 53–59

Sammarco PW, Levinton JS, Ogden JC (1974) Grazing and control of coral reef community structure by *Diadema antillarum* Philippi (Echinodermata: Echinoidea): a preliminary study. J Mar Res 32: 47–53

Sammarco PW, Carleton JH, Risk MJ (1986) Effects of grazing and damsel fish territories on internal bioerosion of dead corals: direct effects. J Exp Mar Biol Ecol 98: 1–9

Scoffin T, Stern CW, Boucher D, Frydl P, Hawkins CM, Hunter IG, MacGeachy JK (1980) Calcium carbonate budget of a fringing reef on the west coast of Barbados II. Erosion, sediments and internal structure. Bull Mar Sci 30: 475–508

Shulman MJ (1990) Aggression among sea urchins on Caribbean coral reefs. J Exp Mar Biol Ecol 140: 197–207

Solandt JL, Campbell AC (2001) Macroalgal feeding characteristics of the sea urchin *Diadema antillarum* Philippi at Discovery Bay, Jamaica. Caribb J Sci 37: 227–238

Stearn CW, Scoffin TP (1977) Carbonate budget of a fringing reef, Barbados. Proc 3rd Int Coral Reef Symp 2: 471–476

Stearn CW, Scoffin TP, Martindale W (1977) Calcium carbonate budget of a fringing reef on the west coast of Barbados. Part 1. Zonation and productivity. Bull Mar Sci 27: 479–510

Stephenson A (1934) The breeding of reef animals. Part I. Invertebrates other than corals. Sci Rep Great Barrier Reef Exped 3(9): 247–272

Stephenson T, Stephenson A, Tandy G, Spender M (1931) The structure and ecology of Low Isles and other reefs. Sci Rep Great Barrier Reef Exped 3: 1–112

Strathmann RR, Fenaux L, Strathmann MF (1992) Heterochronic developmental plasticity in larval sea urchins and its implications for evolution of nonfeeding larvae. Evolution 46: 972–986

The Nature Conservancy (2004) The Diadema Workshop. The Nature Conservancy, National Fish and WIldlife Foundation and Rosenstiel School of Marine & Atmospheric Science, pp 29

Tuason AY, Gomez ED (1979) The reproductive biology of *Tripneustes gratilla* Linnaeus (Echinodermata: Echinoidea) with some notes on *Diadema setosum* Leske. Proc Int Symp Mar Biogeogr Evol South Hem 2: 707–716

Tuya F, Martin JA, Reuss GM, Luque A (2001) Feeding preferences of the sea urchin *Diadema antillarum* (Philippi) in Gran Canaria Island (Central-East Atlantic Ocean). J Mar Biol Ass UK 81: 1–5

Tuya F, Martin JA, Luque A (2004a) Patterns of nocturnal movement of the long-spined sea urchin *Diadema antillarum* (Philippi) in Gran Canaria (the Canary Islands, central East Atlantic Ocean). Helgol Mar Res 58: 26–31

Tuya F, Boyra A, Sanchez-Jerez P, Barbera C, Haroun RJ (2004b) Relationships between rocky-reef fish assemblages, the sea urchin *Diadema antillarum* and macroalgae throughout the Canarian Archipelago. Mar Ecol Prog Ser 278: 157–169

Tuya F, Boyra A, Sanchez-Jerez P, Barbra C, Haroun R (2004c) Can one species determine the structure of the benthic community on a temperate rock reef? The case of the long-spined sea-urchin *Diadema antillarum* (Echinodermata: Echinoidea) in the eastern Atlantic. Hydrobiologia 519: 211–214

Tuya F, Sanchez-Jerez P, Haroun RJ (2005) Influence of fishing and functional group of algae on sea urchin control of algal communities in the eastern Atlantic. Mar Ecol Prog Ser 287: 255–260

Uy FA, Bongalo NCB, Dy DT (2001) Bioerosion potential of three species of sea urchins commonly found in the reef flats of eastern Mactan Island, Cebu, Philippines. Philipp Sci 38: 26–38

Wanders J (1977) The role of benthic algae in the shallow reef of the Curacao (Netherlands, Antilles) III. The significance of grazing. Aquat Bot 3: 357–390

Williams AH (1977) Three-way competition in a patchy back reef environment. Ph.D. Dissertation, University of North Carolina, Chapel Hill

Williams AH (1978) Ecology of the threespot damselfish: social organisation, age, structure, and population stability. J Exp Mar Biol Ecol 34: 197–213

Williams AH (1980) The threespot damshelfish: a noncarnivorous keystone species. Am Nat 116: 138–142

Williams AH (1981) An analysis of the competitive interactions in a patchy back-reef environment. Ecology 62: 1107–1120

Williams SL, Carpenter RC (1988) Nitrogen-limited primary productivity of coral reef algal turfs: potential contribution of ammonium excreted by *Diadema antillarum*. Mar Ecol Prog Ser 47: 145–152

Wirtz P, Martins HM (1993) Notes on some rare and little known marine invertebrates from the Azores, with a discussion of the zoogeography of the region. Arquipe'lago 11A: 55–63

Yoshida, M (1952) Some observation on the maturation of the sea urchin, *Diadema setosum*. Annot Zool Jpn 25: 265–271

Edible Sea Urchins: Biology and Ecology
Editor: John Miller Lawrence

Chapter 12

Ecology of *Loxechinus albus*

Julio A Vásquez

Departamento de Biología Marina, Universidad Católica del Norte, Centros Estudios Avanzados en Zonas Aridas, Coquimbo (Chile).

1. INTRODUCTION

Loxechinus albus Molina 1782 (Plate 2A) is one of the most economically important species in the littoral benthic systems of the southeastern Pacific region of South America. This species has been used as a food source since pre-Colombian times (Jerardino et al. 1992) and by indigenous Chilean coastal populations (Deppe and Viviani 1977; Vásquez et al. 1996). Since *L. albus* is closely associated with *Macrocystis* beds, it is most abundant below 40° S where these beds occur. The first report on landings of *L. albus* (SERNAP 1957) showed 4000 TM extracted from the entire Chilean coastline; since then, landings have increased (SERNAP 1957–1999). This fishery has expanded and is supported mainly from landings from the southern region between 42° and 56° S. *Loxechinus albus* is one of the most important benthic grazers on intertidal and shallow subtidal rocky environments along the Chilean coast (Dayton 1985; Vásquez and Buschmann 1997).

2. MORPHOLOGY

Loxechinus albus has a semispherical shell, is middle- to large-sized, green in color with reddish or purple shades in ambulacra and interambulacra of large individuals. Those living at depth show whitish shades. The ambulacral plates *are* polyparous, with 6–11 pore pairs and a primary tubercle. The ambulacral system is dicyclic with a central anus and numerous large periproctal plates. In adults, primary spines are short and conic. Secondary and miliary spines are numerous. Globiferous pedicellaria have a neck without circular musculature and big valves bearing from one to four lateral teeth on each side (Larraín 1975).

3. DISTRIBUTION

Loxechinus albus is an endemic species of the Peruvian and Chilean coasts, distributed from Isla Lobos de Afuera, Perú (6° 53′ S) to the southern tip of South America (ca. Cape Horn, Chile 56° 70′ S), from tidepools with permanent water circulation through the low intertidal zone to as deep as 340 m (Larraín 1975). Fenucci (1967) collected *L. albus* on the Argentinian continental platform (up to 37° 35′ S–54° 33′ W).

4. HABITAT AND SUBSTRATE PREFERENCES

In northern and central Chile (ca. 18–40° S), larval settlement occurs mainly in rocky intertidal areas (Contreras and Castilla 1987; Castilla 1990; Zamora and Stotz 1992; Vásquez et al. 1998). Postmetamorphic individuals up to about 2 cm diameter are found in intertidal crevices with abundant shell sand, inhabiting intertidal kelp holdfasts, and associated with frondose macroalgal species. Intertidal populations of small *L. albus* between 2 and 4.5 cm diameter live in *intertidal* pools with constant water exchange. Individuals larger than 4.5 cm diameter migrate toward subtidal environments with high cover of subtidal kelp species, forming the population which is exposed to artesanal fisheries (Contreras and Castilla 1987).

In southern Chile, populations of *L. albus* occur mainly associated with kelp beds of *Macrocystis pyrifera*, which grow on rocky substrata. Small *L. albus* are found on and under boulders, and rock crevices. In contrast with coexisting echinoid species, *L. albus* is never *found* inhabiting holdfasts of *Macrocystis* (Vásquez et al. 1984). *Loxechinus albus* are rare in protected areas, contrasting with high densities reported for exposed habitats (Vásquez et al. 1984; Dayton 1985). Hypotheses for such differences in density between habitats include: (1) differential larval availability for both contrasting habitats, (2) possible larval sensitivity to the reduced salinities in many of these habitats, and (3) unsuccessful settlement processes in areas with a high sediment load (Dayton 1985).

5. FOOD

Loxechinus albus is one of the most important benthic grazers in intertidal and shallow subtidal ecosystems, but little information exists on the gut contents of wild populations. Bückle et al. (1980) found 22 taxa (genera) in individuals from Valparaíso (33° S) and Chiloé (ca. 42° S). In southern Chile, Vásquez et al. (1984) identified six algal species in the gut contents of individuals from Puerto Toro (ca. 55° S). Although no direct evidence describes the foraging strategy of *L. albus*, *both* studies indicated no preference in the consumption of prey items, suggesting that large *L. albus* are generalists, consuming the more abundant algal species in each locality (Table 1).

Small individuals of *L. albus* feed on crustose calcareous algae, benthic diatom films, and pieces of drift algae. Large individuals heavily graze intertidal and subtidal kelps, benthic seaweed turf (Dayton 1985), and may also feed on drift algae (Castilla and Moreno 1982; Contreras and Castilla 1987).

Large and small *L. albus* from Iquique in northern Chile (ca. 20° S) do not show significant differences in dietary composition with changes in season (Ojeda et al. 1998). The principal dietary items were the green alga *Ulva* sp. for juveniles and the brown alga *Lessonia* sp. for large individuals. The absorption efficiency for the large individuals,

Table 1 Food items in the gut content of wild population of *Loxechinus albus* from (1) central (Bückle et al. 1980) and (2) southern (Vásquez et al. 1984) Chile.

Items	(1) Relative abundance (%)	(2) Relative abundance (%)
Phaeophyta	26.14	76.9
Macrocystis pyrifera		64.0
Lessonia flavicans		11.7
Lessonia vadosa		1.2
Lessonia spp. & *Durvillaea*	21.6	
Glossophora kunthi	2.5	
Petalonia sp.	2.0	
Scytosiphon sp.	0.02	
Ectocarpus sp.	0.02	
Rhodophyta	68.2	8.6
Gigartina skottsbergii		2.8
Epymenia falklandica		5.0
Callophylis variegata		0.8
Plocamium sp.	19.0	
Polysiphonia sp.	11.1	
Corallina chilensis	9.7	
Gelidium sp.	9.2	
Lithothamnion sp.	14.7	
Centrosceras sp.	3.0	
Iridaea sp.	0.9	
Antithamnion sp.	0.2	
Ceramium sp.	0.2	
Porphyra sp.	0.2	
Chlorophyta	5.64	
Codium sp.	0.3	
Enteromorpha sp.	0.8	
Chaetomorpha sp.	1.9	
Ulva sp.	2.6	
Rhizoclonium	0.02	
Scytosiphon sp.	0.02	
Unidentified items	15.5	

measured as a percentage of the total ingested organic matter, was greatest for individuals fed *Lessonia* sp. (75%), followed by *Ulva* sp. (73.5%) and *Macrocystis* sp. (70.9%). In small individuals it was highest for the chlorophyte *Ulva* sp. (78.7%), followed by *Lessonia* sp. (68.3%) and *Macrocystis* sp. (55.1%). Trophic selectivity experiments in controlled conditions revealed that small and large individuals prefer the most abundant item observed in nature, suggesting that *L. albus* undergoes an ontogenetic diet shift consisting of a differential foraging strategy between small and large individuals. While small individuals appear to behave as time minimizers, large individuals appear to act as energy maximizers (Ojeda et al. 1998).

Strong evidence suggests that sea urchin species in southern South America control neither plant density nor plant diversity (Castilla and Moreno 1982). On the contrary, the grazing of sea urchins in the Northern Hemisphere generates two alternative stable states: one dominated by kelp species and the other dominated by sea urchins and crustose calcareous algae ("barren ground" *sensu* Lawrence 1975). Another factor that can switch states is the abundance and the frequency of drift algae (Harrold and Reed 1985). In this context, besides the abundance and predictability of drift material all along the Chilean coast, the presence of a sucking disc in the terminal end of the oral and aboral podia (Vásquez 1986; Contreras and Castilla 1987) helps *L. albus* capture and retain and manipulate drift algae, decreasing the pressure on marine algae in this ecosystem (Vásquez 1992).

6. REPRODUCTION ECOLOGY

The reproductive period of *L. albus* varies with latitude, showing a temporal displacement of the spawning period (Fig. 1). The spawning period for the 22–24° S region occurs during June (Gutiérrez and Otsu 1975; Zegers et al. 1983), between June and August close to the 30° S (Zamora and Stotz 1992), between August and November on the coast between 32° and 33° S (Bückle et al. 1978; Guisado and Castilla 1987), and between

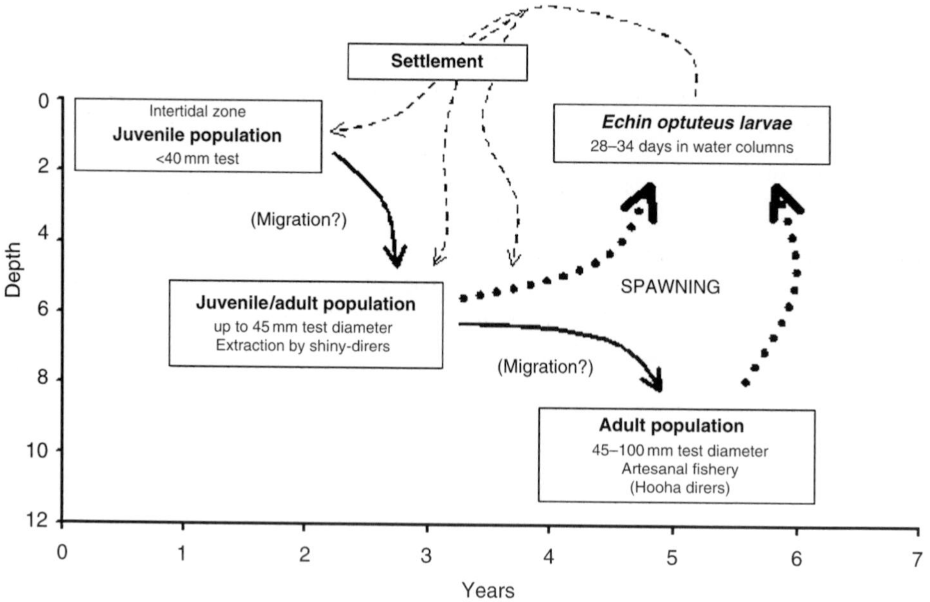

Fig. 1. Life history of *Loxechinus albus*. Modified from Guisado and Castilla (1987) and Castilla (1990).

October and December at Valdivia (ca. 40° S) (Guisado 1985). In the Chiloé and Guaitecas Islands, between 42° and 45° S, Bay-Smith et al. (1981) reported spawning of *L. albus* between November and December.

The southern population of *L. albus* from Punta Arenas (56° S) does not show this tendency; as the main spawning period occurs between September and October (Bay-Smith et al. 1981). Zamora and Stotz (1992) suggested the Punta Arenas populations of *L. albus* are not exposed to the influence of the Humboldt Current which affects *L. albus* between 22° and 45° S. On the contrary, the coast above 48° S is influenced by the Cape Horn Current, which has different oceanographic characteristics (Dayton 1985; Strub et al. 1998). The effect of the Cape Horn Current is an important biogeographical boundary for benthic macroinvertebrate distribution on the southeastern Pacific littoral (Lancellotti and Vásquez 1999).

On the other hand, although mature organisms were present all year around in the Magellan Region of southern Chile, simultaneous spawning of female and males occurred from August to September in Dawson Island (53° 43′ S–72° 00′ W) and between July and September in Cockburn Channel (53° 43′ S–72° 42′ W), suggesting that small-scale variability of spawning period in this region may be explained by differences in food type and availability among areas (Oyarzún et al. 1999). These results also suggest that the Magellan Region is an exception to the latitudinal pattern of spawning period reported for the most of the Chilean coast described by Zamora and Stotz (1992). Oyarzún et al. (1999) suggest that this large-scale variability may be explained by the simultaneous occurrence of low temperatures and short days during late winter and early spring.

In contrast with populations of *L. albus* above 56° S, those between 22° and 45° S have a second period of gonad growth during summer through fall (Bay-Smith 1982). According to Bückle et al. (1978), this second gonadal production may be resorbed, serving as a mode of nutrient storage for the autumn–winter fasting period prior to the single spawning. Absence of echinopluteus larvae in the plankton of southern Chile suggests the second gonadal maturation observed in summer–autumn does not culminate in a spawning. The large number of nutritive phagocytes observed in summer–autumn in individuals observed by Zamora and Stotz (1992) strongly supports this hypothesis.

A comparison of spawning seasons between high and low latitudes with seasonal temperatures over the same gradient (Fonseca 1987) led Zamora and Stotz (1992) to suggest that spawning coincided with temperatures of 14 °C (20° S) and 11 °C (56° S). Other factors suggested to induce spawning of sea urchins are the photoperiod and the availability of food for larvae (Pearse et al. 1986). These factors appear to be of special relevance with respect to the induction of spawning in populations at Punta Arenas in the extreme south of Chile (Zamora and Stotz 1992).

7. POPULATION ECOLOGY

After spawning, the echinopluteus larvae remain in the plankton between 28 and 34 days (Arrau 1958; González et al. 1987), feeding on phytoplankton (Pérez et al. 1995). Settlement occurs primarily in the rocky intertidal, where postmetamorphic juveniles up to 20 mm in diameter occupy crevices containing shelly sand. These juveniles feed principally on benthic diatoms, crustose algae, and propagules or early growth stages of macroalgae (Contreras and Castilla 1987). Guisado and Castilla (1987) suggested that these juveniles migrate to the subtidal, forming populations having test diameters of about 45 mm which feed primarily on drift algae and brown algal thalli. These populations, which have had at least one sexual maturation (Bückle et al. 1978), are heavily fished by shore-based free divers. Individuals greater than 45 mm test diameter migrate to deeper waters, forming populations exposed to commercial fisheries (Guisado and Castilla 1987; Fig. 2). These rocky subtidal ecosystems, 4–15 m in depth, are generally dominated by brown algae *Macrocystis* spp. and *Lessonia* spp (Vásquez 1992; Vásquez and Buschmann 1997).

In contrast to numerous reports concerning mass mortalities of sea urchins in the northern hemisphere (see Pearse et al. 1977; Miller and Colodey 1983; Scheibling and Stephenson 1984), similar reports for the southwestern Pacific coasts do not exist. During the period since the mid-1980s, extensive observations of intertidal and subtidal habitats from 0 to 25 m depth at latitudes from 18° S to 56° S have not reported mass mortalities for any urchin species.

8. COMMUNITY ECOLOGY

The main focus of ecological research in Chile has been the study of biological and physical processes structuring marine communities (see review by Fernandez et al., in press). This information, however, has come from relatively few study sites and types of environments. Although recent reviews summarize much of the information available and present the current view of the understanding of the dynamics of nearshore ecosystems (Santelices 1989; Castilla and Paine 1987; Castilla et al. 1993; MacLachlan and Jaramillo 1995; Vásquez and Buschmann 1997), shallow subtidal habitats which experience the heaviest fisheries are less well studied (Vásquez et al. 1998).

In the intertidal of northern and central Chile, between 18° and 42° S, *Lessonia nigrescens* and *Durvillaea antarctica* provide refuge, food, and substrate for recruitment and nursery areas for juvenile populations of *Loxechinus albus*. *Lessonia trabeculata* and *Macrocystis integrifolia* play a similar ecological role in subtidal habitats at the same latitudinal range. South of 42° S, fiords and semi-enclosed waters are dominated by *Macrocystis pyrifera* which support the most abundant population of *L. albus* (Vásquez 1992; Vásquez and Buschmann 1997).

In northern and central Chile (18°–42° S) *Lessonia trabeculata* and *Macrocystis integrifolia* (up to 35° S) form extensive subtidal kelp beds on rocky bottoms in areas exposed and semi-exposed to heavy surge. At these latitudes, the sea urchin *Tetrapygus niger* and the gastropod *Tegula tridentata* are the most abundant grazers associated with kelp forests. Dense populations of these benthic herbivores impede the settling of macroalgal propagules and maintain extensive barren grounds (Vásquez 1993). Contreras and Castilla (1987) and Vásquez (1992, 1993) suggested that overgrazing of *T. niger* and *T. tridentata* on rocky subtidal and intertidal communities indirectly decreases the abundance of *L. albus* in northern and central Chile. In this context, there is no experimental evidence on competitive interaction between *L. albus* and *T. niger*.

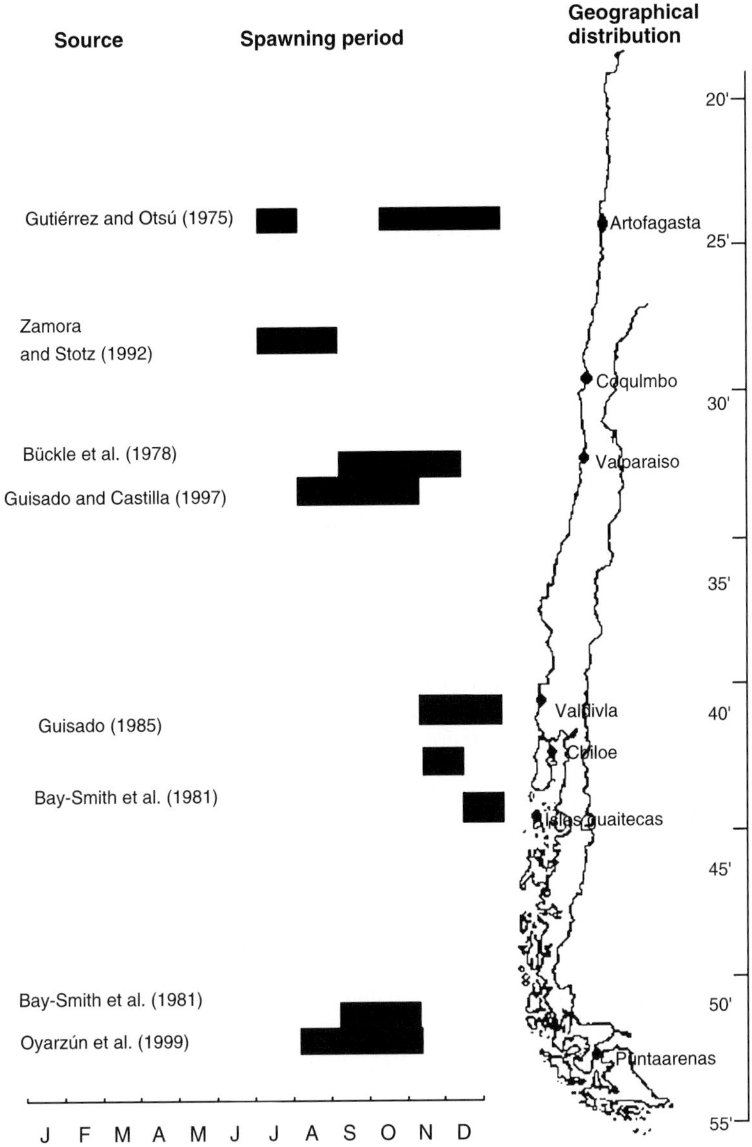

Fig. 2. Geographical variation of the spawning period of *Loxechinus albus* along the Chilean Coast. Modified from Zamora and Stotz (1992).

Viviani (1975) made qualitative observations that suggested that *T. niger* and *L. albus* compete for food and living space in intertidal and subtidal habitats of northern Chile. He concluded that if *Lessonia* spp. were the most important plants in areas where both species occur, *L. albus* would exclude *T. niger*. Field observations and experimental results indicate that adult *Lessonia* spp. strongly decrease sea urchin grazing. Frond movements resulting from water motion may keep populations of sea urchins away from the kelp, especially during low tides and strong bottom surge ("whiplash effect" *sensu* Santelices 1989; Vásquez 1992). Barren areas (*sensu* Lawrence 1975) covered with calcareous crusts and dense aggregations of *T. niger* are especially common in the intertidal–subtidal pools having frequent water change but not directly exposed to wave action. In comparative terms, coastal marine communities dominated by *T. niger* and calcareous crust algae are less diverse than those communities where *L. albus* is more frequent and abundant. If *L. albus* is eliminated as a result of commercial harvesting, *T. niger* typically comes to occupy the open space forming extensive barren grounds. The lack of experimental evidence for biological interaction between sea urchin species has mainly been due to logistical problems involved with making underwater observations, as well as massive commercial harvesting of *L. albus* in northern Chile since the mid-1970s. South of 42° S, where the most populous areas of *L. albus* occur, *T. niger* is much less abundant (Vásquez and Buschmann 1997).

Southeastern Pacific kelp communities, in contrast with similar communities in the Northern Hemisphere, do not have keystone species regulating herbivore densities (Castilla and Paine 1987). Foster and Schiel (1988) strongly suggested that the concept of the sea otter as a keystone species is applicable only to a relatively small number of sites and thus does not constitute a general explanation of kelp community organization in California. In this context, Vásquez (1989) showed that sea star species *Meyenaster gelatinosus*, *Luidia magellanica*, *Stichaster striatus,* and *Heliaster helianthus*, and the carnivorous fishes *Cheilodactylus variegatus*, *Mugiloides chilensis*, and *Semicossiphus maculatus* form a guild that heavily controls the abundance of benthic herbivores associated with kelp communities in northern and central Chile. All these species prey on sea urchins and gastropods (Vásquez 1993; Vásquez et al. 1998).

The most conspicuous habitats for *L. albus* up to 35° S are those dominated by the giant kelp, *Macrocystis pyrifera*. These habitats contain more than 95% of the populations of *L. albus* undergoing fishing. Extensive descriptions of habitat and community structure have been made by Castilla and Moreno (1982), Moreno and Sutherland (1982), Moreno and Jara (1984), Santelices and Ojeda (1984), Ojeda and Santelices (1984), Vásquez et al. (1984), Castilla (1985), and Dayton (1985). Based on a broad geographical survey between 44° and 52° S along southern Chile and Argentina coasts, Dayton (1985) pointed out that the distribution and abundance of *M. pyrifera* were determined by (1) availability of suitable rocky substratum, (2) interspecific competition with *Lessonia vadosa* (shallow waters) and *L. flavicans* (deeper waters), (3) entanglement with drift algae and the heavy settlement of bivalves on the kelp fronds, (4) degree of exposure to waves, (5) grazing by *L. albus*, and (6) indirectly, the effects of human fishing of *L. albus*. He documented that *M. pyrifera* was overgrazed by *L. albus,* in many areas between 44° and 52° S. In other areas *L. albus* existed at lower densities or were absent altogether and did not affect the population of *M. pyrifera*. Furthermore, Dayton (1985) stated that *L. albus* had an important role in the control of abundance of *M. pyrifera* in wave-exposed sites and that its effect on this kelp decreased as the intensity of the wave action decreased. This explanation is the basis for the general belief that sea urchins only play a major role in habitats which are stressed by harsh environmental conditions with severe storms, low nutrients, and/or warm temperatures (Harrold and Reed 1985). For these reasons, wave-protected environments should have long-lived perennial populations of *M. pyrifera* (North 1971; Rosenthal et al. 1974; Gerard 1976; Kirkwood 1977; Dayton et al. 1984; Druehl and Wheeler 1986).

The foraging behavior of *L. albus* is influenced mainly by the degree of wave surge and hunger (Dayton 1985), as in *Strongylocentrotus* spp. (Vadas 1977; Harrold and Reed 1985). As Dayton (1985) indicated, the distribution of *L. albus* is often restricted to areas exposed to strong wave action. The importance of hunger is a well-known factor influencing the behavior of sea urchins and, in some cases, the formation of the sea urchin grazing fronts irrespective of density (Lawrence 1975; Harrold and Reed 1985; Harrold and Pearse 1987). With regard to hunger, drift algae are abundant in all areas with large kelp forests (Castilla and Moreno 1982; Dayton 1985). In these areas *L. albus* was common but usually was not very abundant. It is clear they do not forage far from a source of food (Castilla and Moreno 1982; Santelices and Ojeda 1984). In most protected areas, drift is abundant and *L. albus* is often rare (Vásquez et al. 1984). In most other areas where they are abundant, *L. albus* forages actively (Dayton 1985).

Grazing pressure by *Loxechinus albus* can exert a significant effect on the density of *Macrocystis pyrifera* populations in southern Chile (Buschmann et al. 2003), in contrast to conclusions of previous studies (see Castilla and Moreno 1982). As *M. pyrifera* shows an annual cycle strategy in this area, sea urchins impose a greater grazing pressure on new recruits that appear massively in spring (Fig. 3). With a low grazing pressure, these populations have a high recruitment level but a natural decline in summer mainly due to abiotic factors. The functional model by Buschmann et al. (2003) shows kelp recruitment and abundances in different seasons of the year and the urchin effects at higher densities. With a condition of low urchin abundance, temperature and nutrients seem to be the main environmental factors controlling the kelp abundances. Observation of perennial kelp populations in more exposed locations should not be affected by *Loxechinus albus* as reported by Castilla and Moreno (1982). Dayton (1985) indicated that *L. albus* has a significant effect on the abundance of *Macrocystis pyrifera* in wave-protected environments of southern Chile. Although his conclusions were correct, they were based on a wrong assumption because he observed *Macrocystis* abundance during May when the population was naturally declining (Fig. 3).

Few data are available on the regulation of sea urchin density in southern South America. Dayton (1985) hypothesized that low larval availability was an important factor for explaining the relatively low abundance of population of *L. albus* in Tierra del Fuego. These habitats are influenced by the circumpolar West Wind Drift current, where the only source of larvae of *L. albus* would be from the Cape Horn Archipelago. Assuming larvae of *L. albus* are similar to those of other echinoids in spending four or more weeks in the plankton, the West Wind Drift would carry most of the larvae away and the only recruitment into these habitats would come from eddies or areas where larvae are trapped (Dayton 1985).

The abundance of sea urchins in the Northern Hemisphere is often influenced by predators such as sea otters (Estes and Palmisano 1974; Dayton 1985), fishes and/or lobsters (Mann 1977; Tegner and Dayton 1981; Cowen 1983), crabs (Kitching and Ebling 1961; Bernstain et al. 1983), or asteroids (Mauzey et al. 1968; Paine and Vadas 1969; Rosenthal and Chess 1972). As Castilla and Moreno (1982) and Castilla (1985) indicated, no single efficient sea urchin predator exists in the kelp communities of *Macrocystis*

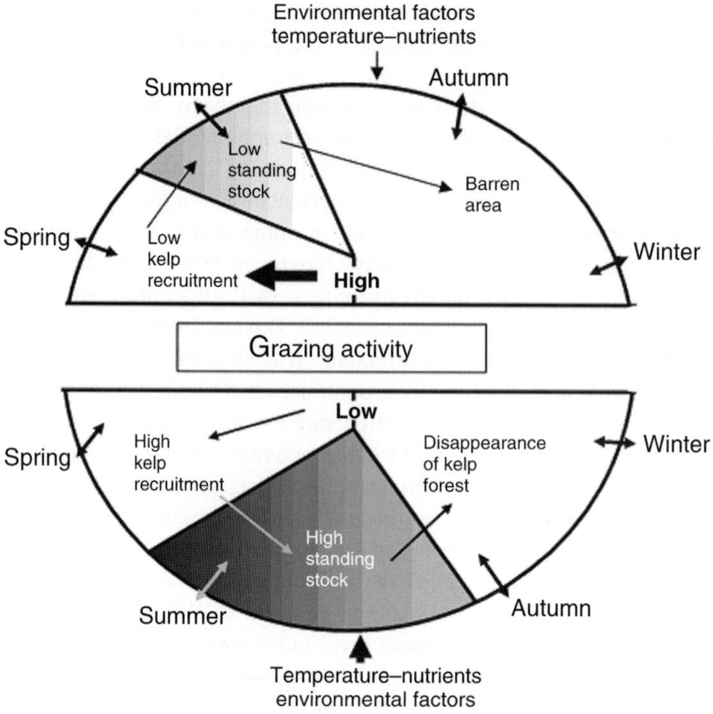

Fig. 3. Qualitative model in conditions of high-urchin abundance (top semicircle) and low-urchin abundance (bottom semicircle). From Buschmann et al. (2003).

pyrifera of southern Chile. Apparently, the most conspicuous predators of sea urchins are two asteroids, *Meyenaster gelatinosus*, which occurs in abundance only north of Golfo de Penas (Dayton 1985) and *Cosmasterias lurida* Philippi, which is the principal predator found in coastal belts of *Macrocystis* at Puerto Toro in southern Chile (Vásquez and Castilla 1984; Castilla 1985). *Loxechinus albus* has effective defense behaviors and *M. gelatinosus* only reduces densities of *L. albus* on isolated boulders from which the sea urchins can be stampeded without immediate immigration (Dayton et al. 1977). When the boulders are sufficiently isolated and large, this stampeding phenomenon can result in the release of *M. pyrifera* from predation by *L. albus* long enough to recruit and survive to reproduce. *Loxechinus albus* and other sea urchin species are rare in the diet of *Cosmasterias lurida* (Vásquez and Castilla 1984).

9. EL NIÑO

Populations of *L. albus* in northern Chile below 30° S are occasionally affected by the El Niño–Southern Oscillation (ENSO), although the magnitude of the effect has not been documented. During the ENSO of 1982–1983 a decrease of populations attached to the substrate was observed between 18° and 20° S (Soto 1985), as well as a decrease of population densities near Antofagasta (ca. 23–24° S) related to massive mortalities of *Lessonia* spp. and *Macrocystis intergrifolia* (Tomicic 1985).

During the ENSO of 1997–1998, superficial sea surface warming was minimized by intense and frequent coastal upwelling (González et al. 1998). As a consequence, no mass mortalities of littoral invertebrates occurred near Antofagasta. After El Niño 1997–1998, and during periods of negative SST anomalies (1999–2000), the mean density of sea urchins increased, tripling their mean density between 1996–1998 (Vega et al. 2005). This change in temporal patterns of abundance of two sea urchin species coincided with the local extinction of the *M. integrifolia* population. An inverse and significant correlation suggested that the density of juvenile and adult of *M. integrifolia* sporophytes decreased with increasing numbers of *T. niger* and *L. albus* in northern Chile. Beginning in 2001, the abundance of sea urchins began to decrease significantly until the end of 2003, giving values similar to those encountered between 1996 and 1998. In this context, it has been noted that areas with intense and permanent offshore transport, such as the area on the Mejillones Peninsula, are typified by high survival, retention, and transport of echinoderm larvae toward the coast, similar to that reported by Ebert and Russell (1988) for *Strongylocentrotus franciscanus* on the west coast of the US. During the ENSO cycle (1997–1998 El Niño and 1998–2000 La Niña), different events favored an increase in the urchin population during the cool phase, including (1) induction of spawning due to increases in SST and persistence of upwelling events, (2) significant reductions in densities of adult individuals of *M. gelatinosus* and *L. magellanica* (Vásquez et al. 1998; Vega et al. 2005), and (3) changes in the feeding behavior of *H. helianthus* (Tokeshi and Romero 1995; Vásquez et al. 1998).

10. CONCLUSIONS

A great amount of additional basic information is necessary to understand the ecology of *L. albus* in the southwestern Pacific coast. Fernández et al. 2000. This is especially true with reference to larval ecology, factors which determine spawning, habitat selection, settlement mechanisms, and the migration of populations of juveniles from intertidal to subtidal habitats. Also, the factors that determine population density in different ontogenetic stages (e.g. predation, competition) need to be studied. These key factors and mechanisms in the population ecology of *L. albus* are critical when considering the broad geographic range of the species. Sampling and evaluation must be carried out over the entire range in order to obtain a complete picture of its ecological potential. For example, although much data have been obtained for populations of the species around 18° S, these data are hardly applicable to populations occurring far to the south where parameters of the habitat may differ significantly.

ACKNOWLEDGMENTS

This contribution was supported by FONDAP O & BM Programa Mayor No. 3, Ecology and Conservation and FONDECYT 1000044. FONDECTY 1040425.

REFERENCES

Arrau L (1958) Desarrollo del erizo comestible de Chile *Loxechinus albus* Molina. Rev Biol Mar (Valparaíso) 7: 39–61

Bay-Smith E (1982) Erizo *Loxechinus albus* (Molina). Echinoidea, Echinoida, Echinidae. Estado actual de las principales pesquerías nacionales. Bases para el Desarrollo Pesquero IFOP (Chile) 9: 1–52

Bay-Smith E, Werlinger C, Silva J (1981) Ciclo anual de reproducción del recurso *Loxechinus albus* entre la X y la XII Región. Informe Final Proyecto de Investigación Subsecretaría de Pesca. Universidad de Concepción, 68 pp

Bernstain BB, Schroeter SC, Mann KH (1983) Sea urchin (*Strongylocentrotus droebachiensis*) aggregating behaviour investigating by a subtidal multifactorial experiment. Can J Fish Aqua Sci 40: 1975–1986

Bückle LF, Guisado C, Tarifeño E, Zuleta A, Córdova L, Serrano C (1978) Biological studies on the Chilean sea urchin *Loxechinus albus* (Molina) (Echinodermata: Echinoidea) IV. Maturation cycle and seasonal changes in the gonad. Ciencias Marinas 5: 1–18

Bückle LF, Alveal K, Tarifeño E, Guisado C, Córdova L, Serrano C, Valenzuela J (1980) Biological studies on the Chilean sea urchin *Loxechinus albus* (Molina) (Echinodermata: Echinoidea). Food analysis and seasonal feeding rate. Anal Cen Cienc Mar Limnol Univ Nac Auton (México) 7: 149–158

Buschmann Att, García C, Espinoza R, Filún L, Vásquez JA (2003) Sea urchin (*Loxechinus albus*) and kelp (*Macrocystis pyrifera*) in protected areas in southern Chile. In: Edible sea urchins: Biology and Ecology Lawrence JM (ed). Developments in Aquaculture and Fisheries Science 32: 161–175

Castilla JC (1985) Food web and functional aspects of the kelp, *Macrocystis pyrifera*, community in the Beagle Channel, Chile. In: Siegfried WR, Condy PR, Laws RM (eds) Antarctic nutrient cycles and food webs. Springer-Verlag, Berlin, pp 407–414

Castilla JC (1990) El erizo chileno *Loxechinus albus*: importancia pesquera, historia de vida, cultivo en el laboratorio y repoblación natural. In: Hernández A (ed.) Cultivo de Moluscos en America Latina. CIID Canadá, Santiago, Chile, pp 83–98

Castilla JC, Moreno CA (1982) Sea urchins and *Macrocystis pyrifera*: an experimental test of their ecological relations in southern Chile. In: Lawrence JM (ed.) Echinoderms: proceedings of the International Echinoderm Conference, Tampa Bay. AA Balkema, Rotterdam, pp 257–263

Castilla JC, Paine RT (1987) Predation and community organization on Eastern Pacific, temperate zone, rocky intertidal shores. Rev Chile Hist Nat 60: 131–151

Castilla JC, Navarrete SA, Lubchenco J (1993) Southeastern Pacific Coastal environments: main features, large-scale perturbations, and global climate changes. In: Mooney H, Fuentes E (eds) Earth systems responses to global change, contrasts between North and South America. Academic Press Inc., San Diego, pp 167–187

Contreras S, Castilla JC (1987) Feeding behaviour and morphological adaptations in two sympatric sea urchin species in central Chile. Mar Ecol Prog Ser 38: 217–224

Cowen RK (1983) The effect of sheephead (*Semicossyphus franciscanus*) populations: an experimental analysis. Oecologia 58: 249–255

Dayton PK (1985) The structure and regulation of some South American kelp communities. Ecol Monogr 55: 447–468

Dayton PK, Rosenthal RJ, Mahan LC, Antezana T (1977) Population structure and foraging biology of the predaceous Chilean asteroid *Meyenaster gelatinosus* and the escape biology of its prey. Mar Biol 39: 361–370

Dayton PK, Currie V, Gerrodette T, Keller BD, Rosentahl R, Van Tresca D (1984) Path dynamics and stability of some Californian kelp communities. Ecol Monogr 54: 253–289

Deppe R, Viviani CA (1977) La pesquería artesanal del erizo comestible *Loxechinus albus* (Molina) (Echinodermata, Echinoidea, Echinidae) en la región de Iquique. Biol Pesq (Chile) 9: 23–41

Druehl LD, Wheeler WN (1986) Population biology of *Macrocystis integrifolia* from British Columbia. Can Mar Biol 90: 173–179

Ebert TA and Rusell MP (1988) Latitudinal variation in size structure of the west coast purple sea urchins: a correlation with headlands. Limnol Oceanog 33: 286–294

Estes JA, Palmisano JF (1974) Sea otters: their role in structuring nearshore communities. Science 185: 1058–1060

Fenucci JL (1967) Contribución al conocimiento del crustáceo decápodo brachiuro *Pinnaxodes chilensis* (M. Edwards) comensal de *Loxechinus albus* (Molina) (Echinodermata: Echinoidea). Physis 27: 125–133

Fernández M, Jaramillo E, Marquet P, Navarrete S, George-Nascimento M, Ojeda FP, Valdovinos C, Vásquez JA (2000) Diversity, dynamics and biogeography of Chilean benthic nearshore ecosystems: an overview and guidelines for conservation. Rev Chile Hist Nat 73: 797–830

Fonseca TR (1987) Anomalías de temperatura y nivel del mar en la costa del Pacifico Sudoriental. Cienc Tecnol Mar CONA 11: 17–43

Foster MS, Schiel DR (1988) Kelp community and sea otters: keystone species or just another brick in the wall? In: VanBlaricom GR, Estes JA (eds) The community ecology of sea otters. Ecol Stud 65: 92–115

Gerard VA (1976) Some aspects of material dynamics and energy flow in a giant kelp forest in Monterrey Bay, California. PhD Thesis, University of California, Santa Cruz, USA

González HE, Daneri G, Figueroa D, Iriarte JL, Lefebre N, Pizarro G, Quiñones R, Sobarzo M, Troncoso A (1998) Producción primaria y su destino en la trama trófico pelágica y océano profundo e intercambio océano-atmósfera de CO_2 en la zona norte de la Corriente de Humboldt (23° S): posibles efectos del evento El Niño 1997–98 en Chile. Rev Chile Hist Nat 71: 429–458

González L, Castilla JC, Guisado C (1987) Effect of larval diet and rearing temperature on metamorphosis and juvenile survival of the sea urchin *Loxechinus albus* (Molina, 1782) (Echinodermata: Echinoidea). J Shellfish Res 6: 109–115

Guisado C (1985) Estrategias de desarrollo larval y ciclo de vida en dos especies de echinoideos regulares del sur de Chile. MSc Thesis, Universidad Austral de Chile, Valdivia, Chile

Guisado C, Castilla JC (1987) Historia de vida, reproducción y avances en el cultivo del erizo comestible chileno *Loxechinus albus* (Molina, 1782) (Echinoidea, Echinidae). In: Arana P (ed.) Manejo y Desarrollo Pesquero, Escuela de Ciencias del Mar. U. Católica de Valparaíso, Valparaíso, Chile, pp 59–68

Gutiérrez J, Otsu I (1975) Periodicidad en las variaciones biométricas de *Loxechinus albus* Molina. Rev Biol Mar (Valparaíso) 15: 179–199

Harrold C, Pearse JS (1987) The ecological role of echinoderms in kelp forest. Echinoderm Studies 2: 137–233

Harrold C, Reed DC (1985) Food availability, sea urchin grazing and kelp forest community structure. Ecology 66: 1160–1169

Jerardino A, Castilla JC, Ramirez JM, Hermosilla N (1992) Early coastal subsistence patterns in central Chile. A systematic study of the marine-invertebrate fauna from the site of Curaumilla-I. Lat Am Antiq 3: 43–62

Kirkwood PD (1977) Seasonal patterns in the growth of giant kelp *Macrocystis pyrifera*. PhD Thesis, California Institute of Technology, Pasadena, California, USA

Kitching JA, Ebling FJ (1961) The ecology of Lough Ine XI. The control of algae *by Paracentrotus lividus* (Echinoidea). J Anim Ecol 30: 373–383

Lancellotti DA, Vásquez JA (1999) Biogeographical patterns of benthic macroinvertebrates in the south-eastern Pacific littoral. J Biogeogr 26: 1001–1006

Larraín A (1975) Los equinodermos regulares fósiles y recientes de Chile. Gayana Zool 35: 1–189

Lawrence JM (1975) On the relationships between marine plants and sea urchins. Ocean Mar Biol Ann Rev 13: 213–286

MacLachlan A, Jaramillo E (1995) Zonation on sandy beaches. Oceanogr Mar Biol Ann Rev 33: 305–335

Mann KH (1977) Destruction of kelp-beds by sea urchins: a cyclical phenomenon or irreversible degradation? Helgol wiss Merres 30: 455–467

Mauzey KP, Birkeland C, Dayton PK (1968) Feeding behaviour of asteroids and escape responses of their prey in the Puget Sound regions. Ecology 49: 603–619

Miller RJ, Colodey AG (1983) Widespread mass mortalities of the green sea urchin in Nova Scotia, Canada. Mar Biol 73: 263–267

Moreno CA, Jara AF (1984) Ecological studies on fish fauna associated with *Macrocystis pyrifera* belts in the north Fuegian Island, Chile. Mar Ecol Progr Ser 15: 99–107

Moreno CA, Sutherland JP (1982) Physical and biological processes in a *Macrocystis pyrifera* community near Valdivia, Chile. Oecologia 55: 1–6

North WJ (1971) Introduction. In: North WJ (ed.) The biology of giant kelp beds (*Macrocystis*) in California. Nova Hedvigia 32: 1–37.

Ojeda FP, Santelices B (1984) Invertebrate communities in holdfast of the kelp *Macrocystis pyrifera* from southern Chile. Mar Ecol Progr Ser 16: 65–73

Ojeda FP, Cáceres CW, González SJ (1998). Experimental feeding ecology of the edible sea urchin, *Loxechinus albus*, off the coast of northern Chile. In: Mooi R, Telford M (eds) Echinoderms: San Francisco. AA Balkema, Rotterdam, Brookfield, p 769

Oyarzún ST, Marín SL, Valladares C, Iriarte JL (1999). Reproductive cycle of Loxechinus albus (Echinodermata: Echinoidea) in two areas of the Magellan Region (53° S, 70–72° W), Chile. Scientia Marina 63: 439–449

Paine RT, Vadas RL (1969) The effects of grazing by sea urchins, *Strongylocentrotus* spp. on benthic algal populations. Limnol Oceanogr 14: 710–719

Pearse JS, Costa DP, Yellin MB, Agegian CR (1977) Localized mass mortality of red sea urchin *Strongylocentrotus franciscanus*, near Santa Cruz, California. Fish Bull US 75: 645–648

Pearse JS, Pearse VB, Davis KK (1986). Photoperiodic regulation of gametogenesis and growth in the sea urchin, *Strongylocentrotus purpuratus*. J Exp Zool 237: 107–118

Pérez MC, González ML, López DA, Zúñiga J (1995) Cultivation of the erizo: an evaluation of eggs and postmetamorphic juveniles size selection. Aquac Internat 3: 364–369

Scheibling RE, Stephenson RL (1984) Mass mortality of *Strongylocentrotus droebachiensis* (Echinodermata: Echinoidea) of Nova Scotia, Canada. Mar Biol 78: 153–164

Rosenthal RJ, Chess JR (1972) A predatory–prey relationship between the leather star (*Dermasterias imbricata*) and the purple urchin (*Strongylocentrotus purpuratus*). Fish Bull 70: 205–216

Rosenthal RJ, Clarck WD, Dayton PK (1974) Ecology and natural history of a stand of giant kelp, *Macrocystis pyrifera* off Corona del Mar. Calif Fish Bull 72: 670–684

Santelices B (1989) Algas Marinas de Chile. Distribución. Ecología, Utilización, Diversidad. Ediciones Universidad Católica de Chile, Santiago 399 pp

Santelices B, Ojeda FP (1984) Population dynamics of coastal forest of *Macrocystis pyrifera* in Puerto Toro, Isla Navarino, Southern Chile. Mar Ecol Progr Ser 14: 175–183

SERNAP (1957–1999) Servicio Nacional de Pesca. Anuario Estadístico de Pesca. Ministerio de Economía Fomento y Reconstrucción. República de Chile

Soto R (1985) Efectos del fenómeno El Niño 1982–1983 en ecosistemas de la I Región. Invest Pesq (Chile) 32: 199–206

Strub PT, Mesias JM, Montecinos V, Rutlant J, Salinas S (1998) Coastal ocean circulation off western South America. In: Brink KH, Robinson AR (eds) The global coastal ocean, The Sea. John Wiley & Sons Inc., New York, Vol. 11, pp 273–313

Tegner MJ, Dayton PK (1981) Population structure, recruitment and mortality of two sea urchins (*Strongylocentrotus franciscanus* and *S. purpuratus*) in a kelp forest. Mar Ecol Prog Ser 5: 255–268

Tokeshi M, Romero L (1995) Quantitative analysis of foraging behaviour in a field population of South American sun–star *Heliaster helianthus*. Mar Biol 122:297–303

Tomicic JJ (1985) Efectos del fenómeno de El Niño 1982–83 en las comunidades litorales de la Península de Mejillones. Invest Pesq (Chile) 32: 209–213

Vadas RL (1977) Preferential feeding: an optimization strategy in sea urchins. Ecol Monogr 47: 196–203

Vásquez JA (1986) Morfología de las estructuras alimentarias como factores en la organización de comunidades submareales. Biota 1: 104

Vásquez JA (1989) Estructura y organización de huirales submareales de *Lessonia trabeculata*. PhD Thesis, Universidad de Chile, Santiago Chile

Vásquez JA (1992) *Lessonia trabeculata*, a subtidal bottom kelp in northern Chile: a case study for a structural and geographical comparison. In: Seeliger U (ed.) Coastal plant communities of Latin America. Academic Press, San Diego, pp 77–89

Vásquez JA (1993) Abundance, distributional patterns and diets of main herbivorous and carnivorous species associated to *Lessonia trabeculata* kelp beds in northern Chile. Serie Ocasional Facultad de Ciencias del Mar, Universidad Católica del Norte (Chile) 2: 213–229

Vásquez JA, Buschmann AH (1997) Herbivore-kelp interactions in Chilean subtidal communities: a review. Rev Chile Hist Nat 70: 41–52

Vásquez JA, Castilla JC (1984) Some aspects of the biology and trophic range of *Cosmasterias lurida* (Asteroidea, Asteriinae) in belts of *Macrocystis pyrifera* at Puerto Toro Chile. Medio Ambiente (Chile) 7: 47–51

Vásquez JA, Castilla JC, Santelices B (1984) Distributional patterns and diets of four species of sea urchins in giant kelp forest (*Macrocystis pyrifera*) of Puerto Toro, Navarino Island, Chile. Mar Ecol Progr Ser 19: 55–63

Vásquez JA, Véliz D, Weisner R (1996) Análisis malacológico de un yacimiento de la cultura Huentelauquen IV Región, Chile. Gayana Oceanol 4: 109–116

Vásquez JA, Camus PA, Ojeda FP (1998) Diversidad, estructura y funcionamiento de ecosistemas rocosos del norte de Chile. Rev Chile Hist Nat 71: 479–499

Vega JMA, Bushmann AH, Vásquez JA (2005) Population biology of the subtidal kelps *Macrocystis integrifolia* and *Lessonia trabeculata* (Laminariales, Pheophyta) in an upwelling ecosystem of northern Chile: interannual variability and El Niño 1997–1998. Rev Chile Hist Nat 78: 33–50

Viviani CA (1975) Las comunidades marinas litorales en el norte grande de Chile. Publ. Ocasional. Lab Ecol Marina, Iquique. Universidad del Norte, 196 pp

Zamora S, Stotz W (1992) Ciclo reproductivo de *Loxechinus albus* (Molina, 1782) (Echinodermata: Echinoidea) en Punta Lagunillas, IV Región, Coquimbo, Chile. Rev Chile Hist Nat 65: 121–133

Zegers J, Oliva M, Hidalgo C, Rodríguez L (1983) Crecimiento de *Loxechinus albus* (Molina, 1782) (Echinodermata: Echinoidea) en sistemas de jaulas suspendidas a media agua. Mem Asoc Latinoam Acuicul 5: 369–378

Edible Sea Urchins: Biology and Ecology
Editor: John Miller Lawrence

Chapter 13

Ecology of *Paracentrotus lividus*

Charles F Boudouresque and Marc Verlaque

Center of Oceanology of Marseilles, Marseilles (France).

1. INTRODUCTION

Paracentrotus lividus (Plate 4A) is a rather large sea urchin. The diameter (horizontal test diameter without spines) of largest individuals can reach 7.5 cm (e.g. Bonnet 1925; Boudouresque et al. 1989; Lozano et al. 1995). In spite of its popular name ('purple sea urchin'), its colour is highly variable: black-purple, purple, red-brown, dark brown, yellow-brown, light brown or olive green. Colour is not related to depth or size (Koehler 1883; Mortensen 1943; Cherbonnier 1956; Tortonese 1965; Gamble 1966–1967).

The analysis of sequence variation in a fragment of the mitochondrial gene cytochrome *c* oxidase suggests that *P. lividus* has two randomly mating populations in the western Mediterranean and in the eastern Atlantic, with panmixis within these two geographical areas (Duran et al. 2004).

In most of its geographical range, in past or present times and on a regular or occasional basis, its gonads have been appreciated as seafood and *P. lividus* has been intensely harvested. Nowadays, the consumption of *P. lividus* is mainly limited to France and Spain, and to a lesser extent to Italy and Greece, although harvesting occurs, or has occurred, over a much larger area (e.g. Ireland, Portugal and Croatia) for export (Régis et al. 1985; Ballesteros and García-Rubies 1987; Ledireac'h 1987; Le Gall 1987; Zavodnik 1987; Byrne 1990; Catoira Gómez 1992; Martínez et al. 2003; Guidetti et al. 2004; Sánchez-España et al. 2004).

2. DISTRIBUTION AND HABITAT

2.1. Habitat

Paracentrotus lividus is distributed throughout the Mediterranean Sea and in the north-eastern Atlantic, from Scotland and Ireland to Southern Morocco and the Canary Islands, including the Azores. It is particularly common in regions where winter water temperatures range from 10 to 15 °C, and summer temperatures from 18 to 25 °C such as western Mediterranean, Portugal and Biscayne Bay. Its northern and southern limits more or less correspond to 8 °C winter and 28 °C summer isotherms, respectively. In the English Channel, lower and upper lethal temperatures are 4 and 29 °C. However, in a Mediterranean lagoon, it can withstand temperatures >30 °C, which suggests a certain physiological diversity (Tortonese 1965; Mortensen 1943; Allain 1975; Kitching and Thain 1983; Bacallado et al. 1985; Le Gall et al. 1990; Fernandez and Caltagirone 1994; Fernandez 1996).

Paracentrotus lividus is typically a subtidal species, living from the mean low water mark down to 10–20 m, and in intertidal rock pools (Gamble 1965; Tortonese 1965; Allain 1975; Régis 1978; Harmelin et al. 1980; Crook et al. 2000). Its upper limit is determined by desiccation. In the Mediterranean, a sea characterised by low amplitude tides, when the sea level drops to unusually low levels due to a particularly high atmospheric pressure, emerged individuals of *P. lividus* usually die quite quickly. Intertidal individuals occur in rock pools. Exceptionally cold winters can eliminate them (Bouxin 1964). Isolated individuals occur at depths of up to 80 m (Cherbonnier 1956; Tortonese 1965). Larvae can be advected downward to depths of several hundred metres; in the Mediterranean, the deep waters are nearly isothermal throughout the year (12–14 °C), such that temperature is not a factor. In addition, larvae can tolerate pressures more than one order of magnitude higher than where the adults occur. Accordingly, absence of *P. lividus* at depth is the result of other factors (Young et al. 1997), possibly biotic factors such as predation.

In the open sea, *P. lividus* occurs mainly on solid rocks, boulders and in meadows of the seagrasses *Posidonia oceanica* and *Zostera marina* (Mortensen 1927; Tortonese 1965; Ebling et al. 1966; Verlaque 1987a). Its surprising absence or relative scarcity in meadows of *Cymodocea nodosa*, though this seagrass is a strongly 'preferred' food, might be due to either the unsuitability of the sand bottom between shoots of *C. nodosa* for locomotion, or a high predation pressure (Traer 1980). Indeed, it is uncommon on sandy and detritic bottoms, bottoms on which individuals cluster on isolated stones, large shells and various refuses (Zavodnik 1980). Shallower individuals, under very exposed conditions and/or in intertidal rock pools, resist dislodgment by waves by burrowing in the substratum (e.g. sandstone, limestone, granite, basalt; not hard slate), creating cup-shaped cavities where they live permanently or temporarily. This behaviour may also provide protection from predators. The burrows are a perfect fit for the animals. They are sometimes so numerous and close together that the substrate is completely honeycombed. The bottoms of the burrows are free of encrusting corallines (rhodobionta) whereas the walls are not. This may be the result of intense feeding, lack of light, or both. This rock-boring behaviour, which occurs both in the Atlantic and the western Mediterranean, and in particular the manner in which *P. lividus* manages to excavate its living quarters, has been the object of great attention and the cause of many controversies. In the nineteenth and early twentieth centuries, certain authors argued that echinoids were not able to burrow into hard rocks (Caillaud 1856; Fischer 1864; Otter 1932; Mortensen 1943; Kempf 1962; Gamble 1965; Neill and Larkum 1966; Goss-Custard et al. 1979; Torunski 1979; Martinell 1981; Lawrence and Sammarco 1982; Cuenca 1987).

In littoral lagoons (e.g. Thau and Urbinu lagoons, Mediterranean, and Arcachon Bay, Atlantic, France), *P. lividus* can live on coarse sand or even on mud substrata (Allain 1975; San Martín 1987; Fernandez et al. 2003). In these lagoons, as in intertidal rock pools, the maximal size of the individuals is always far smaller than in the open sea. Similar dwarf populations occur in the eastern Mediterranean Sea (Koehler 1883; Pérès and Picard 1956; Kempf 1962; Tortonese 1965). Although present in Mediterranean coastal lagoons and Atlantic 'rías' (estuaries), *P. lividus* is sensitive to low and high salinities. Lower and upper lethal salinities are in the order of 15–20 and 39–40 for long-term exposure (Allain 1975; Pastor 1971; Le Gall et al. 1989). In the autumn of 1993, an exceptional rainfall (450 mm in 48 h) occurred on the Urbinu Lagoon, Corsica, and its watershed area, resulting in local salinities as low as 7 and in mass mortality of *P. lividus* (Fernandez et al. 2003).

The species seems insensitive to organic pollution, and in fact such compounds clearly enhance its development (Tortonese 1965; Allain 1975; Zavodnik 1987; Delmas 1992). Dense populations have occurred in the polluted Bay of Brest, Brittany, near the mouth of urban waste discharge at Rabat, Morocco, in the outflow of Marseilles and in the heavily polluted Berre Lagoon, near Marseilles. The density of *P. lividus* increased progressively towards the sewage outlet of Cortiou cove and was most dense a few hundred metres from the discharge point (Koehler 1883; Kempf 1962; Harmelin et al. 1980, 1981; Delmas and Régis 1985; Bayed et al. 2005). Aquarium experiments have shown it is sensitive to ammonia (Lawrence et al. 2003) but at concentrations more likely in aquaculture facilities than in the natural environment. Finally, *P. lividus* withstands high levels of heavy metal pollution, and even accumulates them, although growth rate is reduced (Augier et al. 1989; Delmas 1992; San Martín 1995). In contrast, at least in tide pools, oil spills can cause 100% mortality. It took 3 years for densities of *P. lividus* to recover after the 'Erika' oil spill (Barille-Boyer et al. 2004).

Small individuals (<1–2 cm), which are particularly vulnerable to predation, live permanently in holes, crevices, under pebbles and boulders, within the 'matte' of the seagrass *P. oceanica*, and sometimes under a dense cover of multicellular photosynthetic organisms (MPOs) (Kempf 1962; Gamble 1966–1967; Kitching and Thain 1983; Verlaque 1984, 1987a; Azzolina and Willsie 1987; Azzolina 1988; San Martín 1995). Larger individuals may or may not return to shelters after foraging, depending on their size and density of predatory fishes (Sala 1996; Palacín et al. 1997).

2.2. Densities

Densities of *P. lividus,* ranging from a few to a dozen individuals per square metre, are quite common over its whole geographical and depth ranges (Table 1). Very high densities (>50–100 individuals per square metre) usually occur in shallow habitats, on gently sloping rocks, pebbles or boulders, in intertidal rocky pools and in polluted environments (Table 1; Kempf 1962; Pastor 1971; Crapp and Willis 1975; Torunski 1979; Harmelin et al. 1981; Delmas and Régis 1986; Delmas 1992). Higher densities correspond to localised aggregations (>1600 individuals per square metre) although the basis for this phenomenon remains unclear (Mastaller 1974; Keegan and Könnecker 1980). Such localised aggregations may be a defense strategy against predators, a feeding behaviour or a spawning behaviour.

2.3. Short- and Long-Term Changes in Density

Apparent changes in *Paracentrotus lividus* density can be misleading due to its daily and seasonal behaviour and to the census method. For example, at Lough Hyne, Ireland, the

Table 1 *Paracentrotus lividus*: densities (mean number of individuals per square meter) in different localities, depth and habitats. Small individuals (<2 cm in diameter) are excluded. Depth refers to the upper limit of the subtidal.

Location	Depth (m)	Habitat	Density	References
Elbu Bay, Corsica (Natural Reserve of Scandola)	3–6	Gently sloping rocks and *Posidonia oceanica*	0.1[a]	Boudouresque et al. (1992)
Urbinu Lagoon, Corsica	0–1	Silt	0.2–0.6	Fernandez et al. (2003)
Niolon, near Marseilles, France	3–10	Rocky vertical wall	0.5	Kempf (1962)
Urbinu Lagoon, Corsica	0.5–1	*Cymodocea nodosa* meadow	0.7	Fernandez and Boudouresque (1997)
Plane Island, Marseilles, France	2–5	Subvertical rock with MPOs and gorgonians	0.7	Delmas (1992)
Port-Cros Island, France	0.5	*Posidonia oceanica* barrier-reef	0.9	Azzolina et al. (1985)
Sesimbra, Portugal	0–2	Sloping rock with *Corallina elongata*	<1	Saldanha (1974)
Port-Cros Island, France	15–22	Subhorizontal rocks	1	Harmelin et al. (1980)
Marseilles, France	0–10	Vertical pier	1.5	Kempf (1962)
Tossa de Mar, Catalonia, Spain	4–16	Rocky vertical wall	2	Palacín et al. (1997)
Galeria Bay, Corsica (outside the Natural Reserve of Scandola)	3–6	Gently sloping rocks and *Posidonia oceanica*	3.4	Boudouresque et al. (1992)
Banyuls, French Catalonia	9	*Posidonia oceanica* meadow	3.6	Shepherd (1987)
Tossa de Mar, Catalonia, Spain	13–16	Boulders with turfing and frondose MPOs	3–4	Palacín et al. (1998)
Malta	6–13	Cliff	4[a]	Gamble (1965)
Port-Cros Island, France	5	Subhorizontal rocks	5	Harmelin et al. (1980)
Port-Cros Bay, France	2–3	*Posidonia oceanica* meadow	6	Boudouresque et al. (1980)
Lough Hyne, Cork, Ireland	0.5	Shells, stone, and gravel	7[a]	Kitching and Ebling (1961)
Marseilles, France	11	*Posidonia oceanica* meadow	8	Kirkman and Young (1981)
Niolon, near Marseilles, France	5–8	Gently sloping rock	9	Kempf (1962)

Table 1 Continued

Location	Depth (m)	Habitat	Density	References
Marseilles, France	1–8	Concrete blocks, breakwater	10.5	Kempf (1962)
La Madrague, near Marseilles, France	4–6	Subhorizontal rocks with Corallinaceae	13.5	Kempf (1962)
Gulf of Marseilles, France	5–10	Gently sloping rocks	10–15	Régis (1978)
Banyuls, French Catalonia	5	*Posidonia oceanica* meadow	22	Shepherd (1987)
South of Leghorn, Italy	0	Intertidal rocky pool	24	Benedetti-Cecchi and Cinelli (1995)
Cortiou cove, near Marseilles, France	7–10	Rocks, 500 m from sewage outfall	26	Harmelin et al. (1980)
Sesimbra, Portugal	15	Rocks with Corallinaceae	26	Saldanha (1974)
Port-Cros Bay, France	0.5–1.5	Degraded *Posidonia oceanica* meadow	28	Boudouresque et al. (1980)
Cortiou cove, near Marseilles, France	2–5	Rocks, 500 m from sewage outfall	42	Harmelin et al. (1980)
Banyuls, French Catalonia	1	*Posidonia oceanica* meadow	45	Shepherd (1987)
Loire Atlantique, France	0	Intertidal rocky pools	63	Barille-Boyer et al. (2004)
Ischia Island, near Naples, Italy	3–6	*Posidonia oceanica* meadow	64	Traer (1980)
Cortiou Cove, near Marseilles, France	2.5	Rocks, 1 500 m from sewage outfall	78	Harmelin et al. (1980)
Gulf of Piran, Northern Adriatic	3	Rocks	80	Torunski (1979)
Sausset-les-Pins, near Marseilles, France	4–5	*Posidonia oceanica* meadow	40–110	Verlaque and Nédélec (1983a)
Cabo Mar, Ría de Vigo, Spain	0	Rocky pools	118	Pastor (1971)
Borneira, Ría de Vigo, Spain	0	Rocky pool	125	Pastor (1971)
Gulf of Piran, Northern Adriatic	1	Rocks	180[a]	Torunski (1979)
Cascais, Portugal	Subtidal	Boulders	234	Gago et al. (2003)
Bantry Bay, Ireland	0	Rocky pools	150–300	Crapp and Willis (1975)

MPOs – Multicellular photosynthetic organisms.

[a] Calculated from the authors' figure or table.

apparent population size (visible individuals) differed between day and night (15.8 and 6.7 individuals per square metre, respectively) (Barnes and Crook 2001a).

Populations of *P. lividus* can be relatively stable for several years, e.g. in rocky vertical walls between 3 and 10 m depths, near Tossa de Mar, Catalonia (Turon et al. 1995) and in a meadow of *P. oceanica*, near Marseilles (San Martín 1995). However, this is not common. Conversely, rapid changes in density of large individuals can be observed. For example, ca. 1 km off Cortiou Cove, near Marseilles, the mean density at a 12–15 m depth increased significantly from 12 individuals per square metre in March 1985 to 31 individuals per square metre 2 months later (Delmas 1992). In a shallow boulder habitat of Spanish Catalonia, subject to strong wave action, the mean density varied extensively from one month to the next. The greatest change occurred between August and September 1992, when it dropped from 31 to 5 individuals per square metre (Turon et al. 1995).

Year-to-year changes in density occur very frequently. In intertidal pools, near Leghorn, Italy, the mean density increased from 9 to 24 individuals per square metre between the winter of 1991 and the winter of 1992 (Benedetti-Cecchi and Cinelli 1995). In the Piran Bay, Gulf of Triest, Adriatic Sea, populations exploded between 1972 and 1974, with local densities of up to 350 individuals per square metre (Mastaller 1974; Vukovič 1982; Schneider and Torunski 1983). In Galeria Bay, Corsica, the mean density doubled between 1980 and 1988 (Boudouresque et al. 1989). In contrast, in Port-Cros Bay, Var, France, the density declined drastically between 1979 and 1980, followed by a steady decline until 1984 (Boudouresque et al. 1981; Azzolina et al. 1985; Azzolina 1987, 1988). Such fluctuations can lead to misinterpretations of larger-scale temporal patterns. For example, between 1992 and 1994, *P. lividus* densities were significantly lower in a protected area (a marine reserve of Spanish Catalonia with high fish densities) than in an adjacent unprotected area, supporting the hypothesis that fish predation was the most important factor controlling sea urchin populations (Sala 1996; Sala and Zabala 1996). Subsequently, in 1995 and 1996, *P. lividus* densities in these two areas increased and declined, respectively, becoming identical, indicating that factors other than fish predation actually control sea urchin densities (Sala et al. 1998b).

Long-term changes in density have also been recorded in several areas. At Lough Hyne, Ireland, density increased conspicuously from 1962 to 1965, then declined until 1975. Another peak in density in 1979 was followed by a sharp decline, with no further recovery until 1999. The change was almost four orders of magnitude. Finally, in 2000, the population became extinct (Ebling et al. 1966; Kitching and Thain 1983; Barnes et al. 1999; Barnes and Crook 2001a; Barnes et al. 2002). In the Bay of Biscay, density was high in 1905–1907, very low in 1925, then high again in 1935–1950 (Fischer-Piette 1955). In Brittany, densities have declined greatly since the 1950s or 1960s, as a result of both overharvesting and local proliferation of the predatory starfish *Marthasterias glacialis* (Allain 1975; Le Gall 1987). Near Marseilles, a significant increase occurred from the early 1960s to the late 1970s, possibly related to pollution increases (Kempf 1962; Harmelin et al. 1981). In Port-Cros Bay, Provence, France, an increase occurred from the 1960s to the late 1970s, followed by a decrease in the 1980s and a renewed increase in the 1990s (Azzolina 1987; unpubl. data).

An uneven spawn production (near the limits of its geographical range), losses during larval life, success or failure of recruitment, migrations, natural changes in abundance of sea urchin predators, overfishing of predators (especially crabs and fishes), pollution, exceptional rainfalls, diseases and harvesting may account for these short- and long-term fluctuations (Ebling et al. 1966; Southward and Southward 1975; Le Gall A1987; Byrne 1990; Delmas 1992; Sala 1997; López et al. 1998; Fernandez et al. 2003).

2.4. Co-occurring Species

The burrows of *P. lividus* have encrusting Corallinaceae, especially at the edge, and several animal species. In the northwestern Mediterranean, Kempf (1962) reported the crustacean *Catapaguroides timidus*, the mollusks *Cantharidus striatus*, *C. exasperatus*, *Alvania lineata*, young *Columbella rustica*, *Nassa incrassata* and *Chiton olivaceus*, the annelid *Hormothos spinifera* and sometimes juveniles of the small fish *Lepadogaster* sp.

The rocky substrates with high densities of *P. lividus* are colonised by encrusting Corallinaceae and exhibit a low biomass of primary producers. They have been termed 'barren grounds'. Barren grounds have been observed in most of the geographical range of *P. lividus*: England, Portugal, northwestern Mediterranean, Sardinia and Malta (Kempf 1962; Neill and Larkum 1966; Gamble 1966–1967; Saldanha 1974; Lewis 1976; Verlaque 1987a). In the Atlantic, and in the colder northern parts of the Mediterranean, the main encrusting corallines of the barren grounds are *Lithophyllum incrustans* and *Mesophyllum alternans*, often covered by a turf of the red alga (rhodobionta) *Callithamniella tingitana*. The large green alga (chlorobionta) *Codium fragile* is also usually present. In warmer parts of the Mediterranean, the main encrusting coralline is *Neogoniolithon brassica-florida*, usually associated with an encrusting brown alga (stramenopile), *Pseudolithoderma adriaticum* (Pastor 1971; Kitching and Thain 1983; Verlaque and Nédélec 1983b; Verlaque 1987a, 1987b; Hereu et al. 2005). Like the primary producers, the fauna is very poor in terms of the number of species. In the Mediterranean, the most abundant species are the limpet *Patella caerulea*, the cirriped *Balanus perforatus* and the sea anemones *Anemonia sulcata* and *Aiptasia mutabilis*. Another sea urchin, *Arbacia lixula*, can also occur (Kempf 1962; Verlaque 1987a). In the 1960s and the 1990s, the abundance of this possible competitor of *P. lividus* increased markedly, a phenomenon tentatively attributed to either a warm episode or global warming (Kempf 1962; Francour et al. 1994). In Portugal and Galicia, the dominant species are *Balanus spongicola* and the sea anemones *Anemonia sulcata* and *Corynactis viridis*. The sea urchin *Psammechinus miliaris* (Gmelin) is also sometimes associated with *P. lividus* (Pastor 1971; Saldanha 1974). Finally, in Ireland, the dominant species are the mollusks *Anomia ephippium*, *Chlamys varia*, *Gibbula cineraria* and *Patella aspera*, the polychaete *Pomatoceros* spp. and the stalked colonial tunicate *Morchellium argus* (Kitching and Thain 1983). Most of these species appear to be directly or indirectly favoured by the browsing of *P. lividus*, as this removes their competitors and makes substratum available to those organisms that can resist its browsing or grazing, or for which, although browsed, the cost-benefit balance is beneficial. The densities of these species (e.g. *Patella caerulea* and the chlorobionta *Codium fragile*) consistently vary in the same manner as do the densities of *P. lividus* (Kitching and Thain 1983; Verlaque 1984).

In contrast with barren grounds, which constitute a habitat that is little more than two-dimensional, similar subtidal bottoms with low densities of *P. lividus* are usually occupied by forests of *Laminaria* or *Cystoseira* (stramenopiles), or bush-like MPO assemblages. These three-dimensional habitats provide accommodation and/or food for a rich fauna (polychaetes, amphipods, gastropods) and flora. The biomass of primary producers is usually high (Kitching and Thain 1983).

Juveniles of two clingfishes (*Apletodon incognitos* and *Lepadogaster candollei*) and gobies (*Gobius bucchichi*, *Millerigobius macrocephalus* and *Zebrus zebrus*) hide between the oral spines of *P. lividus* in the western Mediterranean (Patzner 1999).

3. FOOD AND FEEDING

3.1. Food Preferences

For the most part, food preference in *P. lividus* has been determined by means of aquarium experiments, where individuals were exposed to two items in similar and nonlimiting amounts. Further information comes from comparison of item abundance in gut content and in the habitat where individuals were collected (the Ivlev index) (Ivlev 1961). A food is said to be 'preferred' or 'avoided' when its percentage abundance in the sea urchin gut is above or below the amount available, respectively (Table 2). These results must be considered with caution for a number of reasons: (i) *in vitro* results could be artefacts,

Table 2 *Paracentrotus lividus*: food preferences of large individuals: fungi and multicellular photosynthetic organisms (MPOs).

Species	Winter	Spring	Summer	Autumn	Year
Fungi (opisthochontha)					
Corollospora maritima	–	–	–	–	1
Rhodobionta, plantae ("red algae")					
Amphiroa rigida	–	3	–	3	4
Asparagopsis armata (gametophyte)	4	4	4	–	4
Boergeseniella fruticulosa	–	–	–	–	2
Bonnemaisonia hamifera	–	–	–	–	3
Botryocladia botryoides	–	4	–	–	–
Centroceras clavulatum	–	–	4	–	–
Ceramium ciliatum	2–4	–	–	–	–
Ceramium rubrum	–	–	–	–	2–3
Corallina elongata	2–3	2–3	2	–	3
Cryptonemia lomation	–	–	–	–	4
Falkenbergia rufolanosa[a]	–	2	–	–	2
Gelidium spinosum (= *G. latifolium*)	–	–	4	–	4
Gracilaria bursa-pastoris	–	3	–	–	–
Haliptilon virgatum (= *Corallina granifera*)	–	2–3	–	–	–
Halopithys incurva	–	2	2	–	–
Halurus flosculosus	4	–	–	–	–

Table 2 Continued

Species	Winter	Spring	Summer	Autumn	Year
Herposiphonia secunda	–	2	–	–	2
Jania rubens	2	–	–	–	2
Laurencia microcladia	–	3	2	3	–
Laurencia obtusa	–	–	–	–	3
Liagora viscida	–	–	–	–	4
Lithophyllum incrustans	–	–	–	–	4
Neogoniolithon brassica-florida	–	4	4	4	–
Osmundaria volubilis (= *Vidalia volubilis*)	–	4	4	–	3
Osmundea truncata[b]	–	–	–	–	3
Osmundea verlaquei[c]	4	–	–	–	–
Phyllophora nervosa	–	4	–	–	4
Plocamium cartilagineum	1–2	–	–	–	–
Rhodymenia ardissonei	–	4	–	–	4
Rissoella verruculosa	–	1	–	–	1
Rytiphlaea tinctoria	–	–	–	–	4
Solieria chordalis	–	2–3	–	–	–
Sphaerococcus coronopifolius	3	–	4	–	4
Chlorobionta, plantae ("green algae")					
Anadyomene stellata	–	–	4	4	–
Bryopsis muscosa	1	–	–	–	–
Caulerpa prolifera	–	–	4	–	–
Caulerpa racemosa var. *cylindracea*	–	–	–	2	–
Caulerpa taxifolia	2	3–4	4	–	3
Chaetomorpha linum	–	–	–	–	2
Cladophora dalmatica	–	–	–	–	2–3
Cladophora rupestris	–	–	–	–	2
Codium bursa	4	4	–	–	4
Codium fragile	2	2–4	1–3	–	3
Codium vermilara	2	–	2	–	–
Flabellia petiolata	–	–	4	–	4
Halimeda tuna	–	2	3	–	–
Ulva compressa (= *Enteromorpha compressa*)	–	1	–	–	–
Ulva intestinalis (= *Enteromorpha intestinalis*)	–	–	–	–	1
Ulva linza (= *Enteromorpha linza*)	–	–	–	–	3
Ulva lactuca	–	–	–	–	2
Ulva rigida	2–4	–	2	–	2
Valonia utricularis	–	–	–	–	4
Magnoliophyta, plantae ("seagrasses")					
Cymodocea nodosa	–	–	–	–	1
Posidonia oceanica	2–4	1	1–2	1	4
Zostera marina	–	3	–	–	–

(*Continued*)

Table 2 Continued

Species	Winter	Spring	Summer	Autumn	Year
Photosynthetic stramenopiles ("brown algae")					
Cystoseira amentacea var. *stricta*	1	–	–	–	1
Cystoseira barbata	–	–	–	–	1
Cystoseira brachycarpa (= *C. balearica*)	–	1–2	2	–	1
Cystoseira compressa	1–3	1	–	–	1
Cystoseira crinita	1	–	–	–	–
Cystoseira mediterranea	1	–	2	–	1
Cystoseira spinosa	–	–	–	–	1
Dictyota dichotoma	2	1–2	–	1	2
Dictyota fasciola (= *Dilophus fasciola*)	–	–	–	–	1
Dictyota spiralis (= *Dilophus spiralis*)	1–2	1–2	–	–	–
Fucus serratus	–	–	–	–	1
Laminaria japonica	-	2	–	–	–
Padina pavonica	–	1	1	1–2	1
Pseudolithoderma adriaticum	–	4	–	4	–
Sargassum muticum	–	2–3	–	–	–
Scytosiphon lomentaria	2	–	–	–	–
Sphacelaria cirrosa	–	2–3	–	–	2
Stilophora tenella (= *S. rhizodes*)	–	–	–	–	2
Stypocaulon scoparium (= *Halopteris scoparia*)	1–3	2	1	–	2
Undaria pinnatifida	–	1	–	–	–

The use of the term "algae", which is neither a monophyletic clade nor a functional or a morphological group is avoided here. 1: Strongly "preferred" species. 2: Moderately "preferred" species. 3: Indifferent or slightly "avoided" species. 4: Strongly "avoided" species. Presence of two numbers means contrasting results, as a function of authors, methods, test diameter or locality. "Year" means year-round or of unknown season.

[a] Sporophyte of *Asparagopsis armata.*

[b] = *Laurencia pinnatifida* auctores, non (Hudson) Lamouroux.

[c] = *Laurencia undulata* auctores, non Yamada.

Data from Traer (1980, 1984), Cuomo et al. (1982), Nédélec (1982), Kitching and Thain (1983), Verlaque and Nédélec (1983b), Verlaque (1984, 1987b), Knoepffler-Péguy et al. (1987), San Martín (1987), Frantzis et al. (1988), Odile et al. (1988), Fernandez (1989), Rico (1989), Kovitou (1991), Boudouresque et al. (1993), Knoepffler-Péguy and Nattero (1996), Lemée et al. (1996) and Aubin (2004).

although they are usually consistent with in situ experiments or observations (Lemée et al. 1996). However, in the Thau lagoon, France, results from in situ choice experiments and from gut contents (Ivlev index) were not consistent, especially for the chlorobionta *C. fragile* (San Martín 1987). (ii) Food choice is always dependent upon available items (Paine and Vadas 1969). For example, choice experiments show that, in the Thau lagoon, the stramenopile *Undaria pinnatifida* is the most preferred item. However, it was never found in guts of *P. lividus* collected in the field. Indeed, this alga is only found on metallic oyster-culture devices that *P. lividus* cannot reach (San Martín 1987). (iii) A sea urchin might be attracted by one species rather than another, and yet feed more rapidly on the second once feeding has begun. This might be related to the ability of the urchin to manipulate and eat the food (Lawrence 1975). (iv) An item may be preferred in one season and avoided in another (Table 2).

In the field, when the resource is nonlimited, analysis of gut contents indicates that *P. lividus* is basically herbivorous (Mortensen 1943; Kitching and Ebling 1961; Kempf 1962; Ebling et al. 1966; Neill and Larkum 1966; Neill and Pastor 1973; Verlaque and Nédélec 1983b; Verlaque 1987a, 1987b).

A number of species are clearly 'preferred', e.g. the MPOs *Rissoella verruculosa* (rhodobionta, plantae), *Cymodocea nodosa* (seagrass: magnoliophyta, plantae), *Cystoseira amentacea, Padina pavonica* and *Undaria pinnatifida* (photosynthetic stramenopiles). Other MPOs are strongly 'avoided', e.g. *Asparagopsis armata*, *Gelidium spinosum*, *Anadyomene stellata*, *Caulerpa prolifera*, *C. taxifolia* and *Flabellia petiolata* (Table 2).

Paracentrotus lividus consumes all parts of the seagrass *P. oceanica*: living leaves with or without epiphytes, unshed dead leaves, dead leaves of the litter, and even rhizomes and roots. It prefers, in increasing order of preference: green leaves covered by epiphytic MPOs, brown (dead) leaf tips with epiphytes, green leaf tips without epiphytes and, lastly, green leaf bases without epiphytes (Traer 1980, 1984; Verlaque and Nédélec 1983a; Zupi and Fresi 1984; Shepherd 1987; Verlaque 1987a). Authors have ranked brown leaf tips without epiphytes at the top, middle or last level of food choice (Ott and Maurer 1977; Traer 1980; and Shepherd 1987, respectively). Ott (1981) concluded that *P. lividus* feeds on plant parts that are actually debris and defined it as a 'pseudograzer'. The time since leaf death, the nature and abundance of the fungal flora that degrades them, and therefore their chemical composition, may account for these contrasting results. It is of interest to note that the lignicolous ascobionta fungus *Corollospora maritima* is strongly preferred by *P. lividus* (see Table 2), both in pure culture or within dead leaves (Cuomo et al. 1982).

Food choice is frequency-dependent, i.e. dependent upon the relative abundance of available items ('switching' or 'apostatic feeding') (see Harper 1969; Lawrence 1975; Verlaque and Nédélec 1983a; Frantzis et al. 1988). Selectivity is high under conditions of high food supply, but declines as the grazing pressure increases and finally disappears when grazing pressure becomes extreme. Although the chlorobionta *C. fragile* is far from being a strongly preferred food (Table 2), it is quickly and completely eaten when transferred to a barren ground with high densities (50–90 individuals per square metre) of *P. lividus* (Kitching and Thain 1983; see also Gago et al. 2003). Similarly, *P. lividus* on barren grounds ingest large amounts of the incrusting coralline *L. incrustans* (Delmas and Régis 1986). Available items may also include drifting MPOs. In the Mediterranean, leaves of *P. oceanica* can constitute up to 40% of the gut contents of individuals located at hundreds of metres from a meadow (Verlaque and Nédélec 1983b; Maggiore et al. 1987; Verlaque 1987a). In the Gulf of Calvi, Corsica, *P. oceanica* contributes roughly 12% to the carbon content of such individuals (Dauby 1989). At Cascais, Portugal, drift material contribution is particulary important in October and November (Gago et al. 2003).

Food preferences are size dependent. Recently settled sedentary juveniles (1 mm in diameter), graze (*sensu* Ogden 1976) on encrusting and endolithic MPOs: the Corallinaceae *N. brassica-florida* and *Hydrolithon farinosum*, Peyssonneliaceae, *Pseudolithoderma adriaticum*, *Blastophysa rhizopus*, *Phaeophila dendroides*. They also graze on Bacillariophyceae, sponges and foraminifera. Subsequently (3–7 mm), they browse (*sensu* Ogden 1976) on filamentous MPOs: the Ceramiaceae *Spermothamnion repens* and

Antithamnion cruciatum, the Rhodomelaceae *Lophosiphonia scopulorum* and *Dipterosiphonia dendritica*, and Ectocarpaceae; rhodobionta are dominant. Between 7 and 10 mm in diameter, they begin to eat coarsely branched MPOs (e.g. *Stypocaulon scoparium*, *Padina pavonica*, *Corallina elongata*, *Cystoseira* spp.). The adult diet is reached from about 10 mm diameter upwards, at which time photosynthetic stramenopiles ('brown algae') and leaves of *P. oceanica* are dominant (Verlaque and Nédélec 1983b; Verlaque 1984, 1987a, 1987b). However, the frequency of the latter food source increases with size, and the contrast between 'preferred' and 'avoided' species is sharper for small individuals (2.0–2.5 cm in test diameter) than for large individuals (>3.5 cm) (Traer 1980; Nédélec 1982; Verlaque 1987b; Fernandez 1989; Rico 1989).

Food preference or avoidance is often difficult to explain. Some avoided species synthesise toxic or repellent secondary metabolites. This is the case for the introduced chlorobionta *Caulerpa taxifolia,* which produces large amounts of terpenes (Guerriero et al. 1992; Lemée et al. 1996). The rhodobionta *Asparagopsis armata* synthesizes brominated substances (Codomier et al. 1977). In photosynthetic stramenopiles and *P. oceanica*, the chemical defence mechanism appears to be linked to the presence of phenolic compounds (Traer 1980, 1984). However, the amount of these metabolites is not always consistent with preference. For example, the photosynthetic stramenopiles *Cystoseira compressa* and *Stypocaulon scoparium* contain 23% and 2% (in relation to total dry weight) polyphenols, respectively, and yet they present a similar ranking in Table 2 (Frantzis and Grémare 1992). Calcification is also a cause of avoidance. *Paracentrotus lividus* avoids most incrusting and articulated corallines (rhodobionta), e.g. *L. incrustans* and *Amphiroa rigida*, although some tiny articulated corallines, e.g. *Jania rubens*, are moderately preferred (Table 2). Food choice may depend upon the overall morphology and texture of the food, and on the ease with which the sea urchin can catch it. Nitrogen content of food is often assumed to be positively correlated with food choice. The consumption of *P. oceanica* leaves is enhanced when their nitrogen content is increased, such as in polluted habitats (Ruiz-Fernández 2000). In the Mediterranean, however, several strongly 'avoided' species have high nitrogen content, and thus low C/N ratios, such as the rhodobionta *Asparagopsis armata* and *Halurus flosculosa* (Rico 1989). In contrast, the 'preferred' photosynthetic stramenopile *Padina pavonica* has a very low level of both essential and nonessential amino acids (Frantzis and Grémare 1992). Finally, there is no clear relationship between 'preferred' species and their calorific value (Rico 1989).

De Burgh (1975), West et al. (1977) and Régis (1978, 1981) have shown that surface feeding by the spines exists in *P. lividus*. In sites characterised by high domestic pollution, the spines become longer and thinner. This lengthening of the spines, and their more porous internal structure, is considered a morphofunctional adaptation towards a more active and efficient uptake of organic material (Delmas and Régis 1985; Régis 1986). It is consistent with the inverse relation between mean spine length and mean repletion index (Régis and Arfi 1978). This surface feeding, together with the trapping of drifts, may account for very high densities of this species in polluted sites, the presence of individuals imprisoned in burrows, and the populations in pools nearly devoid of MPOs (Table 1). Indeed, it is highly improbable that MPO settlement alone can provide sufficient food, especially when other herbivores, such as limpets, also compete for the same food resources (Mastaller 1974; Crapp and Willis 1975).

Although MPOs obviously constitute the main feeding resource of *P. lividus* in situ, this species appears to be an opportunistic generalist able to exploit any kind of food resource, especially under conditions of limited resources. Indeed, unicellular photosynthetic organisms, sponges, hydrozoa, copepods, etc. can be found in its gut contents (Mortensen 1943; Tortonese 1965; Pastor 1971; Niell and Pastor 1973; Régis 1978; Delmas and Régis 1986; Fernandez 1990; Mazzella et al. 1992). As for MPOs, sharp feeding preferences exist: the sponge *Dysidea avara* is grazed, whereas another sponge, *Crambe crambe* is strongly avoided, in relation with chemical deterrence (Uriz et al. 1996). Harmelin et al. (1981) reported that individuals of *P. lividus* ate dead fish lying on the bottom. In aquaria, they can be fed mussels (Powis de Tenbossche 1978; Haya and Régis 1995). When only strongly 'avoided' species are available, e.g. in a stand of the introduced chlorobionta, *C. taxifolia*, *P. lividus* ingests large amounts of sand (Lemée et al. 1996). Finally, cannibalism can occur as sea urchins remains were found within the gut contents in an exceptionally dense population of *P. lividus*. In aquarium, small individuals, 2–3 cm in diameter, are attacked and consumed by starved large ones (Pastor 1971).

Paracentrotus lividus often covers its aboral side by mean of a variety of items, both in situ and the laboratory: leaves of *P. oceanica*, other MPOs, empty shells, small pebbles, plastic fragments, etc. (Kempf 1962; Dambach and Hentschel 1970; Pastor 1971; Martinell 1981; Rico 1989; Benedetti-Cecchi and Cinelli 1995). This covering behaviour ('heaping behaviour', 'masking behaviour'), which is most intense in summer, has been considered to be a protection against either light (Mortensen 1927, 1943; Sharp and Gray 1962; Barnes and Crook 2001b; Crook and Barnes 2001; Crook 2003), (although it occurs both in the presence and absence of light) (Kempf 1962; Gamble 1965; Dambach and Hentschel 1970), ultraviolet radiation (Verling et al. 2002), item availability (Crook and Barnes 2001) or predation (Mortensen 1927; Pastor 1971). The fact that smaller individuals cover at higher frequencies than larger individuals (Crook et al. 1999; but see Barnes and Crook 2001b) supports the latter hypothesis. Dambach and Henstschel (1970) suggested that the covering process was reflexive. Richner and Milinski (2000) showed that covering protects the delicate apical openings of the water vascular system, which powers all the sea urchin movements, from occlusion by floating sand. Covering would appear to function as an umbrella against floating particles. In addition, this behaviour may play a role in feeding, allowing MPOs to be caught and held for consumption, a process called covering-feeding behaviour (De Ridder and Lawrence 1982; Verlaque and Nédélec 1983b). In fact, covering can play a variety of roles, since data supporting most of these hypotheses are convincing.

3.2. Consumption Rate

Ebert (1968) noted that the consumption rate of sea urchins in the laboratory may not be representative of that in the field. Marked differences between laboratory and field feeding rates occur in *P. lividus* (Table 3) (Mastaller 1974; Shepherd 1987). A rhythmic feeding periodicity of several days (Nédélec 1982; Nédélec et al. 1983; Fernandez 1989; Rico 1989) may account for discrepancies between short-term experiments. For example, the mean periodicity of feeding peaks *in vitro* was 3 days and 3–6 days when fed

Table 3 *Paracentrotus lividus*: consumption rate (mg dry weight per individual per day).

Location	Test diameter (cm)	Food	Consumption rate	References
Villefranche, France	1	*Posidonia oceanica* leaves and epiphytes, in situ	24[a]	Cellario and Fenaux (1990)
Ischia, Italy	1.5	*Posidonia oceanica* leaves, *in vitro*	15[a]	Traer (1980)
Villefranche, France	2	*Posidonia oceanica* leaves and epiphytes, *in vitro*	288[a]	Cellario and Fenaux (1990)
Marseilles, France	3	Artificial food chiefly of animal origin, *in vitro*	340–380[a]	Haya and Régis (1995)
Marseilles, France	3	*Mytilus edulis* (mussel), *in vitro*	210[a]	Haya and Régis (1995)
Marseilles, France	3	*Spirulina* (Cyanobacteria), *in vitro*	210[a]	Haya and Régis (1995)
Marseilles, France	3	Artificial food chiefly of plant origin, *in vitro*	1 480[a]	Haya and Régis (1995)
French Catalonia	3–3.5	*Posidonia oceanica* leaves, in situ	176–269	Shepherd (1987)
Israel	3.5	Artificial food	196–214	Shpigel et al. (2004)
French Catalonia	3.5–4	*Posidonia oceanica* leaves, in situ	538	Shepherd (1987)
Corsica	4	Turf MPOs in situ	25–161	Verlaque (1984, 1987a)
Corsica	4–7	*Posidonia oceanica* and epiphytes, in situ	511	Nédélec et al. (1981)
Corsica	4	*Posidonia oceanica* and epiphytes, in situ	63–171	Nédélec and Verlaque (1984), Verlaque (1987a)
Northern Adriatic Sea	4	Photosynthetic stramenopiles, *in vitro*	66–146	Mastaller (1974)
French Catalonia	4–4.5	*Posidonia oceanica* leaf tips without epiphytes, *in vitro*	181	Shepherd (1987)
French Catalonia	4–4.5	*Posidonia oceanica* dead leaf tips, with epiphytes, *in vitro*	325	Shepherd (1987)

Table 3 Continued

Localities	Test diameter (cm)	Food	Consumption rate	References
Marseilles, France	4–5	Artificial food (mainly seaweed, corn starch, shrimp, fish), In vitro	1 000–2 700[b]	Lawrence et al. (1989)
Ischia, Italy	4.5	*Posidonia oceanica* leaves, *in vitro*	185[a]	Traer (1980)
French Catalonia	4.5–5.8	*Posidonia oceanica* leaves, in situ	496–851	Shepherd (1987)
Corsica	5	Turf MPOs, in situ	50–313	Verlaque (1984, 1987a)
Corsica	5	*Posidonia oceanica* and epiphytes, in situ	124–333	Nédélec and Verlaque (1984), Verlaque (1987a)
French Catalonia	5–5.5	*Posidonia oceanica* young leaves, *in vitro*	58	Fernandez (1989)
French Catalonia	5–5.5	*Codium vermilara, in vitro*	63	Fernandez (1989)
French Catalonia	5–5.5	*Ulva rigida, in vitro*	75	Fernandez (1989)
French Catalonia	5–5.5	*Posidonia oceanica* old leaves, *in vitro*	116	Fernandez (1989)
French Catalonia	5–5.5	*Stypocaulon scoparium, in vitro*	125	Fernandez (1989)
French Catalonia	5–5.5	*Cystoseira mediterranea, in vitro*	246	Fernandez (1989)
Corsica	7	Turf MPOs, in situ	143–833	Verlaque (1984, 1987a)
Corsica	7	*Posidonia oceanica* and epiphytes, in situ	339–914	Nédélec and Verlaque (1984), Verlaque (1987a)

[a] Wet weight.
[b] Probable wet weight.

Ulva rigida and *Stypocaulon scoparium,* respectively. Such variations in the rate of consumption over time, however, were not observed during a 7-day feeding experiment with an artificial diet (Lawrence et al. 1989).

The consumption rate of *P. lividus* is highly variable. Consumption rates for individuals 5 cm in diameter range from 50 to 851 mg dry weight per individual per day (Table 3).

Some of this variability might be due to experimental processes, especially under aquarium conditions. In particular, *in vitro* conditions reduce the sea urchins' food searching time and resting periods due to water turbulence. The absence of density-dependent effects within a meadow of *P. oceanica* (Kirkman and Young 1981) is surprising and requires confirmation. In contrast, the consumption rate of *P. lividus* is strongly dependent upon available food (Table 3). It is higher when fed 'preferred' species than when fed 'avoided' ones (Traer 1984; Fernandez 1989; Rico 1989), especially when the consumption rate is expressed in terms of organic matter (Frantzis and Grémare 1992). When exclusively fed a strongly 'avoided' species, namely the chlorobionta *C. taxifolia,* in summer and autumn, the consumption rate was very low and decreased to nearly zero over time (Boudouresque et al. 1996), which fits the optimisation model. The consumption rate for a food with low organic matter content is higher than for a food with high energy value (Frantzis and Grémare 1992; Haya and Régis 1995; Fernandez and Boudouresque 2000), a result which is consistent with predictions of the compensatory model. *Paracentrotus lividus* seems to fit both the optimisation and the compensatory models.

In the Mediterranean, *P. lividus* mainly feeds at night. Accordingly, the consumption rate is higher under experimental conditions of continual darkness and decreases with an increased number of daylight hours (Shepherd 1987). At Lough Hyne, Ireland, *P. lividus* is diurnal, apparently in response to nocturnal predator activity (Kitching and Thain 1983).

The repletion index (RI), the ratio between the weight of the gut content and that of the test (Lawrence et al. 1965), is usually higher in winter and spring than in summer and autumn (Semroud and Kada 1987; Rico 1989; Semroud 1993; Fernandez and Boudouresque 1997; but see Nédélec 1982). It does not differ significantly between individuals fed 'preferred' or 'avoided' species (Fernandez 1989; Rico 1989; but see Odile et al. 1988), and between individuals living in barren grounds, where food availability is presumed to be limiting, and erect MPO habitats (Régis 1978; Gago et al. 2003; but see Fernandez and Boudouresque 1997). Cyclical changes in the RI, with a periodicity between 3 and 5 days, may express the alternation of feeding and resting phases (Nédélec et al. 1983; Rico 1989). Such an alternation has also been observed in the moving activity of *P. lividus* (Dance 1987).

3.3. Ecological Consequences of Feeding

Paracentrotus lividus clearly appears to be an opportunistic generalist. In the field, the diet of *P. lividus* is selective, whatever the reasons behind these food choices may be. Apostatic feeding ('switching') makes it possible to move from a preferred but rare resource to a less preferred but abundant one. In this way, *P. lividus* plays a major role in determining the organisation of benthic communities, with subtidal MPO assemblages affected by both its feeding preferences and abundance (Barnes et al. 2002).

The density of *P. lividus* shows a general and conspicuous negative correlation with erect MPO coverage and biomass (Kempf 1962; Kitching and Ebling 1967; Pastor 1971; Verlaque and Nédélec 1983b; Verlaque 1987a). Densities ranging from 7 to 20 individuals per square metre (5 cm diameter) may completely remove dense populations of erect MPOs from the substratum in Corsica and subsequently maintain barren grounds on which only encrusting corallines are seen to grow (Verlaque 1984, 1987a). During the

1970s explosion of *P. lividus* in the Northern Adriatic, front lines of feeding sea urchins were observed moving ahead and leaving behind them only stumps (Mastaller 1974; Torunski 1979). Very high densities of sea urchins (>50 individuals per square metre) can even remove the encrusting corallines. They rasp away rock particles while grazing endolithic microscopic photosynthetic organisms and thus become agents of biological erosion. The sea urchin's calcitic teeth are harder than limestone (hardness of 3–4 on the Mohs scale) and the distinct scratches they leave on the substrate surface are easily identifiable. The constant and rapid growth of these teeth (1.0–1.5 mm per week) requires that *P. lividus* graze the substrate to erode them. In the Gulf of Piran (northern Adriatic), limestone bottom erosion has been estimated at 0.3–0.7 mm a^{-1} for sea urchin densities ranging between 50 and 120 individuals per square metre (Torunski 1979; Schneider and Torunski 1983). A bioerosion rate of intertidal limestone reaching 15 mm a^{-1} has been reported from Ireland (Trudgill et al. 1987). Overgrazing may also affect meadows of the seagrass *P. oceanica* (Verlaque and Nédélec 1983a; Shepherd 1987). However, due to the preference of *P. lividus* for leaf epiphytes rather than for *P. oceanica* leaves, the meadow can withstand sea urchin densities of 5–15 individuals per square metre over long periods (Tomas et al. 2005).

Where *P. lividus* shelters in crevices, it causes the formation of haloes ('gardens') with low MPO cover close to the crevices. In the Mediterranean, these crevices are surrounded by a nongrazed 'forest' of the large stramenopile *Cystoseira* spp. (Verlaque 1984; Sala 1996). Where predation pressure is low, 'gardens' occupied by large sea urchins can be observed in places deprived of shelters. The size of the gardens depends upon both the number of sea urchins they harbour and the presence or absence of other sedentary herbivores (Verlaque 1984). Removing the sea urchins induces recolonisation of the gardens by *Cystoseira* and other MPOs. One year after the removal of the sea urchins, the assemblage is no longer distinguishable from that of ungrazed areas (Verlaque 1987a, 1987b; Benedetti-Cecchi and Cinelli 1995). Removal of *P. lividus* from a shallow Mediterranean barren ground characterised by encrusting corallines (*L. incrustans*) resulted in the development of a filamentous MPO cover, mainly *Feldmannia globifera*, *Falkenbergia rufolanosa* and *Polysiphonia subulata* in only 1 month. Subsequently, other filamentous species, such as *Ceramium ciliatum*, developed together with the foliose and coarsely branched *Dictyota dichotoma*, *Colpomenia sinuosa* and *Gelidum spinosum* (Kempf 1962). A similar area cleared of *P. lividus*, in Ireland, changed from an initial (July) MPO coverage of less than 1% to a 50% cover (mainly enteromorpha-like *Ulva*) in September and 100% in July of the next year (Kitching and Ebling 1961). Placing hundreds of *P. lividus* in an *Ulva* stand resulted in a patch free of weeds at the end of 1 year (Kitching and Ebling 1961; see also Palacín et al. 1998).

In the Mediterranean, the recent dramatic regression, or in some regions the disappearance, of the deep subtidal forests of *Cystoseira* described in the first half of the twentieth century has been attributed to increased densities of *P. lividus* (Giaccone 1971; Katzmann 1974; Gros 1978; Verlaque and Nédélec 1983b). Most of these *Cystoseira* species are endemic to the Mediterranean and threats to this natural heritage are a cause of growing concern for conservation biologists.

Several causes have been proposed to explain pullulations of *P. lividus* and subsequent barren grounds: organic pollution (Régis 1978; Harmelin et al. 1981), overfishing of their predatory crabs (Pastor 1971) and overfishing of their predatory fishes (Verlaque 1987a; Boudouresque et al. 1992; Sala et al. 1998a; but see Guidetti et al. 2005). Outside the Natural Reserve of Scandola, Elbu Bay, Corsica, density of *P. lividus* is 34 times higher than within (Table 1) where fish biomass is high (Boudouresque et al. 1992). A high availability of shelters for juveniles may allow them to escape predators and thus to reach the 3–5 cm test diameter which makes it possible to 'escape in size' (Sala et al. 1998a). In southern Italy, the initial denudation of the substrate can occur as a consequence of human activity. Scuba divers break rocks to illegally collect date mussels, *Lithophaga lithophaga*. The exploited rocks remain completely bare and are quickly invaded by *P. lividus* and *Arbacia lixula*, probably hindering the subsequent pattern of recolonisation by erect MPOs. Sea urchin biomass is two- to fourfold greater at the impacted than at control sites (Fanelli et al. 1994; Guidetti et al. 2003). Two other possibilities are to be considered (Sala et al. 1998a), though the data which support them concern other species of sea urchins: first, a reduction in MPO biomass due to other causes such that sea urchin grazing eliminates that which remains and, second, an increase in sea urchin density as a result of successful recruitment by migration (Lawrence 1975).

An established barren ground may persist for an extended period of time. In the Gulf of Piran (northern Adriatic), the explosion of the population of *P. lividus* that occurred in the early 1970s and the formation of barren grounds that ensued lasted, without significant changes, for at least a decade (Vukovič 1982; Schneider and Torunski 1983). The capacity of *P. lividus* to survive on extremely low levels of benthic food, especially when drift MPOs or dissolved and colloidal organic matter are available, obviously facilitates the persistence of these barren grounds. In fact, MPO forests and barren grounds probably constitute two alternative 'stable' states ('switch climaxes'), in the framework of the multiple 'stable' states concept (see Knowlton 2004; Wright et al. 2005). This means that, once established, a barren ground could persist even with relatively low densities of *P. lividus,* due to the fact that corallines are able to reduce settlement of potential competitors and/or to attract grazer recruits. Disturbances would be required to induce the shift from barrens to forest, or from forest to barrens. Bulleri et al. (2002) provide convincing arguments for the occurrence of such multiple 'stable' states in the Mediterranean.

However, stands of large erect MPOs may coexist with high densities of *P. lividus*, despite the fact that such densities are usually associated with barren grounds. The sheltering behaviour of *P. lividus* and its reluctance to move too far away from shelters appear to be key factors allowing such a coexistence (Benedetti-Cecchi and Cinelli 1995). In addition, the presence of predatory fish can reduce the grazing range of *P. lividus* to about 1 m around refuges (Sala 1996).

Most studies have focused on high densities of *P. lividus*, the feeding activity of which can produce barren grounds that are usually very localised, both spatially and temporally. In contrast, very few studies have addressed the impact of sea urchin browsing at low or very low densities, a condition which characterises much larger areas (Lawrence 1975). A removal–reintroduction experiment showed that even low densities of *P. lividus* exert a noticeable effect on habitat structure of Mediterranean rocky substrates, namely the coverage of non-crustose (turfing and frondose) MPOs (Palacín et al. 1998). In the same way, the browsing by sea urchins, even at low densities, can be a factor regulating the MPO biomass on maerl beds of Brittany, France (Guillou et al. 2002).

3.4. Competition with Other Herbivores

Competition between *P. lividus* and other herbivores may affect its abundance and behaviour, and hence its effect on benthic communities. However, very little attention has been directed towards this problem.

In the Mediterranean, *P. lividus* sometimes shares its rocky habitats with another sea urchin, *A. lixula*. The feeding niches of the two species, though overlapping, are nevertheless distinct; the former is mainly a browser (*sensu* Ogden 1976) of coarsely branched erect MPOs and can utilise drift material (MPOs, including seagrass leaves), whereas the latter is a grazer (*sensu* Ogden 1976) of encrusting coralline, unable to trap drift material. However, both browse filamentous turf MPOs. In conditions of limited resources, when erect species are no longer present, such as in autumn or in barren grounds, the feeding niche of *P. lividus* shifts and then largely overlaps that of *A. lixula* (Kempf 1962; Régis 1978; Harmelin et al. 1981; Verlaque and Nédélec 1983a; Maggiore et al. 1987; Frantzis et al. 1988; but see Delmas 1992). The interspecific competition between *P. lividus* and *A. lixula*, and their coevolution, may have been involved in the separation of their feeding niches as suggested by Lawrence (1975). In addition, when present along the same exposed coast, the two species adopt a different vertical zonation, with *A. lixula* dominant at the upper levels and *P. lividus* at lower ones (Ruitton et al. 2000). When experimentally displaced from these lower levels upwards, *P. lividus* has a tendency to move downwards (Chelazzi et al. 1997), which suggests that *A. lixula* is the dominant competitor at the upper levels. When *P. lividus* is removed (e.g. by fishing) the biomass of *A. lixula* increases (Guidetti et al. 2004). In any case, *A. lixula*, which cannot feed on erect MPOs and therefore can only colonize gardens and barren grounds, probably benefits from prior removal of erect species by *P. lividus* (Frantzis et al. 1988).

The limpets *Patella caerulea* and *P. aspera* are common co-dwellers with *P. lividus* in barren grounds and gardens. In the Gulf of Galeria, Corsica, the mean density of *P. caerulea* in these gardens is 8 individuals per square metre (Verlaque 1987b). It grazes on encrusting corallines, endolithic microscopic photosynthetic organisms (e.g. *Phaeophila dendroides* and cyanobacteria), and turf MPOs (e.g. *Lophosiphonia cristata*). In these habitats, the diets of *P. lividus* and *P. caerulea* clearly overlap, such that *P. lividus* must enlarge the surface of the gardens where *P. caerulea* is present. For a given density of *P. lividus*, garden size and limpet density are significantly correlated (Verlaque 1987b).

Other sea urchins are possible competitors of *P. lividus*: *Echinus esculentus* and *Psammechinus miliaris* (in the Atlantic), *Psammechinus microtuberculatus* (in the Mediterranean) and *Sphaerechinus granularis*. Their densities are usually low and/or their habitats and feeding niches only slightly overlap that of *P. lividus,* at least under conditions of nonlimited resource. In meadows of *P. oceanica*, *P. microtuberculatus* and *S. granularis* mainly consume dead seagrass leaves and their epiphytes,probably from the litter. The latter sea urchin also grazes the rhizomes and roots of *P. oceanica* when these are accessible (Verlaque 1981; Campos-Villaça 1984; Paul et al. 1984; Traer 1984). On hard substrates, *S. granularis* and *P. lividus* rarely coexist, the former living in deeper waters.

Moreover, as is true for *A. lixula*, *S. granularis* seems to be mainly an active grazer of encrusting corallines (Sartoretto and Francour 1997).

The most abundant herbivorous fish existing in the geographical range of *P. lividus* is *Sarpa salpa*. Its diet presents some similarities with that of *P. lividus*. Adults mainly browse erect, coarsely branched photosynthetic stramenopiles and the leaves of *P. oceanica* and *Cymodocea nodosa*, the latter being preferred. However, differences in food preferences are apparent. The chlorobionta *Caulerpa prolifera,* which is strongly avoided by *P. lividus*, is browsed by *S. salpa.* In addition, though being a generalist feeder like *P. lividus,* its higher mobility allows it to pass over barren grounds. In the field, its feeding behaviour therefore does not shift from browsing to grazing under conditions of limited resources (Laborel-Deguen and Laborel 1977; Gros 1978; Verlaque 1987b, 1990; Chevaldonné 1990; Sala 1996). Accordingly, competition between *P. lividus* and *S. salpa* is chiefly to be expected in well-developed MPO assemblages at rather low densities of the former species. Such a competition may prove limited. A number of other partially or totally herbivorous fishes, possible competitors of *P. lividus*, have been recorded (Verlaque 1990), but little is known about these with the exception of *Diplodus puntazzo*. This fish is markedly omnivorous, with MPOs representing its most important food source. Two species found to be dominant in their stomachs are *Plocamium cartilagineum* and *Flabellia petiolata* (Sala 1996). The first is 'preferred' and the second 'avoided' by *P. lividus* (Table 2).

4. MOVEMENT AND MIGRATION

Paracentrotus lividus often exhibits daily small-scale migratory movements between shelters (e.g. rock lower surfaces) which provide refuges from predators and areas (e.g. upper surfaces of rocks) which constitute a source of food (Barnes and Crook 2001b). Mediterranean and Atlantic populations usually exhibit a nocturnal activity (Kempf 1962; Gamble 1965; Pastor 1971; Powis de Tenbossche 1978; Shepherd and Boudouresque 1979; Dance 1987; Rico 1989; Sala 1996; Hereu 2005) as is common among echinoids. Locally, however, it can be reversed. For example, at Lough Hyne, Ireland, *P. lividus* has a diurnal activity, inversely correlated with the abundance of its chief predators, crabs and the asteroid *M. glacialis*, which are nocturnal (Muntz et al. 1965; Ebling et al. 1966; Kitching and Thain 1983; Crook et al. 2000; Barnes and Crook 2001a). Differences can be observed even in the same locality. At Medes Islands, Spanish Catalonia, most large *P. lividus* within the marine reserve shelter during the day where high densities of diurnal predator fishes occur, but occupy exposed sites outside the reserve (Sala and Zabala 1996). It is possible that *P. lividus* 'learns', through perception of the body fluids of broken sea urchins, to associate the level of light with the pressures of predation and adopts a pattern of activity opposite to that of its predators. Data from other species of echinoderms support this hypothesis (Lawrence 1975). At Lough Hyne, daily migration between sides, interstices and the upper surface of rocks is size/age dependent. Only three year-classes (2+, 3+, 4+) migrate, whereas younger and older individuals do not (Crook et al. 2000).

During the daily peak of activity, the distance traveled by an individual can reach $40\,cm \cdot h^{-1}$ (Dance 1987; Hereu 2005). On a horizontal surface, speed increases with the test diameter whereas it is size-independent on a vertical surface (Domenici et al. 2003). The direction of movement is usually random, although movement toward deeper water has been observed following a period of water turbulence (Dance 1987; San Martín 1995; Hereu 2005).

The net 24 h movement, i.e. the straight-line distance between the initial and final position of an individual, ranges between 0 and 260 cm (mean: 50 cm). The effective 24 h movement, i.e. the length of the line joining each position sequentially occupied, can be greater and reach up to 480 cm (Kempf 1962; Gamble 1965; Shepherd and Boudouresque 1979; Cuenca 1987; Dance 1987). This distance is not influenced by body size, depth or season. It is affected by the substrate. Individuals in the meadows of *P. oceanica* travel significantly less than those on rocky substrates. Aquarium experiments show that the percentage of time devoted to movement depends upon the nature of the available food, although a distinct relation with food choice is lacking (Rico 1989). Effective movement is affected by water movement as activity decreases under turbulent conditions. Individuals living in very exposed conditions, in burrows and/or in tide pools may not move at all. Also, individuals trapped in their burrows are not uncommon (Bouxin 1964; Gamble 1965; Powis de Tenbossche 1978; Benedetti-Cecchi and Cinelli 1995). Movement is also affected by the abundance of predatory fish. Movement is lower in a no-take marine reserve than in an adjacent area subject to high fishing: 36 and 51 cm per day on average, respectively (Hereu 2005). The distance an individual travels varies greatly from one day to the next. Complete inactivity for 1 or 2 days can occur. The individuals in a population do not synchronise their periods of activity or inactivity, however. Cycles of activity from 2 to 3 (up to 5) days also occur in aquaria (Neill and Larkum 1966; Dance 1987; Rico 1989).

Paracentrotus lividus does not show a clear homing behavior. Nevertheless, large individuals seem to stay in a relatively small area (a few metres) for periods of at least several months (Dance 1987; Fenaux et al. 1987; Hereu 2005). Individuals living in burrows under exposed conditions may return to their holes after 1–2 day foraging excursions (Neill and Larkum 1966).

In contrast with the short-term relative sedentariness of *P. lividus,* long-term migrations apparently occur. In Port-Cros bay, most juveniles occur in a shallow meadow of *P. oceanica*. In the deeper meadows and on rocky substrates, the demographic structure of *P. lividus* and its seasonal changes cannot be explained by growth and survivorship curves alone. This makes it necessary to hypothesise migration in the order of 100 m (Azzolina 1987, 1988). Also, in a *P. oceanica* meadow in Spanish Catalonia, only immigration from adjacent rock walls can account for the presence of *P. lividus* (Tomas et al. 2004). In a coastal lagoon of Corsica, similar observations suggest migration from a shallow pebble habitat (recruitment area) to a seagrass *C. nodosa* meadow (adult habitat), which involves the passing of the sea urchins over the silt and sand bottoms that separate these two habitats (Fernandez et al. 2001). In French Catalonia, growing individuals migrate from shallow recruitment areas to deep (10–20 m) habitats (Lecchini et al. 2002).

Removal of *P. lividus* from populations of high densities, which could lead to overgrazing facies and hence food limitation, shows that recolonisation by migration can occur within 2 months (Kempf 1962). In contrast, removal of individuals from populations of low densities shows very little migration (Palacín et al. 1997). These results, consistent with those of Kitching and Ebling (1961), suggest that migrations are density and food dependent, the abundance of food obviating the need for long foraging trips (Palacín et al. 1997).

5. MORTALITY

5.1. Predators

Predators of *P. lividus* are listed in Table 4. In the Mediterranean Sea, fishes (*Diplodus sargus*, *D. vulgaris*, *Labrus merula* and *Coris julis*), the spider crab *Maja crispata* and the gastropod *Truncularìopsis trunculus* are its major predators. *Diplodus sargus* can successfully attack individuals up to 5 cm, *D. vulgaris* and *M. crispata* up to 3 cm and *Coris julis* less than 1 cm in test diameter (Tertschnig 1989; Sala 1996, 1997; Hereu

Table 4 *Paracentrotus lividus*: main predators.

Species	Location, Country	References
Crustaceans		
Cancer pagurus	Lough Hyne, Cork, Ireland	Muntz et al. (1965), Kitching and Ebling (1967), Barnes and Crook (2001a)
	Ría de Vigo, Spain	Niell and Pastor (1973)
Carcinusmaenas	Lough Hyne, Cork, Ireland	Muntz et al. (1965), Barnes and Crook (2001a)
Eriphia spinifrons	Northern Adriatic	Stirn et al. (1974)
Homarus gammaris (= *H. vulgaris*)	LoughHyne, Cork, Ireland	Muntz et al. (1965)
Macropipus corrugatus	Lough Hyne, Cork, Ireland	Muntz et al. (1965)
Maja crispata	Naples, Italy	Tertschnig (1989)
Maja brachydactyla[a]	Lough Hyne, Cork, Ireland	Muntz et al. (1965)
	Ría de Arousa, Galicia, Spain	Bernárdez et al. (2000)
	Northern Adriatic	Stirn et al. (1974)
Necora puber (= *Portunus puber*)	LoughHyne, Cork, Ireland	Muntz et al. (1965), Kitching and Ebling (1967), Crook et al. (2000), Barnes and Crook (2001a)
Palinurus elephas (= *P. vulgaris*)	Corsica	Campillo and Amadei (1978)
	Roscoff, Brittany, France	Vasserot (1965)
	Adriatic Sea	Zavodnik (1987)
	Ría de Vigo, Spain	Niell and Pastor (1973)
Xantho incisus	Lough Hyne, Cork, Ireland	Muntz et al. (1965)

Table 4 Continued

Species	Location, Country	References
Molluscs		
Truncilariopsis trunculus	Catalonia, Spain	Sala and Zabala (1996)
Octopus vulgaris	Adriatic Sea	Zavodnik (1987)
Starfishes		
Asterias rubens	Atlantic	Cherbonnier (1954)
	Lough Hyne, Cork, Ireland	Barnes and Crook (2001a)
	Vigo, Galicia, Spain	Pastor (1971)
Astropecten auranciacus	Provence, France	Massé (1975)
Luidia ciliaris	Banyuls, France	Vasserot (1964)
	Ligurian Sea	Tortonese (1952)
Marthasterias glacialis	Lough Hyne, Cork, Ireland	Ebling et al. (1966), Kitching and Thain (1983), Crook et al. (2000), Barnes and Crook (2001a)
	Ría de Vigo, Spain	Niell and Pastor (1973)
	Banyuls, France	Prouho (1890), Vasserot (1964)
	Port-Cros, Var, France	Dance and Savy (1987)
	Adriatic Sea	Zavodnik (1987)
Fishes		
Balistes carolinensis	Italy	Bini (1968)
Blennius gattorugina	Italy	Bini (1968)
B. pholis	France	Carvalho (1982)
B. sanguinolentus	Italy	Bini (1968)
B. trigloides	France	Carvalho (1982)
Coris julis	Catalonia, Spain	Sala and Zabala (1996), Sala (1996), Hereu et al. (2005)
	Italy	Bini (1968)
	Western Mediterranean	Sala (1997), Quignard (1966)
Diplodus sargus	Catalonia, Spain	Sala and Zabala (1996), Sala (1996), Hereu et al. (2005)
	Gulf of Lion, France	Rosecchi (1985)
	Port-Cros, Var, France	Khoury (1987)
	Naples, Italy	Ara (1937)
	Western Mediterranean	Sala (1997)
	Northern Adriatic	Stirn et al. (1974), Guidetti et al. (2005)
Diplodus vulgaris	Catalonia, Spain	Sala and Zabala (1996), Sala (1996)
	Gulf of Lion, France	Rosecchi (1985)
	Port-Cros, Var, France	Khoury (1987)
	Naples, Italy	Ara (1937)
	Western Mediterranean	Sala (1997)

(*Continued*)

Table 4 Continued

Species	Location, Country	References
	Adriatic Sea	Zavodnik (1987), Guidetti et al. (2005)
Labrus bergylta	Lough Hyne, Cork, Ireland	Crook et al. (2000)
Labrus merula	Catalonia, Spain	Hereu et al. (2005)
	Port-Cros, Var, France	Khoury (1987)
	Mediterranean	Timon-David (1936)
	Western Mediterranean	Sala (1997), Quignard (1966)
Labrus viridis	Port-Cros, Var, France	Khoury (1987)
	Western Mediterranean	Quignard (1966)
Lithognathus mormyrus		Whitehead et al. (1986)
Sparisoma cretense	Malta	Neill and Larkum (1966)
Sparus aurata	Near Marseilles, France	San Martín (1995)
	Western Mediterranean	Sala (1997)
	Northern Adriatic	Guidetti et al. (2005)
	Mediterranean	Savy (1987)
Symphodus mediterraneus	Port-Cros, Var, France	Khoury (1987)
	Western Mediterranean	Quignard (1966)
S. roissali	Port-Cros, Var, France	Khoury (1987)
	Western Mediterranean	Quignard (1966)
S. tinca	Port-Cros, Var, France	Khoury (1987)
	Western Mediterranean	Quignard (1966)
Thalassoma pavo	Catalonia, Spain	Hereu et al. (2005)
	Italy	Bini (1968)
	Western Mediterranean	Sala (1997), Quignard (1966)
Xyrichthys novacula	Italy	Bini (1968)
	Western Mediterranean	Quignard (1966)
Birds		
Larus spp.	Lough Hyne, Cork, Ireland	Ebling et al. (1966)

[a] Misnamed as *Maja squinado*, a species restricted to the Mediterranean Sea.
Updated from Savy (1987) and Sala (1997).

et al. 2005). In an unprotected coastal area where *D. sargus* and *D. vulgaris* are removed through fishing, predation by fishes and by *T. trunculus* represented 57% and 43% of the total, respectively. In contrast, in a marine protected area characterised by high fish densities, fishes accounted for 100% of the predation (Sala and Zabala 1996). The starfish *M. glacialis* feeds on all size-classes of *P. lividus* (up to 63 mm). In cages, it consumes a sea urchin every 4 or 5 days. As this starfish is rather rare, it should not play a conspicuous role (Dance and Savy 1987). The situation seems to be quite different in the Atlantic Ocean, where starfishes and crustaceans play a major role. The crabs *Cancer pagurus*, *Necora puber*, *Maja brachydactyla* and *Carcinus maenas* consume *P. lividus* up to 6.5, 5, 4.5 and 2 cm in test diameter, respectively. Large *Cancer pagurus* can eat 1–2 large sea urchins per day. Large lobsters *Homarus gammarus* (= *H. vulgaris*) eat individuals measuring up to 6.1 cm in diameter. The starfish *M. glacialis* consumes small (<2 cm) individuals (Muntz et al. 1965; Ebling et al. 1966; Kitching and Ebling 1967; Neill and Pastor 1973; Kitching and Thain 1983; Bernárdez et al. 2000). These predators play a major role, not only in controlling abundance of *P. lividus*, but also in controlling its behaviour, and therefore have a pronounced effect on benthic assemblages dominated by photosynthetic organisms (Sala 1996).

In addition to the predator guild that successfully attacks and breaks open sea urchin tests, the scavenger guild feeds on sea urchin carcasses opened by predators. In the Mediterranean Sea, with the exception of *D. vulgaris,* the ranking of scavenger fishes is the same as that of predators: *D. sargus*, *Thalassoma pavo*, *Coris julis* and *Labrus merula* (Sala 1997). *Serranus cabrilla*, *Diplodus cervinus*, *Oblada melanura*, *Spondyliosoma cantharus*, *Chromis chromis*, *Symphodus melanocercus*, *S. ocellatus* and *Parablennius rouxi* are only scavengers (Sala 1997).

5.2. Diseases and Parasites

A communicable nonspecific disease of *P. lividus* and of several other sympatric species of regular echinoids, was reported in the late 1970s and early 1980s, mainly in the Mediterranean (Spain, France, Italy, Croatia) but also in the Atlantic (Brittany). This disease, known as the 'bald sea urchin disease', produces lesions on the outer body surface. The pathogens are bacteria of the genera *Aeromonas* and *Vibrio* (Höbaus et al. 1981; Jangoux and Maes 1987; Jangoux 1990). As a result of this disease, mass mortalities of *P. lividus* were reported in the northwestern Mediterranean, especially in summer and in shallow waters (Boudouresque et al. 1980, 1981; Azzolina 1987).

Metacercariae of *Macvicaria crassigula,* a parasitic trematode, occur in the water vascular system of *P. lividus*. Adults live in the digestive tract of the sparid fish *D. vulgaris*, a feature which is consistent with its predatory behavior upon *P. lividus* (Bartoli et al. 1989; Jousson et al. 1999). In Corsica, some *P. lividus* harbour thousands of metacercariae. It can be hypothesised that they affect the ability of the sea urchin to move, to grip the substratum and therefore to escape predatory fishes (Pierre Bartoli, pers. comm.). Metacercariae of another species of trematode, *Zoogonus mirus*, occur in various muscles of the Aristotle's lantern, a feature that can affect feeding efficiency. The definitive host is the fish *Labrus merula* (Timon-David 1934).

In Ireland, vermiform parasites have been observed in the gonads, with infestation rates of 4% in sea urchins inhabiting intertidal pools and 38% for those in subtidal pools. No discernable effect on the host's reproduction has been detected (Byrne 1990).

5.3. Other Causes of Mortality

A mass-mortality of *P. lividus,* which occurred in Ireland during the autumn of 1979, was related to a bloom of the dinoflagellate *Gyrodinium aureolum* (Cross and Southgate 1980).

In a shallow exposed habitat of Catalonia, Spain, wave action during storms causes high mortalities, which help explain the changing size structure and month-to-month fluctuations in density at this site (Turon et al. 1995). Mortality also can occur by desiccation during periods of exceptionally low sea levels (unpubl. data).

In rock pools, population structures suggest that more sea urchins die of senescence than are taken by predators, with low levels of mortality during their first 4 years (except perhaps during the first year), and high mortalities between 6 and 9 years of age (Crapp and Willis 1975). Maximum longevity would exceed 13 years (Régis 1978; Delmas 1992).

6. GROWTH

Somatic growth of *P. lividus* in the field appears to be related mainly to water temperature, food quality, and gonadal development (Fernandez 1996). Seasonal variations in the growth rate seem to be related with temperature. According to Le Gall et al. (1990), growth does not occur in English Channel populations between 4 and 7 °C, but increases proportionally between 7 and 18 °C to a maximum from 18 to 22 °C. Above 22 °C, growth declines and stops at 28 °C. In the Mediterranean, maximum growth occurs between 12 and 18 °C in spring, sometimes in autumn, and minimally in winter (Azzolina 1988; Fernandez and Caltagirone 1994; Turon et al. 1995; but see Shpigel et al. 2004). Occurrence of only small individuals in coastal lagoons (Koehler 1883; Kempf 1962) seems due to earlier mortality rather than to a lower growth rate (Fernandez and Caltagirone 1994).

In an exposed habitat with scarce and low quality food, maximal growth and size of *P. lividus* was less than in a deeper habitat with no food limitation (Turon et al. 1995; see also Gago et al. 2003). Tests are thicker in a littoral lagoon of Corsica, a biotope with an abundant and preferred food, which may help prevent fish predation. Conversely, the Aristotle's lantern is relatively larger in individuals living on pebble bottoms with limited food. This probably enables the sea urchin to feed more efficiently (Fernandez 1996; Fernandez and Boudouresque 1997).

Two cm individuals are generally considered to average 2 years old and 4 cm individuals 4–5 years (Table 5). The published growth curves rapidly become flat, and none can account for the largest individuals that are 7 cm in diameter. Either the field data are incomplete or wrong. More probably, these large individuals are several decades old, as suggested by Russell et al. (1998) for *Strongylocentrotus droebachiensis*.

Among an initially homogeneous cohort of juvenile *P. lividus* in aquaria, growth results in a multimodal size distribution, with individuals growing either very fast or very slowly. This variability in growth rate is the consequence of intraspecific competition. Such growth inhibition also in the natural environment could be very efficient in stabilising field aggregative populations by maintaining a protected pool of small individuals with high growth potential but inhibited by the density of larger ones. A decrease in the density of large individuals would remove the inhibition of growth of small sea urchin growth (Grosjean et al. 1996).

Table 5 *Paracentrotus lividus*: relation between test diameter (cm) and age (years).

Localities	Methods	Test diameter and corresponding age							References
		1 cm	2 cm	3 cm	4 cm	5 cm	6 cm	7 cm	
Cork, Ireland	R	–	2–3	3–4	3–5	4–7	–	–	Kitching and Thain (1983)
Bantry Bay, Ireland	R, S	–	1–2	2–3	3–5	6–7	–	–	Crapp and Willis (1975)
Brittany, France	A	1–1.5	1–2	1.5–>2.5	–	–	–	–	Grosjean et al. (1996)
Cascais, Portugal	R	–	–	–	3–7	5–8	–	–	Gago et al. (2003)
Tossa de Mar, Spain	R	1	3	4	5–6	7	–	>12	Turon et al. (1995)
Cubelles, Spain	R	1–2	3–4	5	7	11–12	–	–	Turon et al. (1995)
Marseilles, France	C	2	4	6–7	9	–	–	–	Régis (1978)
	B	–	–	6	9	11	–	–	Delmas (1992)
	A	1–2	2–3	–	–	–	–	–	Régis (1978)
Port-Cros, France	R	–	1–2	2–3	4	7	–	–	Azzolina (1988)
	C	–	2–3	3	5	–	–	–	Azzolina (1988)[a]
	C	–	1–2	2–3	4–5	–	–	–	Azzolina (1988)[b]
Villefranche, France	A	<1	1–2	–	–	–	–	–	Fenaux et al. (1987)
	A	0.5–1	2	–	–	–	–	–	Cellario and Fenaux (1990)
	S	1	1–2	2	3	4	–	–	Fenaux et al. (1987)
Urbinu, Corsica	A	<1	1–2	2–3	3–5	6–10	–	–	Fernandez (1996)[b]
Tunis, Tunisia	R	–	–	1–2	2–3	3–8	–	–	Sellem et al (2000)
Unknown	A	1	1–2	2–3	3	4–6	5–7	–	Grosjean et al (2003)

Method: A – aquarium experiments, B – Bhattacharya method, C – measure of growth within in situ cages, R – from rings in plates, S – size–frequency analysis.

Ages are, for the most part, interpolated from the authors' curves.

[a] Calculation via the Gompertz model.

[b] Calculation via the von Bertalanffy model.

7. REPRODUCTION

7.1. Reproductive Cycles

Sexes are separate in *P. lividus* although hermaphroditism has been observed (Drzewina and Bohn 1924; Neefs 1937; Byrne 1990). The sex ratio seems to change throughout the year and from one year to the next (Neefs 1938; Allain 1975; Crapp and Willis 1975; Guettaf 1997). Maturation *in vitro* occurs at 13–20 mm diameter and 5 months (L Fenaux in Azzolina 1987; Cellario and Fenaux 1990). Sexual maturity may occur later in situ. Unfavourable conditions and limiting food supply can decrease the size at which reproduction begins (Lozano et al. 1995). Overall, the gonad index is higher in 40–70 mm than in 20–40 mm size-classes (Martínez et al. 2003; Sánchez-España et al. 2004).

The annual cycle of the gonad index of *P. lividus* has one or two seasonal peaks (Table 6) and can differ conspicuously between neighbouring localities (Lozano et al. 1995; Guettaf 1997; Sánchez-España et al. 2004). Histological studies of the gonads indicate only a single annual gametogenic cycle (Byrne 1990; Lozano et al. 1995; Guettaf 1997; Spirlet et al. 1998; Martínez et al. 2003) as is prevalent in temperate echinoids, though individuals with mature gonads can be present year-round (Sánchez-España et al. 2004).

Overall, it appears from *in vitro* experiments that both somatic and gonadal production occur in echinoids when food availability is high (Lawrence et al. 1992; Gago et al. 2003), when the amount of organic matter ingested is high (Frantzis and Grémare 1992) and that gonad index is higher when small individuals (not significant for large ones) are fed preferred species (Fernandez 1989). Temperatures between 18 and 22 °C and short days enhance gonadal development (Shpigel et al. 2004). Surprisingly, some contradictory results have been obtained from field data. Larger gonads have been observed in subtidal well-fed populations, both in open-sea and lagoonal habitats (Byrne 1990; Fernandez 1990, 1996; San Martín 1995; Fernandez and Boudouresque 1997). Gonad indices have been found to be higher when the population density is low, despite similar feeding patterns (Guettaf and San Martín 1995). But only weak, or nonsignificant relationships, have been observed between gonad index and repletion index (Régis 1978; Semroud and Kada 1987). In the same way, the gonad index in populations from *P. oceanica* meadows, dense MPO stands on rocky substrates or barren grounds do not differ. It is possible that populations in barren grounds complete their diet with drift MPOs, including dead leaves of *P. oceanica* (Guettaf 1997). In Spanish Catalonia, gonad index values were found to be higher in a shallow habitat subject to high hydrodynamics, with boulders occupied by poorly developed MPOs and high density of *P. lividus*, than in a deeper 'stable' habitat dominated by preferred species and low sea urchin density (Lozano et al. 1995). These authors suggested this implies a higher investment in reproduction under less favourable nutritional conditions, but these results might be due to a high supply in drift MPOs of good nutritional quality.

7.2. Spawning

Male and female individuals aggregate for spawning and simultaneously release their gametes (Cherbonnier 1954). In the Port-Cros Bay, France, spawning aggregations of

Table 6 *Paracentrotus lividus*: period(s) of the year when the gonad index is highest.

Location	Higher gonad index	References
Bantry Bay, Ireland	January and August	Crapp and Willis (1975)
Ballynahown, West Ireland	May–June	Byrne (1990)
Glinsk, West Ireland	March–July	Byrne (1990)
Northern Brittany, France	December–March	Allain (1975)
Western Brittany, France	March	Spirlet et al. (1998)
Rabat-Casablanca, Morocco	February–April	Bayed et al. (2005)
La Herradura, Andalusia, Spain	June–July	Sánchez-España et al. 2004
Palmeral, Andalusia, Spain	November–December	Sánchez-España et al. 2004
Cubelles, Catalonia, Spain	June–October[a]	Lozano et al. (1995)
Tossa de Mar, Catalonia, Spain	February–March and August	Lozano et al. (1995)
Carry-le Rouet, east of Marseilles, France	March–May	Guettaf and San Martín (1995)
Marseilles, France	February–March[b]	Régis (1978)
	April–May and August[c]	Régis (1978)
Maïre, near Marseilles, France	June–July[c]	Régis (1978)
Villefranche, French Riviera	April–May and August	Fenaux (1968)
Urbinu lagoon, Corsica	April[d]	Fernandez (1990)
	April–July[e]	Fernandez (1990)
	February–March[c,e]	Fernandez (1996), Fernandez and Boudouresque (1997)
Aïn Chorb, near Algiers, Algeria	January–March[c]	Semroud and Kada (1987)
Alger-plages, Algiers, Algeria	February–April and June–July	Guettaf (1997)
El Marsa, near Algiers, Algeria	February–March and August	Guettaf (1997)
Aïn-Tagouraït, Tipaza, Algeria	November–January	Guettaf (1997)

[a] Trend unclear.
[b] Test diameter: 30–40 mm.
[c] Test diameter 40–50 mm.
[d] Pebble habitat.
[e] Meadow of the seagrass (magnoliophyta) *Cymodocea nodosa*.

10–20 individuals occur at dusk on prominent stones or at the top of leaves of *P. oceanica*. These aggregations last a few hours (unpubl. obs.). The spawning episodes never involve all individuals of a population (Allain 1975). Suspensions of homogenised ripe testes and ovaries can trigger spawning by ripe females and males, respectively (Kečkeš et al. 1966). The popular belief that spawning episodes occur at full moon has not been confirmed by field observations (Fox 1923; Allain 1975; but see Fol 1879).

Spawning of *P. lividus* has been reported to occur either once or twice in a year (Table 7). An interannual difference in the onset of spawning has been observed (in Ireland), with the start of gamete release differing by as much as 4 weeks between years. Seawater temperature (13–16 °C) may serve as a proximate cue for its induction. When

Table 7 *Paracentrotus lividus*: period(s) of spawning. According to Lozano et al. (1995), data deduced from a drop in gonad index should be considered with caution.

Location	Dates of spawning	References
Bantry Bay, Ireland	January–March and August–September	Crapp and Willis (1975)
Ballynahown, West Ireland	May–July	Byrne (1990)
Glinsk, West Ireland	June–July	Byrne (1990)
Northern Brittany, France	March–September (max.: July)	Allain (1975)
Roscoff, Northern Brittany, France	June–August and September	Cherbonnier (1951)
Western Brittany, France	Late spring–early summer	Spirlet et al. (1998)
Rabat-Casablanca, Morocco	March–June	Bayed et al. (2005)
Barcelona, Catalonia, Spain	Spring–early summer	Lozano et al. (1995)
Marseilles, Provence, France	June and September–November	Régis (1979)
Villefranche, French Riviera	June and late summer	Fenaux (1968)
French Riviera	April–May and September–October	Pedrotti (1993)
Urbinu lagoon, Corsica	March–June and August–October	Fernandez (1996)
	March–June and September–November	Leoni et al. (2003)
Calvi Bay, Corsica	March–April	Leoni et al. (2003)
Algiers, Algeria	March, May, and July	Semroud (1993)
Alger-plages, Algiers, Algeria	April–May and August–September	Guettaf (1997)
El Marsa, near Algiers, Algeria	April and September–October	Guettaf (1997)
Aïn-Tagouraït, Tipaza, Algeria	February–March	Guettaf (1997)

there are two spawning periods, the first would occur as the temperature rises to a critical level and the second when it falls to that level (Fenaux 1968; Byrne 1990; Pedrotti 1993; but see Bayed et al. 2005). The first spawning event may be triggered by day length (about 15 h of daylight) rather than by temperature, whereas the end of the spawning period seems to be influenced by temperature (Spirlet et al. 1998, 2000). The peak in RI occurs just before or during the spawning period (Bayed et al. 2005). The differences described between regions (one or two seasonal peaks) can be observed within a given region, between localities and habitats (Table 7; Guettaf 1997). Lozano et al. (1995) concluded that spawning only occurs during spring and early summer. However, occurrence of larvae in autumn and the presence of benthic post-metamorphic individuals (1 mm diameter) in October suggest a late summer spawning peak.

All things considered (locality, habitat, inter-individual and between-year variability), and regardless of the number of annual peaks, spawning can occur nearly year-round (Table 7), although usually at very low levels. This may represent a strategy whereby the risks of planktonic larval loss, losses which are due to poor trophic conditions or to the transport of larvae by the ocean currents from the coast, are spread out over time.

7.3. Recruitment

In the northwestern Mediterranean, larvae are present in the plankton year-round. One or two peaks occur, the first in May-June, the second, if present, from September–November (Fenaux 1968; Fenaux and Pedrotti 1988; Pedrotti and Fenaux 1992; Pedrotti 1993). Lozano et al. (1995) suggested the autumn peak was either an artefact or the prolonged survival of larvae from a spring-summer spawn occurring under unfavourable environmental conditions. The duration of the planktonic life of larvae of *P. lividus* has been estimated to be 23–29 days in situ (Pedrotti 1993; but see Lozano et al. 1995), but it can drop to 14–19 days *in vitro* with nonlimiting food (George et al. 1989; Fenaux et al. 1992). Larvae from well-fed parents survive better, grow faster and metamorphose earlier than those from starved or poorly fed parents (George et al. 1989).

The test diameter of newly metamorphosed individuals is 0.3–0.4 mm (Fenaux et al. 1987; Cellario and Fenaux 1990; Tomas et al. 2004). In the Mediterranean, benthic settlement occurs once or twice throughout the year (Table 8; Verlaque 1987b; Azzolina 1988; Fenaux et al. 1987; Lozano et al. 1995; Tomas et al. 2004). The main recruitment occurs in spring although some new recruits are found year-round (Lozano et al. 1995; López et al. 1998; Tomas et al. 2004). Azzolina and Willsie (1987) concluded that the possibility of random settlement in a given locality, regardless of depth and habitat, cannot be rejected. In contrast, recruitment may be quite different between localities (Table 9). In 1992 and 1993, recruitment occurred in Tossa de Mar, Spanish Catalonia, and not in nearby Cubelles (Lozano et al. 1995). At Medes Islands, Spanish Catalonia, significant differences were observed between two adjacent sites (Sala et al. 1998b). Unpredictable winds and eddies which affect coastal currents might be responsible.

Some benthic species hinder larval settlement and/or affect postsettlement juvenile survival. Secondary metabolites of sponges, e.g. *C. crambe*, may inhibit recruitment (Becerro et al. 1997). Survival of settlers to a diameter of 2–3 mm is 0.5–0.7% (López et al. 1998) or <10% (Azzolina and Willsie 1987), while 0.04% survive one year and only 0.03% may attain reproductive size. Mortality rates are density-dependent during the early benthic stages (López et al. 1998). Survival of 2–10 mm test diameter juveniles

Table 8 *Paracentrotus lividus*: major benthic settlement periods of young post-metamorphic individuals.

Location	Benthic settlement	References
Catalonia, Spain	June–July	Lozano et al. (1995)
	Late spring–summer	López et al. (1998)
	Spring–early summer and autumn–early winter	Tomas et al. (2004)
Marseilles, France	July–November	Azzolina and Willsie (1987), Azzolina (1988)
Villefranche, France	June and November	Fenaux et al. (1987)
	Late spring and autumn	Pedrotti (1993)
Galeria, Corsica	Late May, early June and early October	Verlaque (1984, 1987a)

Table 9 *Paracentrotus lividus*: mean densities (individuals per square meter) of juveniles.

Location	Depth (m)	Habitat	Density	References
Cubelles, Catalonia, Spain	0.2–0.5	Boulders with poorly developed MPOs	0	Lozano et al. (1995)
Port-Cros, France	11	*Posidonia oceanica* meadow	0	Azzolina and Willsie (1987)
	11	Dead *Posidonia oceanica* “matte”	0	Azzolina and Willsie (1987)
Urbinu lagoon, Corsica	0–1	*Cymodocea nodosa* meadow	0	Fernandez (1990)
Galeria, Corsica	1–5	Gardens with encrusting corallines	13[b]	Verlaque (1984, 1987a)
Urbinu lagoon, Corsica	0–1	Pebbles	17[d]	Fernandez (1990)
Provence, France	1–2	*Cystoseira crinita* forest	19[e]	Bellan-Santini (1969)
Marseilles, France	1	*Corallina elongata*	20[e]	Bitar (1980)
	11	*Posidonia oceanica* meadow	22[c]	Azzolina and Willsie (1987)
Port-Cros, France	1.5	Dead *Posidonia oceanica* “matte”	25[c]	Azzolina and Willsie (1987)
Marseilles, France	11	Dead *Posidonia oceanica* “matte”	37[c]	Azzolina and Willsie (1987)
Sesimbra, Portugal	0–2	*Corallina elongata*	38[e]	Saldanha (1974)
Galeria, Corsica	1–5	Pebbles	52[b]	Verlaque (1984, 1987a)
Sesimbra, Portugal	8–12	*Asparagopsis armata*	121[e]	Saldanha (1974)
Galeria, Corsica	1–5	*Cystoseira brachycarpa* forest	140[b]	Verlaque (1984, 1987a)
Medes Islands, Catalonia, Spain	5–10	Boulders	1160[a]	Sala and Zabala (1996)
Tossa, Catalonia, Spain	4–10	Vertical wall with erect MPOs	12500[b,e,f]	Lozano et al. (1995)

[a] Test diameter <10 mm.
[b] Test diameter <12 mm.
[c] Test diameter <13 mm.
[d] Test diameter <20 mm.
[e] Calculated from the author’s data.
[f] Extrapolation to square meter from smaller surfaces.

decreases with abundance of predatory fishes and increases with structural complexity of the habitat: barren grounds < turf MPOs < erect MPOs < crevices (Hereu et al. 2005).

Interannual variation in recruitment is high, ca. 1 order of magnitude (Tomas et al. 2004). These variations are usually inferred from the demographic structure of the populations, which show over-represented, under-represented or missing cohorts (Delmas 1992). In the Bay of Galeria, Corsica, abundant recruitment occurred in 1986 or 1987 (Boudouresque et al. 1989). In Spanish Catalonia, a heavy recruitment episode occurred in 1992 and a low one the next year (Lozano et al. 1995; López et al. 1998). The conspicuous changes with time in the abundance and demographic structure of the populations that result (Boudouresque et al. 1989; Sala et al. 1998b) obviously modify the ecological impact of the species.

8. CONCLUSIONS

Herbivorous sea urchins are often the determining factor with regard to the abundance and distribution of multicellular photosynthetic organisms (including seagrasses) in shallow-water marine environments, and play a key role in the general functioning of ecosystems (see reviews by Lawrence 1975; Lawrence and Sammarco 1982; Sala et al. 1998a). The ecological role of *P. lividus* is strengthened by the fact that few large and abundant herbivores occur over its geographical range. It is a significant structuring force not only in spatially localized barren grounds with high or very high sea urchin densities, but also in communities with low densities which are representative of larger areas over the geographical range of the species. In addition, it may control the shift between these two possible alternative 'stable' states.

Human impact can noticeably modify the ecological role of *P. lividus*, either by reducing its abundance through harvesting, or by dramatically increasing it directly through pollution or indirectly through overfishing of its predators.

Beyond these generalities, which are well documented, it is clear that our efforts to understand the processes by which *P. lividus* interacts with its biotic environment are thwarted by the often confusing and contradictory information in the literature. *Paracentrotus lividus* is in fact a very irritating species for biologists. Just when we think we have understood a mechanism, when a succession of causes and effects seem to occur in a logical order, and as we attempt to lay down the last piece of the puzzle, it escapes our grasp. This species appears bent on debunking our most appealing and logical theories. In fact, *P. lividus* is an incredibly opportunistic generalist, with a very wide range of adaptive responses to environmental conditions. It can do everything and anything.

ACKNOWLEDGMENTS

We acknowledge with gratitude Pierre Bartoli and Gustavo San Martín for providing valuable information, John M Lawrence, Enric Sala and Catherine Fernandez for stimulating discussions, and finally the assistance of Yolande Bentosella and Michele Perret-Boudouresque for bibliographical research.

REFERENCES

Allain JY (1975) Structure des populations de *Paracentrotus lividus* (Lamarck) (Echinodermata, Echinoidea) soumises à la pêche sur les côtes Nord de Bretagne. Rev Trav Inst Pêches Marit 39 (2): 171–212

Ara L (1937) Contributo allo studio dell'alimentazione dei pesci: *Sargus vulgaris* Geoffr. – *Sargus annularis* L. – *Sargus sargus* L. Bol Pesca Piscic Idrobiol 13: 371–381

Aubin G (2004) Régime alimentaire des deux oursins réguliers *Paracentrotus lividus* et *Sphaerechinus granularis* dans une zone colonisée par *Caulerpa racemosa* var. *cylindracea* (Sonder) Verlaque, Huisman *et* Boudouresque du golfe de Marseille. Rapport stage Maîtrise, Univ Aix-Marseille 2

Augier H, Ramonda G, Rolland J, Santimone M (1989) Teneurs en métaux lourds des oursins comestibles *Paracentrotus lividus* (Lamarck) prélevés dans quatre secteurs tests du littoral de Marseille (Méditerranée, France). Vie Mar HS 10: 226–239

Azzolina JF (1987) Evolution à long terme des populations de l'oursin comestible *Paracentrotus lividus* dans la baie de Port-Cros (Var, France). In: Boudouresque CF (ed.) Colloque international sur *Paracentrotus lividus* et les oursins comestibles. GIS Posidonie, Marseilles, pp 257–269

Azzolina JF (1988) Contribution à l'étude de la dynamique des populations de l'oursin comestible *Paracentrotus lividus* (Lmck.). Croissance, recrutement, mortalité, migrations. Thèse Doct, Univ Aix-Marseille 2

Azzolina JF, Willsie A (1987) Abondance des juvéniles de *Paracentrotus lividus* au sein de l'herbier à *Posidonia oceanica*. In: Boudouresque CF (ed.) Colloque international sur *Paracentrotus lividus* et les oursins comestibles. GIS Posidonie, Marseilles, pp 159–167

Azzolina JF, Boudouresque CF, Nédélec H (1985) Dynamique des populations de *Paracentrotus lividus* dans la baie de Port-Cros (Var): données préliminaires. Sci Rep Port-Cros Nation Park 11: 61–81

Bacallado JJ, Moreno E, Pérez Ruzafa A (1985) Echinodermata (Canary Islands) – provisional checklist. In: Keegan BF, O'Connor BDS (eds) Echinodermata. AA Balkema, Rotterdam, pp 149–151

Ballesteros E, García-Rubies A (1987) La pêche aux oursins en Espagne et plus particulièrement en Catalogne. In: Boudouresque CF (ed.) Colloque international sur *Paracentrotus lividus* et les oursins comestibles. GIS Posidonie, Marseilles, pp 325–328

Barnes DKA, Crook AC (2001a) Implication of temporal and spatial variability in *Paracentrotus lividus* populations to the associated commercial coastal fishery. Hydrobiologia 465: 95–102

Barnes DKA, Crook AC (2001b) Quantifying behavioural determinants of the coastal European sea-urchin *Paracentrotus lividus*. Mar Biol 138: 1205–1212

Barnes DKA, Steele S, Maguire D, Turner J (1999) Population dynamics of the urchin *Paracentrotus lividus* at Lough Hyne, Ireland. In: Candia Carnevali MD, Bonasoro F (eds) Echinoderm research 1998. AA Balkema, Rotterdam, pp 427–431

Barnes DKA, Verling DKA, Crook A, Davidson I, O'Mahoney M (2002) Local population disappearance follows (20 yr after) cycle collapse in a pivotal ecological species. Mar Ecol Progr Ser 226: 311–313

Barille-Boyer AL, Barille L, Grue Y, Harin N (2004) Temporal changes in community structure of tide pools following the "Erika" oil spill. Aq Liv Res 17 (3): 323–328

Bartoli P, Bray RA, Gibson DI (1989) The Opecoelidae (Digenea) of sparid fishes of the western Mediterranean. Syst Parasitol 13: 167–192

Bayed A, Quiniou F, Benrha A, Guillou M (2005) The *Paracentrotus lividus* populations from the northern Moroccan Atlantic coast: growth, reproduction and health condition. J Mar Biol Ass UK 85: 999–1007

Becerro MA, Turon X, Uriz MJ (1997) Multiple functions for secondary metabolites in encrusting marine invertebrates. J Chem Ecol 23 (6): 1527–1547

Bellan-Santini D (1969) Contribution à l'étude des peuplements infralittoraux sur substrats rocheux (étude qualitative et quantitative de la frange supérieure). Rec Trav Stat Mar Endoume 47 (63): 1–294

Bernárdez C, Freire J, González-Gurriarán E (2000) Feeding of the spider crab *Maja squinado* in rocky subtidal areas of the Ría de Arousa (north-west Spain). J Mar Biol Ass UK 80: 95–102

Benedetti-Cecchi L, Cinelli F (1995) Habitat heterogeneity, sea urchin grazing and the distribution of algae in littoral rock pools on the west coast of Italy (western Mediterranean). Mar Ecol Progr Ser 126: 203–212

Bini G (1968) Atlante delle pesci delle coste italiane, vol. IV. Mondo sommerso publ., Italy

Bitar G (1980) Etude de l'impact de la pollution par un émissaire urbain (collecteur Cortiou) sur les peuplements infralittoraux de substrats durs de la côte Sud de Marseilleveyre (Marseille). Thèse Doct 3° cycle, Univ Aix-Marseille 2

Bonnet A (1925) Documents pour servir à l'étude des variations chez les Echinides. Bull Inst Océanogr 462: 1–28

Boudouresque CF, Nédélec H, Shepherd SA (1980) The decline of a population of the sea urchin *Paracentrotus lividus* in the bay of Port-Cros (Var, France). Trav Sci Parc Nation Port-Cros 6: 242–251

Boudouresque CF, Nédélec H, Shepherd SA (1981) The decline of a population of the sea urchin *Paracentrotus lividus* in the bay of Port-Cros (Var, France). Rapp PV Réun Commiss Int Explor Sci Mer Médit 27 (2): 223–224

Boudouresque CF, Verlaque M, Azzolina JF, Meinesz A, Nédélec H, Rico V (1989) Evolution des populations de *Paracentrotus lividus* et d'*Arbacia lixula* (Echinoidea) le long d'un transect permanent à Galeria (Corse). Trav Sci Parc Nat Rég Rés Nat Corse 22: 65–82

Boudouresque CF, Caltagirone A, Lefèvre JR, Rico V, Semroud R (1992) Macrozoobenthos de la réserve naturelle de Scandola (Corse, Méditerranée nord-occidentale). Analyse pluriannuelle de "l'effet réserve". In: Olivier J, Gerardin N, Jeudy de Grissac A (eds) Economic impact of the Mediterranean coastal protected areas. Medpan, Fr., Hyeres, pp 15–20

Boudouresque CF, Rodríguez-Prieto C, Arrighi F (1993) Place de *Caulerpa prolifera* (Chlorophyta) dans les préférences alimentaires de l'oursin *Paracentrotus lividus*. Trav Sci Parc Nat Rég Rés Nat Corse 41: 41–51

Boudouresque CF, Lemée R, Mari X, Meinesz A (1996) The invasive alga *Caulerpa taxifolia* is not a suitable diet for the sea urchin *Paracentrotus lividus*. Aquat Bot 53: 245–250

Bouxin H (1964) Une expérience écologique de 15 années. Evolution des peuplements de *Paracentrotus lividus* Lmk. dans la région de Concarneau. CR Soc Biogéogr 40 (351–355): 94–100

Bulleri F, Bertocci I, Micheli F (2002) Interplay of encrusting coralline algae and sea urchins in maintaining alternative habitats. Mar Ecol Progr Ser 243: 101–109

Byrne M (1990) Annual reproductive cycles of the commercial sea urchin *Paracentrotus lividus* from an exposed intertidal and a sheltered subtidal habitat on the west coast of Ireland. Mar Biol 104: 275–289

Caillaud F (1856) Observations sur les oursins perforants de Bretagne. Rev Mag Zool 4: 3–22

Campillo A, Amadei J (1978) Premières données biologiques sur la langouste de Corse, *Palinurus elephas* Fabricius. Rev Trav Inst Pêches Marit 42 (4): 347–373

Campos-Villaça R (1984) Données préliminaires sur l'éthologie alimentaire de l'oursin *Sphaeechinus granularis* dans l'herbier à *Posidonia oceanica* de la baie de Port-Cros. Dipl Etudes approf Océanol biol, Univ Aix-Marseille 2

Carvalho FP (1982) Ethologie alimentaire de trios poissons Blenniidae de la côte portugaise. Bol Soc Portug Ciênc Nat 21: 31–43

Catoira Gómez JL (1992) La pêche des oursins en Galice, Espagne, pendant la période 1990–1991. In: Scalera-Liaci L, Canicatti C (eds) Echinoderm research 1991. AA Balkema, Rotterdam, pp 199–200

Cellario C, Fenaux L (1990) *Paracentrotus lividus* (Lamarck) in culture (larval and benthic phases): parameters of growth observed two years following metamorphosis. Aquaculture 84: 173–188

Chelazzi G, Serra G, Bucciarelli G (1997) Zonal recovery after experimental displacement in two sea urchins co-occurring in the Mediterranean. J Exp Mar Biol Ecol 212: 1–7

Cherbonnier G (1951) Inventaire de la faune marine de Roscoff. Les Echinodermes. Trav Sci Stat Biol Roscoff, suppl 4: 1–15

Cherbonnier G (1954) Le roman des Echinodermes. Les beaux Livres, Rennes

Cherbonnier G (1956) Les Echinodermes de Tunisie. Bull Stat océanogr Salammbô 53: 1–23

Chevaldonné P (1990) Ciguatera and the saupe, *Sarpa salpa* (L.), in the Mediterranean: a possible misinterpretation. J Fish Biol 37: 503–504

Codomier L, Bruneau Y, Combaut G, Teste J (1977) Etude biologique et chimique d'*Asparagopsis armata* et de *Falkenbergia rufolanosa* (Rhodophycées, Bonnemaisoniales). CR Acad Sci Paris 284 D: 1163–1165

Crapp GB, Willis ME (1975) Age determination in the sea urchin *Paracentrotus lividus* (Lamarck) with notes on the reproductive cycle. J Exp Mar Biol Ecol 20: 157–178

Crook AC (2003) Individual variation in the covering behaviour of the shallow water sea urchin *Paracentrotus lividus*. PSZN: Mar Ecol 24 (4): 275–287

Crook AC, Barnes DKA (2001) Seasonal variation in the covering behaviour of the echinoid *Paracentrotus lividus* (Lamarck). PSZN: Mar Ecol 22 (3): 231–239

Crook AC, Verling E, Barnes DKA (1999) Comparative study of the covering reaction of the purple sea urchin, *Paracentrotus lividus*, under laboratory and field conditions. J Mar Biol Ass UK 79 (6): 1117–1121

Crook AC, Long M, Barnes DKA (2000) Quantifying daily migration in the sea urchin *Paracentrotus lividus*. J Mar Biol Ass UK 80: 177–178

Cross TE, Southgate T (1980) Mortalities of fauna of rocky substrates in south-west Ireland associated with the occurrence of *Gyrodinium aureolum* blooms during autumn 1979. J Mar Biol Ass UK 60: 1071–1073

Cuenca C (1987) Quelques méthodes de marquages des oursins échinidés (Echinodermes). Bull Soc Sci Nat Ouest Fr, NS 9 (1): 26–37

Cuomo V, Vanzanella F, Fresi E, Mazzella L, Scipione MB (1982) Microflora delle fanerogame marine dell'isola d'Ischia: *Posidonia oceanica* (L.) Delile e *Cymodocea nodosa* (Ucria) Aschers. Boll Mus Ist Biol Univ Genova suppl 50: 162–166

Dambach M, Hentschel G (1970) Die Bedeckungsreaktion von Seeigeln. Neue Versuche und Deutungen. Mar Biol 6: 135–141

Dance C (1987) Patterns of activity of the sea urchin *Paracentrotus lividus* in the bay of Port-Cros (Var, France, Mediterranean). PSZNI: Mar Ecol 8 (2): 131–142

Dance C, Savy S (1987) Predation of *Paracentrotus lividus* by *Marthasterias glacialis*: an *in situ* experiment at Port-Cros (France, Mediterranean). Posidonia Newslett 1 (2): 35–41

Dauby P (1989) The stable carbon isotope ratios in benthic food webs of the Gulf of Calvi, Corsica. Cont Shelf Res 9 (2): 181–195

De Burgh ME (1975) Aspects of the absorption of dissolved nutrients by spines of *Paracentrotus lividus* (Lamarck). PhD Thesis, Univ Dublin

Delmas P (1992) Etude des populations de *Paracentrotus lividus* (Lam.) (Echinodermata : Echinoidea) soumises à une pollution complexe en Provence nord-occidentale: densités, structure, processus de détoxication (Zn, Cu, Pb, Cd, Fe). Thèse Doct, Univ Aix-Marseille 3

Delmas P, Régis MB (1985) Impact de la pollution domestique sur la biologie et la morphométrie de l'échinoïde *Paracentrotus lividus* (Lamarck). Données préliminaires. CR Acad Sci Paris 300 (3, 4): 143–146

Delmas P, Régis MB (1986) Données préliminaires sur le contenu digestif de l'oursin comestible *Paracentrotus lividus* (Lamarck) soumis à l'influence d'effluents domestiques. Mar Environ Res 20: 197–220

De Ridder C, Lawrence JM (1982) Food and feeding mechanisms: echinoidea. In: Jangoux M, Lawrence JM (eds) Echinoderm nutrition. AA Balkema, Rotterdam, pp 57–116

Domenici P, González-Calderón D, Ferrari RS (2003) Locomotor performance in the sea urchin *Paracentrotus lividus*. J Mar Biol Ass UK 83 (2): 285–292

Drzewina A, Bohn G (1924) Un nouveau cas d'hermaphrodisme chez l'oursin, *Strongylocentrotus lividus*. CR Acad Sci Paris 178: 662–663

Duran S, Palacín C, Becerro MA, Turon X (2004) Genetic diversity and population structure of the commercially harvested sea urchin *Paracentrotus lividus* (Echinodermata, Echinoidea). Mol Ecol 13: 3317–3328

Ebert TA (1968) Growth rates of the sea urchin *Strongylocentrotus purpuratus* related to food availability and spine abrasion. Ecology 49: 1075–1091

Ebling FJ, Hawkins AD, Kitching JA, Muntz L, Pratt WM (1966) The ecology of Lough Ine. XVI. Predation and diurnal migration in *Paracentrotus* community. J Anim Ecol 35: 559–566

Fanelli G, Piraino S, Belmonte G, Geraci S, Boero F (1994) Human predation along Apulian rocky coasts (SE Italy): desertification caused by *Lithophaga lithophaga* (Mollusca) fisheries. Mar Ecol Progr Ser 110: 1–8

Fenaux L (1968) Maturation des gonades et cycle saisonnier des larves chez *A. lixula, P. lividus* et *P. microtuberculatus* (Echinides) à Villefranche-sur-Mer. Vie Milieu 19 (A1): 1–52

Fenaux L, Pedrotti ML (1988) Métamorphose des larves d'Echinides en pleine eau. PSZNI: Mar Ecol 9 (2): 93–107

Fenaux L, Etienne M, Quelart G (1987) Suivi écologique d'un peuplement de *Paracentrotus lividus* (Lamarck) dans la baie de Villefranche sur Mer. In: Boudouresque CF (ed.) Colloque international sur *Paracentrotus lividus* et les oursins comestibles. GIS Posidonie, Marseilles, pp 187–197

Fenaux L, George SB, Pedrotti ML, Corre MC (1992) Différences dans la succession des stades morphologiques du développement larvaire des échinides en relation avec l'environnement trophique des adultes et des larves. In: Scalera-Liaci L, Canicatti C (eds) Echinoderm research 1991. AA Balkema, Rotterdam, pp 173–180

Fernandez C (1989) Contribution à l'élaboration des bases scientifiques de l'aquaculture de l'oursin *Paracentrotus lividus*. Mém Maîtrise Sci Techn, Univ Corse

Fernandez C (1990) Recherches préliminaires à la mise en place d'un pilote d'aquaculture de l'oursin *Paracentrotus lividus* dans un étang corse. Dipl Etudes approf Océanol, Univ Aix-Marseille 2

Fernandez C (1996) Croissance et nutrition de *Paracentrotus lividus* dans le cadre d'un projet aquacole avec alimentation artificielle. Thèse Doct Océanol, Univ Corse

Fernandez C, Boudouresque CF (1997) Phenotypic plasticity of *Paracentrotus lividus* (Echinodermata: Echinoidea) in a lagoonal environment. Mar Ecol Progr Ser 152: 145–154

Fernandez C, Boudouresque CF (2000) Nutrition of the sea urchin *Paracentrotus lividus* (Echinodermata: Echinoidea) fed different artificial food. Mar Ecol Progr Ser 204: 131–141

Fernandez C, Caltagirone A (1994) Growth rate of adult sea urchins, *Paracentrotus lividus* in a lagoon environment: the effect of different diet types. In: David B, Guille A, Féral JP, Roux M (eds) Echinoderms through time. AA Balkema, Rotterdam, pp 655–660

Fernandez C, Caltagirone A, Johnson M (2001) Demographic structure suggests migration of the sea urchin *Paracentrotus lividus* in a coastal lagoon. J Mar Biol Ass UK 81: 361–362

Fernandez C, Pasqualini V, Johnson M, Ferrat L, Caltagirone A, Boudouresque CF (2003) Stock evaluation of the sea urchin *Paracentrotus lividus* in a lagoonal environment. In: Féral JP, David B (eds) Echinoderm research 2001. AA Balkema, Lisse, pp 319–323

Fischer P (1864) Note sur les perforations de l'*Echinus lividus* Lamk. Ann Sci Nat 5 (1): 321–332

Fischer-Piette E (1955) Répartition, le long des côtes septentrionales de l'Espagne, des principales espèces peuplant les rochers intercotidaux. Ann Inst Océanogr 31 (2): 37–124

Fol H (1879) Recherches sur la fécondation et le début de l'hénologie chez divers animaux. Mém Soc Phys Hist Nat Genève 32 (1): 89–397

Fox HM (1923) Lunar periodicity in reproduction. Proc Roy Soc London B 95: 523–550

Francour P, Boudouresque CF, Harmelin JG, Harmelin-Vivien ML, Quignard JP (1994) Are the Mediterranean waters becoming warmer? Information from biological indicators. Mar Poll Bull 28 (9): 523–526

Frantzis A, Grémare A (1992) Ingestion, absorption and growth rates of *Paracentrotus lividus* (Echinodermata: Echinoidea) fed different macrophytes. Mar Ecol Progr Ser 95: 169–183

Frantzis A, Berthon JF, Maggiore F (1988) Relations trophiques entre les oursins *Arbacia lixula* et *Paracentrotus lividus* (Echinoidea Regularia) et le phytobenthos infralittoral superficiel dans la baie de Port-Cros (Var, France). Sci Rep Port-Cros Nation Park 14: 81–140

Gago J, Range P, Luís O (2003) Growth, reproductive biology and habitat selection of the sea urchin *Paracentrotus lividus* in the coastal waters of Cascais, Portugal. In: Féral JP, David B (eds) Echinoderm research 2001. AA Balkema, Lisse, pp 269–276

Gamble JC (1965) Some observations on the behaviour of two regular echinoids. In: Lythgoe JN, Woods JD (eds) Symposium underwater association. Underwater Ass., Malta, pp 47–50

Gamble JC (1966–1967) Ecological studies on *Paracentrotus lividus* (Lmk.). In: Underwater Association Report. TGW Industrial & Research Promotion, Carshalton, England, pp 85–88

George S, Fenaux L, Lawrence J (1989) Effets du passé alimentaire des parents sur la taille de l'œuf et le développement larvaire de deux échinides *Arbacia lixula* et *Paracentrotus lividus*. Vie Mar HS 10: 258

Giaccone G (1971) Contributo allo studio dei popolamenti algali del basso Tirreno. Ann Univ Ferrara, Bot 4 (2): 17–43

Goss-Custard S, Jones J, Kitching JA, Norton TA (1979) Tide pools of Carrigathorna and Barloge Creek. Phil Trans Roy Soc London 287 (1016): 1–44

Gros C (1978) Le genre *Cystoseira* sur la côte des Albères. Répartition, écologie, morphogénèse. Thèse Doct 3° cycle, Univ Paris 6

Grosjean P, Spirlet C, Jangoux M (1996) Experimental study of growth in the echinoid *Paracentrotus lividus* (Lamarck, 1816) (Echinodermata). J Exp Mar Biol Ecol 201: 173–184

Grosjean P, Spirlet C, Jangoux M (2003) A functional growth model with intraspecific competition applied to a sea urchin, *Paracentrotus lividus*. Can J Fish Aquat Sci 60 (3): 237–246

Guerriero A, Meinesz A, D'Ambrosio M, Pietra F (1992) Isolation of toxic and potentially toxic sesqui- and monoterpenes from the tropical green seaweed *Caulerpa taxifolia* which has invaded the region of Cap Martin and Monaco. Helv Chim Acta 75: 689–695

Guettaf M (1997) Contribution à l'étude de la variabilité du cycle reproductif (indice gonadique et histologie des gonades) chez *Paracentrotus lividus* (Echinodermata: Echinoidea) en Méditerranée sud-occidentale (Algérie). Thèse Doct, Univ Aix-Marseille 2

Guettaf M, San Martín GA (1995) Etude de la variabilité de l'indice gonadique de l'oursin comestible *Paracentrotus lividus* (Echinodermata: Echinidae) en Méditerranée nord-occidentale. Vie Milieu 45 (2): 129–137

Guidetti P, Fraschetti S, Terlizzi A, Boero F (2003) Distribution patterns of sea urchins and barrens in shallow Mediterranean rocky reefs impacted by the illegal fishery of the rock-boring mollusc *Lithophaga lithophaga*. Mar Biol 143: 1135–1142

Guidetti P, Terlizzi A, Boero F (2004) Effects of the edible sea urchin, *Paracentrotus lividus*, fishery along the Apulian rocky coast (SE Italy, Mediterranean Sea). Fish Res 66: 287–297

Guidetti P, Bussotti S, Boero F (2005) Evaluating the effects of protection on fish predators and sea urchins in shallow artificial rocky habitats: a case study in the northern Adriatic Sea. Mar Environ Res 59 (4): 333–348

Guillou M, Grall J, Connan S (2002) Can low sea urchin densities control macro-epiphytic biomass in a nort-east Atlantic maerl bed ecosystem (Bay of Brest, Britanny, France). J Mar Biol Ass UK 82: 867–876

Harmelin JG, Bouchon C, Duval C, Hong JS (1980) Les échinodermes des substrats durs de l'île de Port-Cros, Parc National (Méditerranée Nord-occidentale). Trav Sci Parc Nation Port-Cros 6: 25–38

Harmelin JG, Bouchon C, Hong JS (1981) Impact de la pollution sur la distribution des échinodermes des substrats durs en Provence (Méditerranée Nord-occidentale). Téthys 10 (1): 13–36

Harper JL (1969) The role of predation in vegetational diversity. Brookhaven Symp Biol 22: 48–62

Haya D, Régis MB (1995) Comportement trophique de *Paracentrotus lividus* (Lam.) (Echinodermata: Echinoidea) soumis à six régimes alimentaires dans des conditions expérimentales. Mésogée 54: 35–42

Hereu B (2005) Movement patterns of the sea urchin *Paracentrotus lividus* in a marine reserve and an unprotected area in the NW Mediterranean. Mar Ecol 26: 54–62

Hereu B, Zabala M, Linares C, Sala E (2005) The effect of predator abundance and habitat structural complexity on survival of juvenile sea urchins. Mar Biol 146: 293–299

Höbaus E, Fenaux L, Hignette M (1981) Premières observations sur les lésions provoquées par une maladie affectant le test des oursins en Méditerranée occidentale. Rapp PV Réun Commiss Int Explor Sci Mer Médit 27 (2): 221–222

Ivlev VS (1961) Experimental ecology of the feeding of the fishes. Yale University Press, New Haven

Jangoux M (1990) Diseases of echinodermata. In: Kinne O (ed.) Diseases of marine animals, vol. 3. Biologische Anstalt Helgoland, pp 439–567

Jangoux M, Maes P (1987) Les épizooties chez les oursins réguliers (Echinodermata). In: Boudouresque CF (ed.) Colloque international sur *Paracentrotus lividus* et les oursins comestibles. GIS Posidonie, Marseilles, pp 299–307

Jousson O, Bartoli P, Pawlowski J (1999) Molecular identification of developmental stages in Opecoelidae (Digenea). Int J Parasitol 29: 1853–1858

Katzmann W (1974) Regression for Braunalgenbestande im Mittelmeer. Naturwiss Rundsch 27: 480–481

Kečkeš S, Ozretić B, Lucu Č (1966) About a possible mechanism involved in the shedding of sea-urchins. Experientia 22: 146–147

Keegan BF, Könnecker G (1980) Aggregation in echinoderms on the West coast of Ireland. An ecological perspective. In: Jangoux M (ed.) Echinoderms: present and past. AA Balkema, Rotterdam, p 199

Kempf M (1962) Recherches d'écologie comparée sur *Paracentrotus lividus* (Lmk.) et *Arbacia lixula* (L.). Rec Trav Stat Mar Endoume 25 (39): 47–116

Khoury C (1987) Ichtyofaune des herbiers de posidonies du Parc national de Port-Cros: composition, éthologie alimentaire et rôle dans le réseau trophique. Thèse Doct 3° cycle, Univ Aix-Marseille 2

Kirkman H, Young PC (1981) Measurement of health, and echinoderm grazing on *Posidonia oceanica* (L.) Delile. Aquat Bot 10 (4): 329–338

Kitching JA, Ebling FJ (1961) The ecology of Lough Ine. CI. The control of algae by *Paracentrotus lividus* (Echinoidea). J Anim Ecol 30: 373–383

Kitching JA, Ebling FJ (1967) Ecological studies at Lough Ine. Adv Ecol Res 4: 197–291

Kitching JA, Thain VM (1983) The ecological impact of the sea urchin *Paracentrotus lividus* (Lamarck) in Lough Ine, Ireland. Phil Trans Roy Soc London B 300: 513–552

Knoepffler-Péguy M, Maggiore F, Boudouresque CF, Dance C (1987) Compte Rendu d'une expérience sur les preferenda alimentaires de *Paracentrotus lividus* (Echinoidea) à Banyuls-sur-Mer. In: Boudouresque CF (ed.) Colloque international sur *Paracentrotus lividus* et les oursins comestibles. GIS Posidonie, Marseilles, pp 59–64

Knoepffler-Péguy M, Nattero MJ (1996) Comportement de *Paracentrotus lividus* en présence de la souche de *Caulerpa taxifolia* du port de St Cyprien (Fr. Pyr. or.). In: Ribera MA, Ballesteros E, Boudouresque CF, Gómez A, Gravez V (eds) Second international workshop on *Caulerpa taxifolia*. Publ Univ Barcelona (Spain), pp 309–314

Knowlton N (2004) Multiple "stable" states and the conservation of marine ecosystems. Progr Oceanogr 60: 387–396

Koehler R (1883) Recherches sur les Echinides des côtes de Provence. Ann Mus Hist Nat Marseille, Zool 1 (3): 5–167, pl. 1–7

Kovitou M (1991) Feeding preferenda of the sea-urchin *Paracentrotus lividus* (in Greek, English abstract). Res Dipl Univ Thessaloniki

Laborel-Deguen F, Laborel J (1977) Broutage des posidonies à la plage du Sud. Trav Sci Parc Nation Port-Cros 3: 213–214

Lawrence J, Régis MB, Delmas P, Gras G, Klinger T (1989) The effect of quality of food on feeding and digestion in *Paracentrotus lividus* (Lamarck) (Echinodermata: Echinoidea). Mar Behav Physiol 15: 137–144

Lawrence J, Fenaux L, Corre MC, Lawrence A (1992) The effect of quantity and quality of prepared diets on production in *Paracentrotus lividus* (Echinodermata: Echinoidea). In: Scalera-Liaci L, Canicatti C (eds). Echinoderm research 1991. AA Balkema, Rotterdam, pp 107–110

Lawrence JM (1975) On the relationships between marine plants and sea urchins. Oceanogr Mar Biol Annu Rev 13: 213–286

Lawrence JM, Sammarco PW (1982) Effects of feeding on the environment: Echinoidea. In: Jangoux M and Lawrence JM (eds) Echinoderm nutrition. AA Balkema, Rotterdam, pp 499–519

Lawrence JM, Lawrence AL, Holland MO (1965) Annual cycle in the size of the gut of the purple sea urchin, *Strongylocentrotus purpuratus* (Stimpson). Nature 205 (4977): 1238–1239

Lawrence JM, McBride SC, Plank LR, Shpigel M (2003) Ammonia tolerance of the sea urchins *Lytechinus variegatus, Arbacia punctulata, Strongylocentrotus franciscanus,* and *Paracentrotus lividus*. In: Féral JP, David B (eds) Echinoderm reserch 2001. AA Balkema, Lisse, pp 233–236

Lecchini D, Lenfant P, Planes S (2002) Variation in abundance and population dynamics of the sea-urchin *Paracentrotus lividus* on the Catalan coast (north-western Mediterranean Sea) in relation to havitat and marine reserve. Vie Milieu 52 (2–3): 111–118

Ledireac'h JP (1987) La pêche des oursins en Méditerranée: historique, techniques, législation, production. In: Boudouresque CF (ed.) Colloque international sur *Paracentrotus lividus* et les oursins comestibles. GIS Posidonie, Marseilles, pp 335–362

Le Gall P (1987) La pêche des oursins en Bretagne. In: Boudouresque CF (ed.) Colloque international sur *Paracentrotus lividus* et les oursins comestibles. GIS Posidonie, Marseilles, pp 311–324

Le Gall P, Bucaille D, Dutot P (1989) Résistance aux variations de salinité chez *Paracentrotus* et *Psammechinus*. Vie Mar HS 10: 83–84

Le Gall P, Bucaille D, Grassin JB (1990) Influence de la température sur la croissance de deux oursins comestibles, *Paracentrotus lividus* et *Psammechinus miliaris*. In: De Ridder C, Dubois P, Lahaye MC, Jangoux M (eds) Echinoderm research. AA Balkema, Rotterdam, pp 183–188

Lemée R, Boudouresque CF, Gobert J, Malestroit P, Mari X, Meinesz A, Ménager V, Ruitton S (1996) Feeding behaviour of *Paracentrotus lividus* in the presence of *Caulerpa taxifolia* introduced in the Mediterranean Sea. Oceanol Acta 19 (3–4): 245–253

Leoni V, Fernandez C, Johnson M, Ferrat L, Pergent-Martini C (2003) Preliminary study on spawning periods in the sea urchin *Paracentrotus lividus* from lagoon and marine environments. In: Feral JP, David B (eds) Echinoderm research 2001. AA Balkema, Lisse, pp 277–280

Lewis JR (1976) The ecology of rocky shores. Hodder and Stoughton, London

López S, Turon X, Montero E, Palacín C, Duarte CM, Tarjuelo I (1998) Larval abundance, recruitment and early mortality in *Paracentrotus lividus* (Echinoidea). Interannual variability and plankton-benthos coupling. Mar Ecol Progr Ser 172: 239–251

Lozano J, Galera J, López S, Turon X, Palacín C, Morera G (1995) Biological cycles and recruitment of *Paracentrotus lividus* (Echinodermata: Echinoidea) in two contrasting habitats. Mar Ecol Progr Ser 122: 179–191

Maggiore F, Berthon JF, Boudouresque CF, Lawrence J (1987) Données préliminaires sur les relations entre *Paracentrotus lividus, Arbacia lixula* et le phytobenthos dans la baie de Port-Cros (Var, France, Méditerranée). In: Boudouresque CF (ed.) Colloque international sur *Paracentrotus lividus* et les oursins comestibles. GIS Posidonie, Marseilles, pp 65–82

Martinell J (1981) Actividad erosiva de *Paracentrotus lividus* (Lmk.) (Echinodermata, Echinoidea) en el litoral gerundense. Oecol Aquat 5: 219–225

Martínez I, García FJ, Sánchez AI, Daza JL, del Castillo F (2003) Biometric parameters and reproductive cycle of *Paracentrotus lividus* (Lamarck) in three habitats of southern Spain. In: Féral JP, David B (eds) Echinoderm research 2001. AA Balkema, Lisse, pp 281–287

Massé H (1975) Etude de l'alimentation de *Astropecten aranciacus.* Cah Biol Mar 16: 495–510

Mastaller M (1974) Zerstörung des Makrophytals an der nordadriatischen Küste durch intensives Abweiden durch Seeigel; eine Untersuchung über pipulationsstrukturen, nahrung und Fressverhalten bei dem Echinoiden *Paracentrotus lividus* (Lmrck). Dipl Arbeit, Zool Inst Univ München

Mazzella L, Buia MC, Gambi MC, Lorenti M, Russo GF, Scipione MB, Zupo V (1992) Plant-animal trophic relationships in the *Posidonia oceanica* ecosystem of the Mediterranean: a review. In: John DM, Hawkins SJ, Price JH (eds) Plant–animal interactions in the marine benthos. Oxford Science, Oxford, pp 165–187

Mortensen T (1927) Handbook of the echinoderms of the British Isles. University Press, Oxford

Mortensen T (1943) A monograph of the Echinoidea, III.3 Camarodonta. II. Echinidae, Strongylocentrotidae, Parasaleniidae, Echinometridae. CA Reitzel, Copenhagen

Muntz L, Ebling FJ, Kitching JA (1965) The ecology of Lough Ine. XIV. Predatory activity of large crabs. J Anim Ecol 34: 315–329

Nédélec H (1982) Ethologie alimentaire de *Paracentrotus lividus* dans la baie de Galeria (Corse) et son impact sur les peuplements phytobenthiques. Thèse Doct 3° cycle, Univ Aix-Marseille 2

Nédélec H, Verlaque M (1984) Alimentation de l'oursin *Paracentrotus lividus* (Lamarck) dans un herbier à *Posidonia oceanica* (L.) Delile en Corse (Méditerranée – France). In: Boudouresque CF, Jeudy de Grissac A, Olivier J (eds) International workshop on *Posidonia oceanica* beds. GIS Posidonie, Marseilles, pp 349–364

Nédélec H, Verlaque M, Diapoulis A (1981) Preliminary data on *Posidonia* consumption by *Paracentrotus lividus* in Corsica (France). Rapp PV Réun Commiss Int Explor Sci Mer Médit 27 (2): 203–204

Nédélec H, Verlaque M, Dallot S (1983) Note préliminaire sur les fluctuations de l'activité trophique de *Paracentrotus lividus* dans l'herbier de Posidonies. Rapp PV Réun Commiss Int Explor Sci Mer Médit 28 (3): 153–155

Neefs Y (1937) Sur divers cas d'hermaphrodisme fonctionnel chez l'oursin *Strongylocentrotus lividus.* CR Acad Sci Paris 204 (11): 901–902

Neefs Y (1938) Remarques sur le cycle sexuel de l'oursin, *Strongylocentrotus lividus,* dans la région de Roscoff. CR Acad Sci Paris 206 775–777

Neill SRSJ, Larkum H (1966) Ecology of some echinoderms in Maltese waters. In: Lythgoe JN, Woods JD (eds) Symposium underwater association. Underwater Ass, Malta, pp 51–55

Neill FX, Pastor R (1973) Relaciones tróficas de *Paracentrotus lividus* (Lmk.) en la zona litoral. Invest Pesq 37 (1): 1–7

Odile F, Boudouresque CF, Knoepffler-Péguy M (1988) Etude expérimentale des preferenda alimentaires de *Paracentrotus lividus* (Echinoidea). In: Régis MB (ed.) Actes du 6° Séminaire international sur les Echinodermes, Ile des Embiez (Var – France), p 85

Ogden JC (1976) Some aspects of herbivore-plant relationships on Caribbean reefs and seagrass beds. Aquat Bot 2 (2): 103–116

Ott J, Maurer L (1977) Strategies of energy transfer from marine macrophytes to consumer levels: the *Posidonia oceanica* example. In: Keegan BF, Céidigh PO, Boaden PJS (eds) Biology of benthic organisms. Pergamon Press, Oxford, pp 493–502

Ott JA (1981) Adaptative strategies at the ecosystem level: examples from two benthic marine systems. PSZNI: Mar Ecol 2 (2): 113–158

Otter GW (1932) Rock-burrowing echinoderms. Biol Rev 7 (2): 89–107

Paine RT, Vadas RL (1969) The effects of grazing by sea urchins *Stongylocentrotus* spp. on benthic algal populations. Limnol Oceanogr 14: 710–719

Palacín C, Giribet G, Turon X (1997) Patch recolonization through migration by the echinoid *Paracentrotus lividus* in communities with high algal cover and low echinoid densities. Cah Biol Mar 38: 267–271

Palacín C, Giribet G, Carner S, Dantart L, Turon X (1998) Low densities of sea urchins influence the structure of algal assemblages in the western Mediterranean. J Sea Res 39: 281–290

Pastor R (1971) Distribución del erizo de mar, *Paracentrotus lividus* (Lmk), en la Ría de Vigo. Publ Técn Dir Gen Pesca Marít, Spain 9: 255–270

Patzner RA (1999) Sea urchins as hiding-place for juvenile benthic teleosts (Gobiidae and Gobiesocidae) in the Mediterranean Sea. Cybium 23 (1): 93–97

Paul O, Verlaque M, Boudouresque CF (1984) Etude du contenu digestif de l'oursin *Psammechinus microtuberculatus* dans l'herbier à *Posidonia oceanica* de la baie de Port-Cros (Var – France). In: Boudouresque CF, Jeudy de Grissac A, Olivier J (eds) First international workshop on *Posidonia oceanica* beds. GIS Posidonie, Marseilles, pp 365–371

Pedrotti ML (1993) Spatial and temporal distribution and recruitment of echinoderm larvae in the Ligurian Sea. J Mar Biol Ass UK 73: 513–530

Pedrotti ML and Fenaux L (1992) Dispersal of echinoderm larvae in a geographical area marked by upwelling (Ligurian Sea, NW Mediterranean). Mar Ecol Progr Ser 86: 217–227

Pérès JM, Picard J (1956) Note préliminaire sur la campagne de recherches benthiques effectuées par la "Calypso" sur le seuil siculo-tunisien. Rapp PV Réun Commiss Int Explor Sci Mer Médit 13: 219–224

Powis de Tenbossche T (1978) Comportement alimentaire et structures digestives de *Paracentrotus lividus* (Lamarck) (Echinodermata – Echinoidea). Mém Licence Sci biol, Univ libre Bruxelles

Prouho H (1890) Du rôle des pédicellaires gemmiformes des oursins. CR Acad Sci Paris 111: 62–64

Quignard JP (1966) Recherches sur les Labridae (poissons téléostéens Perciformes) des côtes européennes. Systématique et biologie. Naturalia Monspeliensis, Sér Zool 5: 1–247

Régis MB (1978) Croissance de deux échinoïdes du golfe de Marseille (*Paracentrotus lividus* (Lmk) et *Arbacia lixula* L.). Aspects écologiques de la microstructure du squelette et de l'évolution des indices physiologiques. Thèse Doct Sci, Univ Aix-Marseille 3

Régis MB (1979) Analyse des fluctuations des indices physiologiques chez deux échinoïdes (*Paracentrotus lividus* (Lmk) et *Arbacia lixula* L.) du golfe de Marseille. Téthys 9 (2): 167–181

Régis MB (1981) Aspects morphométriques de la croissance de deux échinoïdes du golfe de Marseille, *Paracentrotus lividus* (Lmk) et *Arbacia lixula* L. Cah Biol Mar 22: 349–370

Régis MB (1986) Microstructure adaptative des radioles de *Paracentrotus lividus* (Echinodermata: Echinoidea) en milieu eutrophisé par des eaux usées. Mar Biol 90: 271–277

Régis MB, Arfi R (1978) Etude comparée de la croissance de trios populations de *Paracentrotus lividus* (Lamarck), occupant des biotopes différents, dans le golfe de Marseille CR Acad Sci Paris, Sér D 286: 1211–1214

Régis MB, Pérès JM, Gras G (1985) Données préliminaires sur l'exploitation de la ressource: *Paracentrotus lividus* (Lmck) dans le Quartier Maritime de Marseille. Vie Mar 7: 41–60

Richner H, Milinski M (2000) On the functional significance of masking behaviour in sea urchins – an experiment with *Paracentrotus lividus*. Mar Ecol Progr Ser 205: 307–308

Rico V (1989) Contribution à l'étude des preferenda alimentaires et du comportement moteur de l'oursin régulier *Paracentrotus lividus*. Dipl Etudes Approf Océanogr, Univ Aix-Marseille 2

Rosecchi E (1985) Ethologie alimentaire des Sparidae *Diplodus annularis, Diplodus vulgaris, Diplodus sargus, Pagellus erythrinus, Sparus aurata* du golfe du Lion et des étangs palavasiens. Thèse Doct 3° cycle, Univ Montpellier

Ruitton S, Francour P, Boudouresque CF (2000) Relationships between algae, benthic herbivorous invertebrates and fishes in rocky sublittoral communities of a temperate sea (Mediterranean). Estuar Coast Shelf Sci 50: 217–230

Ruiz-Fernández JM (2000) Respusta de la fanerógama marina *Posidonia oceanica* (L.) Delile a perturbaciones antrópicas. Tesis Biología Univ Murcia, Spain

Russell MP, Ebert TA, Petraitis PS (1998) Field estimates of growth and mortality of the green sea urchin *Strongylocentrotus droebachiensis*. Ophelia 48 (2): 137–153

Sala E (1996) The role of fishes in the organization of a Mediterranean sublittoral community. Thèse Doct, Univ Aix-Marseille 2

Sala E (1997) Fish predators and scavengers of the sea urchin *Paracentrotus lividus* in protected areas of the north-west Mediterranean Sea. Mar Biol 129: 531–539

Sala E, Zabala M (1996) Fish predation and the structure of the sea urchin *Paracentrotus lividus* populations in the NW Mediterranean. Mar Ecol Progr Ser 140: 71–81

Sala E, Boudouresque CF, Harmelin-Vivien M (1998a) Fishing, trophic cascades, and the structure of algal assemblages: evaluation of an old but untested paradigm. Oikos 82: 425–439

Sala E, Ribes M, Hereu B, Zabala M, Alvà V, Coma R, Garrabou J (1998b) Temporal variability in abundance of the sea urchins *Paracentrotus lividus* and *Arbacia lixula* in the northwestern Mediterranean: comparison between a marine reserve and un unprotected area. Mar Ecol Progr Ser 168: 135–145

Saldanha L (1974) Estudo do povoamento dos horizontes superiores da rocha litoral da costa da Arrábida (Portugal). Arq Mus Bocage, Lisbon, 2° Sér, 5 (1): 1–382 + 28 pl

Sánchez-España AI, Martínez-Pita I, García FJ (2004) Gonadal growth and reproduction in the commercial sea urchin Paracentrotus lividus (Lamarck, 1816) (Echinodermata: Echinoidea) from souther Spain. Hydrobiologia 519: 61–72

San Martín GA (1987) Comportement alimentaire de *Paracentrotus lividus* (Lmk) (Echinodermata: Echinidae) dans l'étang de Thau (Hérault, France). In: Boudouresque CF (ed.) Colloque international sur *Paracentrotus lividus* et les oursins comestibles. GIS Posidonie, Marseilles, pp 37–57

San Martín GA (1995) Contribution à la gestion des stocks d'oursins: étude des populations et transplantations de *Paracentrotus lividus* à Marseille (France, Méditerranée) et production de *Loxechinus albus* à Chiloe (Chili, Pacifique). Thèse Doct, Univ Aix-Marseille 2

Sartoretto S, Francour P (1997) Quantification of bioerosion by *Sphaerechinus granularis* on "coralligene" concretions of the western Mediterranean. J Mar biol Ass UK 77: 565–568

Savy S (1987) Les prédateurs de *Paracentrotus lividus* (Echinodermata). In: Boudouresque CF (ed.) Colloque international sur *Paracentrotus lividus* et les oursins comestibles. GIS Posidonie, Marseilles, pp 413–423

Schneider J, Torunski H (1983) Biokarst on limestone coasts, morphogenesis and sediment production. PSZNI: Mar Ecol 4 (1): 45–63

Sellem F, Langar H, Pesando D (2000) Age et croissance de l'oursin *Paracentrotus lividus* Lamarck, 1816 (Echinodermata-Echinoidea) dans le golfe de Tunis (Méditerranée). Oceanol Acta 23 (5): 607–613

Semroud R (1993) Contribution à la connaissance de l'écosystème à *Posidonia oceanica* (L.) Delile dans la région d'Alger (Algérie): étude de quelques compartiments. Thèse Doct, Univ Alger

Semroud R, Kada H (1987) Contribution à l'étude de l'oursin *Paracentrotus lividus* (Lamarck) dans la région d'Alger (Algérie): indice de réplétion et indice gonadique. In: Boudouresque CF (ed.) Colloque international sur *Paracentrotus lividus* et les oursins comestibles. GIS Posidonie, Marseilles, pp 117–124

Sharp DT, Gray JE (1962) Studies on factors affecting the local distribution of two sea urchins *Arbacia punctulata* and *Lytechinus variegates*. Ecology 43 (2): 309–313

Shepherd SA (1987) Grazing by the sea-urchin *Paracentrotus lividus* in *Posidonia* beds at Banyuls, France. In: Boudouresque CF (ed.) Colloque international sur *Paracentrotus lividus* et les oursins comestibles. GIS Posidonie, Marseilles, pp 83–96

Shepherd SA, Boudouresque CF (1979) A preliminary note on the movement of the sea urchin *Paracentrotus lividus*. Trav Sci Parc Nation Port-Cros 5: 155–158

Shpigel M, McBride SC, Marciano S, Lupatsch I (2004) The effect of photoperiod and temperature on the reproduction of European sea urchin *Paracentrotus lividus*. Aquaculture 232: 342–355

Southward A, Southward E (1975) Endangered urchins. New Sci 66 (944): 70–72

Spirlet C, Grosjean P, Jangoux M (1998) Reproductive cycle of the echinoid *Paracentrotus lividus*: analysis by means of maturity index. Invert Reprod Develop 34 (1): 69–81

Spirlet C, Grosjean P, Jangoux M (2000) Optimization of gonad growth by manipulation of temperature and photoperiod in cultivated sea urchins, *Paracentrotus lividus* (Lamarck) (Echinodermata). Aquaculture 185: 85–99

Stirn J, Avcin A, Cencel J, Dorer M, Gomiscek S, Kveder S, Malej A, Meischner D, Nozina I, Paul J, Tusnik P (1974) Pollution problems of the Adriatic Sea. An interdisciplinary approach. Rev Int Océanogr Méd 35–36: 21–78

Tertschnig WP (1989) Predation on the sea urchin *Paracentrotus lividus* by the spider crab *Maia crispata*. Vie Mar HS 10 : 95–103

Timon-David J (1934) Recherches sur les Trématodes parasites des oursins en Méditerranée. Bull Inst Océanogr 652: 1–16

Tomas F, Romero J, Turon X (2004) Settlement and recruitment of the sea urchin *Paracentrotus lividus* in two contrasting habitats in the Mediterranean. Mar Ecol Progr Ser 282: 173–184

Tomas F, Turon X, Romero J (2005) Effects of herbivores on a *Posidonia oceanica* seagrass meadow: importance of epiphytes. Mar Ecol Progr Ser 287: 115–125

Tortonese E (1952) Gli echinodermi del Mar Ligure e delle zone vicine. Atti Accad Lig Sci Lett 8: 163–242

Tortonese E (1965) Fauna d'Italia, Echinodermata. Edizioni Calderini, Bologna

Torunski H (1979) Biological erosion and its significance for the morphogenesis of limestone coasts and for nearshore sedimentation (northern Adriatic). Senckenberg Marit 11 (3–6): 193–265

Traer K (1980) The consumption of *Posidonia oceanica* Delile by Echinoids at the Isle of Ischia. In: Jangoux M (ed.) Echinoderms: present and past. AA Balkema, Rotterdam, pp 241–244

Traer K (1984) Ernährung und Energetik regulärer Seeigel in Beständen des Mediterranen Seegrases *Posidonia oceanica*. Thesis Doct, Univ Wien

Trudgill ST, Smart PL, Friederich H, Crabtree RW (1987) Bioerosion of intertidal limestone, Co. Clare, Eire.1. Mar Geol 74: 85–98

Turon X, Giribet G, López S, Palacín C (1995) Growth and population structure of *Paracentrotus lividus* (Echinodermata: Echinoidea) in two contrasting habitats. Mar Ecol Progr Ser 122: 193–204

Uriz MJ, Turon X, Becerro MA, Galera J (1996) Feeding deterrence in sponges. The role of toxicity, physical defenses, energetic contents, and life-history stages. J Exp Mar Biol Ecol 205: 187–204

Vasserot J (1964) "Défense passive" de l'oursin *Arbacia lixula* contre *Marthasterias glacialis*. Vie Milieu 17: 173–176

Vasserot J (1965) Un prédateur d'échinodermes s'attaquant particulièrement aux ophiures: la langouste *Palinurus vulgaris*. Bull Soc Zool Fr 90 (2–3): 365–384

Verlaque M (1981) Preliminary data on some *Posidonia* feeders. Rapp P. Réun Commiss Int Explor Sci Mer Médit 27 (2): 201–202

Verlaque M (1984) Biologie des juvéniles de l'oursin herbivore *Paracentrotus lividus* (Lamarck): sélectivité du broutage et impact de l'espèce sur les communautés algales de substrat rocheux en Corse (Méditerranée, France). Bot Mar 27 (9): 401–424

Verlaque M (1987a) Relations entre *Paracentrotus lividus* (Lamarck) et le phytobenthos de Méditerranée occidentale. In: Boudouresque CF (ed.) Colloque international sur *Paracentrotus lividus* et les oursins comestibles. GIS Posidonie, Marseilles, pp 5–36

Verlaque M (1987b) Contribution à l'étude du phytobenthos d'un écosystème photophile thermophile marin en Méditerranée occidentale. Etude structurale et dynamique du phytobenthos et analyse des relations faune-flore. Thèse Doct, Univ Aix-Marseille 2

Verlaque M (1990) Relations entre *Sarpa salpa* (Linnaeus, 1758) (Téléostéen, Sparidae), les autres poissons brouteurs et le phytobenthos algal méditerranéen. Oceanol Acta 13 (3): 373–388

Verlaque M, Nédélec H (1983a) Note préliminaire sur les relations biotiques *Paracentrotus lividus* (Lmk.) et herbier de posidonies. Rapp PV Réun Commiss Int Explor Sci Mer Médit 28 (3): 157–158

Verlaque M, Nédélec H (1983b) Biologie de *Paracentrotus lividus* (Lamarck) sur substrat rocheux en Corse (Méditerranée, France): alimentation des adultes. Vie Milieu 33 (3–4): 191–201

Verling E, Crook AC, Barnes DKA (2002) Covering behaviour in *Paracentrotus lividus*: is light important? Mar Biol 140: 391–396

Vukovič A (1982) Florofavnistične spremembe infralitorala po populacijski eksploziji *Paracentrotus lividus* (L.). Acta Adriat 23 (1–2): 237–241

West B, De Burgh M, Jeal F (1977) Dissolved organics in the nutrition of benthic invertebrates. In: Keegan BF, Céidigh PO, Boaden PJS (eds) Biology of benthic organisms. Pergamon Press, Oxford, pp 587–593

Whitehead PJP, Bauchot ML, Hureau JC, Nielsen J, Tortonese E (1986) Fishes of the North-Eastern Atlantic and the Mediterranean. UNESCO, Paris

Wright JT, Dworjanyn SA, Rogers CN, Steinberg PD, Williamson JE, Poore AGB (2005) Density-dependent sea urchin grazing: differential removal of species, changes in community composition and alternative community states. Mar Ecol Progr Ser 298: 143–156

Young CM, Tyler PA, Fenaux L (1997) Potential for deep sea invasion by Mediterranean shallow water echinoids: pressure and temperature as stage-specific dispersal barriers. Mar Ecol Progr Ser 154: 197–209

Zavodnik D (1980) Distribution of Echinodermata in the north Adriatic insular region. Acta Adriat 21 (2): 437–468

Zavodnik D (1987) Synopsis of the sea urchin *Paracentrotus lividus* (Lamarck, 1816) in the Adriatic Sea. In: Boudouresque CF (ed.) Colloque international sur *Paracentrotus lividus* et les oursins comestibles. GIS Posidonie, Marseilles, pp 221–240

Zupi V, Fresi E (1984) A study of the food web of the *Posidonia oceanica* ecosystem: analysis of the gut contents of echinoderms. In: Boudouresque CF, Jeudy de Grissac A, Olivier J (eds) International workshop on *Posidonia oceanica* beds. GIS Posidonie, Marseilles, pp 373–379

Edible Sea Urchins: Biology and Ecology
Editor: John Miller Lawrence

Chapter 14

Ecology of *Psammechinus miliaris*

Maeve S Kelly, Adam D Hughes and Elizabeth J Cook

Scottish Association for Marine Science, Oban (UK).

1. APPEARANCE

Psammechinus miliaris (Gmelin) (Plate 3B) is a regular echinoid with some dorso-ventral flattening of the test. The colour varies with habitat (Bull 1939; Lindahl and Runnström 1929; Comely 1979). Shallow or littoral individuals are a deep purplish-brown and show no difference between the colour of the test and the spines. Those from deeper water tend to be much paler in colour, with a light green test and vivid purple spine tips. The external morphology does not change when the deep-water morphotype is maintained in shallow water or vice versa (Gezelius 1962). The horizontal test diameter is typically no greater than 40 mm although larger individuals have been recorded (Mortensen 1943; Allain 1978).

2. DISTRIBUTION

Psammechinus miliaris is found all around the British Isles, as far north as Scandinavia and south to Morocco, but not in the Mediterranean (Mortensen 1943). Its distribution is mainly confined to the southern and eastern parts of the North Sea (Süssbach and Breckner 1911, in Ursin 1960) particularly the Dogger Bank, German and Danish Waddensea and the Southern Bight (Hagmeier and Kändler 1927, in Ursin 1960). Its bathymetrical distribution extends from the littoral zone where populations can be found exposed on boulder shores (Lindahl and Runnström 1929) to depths of 100 m (Mortensen 1943). Ursin (1960) reported *P. miliaris* to have a high tolerance for low temperatures and that it is found in Limfjord, Denmark, where winter temperatures are just above zero. It reproduces in waters around the Faroes where the summer temperatures seldom exceed 11 °C (Ursin 1960).

3. HABITAT

In Scotland *P. miliaris* typically occurs in dense, localised populations in sheltered areas of sea lochs on the west coast (Davies 1989; Holt 1991). Its distribution frequently coincides with that of the brown macroalga *Laminaria saccharina*. *Psammechinus miliaris* is found on the fronds as well as on rock surfaces below the fronds. Some populations are exposed to air at low spring tides. They are found attached to the undersides of rocks, boulders and seaweed and shallowly buried under gravel on the foreshore (Kelly 2000). Individuals from these littoral and subtidal habitat types on the west coast of Sweden were termed, respectively, the Z and S forms by Lindahl and Runnström (1929). The Z forms lived in the 'seaweed region' and were larger and darker than the S forms found at greater depths.

Larsson (1968) reported a clumped distribution pattern for *P. miliaris* in the Gullmar and Saltkälle Fjords, Sweden, but no evidence of a size–depth relationship. Smidt (1951) and Hagmeier and Kändler (1927, in Ursin 1960) as cited in Ursin (1960) found *P. miliaris* to be abundant on rough ground such as oyster banks in the German and Danish Waddensea. Comely (1979) described a population of *P. miliaris* from a bed of *Zostera marina* in a stable-salinity, shallow inlet in Loch Sween, Scotland. These individuals were found at depths of 1–2 m on the bottom mud and attached to the rhizomes. Very young individuals have been found in the holdfasts of *Laminaria* from the Clyde sea, Scotland (Moore 1971). *Psammechinus miliaris* has also been found to settle in large numbers on artificial structures associated with aquaculture such as suspended rope cultures of mussels in Killary Harbour and Bantry Bay in Ireland (Leighton 1995), on suspended scallop cultures in Loch Fyne, Scotland (Cook 1999) and on artificial structures deployed adjacent to salmon cages in the Lynne of Lorne, Scotland (Cook et al. 2006).

Lindahl and Runnström (1929) showed through acclimatisation experiments that the Z (littoral) and S (deeper water) forms have different salinity optima. Gezelius (1962) showed that the eggs from the Z and S forms of *P. miliaris* differed in the range of salinities they would tolerate and over which fertilisation would successfully occur (20–32 and 26–38 ppt at 17–19 °C, respectively). Cook (1999) found that individuals from littoral populations were more tolerant of an increased temperature and decreased relative humidity during aerial exposure than those from subtidal populations. A more recent study from a Scottish sea loch has provided further evidence that there is an ecological basis for the distinction of intertidal and subtidal populations (Hughes et al. 2005). However, the degree of genetic as opposed to environmentally induced adaptation to these conditions is yet to be determined.

4. DENSITY

Population densities of *P. miliaris* on the west coast of Scotland vary with habitat. Densities have been recorded of 18 individuals per 100 g dry weight of sea weed (Bedford and Moore 1985); 34 individuals per square metre in a subtidal *L. saccharina* bed, 182.4 individuals per square metre in a shallow bed of *Zostera* and 28.4 individuals per square metre on adjacent mud surfaces (Comely 1979) and 352 individuals per square metre for littoral populations (Kelly 2000) where individuals in one 0.25 m^2 quadrat ranged from 3.7 to 24.2 mm horizontal test diameter. In the German Wadden Sea Hagmeier and Kandler (1927, in Ursin 1960) found densities of several individuals per square metre. Larsson (1968) found up to 10 individuals per square metre in Saltkälle Fjord, Sweden. A fuller comparison of bottom type would enable an assessment of the influence of refuge availability on population density.

5. POPULATION STRUCTURE

Fjordic and littoral populations of *P. miliaris* range in size from 5 to 35 mm (Larsson 1968; Bedford and Moore 1985). The structure of two adjacent littoral populations of *P. miliaris* (A and B) on the west coast of Scotland was examined over a 2 year period (Kelly 2000). Site A was characterised by a thin layer of pebbles overlying finer sediments with occasional small boulders (<50 cm diameter exposed) and mostly covered by intertidal macroalgae with a narrow zone of *L. saccharina* only exposed on low spring tides. Boulder size and the amount of algal cover were less at site B. Individuals with a test diameter <5 mm in midwinter were assumed to represent the last season's recruits. The population at site A had a high proportion of large individuals (22–38 mm test diameter), whereas the population at site B had no individuals with a test diameter >30 mm. This was attributed to the smaller boulders at this site, offering fewer large refuges. There was evidence of bimodality at both sites with proportionally low representation of the 12–17 mm size classes. This was attributed to a recruitment failure in a previous season. Recruitment occurred at both sites in both seasons studied. The largest number of recruits (<5 mm) collected on any date represented 26% of the population at that site. On every sample date some individuals <5 mm test diameters were recorded, indicating either slow growth rates or a succession of late recruitment.

6. FOOD AND TROPHIC ECOLOGY

Analysis of gut contents or direct observation has shown that *P. miliaris* feeds upon hydroids, worms, echinoderms, crustaceans, diatoms, molluscs and bottom material (Eichelbaum 1909); on fresh plants, dead fragments of *Zostera*, animals on plants especially bryozoans, sponges and hydroids (Blegvad 1914); on small attached algae and *Corallina* (Lindahl and Runnström 1929) and *Rhodymenia, Chondrus, Fucus, Laminaria* (authors as cited in Lawrence 1975). Hancock (1975) reported that *P. miliaris* fed voraciously on cockles, barnacles and ascidians and also grazed boring sponges and worms from oyster shells. Jensen (1969) reported better growth rates in *P. miliaris* feeding on *L. saccharina* with epizoic animals than on clean *L. saccharina*. Faller-Fritsch and Emson (1972, in Lawrence 1975) found the principal food of *P. miliaris* varied markedly with locality and included *Spirorbis, Mytilus* and algae. They concluded that the distribution of *P. miliaris* was more likely to be limited by physical factors than food availability.

Bedford and Moore (1985) found that *P. miliaris* from the Clyde Sea area, Scotland, grazed on sublittoral beds of detached *L. saccharina*. The sea urchins' response to fresh and rotting weed suggested that large *P. miliaris* have different strategies for exploiting these weed types. Fresh weed is relatively difficult to digest. When it is consumed, gut retention times are long which results in high protein absorption efficiencies. When rotting weed is consumed, gut retention time is short and more food passes through the gut. Thus the growth rates of individuals on the two weed types remain equivalent.

Maeaettea and Roedstroem (1989) found grazing *P. miliaris* cleaned oyster trays of fouling organisms so effectively that they needed no further antifouling treatments. Similarly Kelly et al. (1998a) and Cook (1999) found *P. miliaris* effectively grazed a range of fouling organisms both from salmon cage netting and from corrugated PVC collector plates. The grazing activity of *P. miliaris* can profoundly affect benthic community structure. Over a 4 month period when *P. miliaris* were removed from defined intertidal areas, reducing grazing pressure, a significantly different community of encrusting organisms developed on the substrate, compared to sites where there was no reduction in grazing pressure (Hughes et al. 2004). The most notable changes were an increase in the tube worm *Pomatoceros* and ephemeral algae in areas of reduced urchin density. More specifically, controlled studies of foraging rate and preference have shown that when foraging on barnacles, *Semibalanus balanoides, P. miliaris* consistently exerted a preference for small (<7.5 mm basal plate length) barnacles. However, barnacles of basal plate length >12.5 mm were also consumed by *P. miliaris* of test diameter 14–22 mm. On average, *P. miliaris* from two different size classes consumed 8–15 barnacles each day (Otero-Villanueva et al. 2006). Large urchins of 38–46 mm test diameter showed a preference for intermediate sizes (>13.4–17.8 mm shell length) of mussel, ingesting a maximum of six mussels a day and increasing their consumption in a density-dependent manner until reaching a plateau, once more than 10 mussels were offered (Otero-Villanueva et al. in press).

Psammechinus miliaris fed prepared salmon feed had enhanced gonadal and somatic growth rates indicating it can utilise diets rich in animal lipids (Cook et al. 1998). Cook et al. (2000) compared essential fatty acid profiles of gonads of *P. miliaris* fed controlled diets and from wild populations. The proportion of docosahexaenoic acid (DHA), 22:6(n-3), was significantly higher in the gonads of individuals fed salmon feed or collected from suspended rope culture heavily fouled with spat of *Mytilus edulis*. Both salmon feed and mussel tissue contain high proportions of DHA. The proportions of stearidonic acid, 18:4(n-3) and arachidonic acid 20:4(n-6) were proportionately higher in the gonads of individuals fed *L. saccharina* and from littoral locations. *Laminaria saccharina* similarly contained a high proportion of stearidonic and arachidonic acids.

Fatty acid signature has similarly been used as a tool to determine if there were differences in the diets of intertidal and subtidal populations of *P. miliaris*, from the west coast of Scotland. Contrasting populations of *P. miliaris* exhibit significantly different fatty acid profiles (Hughes et al. 2005). Both 20:4(n-6) and 20:5(n-6), associated with brown algae were found in higher levels in subtidal populations, while levels of 22:6(n-3) and 18:4(n-3) were higher in the intertidal urchins. Of these acids 22:6(n-3) is associated with suspension feeding invertebrates, and 18:4(n-3) with green algae. The data show the observed differences in gondal index between intertidal and subtidal *P. miliaris* result from a higher tendency to omnivory in the intertidal.

In a study of the biosynthetic pathways of *P. miliaris* Bell et al. (2001) showed that it has the capability to produce C-20 polyunsaturated fatty acids (PUFAs) from shorter-chain precursor fatty acids. While the urchins were shown to convert 18:3(n-3) to 20:5(n-3), the conversion rate was slow, and no further elongation to 22:6(n-3) was found. It is perhaps surprising that these animals have the ability to synthesise C-20 PUFA when their marine diet typically contains them in abundance, and this suggests an essential role for C-20 PUFA in this species.

7. GROWTH RATES, AGEING AND ENERGY PARTITIONING

From her examination of growth bands in the test, Jensen (1969) concluded that *P. miliaris* completed most of its test growth in 3 years. She reported 4- and 6-year-old *P. miliaris* from Norway and Denmark had 16 and 15 mm test diameters, respectively. Allain (1978) estimated the age of large *P. miliaris* (>45 mm) from Brittany, France as 3 and 4 years and suggested individuals may reach 10 years of age. Gage (1991) reported some *P. miliaris* from Scottish waters of test diameter <30 mm may be 10 years old. However, accurate ageing of older specimens was difficult because of the closely spaced bands on the plate margins (Gage 1991).

Gage (1991) suggested that, contrary to the findings of Jensen (1969), somatic growth is maximal in early spring/summer months and little or no growth occurs after August. The data of Cook et al. (1998), from a laboratory study, support the observation that growth rates slow over winter.

Psammechinus miliaris on a high energy diet maintain high gonad indices out of the reproductive season but have gonads which contain a predominance of nutritive phagocytes rather than developed gametes (Cook et al. 1998; Kelly et al. 1998b). The wild individuals described by Kelly (2000) appear to be food-limited as individuals had very low over-winter gonad indices and gonads that contain a loose network of empty nutritive phagocytes. For such food-limited *P. miliaris*, it is possible to correlate gonad indices and reproductive state with prespawned urchins having higher gonadal indices than postspawned urchins (Hughes et al. 2006). These food-limited individuals likely channel energy to reproductive effort at the expense of somatic growth during the breeding season as postulated for *Paracentrotus lividus* and *Sphaerechinus granularis* (Lozano et al. 1995; Guillou and Lumingas 1999). In a detailed study on the effect of diet, *P. miliaris* were fed a range of diets including one of mussel flesh (*M. edulis*) which was high in protein and carbohydrate, and one of macroalgae (*L. saccharina*) which was low in protein and high in carbohydrate, and showed different strategies of energy partitioning with both diet composition and the size class of urchin (Otero-Villanueva et al. 2004). *Psammechinus miliaris* fed the algal diet showed poorer absorption efficiency and had low assimilation, gonadal and somatic growth rates compared with those fed the diet of mussel flesh. A diet of salmon feed (high protein and high lipid) promoted greater gonadal growth than the other diets. A negative energy balance resulted when large urchins (29–37 mm test diameter) were fed the algal diet. The data support the theory *P. miliaris* is well adapted to feed on other invertebrates, utilising the high-energy diet for greater reproductive success.

Kelly et al. (1998b) showed *P. miliaris* with access to salmon feed pellets, a nutritious food, had higher alimentary indices than local wild individuals suggesting the use of the gut as a storage organ. The alimentary index of *P. miliaris* populations shows a decrease during the spawning period (Kelly 2000) which may reflect the mobilisation of nutrients from the gut wall to help meet the energetic demands of reproduction.

8. REPRODUCTION

The spawning period for *P. miliaris* has been reported to be June to August in the Clyde sea area (Elmhirst 1922); June to October near Bergen, Norway (Runnström 1925); June to October and May to October in West Norway and Denmark (Jensen 1969); and July and August on the west coast of Scotland (Comely 1979). Both Elmhirst (1922) and Brattström (1941) found sexually mature individuals were in their first year. Brattström (1941) reported they had a test diameter of 6–7 mm.

Psammechinus miliaris from two typical but contrasting habitats (littoral and subtidal) on the west coast of Scotland had a defined annual cycle of gametogenesis with a single spawning period (Kelly 2000). The annual cycle was similar to that described for sympatric species, e.g. *Echinus esculentus* and *P. lividus* (Comely and Ansell 1989; Byrne 1990). No secondary spawning events as described for *P. lividus* and Mediterranean *S. granularis* (Crapp and Willis 1975; Guillou and Michel 1993) occurred. The gonad index varies between habitat types, replicate sites within a habitat type and years (Kelly 2000). Gonad indices are maximal prior to the onset of spawning in June and July. High summer gonad indices at littoral sites may result from a seasonal influx of a more nutritious or more abundant food resource. This food resource could be encrusting invertebrates on which *P. miliaris* feeds.

As female *P. miliaris* maintained under short photoperiods do not complete gametogenesis (Kelly 2001), it is likely that *P. miliaris* is a 'lengthening day' species and that photoperiod and not temperature is the primary stimulus for gametogenesis. The gonad fatty acid signature of mature male and female *P. miliaris* is significantly different, ovaries having higher proportions of 14:0, 16:1(n-7) and 18:1(n-9), and testes being higher in 20:4(n-6) and 22:6(n-3). The relative proportions of these fatty acids change over the reproductive cycle of the urchin, suggesting a functional significance, as gametes mature and are shed. (Hughes et al. 2005).

The gonad biomass in wild individuals of *P. miliaris* in Scotland is generally low (Kelly, 2000). Gonad colour varies both with season and with diet. The highest percentages of individuals with gonads of a bright orange occur over the summer months when the gametes mature. The carotenoids responsible for colouring the gonad are likely synthesed from precursors in algal materials (McLaughlin and Kelly 2001).

9. LARVAL BIOLOGY

Larvae of *P. miliaris* are present in the plankton from early spring to December off Denmark (Thorson, 1946) and from July to September in Scotland (Bruce et al. 1963). In culture, the larval phase of *P. miliaris* can be as short as 20 days (Kelly et al. 2000). Planktotrophy begins after approximately 48 h when the stomach forms. With an optimal food ration the rudiment forms on day 14 and the larvae are competent to settle on day 20–21. Food ration affects the morphology of the developing larvae (Kelly et al. 2000). Larvae on a high food ration of the microalgae *Pleurocrysis elongata* reduce larval arm length and are unable to maintain their position in the water column. Larvae fed a low ration fail to develop to metamorphosis. Larval diet type also affects larval morphology, survivorship at metamorphosis and juvenile size (Kelly et al. 2000). Borei and Wernstedt (1939) speculated that the Z (littoral) and S (deeper water) forms were so distinct physiologically and ecologically that recruitment may be restricted to their specific depths.

Immediately postmetamorphosis the young individual consists of the ventral half of the test and an undifferentiated mass of soft tissue (Hinegardner 1969). At this stage it lacks a digestive tract and is more properly referred to as a postlarva. In *P. miliaris* the digestive tract is formed after 5–7 days (Leighton 1995) and the juvenile is then complete and ready to feed. In the laboratory *P. miliaris* will freely hybridise with *E. esculentus*, at fertilisation, and continue to develop to early larval life-history stages. It is possible to raise the progeny (of both crosses; male *P. miliaris* × female *E. esculentus* and female *P. miliaris* x male *E. esculentus*) through metamorphosis to settlement, although in very reduced numbers compared with single species cultures. The resulting offspring have a mixture of the physical attributes (spine length, test shape and colouring) of both parents, however, no such hybrid forms have been observed in wild populations (Kelly, unpubl. obs.).

10. CONCLUSIONS

Psammechinus miliaris occurs in a diverse range of habitats, frequently at high densities, particularly in shallow or littoral locations. Its omnivory is well documented and the observation that a diet rich in encrusting invertebrates leads to high gonad indices in this species has been supported by examination of the fatty acid profile of the gonad tissue. The consistent differences in fatty acid profile between males and females suggest a functional significance which has yet to be elucidated. Similarly, the transformations of ingested carotenoids and the synthesis of those expressed as gonad colour are poorly understood. It is likely the grazing activity of *P. miliaris* has a profound impact on the biodiversity and distribution of subtidal and intertidal encrusting invertebrates as well as the flora. However, field studies that demonstrate the full extent to which *P. miliaris* regulates the structure of these communities are lacking.

REFERENCES

Allain JY (1978) Age et croissance de *Paracentrotus lividus* (Lamarck) et de *Psammechinus miliaris* (Gmelin) des cotes nord de Bretagne (Echinoidea). Cah Biol Mar 19: 11–21

Bedford AP, Moore PG (1985) Macrofaunal involvement in the sublittoral decay of kelp debris: The detritivore community and species interactions. Estuar Coast Shelf Sci 18: 97–111

Bell MV, Dick JR, Kelly MS (2001) Biosynthesis of eicosapentaenoic acid in the sea urchin *Psammechinus miliaris* (Gmelin). Lipids 36: 79–82

Borei H, Wernstedt C (1939) Zur Ökologie und Variation von *Psammechinus miliaris*. Arkiv Zool 28: 1–15

Brattström H (1941) Studien über die Echinodermen des Gebietes zwischen Skagerak and Ostsee, besonders des Oresundes, mit einer Ubersicht über die physische Geographie. Unders över Öresund 27

Bruce JR, Colman JS, Jones NS (1963) Marine fauna of the Isle of Man. LMBC Mem Typ Br Mar Pl Anim 36: 1–307

Bull HO (1939) The growth of *Psammechinus miliaris* (Gmelin) under aquarium conditions. Rep Dove Mar Lab 3 Ser 6: 39–41

Byrne M (1990) Annual reproductive cycles of the commercial sea urchin *Paracentrotus lividus* from an exposed intertidal and a sheltered subtidal habitat on the West Coast of Ireland. Mar Biol 104: 275–289

Comely CA (1979) Observation on two Scottish West coast populations of *Psammechinus miliaris*. SMBA Internal Report, Oban

Comely CA, Ansell AD (1989) The reproductive cycle of *Echinus esculentus* L. on the Scottish West Coast. Estuar Coastal Shelf Sci 29: 385–407

Cook EJ (1999) *Psammechinus miliaris* (Gmelin) (Echinodermata: Echinoidea): Factors effecting its somatic growth and gonadal growth and development. PhD Thesis, Napier University, Edinburgh

Cook EJ, Kelly MS, McKenzie JD (1998) Gonad development and somatic growth in the green sea urchin, *Psammechinus miliaris* fed artificial and natural diets. J Shellfish Res 17: 1549–1555

Cook EJ, Bell MV, Black KD, Kelly MS (2000) Fatty acid compositions of gonadal material and diets of the sea urchin *Psammechinus miliaris*: Trophic and nutritional implications. J Exp Mar Biol Ecol 255: 261–274

Cook EJ, Black KD, Sayer MDJ, Cromey C, Angel D, Katz T, Eden N, Spanier E, Karakassis I, Tsapakis M, Malej A (2006) Pan-European study on the influence of caged mariculture on the development of sub-littoral fouling communities. ICES J Mar Sci 63: 637–649

Crapp GB, Willis ME (1975) Age determination in the sea urchin, *Paracentrotus lividus* (Lamarck), with notes on the reproductive cycle. J Exp Mar Biol Ecol 20: 157–178

Davies M (1989) Surveys of Scottish Sealochs, Lochs A'Chairn Bhain, Glendhu and Glencoul. Marine Nature Conservation Review Report no. 983. JNCC, Peterborough, UK.

Elmhirst R (1922) Notes on the breeding and growth of marine animals in the Clyde sea area (Millport, Marine Biol. St.). Ann Rep Scot Mar Biol Ass 1–47. Appendix to Council Report. Scottish Association for Marine Science, Oban, Argyll, PA34 1QA.

Gage JD (1991) Skeletal growth zones as age markers in the sea urchin *Psammechinus miliaris*. Mar Biol 110: 217–228

Gezelius G (1962) Adaptation of the sea urchin *Psammechinus miliaris* to different salinities. Zool Bildrag Fran Uppsala 35: 329–337

Guillou M, Lumingas LJL (1999) Variation in the reproductive strategy of the sea urchin *Sphaerechinus granularis* (Echinodermata: Echinoidea) related to food availability. J Exp Mar Biol Ecol 79: 131–136

Guillou M, Michel C (1993) Reproduction and growth of *Sphaerechinus granularis* (Echinodermata: Echinoidea) in Southern Brittany. J Exp Mar Biol Ecol 73: 179–192

Hancock DA (1975) The feeding behaviour of the sea urchin *Psammechinus miliaris* (Gmelin) in the laboratory. Proc Zool Soc Lond 129: 255–263

Hinegardner R (1969) Growth and development of the laboratory cultured sea urchin. Biol Bull 137: 465–475

Holt R (1991) Surveys of Scottish Sealochs, Lochs Laxford, Inchard, Broom and Little Loch Broom. Marine Nature Conservation Review Report no. 16. JNCC, Peterborough, UK.

Hughes AD, Kelly MS, Barnes DKA (2004) Sea bed scrapers and shapers: The role of urchins in regulating hard substrate communities of Scottish Sea Lochs. Biol Mar Mediter 11: 100

Hughes AD, Catarino AI, Kelly MS, Barnes DKA, Black KD (2005) Gonad fatty acids and trophic interactions of the echinoid *Psammechinus miliaris*. Mar Ecol Prog Ser 305: 101–111

Hughes AD, Kelly MS, Barnes DKA, Catarino, AI, Black KD (in press) The dual function of sea urchin gonads are reflected in the temporal variation of their biochemistry. Mar Biol 48(4): 789–798

Jensen M (1969) Breeding and growth of *Psammechinus miliaris*. Ophelia 7: 65–78

Kelly MS (2000) The reproductive cycle of the sea urchin *Psammechinus miliaris* (Echinodermata: Echinoidea) in a Scottish sea loch. J Mar Biol Ass UK 80: 909–919

Kelly MS (2001) Environmental parameters controlling gametogenesis in the echinoid *Psammechinus miliaris*. J Exp Mar Biol Ecol 266 (1) 67–80

Kelly MS, Brodie CC, McKenzie JD (1998a) Sea urchins in polyculture: The way to enhanced gonad growth. In: Mooi R, Telford M (eds) Echinoderms: San Francisco. AA Balkema, Rotterdam, pp 707–711

Kelly MS, Brodie CC, McKenzie JD (1998b) Somatic and gonadal growth of the sea urchin *Psammechinus miliaris* maintained in polyculture with the Atlantic salmon. J Shellfish Res 17: 1557–1562

Kelly MS, Hunter AJ, Scholfield C, McKenzie JD (2000) Morphology and survivorship of larval *Psammechinus miliaris* (Gmelin) (Echinoidea: Echinodermata) in response to varying food quantity and quality. Aquaculture 183: 223–240

Larsson BAS (1968) SCUBA-studies on vertical distribution of Swedish rocky-bottom echinoderms. A methodological study. Ophelia 5: 137–156

Lawrence J (1975) On the relationships between marine plants and sea urchins. Oceanogr Mar Biol Ann Rev 13: 213–286

Leighton P (1995) Contributions towards the development of echinoculture in North Atlantic waters with particular reference to *Paracentrotus lividus* (Lamarck). PhD Thesis, National University of Ireland, Galway

Lindahl PE, Runnström J (1929) Variation und Ökologie von *Psammechinus miliaris* (Gmelin). Acta Zool 10: 401–484

Lozano J, Galera J, Lopez S, Turon X, Palacin C, Morera G (1995) Biological cycles and recruitment of *Paracentrotus lividus* (Lamarck) (Echinodermata: Echinoidea) in two contrasting habitats. Mar Ecol Prog Ser 122: 179–191

Maeaettea M, Roedstroem EM (1989) Biological control of fouling organisms in suspended oyster culture on the Swedish west coast. International Aquaculture Conference, Bordeaux

McLaughlin G, Kelly MS (2001) Effect of artificial diets containing carotenoid-rich microalgae on gonad growth and colour in the sea urchin *Psammechinus miliaris* (Gmelin). J Shellfish Res 20: 377–382

Moore PG (1971) A study of pollution in certain marine ecosystems. PhD Thesis, University of Leeds, Leeds

Mortensen TH (1943) A monograph of the Echinoidea. Camarodonta II. CA Reitzel, Copenhagan

Otero-Villanueva M, Kelly MS, Burnell G (2004) How diet influences energy partitioning in the regular echinoid *Psammechinus miliaris*; constructing an energy budget. J Exp Mar Biol Ecol 304: 159–181

Otero-Villanueva M, Kelly MS, Burnell G (2006) Foraging decisions by the regular echinoid *Psammechinus miliaris*. J Mar Biol Ass UK 86: 773–781

Runnström S (1925) Temperatur och utbredning, nagra experimentellt biologiska undersökningar. Naturen, Vol. 49. Bergen, pp 268–288

Smidt ELB (1951) Animal production in the Danish Waddensea. Medd Komm Danm Fisk-og Havunders Ser Fisk 11(6) 1–151

Thorson G (1946) Reproduction and larval development of Danish marine bottom invertebrates. Medd Komm Danm Fisk-og Havunders Ser Plankton 4(1): 1–523

Ursin E (1960) A quantitative investigation of the Echinoderm fauna of the central North Sea. Medd Komm Danm Fisk-og Havunders Ser Fisk 24: 204 pp

Edible Sea Urchins: Biology and Ecology
Editor: John Miller Lawrence

Chapter 15

Ecology of *Echinometra*

TR McClanahan and NA Muthiga

Wildlife Conservation Society, Mombasa (Kenya).

1. INTRODUCTION

1.1. Species of *Echinometra*

The total number of species of *Echinometra* and a species-level taxonomy has yet to be completed, particularly for the central Indo-Pacific (Palumbi 1996). *Echinometra* forms a monophyletic group composed of two clades – the Indo-West Pacific species and eastern Pacific to Atlantic species (Kinjo et al. 2004). Species diversity is likely to show a pattern similar to other tropical taxa, with the highest genetic and species diversity in the west Indo-Pacific and decreasing with the distance from a center somewhere in Indonesia (Palumbi et al. 1997). The existing nomenclature does not coincide well with diversity pattern as only one species of *Echinometra, E. mathaei* (Plate 4C), is reported from the Indo-Pacific (Clark and Rowe 1971). Edmonson (1935) reported two species from Hawaii, *E. mathaei* and *E. oblonga* but, Mortensen (1943) argued, these were morphs of a single species, and Clark and Rowe (1971) did not recognize *E. oblonga*.

Genetic, morphological, biochemical, ecological, and reproductive studies suggest at least four sympatric species of *Echinometra*, currently referred to as *Echinometra* species A, B, C, and D occur in Okinawa (Uehara et al. 1990, 1991; Arakaki and Uehara 1991; Matsuoka and Hatanaka 1991; Nishihira et al. 1991; Palumbi and Metz 1991; Palumbi et al. 1997) and in Fiji (Appana et al. 2004). Additionally, *E. oblonga* is distinct morphologically and gametically incompatible with sympatric *E. mathaei* (Metz et al. 1994). Studies on morphological characteristics (Arakaki et al. 1998; Arakaki and Uehara 1999) and genetics (Palumbi 1996) suggest that *Echinometra* sp. B and *Echinometra* sp. D in Okinawa are *E. mathaei* and *E. oblonga*, respectively. Nonetheless, as various species of allopatric and sympatric *Echinometra* do hybridize (Aslan and Uehara 1997; Rahman et al. 2001, 2002, 2004; Rahman and Uehara 2003; Rahman and Uehara 2004), reproductive isolation is maintained by spatial and temporal segregation of gametes as well as genetic incompatibility, depending on the species.

For example, among the Okinawan *Echinometra*, *Echinometra* sp. A and *Echinometra* sp. D are genetically very divergent species that are maintained by prezygotic isolation. *Echinometra* sp. A and *Echinometra* sp C are most closely related to each other genetically (Matsuoka and Hatanaka 1991), but can be distinguished from each other by differences in adult morphology and microhabitat preferences. Although A and B species produce hybrids that grow to sexual maturity in the lab, intensive surveys have failed to find hybrids in the field, suggesting that other prezygotic isolating mechanisms occur between these two species (Rahman et al. 2001, 2002; Rahman and Uehara 2003). Similarly, prezygotic isolating mechanisms such as habitat segregation as well as species-specific sperm chemotaxis and/or gamete compatibility play a role in maintaining reproductive isolation between *E. mathaei* and *E. oblonga* in Okinawa (Rahman et al. 2004). *E. mathaei* and *Echinometra* sp. A hybrids exhibit hybrid vigor and are potential candidates for aquaculture (Rahman et al. 2005). Studies of the two Caribbean species that live sympatrically show no hybridization (Lessios and Cunningham 1990; McCartney et al. 2000).

Both Caribbean and Indo-Pacific species appeared to have formed during the Pleistocene with speciation associated with sea-level drops (Palumbi 1996; McCartney et al. 2000). The two sympatric Caribbean species of *Echinometra* formed 1.27–1.62 mya (McCartney et al. 2000) and Indo-Pacific *Echinometra* appear to have formed from 1.5 to 3 mya (Palumbi 1996). Indo-Pacific and Atlantic lineages of *Echinometra* split between 3.3 and 4.2 mya, shortly before the rise of the Isthmus of Panama (McCartney et al. 2000). The more broadly distributed Caribbean species, *E. lucunter*, has an Atlantic and Caribbean subspecies (*polypora* and *lucunter* respectively), which were estimated to have separated between 200 000 and 250 000 years (McCartney et al. 2000).

Two other unnamed species of *Echinometra* have not been reported in Hawaii or the western Indian Ocean and are probably restricted to the west Indo-Pacific between Japan (found in Okinawa at 26 °N) and northern Australia (Palumbi 1996). Two species are reported from the Atlantic-Caribbean, *E. lucunter* and *E. viridis,* and one species form the Eastern Pacific, *E. vanbrunti*. *Echinometra insularis* has been reported from the remote Easter Island (Fell 1974, Osorio and Atan 1993). Consequently, we estimate eight common species of *Echinometra* at the global level. Ecological comparisons of these eight (Kelso 1971; Russo 1977; Tsuchiya and Nishihira 1984, 1985, 1986; Neill 1988; Arakaki and Uehara 1991) suggest differences in morphology, behavior, diets, tolerance to environmental stress, and reproduction between and within species based on population density, predators, habitat, particularly water movement, and other local environmental conditions.

1.2. Biogeography and Large-scale Distribution

Echinometra are commonly reported from shallow waters between the average low tide and depths of 10 m. Most species are commonly found in the first few meters of water, but the Caribbean species, *E. viridis*, can be common in submerged patch reefs to depths of 20 m. Most of the ecological scientific literature does little more than describe *Echinometra* as a member of the sub-to-intertidal fauna. Consequently, it is probable the genus exists throughout most of the tropics including the west African coast, St Helena, and Sao Tome (McCartney et al. 2000). Clark and Rowe (1971) did not report *Echinometra* from the western Indian and Pakistan coastline, so this may be one of the very few unsuitable locations.

1.3. Local Distribution and Abundance Patterns

The upper limits to the distribution of *Echinometra*, mean low water, is likely to be caused by desiccation or high water temperatures. The four Okinawa species were shown to differ in their tolerance to environmental conditions (Tsuchiya et al. 1987; Arakaki and Uehara 1991); the two Hawaiian species *E. mathaei* and *E. oblonga* did not (Kelso 1971). Temperatures >40 °C are lethal for *Echinometra* (Tsuchiya et al. 1987), although this probably varies with the local water temperature environment. For example, mass mortality of macrobenthos including *Echinometra* occurred at Hikueru atoll (French Polynesia) due to hydrographic conditions that drastically reduced water exchange in the lagoon (Adjeroud et al. 2001). *Echinometra* burrows are a common feature of the upper margin of rocky intertidal and reef flat zones where they form a distinct zone at mean low water (Schoppe and Werdings 1996). The lower depth limit, in contrast, is likely to be primarily caused by predation (McClanahan and Muthiga 1989). Food is of secondary importance, but may be important for *Echinometra* that feed principally on drift algae in the intertidal zone (Russo 1977).

Echinometra distributions can be patchy with densities varying from 0.1 to 100 individuals per square meter over very short distances. This patchiness is not always easy to explain and generally increases with the spatial scale of the comparison, being less than 50% within reef sites, but being over 150% between reefs. Spatial distribution and abundance patterns also differ among the ecomorphs of *Echinometra*. For example, on a Fijian reef, *E*. sp. A and *E*. sp. C not only preferred different zones (calmer reef flats and high energy crests, respectively), but also demonstrated different degrees of agnostic behavior (marginal to strong cluster tendencies, respectively) (Appana et al. 2004). Patchiness of settling larvae, postsettlement environmental factors such as desiccation, food and crevice availability, and predation, may be responsible for this variability. The degree of agonistic or aggregating behavior of *Echinometra* (McClanahan and Kurtis 1991) may also be involved.

Settlement of small recruits (<15 mm) on shallow coral reefs in Kenya occurs largely on coral rubble and can vary between 0.4 and 42 recruits per square meter per year (Muthiga 1996). Low recruitment densities of around 1 individual per square meter per year are typical as the distribution of these data is not normal, but leptokurtic with high recruit levels as rare events. Muthiga (1996) concluded that, although rare high recruitment events could temporarily elevate the population densities of juveniles, adult *Echinometra* (>15 mm) populations were caused primarily by levels of predation or post–recruitment mortality. A long-term monitoring study of *E. mathaei* on coral reefs in Kenya found that spatial variability in population densities is strongly and positively related to predation levels (McClanahan 1998). In contrast, studies on the intertidal rocky shore of Hawaii suggest that water movement and the availability of algal drift positively influences the density of *Echinometra* as well as the ratio of *E. mathaei* to *E. oblonga* (Russo 1977). The sturdier *E. oblonga* is more dominant with increasing water movement. Chiappone et al. (2002) reported very low densities of *E. viridis* (0.0013 individuals per square meter) and no significant changes in density and distribution in the Florida keys compared to historical records.

Some *Echinometra* exhibit agonistic behavior with individuals defending their burrows or cervices against intruders (Grunbaum et al. 1978; Tsuchiya and Nishihira 1985; Shulman 1990; McClanahan and Kurtis 1991). Others are less aggressive and, at times, form social aggregations (Tsuchiya and Nishihira 1985; McClanahan 1999). As some species can be aggressive in one habitat, environment or population density, but not in another, it is difficult to generalize about behavior based on species alone (McClanahan and Kurtis 1991). At the small scale of habitats, the agonistic species or types have more uniform distributions than the less agonistic species. Recolonization of disturbed areas is slower for the territorial than nonterritorial species (Tsuchiya and Nishihira 1986). This agonistic behavior is primarily defense for predator-free space rather than a defense of food resources and is flexible depending on the abundance of predators (McClanahan and Kurtis 1991). Consequently, both genes and environment appear to interact and influence this behavior and associated distribution patterns in *Echinometra*.

2. POPULATION BIOLOGY AND ECOLOGY

2.1. Reproductive Biology and Ecology

Echinometra reproduces sexually, however, there is no obvious external sexual dimorphism. Genital papillae in *E. mathaei* can be distinguished at sexual maturity when male papillae become long and the female papillae remain stumpy protuberances (Tahara and Okada 1968). In general, the sex ratio of *Echinometra* populations is 1:1 (Pearse and Phillips 1968; Muthiga 1996; Muthiga and Jaccarini 2005) exceptions include a predominance of males of *E. mathaei* in the Gulf of Suez (Pearse 1969) and of females in the west tropical Pacific (Pearse 1968). The mechanism that determines gender in echinoids is not known (Pearse and Cameron 1991), although sex chromosomes have been reported for *E. mathaei* (Uehara and Taira 1987).

In *Echinometra*, the protein bindin influences the attachment of sperm to eggs, and eggs show strong selection to sperm with a bindin genotype similar to their own, suggesting a strong linkage between female choice and male trait loci (Palumbi 1999). Similarly, comparisons between the bindin of sympatric and allopatric populations of *E. oblonga* and *E.* sp. C showed reproductive character displacement between these two species that is driven by selection (Geyer and Palumbi 2003). However, a comparison of bindin polymorphism between *E. lucunter* and *E. viridis* from the Caribbean and *E. vanbrunti* from the eastern Pacific showed that the eggs of *E. lucunter* have developed a strong block to fertilization by sperm of the other two species, whereas eggs of the other two species have not, suggesting that different selection processes act to varying degrees within species (McCartney and Lessios 2004). Additional reproductive selective pressures include shear stress and strain forces that eggs are subjected to as they pass through the oviduct-gonopore that can damage eggs during spawning. Unlike *E. lucunter*, the eggs of *E. vanbrunti* do not experience strain during spawning because the diameter of its eggs is less than the diameter of the oviduct-gonopore at all stages of adult growth (Bolton and Thomas 2002).

Pearse (1969) presents a detailed description of events that occur during reproduction of *E. mathaei* that is similar to other echinoids (Pearse and Cameron 1991). The factors that influence the timing of the reproductive cycle of *Echinometra*, however, are not well understood (Muthiga 1996; Muthiga and Jaccarini 2005). Of the 21 studies carried out on the reproduction of *Echinometra* (Table 1), only three indicate continuous reproduction; *E. mathaei* in Hawaii (Kelso 1971), Rottnest Island (Pearse and Phillips 1968), and in the NW Red Sea (Pearse 1969). However, even these populations have peak periods of spawning.

Pearse (1969, 1974) suggested that *E. mathaei* might have a restricted spawning period in the higher latitudes and continuous spawning throughout the year closer to the equator where the environmental factors, especially temperature, are presumed more stable. However, *E mathaei* at Diani, Kanamai, and Vipingo on the Kenyan coast (latitude 04 °S) has an annual reproductive cycle with gametogenesis commencing in July and spawning peaking in February–March of each year (Fig. 1; Muthiga 1996; Muthiga and Jaccarini 2005). The length of the reproductive cycle from beginning of ova or sperm accumulation to phagocytosis of relic ova or sperm is approximately 6 months, where a similar period has been found for *E. mathaei* at Wadi el Dom in the Gulf of Suez (Pearse 1969) and the eastern coast of South Africa (Drummond 1995).

The seasonal reproductive pattern of *E. mathaei* on the Kenyan coast is closely correlated to seawater temperatures as well as light, which are influenced by the monsoons

Table 1 *Echinometra*: spawning seasons.

Species	Location	Spawning season	References
E. mathaei	S.E. Honshu, Japan	Jan–Sep	Kobayashi (1969)
	Minatogawa (Okinawa Island) Japan	May–Dec	Fujisawa and Shigei (1990)
	Sakurajima (Kagoshima Bay) Japan	Jun–Sep	Fujisawa and Shigei (1990)
	Shirahama (Kii Pen) Japan	Jul–Aug	Fujisawa and Shigei (1990)
	Seto, Japan	Jul–Aug	Onada (1936)
	Sesoko Island, Japan	Sep–Oct	Arakaki and Uehara (1991)
	Hawaii	Continuous	Kelso (1971)
	Gulf of Suez	Jul–Sep	Pearse (1969)
	NW Red Sea	Continuous	Pearse (1969)
	Diani, Kanamai and Vipingo, Kenya	Nov–Mar	Muthiga (1996), Muthiga and Jaccarini (2005)
	Eastern coast South Africa	Jan–Mar	Drummond (1995)
	Rottnest Island	Continuous	Pearse and Phillips (1968)
E. lucunter	Bermuda	Jul–Oct	Harvey (1947)
	South Florida	Jun–Aug	McPherson (1969)
	Puerto Rico	Aug–Oct	Cameron (1986)
	Barbados	Jul–Oct	Lewis and Storey (1984)
	Panama	Continuous	Lessios (1981)
	Brazil	Dec–May	Ventura et al. (2003)
E. vanbrunti	Panama	Sep–Nov	Lessois (1981)
E. viridis	South Florida	Jun–Aug	McPherson (1969)
	Puerto Rico	Sep–Nov	Cameron (1986)
	Panama	Apr–Dec	Lessios (1981)

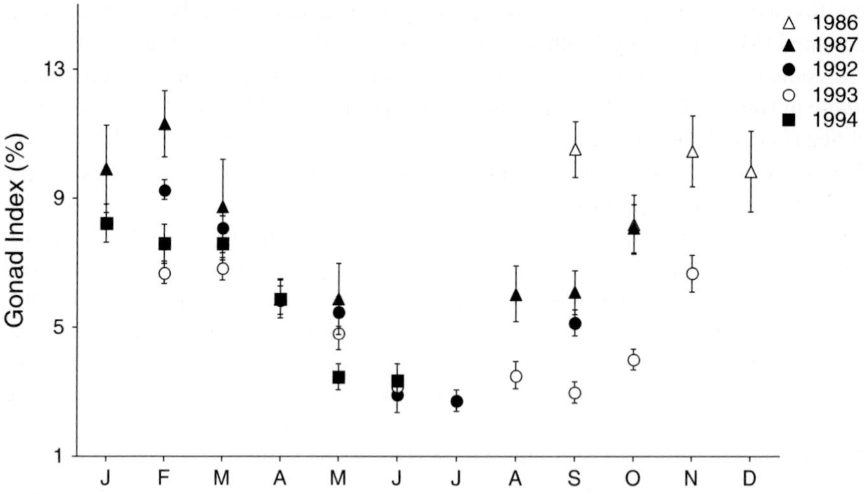

Fig. 1. Seasonal changes in the gonad indices of *Echinometra mathaei* at Kanamai, Kenya (Muthiga and Jaccarini 2005).

(McClanahan 1988). Interestingly, *E. mathaei* populations at Rottnest Island, western Australia, spawn continuously throughout the year in seawater temperatures that are cooler than on the Kenyan coast. Environmental variation may be the key factor in *E. mathaei* rather than an absolute temperature minimum or threshold for spawning. Availability of food for the larvae may be an important factor in controlling spawning of *E. mathaei* along the Kenyan coast as spawning occurs just prior to the peak in phytoplankton concentration. Availability of food for newly metamorphosed young is an important factor in *E. vanbrunti*, which spawns during the dry season upwelling (Lessios 1981). Other factors that play a modifying role include population density and the physical environment. Lessios (1981) showed that spawning synchrony is tighter in sparsely populated populations of *E. viridis* and less important in densely populated populations of *E. lucunter*. He postulated that salinity could act as a proximate cue for controlling the timing of spawning as it coincided with increased salinity for *E. lucunter* and *E. viridis* at Fort Randolph, Caribbean. Spawning also occurs in *E. lucunter* during a discrete period from July to October in wave-swept areas, and occurred during several periods in populations living in calm waters (Lewis and Storey 1984). In Brazilian populations of *E. lucunter*, spawning occurred in the late spring and early summer although differences occurred in the length of the spawning period, being longer in populations on coral reefs than on rocky shores (Ventura et al. 2003). Lunar periodicity in spawning has not been found in *E. mathaei* (Muthiga, 1996; Muthiga and Jaccarini 2005) and *E. viridis* (Lessios 1991).

2.2. Feeding Ecology

Echinometra has two basic feeding modes that vary with species and environment: (1) catching algal drift and (2) benthic grazing. Probably most individuals of all species use both modes at any time. It is not always easy to discern the source of food for

Echinometra, as guts are often full of masticated algae and sediments. *Echinometra* with little sediment in the gut are probably feeding on drift algae, but may be eating abundant fleshy algae in areas with low sedimentation. Feeding on drift algae is most common in shallow waters along shorelines, but also occurs in areas with currents such as reef channels, or shallow tops of patch reefs, or reef flats. Additional work is needed to determine when and where the various species feed on benthic versus drift algae.

Calcium carbonate sediments are usually the largest fraction of the gut content of *E. mathaei*, being between 65% and 95% of the gut (Black et al. 1984; Downing and El-Zahr 1987; McClanahan and Kurtis 1991). These measurements support the conclusion that benthic epi- and endolithic algae are the major source of food for *Echinometra* (Odum and Odum 1955). These algae are mostly small, fast-growing filamentous turf-forming algae, generally green and blue-green algae that grow in and on the surfaces of coral rock. Many herbivorous sea urchins will feed on animals if given the choice. This is true for *Echinometra*, but it has lesser preference for animal food compared to other sea urchins (McClintock et al. 1982). Nonetheless, *Echinometra* is a generalized herbivore, feeding on a variety of macrophytes, including seagrass, but occasionally consumes benthic animals, such as sponges and corals, as well as algae.

2.3. Energetics

Echinometra has low energy requirements that can be lowered further during starvation conditions (McClanahan and Kurtis 1991). One reason for the low energy requirement is the low level of organic matter in the body. *Echinometra mathaei* are reported to be 55% water and 40% spines and test or largely calcium carbonate, with only 5% of its body being organic matter (McClanahan and Kurtis 1991). *Echinometra* respire only about 0.1% of their wet weight per day and can survive up to 170 days without food (McClanahan and Kurtis 1991; Lawrence and Bazhin 1998). Daily consumption of organic matter is between 2.5% and 4% of the wet body weight (McClanahan and Kurtis 1991; McClanahan 1992). Much of the ingested organic matter, around 50%, is defecated and not absorbed (Mills et al. 2000). Consequently, around 25% of the body's organic matter is absorbed daily, although this may vary with diet and species. This aspect of their energetics has yet to be investigated.

Since respiration is a small part of the energetic demands, other biological processes, largely reproduction, is a major part of *Echinometra* energy expenditures. In the absence of food, reduced reproduction is likely to be a major adaptation. However, Muthiga (1996) found that gonads are often conserved relative to other body parts. Increased jaw size (Black et al. 1984; McClanahan and Kurtis 1991; Muthiga 1996; Ventura et al. 2003), reduced body size (Muthiga 1996), lowered respiration (McClanahan and Kurtis 1991), and reduced lipid content of the gut (Lawrence 1970) are frequently observed responses to reduced food. Consequently, reproduction is one of the most important activities of *Echinometra*, and is conserved in the face of reduced energy inputs.

Echinometra can reach a wet biomass level of around $1.1\,\mathrm{kg\,m^{-2}}$, which is among the highest reported values for an herbivore (McClanahan 1992). Experimental (McClanahan and Kurtis 1991) and energy-based simulation model studies (McClanahan 1992, 1995) suggest that the high net production of shallow-water turf algae and low per capita consumption rates of *Echinometra* are responsible. Modelling studies suggest that low resource consumption rates make populations of *Echinometra* tolerant of starvation and resource competition. Low consumption is associated with slow growth, particularly if growth is channelled toward reproduction rather than somatic growth, which is the case for *Echinometra*. *Echinometra* tolerates resource levels that are lower than many other herbivores (McClanahan et al. 1994), particularly bony fish, which have consumption rates ~3–5 times higher (McClanahan 1992). This makes *Echinometra* tolerant of environmental stress and competition, but intolerant of predation and harvesting. High levels of reproduction and larval recruitment allow *Echinometra* to quickly colonize disturbed areas, but slow growth decreases the time to reach a maximum biomass. These energetic conditions enable *Echinometra* to become dominant in environmentally stressful conditions with low predation. The same life history and energetic characteristics would, however, cause a rapid overharvesting such that short intense harvesting periods would require long periods for their biomass to recover.

2.4. Growth and Longevity

Growth estimates of *Echinometra* have been reported from a variety of sources including estimates from size distributions (Kelso 1971; Drummond 1994; Muthiga, 1996), from natural growth lines on test plates (Ebert 1988), from tagging (Ebert 1975), and from observations of growth in aquaria (Lawrence and Bazhin 1998). In general, *Echinometra* has a moderate, although variable, growth relative to other species of sea urchins (Ebert 1975). Estimates of growth for adults range from constant K of 0.19–0.46 with maximum lengths between 30 and 85 mm (Table 2). Few studies have been done for different sizes or growth stages of *Echinometra*. Differences in the size of individuals used in the reported studies partly explain the high variation in reported growth constants. Like many echinoids, however, growth is relatively rapid during the juvenile stages and slows down as the individual approaches the asymptotic size.

Table 2 *Echinometra*: Number of individuals sampled (N), growth (K), maximum lengths (L_{inf}) in mm, and mortality estimates (M).

Species	Location	N	K	L_{inf} (mm)	M	Reference
E. mathaei	Lani Overlook, Hawaii		0.46	38.1		Ebert (1975)
		77	0.29	40.1		Ebert (1982)
	Diani Kenya	248	0.32	62.0		Muthiga (1996)
	Kanamai, Kenya	175	0.25	71.1		Muthiga (1996)
	Vipingo, Kenya	189	0.19	84.9		Muthiga (1996)
	Kapapa Island Oahu				0.05	Kelso (1971)
				55.0	0.20	Ebert (1975)
	South Africa		0.32			Drummond (1994)
E. oblonga	Lani overlook, Hawaii		0.40	32.2		Ebert (1975)
		35	0.36	30.9		Ebert (1982)
	Kapapa Island, Hawaii				0.07	Kelso (1971)
	Kapapa Island Hawaii			50.0	0.18	Ebert (1975)

Juvenile *E. mathaei* on the Kenyan coast increased by 325% in length from 4 to 17 mm (horizontal test diameter) in 8 months (Muthiga 1996), while recruits fed algal turf in aquaria increased by 896% in weight from 118 to 1175 mg in 6 months (Lawrence and Bazhin 1998). Muthiga (1996) found that *E. mathaei* approached an asymptotic growth at around 5–7 years. *E. mathaei* is reported to live up to 8–10 years along the east African coast (Drummond 1994; Muthiga 1996) and reach a maximum size of 85 mm diameter in Vipingo, Kenya (Muthiga 1996). The size at first reproduction has been reported for *E. mathaei* as 12 mm in South Africa (Drummond 1994).

Growth in *Echinometra* is highly dependent on food availability (Muthiga 1996). Comparisons of three populations of *E. mathaei* on the Kenyan coast with differing levels of food availability support the allocation model where resources are appropriately allocated to growth, reproduction, and maintenance depending on food availability (Gadgil and Bossert 1970). Individuals from Vipingo, the reef with highest food availability, grew faster, were larger, and had larger gonads than those at Kanamai or Diani, which were food-limited reefs (Muthiga 1996). Diani reef, the most food-limited, had the smallest individuals and maximum sizes, but these individuals had larger gonads than those at Kanamai indicating that sea urchins at Diani allocate resources to reproduction at the expense of body growth. This study also showed that *E. mathaei* attains sexual maturity at a larger size and has larger gonads when food is not limiting, compared to conspecifics in food-limited environments (Muthiga 1996).

Growth and allocation of resources to key functions are also affected by population density, as the amount of resources available for growth will be constrained by competition with conspecifics. Individuals at Diani, the reef with the highest population density not only had the smallest body size but the largest jaws relative to body size, an indication of food limitation (Ebert 1980). Large jaws relative to body size were also reported for *E. mathaei* on Rottnest Island in Australia (Black et al. 1984). *E. mathaei*, like other echinoids (Ebert 1980), has morphological plasticity with individuals allocating more resources to various body parts depending on prevailing conditions. For example, more resources are allocated to strengthen body wall under rough hydrodynamic conditions thereby enhancing individual survivability (Drummond 1994).

2.5. Pelagic Larval Dynamics

Pelagic larvae of the family Echinometridae can be distinguished from all other urchins by the presence of a double recurrent rod (Pearse and Cameron 1991). Larval structure and development of *E. lucunter and E. mathaei* were described by Mortensen (1921, 1937) and Onada (1936). Several factors including temperature, salinity, and food availability may affect the rate of larval development, but few field data are available. Although the eggs of *E. mathaei* can be fertilized normally at 28–36 °C normal development only occurs at temperatures below 34 °C (Rupp 1973). Estimates of the larval period have been reported for *E. lucunter* (23 days), *E. vanbrunti* (18 days) ,and *E. viridis* (30 days) (Cameron 1986; Emlet et al. 1987).

The rate of larval development in laboratory studies on *Echinometra* is strongly affected by diet (Emlet et al. 1987). Larvae of *E. lucunter* increased growth with increased food ration (Metaxas and Young 1998). Poor dietary conditions affect the shape, the densities, and locomotory abilities of echinoid larvae eventually affecting recruitment (Strathmann 1987; Metaxas and Young 1998). *Echinometra lucunter* and *E. mathaei* (Cameron and Hinegardner 1974) and *E. viridis* (Cameron 1986) have been reared in laboratory to metamorphosis.

Events that occur at or near the time of echinoid settlement and the first few days of benthic life may cause significant mortality, but the causes are not well understood. Recruitment is usually reported for juvenile *Echinometra* from year–class determination of size frequencies (Ebert 1975), and also from the repeated census of small juveniles (2–4 mm test diameter) (Muthiga 1996). The first visible recruits of *Echinometra* are usually closely correlated, although offset by 1–2 months, after their peak spawning period. Caribbean species, *E. lucunter* and *E. viridis,* show seasonality, spawning in August–September and recruiting in September–October (Cameron 1986). Although no recruitment pulse was observed for *E. lucunter* in Puerto Rico, *E. viridis* recruited in several seasons (Pearse and Cameron 1991). Interestingly, at Rottnest Island, western Australia, *E. mathaei* recruit seasonally despite continuous reproduction indicating that not all months have the right environmental conditions for successful larval development and recruitment (Watts et al. 1990).

Echinometra shows temporal and spatial variability in recruitment. Comparisons of recruitment on three different reefs between 1992 and 1994 on the Kenyan coast showed high spatial and temporal variability for *E. mathaei* (Muthiga 1996). There was no significant relationship between population density and recruitment, as the reef with the highest population density showed similar recruitment levels to the reef with the lowest population density. However, there was a positive relationship between recruitment and annual seawater chlorophyll concentration in this study. Interannual variation in water-column productivity should, therefore, be expected to increase the success of larvae and the numbers of benthic recruits. As highly successful recruitment years and subsequent population density of adults were not related, mortality in the early postsettlement stage was a better predictor of adult populations (Muthiga 1996).

2.6. Benthic Population Dynamics

Long-term population dynamic studies have been undertaken in the Persian Gulf (K Allen, pers. comm.) and the fringing-reef lagoon of southern Kenya (McClanahan and Kurtis 1991). *Echinometra* population densities on patch reefs of the Persian Gulf have shown a steady rise since 1987, most pronounced after the 1998 El Niño loss of coral and increased turf algae, probably explain the increase in populations of *E. mathaei* in this environmentally stressful and predator-depauperate environment.

Populations on Kenyan reefs show marked differences between shallow reef lagoons, ranging from 0 to 300 individuals per 10 m^2. This has been largely attributed to varying levels of postsettlement predation in these reefs (McClanahan and Muthiga 1989; Muthiga 1996). Population on most reefs has been fairly stable (<10% annual variation) in reefs with moderate population densities (10–70 individuals per 10 m^2), but at the low and the high ends, population densities are more variable (McClanahan 1998). At low population densities (Malindi), high predation and variable recruitment result in high total variation of population estimates (188%) of which nearly 140% is annual variation. As almost no *E. mathaei* >2 cm in diameter occurred in this reef , this variation was largely due to larval recruitment variability and rapid rates of predation on the recruits. At the high end of population densities (Diani) the numbers increased steadily from 50 to 300 individuals per 10 m^2 from 1970 (Khamala 1971) to 1992 and then a subsequent decline stabilizing at 150 individuals per 10 m^2 (~5000 kg ha^{-1}) (McClanahan and Kurtis 1991). This change is most likely explained as a response to reduced predation on these reefs over the early part of this period, caused by high levels of fishing, followed by a compensation and stabilization at the energetic maximum (~5000 kg ha^{-1}) of *E. mathaei* (McClanahan 1992).

The Kenyan study indicates that *E. mathaei* has more stable and predictable populations, both in space and time, than other sea urchin species (McClanahan 1999). Both life history and community ecology forces are probably responsible. These include continuous reproduction during the optimal spawning and larval growth period, relatively high levels of recruitment to the benthos for a tropical species, flexible diet, low energetic requirements and the ability to resist starvation, and defense of burrow space. McClanahan (1999) suggested that *E. mathaei* may also be the top competitor in the sea urchin guild in shallow low-predation environments and that its populations are, therefore, less affected by other sea urchin competitors. Nonetheless, this stability does not imply an inability to respond to changes in predation and food availability. High reproduction, settlement, and flexible diet should make it quite responsive to changes in resources. As evidence for this is largely circumstantial, such as the population surges and stabilization described above, more experimental studies at the appropriate scale (>1 year and >1 m^{-2}) of population dynamics are required.

3. COMMUNITY ECOLOGY AND COEXISTENCE

Echinometra can achieve high population densities, wet weights temporarily reaching 11 000 kg ha^{-1}, and consume a large fraction of the benthic primary production. Because *Echinometra* are only seldom harvested in sufficient quantities to influence their populations, they can be considered pests at times of high population abundance. *Echinometra* consume much of the net benthic production that might otherwise be consumed by other commercial species such as herbivorous rabbitfish, parrotfish, and surgeonfishes. They can erode reef substratum, reducing its complexity and suitability to other coral reef species. Consequently, the factors that control their populations, allow them to coexist with other herbivorous species, and keep them as a more minor part of the food web are important for marine habitat and fisheries management.

3.1. Competitive Interactions with Other Herbivores

Large-scale population reduction experiments (50 m × 50 m) that reduced *Echinometra* numbers in Diani, Kenya, resulted in increased abundance of algae, reduced coral abundance, and increased herbivorous and invertivorous fish numbers (McClanahan et al. 1996). Small-scale reductions (5 m × 5 m) of *Echinometra* did not produce large changes in the benthos other than increased nonencrusting algae in one reef flat site with initially low (~10%) erect algal cover, but less change on a reef that already had high (~70%) erect algal cover (Prince 1995). Consequently, the influence of *Echinometra* on the local benthic and associated species may depend on the levels of food and whether they are feeding on drift or benthic algae (McClanahan 1999). Under most conditions of low population densities and with low energy requirements, *Echinometra* is probably not competing strongly with other large herbivores. At high population densities it may be dominant in the consumption of benthic algae.

3.2. Competitive Interactions with Other Sea Urchins

Most species of *Echinometra* defend their burrows against other *Echinometra* and other species of sea urchin such as *Diadema* (Grunbaum et al. 1978; Neill 1988; McClanahan 1988; Shulman 1990). Usually the inhabitant expels the intruder, but not necessarily if differences in body size are large (Grunbaum et al. 1978; Neill 1988; McClanahan and Kurtis 1991). *Echinometra* are able to do this because of spatial vision that is oriented toward a target with an angular width of 33° allowing the animal to leave and return to small dark shelters (Blevins and Johnsen 2004). Removal of *Echinometra* also results in recolonization by neighboring conspecifics and is faster for the less agonistic forms (Neill 1988). Small-scale reductions of *E. mathaei* on the tops of coral heads resulted in a rapid, but partial recolonization of the heads by smaller individuals (McClanahan and Kurtis 1991). This suggests that smaller individuals hide and occupy suboptimal locations as a result of intraspecific competition with larger individuals. These intraspecific competition studies may be less applicable to *Echinometra* that have less agonistic behavior, or feed on drift algae.

Interspecific competitive interactions occur between *Echinometra* and other genera, largely *Diadema* (Williams 1981; McClanahan 1988; Shulman 1990). Studies of resource competition are fewer and less informative about interspecific competition (Lessios 1988). Williams (1981) found that *Diadema* and *Echinometra* responded to changes in their numbers by either increasing or decreasing to compensate for the direction of the manipulation. In general, *Echinometra* manipulations had stronger effects than the *Diadema* manipulations. The three-spot damselfish (*Eupomacentrus planifrons*) may mediate this interspecific competition because it is aggressive toward *D. antillarum,* but not *Echinometra* (Williams 1979, 1981).

Williams (1980) and Shulman (1990) suggested that *D. antillarum* is the superior resource competitor over *Echinometra* in the Caribbean. Williams (1980) suggested that preferential attacks by the three-spot damselfish on *D. antillarum,* prevents *Echinometra*, from being competitively excluded. This hypothesis, while intriguing, is probably not correct. Lessios (1988) did not find population increases in other sea urchins, including *Echinometra*, after the mass mortality of *D. antillarum* in the early 1980s. *Diadema antillarum* is unlikely to competitively exclude *Echinometra* because it has a larger body and spine length that will prevent it from occupying *Echinometra* crevices. *Diadema antillarum* may bite the spines of intruding *Echinometra* but it does not push competitors (Shulman 1990) and is unlikely to exclude the agonistic forms of *Echinometra*. At most, *D. antillarum* may restrict *Echinometra* to their crevices. High predation on *Echinometra* outside the crevices (McClanahan 1988, 1999) is more likely to restrict *Echinometra* to crevices than competitive interactions with *Diadema*.

In contrast to the Caribbean studies, crevice competition studies in Kenya found that *E. mathaei* is a superior interference competitor over both *D. setosum* and *D. savignyi* (McClanahan 1988; McClanahan and Shafir 1990). For shallow fringing reefs with low levels of predators, *E. mathaei* lives outside of cervices and other species are found only at very low population densities (McClanahan 1998). This suggests that *E. mathaei* may be the superior resource competitor. Experimental reduction of *Echinometra* populations in Diani, Kenya resulted in an increase in the other species, particularly *Tripneustes gratilla* (McClanahan et al. 1996; unpubl. data). A survey of sea urchins on patch reefs of northern Tanzania found that *E. mathaei* was uncommon and the large sea urchin, *Echinothrix diadema,* was often numerically dominant when predation was low (McClanahan et al. 1999). These studies suggest that *Echinometra* may be the competitive dominant in some of the habitats such as shallow fringing reefs and rocky shores, but that *E. diadema* may dominate in other habitats like patch reefs or reefs with unstable rubble substratum where *Echinometra* burrows are difficult to create.

Competition between *Echinometra* and other grazing invertebrates is likely but not well studied. Prince (1995) found an increase in the herbivorous snail, *Turbo intercostalis*, in plots from which *Echinometra* had been removed and suggested competition, but the experimental design lacked statistical power to be certain about this interaction. Positive associations between *Echinometra* and other invertebrates may occur. Crabs, fish, and brittlestars are associated with *Echinometra* burrows (Schoppe and Werdings 1996).

3.3. Competitive Interactions with Fish

No interspecific competition between fish and *Echinometra* has been reported other than the studies of Williams (1979, 1980, 1981), which found low levels of aggression toward *Echinometra*. Consequently, interference competition between *Echinometra* and fish is probably not very common. Resource competition may be common when *Echinometra* reaches high population levels as indicated by slightly increased herbivore fish abundance (damselfish and parrotfish) and grazing rates in areas where *Echinometra* populations were reduced (McClanahan et al. 1994, 1996).

3.4. Predation and Predators

A number of marine fish and invertebrates as well as shorebirds feed on *Echinometra* in the intertidal zone (Schneider 1985). Fish predators have been identified from the gut contents (Randall 1967) and from direct observations of baited sites (McClanahan 1995, 1999, 2000). It is difficult to distinguish predators of already opened carcasses from analysis of gut contents partly because the number of scavenger species is considerably larger than the number of predatory species (McClanahan 1995, 1999, 2000). The main fish predators include triggerfish, wrasses, emperors, porgies, pufferfish, and grunts. Only a few species in each family are likely to make *Echinometra* a significant part of their diet. Many species may feed on small individuals, although this has been poorly studied, but far fewer feed on adults. A study of predators on large *E. viridis* at an atoll in Belize, found only four common predators, namely the jolthead porgy, the ocean and queen triggerfishes, and the hogfish (McClanahan 1999). The Spanish grunt and porcupinefish are also likely to be predators in the Caribbean (Randall 1967). Ten fish species acted primarily as scavengers (McClanahan 1999), but studies of gut contents suggest a much larger number (Randall 1967).

Eight common predators of *E. mathaei* occur in Kenya, of which one, the red-lined triggerfish (*Balistapus undulatus*), is responsible for nearly 80% of the observed predation (McClanahan 2000). The other predators were mostly *Coris* and *Cheilinus* wrasses, two other less common triggerfish, and an emperor. The red-lined triggerfish is aggressive and dominates most other predators with the exception of the larger titan triggerfish (*Balistiodes viridescens*). Dominance of predation by the red-lined triggerfish increased with time since fishing was eliminated in marine parks in Kenya (McClanahan 2000). Consequently, it is likely that the red-lined triggerfish is the dominant predator in the unfished reefs. Both studies suggest the numbers of predators is considerably fewer than the scavengers, and that gut-content studies can overestimate species that kill and have significant effects on *Echinometra* populations.

Invertebrates, mostly snails and starfish, also prey on *Echinometra*. Their effect on *Echinometra* populations is probably much less than the predatory fishes (McClanahan and Muthiga 1989; McClanahan 1999). Less than 10% of the predation on tethered *Echinometra* in Kenya was attributable to snails and starfish, with each group contributing an estimated 5% of the predation in fished and a negligible amount in unfished reefs (McClanahan and Muthiga 1989). Consequently, heavy fishing may cause invertebrates to play a more important role as predators.

In the western Indian Ocean, the bullmouth helmet shell, *Cypraecassis rufa*, is the most dominant snail predator of *Echinometra* (McClanahan and Muthiga 1989). About 7% of the predation in the Belize study could be attributed to a helmet shell, probably *Cassis madagascariensis* (McClanahan 1999). Observation that large generalist starfish such as species in the genera *Culcita* and *Protoreaster,* and other invertebrates including lobsters, crabs, and sea anemones opportunistically feed on *Echinometra*, are anecdotal and probably do not control the abundance of *Echinometra*. The proliferation of sea urchins and *Echinometra* on many fished reefs can largely be attributed to the loss of bony fish, particularly benthic-feeding triggerfish, rather than the collection of ornamental snails or invertebrates (McClanahan and Muthiga 1989; McClanahan 2000). Triggerfish, although seldom a target species, appear to be aggressive in taking bait and are, therefore, highly vulnerable to fishing and have slow recovery rates when fishing ceases (McClanahan 2000).

4. HERBIVORY AND GRAZING EFFECTS

4.1. Herbivory

Echinometra is often the dominant herbivore of benthic algae in shallow intertidal areas (McClanahan and Kurtis 1991; Prince 1995), but its effects on algae and plants is likely to be minimal when it feeds on drift algae (Prince 1995; McClanahan 1999). *Echinometra* like other reef herbivores (Choat 1991), probably feeds preferentially on early successional turf algae, avoiding late successional and less palatable forms of fleshy algae. In late successional algal communities, *Echinometra* may turn to drift algae as the main diet and, therefore, may vary in its dependence on algae based on the state of the algal ecological succession. Early successional algae dominate most reefs and rocky shores, so the benthic herbivore mode is likely to be its most common ecological role. As a generalized herbivore, *Echinometra* may feed on many species and ecological types of algae, but its ability to influence the abundance of specific algal or plant species is only partially understood.

Coral reefs are estimated to have a net benthic production of around 2 g carbon per square meter per day (Larkum 1983) or ~20 g wet organic matter per square meter per day (Atkinson and Grigg 1984). As individual *Echinometra* consume about 1 g organic matter per day, it is possible that the grazing influence of *Echinometra* is great. At a density of around 20 *Echinometra* per square meter all net production would be consumed by *Echinometra.* About half of this would be defecated. Some of the faeces would be recycled by grazers and some lost to the water column. *Echinometra* populations can be more than 100 m^{-2} in some reef flats. Allochtonous production of drift algae is likely to be a major resource for *Echinometra* at this density. The effects of *Echinometra* on algae and plants will therefore be influenced by population numbers and mode of feeding.

Experimental reduction of *Echinometra* populations at around 25 individuals per square meter were undertaken in an area of Diani Beach, Kenya and resulted in a many-fold increase in the brown algae *Sargassum, Padina,* and *Turbinaria* on rocky coral rubble bottoms and the sea grass *Thalassia hemprichii* on sandy bottoms (McClanahan et al. 1996). The effect on sea grass was more surprising than the expected effects on the rocky bottoms as sea grass is only a small portion of the *Echinometra* diet in this reef (McClanahan, pers. obs.). Sediment accumulated and filled in the spaces between rock and rubble areas, which in the absence of *Echinometra,* probably improved conditions for sea grass colonization. This suggests that benthic grazing *Echinometra* can cycle sediments, reduce their accumulation rates, and keep rocky or rubble surfaces from being buried in sand and colonized by sea grass. Consequently, *Echinometra* effects on plants are not always directly related to feeding preferences. This study of the high-population-density benthic grazing mode of *Echinometra* suggest that a number of mid to late successional brown algae are suppressed by *Echinometra,* as would be expected from their energy demands.

4.2. Erosion of Calcium Carbonate

The observation that 60–95% of the gut of *Echinometra* that lives in excavated burrows is calcium carbonate sediments, has prompted study of the role of *Echinometra* in the inorganic carbon cycle of coral reefs (Ogden 1977; Russo 1980; Bak 1990; Downing and El-Zahr 1987; McClanahan and Kurtis 1991; Peyrot-Clausade 1995; Mokady et al. 1996; Conand et al. 1997; Peyrot-Clausade et al. 2000). Pennings and Svedberg (1993) suggested that calcium carbonate does not deter the feeding of *Echinometra* but is detrimental to other coral reef grazers. Some erosion by *Echinometra* is done through scraping coral rock with teeth and spines. However, some of the sediments in their guts are from already

Table 3 Rates of bioerosion by *Echinometra* (kg m^{-2} yr^{-1}) in Pacific and Caribbean reefs.

Species	Location	Density (individuals per square meter)	$CaCO_3$ erosion rates (kg m^{-2} yr^{-1})	References
E. mathaei	Diani, Kenya	1.7–14	3.8	McClanahan and Muthiga (1988)
	Marshall Islands	2.8	3.2	Russo (1980)
	Reunion	3.8–73.6	0.4–8.3	Conand et al. (1997)
	French Polynesia	7.38	0.4	Bak (1990)
		7–10	0.6–7.5	Peyrot–Clausade et al. (2000)
E. lucunter	Virgin Islands	92–176	3.9	Ogden (1977)
E. viridis	Puerto Rico	0.8–62.6	1.1–4.1	Griffin et al. (2003)

eroded sediments. The ratio of new to previously eroded sediments has been a problem for estimating the amount of actual or new erosion rates.

Estimates of rates of erosion of coral rock by individual *E. mathaei* vary from 0.1 to 0.4 g per individual per day (McClanahan and Muthiga 1988; McClanahan and Kurtis 1991; Mokady et al. 1996), although, Downing and El-Zahr (1987) reported values up to 1.4 g per individual per day. Consequently *E. mathaei* may erode between 1 and 8 kg calcium carbonate per square meter per year at some high population densities (Table 3), or considerably more than typical calcium carbonate deposition rates for coral reefs of between 1 and 4 kg per square meter per year (Smith 1983). The bioerosion rates for the Caribbean species *E. lucunter* were within similar ranges to *E. mathaei* (Ogden 1977), but the rates for *E. viridis* were low with individual rates of 0.2 g per individual per day and an annual rate of 0.1–4 kg calcium carbonate per square meter per year (Griffin et al. 2003). The complexity of reef environments can be increased by bioerosion, but it may be greatly reduced when erosion exceeds accretion. Consequently, the effects of *Echinometra* on the inorganic carbon cycle can be large as well as their influence on the topography of the substratum.

5. CONCLUSION

Echinometra are the most ubiquitous and abundant shallow-water sea urchins in the tropics. Their small body size, oblong body shape, flexible behavior and diet, high reproduction and recruitment rates, and low resource requirements are life-history properties that allow them to survive in cryptic and environmentally stressful environments. *Echinometra* are flexible in their allocation of resources to different body parts and in reproduction and feeding. This allows them to adapt to variable environmental conditions. They are conservative in the maintenance of gonads and reproduction, lack a lunar cycle, and are variable for seasonal cycles. Behavior and feeding vary with and between species.

Physical isolation mediated by sea-level changes, rather than differences in reproductive behavior, is probably the main mechanism for speciation.

The small body size and spine length of *Echinometra* make them highly susceptible to predators and they have a number of fish and invertebrate predators that feed on them when exposed, particularly bottom-feeding durophagous triggerfish and wrasses. Opened carcasses are fed on by a wide variety of fishes. *Echinometra* can become very abundant, dominant herbivores and major eroders of reef–rock substrata in reefs and rocky shores with stable substratum and few fish predators. About half of the food consumed will not be absorbed and becomes detritus. Most of the consumed energy will be released as gametes. At the highest population densities feeding, scraping, and spine abrasion by *Echinometra* will cause reef erosion to exceed reef growth rates. The success of *Echinometra* in areas with few predators makes them a good indicator species for environmentally stressed or overfished reefs. When unpalatable fleshy algae and associated algal drift are abundant, *Echinometra* will feed primarily on drift algae and can reach biomass levels beyond what could be maintained by local primary production.

REFERENCES

Adjeroud M, Andrefouet S, Payri C (2001) Mass mortality of macrobenthic communities in the lagoon of the Hikueru atoll (French Polynesia). Coral Reefs 19: 287–291

Appana SD, Vuki VC, Cumming RL (2004) Variation and spatial distribution of ecomorphs of the sea urchins, *Echinometra* sp. nov A and *E.* sp. nov. C on a Fijian reef. Hydrobiologia 518: 105–110

Arakaki Y, Uehara T (1991) Physiological adaptations and reproduction of the four types of *Echinometra mathaei* (Blainville). In: Yanagisawa T, Yasumasu I, Oguro C, Suzuki N, Motokawa T (eds) Biology of echinodermata. AA Balkema, Rotterdam, pp 105–112

Arakaki Y, Uehara T (1999) Morphological comparison of black *Echinometra* individuals among those in the Indo-West Pacific. Zool Sci 16: 551–558

Arakaki Y, Uehara T, Fagoone I (1998) Comparative studies of the genus *Echinometra* from Okinawa and Mauritius. Zool Sci 15(1): 159–168

Aslan LM, Uehara T (1997) Hybridization and FI backcrosses between two closely related tropical species of sea urchins (genus *Echinometra*) in Okinawa. Inver Reprod Dev 31: 319–324

Atkinson MJ, Grigg RW (1984) Model of a coral reef ecosystem II. Gross and net benthic primary production at French frigate shoals, Hawaii. Coral Reefs 3: 13–22

Bak RPM (1990) Patterns of echnoid bioerosion in two Pacific coral reef lagoons. Mar Ecol Prog Ser 66: 267–272

Black R, Codd C, Hebbert D, Vink S, Burt J (1984) The functional significance of the relative size of Aristotle's lantern in the sea urchin *Echinometra mathaei* (de Blainville). J Exp Mar Bio Ecol 77: 81–97

Blevins E, Johnsen S (2004) Spatial vision in the echinoid genus *Echinomatra.* J Exp Biol 207: 4249–4253

Bolton TF, Thomas FIM (2002) Physical forces experienced by echinoid eggs in the oviduct during spawning: comparison of the geminate pair *Echinometra vanbrunti* and *Echinometra lucunter.* J Exp Mar Biol Ecol 267: 123–137

Cameron RA (1986) Reproduction larval occurrence and recruitment in Caribbean sea urchins. Bull Mar Sci 39: 332–346

Cameron RA, Hinegardner RT (1974) Initiation of metamorphosis in laboratory-cultured sea urchins. Biol Bull 146: 335–342

Chiappone M, Swanson DW, Miller SL (2002) Density, spatial distribution and size structure of sea urchins in Florida Keys coral reef and hard-bottom habitats. Mar Ecol Prog Ser 235: 117–126

Choat JH (1991) The biology of herbivorous fishes on coral reefs. In: Sale PF (ed.) The ecology of fishes on coral reefs. Academic Press, New York, pp 120–155

Clark AM, Rowe FEW (1971) Monograph of the shallow-water Indo-West Pacific echinoderms. Pitman Press, Bath

Conand C, Chabanet P, Cuet P, Letourner Y (1997) The carbonate budget of fringing reef in La Reunion Island (Indian Ocean): sea urchin and fish bioerosion and net calcification. Proc 8th Int Coral Reef Symp 1: 953–958

Downing N, El-Zahr CR (1987) Gut evacuation and filling rates in the rock-boring sea urchin, *Echinometra mathaei.* Bull Mar Sci 41: 579–584

Drummond AE (1994) Aspects of the life history biology of three species of sea urchin on the South African East coast. In: David B, Guille A, Feral JP, Roux M (eds) Echinoderms through time. AA Balkema, Rotterdam, pp 637–641

Drummond AE (1995) Reproduction of the sea urchins *Echinometra mathaei* and *Diadema savignyi* on the South African coast. Mar Freshwater Res 46: 751–757

Ebert TA (1975) Growth and mortality of post-larval echinoids. Amer Zool 15: 755–775

Ebert TA (1980) Relative growth of sea urchin jaws: an example of plastic resource allocation. Bull Mar Sci 30: 467–474

Ebert TA (1982) Longevity, life history, and relative body wall size in sea urchins. Ecol Monogr 52: 353–394

Ebert TA (1988) Calibration of natural growth lines in the ossicles of two sea urchins *Strongylocentrotus purpuratus* and *Echinometra mathaei* using tetracycline. In: Burke RD Maldenov PV, Lambert P, Parsley RL (eds) Echinoderm biology. AA Balkema, Roterdam, pp 435–443

Edmonson CH (1935) Hawaiin reef and shore fauna. Bishop Museum Press, Honolulu

Emlet RB, McEDward LR, Strathman RR (1987) Echinoderm larval ecology viewed from the egg. Echinoderm Stud 2: 55–136

Fell FJ (1974) The echinoids of Easter Island (Rapa Nui). Pac Sci 28: 147–158

Fujisawa H, Shigei M (1990) Correlation of embryonic temperatures sensitivity of sea urchins with spawning season. J Exp Mar Biol Ecol 136: 123–139

Gadgil G, Bossert WH (1970) Life historical consequences of natural selection. Am Nat 104: 1–24

Geyer LB, Palumbi SR (2003) Reproductive character displacement and the genetics of gamete recognition in tropical sea urchins. Evolution 57(5): 1049–1060

Griffin SP, Garcia RP, Weil E (2003) Bioerosion in coral reef communities in southwest Puerto Rico by the sea urchin *Echinometra viridis*. Mar Biol 143: 79–84

Grunbaum HG, Bergman G, Abbott DP, Ogden JC (1978) Intraspecific agonistic behavior in the rock boring sea urchin *Echinometra lucunter* (L.). Bull Mar Sci 28: 181–188

Harvey EB (1947) Bermuda sea urchins and their eggs. Biol Bull 93: 217–218

Kelso DP (1971) Morphological variation, reproductive periodicity, gamete compatibility and habitat specialization in two species of the sea urchin *Echinometra* in Hawaii. PhD Thesis, University of Hawaii

Khamala CPM (1971) Ecology of *Echinometra mathaei* (Echinoidea: Echinodermata) at Diani beach, Kenya. Mar Biol 2: 167–172

Kinjo S, Shirayama Y, Wada H (2004) Phylogenetic relationships and morphological diversity in the family Echinometridae (Echinoida, Echinodermata) In: Heinzeller T, Nebelsick JH (eds) Echinoderms. München. AA Balkema Publishers, Leiden, pp 527–530

Kobayashi N (1969) Spawning periodicity of sea urchins at Setto III. *Tripneustes gratilla, Echinometra mathaei, Anthocidaris crassipina* and *Echinostrephus aciculatus*. Sci Eng Rev Doshisha Uni 9: 254–269

Larkum WD (1983) The primary productivity of plant communities on coral reefs. In: Barnes DJ (ed.) Perspectives on coral reefs. Brian Clouston Publisher, Manuka, pp 221–230

Lawrence JM (1970) The effect starvation on the lipid and carbohydrate levels of the gut of the tropical sea urchin *Echinometra mathaei* (de Blainville). Pac Sci XXIV: 4487–4489

Lawrence JM, Bazhin A (1998) Life history strategies and the potential of sea urchins for aquaculture. J Shellfish Res 17: 1515–1522

Lessios HA (1981) Reproductive periodicity of the Echinoids *Diadema* and *Echinometra* on the two Coasts of Panama. J Exp Mar Biol Ecol 50: 47–61

Lessios HA (1988) Population dynamics of *Diadema antillarum* (Echinodermata: Echinoidea) following mass mortality in Panama. Mar Biol 99: 115–526

Lessios HA (1991) Presence and absence of monthly reproductive rhythms among eight Caribbean echinoids off the coast of Panama. J Exp Mar Biol Ecol 153: 27–47

Lessios HA, Cunningham CW (1990) Gametic incompatibility between species of the sea urchin *Echinometra* on the two sides of the Isthmus of Panama. Evol 44: 933–941

Lewis JB, Storey GS (1984) Differences in morphology and life history traits of the echinoid *Echinometra lucunter* from different habitats. Mar Ecol Prog Ser 15: 207–211

Matsuoka N, Hatanaka T (1991) Molecular evidence for the existence of four sibling species within the sea-urchin, *Echinometra mathaei* in Japanese waters and their evolutionary relationships. Zool Sci 8(1): 121–133

McCartney MA, Lessios HA (2004) Adaptive evolution of sperm bindin tracks egg incompatibility in neotropical sea urchins of the genus *Echinometra*. Mol Biol Evol 21(4): 732–745

McCartney MA, Keller G, Lessois HA (2000) Dispersal barriers in tropical oceans and speciation in Atlantic and eastern Pacific sea urchins of the genus *Echinometra*. Mol Ecol 9: 1391–1400

McClanahan TR (1988) Coexistence in a sea urchin guild and its implications to coral reef diversity and degradation. Oecologia 77: 210–218

McClanahan TR (1992) Resource utilization, competition and predation: a model and example from coral reef grazers. Ecol Mod 61: 195–215

McClanahan TR (1995) Fish predators and scavengers of the sea urchin *Echinometra mathaei* in Kenyan coral-reef marine parks. Env Biol Fish 43: 187–193

McClanahan TR (1998) Predation and the distribution and abundance of tropical sea urchin populations. J Exp Mar Biol Ecol 221: 231–255

McClanahan TR (1999) Predation and the control of the sea urchin *Echinometra viridis* and fleshy algae in the patch reefs of Glovers Reef, Belize. Ecosystems 2: 511–523

McClanahan TR (2000) Recovery of the coral reef key stone predator, *Balistapus undulatus*, in east African marine parks. Biol Cons 94: 191–198

McClanahan TR, Kurtis JD (1991) Population regulation of the rock-boring sea urchin *Echinometra mathaei* (de Blainville). J Exp Mar Biol Ecol 147: 121–146

McClanahan TR, Muthiga NA (1988) Changes in Kenyan coral reef community structure and function due to exploitation. Hydrobiologia 166: 269–276

McClanahan TR, Muthiga NA (1989) Patterns of predation on a sea urchin, *Echinometra mathaei* (de Blainville), on Kenyan coral reefs. J Exp Mar Biol Ecol 126: 77–94

McClanahan TR, Shafir SH (1990) Causes and consequences of sea urchin abundance and diversity in Kenyan coral reef lagoons. Oecologia 83: 362–370

McClanahan TR, Nugues M, Mwachireya S (1994) Fish and sea urchin herbivory and competition in Kenyan coral reef lagoons: the role of reef management. J Exp Mar Biol Ecol 184: 237–254

McClanahan TR, Kamkuru AT, Muthiga NA, Gilagabher Yeboi M, Obura D (1996) Effect of sea urchin reductions on algae, corals and fish populations. Cons Biol 10: 136–154

McClanahan TR, Muthiga NA, Kamakuru AT, Machano H, Kiambo R (1999) The effects of marine parks and fishing on the coral reefs of northern Tanzania. Biol Cons 89: 161–182

McClintock JB, Klinger TS, Lawrence JM (1982) Feeding preferences of echinoids for plant and animal food models. Bull Mar Sci 32: 365–369

McPherson BF (1969) Studies of the biology of the tropical sea urchins *Echinometra lucunter* and *Echinometra viridis*. Bul Mar Sci 19: 194–213

Metaxas A, Young CM (1998) Response of echinoid larvae to food patches of different algal densities. Mar Biol 130: 433–445

Metz EC, Kane RE, Yanagimachi H, Palumbi SR (1994) Fertilization between closely related sea urchins is blocked by incompatibilities during sperm–egg attachment and early stages of fusion. Biol Bull 187: 23–34

Mills S, Peyrot-Clausade M, Fontaine FM (2000) Ingestion and transformation of talgal turf by *Echinometra mathaei* on Tiahura fringing reef (French Polynesia) J Exp Mar Biol Ecol 254(1): 71–84

Mokady O, Lazar B, Loya Y (1996) Echinoid bioerosion as a major structuring force of Red Sea coral reefs. Biol Bull 190: 367–372

Mortensen T (1921) Studies of the development and larval forms of echinoderms. GEC Gad, Copenhagen, 266pp

Mortensen T (1937) Contributions to the study of the development of the larval forms of the echinoderms III. Mem de l'Academie Royale des Sciences et des Lettres de Danemark, Copenhague, Section des Sciences, 9me serie t. VII. No. 1: 1–65

Mortensen TH (1943) Monograph of the echinoidea: camarodonta. CA Reitzel, Copenhagen

Muthiga NA (1996) The role of early life history strategies on the population dynamics of the sea urchin *Echinometra Mathaei* (de Blainville) on reefs in Kenya. PhD Thesis, University of Nairobi

Muthiga NA, Jaccarini V (2005) Effects of seasonality and population density on the reproduction of the Indo-Pacific echinoid *Echinometra mathaei* in Kenyan coral reef lagoons. Mar Biol 146: 445–453

Neill JB (1988) Experimental analysis of burrow defence in *Echinometra mathaei* (de Blainville) on an Indo-West Pacific reef flat. J Exp Mar Biol Ecol 115: 127–136

Nishihira M, Sato Y, Arakaki Y, Tsushiya M (1991) Ecological distribution and habitat preference of four types of the sea urchin *Echinometra mathaei* on the Okinawan coral reefs. In: Yanagisawa T, Yasumasu I, Oguro C, Suzuki N, Motokawa T (eds) Biology of echinodermata. AA Balkema, Rotterdam, pp 91–104

Odum HT, Odum EP (1955) Trophic structure and productivity of a windward coral reef community in Eniwetok Atoll. Ecol Monogr 25: 291–320

Ogden JC (1977) Carbonate-sediment production by parrot fish and sea urchins on Caribbean reefs. In: Frost S, Weiss M (eds) Caribbean reef systems: Holocene and ancient. American Association of Petroleum Geologists Special Paper 4, pp 281–288

Onada K (1936) Notes on the development of some Japanese echinoids with special reference to the structure of the larval body. Jap J Zool 6: 635–654

Osorio RC, Atan HH (1993) Biological relationships between *Luetzenia goodingi* Rehder 1980 (Gastropoda, Stiliferidae) parasite of *Echinometra insularis* Clark 1972 (Echinoidea) of Easter Island. Rev Biol Mar 28: 99–109

Palumbi SR (1996) Macrospatial genetic structure and speciation in marine taxa with high dispersal abilities. In: Ferraris JD, Palumbi SR (eds) Molecular zoology: advances, strategies, and protocols. Wiley-Liss Inc., New York, pp 101–117

Palumbi SR (1999) All males are not created equal: fertility differences depend on gametic polymorphisms in sea urchins. Evol 96 (22): 12632–12637

Palumbi SR, Metz EC (1991) Strong reproductive isolation between closely related tropical sea urchins (genus *Echinometra*). Mol Biol Evol 8: 227–239

Palumbi SR, Grabowsky G, Duda T, Geyer L, Tachino N (1997) Speciation and population genetic structure in tropical Pacific sea urchins. Evolution 51: 1506–1517

Pearse JS (1968) Pattens of reproductive periodicities in four species of Indo-Pacific echinoderms. Proc Indian Acad Sci 68: 247–279

Pearse JS (1969) Reproductive periodicities of Indo-Pacific invertebrates in the Gulf of Suez II. The echinoid *Echinometra mathaei* (de Blainville). Bull Mar Sci 19: 580–613

Pearse JS (1974) Reproductive patterns of tropical reef animals: three species of sea urchins. Proc 2nd Int Coral Reef Symp 1: 235–240

Pearse JS, Cameron R (1991) Echinodermata: echinoidea. In Giese AC, Pearse JS, Pearse VB (eds) Reproduction of marine invertebrates, Volume VI, Echinoderms and Lophophorates. Boxwood Press, Pacific Grove, pp 513–662

Pearse JS, Phillips BF (1968) Continuous reproduction in the Indo-Pacific sea urchin *Echinometra mathaei* at Rottnest Island, Western Australia. Aust J Mar Freshwater Res 19: 161–172

Pennings SC, Svedberg JM (1993) Does $CaCO_3$ in food deter feeding in sea urchins? Mar Ecol Prog Ser 101: 163–167

Peyrot-Clausade M (1995) Initial bioerosion and bioaccretion on experimental substrates in high island and atoll lagoons (French Polynesia). Oceanolog Acta 18: 531–541

Peyrot M, Chabanet P, Conand C, Fontaine MF, Letourneur Y, Harmelin-Vivien M (2000) Sea urchin and fish bioerosion on La Reunion and Moorea reefs. Bull Mar Sci 66: 477–485

Prince J (1995) Limited effects of the sea urchin *Echinometra mathaei* (de Blainville) on the recruitment of benthic algae and macro-invertebrates into intertidal rock platforms at Rottnest Island, Western Australia. J Exp Mar Biol Ecol 186: 237–258

Rahman MA, Uehara T (2003) F1 and F2 backcrosses in the hybrids between two unnamed genetically distinct species of tropical sea urchins, *Echinometra* sp. A and *Echinometra* sp. C. Pakistan J Biol Sci 6(13): 1163–1175

Rahman MA, Uehara T (2004) Interspecific hybridization and backcrosses between two sibling species of Pacific sea urchins (Genus *Echinometra*) on Okinawan intertidal reefs. Zool Stud 43: 93–111

Rahman MA, Uehara T, Pearse JS (2001) Hybrids of two closely related tropical sea urchins (Genus *Echinometra*): evidence against postzygotic isolating mechanisms. Biol Bull 200: 97–106

Rahman MA, Uehara T, Rahman SM (2002) Effects of egg size on fertilization, fecundity and offspring performance: a comparative study between two sibling species of tropical sea urchins (Genus *Echinometra*). Pak J Biol Sci 5(1): 114–121

Rahman MA, Uehara T, Pearse JS (2004) Experimental hybridization between two recently diverged species of tropical sea urchins, *Echinometra mathaei* and *Echinometra oblonga*. Inver Reprod Dev 45: 1–14

Rahman MA, Uehara T, Lawrence JM (2005) Growth and heterosis of hybrids of two closely related species of Pacific sea urchins (genus *Echinometra* in Okinawa). Aquaculture 245: 121–133

Randall JE (1967) Food habits of reef fishes of the West Indies. Stud Trop Oceanog 5: 665–847

Rupp JH (1973) Effects of temperature and fertilization on early cleavage of some tropical echinoderms with emphasis on *Echinometra mathaei*. Mar Biol 23: 183–189

Russo AR (1977) Water flow and the distribution and abundance of echinoids (genus *Echinometra*) on an Hawaiian reef. Aust J Mar Freshwater Res 28: 693–702

Russo AR (1980) Bioerosion by two rock boring echinoids (*Echinometra mathaei* and *Echinostrephus aciculatus*) on Enewetak Atoll, Marshall Islands. J Mar Res 38: 99–110

Schneider DC (1985) Predation on the urchins *Echinometra lucunter* (Linnaeus) by migratory shorebirds, on a tropical reef flat. J Exp Mar Biol Ecol 92: 19–27

Schoppe S, Werdings B, (1996) The boreholes of the sea urchin genus *Echinometra* (*Echinodermata*: *Echinoidea*: *Echinometridae*) as a microhabitat in tropical South America. Mar Ecol 17: 181–186

Shulman MJ (1990) Aggression among sea urchins on Caribbean coral reefs. J Exp Mar Biol Ecol 17: 181–186

Smith SV (1983) Coral reef calcification. In: Barnes DJ (ed.) Perspectives on coral reefs. Brian Clouster Publisher, Manuka, pp 240–247

Strathman RR (1987) Larval feeding. In: Giess AC, Pearse JS, Pearse VB (eds) Reproduction of marine invertebrates, Vol. 9. Blackwell Science Publications, Boxwood Press, Pacific Grove, pp 465–550

Tahara Y, Okada M (1968) Normal development of secondary sexual characters in the sea urchin *Echinometra mathaei*. Publ Seto Mar Biol Lab 16: 41–50

Tsuchiya M, Nishihira M (1984) Ecological distribution of two types of the sea urchin, *Echinometra mathaei* (Blainville), on Okinawan reef flat. Galaxea 3: 131–143

Tsuchiya M, Nishihira M (1985) Agonistic behavior and its effect on the dispersion pattern in two types of sea urchins, *Echinometra mathaei* (Blainville). Galaxea 4: 37–48

Tsuchiya M, Nishihira M (1986) Re-colonization process of two years of the sea urchin, *Echinometra mathaei* (Blainville), on the Okinawan reef flat. Galaxea 5: 283–294

Tsuchiya M, Yanagiya K, Nishihira M (1987) Mass mortality of the sea urchin *Echinometra mathaei* (Blainville) caused by high water temperature on the reef flats in Okinawa, Japan. Galaxea 6: 375–385

Uehara T, Taira K(1987) Heteromorphic chromosomes in sea urchins, Type B of *Echinometra mathaei* from Okinawa. Zool Sci 4: 1001

Uehara T, Asakura H, Arakaki Y (1990) Fertilization blockage and hybridization among species of sea urchins. In: Hoshi M, Yamashita O (eds) Advances in invertebrate reproduction. Elsevier, Amsterdam, pp 305–310

Uehara T, Shingaki M, Tairo K, Arakaki Y, Nakatomi H (1991) Chromosome studies in eleven Okinawan species of sea urchins, with special reference to four species of the Indo-Pacific *Echinometra*. In: Yanagisawa T, Yasumasu I, Oguro C, Suzuki N, Motokawa T (eds) Biology of echinodermata. AA Balkema, Rotterdam, pp 119—129

Ventura CRR, Varotto RS, Carvalho ALPS, Pereira AD, Alves SLS, MacCord FS (2003) Interpopulation comparison of the reproductive and morphological traits of *Echinometra lucunter* (Echinodermata: Echinoidea) from two different habitats on Brazilian coast. In: Féral JP, David B (eds) Echinoderm research 2001. AA Balkema Publishers, Lisse, pp 289–293

Watts RJ, Johnson MS, Black R (1990) Effects of recruitment on genetic patchiness in the urchin *Echinometra mathaei* in Western Australia. Mar Biol 105: 145–151

Williams AH (1979) Interference behaviour and ecology of threespot damselfish (*Eupomacentrus planifrons*). Oecologia 38: 223–230

Williams AH (1980) The threespot damselfish: a noncarnivorous keystone species. Am Nat 116: 138–142

Williams AH (1981) An analysis of competitive interactions in a patchy back-reef environment. Ecology 62: 1107–1120

Edible Sea Urchins: Biology and Ecology
Editor: John Miller Lawrence

Chapter 16

Ecology of *Evechinus chloroticus*

Michael F Barker

Department of Marine Science, University of Otago, Dunedin (New Zealand).

1. INTRODUCTION

Evechinus chloroticus (Valenciennes) (Plate 3A) known locally as *kina* is a ubiquitous echinometrid sea urchin endemic to New Zealand. It is one of the largest sea urchins known, with a maximum test diameter (TD) of 16–17 cm (pers. obs.). It was harvested by Maori people before the arrival of Europeans in New Zealand and more recently has been exploited in commercial fisheries and in small quantities in restricted areas around New Zealand.

The morphology and taxonomy of *E. chloroticus* are well described by (McRae 1959) and Mortensen (1922). Morphology and distribution are also described by Cook (in press). Northern populations of *E. chloroticus* have been the subject of a number of ecological studies (reviewed by Andrew 1988). This chapter describes the ecology of populations throughout New Zealand.

2. GEOGRAPHIC DISTRIBUTION

Evechinus chloroticus is widely distributed around the main islands of New Zealand and is also found at the Chatham Islands to the west, the Snares Islands to the south (Pawson 1965) and from the Three Kings (Schiel et al. 1986) and Kermadec Islands to the north (Pawson 1961). This latter record has been questioned (Dix 1970a) and *E. chloroticus* was not seen during a recent collecting trip to the Kermadec Islands (Gardner, pers. comm.).

3. HABITAT

Evechinus chloroticus is generally found in water less than 12–14 m deep although individuals have been collected from at least 60 m (Cook in press). Intertidal populations also occur, mainly in the north of the North Island (Dix 1970a; McRae 1959). However, Dix (1970a) also notes intertidal populations at Kaiteriteri in Golden Bay at the north of the South Island. Although *E. chloroticus* is occasionally found on sandy or shingle areas (Dix 1970a), this is not a common habitat and urchins are very seldom seen on fine sediments such as silt or mud (pers. obs.). Adults are normally found in areas with moderate currents and wave action, and are seldom found in areas of extreme exposure such as the surf shores that occur along much of New Zealand's west coast. *Evechinus chloroticus* is found in somewhat different patterns of distribution and relationships with kelp and other invertebrates throughout New Zealand and it is difficult to make broad generalisations about its habitat. In northern parts of New Zealand *E. chloroticus* is most commonly found on shallow rocky reefs, dominated by encrusting coralline algae. In such areas urchins may be extremely abundant, with maximum densities of 40 individuals per square metre reported by Choat and Schiel (1982) at Poor Knights Islands. When urchins are closely packed density is strongly influenced by behaviour and substrate availability and density is less useful than biomass per square metre although this fact is hardly ever included in population statistics. Coralline dominated reef flats are often interspersed with mature stands of laminarian algae. Within rocky reefs, *E. chloroticus* is generally clumped, and there may be patches dominated by urchins which are almost devoid of kelp alongside areas where kelp is dense (Choat and Schiel 1982). The borders between such areas appear to be stable. This suggests a balance between echinoid grazing and algal colonisation, a similar situation to that described for *Strongylocentrotus* spp. and *Macrocystis pyrifera* on the west coast of North America (Harrold and Reed 1985). In the Chatham Islands, Schiel et al. (1995) found *E. chloroticus* at almost every site sampled at mean densities of 5–15 individuals per square metre, but at maximum densities of 40 large individuals per square metre. Urchins tended to be abundant in sites where the deep water fucalean *Carpophyllum flexuosum* was common. It was not seen in the characteristically extensive deforested areas described above for northern New Zealand (Choat and Schiel 1982) and deforested patches were never seen larger than $\sim 25\,m^2$. In the South Island *E. chloroticus* in open coast habitats are less often seen on extensive rock flats as in the north. More commonly, urchins are in aggregations, either between individual or small groups of kelp, or in barren areas of approximately 5–6 m diameter. On the Otago coast *E. chloroticus* is uncommon although isolated aggregations of extremely large (120–160 mm diameter) urchins do occur. The apparent rarity of *E. chloroticus* on this area of coast is anomalous, as the habitat would appear to be very suitable over much of the region. This may reflect very sporadic and low levels of recruitment. On the most southern coasts of the South Island, *E. chloroticus* increases in abundance and is very common in the sheltered inlets of Stewart Island. On the south western coast of Fiordland, *E. chloroticus* is common on the open coast and within the Fiords. For example, in the outer reaches of Dusky Sound, urchins occur in highest densities of 2.6 individuals per square metre at depths of 3.5–6.5 m (Villouta et al. 2001) but are also abundant throughout the inner fiords which indent this coastline. In each fiord steep cliffs covered with luxuriant rainforest overhang steep fiord walls that drop into deep water. This area of New Zealand is subjected to extremely high rainfall and many of the fiords have a surface low salinity layer (LSL) that may be up to 10–12 m in depth after several days of heavy rain in the local catchment. This fresh water layer is particularly deep in Doubtful Sound, because of additional discharge of freshwater from a hydroelectric power station. In these fiords *E. chloroticus* is abundant, often in aggregations of up to

20–30 individuals per square metre from immediately below the LSL, clinging to the steep rock faces, sometimes on bare or coralline encrusted rock, sometimes on sunken trees caught on ledges, or in shallower areas. Laminarians are only found towards the entrance of the fiord, although they can be quite luxuriant in such areas (Villouta et al. 2001). Juvenile *E. chloroticus* are cryptic throughout New Zealand. Dix (1970a) noted that individuals less than 10 mm TD were generally attached beneath both intertidal and subtidal rocks. Individuals between 10 and 40 mm TD occurred in the intertidal and subtidal areas under rocks, or within small crevices or depressions in rocks and typically covered in debris. At around 40 mm TD, individuals migrated into open habitats.

The size structures of *E. chloroticus* populations vary over its geographic range. Populations of *E. chloroticus* typically have a unimodal size distribution and are dominated by larger individuals. Such populations have been documented for the Otago coast (pers. obs.), Kaikoura and Kaiteriteri (Dix 1972), Tory Channel (Lamare and Barker 2001) and Dusky Sound (McShane 1992). The mean size of the urchins in these populations ranges from ≈40 to 50 mm TD (e.g. Kaiteriteri, Dix 1972), to 112 mm TD, Dusky Sound (McShane et al. 1996). Populations with a bimodal or polymodal distribution are less common but have been described for Kaikoura (Dix 1972), Dusky Sound (McShane et al. 1994) and Doubtful Sound (Lamare and Barker 2001). There can also be distinct differences in the size structures of *E. chloroticus* populations over relatively short distances. Along the Kaikoura Peninsula for example, populations vary in mean body size and abundance of juveniles over distances of less than 5 km (Dix 1972).

4. ASSOCIATED SPECIES

4.1. Kelp

In northern New Zealand *E. chloroticus* is strongly associated with kelp forests often dominated by the laminarian *Ecklonia radiata* (Choat and Schiel 1982; Schiel 1982; Andrew and Stocker 1986). In very shallow water urchins can be found in association with the fucaleans *Carpophyllum mashalocarpum* and *Carpophyllum angustifolium* (Schiel 1982). In the South Island in the Marlborough Sounds urchins are mostly found with *C. mashalocarpum* and in areas subjected to high current flows, *M. pyrifera* (Dix 1970a; Brewin et al. 2000). Further south on the Otago coast and at Stewart Island *E. radiata* only occurs as occasional isolated plants, and *M. pyrifera* is often the dominant kelp forest forming laminarian species (pers. obs.). In Dusky Sound in southern Fiordland (McShane et al. 1994; Villouta et al. 2001) and further north at the entrance to Doubtful Sound (pers. obs.) *E. chloroticus* is commonly found with the laminarians *E. radiata, C. flexuosum, Sargassum sinclairiii,* and *Lessonia variagata,* although there are a number of red and green seaweeds varying seasonally in abundance.

4.2. Gastropods

Evechinus chloroticus is often found in association with grazing gastropod species. Andrew and Choat (1982) report a positive correlation between the abundance of herbivorous gastropods at depths less than 12 m with the limpet *Cellana stellifera*, the topshell *Trochus viridis* and the turbinid *Cookia sulcata*, abundance being significantly lower in the absence of *E. chloroticus*. These observations were explored in more detail by caging specific gastropods with *E. chloroticus*. The results further supported the suggestion of a positive relationship between *E. chloroticus* and *C. sulcata* with an increase in the mean dry weight of the sea urchins in the presence of the turbinid (Choat and Andrew 1986). In contrast (Shears and Babcock 2003) found *C. sulcata* in northern New Zealand had densities 3 times higher in marine reserve areas with lower urchin densities and higher amounts of coralline turf compared to nonprotected areas where urchin numbers were higher. *C. stellifera* abundance, however, was 2.5 times lower in the marine reserves areas. They suggest *E. chloroticus* maintains a crustose coralline algal substrate suitable for the grazing of *C. stellifera*. Although both *T. viridis*, *C. sulcata* and a range of subtidal limpets occur in the South Island, detailed relationships between *E. chloroticus* and other grazers have not been investigated. In some southern open-coast situations *E. chloroticus* is often found in association with the abalone *Haliotis iris* especially in depths of 5–10 m. Wing (pers. comm.) completed a 3 year stratified random sampling programme in both Paterson Inlet and Port Pegasus. He found that both *E. chloroticus* and *H. iris* are found on shallow rocky reefs and the peak abundances overlap in depth. However, they are rarely found together in the same quadrat (2 m^2) so that at a relatively small scale their distributions are negatively correlated.

5. FEEDING

5.1. Diet

By examining the mouths of urchins in situ and noting the type of food material present (Dix 1970a), was able to observe feeding of *E. chloroticus* directly at both Kaikoura and Kaiteriteri. At least 11 species of identifiable algae were eaten at Kaikoura while only two species, *C. maschallocarpum* and *Hormosira banksii,* were eaten at Kaiteriteri. Additional laboratory observations confirmed that *E. chloroticus* is chiefly herbivorous but will eat a variety of food when algae are scarce. While *E. chloroticus* feeds on a wide variety of algal species, it shows a preference for particular species. Don (1975) attempted to examine this question in the laboratory and concluded that *E. chloroticus* had a preference for *E. radiata,* but that when large brown algae were unavailable the urchins grazed indiscriminately on all encrusting organisms. The degree to which food choice experiments conducted in the laboratory using choice chambers can be extrapolated to the field is problematic, however, and arguably such experiments should be conducted in natural conditions whenever possible. The results of Schiel (1982) highlight this problem. He examined feeding in a series of laboratory and field experiments. In the first, seven species of algae were presented to urchins in the laboratory and in a second experiment replicates of the same algal species were randomly placed subtidally on a rocky reef where *E. chloroticus* were relatively abundant (2.5 individuals per square meter) and randomly distributed. The results showed that there were differences in the amount of each alga grazed by the urchins. However, the ranking of the algal species from the field experiment did not correlate with the rankings established by the laboratory choice experiment.

It is clear that *E. chloroticus* do exert a major influence on algal stands in the subtidal. Andrew and Choat (1982) used cages to exclude *E. chloroticus* from a 1000 m^2 subtidal coralline flat area in northern New Zealand and found a large increase in biomass in a range of kelp species followed, especially *E. radiata.* Shears and Babcock (2002) removed urchins from barrens outside of a marine reserve in the north of New Zealand, which resulted in a change from a crustose coralline algal to a macroalgal dominated habitat after 12 months. They showed that this was largely a consequence of much higher predation on urchins within marine reserves (see later section on predation). In a subsequent study Shears and Babcock (2003) examined changes in community structure in the Leigh Marine Reserve between 1996 and 2000. During this period benthic communities changed from being dominated by *E. chloroticus* to being dominated by macroalgae with an ongoing reduction in urchin numbers. Villouta et al. 2001) experimentally removed urchins from strata at three depths in the outer reaches of Dusky Sound, Fiordland. After 2 years of continuous removal there were marked changes in the algal communities at all three depths with increases in the density of *E. radiata, Carpophyllum* spp., *Sargassum* spp., *Cystophora* spp. and *Landsburgia quercifolia*. Barrens habitat was largely replaced by algal assemblages. The grazing activities of *E. chloroticus* have been shown to influence the structure of encrusting communities as well as algal stands. Ayling (1978) investigated the grazing of *E. chloroticus* on encrusting communities dominated by sponges in both the field and the laboratory. In the field, urchins grazed on sponge species according to abundance but preferences in the laboratory for particular sponge species did not relate to the diet of urchins in the field. Ayling (1981) suggests that in benthic communities made up of encrusting organisms, *E. chloroticus* will readily graze all encrusting species except a few of the more massive sponges, and even these are grazed when food is in short supply. Although it has been shown that many encrusting animals produce chemical toxins or large quantities of mucus, the feeding preferences found by Ayling (1978) were not related to toxicity. She does suggest that toxins of some species of sponges may have some role in influencing feeding preferences. Algae contain phlorotannins (polyphenolic compounds that are the dominant secondary metabolites in temperate brown seaweeds) which are known to deter feeding by some invertebrate grazers (Van Alstyne 1988). Steinberg and Altena (1992) tested the tolerance of *E. chloroticus* to brown algal phlorotannins in six algal species and found no correlation between feeding preferences and algal phenolic levels. Cole and Haggitt (2000), however, found some evidence that high levels of phlorotannins influenced *E. chloroticus* grazing on certain species, and particularly *C. flexuosum* which has two growth forms. One, which had high phlorotannin levels, was not grazed by *E. chloroticus* in the field or in the laboratory. A more detailed investigation of the relationship of diet preferences and metabolites that deter grazers is required.

Aside from the early studies by (Dix 1970a) at Kaikoura and Kaiteriteri, all the studies mentioned so far are for northern *E. chloroticus* populations and there have been no field experiments on diet or feeding preferences in southern populations. Research on morphometric variation and calorific content of *E. chloroticus* gut contents has been undertaken in Doubtful Sound (Wing et al. 2001). They found that sea urchins near the entrance of fiords had a larger percentage of algae, smaller Aristotle's lantern indices and higher calorific content of the gut contents than populations in the inner sounds strongly suggesting nutritional limitations in inner fiord sites. The entrance and outer fiords are characterised by assemblages of laminarian kelps dominated by *E. radiata*, while the inner fiords with steeper rock walls only support filamentous chlorophytes and small amounts of rhodophytes. *Evechinus chloroticus* there are often found grazing sunken logs (pers. obs.). Andrew and Stocker (1986) investigated the relationship between diet and reproductive output and found that *E. chloroticus* caged at low densities and fed the more preferred algae *E. radiata* had a higher reproductive output than urchins fed less preferred algae (*Carpophyllum* sp.). This relationship broke down when urchins were caged at higher densities. Laboratory experiments using prepared artificial diets (Barker et al. 1998) also show that diet quality has a significant influence on reproductive output in *E. chloroticus.* A great deal more information is required from field studies, however, before we will understand the relationship between food selection and other metabolic requirements, i.e. whether urchins selectively graze on algae or other organisms in order to maximize reproductive output or enhance other physiological processes.

5.2. Feeding Rate

In a laboratory study Barker et al. (1998) investigated the seasonal changes in feeding rate of 3 sizes of *E. chloroticus* fed either one of a number of prepared artificial diets or the algae *M. pyrifera* and *Ulva lactuca* (Table 1). Feeding rates differed significantly between diets, urchins eating more prepared feeds than algae. For all diets, feeding rates showed a clear seasonal trend directly correlated with water temperature. Of the two algal diets, *M. pyrifera* was clearly preferred over *U. lactuca* for all urchin sizes. Schiel (1982) fed *E. chloroticus,* similar in size to the medium-size urchins used by Barker et al. (1998) (60 mm TD), a range of algal species in the laboratory. Although both, the range of algal species and the experimental design were different, the mean daily feeding rate per urchin

Table 1 Mean feeding rate on algae (g ind^{-1} day^{-1}) over a 10 month period in the laboratory of small (30–40 mm), medium (50–50 mm) and large (80–90 mm TD) *E. chloroticus* held in individual containers with running seawater at ambient temperatures.

	Feeding rate	**s.d.**
Macrocystis pyrifera		
Small	0.58	0.25
Medium	0.88	0.46
Large	1.08	0.38
Ulvalactuca		
Small	0.15	0.09
Medium	0.31	0.17
Large	0.64	0.41

calculated from Schiel's 1982 data is 0.69 g per individual per day for the most preferred alga (*Cystophora torulosa*). This is similar to 0.88 g per individual per day of *M. pyrifera* eaten by *E. chloroticus*, suggesting that feeding rates for urchins may be similar even though they were collected from widely different latitudinal locations.

6. MOVEMENT

The feeding activity of grazing species is markedly affected by their foraging behaviour. Drift algae is known to make up an important component in the diet of *Strongylocentrotus purpuratus* and *Strongylocentrotus fransciscanus* (Harrold and Reed 1985). In regions where drift algae is plentiful, active foraging is largely unnecessary because currents will move adequate quantities of kelp for sedentary urchins. This markedly affects the behaviour of urchins which tend to be more cryptic when drift algae is abundant (Harrold and Reed 1985). In contrast, urchins tend to actively forage when drift kelp is in short supply and it is in these circumstances that feeding aggregations develop (Harrold and Reed 1985). A similar relationship between *E. chloroticus* and abundance of drift algae has yet to be established and active feeding fronts of urchins have yet to be documented.

Movement in *E. chloroticus*, especially in relationship to levels and quality of food, is poorly understood. Dix (1970b) investigated movement in aggregated populations of *E. chloroticus* at Kaikoura and at Kaiteriteri by inserting anchor tags through the test and searching for tagged urchins after 3, 6 and 9 months. At Kaikoura he found only small numbers (6% at 3 months, 10% at 6 months) had moved, with a maximum distance of 4.8 m. At Kaiteriteri, a site where brown algae are less abundant, movement was slightly greater. Tagged urchins, however, had lower gonad indices and less food in the gut than untagged animals. This suggests that tagging affects feeding and is likely to have influenced movement. Tag recapture studies by Lamare and Mladenov (2000) indicate that a high recovery rate of tagged individuals at 1 year (30–37%) and even after 4 years (23.5%), is possible even with no physical barriers to movement. In laboratory studies Dix (1970) found that *E. chloroticus* moved more at night than during the day in tanks with running seawater. In a detailed set of field experiments Andrew and Stocker (1986) examined the movement of *E. chloroticus* in relationship to the availability of *E. radiata* at two subtidal sites in northern New Zealand. Their results confirm Dix's suggestion that movement is related to the availability of food. Movement was not directional and in the presence of drift algae urchins doubled their overnight movements, suggesting a chemosensory response to the presence of damaged kelp although they did not appear capable of responding to chemical stimuli in a directional manner. Movement may also be affected by the presence of predators. Andrew and MacDiarmid (1991) report that *E. chloroticus* moves less in habitats in which the spiny lobster *Jasus edwardsii* (a potential predator, see section on mortality) is abundant than in areas where lobsters are scarce. There is a need for more studies on movement and behaviour in relation to food availability, especially in southern populations in fiordland. Here the fluctuating levels of the low salinity layer causes vertical migration of urchin populations (pers. obs.) as animals maintain their distribution within high salinitiy strata of the halocline, a physical requirement which must impact markedly on both the type and quantity of available food.

7. REPRODUCTION

Like most species of temperate echinoids *E. chloroticus* is gonochoric with a 1:1 sex ratio and an annual breeding cycle (Dix 1970c), the reproductive effort varying with diet quality (Barker et al. 1998) and population density (Andrew 1986).

7.1. Gametogenesis

The gametogenic cycle has been described in detail by several authors (Dix 1970b; Walker 1982; Brewin et al. 2000; Lamare et al. 2002). The developmental sequence of stages differs little in these studies and the following description can be regarded as applying to all populations: (1) spawning occurs during the austral summer, generally being completed by March in most populations; (2) a build up of nutrient reserves in the form of nutritive phagocytes occurs during autumn and early winter (March–May) followed by gametogenesis during mid-late winter and into spring (June to October); (3) gonads are ripe during spring October and November although some individuals collected from the field can be spawned in the laboratory from October to March (pers. obs.).

7.2. Reproductive Cycle

The annual reproductive cycle has been described in detail by the calculation of monthly gonad size indices for populations from the north of New Zealand (Hauraki Gulf, Walker 1982), central regions (Wellington and Northern South Island (Dix 1970c; McShane et al. 1996) and Brewin et al. (2000) and for the south west of the South Island, Dusky Sound (McShane et al. 1996; Doubtful Sound, Lamare 1998; Lamare and Barker 2001; Lamare et al. 2002, 2004; Wing et al. 2003; Table 2). Reproductive cycles of males and females are generally synchronous in most populations where they have been described. Surprisingly, considering the wide latitudinal range covered by these studies (36° to 45° S) and the marked differences in seasonal temperatures (13–22 °C in the north, 8–15 °C in the south), there is no clear latitudinal trend in the breeding season. Gamete release occurs from November to February in most populations. The broadest range of times when ripe individuals can be found is at Ranson Head (October to May) near the entrance to Doubtful Sound (Table 2). Different methods have been used by different authors to calculate gonad indices, making comparison among all populations difficult. However, the maximum index reported is approximately 25–29 from populations in both the Marlborough Sounds and Doubtful Sound, respectively (Table 2). There are marked differences in reproductive seasonality between populations that are often only a few kilometres apart. For example, Walker (1982) found significant differences in gonad volume in the Hauraki Gulf (three populations, 10 km apart) during most of the year but most pronounced during summer. Similarly, Brewin et al. (2000) found temporal and spatial differences over less than 10 km

Table 2 Spawning period and maximum gonad index (GI) of *Evechinus chloroticus* arranged latitudinally from the north to south of New Zealand.

Region	Sampling site	Year	Spawning period	Max GI	GI formula	Author
Hauraki Gulf	Noises Island	1975–1976	Nov–Jan	N/A	1	Walker (1982)
Hauraki Gulf	Rangitoto Island	1975–1976	Nov–Jan	N/A	1	Walker (1982)
Hauraki Gulf	Crusoe Island	1975–1976	Jan–Feb	N/A	1	Walker (1982)
Wellington	Reef Bay	1993–1994	Dec–Feb	~20	2	McShane et al. (1996)
Marlborough Sounds	Perano Heads	1990–1992	Nov–Jan	19.3–22.95	3	Brewin et al. (2000)
Marlborough Sounds	Titi Bay	1990–1992	Nov–Feb	15.8–26.99	3	Brewin et al. (2000)
Marlborough Sounds	Dieffenbach Point	1990–1992	Jan–Mar	11.87–20.9	3	Brewin et al. (2000)
Golden Bay	Kaiteriteri	1968–1969	Dec–Apr	N/A	4	Dix (1970c)
Kaikoura		1968–1969	Jan–Feb	N/A	4	Dix (1970c)
Doubtful Sound (DS)	Causet Cove	1993–1995	Jan–April	19.21–22.47	3	Lamare et al. (2002)
DS	Causet Cove	1999–2000	Nov–May	15–16	3	Knapp (pers. comm.)
DS	Espinosa Point	1992–1995	Jan–April	15.26–21.12	3	Lamare et al. (2002)
DS	Deep Cove	1993–1995	March–April	13.74–19.1	3	Lamare et al. (2002)
DS	Seymore Island	1999–2000	Sept–Jan	28–29	3	Knapp (pers. comm.)
DS	Ranson Head	1999–2000	Oct–May	19–21	3	Knapp (pers. comm.)
Dusky Sound	Anchor Island	1992–1993	Dec–Feb	~8–15	2	McShane et al. (1996)

1. Gonad volume/test diameter; 2. Gonad volume/drained wet weight × 100; 3. Gonad wet weight/drained wet weight × 100; 4. Gonad volume/test volume × 10.

in three populations in the Marlborough Sounds and Lamare et al. (2002) throughout the length of Doubtful Sound. In these populations there were also differences in rates of gametogenesis and recovery of individuals and gametogenic differences between sexes, although these were minor in Doubtful Sound (Lamare et al. 2002).

7.3. Reproductive Output

While gonad index (GI), which is commonly used by many authors to describe the reproductive cycle, provides a measure of reproductive potential, reproductive output (difference in pre- and postspawning GI) is a useful measure of actual gamete production into the pool of larvae available for recruitment. While this is easily calculated, few authors provide this information. Brewin et al. (2000) found that annual mean gamete output (AGO) for three populations in the Marlborough Sounds varied from 0.068 to 0.108 g per gram individual per year with no direct relationship between output and length of spawning period. Lamare et al. (2002) report that in the three sites monitored in Doubtful Sound, Causet Cove (outer) Espinosa Point (mid) and Deep Cove (inner), the AGO was 0.108, 0.092 and 0.058 g per gram individual per year, respectively, probably reflecting differences in food availability. This conclusion is similar to that drawn by Wing et al. (2003).

7.4. Size at Sexual Maturity

The size at which *E. chloroticus* becomes sexually mature differs between populations, being 35–45 mm TD at Kaiteriteri, and 55–75 mm at Kaikoura (Dix 1970c). In Dusky Sound *E. chloroticus* reached sexual maturity at 50–60 mm TD with little difference between populations (McShane et al. 1996). In the laboratory Barker et al. (1998) found that *E. chloroticus* as small as 30 mm TD will develop a gonad when fed prepared diets. Although the gonads were largely composed of nutritive phagocytes, they also contained either oocytes or spermatocytes confirming precocious development. This suggests that sexual maturity may be determined by nutritive input and not urchin size.

7.5. Spawning

Spawning cues are poorly understood in echinoids and there have been few field observations of spawning. Lamare and Stewart (1998) observed a mass in situ synchronous spawning of *E. chloroticus* in Doubtful Sound, on 27 January 1994. Spawning occurred between 17:30 and 18:30 h, 20 min before low tide and coincided with a full moon, spring tides and a period of decreasing sea temperatures. More than 90% of both males and females spawned with gametes clouding the water. A number of gametes were eaten by small labrid fish species. The spawning, which the authors suggest may have occurred over the whole fiord, was followed by a 42–50% decrease in gonad indices and resulted in a widespread, dense cohort of larvae within the fjord (see Section 8). Wing et al. (2003) observed a similar mass spawning with 60–70% of adult urchins spawning at three separate sites in the same general area (Doubtful and Thompson Sounds) on 13 November 1999 during a low spring tide.

Such synchronous spawning may be quite unusual and perhaps reflects the unique hydological conditions within a fiord, open coast populations could be expected to show less synchrony. Certainly even though gonad indices generally indicate that gametogenesis may be synchronous, spawning is usually not (Walker 1982) and is probably induced by very local cues.

8. LARVAL DEVELOPMENT

Fertilized eggs develop through typical four-armed pluteus and eight-armed echinopluteus stages (Hyman 1955) to a competent larva with a juvenile rudiment able to selectively settle and complete metamorphosis. Early development was first described by (Mortensen 1921) through the pluteus stage only. Dix (1969) and Walker (1984) succeeded in rearing larvae to metamorphosis in a minimum time of approximately 30 days. Both authors provide a cursory description of larval development and metamorphosis. The most rapid larval development in culture, in which larvae reached competency at 22 days postfertilization, was obtained by Lamare and Barker (1999). Although these studies provide useful information on the developmental timetable, it is difficult to extrapolate laboratory developmental rates to the field. The only comprehensive investigation on larval ecology of *E. chloroticus* was done by Lamare (1998), who made repetitive plankton samplings at five permanent stations located along a transect running from the head of Doubtful Sound to the entrance. He was able to follow the transport and development of larvae in Doubtful Sound following the mass, synchronous spawning of *E. chloroticus* during the summer of 1993/1994 and also during the following summer (1994/1995) when no mass spawning was observed. Larval densities were approximately ten times higher in 1993/1994 (2743.0 larvae per tow) than in 1994/1995 (155.3 larvae per tow) reflecting the marked seasonal differences in reproductive output mentioned earlier. Larval distribution in the water column suggests that there was a high level of larval retention within the fiord, presumably by entrainment of larvae within the estuarine circulation that occurs within Doubtful Sound (Lamare 1998). Larvae completed development between 4 and 6 weeks within Doubtful Sound, a longer period than predicted by Dix (1969) and Walker (1984). This suggests that development in the plankton was slower than potentially possible. Food limitation of larvae was examined by comparison of larval morphometrics. The larvae sampled exhibited a food-limited morphology (Fenaux et al. 1994). These results strongly suggest that recruitment of *E. chloroticus* in Doubtful Sound is from larvae that originated, were retained and completed development within the fiord. Lamare and Barker (1999) followed the cohort of larvae resulting from the mass spawning of adults mentioned above by sampling with vertical plankton hauls 7 days after this spawning (3 February 1994) and subsequently every two days until no larvae were present in the plankton. By determining the reduction in numbers of each larval stage instantaneous mortality rates (M) for *E. chloroticus* larvae were calculated using three different mathematical models. Instantaneous mortality rate was found to be constant and estimated at 0.164, 0.173 and 0.085 larvae per day for the 3 models, the most accurate estimate of mortality probably being $M = 0.16$ per day. Larvae reach competency in Doubtful Sound between 18 and 31 days, which is 1.05–1.82-fold slower than maximum growth rates recorded in laboratory cultures and further evidence of food limitation in the plankton. Very few estimates of larval mortality in the plankton have ever been calculated and interestingly the mortality estimates determined by (Lamare and Barker 1999) are very similar to those of Rumrill (1987) of 0.156 larvae per day for *Strongylocentrotus droebachiensis*.

9. RECRUITMENT

Recruitment processes in marine invertebrates are complex and involve the interaction between reproduction, larval supply (affected by length of larval development, larval dispersal and larval mortality from multiple sources), settlement and postsettlement survival. Examining recruitment processes is, however, necessary in order to understand the dynamics of marine populations and the identification of metapopulations. It is essential for the effective long-term management of exploited species such as sea urchins. Early juvenile echinoderms are microscopic and tend to be cryptic, and examining patterns of recruitment has proven difficult in many species. They are poorly understood in most populations. Dix (1972) noted populations were often made up of single cohorts and suggested that settlement is irregular from year to year. Walker (1984) examined scrapings from $10\,cm^2$ areas of rock and associated algae collected subtidally from Goat Island Bay, North Auckland and found a very small number (13) of recently settled (1.3 ± 0.7 mm) juveniles in substrata containing the turfing coralline species *Corallina officinalis*. When these juveniles were maintained in laboratory aquaria for 2 months they increased approximately 1 mm in test diameter per month. Andrew and Choat (1985) surveyed numbers of juveniles (21–30 mm TD) in three habitats in northeastern New Zealand and found numbers were high and variable in coralline flat habitats (areas devoid of large brown algae) and low in *E. radiata* forests and deep reef habitats. Juveniles transplanted into deep reef habitats suffered a high mortality rate with a mean of 70% being dead after 18 weeks compared to only 3% in coralline flat habitats. Mortality in *E. radiata* forest treatments was intermediate, with a mean of 37%. Presence of conspecific adults did not significantly influence survivorship of juveniles in any habitat and the authors were unable to ascertain the causes of mortality. They concluded that abundance of juvenile *E. chloroticus* is strongly linked to habitat and argued that successful recruitment into canopy forming algal stands of *E. radiata* is unlikely. These conclusions are based on data generated by manipulations of small urchins of a size which suggests that they are in at least their second or third year. Abundance patterns are likely to have been profoundly influenced by earlier postsettlement processes and may well be very different to patterns of recruitment. In contrast Barker (unpubl. data) transplanted recently settled (0.37 mm diameter) juvenile *E. chloroticus* settled on corallina alga encrusted pebbles in both caged and uncaged treatments into a shallow (12 m) site outside of an *E. radiata* kelp forest and a deeper (16 m) site within the kelp forest in Doubtful Sound. Survival was variable but significantly greater in the deeper kelp forest than the shallow site. Although survival was always greater in caged treatments, the difference between caged and uncaged treatments was not statistically significant. This suggests that if the hypothesis of Andrew and Choat (1985) is correct, patterns of recruitment in Doubtful Sound are clearly different than in northern New Zealand.

The most detailed study on recruitment in *E. chloroticus* is that of Lamare and Barker (2001), where settlement and recruitment was determined by a series of field and laboratory experiments in Doubtful Sound and in Tory Channel, Marlborough Sounds. Settlement samplers based on the design of Harrold et al. (1991) were used to monitor settlement rates of larvae at 1–2 month intervals during 1992, 1993 and 1994. Recruitment rates over the same period were monitored from 3 to 8 month intervals by quantifying both the population size distribution and density of juveniles (individuals <20 mm TD). Rates of settlement and recruitment were higher in Doubtful Sound than in Tory Channel. For Doubtful Sound, a large settlement between August 1992 and February 1993 (up to 1.14 recruits per sampler per day) was followed by an increase in juvenile density, from 2.1 up to 13.8 juveniles per 20 m^2 during the subsequent 9 months. Settlement during the following 2 years was comparatively poor (<0.12 recruits per sampler per day), during which time the density of juveniles decreased from 13.8 down to 2.1 juveniles per 20 m^2. A similar pattern was found in Tory Channel where a small rate of settlement was observed in 1992 (<0.05 recruits per sampler per day) and the density of juveniles over the following year was less than 0.6 individuals per 20 m^2. In 1993, settlement was up to 0.54 recruits per sampler per day, and the density of juveniles increased from 0.3 to 5.0 juveniles per 20 m^2 during the following 5 months. Settlement and recruitment were highly correlated.

In the laboratory, larvae exhibited selective settlement on a range of natural (coralline encrusted rocks and weathered bivalve shells) and artificial substrata (weathered plastic plates as used in the settlement samplers) with settlement approaching 100% on coralline algae (97%) but significantly less on the artificial substrates (58%). Survival of laboratory reared new juveniles settled onto coralline encrusted rocks transplanted into Doubtful Sound was 95% after 21 days in the laboratory and 10% and 24% at depths of 12 and 16 m in the field. Using these data, Lamare and Barker (2001) calculated that the settlement samplers underestimate settlement by 41–100% over the average deployment period (45 days). Even though settlement samplers underestimate settlement, they do appear to provide a powerful yet relatively simple method of measuring recruitment. While much more information is needed to comment more specifically on patterns of recruitment by *E. chloroticus*, it does appear that spatial and temporal variation in recruitment in the Tory Channel is linked to the breeding and spawning periodicity discussed earlier. It is almost certainly accentuated by the relatively open-coast situation compared to the more predictable recruitment in a fiord, a consequence of the closed nature of the Doubtful Sound population (Lamare and Barker 2001). Wing et al. (2003) also monitored settlement within Doubtful and Thomson Sounds and found settlement was highest in mid sound sites and decreased at the entrance and arms of the sounds. They suggest that the most intense settlement resulted from a single mass spawning of urchins within the Sound seen at three separate sites in November 1999 and that this cohort of juveniles settled in late December of that year.

10. POPULATION BIOLOGY

10.1. Growth

Information on the growth rate of any species is crucial to understanding population biology and also critical in estimating rates of mortality (and other life-history characteristics). Growth rates have been modeled for a wide range of echinoids from around the world and a variety of growth models, applied (Ebert and Russell 1993). Less is known about growth in *E. chloroticus* than for other sea urchin species of *Strongylocentrotus* sp. Nevertheless we have a better understanding of growth than for some other life-history characteristics such as recruitment. The earliest study was by Dix (1972), who estimated the age of urchins collected at Kaikoura and at Kaiteriteri, ranging in diameter from 40 to 140 mm, by the use of growth lines in apical plates. He used this and growth of laboratory held juveniles (18 mm test diameter) to estimate annual growth. Walker (1981) also used annual lines in combination with external tags threaded through holes drilled in the test in a tag-recapture experiment to investigate seasonal growth in urchins at Goat Island, Leigh in the north of New Zealand.

Growth estimated from growth lines determined from etched genital plates is confounded by various problems. Firstly, while it is generally valid to assume bands in test plates are laid down annually, other factors such as prolonged shortages of food, or periods of abnormally cold water can also create banding patterns. There is also the general difficulty of validating the age at which the first line is laid down, as this is the oldest and often most difficult to see. These problems have been reviewed in detail by Gage (1991). The most successful method of estimating growth is by the injection of a fluorescent marker such as tetracycline hydrochloride or calcein into the body cavity which is subsequently incorporated into the calcium carbonate skeleton at sites of active calcification and can be seen under ultraviolet light. Urchins can be tagged in situ by injection, and later collected after growth has occurred. Growth estimates can be determined from the increase in size of a skeletal element (e.g. a skeletal plate or demi-pyramid from Aristotle's lantern) from the point where the florescent marker is visible. If the relationship between the skeletal structure and test diameter is known, and provided a range of animals of different sizes are tagged, growth estimates for the population can be determined. This method was used by Lamare and Mladenov (2000) in conjunction with measurements of laboratory reared smaller animals (<10 mm TD) and data gathered from newly settled cohorts to estimate growth of two populations of *E. chloroticus* from Espinosa Point, Doubtful Sound and the Tory Channel. A small sample (5) of recently settled juveniles (0.37 mm TD at settlement) had grown to 8.05 mm (laboratory) and 10.5 mm (field) TD after 1 year. Growth was slow for the first 200 days after settlement and then increased. Calcein tagged adult urchins were sampled at 1 (37.7% recovered) and 3.9 (30.12% recovered) years after tagging and their growth modelled using one nonasymptotic (Tanaka) and three asymptopic growth functions (Brody-Bertalanffy, Richards and Jolicoeur). All models predicted faster growth over the first 2–4 years in the Tory Channel population but a larger maximum size in Doubtful Sound. Lamare and Mladenov (2000) concluded that overall the Richards model best describes growth in *E. chloroticus* because it most accurately predicts the size at age 1 year and also predicts a noted decrease in somatic growth at reproductive maturity (30–40 mm TD). Growth curves using other models do not meet these criteria. The Richards model predicts a decrease in somatic growth in the Tory Channel after an age of 1.76 years (22 mm TD) and 2.31 years (30 mm TD) in Doubtful Sound with growth decreasing rapidly after ages 3–4 years (40–50 mm TD). Wing et al. (2003) used calcein tagging to examine *E. chloroticus* growth throughout the Doubtful–Thompson Sound complex and found adult growth rates differed markedly between sites. Fastest growth and larger overall sizes were found at outer sound sites where there was an abundance of laminarian algae, while inner sound sites with poorer food quality had slower growth and smaller mean size suggesting that growth may be energetically limited in this fiord complex. The growth curves calculated by Dix (1972) show considerably faster growth for two of the three populations studied. While it is possible the suggested sizes at age are correct, the problem mentioned above makes it highly likely that Dix (1972) overestimated the growth rate. McShane et al. (1997) also used calcein tagging to model growth in *E. chloroticus* from populations in Dusky Sound and from Arapawa Island in the Marlborough Sounds. They concluded that urchins in Dusky Sound reached a larger size than individuals elsewhere but no growth curve is provided making size at age comparisons with other populations difficult.

10.2. Mortality

Mortality is most accurately determined from tagging and recapture or from population size–frequency analysis. However, there have been very few studies in which mortality has been calculated for *E. chloroticus*. Lamare (1997) applied Ebert's (1973) model, based on analysis of population size–structure, to determine instantaneous mortality (Z) in Doubtful Sound and Tory Channel populations. The results of these analyses suggest mortality is higher in Doubtful Sound compared to Tory Channel. Annual mortality and mean longevity were calculated as 9.21% and 10.38 years in the Doubtful Sound populations and 5.01% and 19.44 years for Tory Channel.

Mortality could result from predation (natural and human), disease, morbidity or from physical damage due to environmental perturbations or even ultraviolet radiation (Lamare et al. 2004). Lamare (1997) ascribes higher mortality in Doubtful Sound to predation by a number of potential predators including a number of benthic feeding fishes, asteroids, molluscs and lobsters. The most conspicuous and probably the most important predators are the large asteroids, *Coscinasterias muricata* and *Astrostole scabra.* These were often observed feeding on *E. chloroticus*. In comparison, Tory Channel has lower densities of most of these predators, particularly benthic invertebrate predators. Periodic rapid intrusions of low-salinity water could also contribute to mortality in Doubtful Sound, particularly for juveniles which commonly occur at shallower depths than adults. Some commercial harvesting of *E. chloroticus* also occurs in Tory Channel. However, harvesting is very low or absent in the Doubtful Sound and probably only accounts for a minor proportion of total mortality in this population Lamare (1997).

Although such estimates of mortality based on population data are rare, there have been several other studies where predation on *E. chloroticus* has been the focus. Andrew and

Choat (1982) used exclusion cages over a 16 month period to evaluate fish predation on juvenile *E. chloroticus* at a subtidal coralline flat site where invertebrate predators occurred in low numbers. The abundance of juveniles within the cages increased significantly over the experimental period, indicating that fish are important predators on juvenile *E. chloroticus*. A number of studies have suggested that the clawed lobster *Homarus americanus* is important in controlling the abundance of *S. droebachiensis* in the east coast of Canada and USA (Bernstein et al. 1981; Breen and Mann 1976; Garnick 1989), although some authors disagree that lobsters play a keystone role in urchin populations (Miller 1985). On rocky reefs throughout New Zealand *E. chloroticus* and the rock lobster, *J. edwardsii,* have a loose association although in southern New Zealand they are seldom seen in close proximity (pers. obs.). Andrew and MacDiarmid (1991) investigated the interaction between *E. chloroticus* and *J. edwardsii* in the shallow subtidal zone of northern rocky reefs. During the day, spiny lobsters were cryptic and spatially segregated on a small scale from sea urchins which were generally exposed. However, during the night both species were active and lobsters moved considerable distances and were likely to have preyed on sea urchins. Laboratory experiments demonstrated that large lobsters ate all sizes of sea urchins and all sizes of lobsters ate small sea urchins (<50 mm TD) in preference to larger sea urchins. Sea urchins have also been found in the guts of lobsters and broken tests in lobster dens (Andrew and McDiamid 1991). Clearly, lobsters prey on sea urchins. However, as lobsters are rare in areas of barrens where *E. chloroticus* is most abundant, and urchins are absent from reefs deeper than 12 m where spiny lobsters are abundant, Andrew and McDiamid (1991) doubted spiny lobsters regulate urchin numbers. If regulation does occur they believe it is mainly on urchins of 40–50 mm TD when they become less cryptic and move from crevices into open rock habitat. Shears and Babcock (2003) used tethering experiments to examine predation on *E. chloroticus* in two marine reserves and in unprotected sites in northern New Zealand. They found that predation in both reserves was 7 times higher than outside the reserve. Predation was highest on small urchins (30–40 mm TD) and, in contrast to the conclusions drawn by Andrew and McDiamid (1991), most predation on larger urchins (55–80 mm) within the marine reserves could be attributed to the spiny lobster, perhaps as a result of a gradual increase in lobster numbers within the marine protected areas. Predation on smaller urchins was by both lobsters and the fish *Pagrus auratus*. Within nonprotected areas predation of tethered urchins was from different predators, principally the starfish *C. muricata* and the gastropod *Charonia lampas*.

Although disease decimates urchin populations of *S. droebachiensis* in Nova Scotia (Scheibling and Stephenson 1984, Scheibling and Raymond 1990) it does not appear to be such an important cause of mortality in other sea urchin species, although there are some reports of disease affecting *E. chloroticus*. Lamare (1997) observed a few individuals of *E. chloroticus* with spine loss resembling 'bald-sea-urchin disease' (Maes et al. 1986) in both Doubtful Sound and the Tory Channel. During the summer of 1999/2000 diseased urchins in northern New Zealand were also reported to be dying of 'bald-sea-urchin disease'. The incidents of infection were widespread, from Cape Reinga, Mayor Island, Leigh and the Bay of Plenty with deaths higher in areas where urchins were present in higher densities (Babcock, pers. comm.). Babcock also established that apparently normal urchins were infected when held in a seawater tank with infected individuals.

10.3. Population Genetics

With a latitudinal distribution of over 1500 km on east and west coasts of New Zealand and the established differences between populations already mentioned (size, recruitment, habitat preferences, etc.) it might be expected that *E. chloroticus* populations would show some genetic variation over their range. Using gel electrphoresis, Mladenov et al. (1997) examined genetic differences at five polymorphic enzyme loci in six widely separated urchin populations (Leigh in the north to Stewart Island in the south). They found little genetic variation between populations and suggested the long larval life (probably 4–6 weeks, see Section 8) allowed high gene flow between populations. Urchins collected from Doubtful Sound showed slight evidence of genetic differentiation ($D = 0.011–0.019$), possibly the result of larval retention within the fiord discussed earlier (see Section 8). In contrast, Perrin (2002) and Perrin et al. (2003) used six microsatellite loci to examine genetic variation in *E. chloroticus* populations in both southern fiords and other sites around New Zealand. At a scale of >1000 km, restricted gene flow between North and South Island was observed. On a scale of 10–200 km, significant differences were found within fiords, among fiords, and between fiords and the open coast. Partial reproductive isolation was observed between two habitats: (1) the open coast and outer fiord populations and (2) the inner fiords. Perrin (2002) indicates that the divergence between the two ecotypes is maintained by local adaptation. Furthermore, the study suggests that genetic differentiation among some inner fiord populations is a result of both historic recruitment events and subsequent restricted larval dispersal within the fiords (see Section 8).

REFERENCES

Andrew NL (1988) Ecological aspects of the common sea urchin, *Evechinus chloroticus*, in northern New Zealand: A review. NZ J Mar Freshwater Res 22: 415–426

Andrew NL, Choat JH (1982) The influence of predation and conspecific adults on the abundance of juvenile *Evechinus chloroticus* (Echinoidea: Echinometridae). Oecologia 54: 80–87

Andrew NL, Choat JH (1985) Habitat related differences in the survivorship and growth of juvenile sea urchins. Mar Ecol Prog Ser 27: 155–161

Andrew NL, MacDiarmid AB (1991) Interrelations between sea urchins and spiny lobsters in northeastern New Zealand. Mar Ecol Prog Ser 70: 211–222

Andrew NL, Stocker LJ (1986) Dispersion and phagokinesis in the echinoid *Evechinus chloroticus* (Val.). J Exp Mar Biol Ecol 100: 1–3

Ayling AL (1978) The relation of food availability and food preferences to the field diet of an echinoid *Evechinus chloroticus* (Valenciennes). J Exp Mar Biol Ecol 33: 223–235

Ayling AM (1981) The role of biological disturbance in temperate subtidal encrusting communities. Ecology 62: 830–847

Barker MF, Keogh JA, Lawrence JM, Lawrence AL (1998) Feeding rate, absorption efficiencies, growth, and enhancement of gonad production in the New Zealand sea urchin *Evechinus chloroticus* Valenciennes (Echinoidea: Echinometridae) fed prepared and natural diets. J Shellfish Res 17: 1583–1590

Bernstein BB, Williams, BE, Mann, KH (1981) The role of behavioural responses to predators in modifying urchins (*Strongylocentrotus droebachiensis*) destructive grazing and seasonal foraging patterns. Mar Biol 63: 39–49

Breen PA, Mann, KH (1976) Changing lobster abundance and the destruction of kelp beds by sea urchins. Mar Biol 34: 137–142

Brewin PE, Lamare MD, Keogh JA, Mladenov PV (2000) Reproductive variability over a four year period in the sea urchin, *Evechinus chloroticus* (Echinoidea: Echinodermata) from differing habitats in New Zealand. Mar Biol 137: 543–557

Choat JH, Andrew NL (1986) Interactions amongst species in a guild of subtidal benthic herbivores. Oecologia 68: 387–394

Choat JH, Schiel DR (1982) Patterns of distribution and abundance of large brown algae and invertebrate herbivores in subtidal regions of northern New Zealand. J Exp Mar Biol Ecol 60: 129–162

Cole RG, Haggitt T (2000) Dietary preferences of *Evechinus chloroticus*. In: Barker MF (ed.) 10th International Echinoderm Conference, pp 425–430

Cook S (in press) Echinoids. In: Cook S (ed.) New Zealand coastal invertebrates. Canterbury University Press, Christchurch

Dix TG (1969) Larval life span of the echinoid *Evechinus chloroticus* (VAL.). NZ J Mar Freshwater Res 3: 13–16

Dix TG (1970a) Biology of *Evechinus chloroticus* (Echinodermata: Echinometridae) from different localities. 1. General. NZ J Mar Freshwater Res 4: 91–116

Dix TG (1970b) Biology of *Evechinus chloroticus* (Echinodermata: Echinometridae) from different localities. 2. Movement. NZ J Mar Freshwater Res 4: 267–277

Dix TG (1970c) Biology of *Evechinus chloroticus* (Echinodermata: Echinometridae) from different localities. 3. Reproduction. NZ J Mar Freshwater Res 4: 385–405

Dix TG (1972) Biology of *Evechinus chloroticus* (Echinodermata: Echinometridae) from different localities. 4. Age, growth and size. NZ J Mar Freshwater Res 6: 48–68

Don GL (1975) The effects of grazing by *Evechinus chloroticus* (Val.) on populations of Ecklonia radiata (Ag.). MSc Thesis, University of Auckland

Ebert TA (1973) Estimating growth and mortality rates from size data. Oecologia 11: 281–298

Ebert TA, Russell MP (1993) Growth and mortality of subtidal red sea urchins (*Strongylocentrotus franciscanus*) at San Nicolas Island, California, USA: Problems with models. Mar Biol 117: 79–89

Fenaux L, Strathmann MF, Strathmann RR (1994) Five tests of food-limited growth of larvae in coastal waters by comparisons of rates of development and form of echinoplutei. Limnol Oceanogr 39: 84–98

Gage JD (1991) Skeletal growth zones as age-markers in the sea urchin *Psammechinus miliaris*. Mar Biol 110: 217–228

Garnick E (1989) Lobster (*Homarus americanus*) population declines, sea urchins, and barren grounds': A space-mediated competition hypothesis. Mar Ecol Prog Ser 58: 23–28

Harrold C, Reed DC (1985) Food availability, sea urchin grazing, and kelp forest community structure. Ecology 66: 1160–1169

Harrold C, Lisin S, Light KH, Tudor S (1991) Isolating settlement from recruitment of sea urchins. J Exp Mar Biol Ecol 147: 81–94

Hyman LH (1955) The Invertebrata, Vol. 4. Echinodermata. McGraw Hill, New York

Lamare MD (1997) Population biology, pre-settlement processes and recruitment in the New Zealand sea urchin *Evechinus chloroticus* Valenciennes (Echinoidea: Echinometridae). PhD Thesis, University of Otago, Otago

Lamare MD (1998) Origin and transport of larvae of the sea urchin *Evechinus chloroticus* (Echinodermata: Echinoidea) in a New Zealand fiord. Mar Ecol Prog Ser 174: 107–121

Lamare MD, Mladenov PV (2000) Modelling somatic growth in the sea urchin *Evechinus chloroticus* (Echinoidea: Echinometridae). J Exp Mar Biol Ecol 243: 17–43

Lamare MD, Barker MF (1999) In situ estimates of larval development and mortality in the New Zealand sea urchin *Evechinus chloroticus* (Echinodermata: Echinoidea). Mar Ecol Prog Ser 180: 197–211

Lamare MD, Barker MF (2001) Settlement and recruitment in the New Zealand sea urchin *Evechinus chloroticus* Valenciennes (Echinoidea: Echinometridae) over a three year period. Mar Ecol Prog Ser 218: 153–166

Lamare MD, Stewart BG (1998) Mass spawning by the sea urchin *Evechinus chloroticus* (Echinodermata: Echinoidea) in a New Zealand fiord. Mar Biol 132: 135–140

Lamare MD, Brewin PE, Barker MF, Wing SR (2002) Reproduction of the sea urchin *Evechinus chloroticus* (Echinodermata: Echinoidea) in a New Zealand fiord. NZ J Mar Freshwater Res 36(4): 719–732

Lamare MD, Lesser MP, Barker MF, Schimanski KB (2004) Variation in sunscreen compounds (mycosporine-like amino acids) for marine species along a gradient of ultraviolet radiation transmission within Doubtful Sound, New Zealand. NZ J Mar Freshwater Res 38(5): 775–793

Maes P, Jangoux M, Fenaux L (1986) The "bald-sea-urchin" disease: Ultrastructure of the lesions and nature of their pigmentation. Ann Inst Oceanogr, Paris (Nouv Ser) 62: 37–45

McRae A (1959) *Evechinus chloroticus* (Val.), an endemic New Zealand echinoid. Trans Roy Soc NZ 86: 205–267

McShane (1992) Sea urchin fisheries of the world-prospects for development of a kina fishery in New Zealand. NZ Professional Fisherman June 92: 27–40

McShane PE, Anderson O, Gerring PK, Stewart RA, Naylor JR (1994) Fisheries biology of kina (*Evechinus chloroticus*). MAF Fisheries, New Zealand Fisheries Assessment Research Document 94/17, 34pp

McShane PE, Gerring PK, Owen AA, Stewart RA (1996) Population differences in the reproductive biology of *Evechinus chloroticus* (Echinoidea: Echinodermetridae). NZ J Mar Freshwater Res 30: 333–339

Miller RJ (1985) Seaweeds, sea urchins, and lobsters: A reappraisal. Can J Fish Aquat Sci 42: 2061–2072

Mladenov PV, Allibone RM, Wallis GP (1997) Genetic differentiation in the New Zealand sea urchin *Evechinus chloroticus* (Echinodermata: Echinoidea). NZ J Mar Freshwater Res 31: 261–269

Mortensen T (1921) Studies on the development and larval forms of echinoderms. GEC Gad, Copenhagen, 261pp

Mortensen T (1922) Echinoderms of New Zealand and the Auckland Campbell Islands. I. Echinoidea. Vidensk Meddr dansk naturh Foren 73: 139–198

Pawson DL (1961) Distribution patterns of New Zealand echinoderms. Tuatara 9: 9–18

Pawson DL (1965) New records of echinoderms from the Snares Islands to the south of New Zealand. Tuatara 6: 9–18

Perrin C (2002) The effects of fiord hydrography and environment on the population genetic structures of the sea urchin *Evechinus chloroticus* and the sea star *Coscinasterias muricata* in New Zealand. PhD Thesis, University of Otago, New Zealand, p 206

Perrin C, Roy MS, Wing SR (2003) Genetic differentiation amongst populations of the sea urchin *Evechinus chloroticus* and the sea star *Coscinasterias muricata* in New Zealand's fiords. In: Féral JP, David B (eds) Echinoderm research 2001. AA Balkema, Rotterdam, pp 7–13

Rumrill SS (1987) Differential predation upon embryos and larvae of temperate Pacific echinoderms. PhD Thesis, University of Alberta, Edmonton

Scheibling RE, Raymond BG (1990) Community dynamics on a subtidal cobble bed following mass mortalities of sea urchins. Mar Ecol Prog Ser 63: 127–145

Scheibling RE, Stephenson RL (1984) Mass mortality of *Strongylocentrotus droebachiensis* (Echinodermata: Echinoidea) off Nova Scotia, Canada. Mar Biol 78: 153–164

Schiel DR (1982) Selective feeding by the echinoid, *Evechinus chloroticus*, and the removal of plants from subtidal algal stands in northern New Zealand. Oecologia 54: 379–388

Schiel DR, Kingsford MJ, Choat JH (1986) Depth distribution and abundance of benthic organisms and fishes at the subtropical Kermadec Islands. NZ J Mar Freshwater Res 20: 521–535

Schiel DR, Andrew NL, Foster MS (1995) The structure of subtidal algal and invertebrate assemblages at the Chatham Islands, New Zealand. Mar Biol 123: 355—367

Shears NI, Babcock RI (2002) Marine reserves demonstrate top-down control of community structure on temperate reefs. Oecologia 132: 131–142

Shears NT, Babcock RC (2003) Continuing trophic cascade effects after 25 years of no-take marine reserve protection. Mar Ecol Prog Ser 246: 1–16

Steinberg PD, Altena I (1992) Tolerance of marine invertebrate herbivores to brown algal phlorotannins in temperate Australasia. Ecol Monogr 62: 189–222

Van Alstyne KL (1988) Herbivore grazing increases polyphenolic defenses in the intertidal brown alga *Fucus distichus*. Ecology 69: 655–663

Villouta E, Chadderton WL, Pugsley CW, Hay CH (2001) Effects of sea urchin (*Evechinus chloroticus*) grazing in Dusky Sound, Fiordland, New Zealand. NZ J Mar Freshwater Res 35: 1007–1024

Walker MM (1981) Influence of season on growth of the sea urchin *Evechinus chloroticus*. NZ J Mar Freshwater Res 15: 201–205

Walker MM (1982) Reproductive periodicity in *Evechinus chloroticus* in the Hauraki Gulf. NZ J Mar Freshwater Res 16: 19–26

Walker MM (1984) Larval life span, larval settlement, and early growth of *Evechinus chloroticus* (Valenciennes). NZ J Mar Freshwater Res 18: 393–397

Wing SR, Vasques J, Lamare MD (2001) Population structure of sea urchins (*Evechinus chloroticus*) along gradients in bentic productivity in the New Zealand fjords. Proceeding of the 10th International Echinoderm Conference, Dunedin, New Zealand, pp. 569–575

Wing SR, Gibbs MT, Lamare MD (2003) Reproductive sources and sinks within a sea urchin, *Evechinus chloroticus*, population of a New Zealand fjord. Mar Ecol Prog Ser 248: 109–123

Edible Sea Urchins: Biology and Ecology
Editor: John Miller Lawrence

Chapter 17

Ecology of *Heliocidaris erythrogramma*

John K Keesing

Commonwealth Scientific and Industrial Research Organization, Wembley, WA (Australia).

1. INTRODUCTION

Commonly known as the purple sea urchin, *Heliocidaris erythrogramma* (Valenciennes 1846) is endemic to Australia, being distributed on rocky reefs from the intertidal zone down to about 35 m along the east, west and south Australian coasts, south of ~25° S (Endean 1957; Dix 1977a; Kailola et al. 1993). The species is common throughout its distribution and may be abundant (ca. 20–80 individuals per square metre) (Andrew 1999). While usually 60–90 mm in diameter, it can reach 125 mm in Tasmania (Growns and Ritz 1994).

Heliocidaris erythrogramma occurs in a variety of habitats throughout its range, from moderate energy, open rocky shores to estuaries. In moderate wave-energy sites it occupies crevices in the reef. In more sheltered areas such as embayments, it aggregates on rocks and also on muddy bottoms and seagrass areas (Kailola et al. 1993). Dix (1977b) recorded *H. erythrogramma* in sheltered to moderately exposed sites in Tasmania amongst boulders, rubble and ledges, most abundantly in depths less than 10 m. Dix (1977b) noted that *H. erythrogramma* is not found in very exposed environments along the southern and south west coast of Tasmania where the bull kelp, *Durvillea potatorum,* is the dominant flora and the abalone *Haliotis rubra*, a dominant herbivore. Sanderson et al. (1996) describe the algal flora of typical habitat of *H. erythrogramma*, noting that in many areas urchin barrens have resulted in macroalgae only occurring in refuges in very shallow waters.

Underwood et al. (1991) characterised assemblages of marine communities in New South Wales and found *H. erythrogramma* most abundant in the immediate subtidal fringe of reefs adjacent to the intertidal zone down to about 3 m depth. In an earlier study in NSW, Shepherd (1973) found that *H. erythrogramma* preferred shallow (ca. 1–5 m) sheltered habitat dominated by *Durvillea,* but in more exposed conditions *H. erythrogramma* extended down to about 20 m depth amongst both *Durvillea* and *Ecklonia* kelp dominated communities where the other common sea urchin species *Centrostephanus rodgersii* also occurred. Wright et al. (2005) and Wright and Steinberg (2001) described high-density (80–192 individuals per square metre) feeding fronts of *H. erythogramma* in 3–4 m depth in a diverse subtidal macroalgal community near Sydney, New South Wales.

Constable (1989) described the habitats of *H. erythrogramma* in Port Philip Bay, Victoria. They varied between sites from well-vegetated reefs dominated by kelps to seagrass beds and barren areas. On the mostly limestone and sandstone reefs, sea urchins occupied crevices and burrows created from a long history of grazing. In other locations they occupied boulder areas. Constable (1989) also found individuals in barren areas separated from kelp beds by short stretches of sand, which precluded their movement between the barrens and the kelp. Sea urchins in barrens occurred all over the boulders while those in the kelp areas occupied burrows. The mean densities of individuals in the seagrass areas were over 50 individuals per square metre compared to 2–6 individuals per square metre in the kelp and barrens habitats.

Constable (1989) found *H. erythrogramma* forming a dense front 3–10 m wide at the edge of a bed of the seagrass *Heterozostera tasmanica*. Few individuals were found actually within the bed. In these areas the lack of hard substrate or vertical relief probably had a strong influence on their distribution. High-density feeding fronts in seagrass beds have also been recorded in Botany Bay near Sydney, New South Wales (Buchner 1986; Roberts et al. 1986).

In South Australia, *H. erythrogramma* is common on shallow subtidal limestone and granite reefs among rocks, boulders or in crevices (Connolly 1986), often being quite abundant and in association with the abalone *H. laevigata* (Keesing, pers.obs.). Connolly (1986) described the physical characteristics and vegetation of habitats of *H. erythrogramma* at coastal sites in South Australia where habitat variability was similar to that of Victorian studies. Shepherd (1974) described the boulder habitat of a large population of *H. erythrogramma* in northern Spencer Gulf in South Australia. There sea urchins occupied boulder habitat down to about 3.5 m at densities of 3 to 22 individuals per square metre and barrens had developed where urchin densities exceeded 10 individuals per square metre. Beyond this depth *H. erythrogramma* occurred amongst seagrass beds where it lived on the live shells of the large bivalve *Pinna*, which protrudes from the sediment in which it is buried.

Vanderklift and Kendrick (2004) examined distribution and abundance of temperate sea urchins including *H. erythrogramma* in south-western Australia. Their studies (see also Vanderklift 2002) over a 250 km region showed extensive patchiness in sea urchin abundance with densities up to 8 individuals per square metre. They found *H. erythrogramma* was equally abundant on both reef flat and rock face habitats unlike two other sympatric species (*Phyllacanthus irregularis* and *Centrostephanus tenuispinus*) which principally occupied the more structurally complex reef face habitat.

The congeneric species, *Heliocidaris tuberculata* (Lamarck 1816) also occurs within a narrow distribution in Australia between southern NSW and southern Queensland (Edgar 1997). It also occurs on the Kermadec Islands off New Zealand (Edgar 1997) and on Lord Howe Island where it is particularly abundant and the dominant sea urchin species (Dakin 1980). In New South Wales *H. tuberculata* occurs in the sublittoral in contrast to *H. erythrogramma* which occurs in the intertidal and shallow subtidal areas. The two species can occur sympatrically over the distribution of *H. tuberculata* in New South Wales where it grows to a much larger size (106 mm test diameter) than *H. erythrogramma* (Clark 1946).

2. POPULATION GENETICS AND COLOUR VARIABILITY

As might be expected of a species with a very short planktonic larval stage, *H. erythrogramma* has significant levels of genetic variation geographically. Two clades are recognised, one on the western coast of Australia and another in the east and south. Significant population differentiation also occurs between eastern and southern populations and within southern populations separated by Bass Strait (McMillan 1991; McMillan et al. 1992).

Although known as the purple sea urchin, *H. erythrogramma* occurs in a broad range of colours from violet to green and white. Growns and Ritz (1994) examined and defined the spinochrome pigments responsible for colour expression in *H. erythrogramma*. They found colour variation was not random but varied between populations and between habitats, suggesting that colour was environmentally and/or genetically determined. Growns (1991, in Growns and Ritz 1994) showed some correlation between habitat type, including level of wave exposure and sea urchin colour. Growns and Ritz (1994) postulated that gene flow between populations will be low because of the short planktonic larval phase and that selective pressures may vary between locations and habitats giving rise to variable colour expressions between populations.

3. REPRODUCTION, DEVELOPMENT, SETTLEMENT AND RECRUITMENT

The lecithotrophic development of *H. erythrogramma* is unusual. For part of its range, *H. erythrogramma* occurs sympatrically with the much larger, congeneric *H. tuberculata*, which exhibits more typical planktotrophic larval development (Laegdsgaard et al. 1991). The large (400 μm) buoyant eggs develop as fully planktonic vitellaria after 36 h metamorphosing after about five days remaining as a lecithotroph for about 3 weeks (Williams and Anderson, 1975). This compares with an egg size of 95 μm and a larval development period of 21–30 days in *H. tuberculata* (Hoegh-Guldberg and Emlet 1997).

Lawrence and Byrne (1994) contrasted various structural and soft organ body components in *H. erythrogramma* with *H. tuberculata* and related these to the different life history strategies. The larger size of *H. tuberculata* results in a greater reproductive output. *Heliocidaris erythrogramma* has ovarian lipid concentration double that of *H. tuberculata*, providing nutrients for the large nonfeeding larvae. Emlet and Hoegh-Guldberg (1997) have shown the importance of the lipid-rich material in eggs of *H. erythrogramma* to growth and survivorship of postlarval juveniles as it enables postmetamorphic individuals to outgrow some predators and assimilate with their new habitat prior to having to feed. Spawning occurs in early summer to autumn (December to March) over a large part of the species' eastern Australian distribution, including Sydney (Williams and Anderson 1975; Laegdsgaard et al. 1991), Hobart (Dix 1977a) and Melbourne (Constable 1989). The most distance sites are separated by about 10° of latitude. Reproductive periodicity has not been studied in South Australia or Western Australia.

Styan (1997) examined the fertilisation success of *H. erythrogramma* in the field by measuring fertilisation rates of eggs from spawning females upstream of a spawning male. He found that the fertilisation rate of its eggs is diminished greatly if spawning does not occur in close proximity to males. Fertilisation success of 48% occurred when males and females were separated by more than 1 m and only 7% when separated by 10 m. This indicates the importance of high-density aggregations in ensuring fertilisation success, especially when the larvae have very limited capacity for dispersal (McMillan et al. 1992) and particularly in asynchronous spawners (*sensu* Dix 1977a) such as *H. erythrogramma*. More recently, Marshall et al. (2004) also showed that timing of spawning (mating order) enhanced fertilisation success and offspring size in *H. erythrogramma* with fertilisation success rates favouring males whose sperm had first access to spawned eggs and hence fertilised the largest and most viable eggs in a spawned batch.

Dix (1977a) found that *H. erythrogramma* matures at 2.3 cm diameter, but that spawning may not occur until it is 4–5 cm. Dix also found that 56% of the individuals sampled >6 cm were female and noted that this was significantly different from a 1:1 ratio. An increase in preponderance of females with size would suggest that females have a faster growth rate or are subject to differential mortality rates than males. Age at maturity is not well defined and the data that are available are conflicting. Constable (1989) provided the best data. He found that gametes are present in *H. erythrogramma* at about 2 years of age with size at maturity differing between sites, 30 mm in faster-growth sites and 25 mm in slower-growth sites. Sanderson et al. (1996) suggested that maturity occurred at 5–10 years based on their tetracycline tagging data, but acknowledged that this estimate was ambiguous. In addition to good data on size and age at maturity being lacking for *H. erythrogramma* across its distribution, there are no data available on fecundity.

There are also almost no data on levels of population replacement in *H. erythrogramma*. Ebert's (1982) study of survival rates among populations of *H. erythrogramma* found very high rates of survivorship ($M < 0.1$) which is consistent with very low levels of recruitment. There is some evidence of this from Sanderson et al. (1996), who found recruitment rates of between 0.3 and 1.0 individuals per square metre per year. Vanderklift and Kendrick (2004) observed little change in abundance of populations of *H. erythrogramma* on six reefs over 26 months and also concluded that recruitment rates were either low or sporadic.

There are few records of juvenile *H. erythrogramma* with the exception of Constable (1989). Connolly (1986) found no individuals <2 cm at a number of sites in South Australia. Sanderson et al. (1996) made no mention of small individuals. Dix (1977a) collected six individuals <2 cm but did not comment on the habitat or the general occurrence or abundance of small *H. erythrogramma*. The only detailed report on the recruitment of small *H. erythrogramma* is that of Constable (1989), who found individuals >10 mm about 6 months after spawning. Constable (1989) found small individuals in seagrass, barrens and kelp habitats. The smallest ones (<10 mm) occurred buried among shell fragments in seagrass habitats. It seems logical that small juveniles are likely to be very cryptic given their vulnerability to predation, although Connolly (1986) noted that small individuals (2–3 cm) were only found on the rocks where macroalgal cover was absent.

4. GROWTH AND AGE

Ebert (1982) concluded that *H. erythrogramma* is a long-lived, slow-growing species with low levels of natural mortality. Sanderson et al. (1996) used three methods to determine the growth rate of *H. erythrogramma* in Tasmania. Extensive size–frequency analysis proved inconclusive except in one area where freshwater runoff during the wettest summer for 25 years killed the majority of the individuals in 1993/1994. This enabled a cohort with a modal size of 27.5 mm diameter (assumed to be 18 months old in the winter of 1995) to be followed for 10 months, during which time they grew to 39 mm or about 14 mm per annum. Constable (1989) gave the only assessment of growth in first year cohorts of *H. erythrogramma*. He examined the growth of a cohort of individuals found at approximately 10 mm diameter. From knowledge of the spawning period, he established that growth in the first year was to 20–25 mm with much slower growth in larger individuals, those of 40 mm increasing only about 5 mm in a year.

Sanderson et al. (1996) also examined the number of growth ridges on Aristotle's lantern teeth and growth of teeth using the tetracycline marking method of (Ebert 1982). The former method gave much faster-growth characteristics (growth coefficient k averaging 0.42 over a number of sites with an average maximum size (L_{inf}) of 80 mm) than the latter at two sites ($k = 0.2$–0.21, $L_{\mathrm{inf}} = 64$–85 mm). Both methods have problems in that the number of growth ridges laid down per year varies between 1 and 3 per year at different locations. The tetracycline tagging method is perhaps more reliable, but is invasive and subject to underestimating k where sample size is small and if insufficient young animals are included in the sample as in this case. The cohort analysis data obtained by Sanderson et al. (1996) suggests k in the range of 2.8–3.5 although maximum size for this site is not known. Ebert (1982) used the tetracycline tagging method at two locations geographically remote from each other and the Tasmanian sites studied by Sanderson et al. (1996). In Western Australia Ebert (1982) found a k value for *H. erythrogramma* of 0.191 ($L_{\mathrm{inf}} = 72$ mm) and in New South Wales a faster rate of 0.356 ($L_{\mathrm{inf}} = 76$ mm). Ebert (1982) estimated that *H. erythrogramma* might grow to 4 cm in its first year in Western Australia. This is probably an over estimation based on the results of Constable (1989).

Sanderson et al. (1996) found the maximum size attained by *H. erythrogramma* in areas studied in Tasmania was about 100 mm but individuals >80 mm were not abundant at most sites. The species has been recorded to grow to 125 mm in test diameter in Tasmania (Growns and Ritz, 1994). At the mainland sites studied by Ebert (1982) *H. erythrogramma* grew only to about 90 mm but individuals >75 mm (New South Wales) and >80 mm (Western Australia) were uncommon. There remains a great need to examine the growth and survivorship of *H. erythrogramma* throughout its range using a suite of complementary methods to obtain reliable and representative estimates of size at age.

Constable (1993) used *H. erythrogramma* to demonstrate that shrinkage in test diameter and volume in starved sea urchins was explained by reduction in suture width. This phenomenon of negative growth is often encountered in studies of sea urchins where seasonal food limitation is encountered and compromises estimates of size at age. These conditions certainly occur in *H. erythrogramma*. Constable (1989) found that the size attained in *H. erythrogramma* is dependent on food supply and probably exposure to wave action in his comparison of a number of populations of *H. erythrogramma* in different habitat types.

5. MOVEMENT AND FEEDING

Heliocidaris erythrogramma moves mostly at night and unlike many sea urchins, shows little fidelity to individual crevices and shelters (Andrew 1999). Connolly (1986) found negligible diel movement by *H. erythrogramma* in South Australia. Sanderson et al. (1996) found minimal movement by *H. erythrogramma* over periods of up to 1 year in Tasmania in cleared $100\,m^2$ quadrats. While their data support the generalisation that *H. erythrogramma* moves little, other conclusions drawn about movement into cleared areas being minimal are not necessarily supported by the data. Sanderson et al. (1996) concluded that movement of sea urchins into cleared areas was low because of the shift in size structure from large individuals before clearing to small ones over a period of 1 year. They concluded that small sea urchins found in the cleared sites were new recruits and dismissed their appearance being related to migration as unlikely. However, the size structure of these small individuals when examined against growth data presented elsewhere in the report by Sanderson et al. (1996) suggests that even these small individuals (<50 mm) would be several years old. Firm conclusions on movement rates (particularly of small sea urchins) are not possible given the possibility that some small individuals could have been missed when the areas were cleared and the uncertainty about growth rates.

Connolly (1986) examined the factors that determine distribution of *H. erythrogramma* among rocks and boulders in South Australia. He tagged the spines of *H. erythrogramma* to evaluate movement in relation to vegetation type in an attempt to explain the nonrandom clumped distribution of urchins. His results were inconclusive with a high incidence of movement of tagged animals off rocks on which they were placed, probably as a result of disturbance. Others remained practically stationary for as long as 2 months. Smoothey (2003) found a nonrandom distribution of *H. erythrogramma* in boulder habitat in New South Wales, some boulders were always more likely to host urchins than others. Depth and type of biotic cover on boulders influenced which boulders were more likely to be used as shelter by urchins. Although urchins moved between boulders, Smoothey (2003) found that the nonrandom distribution of urchins was maintained by urchins moving predominantly towards 'favoured' boulders and away from 'nonfavoured' boulders.

Heliocidaris erythrogramma feeds both by grazing or scraping on the substrate and capturing drift algae (Andrew 1999). The work of both Constable (1989) and Connolly (1986) points strongly to the importance of drift algae as the predominant nutritive source of *H. erythrogramma*. However, there are surprisingly few studies of diet and feeding in *H. erythrogramma* considering it is a dominant herbivore in many southern Australian subtidal habitats (Sanderson et al. 1996). Connolly (1986) found that the diet of *H. erythrogramma* consists of grazed filamentous and encrusting algae, and drift-caught algae and seagrass. The seagrass is an important part of the food captured as drift. Experiments conducted to establish dietary preferences were inconclusive. In Tasmania the preferred habitat and diet of *H. erythrogramma* is the kelp *Macrocystis pyrifera* (Sanderson et al. 1996). More recently Vanderklift et al. (2006) compared the diets of three sympatric species of sea urchins, including *H. erythrogramma*, from multiple sites over a 250 km range of coastline. They confirmed, on the basis of gut content analysis and stable isotope analysis of muscle tissue, that *H. erythrogramma* is primarily an algivorous herbivore. They found *H. erythrogramma* gut contents contained macroalgae in 98% of cases (seagrass 2.4%) and only 1% contained animal foods. Almost 60% of the macroalgae consumed was brown algae (35% red alga) and less than 1% sand or rock indicating that drift feeding on brown algae was the main mode of feeding. Another species (*C. tenuispinus*) examined by this method contained animal foods in 10% of cases and macroalgae in 82% of cases. About half the algae consumed was red alga and almost 10% of gut contents contained sand or rock indicating that grazing as a feeding mode and red alga as a diet, is more important to *C. tenuispinus* than *H. erythrogramma*. Other research by Vanderklift and Kendrick (2005) has confirmed that *H. erythrogramma* preferentially traps and retains the kelp *Ecklonia radiata* in south-western Australia.

In monitoring feeding and movement of *H. erythrogramma*, Connolly (1986) found minimal movement in relation to feeding and no diel pattern in feeding behaviour. Connolly (1986) concluded that sea urchins position themselves in locations favourable for both grazing and catching drift algae.

Constable (1989) examined resource allocation and reproductive output in *H. erythrogramma* in three habitat types – seagrass, barrens and kelp – and compared relative food availability in each habitat type. He found sea urchins in the kelp sites and one of the seagrass sites had higher organic contents in their guts than those from barrens sites. Constable (1989) found diet to correlate with habitat type, with sea urchins in seagrass beds having diets dominated by seagrass and those from the kelp and barrens sites to be dominated by red, green and brown algae. Sea urchins in sites with highest supply of drift algae tended to have lower amounts of solid material such as sand and grit in their guts, suggesting they were able to rely more consistently on a supply of drift algae. Constable (1989) noted that seagrass was a poorer nutritional source than algae and found a food availability hierarchy of 'kelp habitats' > 'barrens habitats' > 'seagrass habitats'.

Constable (1989) found that food availability and possibly exposure as inferred from the comparisons between habitats did affect resource allocation within sea urchins and found a range of variation in structural and soft tissues related to nutrition. Constable (1989) suggested that the hierarchy of resource allocation was: gut and lantern > test and spines > gonads. Indeed, gonad resorption and test shrinkage occurred in individuals with low food availability. Of particular note is his finding that gonad growth and hence reproductive output was correlated with food availability and thus habitat type. This phenomenon is likely to influence recruitment rates and be a key selective agent between populations given short dispersal distances established for this species.

The record of feeding behaviour and diet of *H. erythrogramma* in Tanzania by Shunula and Ndibalema (1986) is obviously a misidentification as *H. erythrogramma* is endemic to Australia.

6. INFLUENCE ON BENTHIC PLANTS, OCCURRENCE OF URCHIN BARRENS AND FEEDING FRONTS

The role of *H. erythrogramma* in structuring communities of benthic flora is apparent from a number of studies. However, this is a complex interaction and even where abundant *H. erythrogramma* does not necessarily impose a recognisable influence on its habitat.

In some locations *H. erythrogramma* is predominantly a drift algal feeder relying on a food source translocated by currents and having little or no impact on benthic macroalgae via direct grazing (e.g. Vanderklift and Kendrick, 2005). However, at higher densities, perhaps in the absence of sufficient drift algal food, *H. erythrogramma* may become a more aggressive grazer and then exert a measurable or even destructive influence on benthic flora (e.g. Sanderson et al. 1996). This is known for other species of sea urchins (e.g. Harrold and Reed 1985).

There have been a number of studies on the influence of grazing of *H. erythogramma* on algal community assemblages. Connolly (1986) observed that the flora on rocks where *H. erythrogramma* occurred consisted of primarily short brown turf algae (mostly *Sphacelaria*) and that macroalgae were absent from these rocks. In contrast, rocks where *H. erythrogramma* were absent supported dense stands of red and brown algae (including *Scaberia, Sargassum* and *Jeanerettia*) to 40 cm height. In a study to determine the effects of grazing on algal community structure before, during and after high-density feeding fronts of urchins in New South Wales, Wright et al. (2005) examined whether *H. erythrogramma* selectively consumed different algal species and if this selectivity changed at high experimental densities maintained for several months. They found that *H. erythrogramma* in feeding fronts with average densities of 80 individuals per square metre (as high as 192 individuals per square metre) selectively grazed some species over others. This affected the survivorship of remaining algal species after the feeding front had passed. Wright et al. (2005) also found that, at high experimentally manipulated densities of 40 and 100 indiduals per square metre, algal community assemblage was affected. This indicates normal patterns of selectivity can be affected by high-density competitive grazing. Wright and Steinberg (2001) and Wright et al. (2005) also found even a chemically defend alga, normally avoided, was consumed at densities >80 individuals per square metre to the extent that very high mortalities of the alga occurred.

As in other temperate areas, urchin barrens occur where kelp beds, in this case *M. pyrifera*, might otherwise prevail. Sanderson et al. (1996) recorded urchin barrens in many parts of Tasmania, noting that barrens may make up to 25% of reef areas in sheltered waters over a significant length of the Tasmanian coastline. Sanderson et al. (1996) noted the ecological impacts of such barrens such as reduced productivity and biodiversity. However, their study was principally concerned with impact of the urchin barrens on fisheries. As a result of the high density of sea urchins and the propensity for dominance of large and hence probably older individuals, roe recovery rates both in terms of quantity and quality were low. Sanderson et al. (1996) saw the urchin barrens as being a factor responsible for limited opportunity for expansion of the Tasmanian sea urchin fishery. Roe quantity and quality were very significantly inversely related to density. This raises the interesting question of the ecological impact of harvesting sea urchins at low densities, given that they are still likely to be the dominant invertebrate herbivore in those environments (Sanderson et al. 1996).

Sanderson et al. (1996) found that removal of the urchins had the potential to rehabilitate urchin barrens. While plant regeneration was variable, roe condition (quality) increased and roe recovery (quantity) increased by up to twofold in areas where sea urchin densities were reduced. On the other hand, Valentine and Johnson (2005a) found removing *H. erythrogramma* from sea urchin barrens resulted only in very slow recovery of native algal species and that recovery of the native algal canopy was very low. Sedimentation was thought to be a factor in this slow recovery.

There are two records of high-density populations of *H. erythrogramma* forming feeding fronts in seagrass habitats. Constable (1989) found *H. erythrogramma* in densities up to 180 individuals per square metre at the edge of a seagrass habitat in southern Victoria. The fronts moved little and persisted for many years (Andrew 1999). In contrast, high-density (120 individuals per square metre) feeding fronts of *H. erythrogramma* near Sydney in New South Wales reportedly devastated more than 75 ha of seagrass meadows over a 5 year period (Buchner 1986, Roberts et al. 1986) and prompted attempts to control them by removal. Roberts et al. (1986) estimated 500 000 individuals occurred in a 2 km front only 1.5 m wide. Over 150 000 individuals were removed (Roberts et al. 1986) but this was regarded as having little effect (Andrew 1999). The phenomenon has had little documentation in New South Wales.

In contrast to detailed studies on the displacement of herbivorous gastropods such as abalone by other sea urchins, including species in Australia (Andrew et al. 1998), the interaction between *H. erythrogramma* and abalone is poorly known. Shepherd (1973) postulated that overfishing of abalone could give rise to expansion of numbers of *H. erythrogramma* leading to sea urchin barrens forming in otherwise productive abalone areas.

A series of studies in Tasmania have sought to examine the role of *H. erythrogramma* in a range of complex interactions involving sea urchins, barrens, an introduced, highly invasive algae and fishing (Johnson et al. 2004). These have provided a large amount of new information on ecological interactions involving *H. erythrogramma.* Valentine and Johnson (2003) described the invasion of the Asian kelp, *Undaria pinnatifida* to Tasmania and demonstrated how destructive grazing by *H. erythrogramma* leading to barrens created the disturbance and space that enabled *U. pinnatifida* to recruit at high densities and form dominating canopies which restricted recovery of and displaced native algae. Johnson et al. (2004) and Valentine and Johnson (2003) proposed that overfishing of rock lobsters could have led to the ability of the rock lobster's prey, *H. erythrogramma,* to increase levels of destructive grazing and the development of urchin barrens which subsequently facilitated the successful establishment of the introduced algae. Valentine and Johnson (2004) also showed that forms of disturbance other than urchin grazing could also create the ecological space required for *U. pinnatifida* to establish. Valentine and Johnson (2005a) demonstrated, using removal experiments, that once established, urchin barrens were often maintained in the absence of sea urchin grazing by factors such as increased sedimentation. Valentine and Johnson (2005b) further described how grazing by *H. erythrogramma* on sporophytes of *U. pinnatifida* could have a significant effect on the ability of *U. pinnatifida* to establish a canopy except in years of very heavy recruitment by the algae.

Also in Tasmania, Edgar et al. (2004) proposed the use of marine reserves as a mechanism to increase predation pressure on *H. erythrogramma*, thereby reducing their ability to create urchin barrens which facilitate the successful recruitment of *U. pinnatifida.* The effectiveness of this measure, however, is not likely to be straightforward. Edgar and Barrett (1999) compared the abundance of a range of major taxa inside and outside marine reserves that had been in place for 6 years. They also had historical data for the reserves and they found that the reserves had no impact on abundance of *H. erythrogramma*. This may have been because the abundance and size structure of major predators or sea urchins such as lobsters and fish had increased very markedly in the reserves.

7. PREDATORS, PARASITES, COMMENSALS AND OTHER ECOLOGICAL INTERACTIONS

Predators, commensal and parasitic associates of *H. erythrogramma* are not well known. Connolly (1986) recorded eight known and 10 putative predators of adult and juvenile *H. erythrogramma*. Known predators were mostly fish species including wrasses and stingrays and a bird, the Pacific Gull. Of these, Connolly (1986) concluded the most important predator at the site he studied was the common *Octopus australis*. He believed it was responsible for all the dead tests in the area as these carried drill holes typical of octopus predation. Andrew (1999) notes that the sea star *Coscinasterias calamaria* is an important predator of sea urchins in exposed habitats. Shepherd and Clarkson (2001) studied the feeding behaviour of the wrasse *Notolabrus tetricus* in South Australia and found that while *H. erythrogramma* formed part of the diet of wrasse, particularly larger fish, it was not a preferred prey item relative to crabs and molluscs, particularly abalone. On the other hand, Johnson et al. (2004) found that the rock lobster *Jasus edwardsi* is a more important predator of *H. erythrogramma* than fishes. Griffiths (2003) examined fish distribution and abundance in rock pools in New South Wales and found that the goby *Aspasmogaster costatus* sought refuge among the spines of sea urchins and was most abundant in habitats with large numbers of *H. tuberculata* and *C. rodgersii*.

Shepherd (1974) described an unusual association between the large bivalve *Pinna*, which lives partly emerged from sediment and *H. erythrogramma*. In some seagrass beds *Pinna* are the only large hard substrate available and are colonised by *H. erythrogramma*. Halos of grazed seagrass occur around the *Pinna*. Shepherd (1974) found an inverse relationship between abundance of *Pinna* (and sea urchins) and density of seagrass. He postulated that the impact that the sea urchins had on seagrass beds was a direct function of the variation in, and success of, *Pinna* recruitment rates.

Constable (1989) recorded the occurrence of sick and dying sea urchins among the high-density aggregations he recorded in Victoria. He suggested this could have been due to either disease or poor nutrition leading to disease. Connolly (1986) noted that the parasitic turbellarian *Syndisyrinx punicea* occurs in the gut of most *H. erythrogramma*. Laegdsgaard et al. (1991) recorded the presence of trematodes in the gonads of 1–3% of *H. erythrogramma* in their study sites near Sydney.

Heliocidaris erythrogramma, like other nearshore attached or slow-moving animals, is subject to catastrophic mortality events. Sanderson et al. (1996) recorded a mass mortality resulting from very heavy unseasonal freshwater runoff and Wells and Keesing (1986) reported extensive mortalities of marine invertebrates including *H. erythrogramma* after unusually long and very low tides during daylight hours coinciding with very hot still conditions for several days. Constable (1989) recorded mortality events involving *H. erythrogramma* which resulted from storms. He noted that the occurrence of poor health in some urchins may have been due to high temperatures or low salinity following extremes of temperature or rainfall, respectively.

8. CONCLUSIONS

The ecology of the purple sea urchin *H. erythrogramma* remains little studied and poorly known. Research has centred mainly on its reproductive biology and on the unusual lecithotrophic type of development and its evolutionary significance (Jackson and Cheetham 1999; Lawrence and Herrera 2000). Studies that have provided a good insight into the ecology of *H. erythrogramma* include those on physiology in relation to nutrition, population genetics and some aspects of feeding. There have been some studies of the urchin's distribution and abundance but these are largely incidental and most aspects of the biology and ecology of *H. erythrogramma* is inadequately or not at all researched. Recent studies on the experimental assessment of the influence of *H. erythrogramma* on marine macroalgal communities and in structuring subtidal communities have added considerably to knowledge of the ecology of *H. erythrogramma*.

As the important role of *H. erythrogramma* in the complex interactions that structure Australian temperate macroalgal reef systems is increasingly recognised, this will ultimately lead to more studies which take a modelling approach incorporating population dynamics which attempt to represent the interactions and predict ecosystem dynamics. Such approaches will be greatly hampered by the lack of information on the fecundity, movement, growth, ageing, settlement and recruitment of *H. erythrogramma*. Studies which use a comparative ecology approach among sympatric sea urchins (e.g. Vanderklift 2002) have the potential to add greatly to our knowledge of the biology and ecology of sea urchins than studies of single species, and should be encouraged.

There are also few studies that contribute anything to knowledge of the ecology of the sympatric *H. tuberculata*. Although it is morphologically very similar to *H. erythrogramma*, the two species differ in size and in habitat (Dakin 1980; Lawrence and Byrne 1994). The difference in development, lecithotrophic in *H. erythrogramma* and planktotrophic in *H. tuberculata* (Raff 1996), has received most attention. *H. tuberculata* deserves more attention by ecologists.

REFERENCES

Andrew NL (1999) Under Southern Seas: the ecology of Australia's rocky reefs. University of New South Wales Press, Sydney

Andrew NL, Worthington DG, Brett PA, Bentley N, Chick RC and Blount C (1998) Interactions between the abalone fishery and sea urchins in New South Wales. Final Report to the Fisheries Research and Development Corporation for Project 93/102

Buchner KA (1986) Population density of the sea urchin (*Heliocidaris erythrogramma*) within the seagrass beds of Botany Bay, NSW. Proceedings of the Australian Marine Sciences Association Annual Conference (abstracts only), p 50

Clark HL (1946) The echinoderm fauna of Australia: its composition and origin. Carnegie Inst Wash Publ 566, p 329

Connolly R (1986) Behaviour and ecology of the sea urchin *Heliocidaris erythrogramma* (Valenciennes). Honours thesis, University of Adelaide

Constable AJ (1989) An investigation of resource allocation in the sea urchin *Heliocidaris erythrogramma* (Valenciennes). PhD thesis, University of Melbourne

Constable AJ (1993) The role of sutures in shrinking of the test in *Heliocidaris erythrogramma* (Echinodermata: Echinometridae) Mar Biol 117: 423–430

Dakin WJ (1980) Australian Seashores. Rev. ed. Angus and Robertson, Sydney, p 332

Dix TG (1977a) reproduction in Tasmanian populations of *Heliocidaris erythrogramma* (Echinodermata: Echinometridae) Aust J Mar Freshwater Res 28: 509–520

Dix TG (1977b) Survey of Tasmanian sea urchin resources. Tasmanian Fish Res 21: 1–14

Ebert TA (1982) Longevity, life history, and relative body wall size in sea urchins. Ecol Monogr 52: 353–394

Edgar GJ (1997) Australian Marine Life. Reed Books, Melbourne, p 366

Edgar GJ and Barrett NS (1999) Effects of the declaration of marine reserves on Tasmanian reef fishes, invertebrates and plants. J Exp Mar Biol Ecol 242: 107–144

Edgar GJ, Barrett NS, Morton AJ and Samson CR (2004) Effects of algal canopy clearance on plant, fish and macroinvertebrate communities on eastern Tasmanian reefs. J Exp Mar Biol Ecol 312: 67–87

Emlet R and Hoegh-Guldberg O (1997) Effects of egg size on postlarval performance: experimental evidence from a sea urchin. Evolution 51: 141–152

Endean R (1957) The biogeography of Queensland's shallow-water echinoderm fauna (excluding Crinoidea), with a rearrangement of the faunistic provinces of tropical Australia. Aust J Mar Freshwater Res 8: 233–273

Griffiths SP (2003) Spatial and temporal dynamics of temperate Australian rockpool icthyofaunas. Mar Freshwater Res 54: 163–176

Growns JE (1991) Some evolutionary and ecological implications of colour variation in the sea urchin *Heliocidaris erythrogramma*. PhD thesis, University of Tasmania

Growns JE and Ritz DA (1994) Colour variation in southern Tasmania populations of *Heliocidaris erythrogramma* (Echinodermata: Echinoidea). Aust J Mar Freshwater Res 45: 233–242

Harrold C and Reed DC (1985) Food availability, sea urchin grazing, and kelp forest community structure. Ecology 1985: 1160–1169

Hoegh-Guldberg O and Emlet RB (1997) Energy use during the development of a lecithotrophic and a planktotrophic echinoid. Biol Bull 192: 27–40

Jackson JBC and Cheetham AH (1999) Tempo and mode of speciation in the sea. Trends Ecol Evol 14: 72–77

Johnson CR, Valentine, JP and Pederson HG (2004) A most unusual barrens: Complex interactions between lobsters, sea urchins and algae facilitates spread of an exotic kelp in eastern Tasmania. In: Heinzeller T, Nebelsick JH (eds) Echinoderms: München. Balkema, Leiden, pp 213–220

Kailola PJ, Williams MJ, Stewart PC, Reichelt RE, McNee A and Grieve C (1993) Australian fisheries resources. Bureau of Resource Sciences, Canberra, Australia. pp 180–182

Laegdsgaard P, Byrne M and Anderson DT (1991) Reproduction of sympatric populations of *Heliocidaris erythrogramma* (Echinoidea) in New South Wales. Mar Biol 110: 359–374

Lawrence JM and Byrne M (1994) Allocation of resources to body components in *Heliocidaris erythrogramma* and *Heliocidaris tuberculata* (Echinodermata: Echinoidea). Zool Sci 11: 133–137

Lawrence JM and Herrera J (2000) Stress and deviant reproduction in echinoderms. Zool Stud 39: 151–171

Marshall DJ, Steinberg PD and Evans JP (2004) The early sperm gets the good egg: mating order effects in free spawners. Proc Roy Soc Lond B 271: 1585–1589

McMillan WO (1991) Larval life history and population subdivision within two temperate sea urchins (*Heliocidaris*). Pac Sci 45: 98–99

McMillan WO, Raff RA and Palumbi SR (1992) Population genetic consequences of developmental evolution in sea urchins (Genus *Heliocidaris*) Evolution 46: 1299–1312

Raff RA (1996) The shape of life. Univ Chicago Press, Chicago

Roberts DE, Jacobs NE and Anink PJ (1986) management of a sea urchin (*Heliocidaris erythrogramma*) infestation in seagrass beds in Botany Bay, NSW. Proceedings of the Australian Marine Sciences Association Annual Conference (abstracts only), p 54

Sanderson JC, Le Rossingnol M and James W (1996) A pilot program to maximise Tasmania's sea urchin (*Heliocidaris erythrogramma*) resource. Final Report to the Fisheries Research and Development Corporation for Project 93/221

Shepherd SA (1973) Competition between sea urchins and abalone. Aust Fish 32: 4–7

Shepherd SA (1974) An underwater survey near Crag Point in Upper Spencer Gulf. South Australian Dept of Fisheries Technical Report 1: 1–29

Shepherd SA and Clarkson PS (2001) Diet, feeding, behaviour, activity and predation of the temperate blue-throated wrasse, *Notolabrus tetricus*. Mar Freshwater Res 52: 311–322.

Shunula JP and Ndibalema V (1986) Grazing preferences of *Diadema setosum* and *Heliocidaris erythrogramma* (Echinoderms) on an assortment of marine algae. Aquat Bot 25: 91–95

Smoothey A. (2003) Intertidal boulder-fields as habitats for the sea urchin, *Heliocidaris erythrogramma.* Honours thesis, University of Sydney

Styan CA (1997) Inexpensive and portable sampler for collecting eggs of free spawning marine invertebrates underwater. Mar Ecol Prog Ser 150: 293–296

Underwood AJ, Kingsford MJ and Andrew NL (1991) Patterns in shallow subtidal marine assemblages along the coast of New South Wales. Aust J Ecol 6: 231–249

Valentine JP and Johnson CR (2003) Establishment of the introduced kelp *Undaria pinnatfida* in Tasmania depends on disturbance to native algal assemblages. J Exp Mar Biol Ecol 295: 63–90

Valentine JP and Johnson CR (2004) Establishment of the introduced kelp *Undaria pinnatfida* following dieback of the native macroalga *Phyllospora comosa* in Tasmania, Australia. Mar Freshwater Res 55: 223–230

Valentine JP and Johnson CR (2005a) Persistence of sea urchin (*Heliocidaris erythrogramma*) barrens on the east coast of Tasmania: inhibition of macroalgal recovery in the absence of high densities of sea urchins. Botanica Marina 48: 106–115

Valentine JP and Johnson CR (2005b) Persistence of the exotic kelp *Undaria pinnatfida* does not depend on sea urchin grazing. Mar Ecol Prog Ser 285: 43–55

Vanderklift MA (2002) Interactions between sea urchins and macroalgae in south-western Australia: testing general predictions in a local context. PhD thesis, University of Western Australia

Vanderklift MA and Kendrick GA (2004) Variation in abundances of herbivorous invertebrates in temperate subtidal rocky reef habitats. Mar Freshwater Res 55:93–103

Vanderklift MA and Kendrick GA (2005) Contrasting influence of sea urchins on attached and drift macroalgae. Mar Ecol Prog Ser 299:101–110

Vanderklift MA, Kendrick GA and Smit AJ (2006) Differences in trophic position among sympatric sea urchin species. Estuar Coast Shelf Sci 66: 291–297

Wells FE and Keesing JK (1986) An investigation of the mollusc assemblages on intertidal beachrock platforms in the Perth metropolitan area, with particular emphasis on the abalone *Haliotis roei.* Volume 1, Platform Molluscs, unpublished report to the Department of Fisheries, Western Australia.

Williams DHC and Anderson DT (1975) The reproductive system, embryonic development, larval development and metamorphosis of the sea urchin *Heliocidaris erythrogramma* (Val.) (Echinodermata: Echinoidea). Aust J Zool 23: 371–403

Wright TJ and Steinberg PD (2001) Effects of variable recruitment and post-recruitment herbivory on local population size of a marine alga. Ecology 82: 2200–2215

Wright TJ, Dworjanyn SA, Rogers CN, Steinberg PD, Williamson JE and Poore AGB (2005) Density dependent sea urchin grazing: differential removal of species, changes in community composition and alternative community states. Mar Ecol Prog Ser 298: 143–156

Edible Sea Urchins: Biology and Ecology
Editor: John Miller Lawrence

Chapter 18

Ecology of *Strongylocentrotus droebachiensis*

RE Scheibling[a] and BG Hatcher[b]

[a]*Department of Biology, Dalhousie University, Halifax, NS (Canada).*
[b]*Bras d'Or Institute for Ecosystem Research, Cape Breton University, Sydney, NS (Canada).*

1. DISTRIBUTION AND ABUNDANCE

Strongylocentrotus droebachiensis (Müller) (Plate 4B) is the most widely distributed member of the family Strongylocentrotidae (Mortensen 1943). Throughout its range, it plays a key ecological role in determining the distribution and abundance of benthic macroalgae, particularly kelps (Chapman and Johnson 1990). Consequently, it has been considered a pest in areas where intensive grazing destroys kelp habitat and limits production available to commercial species such as lobster (Wharton and Mann 1981). *Strongylocentrotus droebachiensis* has become a valued commodity and it is now extensively fished and cultured for its roe in Northwest Atlantic (Hatcher and Hatcher 1997; Miller and Nolan 2000; Chen et al. 2003; Grabowski and Chen 2004) and Northeast Pacific (Perry et al. 2002). All of these factors have combined to make *S. droebachiensis* one of the most studied, if not the most well known, sea urchin species.

1.1. Geographic Range

Strongylocentrotus droebachiensis has a broad arctic-boreal distribution (Mortensen 1943; Jensen 1974; Bazhin 1998). In the North Atlantic, it extends from the Canadian Archipelago and Greenland down the east coast of North America to Cape Cod, USA, and across to Iceland, the Shetland Islands and northern Scotland, Norway, Denmark, and the tip of Sweden. It also occurs in Barents Sea, the White Sea (but very rarely), and the Kara Sea as far east as the coast of Taymyr. It is not found in the Laptev Sea or Eastern Siberian Sea, but occurs in the Chukchi Sea (only near Wrangel Island). In the North Pacific, it extends along the east coast of Siberia to the middle of the Kuril Island chain and east coast of Sakhalin Island, and from the Aleutian Islands and Alaska down the west coast of North America to Oregon, USA. Although genetic differences exist between Atlantic and Pacific populations of *S. droebachiensis*, overall genetic divergence is low across the species' range (Palumbi and Wilson 1990). Addison and Hart (2004) found little evidence of trans-Atlantic gene flow: populations in the Northwest Atlantic were genetically homogenous and more similar to populations in the Pacific than to those in the Northeast Atlantic.

1.2. Population Density and Spatial Distribution

Strongylocentrotus droebachiensis ranges from 0 to 300 m in depth, but is most common in the shallow subtidal zone from 0 to 50 m (Jensen 1974). It is also found in tide pools in the low intertidal zone (Paine and Vadas 1969; Russell et al. 1998). It generally occurs on rocky substrata such as bedrock outcrops, boulders, and cobbles (Himmelman 1986; Scheibling and Raymond 1990), but is also found on gravel bottoms in deeper waters and less frequently on sand (Sivertsen and Hopkins 1995; Brady and Scheibling 2005). Upper depth limits vary seasonally with wave action that can dislodge or abrade sea urchins, or limit their ability to graze macroalgae (Chapman and Johnson 1990; Siddon and Witman 2003; Gagnon et al. 2004). Ice scouring can also seasonally restrict the distribution of individuals to deeper zones (Himmelman and Lavergne 1985; Gagnon et al. 2004). Sea urchin density generally decreases with depth to about 20–30 m, which in many areas corresponds to the lower limits of the rocky subtidal zone and distribution of kelps and other macroalgae upon which sea urchins feed (Propp 1977; Moore and Miller 1983; Himmelman 1986; Ojeda and Dearborn 1989; Sivertsen and Hopkins 1995).

Throughout its range, *S. droebachiensis* is commonly associated with laminarian kelps. At high population densities, the sea urchins destructively graze kelp beds, forming extensive "barrens" dominated by encrusting coralline red algae. They can persist in these barrens indefinitely, precluding the regrowth of erect macroalgae. Spatial and temporal variations in population density and size structure of *S. droebachiensis* have been recorded along the Atlantic coast of Nova Scotia since the 1970s. Densities differ by an order of magnitude between subpopulations in kelp beds (mean ±SE, $n = 11$ studies: 14 ± 5.3 individuals per square meter), destructive grazing aggregations or fronts (136 ± 46.4 individuals per square meter), and recently formed (41 ± 10.1 individuals per square meter), or older barrens (71 ± 6.4 individuals per square meter) (Meidel and Scheibling 2001). Because populations in kelp beds are composed primarily of juveniles, and the proportion of large adults is much greater in grazing fronts than in barrens (Meidel and Scheibling 2001), biomass differences among these subpopulations are even greater. For example, Scheibling et al. (1999) found biomass in a grazing front ($\sim$10 kg m^{-2}) was two orders of magnitude greater than that in the kelp bed directly ahead of the front ($\sim$0.1 kg m^{-2}), and one order of magnitude greater than that in recently formed barrens ($\sim$1 kg m^{-2}). Similar patterns of abundance have been observed for *S. droebachiensis* in other geographic areas: Barents Sea (Propp 1977), Gulf of St. Lawrence (Himmelman et al. 1983b; Gagnon et al. 2004), Gulf of Maine (Witman 1985; Martin et al. 1988; Ojeda and Dearborn 1989), Newfoundland (Himmelman 1984, 1986), Norway (Hagen 1987; Sivertsen 1997), and Iceland (Hjörleifsson et al. 1995).

Multivariate analysis based on an extensive survey of the Norwegian coast showed that the distribution of *S. droebachiensis* and its associated biota was influenced by a suite of geographic and environmental factors (including depth, substratum type, wave exposure, proximity to mainland, latitude, and year of sampling), and parasite prevalence in sea urchins (Sivertsen 1997). Density and size of *S. droebachiensis* is limited by intraspecific competition on barrens where food is in short supply (Lang and Mann 1976; Himmelman 1984, 1986; Sivertsen 1997). Likewise, interspecific competition may limit the distribution of *S. droebachiensis* where it overlaps with other species of sea urchins. Sivertsen (1997) suggested that competition accounts for an inverse relationship between densities of *S. droebachiensis* and *E. esculentus* in barrens in Norway. Foreman (1977) and Duggins (1983) found little mixing between populations of *S. droebachiensis* and *S. franciscanus* in the Northeast Pacific, which may also reflect a competitive interaction. However, experimental tests to evaluate the importance of intra- or interspecific competition in determining density or size of *S. droebachiensis* are lacking.

1.3. Physiological Tolerance Limits

Strongylocentrotus droebachiensis adjusts its metabolism and activity level to compensate for seasonal variations in sea temperature (Percy 1972, 1973, 1974b) which can range from −1 to 20 °C (Scheibling and Hennigar 1997). This seasonal acclimatization enables sea urchins to feed at relatively high rates in winter and support gonadal growth and maturation. The upper thermal tolerance limit for warm-acclimated sea urchins in the laboratory is about 22 °C (Percy 1973; Scheibling and Stephenson 1984; Pearce et al. 2005). Although mass mortalities of *S. droebachiensis* occur during periods of peak sea temperature off Nova Scotia, they are caused by an amoebic pathogen that is facilitated (and possibly introduced) by warm-water masses.

Strongylocentrotus droebachiensis is an osmoconformer with a limited capacity for osmoregulation (Lange 1964; Stickle and Denoux 1976). Thus, its distribution within estuaries (Himmelman et al. 1983b; Drouin et al. 1985), or coastal areas with periodic freshwater-water runoff (Stickle and Denoux 1976; Hooper 1980), is constrained by the salinity regime. However, the salinity tolerance of *S. droebachiensis* varies considerably among populations and size classes, and is dependent on acclimation history (Stickle and Denoux 1976; Himmelman et al. 1983b). Lange (1964) reported a lower lethal limit of 21.5 practical salinity units (psu) for *S. droebachiensis* in Droebak (its namesake), Norway. In contrast, Stickle and Denoux (1976) found that sea urchins at an Alaskan site, influenced by a large melt-water lens, remained active through tidal fluctuations from 14 to 28 psu. Stickle and Ahokas (1974) found that *S. droebachiensis* survived a 30–10–30 psu semidiurnal tidal cycle that simulated conditions in a stratified estuary. Sabourin and Stickle (1981) estimated a 14-day median tolerance limit (salinity at which 50% survive for 14 days) of 12.8 psu for individuals transferred from 30 psu to salinities down to 10 psu.

Himmelman et al. (1984) attributed marked differences in population structure and abundance of *S. droebachiensis* along the salinity gradient of the St. Lawrence Estuary, Quebec, to osmotic stress. In the outer marine region of the estuary, sea urchins are extremely abundant and generally small (<30 mm in test diameter). However, in the upper region, where mean monthly salinities can fall below 20 psu, densities are much lower and most individuals are large (40–80 mm) (Himmelman et al. 1983b; Drouin et al. 1985). In a field experiment, Himmelman et al. (1984) found that sea urchins >10 mm generally survived brief exposures to salinities as low as 3.5 psu at 0 m (the level of lowest water spring tides), but smaller individuals died. At 0.5 m, all size classes generally survived salinities as low as 14.3 psu. Sea urchins from the estuary were more tolerant of low salinities than those from the Atlantic coast of Nova Scotia (Himmelman et al.

1984), suggesting that the two populations represent different physiological (and perhaps genetic) races adapted to local conditions.

Certain wavelengths of UVR (UVA and UVB) can induce cellular damage in aquatic organisms (Lesser and Barry 2003). Adams (2001) showed that adults of *S. droebachiensis* exhibit a shade-seeking or covering response when exposed to UVR (particularly UVB), which may confer protection from damage. However, Lesser and Barry (2003) found significant delays in development (cell division), increased DNA damage, and decreased survival of embryos and larvae of *S. droebachiensis* with increased exposure to UVR. They concluded that adverse effects of UVR on early life-history stages may ultimately affect the distribution, health, and abundance of this species.

2. FOOD AND FEEDING

Strongylocentrotus droebachiensis is an omnivore, feeding on a great range of algae, invertebrates, and microbes (Duggins 1981; Johnson and Mann 1982; Briscoe and Sebens 1988; Nestler and Harris 1994). Nutritional latitude is enhanced by the ability to selectively assimilate different forms of dissolved organic material (Propp 1977), and by microbial associations that allow cellulose digestion (Fong and Mann 1980) and nitrogen fixation (Guerinot and Patriquin 1981). An ecologically and evolutionarily important implication of generalist feeding is that *S. droebachiensis* can survive extended periods when preferred or optimal diets are unavailable, albeit with corresponding decreases in growth and increases in mortality, dispersion, and even cannibalism (Himmelman and Steele 1971; Bernstein et al. 1983).

Despite its breadth of diet, *S. droebachiensis* exhibits clear preferences for foods, and marked differences in its abilities to consume and assimilate them (Vadas 1977; Himmelman and Nedelec 1990). Individuals detect food from a distance and aggregate around it (Garnick 1978; Mann et al. 1984; Vadas et al. 1986), and select specific items from natural assemblages (Larson et al. 1980; Himmelman 1984). Somatic and gonad growth, time to maturity, reproductive output, settlement, survivorship, longevity, and mobility of *S. droebachiensis* have all been demonstrated to be influenced strongly by food availability, quality, and quantity. On this basis, highest rates of population production are predicted to occur where an abundance of highly preferred foods is available under environmental conditions that foster rapid feeding and efficient assimilation. Feeding capabilities and preferences of *S. droebachiensis* in turn have profound impacts on the dynamics of benthic communities (Miller 1985b; Elner and Vadas 1990; Scheibling et al. 1999; Sumi and Scheibling 2005).

2.1. Food Preferences and Nutrition

Foraging by *S. droebachiensis* has been viewed as an energy or nutrient optimization process, in which various foods are selected and consumed in proportion to their net nutritional value (interaction of calorific or protein content and digestibility), rather than simply their availability (Propp 1977; Vadas 1977; Vadas and Elner 1992). This view is largely the result of experiments in which individuals are presented with discrete arrays of foods under controlled (usually laboratory) conditions (MacKay 1976; Larson et al. 1980; Himmelman 1984; Lemire and Himmelman 1996; Pelletreau and Muller-Parker 2002). Natural environmental factors such as temperature, seabed topography, wave action, competition, and predation can strongly influence the expression of such preferences by field populations (Himmelman and Nedelec 1990; Vadas 1990; Scheibling 1996) such that the actual food eaten reflects the availability of foods regardless of their palatability. While *S. droebachiensis* has been considered a facultative specialist (e.g. Briscoe and Sebens 1988), a better generalization is that it is a generalist that eats the best food accessible within that available in a given habitat (Lawrence 1975).

Most of the diet of *S. droebachiensis* is macroalgae. Laboratory experiments have demonstrated a marked preference for various species, rather consistently for large brown algae (i.e. kelps and rock weeds: Kuznetzov 1946; Mackay 1976; Vadas 1977; Himmelman 1984; Briscoe and Sebens 1988; Himmelman and Nedelec 1990; Scheibling and Anthony 2001). This is not universal, as foliose and articulated calcareous red macroalgae (or even microalgae) may be equally or more preferred in certain populations, habitats, or seasons (Larson et al. 1980; Duggins 1981; Himmelman and Nedelec 1990; Lemire and Himmelman 1996).

Of the common macroalgae available, *S. droebachiensis* usually strongly prefers brown algae of the genera Alaria, Chordaria, Laminaria, Nereocystis, Petalonia and Ulvaria, while the brown algal genus Agarum, and many (but not all, e.g. *Chondrus crispus*, *Corallina officinalis*) of the red algal genera (e.g. Ptilota, Palmaria, Porphyra) are rarely consumed or even actively avoided (Vadas 1977; Larson et al. 1980; Himmelman 1984; Himmelman and Nedelec 1990; Gagnon et al. 2005a). The genus *Desmarestia* and the fucoid algae (e.g. the common intertidal genera Fucus and Ascophyllum) generally fall in the intermediate to low-preference categories. Mackay (1976) found that *S. droebachiensis* in the Bay of Fundy are an exception to this generalization, indicating that preferences for macroalgal species can be habitat-specific. The most abundant alga there, *Ascophyllum nodosum*, was the first choice of sea urchins in feeding experiments. Population or habitat-specific preference lists vary somewhat with season and adult size, but the differences are neither large nor consistent (Himmelman 1984; Himmelman and Nedelec 1990). Sea urchins generally have little access to macrophytic green algae in subtidal environments, but recent introductions of the siphonaceous green macroalga *Codium fragile* ssp. *tomentosoides* have prompted feeding preference trials. These demonstrate a low preference for this alga, and lower somatic and gonadal growth in urchins feeding on it (Scheibling and Anthony 2001; Sumi and Scheibling 2005).

Feeding preferences of *S. droebachiensis* are inversely correlated with the calorific (energy) content of algal foods and positively correlated with absorption efficiencies, growth rates and gonad quality (Mackay 1976; Vadas 1977; Larson et al. 1980; Himmelman 1984; Keats et al. 1984; Walker and Lesser 1998). The weak inverse correlation with calorific content challenges the model of energy-maximizing feeding strategies, suggesting that other factors, such as nitrogen and phosphorus content, affect food selection (e.g. Propp 1977). Himmelman and Nedelec (1990) concluded that algae avoided by *S. droebachiensis* have attributes that make them resistant to grazing. However, they found that calorific value, dry weight, ash content, and phenolic content individually are not correlated well with metrics of attraction or feeding, indicating that expressed feeding preferences are a complex behavioral response to a suite of physiological and environmental factors. Pelletreau and Muller-Parker (2002) also found that nutritional content and food preference were uncorrelated, but showed that *S. droebachiensis* avoided algae having high sulfuric acid and polyphenolic concentrations. McClintock (1980) suggested that the amount of energy in a food that can be digested and absorbed, rather than the total amount of energy in the food, must be used to test optimal foraging theories.

The laminarian kelps, whether consumed as attached fronds or detritus, form the primary diet for most populations of *S. droebachiensis* because of their abundance, ease of chemical detection, nutritional content, palatability and digestibility. The ability to feed effectively on drifting fronds and detrital fragments of kelp confers a particular advantage, as it allows populations to thrive in habitats where their preferred food does not grow. When macroalgal food is not available, the sea urchins can survive (albeit at reduced rates of growth and reproduction) on a remarkable array of foodstuffs, ranging from microbial films to conspecifics.

Strongylocentrotus droebachiensis consumes animal food if it is available (Himmelman and Steele 1971; Arnold 1976; Mackay 1976; Duggins 1981; Thompson 1984; Briscoe and Sebens 1988; Nestler and Harris 1994; Witman 1985; Meidel and Scheibling 1999). The most common animal prey are mussels, barnacles, ectoprocts, bryozoans, and crustaceans, but hydroids, gastropods, echinoids, asteroids, polychaetes, sponges, ascidians, and fish are also consumed. Sea urchins may consume animal tissue as predators, cannibals and scavengers (detritivores), or they may ingest animal food inadvertently when consuming plant material. Positive, size-selective feeding on mussels by *S. droebachiensis* apparently reflects prey-handling capability (Briscoe and Sebens 1988), but mussels cannot be consumed effectively at temperatures <1.8 °C.

Entirely animal diets appear to have insufficient carbohydrates or energy to support growth by *S. droebachiensis*, but animal foods are consumed opportunistically and if macroalgae are unavailable or the plant food is of very poor quality (Briscoe and Sebens 1988). The high C:N:P and carbohydrate: protein ratios of marine macroalgae relative to that of sea urchin tissues suggest that certain components of growth may be nitrogen or protein limited even when abundant macroalgal food is available (Miller and Mann 1973; Propp 1977; Takagi et al. 1980; Thompson 1982). A mixed diet of macroalgae and animal tissue enhances growth (Thompson 1982; Briscoe and Sebens 1988; Nestler and Harris 1994; Meidel and Scheibling 1999). For example, Nestler and Harris (1994) measured 54–297% higher growth of *S. droebachiensis* fed kelp covered with bryozoans than those fed kelp alone. This result extends to juvenile urchins, which grew as fast on this animal-coated plant diet as on protein-enhanced artificial diets (Williams and Harris 1998).

The presence of a diverse gut microflora in *S. droebachiensis* (Fong and Mann 1980; Guerinot and Patriquin 1981; Johnson and Mann 1982) allows it to meet basic requirements for carbon, nitrogen, and amino acids in protein-deficient feeding situations. This is sufficient for survival, but it appears that additional animal protein is necessary in the diet for maximal somatic and gonad growth. Certainly *S. droebachiensis* can switch its feeding habitats to consume animals (Himmelman and Steele1971; Duggins 1981;

Briscoe and Sebens 1988; Meidel and Scheibling 1999), a flexibility that would increase fitness (Himmelman and Nedelec 1990).

Food preferences may profoundly affect growth and reproduction. Somatic and gonadal growth in *S. droebachiensis* has been measured in individuals fed diets that differed in degree of preference (Kuznetzov 1946; Mackay 1976; Lang and Mann 1976; Vadas 1977; Larson et al. 1980; Himmelman 1984; Keats et al. 1983, 1984; Raymond and Scheibling 1987; Scheibling and Hamm 1991; Briscoe and Sebens 1988; Munk 1992; Cuthbert et al. 1995; Hooper et al. 1996; Russell 1998; Meidel and Scheibling 1999; Scheibling and Anthony 2001; Pearce et al. 2002a,b, 2004; Daggett et al. 2005). The differences in rates of increase in test diameter or body mass attributable to diet may be as great as a factor of 3.5 over periods of 2–5 months, but increases of 50–85% over a year are characteristic of *S. droebachiensis* when shifted from a poor diet (e.g. filamentous and encrusting algae) to a good one (e.g. kelp and mussels). Growth rates of juvenile *S. droebachiensis* over 22 months were almost twice as high on a diet of unlimited kelp and mussels than those on a limited diet of kelp provided weekly (Meidel and Scheibling 1999). Individuals fed only coralline algae failed to develop gonads. Populations may adapt to available foods. Growth and gonad development were best on a diet of Ascophyllum in the Bay of Fundy (Mackay 1976) but not in the Gulf of St. Lawrence (Himmelman and Nedelec 1990). Diets of three species of kelp produced no significant differences in the growth of juvenile urchins (Pearce et al. 2004).

The ranking of foods in terms of gonad production by *S. droebachiensis* approximates that obtained in feeding preference trials, but the relative values of different macroalgae to gonad and somatic production can differ. For example, individuals fed red algae produced the largest gonads although brown algae were preferred (MacKay 1976). Highest gonad indexes resulted from a diet of *Fucus edentatus* in laboratory experiments but from *Laminaria digitata* in field experiments in Newfoundland (Keats et al. 1983). In contrast, Cuthbert et al. (1995) observed no increase in gonad yields over 6 weeks in field trials on a diet of *F. vesiculosus*. Even diets of similar species of laminarian kelps may support significantly different rates of gonad growth (Cuthbert et al. 1995; Russell 1998), with *S. droebachiensis* showing an apparent preference for the better species (Hooper et al. 1996).

The amount of food eaten affects growth and gonad development of *S. droebachiensis*. Kelp or kelp and mussel diets at 14–50% of the *ad libitum* feeding rate (equivalent to about 0.8–2.5% of body weight per day) result in modest reductions of somatic growth rates (4–36%), but larger reductions in gonad growth rates (35–64%) over 20–40 week periods (Thompson 1982; Hooper et al. 1996; Minor and Scheibling 1997; Meidel and Scheibling 1999). Food rations within this range appear to have insignificant effects on gametogenesis and maturation of gametes (including egg size), but can have significant effects (albeit always less than a factor of 2) on the abundance of nutritive phagocytes, numbers of sperm and eggs, egg size, and egg quality (i.e. lipid and energy content), as well as sex-specific effects on gonad size (Thompson 1982; Minor and Scheibling 1997; Meidel and Scheibling 1999; Garrido and Barber 2001). Hooper et al. (1996) concluded that a ration of 75% *ad libitum* consumption (3.8% body weight per day) represents optimal use of food resources. Pearce et al. (2002b) found that considerably lower rations (0.5–1.0% of body weight per day) of a prepared diet were sufficient to maintain gonad yields above those of control animals fed kelp *ad libitum*. The fact that *S. droebachiensis* can grow and reproduce on greatly reduced rations of preferred foods and significantly increase growth and reproductive output when food is abundant suggests mechanisms that allow the species' persistence despite large fluctuations in the biomass of macroalgal-dominated benthic communities (Scheibling et al. 1999). The data available on food rations for this species are adequate for the construction of ration-based models of community dynamics (Mohn and Miller 1987).

2.2. Feeding Behavior

Two general processes interact to control the degree and pattern of aggregation of *S. droebachiensis* on natural seabeds: food availability and predation. Either process may dominate in a particular habitat or season, and they may be modified by physical environmental factors such as temperature, salinity, and wave action. Such basic factors are unlikely to operate independently, however, and much of the apparent contradiction and controversy in the literature concerning sea urchin behavior results from attempts to ascribe causality to a single factor (Pringle 1986; Elner and Vadas 1990). In the simplest sense, the degree of aggregation can be interpreted as a function of the balance between minimizing the risk of predation while maximizing the intake of food. Mann (1985) attributed large qualitative shifts between different modes of feeding (and their attendant patterns of aggregation) by *S. droebachiensis* to its "nonlinear, adaptive behavioral responses" to food and predation.

As *S. droebachiensis* increase in size, the availability of suitable shelter sites decreases and individuals larger than about 20 cm are increasingly found on the exposed surfaces of the seabed (Scheibling and Hamm 1991; Dumont et al. 2004). Food detection and aggregation behavior as adults is of primary relevance to feeding. Adults can sense food several metres distant (Vadas 1977; Larson et al. 1980; Mann et al. 1984) and can move relatively rapidly over the seabed at rates of 0.5–3.0 m per day (Propp 1977; Garnick 1978; Lauzon-Guay et al. 2006). Thus, *S. droebachiensis* may disperse over substantial distances in search of food or may remain static for extended periods at low densities, presumably minimizing their exposure to predators (Bernstein et al. 1983). Most significantly, *S. droebachiensis* form large feeding aggregations that move through kelp beds, destructively grazing at rates of up to 4.0 m per month (Breen and Mann 1976b; Foreman 1977; Mann 1977, 1985; Scheibling et al. 1999; Gagnon et al. 2004). Large *S. droebachiensis* can shift among different forms of aggregations at timescales of days to decades, and the factors which cause such shifts are the subject of debate (Pringle 1986; Miller 1985a; Elner and Vadas 1990; Hagen and Mann 1994; Sivertsen and Hopkins 1995; Scheibling et al. 1999), in part because the patterns of feeding behavior differ among geographic habitat zones (e.g. the Atlantic coast of Nova Scotia vs. the Fundy coast of New Brunswick vs. the outer coast of Norway).

The availability of good quality food appears to be the primary determinant of aggregation by adult *S. droebachiensis* (Hagen and Mann 1994).The presence of food can override predator avoidance behavior (Vadas et al. 1986), but sea urchins may aggregate for defensive purposes as well, even in the presence of food (Bernstein et al. 1983). Mann (1985) recognized three modes of feeding behavior in populations of *S. droebachiensis*:

1. passive detritivory: stationary, small aggregations in topographic refuges, depending on food coming to them in the form of drift algae;
2. dispersed browsing: mobile, nonaggregated small individuals in low densities, depending on small algae and alternative forms of nutrition to survive in sea urchin-dominated barren grounds (Johnson and Mann 1982); and
3. aggressive herbivory: slow-moving, massive aggregations ("grazing fronts") of large individuals at high density on the margins of kelp beds, which graze rapidly and continuously on attached algae and create barren grounds in their wake (Breen and Mann 1976b; Lang and Mann 1976; Miller 1985a; Hagen and Mann 1992, 1994; Scheibling et al. 1999; Gagnon et al. 2004).

Well-nourished individuals in kelp grazing aggregations have high somatic growth rates and gonad indices (Meidel and Scheibling 1998a; Scheibling et al. 1999; Wahle and Peckham 1999). In certain regions, such as near the Island of Grand Mannan in the Bay of Fundy, large aggregations of well-fed *S. droebachiensis* may also occur on soft sediments, distant from hard substratum supporting kelp beds (Mackay 1976). Presumably these sites are hydrodynamically defined deposition areas for macroalgal detritus, but it is not known whether the feeding rates are similar to those in grazing fronts.

2.3. Feeding Rates

Strongylocentrotus droebachiensis feeds more efficiently on fixed food items than on those that are unattached and mobile (Himmelman 1984). Many inter-habitat differences in feeding and growth rates have been attributed to variation in the accessibility rather than the absolute abundance of algal food items, with water movement often invoked as a factor limiting access (Himmelman and Steele 1971; Himmelman et al. 1983a; Vadas 1990; Siddon and Witman 2003). However, there are no in situ comparisons of consumption or growth rates of *S. droebachiensis* feeding on attached versus unattached (drift) algae. At sufficiently high densities (i.e. >150 individuals per square meter, such as found at grazing fronts) sea urchins are capable of anchoring and consuming even large algal fronds in hydrodynamically energetic conditions (Breen and Mann 1976b).

Laboratory studies provide information on feeding rates that may occur in the field (Vadas 1977; Larson et al. 1980; Keats et al. 1983; Himmelman 1984; Briscoe and Sebens 1988; Himmelman and Nedelec 1990; Cuthbert et al. 1995; Minor and Scheibling 1997; Meidel and Scheibling 1999). Feeding rates vary primarily with food type (i.e. algal species and availability of animal food), and to a lesser extent with body size, temperature, reproductive stage, and season. The last three factors are often confounded in feeding experiments. Consumption of the most preferred kelps by *S. droebachiensis* ranged from 0.026 g individual per day (*L. longicruris*, 15 mm diameter, 3.5 °C, spring; Meidel and Scheibling 1999) to 4.97 g individual per day (*Neoreocystis luetkeana*, 72 mm diameter, 9.5 °C, winter; Vadas 1977). The mean consumption rate of four species of kelp (by sea urchins ranging from 13–82 mm diameter at temperatures of 0–18 °C in all seasons) was 1.301 g individual per day, whereas that of 8 species of the least preferred macroalgae (Agarum spp. and several red algae) was 0.280 g individual per day. The lowest feeding rates are generally associated with the smallest sea urchins (Larson et al. 1980; Himmelman 1984). The feeding rate of individuals <15 mm diameter is typically an order of magnitude slower than that of large individuals (Minor and Scheibling 1997; Meidel and Scheibling 1999). The difference is much smaller when feeding rates are expressed per unit of sea urchin mass. The ration of food supplied also affects sea urchin feeding rates in the laboratory. *Strongylocentrotus droebachiensis* feeds 1.6–3.5 times faster when macroalgae is available in limited amounts or infrequently rather than *ad libitum* (Hooper et al. 1996; Minor and Scheibling 1997; Meidel and Scheibling 1999).

Feeding of *S. droebachiensis* on animal food (i.e. mussels) is much slower, ranging from 0.002 to 0.023 g individual per day (Briscoe and Sebens 1988). Animal food in the diet reduces the feeding rates of sea urchins on preferred macroalgae by factors of 1.8–5.7 (Briscoe and Sebens 1988, Meidel and Scheibling 1999), presumably because the animal protein reduces the requirement for plant protein.

Strongylocentrotus droebachiensis feeding rates are not linearly related to temperature, but generally show a strong relationship with the reproductive cycle, which itself varies seasonally with temperature. Feeding rates on preferred food (*L. digitata*) in the laboratory over a 6-month period were about 50% higher at an average temperature of 8.9 °C than at 3.0 °C (Cuthbert et al. 1995). The difference was less or not significant for less preferred algae, and smaller amounts were eaten. Larson et al. (1980) found that feeding rates peaked in summer and dropped ca. 30% in winter (January to March) in the laboratory in Maine. The relevance of this pattern to natural populations is questionable as feeding rates in the field in Newfoundland increased during and after spawning in March to May (coldest ambient temperatures), but were depressed during the period of gonad growth in August to November despite high water temperatures (Himmelman 1984). The ability of *S. droebachiensis* to maintain high feeding rates at low temperatures is likely related to the distinct thermal acclimatization of respiration and activity exhibited by the species (Percy 1972, 1973).

Features of the habitat or location apparently influence the feeding dynamics of *S. droebachiensis* populations, but causality is hard to demonstrate. *Strongylocentrotus droebachiensis* from Atlantic Canada appear to have higher feeding rates on preferred macroalgae than those in the Gulf of St. Lawrence, but slower than those on the North American west coast (cf., e.g. Himmelman 1984; Larson et al. 1980, with Himmelman and Nedelec 1990; Vadas 1977), but these differences may simply be experimental artifacts. The ability of *S. droebachiensis* to rapidly change its feeding rate and the allocation of ingested resources in response to seasonal and habitat-related differences in food type and availability contributes greatly to its fitness (Russell 1998; Meidel and Scheibling 1999).

3. GROWTH

3.1. Determinants of Growth

Gross absorption efficiencies of preferred algal foods (kelps) by *S. droebachiensis* have been estimated at 9–91% of the ingested food (although the higher values appear anomalous on theoretical grounds), with both body size and season accounting for significant components of the variance (Larson et al. 1980; Thompson 1982). Seasonal variations in metabolic rates are generally less than a factor of 2 (e.g. Percy 1974a; Miller and Mann1973; Larson et al. 1980), while differences with size may span an order of magnitude, decreasing with increasing size (e.g. Vadas 1977). Realistic assimilation efficiency for a 50 mm *S. droebachiensis* fed *ad libitum* on kelp is (60+10)%. Propp (1977) estimated an absorption efficiency of 80%, 60%, and 30% for N, P, and C respectively. Enhanced use of nutrients in its food suggests that the protein concentration may limit the growth of some organs in *S. droebachiensis* (Miller and Mann 1973; Propp 1977; de Jong-Westman et al. 1995a). Measures of dissolved organic carbon (DOC) excretion (Field 1972; Propp 1977) and the growth of test and gonad on protein rich diets support this hypothesis (e.g. Briscoe and Sebens 1988; Nestler and Harris 1994; de Jong-Westman et al. 1995a; Meidel and Scheibling 1999).

Typical net assimilation efficiencies measured for Strongylocentrotus species feeding on macroalgae range from about 4% to 32%, depending on the type of food and body size (Lawrence 1975). Propp (1977) and Thompson (1982) calculated net assimilation efficiencies of *S. droebachiensis* ranging from 14% to 43%, while Miller and Mann (1973) estimated efficiencies of 6–22% for a population in Nova Scotia. The discrepancies may reflect different methodologies.

In a population of mature *S. droebachiensis* in the Barent's Sea for example, young (small) animals (<3-year old) allocated less than 20% of their total production to gonad growth, while older (larger) individuals (6-year old) allocated 45%, and old animals (>8-year old) allocated less than 33% (Propp 1977) At peak reproductive size (and age), almost 7 times more energy was allocated to gonad growth than to somatic growth. While these size or age-related patterns may be consistent among different populations, the absolute values vary markedly with population and habitat. For example, Miller and Mann (1973) found that gonad growth was only about 25% of somatic growth in a population of sea urchins in St. Margaret's Bay, Nova Scotia. Thompson (1984) found that gonad production approximated the production of soft tissues in 4–5-year-old individuals, and then exceeded soft tissue production by as much as fivefold in 8-year-old individuals from three populations in Newfoundland. Keats et al. (1984) calculated that the gonad output of *S. droebachiensis* in Newfoundland increased by a factor of 2–3 along a gradient from about 15 to 1 m depth.

Such differences in the absolute and relative importance of somatic and gonadal growth among populations of *S. droebachiensis* are generally attributed to locale-specific differences in their diets. This conclusion is supported by Thompson's (1982) comparison of growth allocation at three different rations of combined algal and animal foods. The best-fed sea urchins produced the most gonad and somatic tissue, but as the food supply was limited, individuals maintained gonad production (nutritive phagocytes and gametes) while somatic production (as measured by change in the mass of soft tissues other than gonad and tube feet) decreased. The results of Thompson's (1982, 1984) physiological studies suggest that gonad production may be sustained at the expense of somatic growth if sea urchins eat high-quality foods at rations somewhat less than those consumed in *ad libitum* feeding. The results of rationing experiments by Minor and Scheibling (1997) and Meidel and Scheibling (1999) support this model. The effect may not be large, however, as Hooper et al. (1996) observed no significant change in gonad indices until rations were decreased below 50% *ad libitum*. The mechanisms of nutritional limitation that could lead to the apparent conflict between somatic and gonad growth have yet to be elucidated, however, and the hypothesis that the two growth processes proceed in isolation cannot yet be rejected (Lawrence 2000).

3.2. Growth Rates

Strongylocentrotus droebachiensis is a long-lived, slow-growing species that exhibits great phenotypic plasticity of growth in response to environmental factors (Sivertsen and Hopkins 1995; Russell 1998; Russell et al. 1998). Juvenile *S. droebachiensis* in the field increase in size (test diameter) at rates of 2.6–17.0 mm yr^{-1} during their first two years after settlement, at which age they range from 6 to 26 mm in diameter (Himmelman 1986; Raymond and Scheibling 1987; Meidel and Scheibling 1998a). The great variability primarily reflects habitat differences in food availability and quality, and sea urchin density (Himmelman 1986; Meidel and Scheibling 1998a). At sizes greater than about 25 mm, growth of *S. droebachiensis* under ideal conditions may exceed 2.0 mm per month (up to 27.5 mm yr^{-1}, Swan 1961; Miller and Mann 1973; Vadas 1977; Raymond and Scheibling 1987; Nestler and Harris 1994; Minor and Scheibling 1997; Meidel and Scheibling 1999), although rates of about 1.0 mm per month (ca. 12.0 mm yr^{-1}) are more realistic for natural populations (Swan 1958; Propp 1977; Lang and Mann 1976; Himmelman et al. 1983b; Sivertsen and Hopkins 1995). Individuals in tide pools can grow less than 0.25 mm yr^{-1} (Russell et al. 1998). As diameter and body weight are highly correlated for a population (Ebert 1988; Meidel and Scheibling 1999), measured increase in test diameter can be readily converted to tissue growth.

The rate of growth of *S. droebachiensis* increases with size towards an asymptotic level (Swan 1961; Propp1977; Thompson 1984; Sivertsen and Hopkins 1995; Russell et al. 1998). This is not always true during the first few years post settlement, when there is a lag phase in growth followed by an acceleration to intermediate size (e.g. Himmelman 1986; Raymond and Scheibling 1987; Russell et al. 1998). Typical von Bertalanffy growth parameter values derived from measurements of natural populations range greatly from $K = 0.11$–0.25, and $L_{inf} = 45$–92.4 mm (Munk 1992; Sivertsen and Hopkins 1995; Meidel and Scheibling 1998a). Russell et al. (1998) used the Tanaka function to obtain growth parameter values ranging from $a = 15.1$ to 36.4, $d = -0.40$ to -0.16, and $f = 19.2$ to 38.9. Translating growth rates into size at age for comparison reveals size estimates ranging from 6 to 17 mm at 1 year, and 6 to 26 mm at 2 years (Raymond and Scheibling 1987). Converting age at size for larger individuals yields estimates ranging from 1.7 to 18.6 years for a 50 mm diameter individual (Russell et al. 1998).

Russell and Meredith (2000) advised caution in using age estimates that assume growth bands in ossicles are deposited annually. Using fluorescent markers to indicate growth bands on ossicles, they found only 23% of *S. droebachiensis* consistently added a complete band after 1 year, 20% added more than one band in some ossicles, while many of the largest urchins had no complete bands. They suggest that growth of urchins is seasonally episodic and that growth curves based on this aging technique underestimate urchin age and bias population parameters inferred from age estimates. However, their study was conducted in tide pools, which may differ markedly in environmental conditions compared to subtidal habitats. Chen et al. (2003) constructed a growth-transition matrix using von Bertalanffy growth parameters estimated for *S. droebachiensis* along the coast of Maine, which incorporated large size-specific variation in annual growth increments. Even gross morphology, as characterized by change in test diameter and shape during growth, is variable in sea urchins as a function of feeding, because of skeletal flexibility (Johnson et al. 2002).

The great variability of individual growth parameters in natural populations of *S. droebachiensis* results in part from the errors of size-based ageing or tagging methods, in part from the phenotypic plasticity of the species' growth response, but primarily from the differences in food quality and availability in their habitats. Of these, the main determinant of growth performance is whether the population is in a kelp bed or a barren ground. Typically, the growth rates of both somatic and gonad tissue is greater by factors of 1.1–2.8 in kelp-rich habitats than in sea urchin-dominated barren grounds receiving little macroalgal detritus (Lang and Mann 1976; Hagen 1983; Keats et al. 1984; Himmelman et al. 1983b; Sivertsen and Hopkins 1995; Leinaas and Christie 1996; Meidel and Scheibling 1998a; Brady and Scheibling 2006). These habitat differences are reflected in the size and age frequency distributions of sea urchins in natural populations (e.g. Lang and Mann 1976; Sivertsen and Hopkins 1995; Leinaas and Christie 1996; Meidel and Scheibling 1998a), although Vadas et al. (2002) reported two sympatric growth morphs of *S. droebachiensis* in the Gulf of Maine, characterized by differences in von Bertalanffy growth parameters and observed maximum size and age. Size-related differences in the mobility and ability of sea urchins of different sizes to graze macroalgal fronds and capture drift algae generally have been postulated to explain cases of increased growth rates with size, as urchins shift from cryptic, juvenile behaviors to adult foraging behavior (Himmelman 1986; Meidel and Scheibling 1998a; Dumont et al. 2004).

Experiments with prepared foods of varying protein concentration and source shed light on the role of food composition in determining sea urchin growth rates in terms of size, mass, or gonad tissue (Kennedy et al. 2001; Pearce et al 2002a, 2002b, 2002c; Daggett et al. 2005). Food enriched in protein of readily assimilable composition compared with natural foods always results in higher growth rates, although the degree of enhancement over natural mixed algal-animal diets is generally of the order of a few percent. Limited food availability and added metabolic costs of optimizing diet in natural marine communities compared to culture conditions presumably explains this incremental effect.

Growth rates in natural populations are not clearly density-dependent because of confounding factors of food availability and feeding behavior. Indeed, growth is often faster in high-density aggregations of sea urchins at grazing fronts than at much lower densities in adjacent barren grounds (Lang and Mann 1976; Sivertsen and Hopkins 1995; Meidel and Scheibling 1998a). When densities of small individuals (with limited mobility) reach levels that result in food limitation, however, growth rates may become extremely low (Himmelman et al. 1983b; Himmelman 1986; Raymond and Scheibling 1987). The role of recruitment in determining density (and possibly growth of small individuals) is linked also to habitat differences.

Growth rates of *S. droebachiensis* are not strongly temperature dependent (e.g. $Q_{10} = 1.2$, Raymond and Scheibling 1987), although few controlled studies have been undertaken. Pearce et al. (2005) found the growth of recently settled juveniles to differ little between 9 and 13 °C, but to be slightly lower at 4.7 °C, and significantly lower at 19.7 °C. The maintenance of near-linear growth rates across large seasonal changes in temperature may reflect both the thermal insensitivity of respiration, feeding and aggregating behavior (Percy 1973; Garnick 1978; Bernstein et al. 1983; Scheibling et al. 1999), and the ability of *S. droebachiensis* to acclimatize its metabolic rates to temperature (Percy 1972, 1974b).

4. REPRODUCTION

The process of gonadogenesis and spawning in sea urchins is an interactive, multistage sequence with two major phases: storage of nutrients and production of gametes (eggs and sperm). The timing of the reproductive cycle is regulated by the complex interaction of endogenous and exogenous controls, such that considerable variability is observed within and among natural populations (Pearse 1991; Walker et al., this volume). Food availability and quality, temperature, photoperiod, and water-born chemicals influence the frequency of events and rates of reproductive processes in *S. droebachiensis* (Cocanour and Allen 1967; Stephens 1972; Himmelman 1975, 1978; Thompson 1979, 1984; Falk-Petersen and Lönning 1983; Keats et al. 1984, 1987; Munk 1992; Starr et al. 1992, 1994; Lemire and Himmelman 1996; Hagen 1997, 1998; Minor and Scheibling 1997; Meidel and Scheibling 1998b, 1999; Wahle and Peckham 1999). For example, spawning by *S. droebachiensis* in response to a chemical exuded by phytoplankton during the spring bloom in the Gulf of St. Lawrence (Starr et al. 1992, 1994) provides clear evidence of the link between food availability for larvae and fitness that is a dominant theme in the biology of *S. droebachiensis*.

4.1. Reproductive Timing

The annual reproductive cycle of *S. droebachiensis* has been well described in generalized and locale-specific terms (Boolootian and Turner 1965; Cocanour and Allen 1967; Himmelman 1978; Keats et al. 1984; Falk-Petersen and Lönning 1983; Meidel and Scheibling 1998b). The basic cycle appears to be similar throughout the species' range in the Atlantic ocean. In general, well-fed *S. droebachiensis* develop macroscopic gonads by the spring of their third year of life (spawning as early as 2.75 years after settling, Raymond and Scheibling 1987). Following the phagocytosis of the relict gametes after spawning, gonad growth accelerates through the summer first through the accumulation of nutritive phagocytes, reaching maximum rates in autumn when the proliferation of primary oocytes and initiation of vitellogenesis (spermatogenesis in males) begins. The maturation and storage of ova and sperm (at the expense of nutritive cell mass) proceeds through the winter as gonad mass continues to increase at a decelerating rate into early spring, when spawning begins. Near-synchronous spawning in intermittent pulses or mass spawning events proceeds over one to two months in spring, and the cycle starts anew.

The periodicity implicit in this generalized description belies a great deal of variability in the timing and duration of every one of the six phases of the reproductive cycle (Meidel and Scheibling 1998b). The causes of this variability are manifold and the details poorly understood, but they are clearly related to differences in the local environments of subpopulations. For example, the onset of sexual maturity can be delayed for years in dense populations of *S. droebachiensis* with little food (e.g. Propp 1977; Himmelman 1986). The age of sexual maturity can be more than a year older in populations in barrens than in kelp beds (Sivertsen and Hopkins 1995). Walker and Lesser (1998) found that male gonads of both experimentally treated and field-collected *S. droebachiensis* from the coast of New Hampshire matured more rapidly than female gonads (although Meidel and Scheibling, 1998b, observed no difference between the sexes) and suggested that cold water temperatures may be needed for the completion of vitellogenesis in females. Likewise, a study by Garrido and Barber (2001), which examined the effects of temperature and food ration on gonadal growth, concluded that low temperatures are required for continued growth and maturation of primary oocytes during the later stages of oogenesis, and suggested that temperature influences the duration of the reproductive cycle. Thus, the oogenic cycle generally determines the period in which spawning and fertilization may occur, and the stimulus necessary to induce spawning gets progressively weaker as female gonad maturation proceeds (Stephens 1972; Himmelman 1978).

The timings of peak gonad index and spawning exhibit considerable interhabitat and interannual variability (Himmelman 1978; Keats et al. 1984; Munk 1992). Gonad growth is more rapid, and gamete production greater in individuals inhabiting food-rich habitats (Meidel and Scheibling 1999; Wahle and Peckham 1999). The timing of the major gametogenic stages, however, is similar (in synchrony) across barren to kelp bed habitats in Nova Scotia, even though different individuals in a population can be in two or even three gametogenic stages at any one time during the spawning season (Meidel and Scheibling 1998b). Population-scale synchrony suggests larger-scale environmental factors such as temperature, photoperiod, and current regimes interact with habitat-specific food quality and availability to determine the timing of spawning. For example, Walker and Lesser (1998) found the initiation of mitosis in sea urchins off New Hampshire to be correlated with the occurrence of autumn photoperiod, but that cold temperatures were required for the completion of vitellogenesis.

Actual spawning (at least in the Northwest Atlantic) is triggered by phytoplankton blooms (Starr et al. 1992, 1994) and can be expected to be correlated with meteorological and oceanographic patterns and features. In a few cases, a second spawning event occurs in populations inhabiting food-rich habitats (Keats et al. 1987; Meidel and Scheibling 1998b).

4.2. Gonad Growth and Gamete Production

Gonad growth in *S. droebachiensis* is highly dependent on feeding and absorption of nutrients. Food supply and absorption are largely under environmental control, and gonad mass (which varies logistically with test diameter, Russell 1998; Meidel and Scheibling 1998b) is strongly correlated with the availability of high-quality food in the habitat during the postspawning period. Walker and Lesser (1998) found that the changing autumn photoperiod may activate oogonial or spermatogonial mitosis directly, or mobilize energy stores from nutritive phagocytes, which in turn stimulates gonial cell mitosis. In controlled temperature and feeding experiments, Garrido and Barber (2001) found that food availability was the most important factor regulating energy storage, cellular composition, and relative size of gonads throughout the year. Urchins consumed more food at higher temperatures, suggesting that temperature directly affects production of nutritive phagocytes. However, once gametogenesis was initiated (by low temperatures) mature ova were produced, regardless of food availability. Based on these results, Garrido and Barber (2001) concluded that high food availability and low temperatures in the winter favor reproductive output in *S. droebachiensis*.

The total mass of the gonad increases at a decelerating rate until spawning occurs. Just prior to spawning, the mass of the ripe gonad is 2.5–10.0 times greater than that of the residual gonad tissue remaining after spawning in the previous year (Cocanour and Allen 1967; Himmelman 1978; Falk-Petersen and Lönning 1983; Keats et al. 1984; Meidel and Scheibling 1998b). The gonad index is slightly higher in females than males (Munk 1992; Minor and Scheibling 1997; Meidel and Scheibling 1998b), and the development of nutritive phagocytes and oocytes in females is more sensitive to diet and ration than in males (Minor and Scheibling 1997; Meidel and Scheibling 1999). It follows that females release approximately 20% more of their body mass as gametes than do males (Meidel and Scheibling 1998b).

4.3. Fertilization Rates

Understanding reproduction in sea urchin populations has contributed to models of evolutionary biology. The production of zygotes is the ultimate measure of fitness, and depends on both fecundity and the fertilization of spawned eggs. Fecundity (i.e. reproductive output, as inferred from gonad mass) in *S. droebachiensis* varies primarily with habitat correlates such as depth, topography, and exposure that affect the type and availability of food. Wahle and Peckham (1999) found that reproductive output in Maine populations was highest in shallow, food-rich habitats with low density, and declined very rapidly with increasing density. Gonad mass was essentially constant at densities of large individuals greater than about 25 m^{-2}. The relationship was weak and nonlinear, as a 1000-fold decrease in density corresponded to only a twofold increase in gonad mass. Estimated fertilization rates, however, tripled from 18% to 62% over a 64-fold increase in density (ca. 2.5–160 m^{-2}). Wahle and Peckham (1999) predicted a sevenfold increase in zygote production across the observed density range. They concluded that the reproductive benefits of aggregation on fertilization success greatly outweigh possible costs associated with reduced fecundity due to lower individual ration at high densities. It is possible to extend these results to natural populations of *S. droebachiensis* if the effects of population size are considered (Meidel and Scheibling 2001). Aggregations at grazing fronts may maximize both components of reproductive success, making these subpopulations disproportionately important contributors to the larval pool (Meidel and Scheibling 1998b, 2001). However, a simulation model of fertilization dynamics of *S. droebachiensis* did not indicate a strong decline in mean fertilization with decreasing density (Lundquist and Botsford 2004). At high densities, aggregation had little effect on fertilization success, possibly due to an upper density threshold above which fertilization is maximal.

Numerous field studies of fertilization rate in sea urchins suggest severe sperm limitation in nature, requiring ideal flow conditions and close proximity to other spawning individuals for successful fertilization (reviewed by Yund and Meidel 2003). These authors pointed out, however, that gamete longevity and release in viscous fluids at the surface of the test may result in lower threshold densities for fertilization failure than suggested by short-term studies. Meidel and Yund (2001) found that more than 75% of eggs of *S. droebachiensis* maintained in aged seawater in the laboratory were viable for 48 h and that viability exceeded 50% at 72 h, greatly surpassing previous estimates (Wahle and Peckham 1999). In field experiments in which egg samples were exposed to slowly diffusing sperm from a spawning male, Meidel and Yund (2001) showed that fertilization levels increased significantly with exposures ranging from 15 min to 3 h. Yund and Meidel (2003) spawned individuals in a laboratory flume and observed that gametes spawned in the benthic boundary layer were advected slowly and continuously from the aboral test surface of *S. droebachiensis* for several hours, depending on flow velocity (free-stream velocities ranged from 2.5 to 8.5 cm s^{-1}). They found that about half of the fertilization events took place before eggs were advected from the test, and that fertilization levels slightly decreased at the higher flow velocity. Wahle and Gilbert (2002) documented the time course and spatial extent of sperm availability adjacent to an aggregation of *S. droebachiensis* over the spawning season. Using time-integrated fertilization assays they found fertilization rates to decline sharply over a distance of less than 1 m, and sperm availability to range from brief periods of intense supply to a fairly continuous trickle.

Levitan (2002a) compared fertilization success among three species of Strongylocentrotus in their natural habitats and found that each experienced density-dependent species-specific selection on gamete traits. *Strongylocentrotus droebachiensis* occurring at low densities off Vancouver Island, British Columbia, had the lowest levels of fertilization. Gametes of this species exhibited traits (e.g., larger eggs, receptive egg surfaces, and slow but long-lived sperm) that enabled it to outperform congenerics (*S. franciscanus* and *S. purpuratus*) under conditions of sperm limitation (Levitan 2002a, 2002b). The cost of specialization for a particular demographic condition however, is that it may lead to Allee effects when environmental conditions change, especially when female fertilization success is sensitive to variations in population density. One drawback of increasing the likelihood of fertilization under sperm-limiting conditions is an increased risk of hybridization (Levitan 2002a). *Strongylocentrotus droebachiensis* is highly susceptible to heterospecific fertilization, which decreases rates of conspecific fertilization and produces hybrid larvae that have low survival (Levitan 2002b). Nevertheless, Levitan (2002b) concluded that benefits from production of promiscuous eggs by *S. droebachiensis* actually outweigh the cost of hybrid fertilization under conditions of sperm limitation.

5. SETTLEMENT AND RECRUITMENT

5.1. Larval Development and Settlement Behavior

Strongylocentrotus droebachiensis exhibits typical planktotrophic larval development with a larval period that ranges from 4 to 21 weeks, depending on temperature (Strathmann 1978; Hart and Scheibling 1988). Stephens (1972) found that larvae of *S. droebachiensis* develop normally at temperatures as low −1 °C, but observed gross asynchrony in embryonic development above 10 °C and irreversible arrest of cell division above 12 °C. He suggested that larval supply to adult populations is restricted to areas where temperatures in spring (when larvae are in the plankton) do not exceed 10 °C. However, Hart and Scheibling (1988) observed normal and relatively rapid growth of plutei between 12 and 18 °C, exceeding the lethal limit (14–15 °C) proposed by Stephens (1972). Roller and Stickle (1985) found that larvae of *S. droebachiensis* develop normally at salinities as low as 20 psu, in contrast to *Strongylocentrotus pallidus* that did not survive 25 psu. Larvae from a hybrid-cross of *S. droebachiensis* and *S. pallidus* showed developmental patterns and survival rates intermediate to those of the parental species, indicating a genetic component to the euryhalinity of *S. droebachiensis*.

Studies of larval nutrition have shown that effects of maternal condition on development rate (through investment in eggs) are small relative to the effects of larval food ration (de Jong-Westman et al. 1995b; Bertram and Strathmann 1998; Meidel et al. 1999). Burdett-Coutts and Metaxas (2004) showed that larvae of *S. droebachiensis* can actively aggregate at food patches (microalgae) in experimental water columns and proposed a chemosensory mechanism for patch detection. The lengthy developmental period of *S. droebachiensis* suggests that larvae may be transported to settlement sites far from source populations, although information on larval dispersal is lacking for this species. The tolerance of larvae to temperatures and salinities normally encountered throughout the natural range of the sea urchin indicates that the simple effects of these factors probably do not restrict its distribution. However, environmental stressors may have interactive effects that affect development and survival, particularly under low nutritional conditions in the plankton (Hart and Scheibling 1988).

Settlement behavior of *S. droebachiensis* has been examined in laboratory studies. Burke (1980) described a substratum-testing behavior in competent larvae that may detect metamorphosis-inducing stimuli. Pearce and Scheibling (1990) showed that coralline red algae strongly induce settlement and metamorphosis through contact chemoreception, and identified GABA-mimetic molecules, present in coralline algae, as possible inducers. Settlement also is induced, to varying degrees, by a variety of macroalgal species and microbial films (Pearce and Scheibling 1991). Thus, cues for settlement are not highly substratum specific and there is no evidence that conspecific cues are used in settlement (Pearce and Scheibling 1991). McNaught (1999) found that spatial patterns of settlement on artificial turf mirrored those on various natural substrata, and concluded that regional differences in substrate type and availability had little influence on settlement rate of *S. droebachiensis* in the Gulf of Maine. Lambert and Harris (2000) found no preference for substrate type during peak settlement in the southern the Gulf of Maine, although there was a preference for substrate covered with live coralline algae during the rest of the settlement period.

5.2. Temporal and Spatial Patterns of Settlement

Artificial collectors (plastic turf) have been used to document variation in settlement of *S. droebachiensis* at various temporal and spatial scales in the Northwest Atlantic. Settlement occurs mainly in June and July in the Gulf of Maine and Nova Scotia, respectively, although low numbers of settlers have been found as late as October in Nova Scotia (Harris et al. 1994; Balch and Scheibling 2000; Lambert and Harris 2000). Interannual differences in settlement rate, ranging an order of magnitude, were recorded in the early 1990s in Nova Scotia (Balch and Scheibling 2000) and the southwestern Gulf of Maine (Harris et al. 1994). Foreman (1977) found that settlement of *S. droebachiensis* in British Columbia was greatest in years with low spring temperatures, which favored larval survival according to Stephens (1972). In contrast, Hart and Scheibling (1988) argued that settlement pulses of sea urchins in Nova Scotia occur during unusually warm years because of accelerated larval development that reduces exposure to planktonic predators. Balch and Scheibling (2000) found that the interannual pattern of settlement of *S. droebachiensis* was not consistent with those of other echinoderm species with similar larval types and timing of settlement. They suggested that interspecific differences in reproductive output or larval survival, rather than environmental factors such as temperature, are responsible for large variations in interannual settlement.

Balch et al. (1998) found order of magnitude differences in settlement of *S. droebachiensis* among three geographic regions, the Gulf of Maine (103–104 settlers per square meter), the Atlantic coast of Nova Scotia (102 settlers per square meter), and the Bay of Fundy (101 settlers per square meter), that are likely caused by large-scale oceanographic features influencing larval supply. Within the Gulf of Maine, there is a strong gradient of decreasing settlement from the northeastern to southwestern subregions, which has been linked to coastal circulation patterns (Harris and Chester 1996; Balch et al. 1998; McNaught 1999). At smaller spatial scales, variability among sites has been attributed to local hydrodynamic features that may retain larvae or transport them away (Harris et al. 1994; McNaught 1999; Balch and Scheibling 2000; Lambert and Harris 2000). Harris et al. (1994) found that settlement varied strongly with depth at a site in the southwestern Gulf of Maine, with highest settlement at 6–9 m (>15 000 settlers per square meter) and lower settlement (by 1–2 orders of magnitude) below thermoclines at 12 and 25 m.

In the Gulf of Maine, Harris and Chester (1996) and McNaught (1999) found that settlement of *S. droebachiensis* was generally greater in natural or artificial kelp beds than in adjacent barrens or areas cleared of macroalgae. In contrast, Balch et al. (1998) found greater settlement in barrens than in kelp beds at other sites in Maine and Nova Scotia (see also Balch and Scheibling 2000), and studies of *S. purpuratus* and *S. franciscanus* in California found no effect of kelp on settlement (Rowley 1989; Schroeter et al. 1996).

Regional disparities in interhabitat settlement patterns may be related to the physical and biological characteristics of kelp beds, or the surrounding environment, which affect larval supply (Balch and Scheibling 2001). Although artificial collectors facilitate settlement sampling by providing standardized substrates that are easily replicated, differences in collector type or sampling frequency may bias results and limit comparability between studies, or locations within studies (Balch and Scheibling 2001).

5.3. Temporal and Spatial Patterns of Recruitment

Most studies have found that recruitment of *S. droebachiensis*, like that of other sea urchins with broadly dispersing larvae, is patchy in time and space (reviewed by Ebert 1983; Balch and Scheibling 2001). Episodic recruitment of *S. droebachiensis* has been reported in southern Alaska (Estes and Duggins 1995), British Columbia (Foreman 1977), the St. Lawrence Estuary (Himmelman et al. 1983b), Nova Scotia (Scheibling 1986; Raymond and Scheibling 1987; Balch and Scheibling 2000), and Newfoundland (Keats et al. 1985). In contrast, substantial annual recruitment has been recorded in several successive years in the southern Gulf of Maine (Harris et al. 1985, 1994; Martin et al. 1988; Harris and Chester 1996). Interannual variation in recruitment success has been attributed to factors affecting larval production (Harris and Chester 1996; Meidel 1998), survival (Foreman 1977; Hart and Scheibling 1988), and transport by physical oceanographic processes (Estes and Duggins 1995; McNaught 1999).

At the habitat scale, lower rates of recruitment of *S. droebachiensis* have been recorded in kelp beds than in barrens in Nova Scotia (Scheibling 1986; Balch and Scheibling 2000), Gulf of Maine (McNaught 1999), and Norway (Leinaas and Christae 1996). Similar patterns have been observed for other strongylocentrotids in kelp forests in California (Tegner and Dayton 1981). Various mechanisms that reduce larval supply, settlement or early postmetamorphic survival in kelp beds/forests have been proposed to account for these differences in recruitment, but none have been unequivocally demonstrated (Balch and Scheibling 2001). In Nova Scotia, Balch and Scheibling (2000) found a strong positive relationship between recruitment of *S. droebachiensis* and settlement (on artificial collectors) in the previous year. Thus, patterns of settlement pooled across habitats, sites, and years were not obscured by differences in postsettlement survival between kelp beds and barrens. In the Gulf of Maine, McNaught (1999) also found a positive relationship between settlement and subsequent recruitment of *S. droebachiensis* in coralline-dominated sites but no relationship in sites dominated by fleshy macroalgae, which he attributed to greater postsettlement mortality in algal beds.

In Norway, Sivertsen and Hopkins (1995) characterized loose substrata (gravel, shells, corallines), usually at 8–30 m depth, as "recruiting areas" for *S. droebachiensis* because these habitats are dominated by small individuals (5–20 mm). They proposed that juveniles migrate from these areas, as they grow, to join populations of adults on shallow rocky bottoms. High densities of recruits also have been recorded in coralline-encrusted cobble beds in Newfoundland (Himmelman 1986) and Nova Scotia (Scheibling and Raymond 1990).

5.4. Early Postsettlement Mortality

Little is known about the causes of early postsettlement mortality, which influences rates and patterns of recruitment in sea urchins. Himmelman et al. (1983b) found recruitment of *S. droebachiensis* in the upper St. Lawrence Estuary was limited to below 2 m due to the inability of small individuals to tolerate periodic low salinities in the surface layer; high recruitment occurred at all depths near the mouth of the estuary. Raymond and Scheibling (1987) found that both food and temperature affected juvenile survival in laboratory experiments. Survival was >95% after 1 year in well-fed treatments, regardless of temperature regime, or at low temperature (5 °C) regardless of feeding regime. However, survival decreased significantly under low food supply and high temperature (16 °C). Raymond and Scheibling concluded that the interaction of food and temperature reflects the temperature dependence of metabolic rate, which determines nutritional requirements, and found evidence of cannibalism and/or scavenging by conspecifics under the most stressful conditions. Pearce et al. (2005) also reported high survival rates between 5 and 16 °C for small juveniles fed a constant diet, but a significant decrease in survival (26%) at 19 °C. Attempts to acclimate individuals to 24 °C resulted in death within a few days. Meidel et al. (1999) showed that rates of metamorphosis and settler size are positively related to larval food ration, suggesting that nutritional conditions during the planktotrophic larval stage could influence early postsettlement survival.

Predation often is invoked as a determinant of sea urchin recruitment but few studies have recorded predation of early juveniles in the field or demonstrated its importance as a regulatory mechanism (reviewed by Scheibling 1996). Recent settlers of *S. droebachiensis* tend to be cryptic, living in crevices, under rocks, or in biogenic microhabitats such as branched or undercut coralline crusts, macroalgal turfs, and mussel patches (Keats et al. 1985; Witman 1985; Himmelman 1986; Scheibling and Raymond 1990). Although, these microhabitats afford shelter from demersal fish and macroinvertebrate predators (Scheibling and Hamm 1991; Ojeda and Dearborn 1991), they also harbor small predators that could consume small individuals. It is not known whether the occurrence of early juveniles in cryptic microhabitats results from settlement preferences, early postsettlement migration, or differential mortality of randomly distributed settlers.

6. PREDATION

6.1. Predators

Strongylocentrotus droebachiensis is prey to a wide range of fish and invertebrate predators throughout its life history (Scheibling 1996). Information about predators of the planktonic larvae is limited to anecdotal observations by Hooper (1980), who reported that scyphomedusae (notably *Cyanea capillata*), sponges, tunicates, and larval fish "can eat vast numbers of larvae." He also listed ophiuroids and ctenophores as predators of larval *S. droebachiensis* in Newfoundland waters. Predators of recently settled sea urchins (<5 mm) are also poorly known, but probably include juvenile rock crabs, lobsters and sea stars, amphipods, hermit crabs, and small fish (Hooper 1980; Scheibling and Hamm 1991; McNaught 1999; Robinson and Scheibling, unpubl.). McNaught (1999) found that fleshy macroalgal beds harbor more potential micropredators of sea urchins than barrens, and conducted caging studies which suggested that predation in algal habitats may be important in regulating early postsettlement survival of *S. droebachiensis*.

Larger juvenile and small adult (5–25 mm) *S. droebachiensis* are prey to a variety of small-mouthed fish such as cunner (*Tautogolabrus adsperses*), winter flounder (*Pseudopleuronectes americanus*), American plaice (*Hippoglossoides platessoides*), and ocean pout (*Macrozoarces americanus*), as well as crabs (*Cancer irroratus*, *Cancer borealis*, *Cancer pagurus* lobsters (*Homarus americanus*), and starfish (*Asterias vulgaris*) (Himmelman and Steele 1971; Bernstein et al. 1981; Green et al. 1984; Keats et al. 1985, 1987; Witman 1985; Keats 1990, 1991a; Miller 1985a; Ojeda and Dearborn 1991; Scheibling and Hamm 1991; Vadas and Steneck 1995; Hagen et al. 2002; Siddon and Witman 2004). Mobile predators of *S. droebachiensis* generally are size-selective and larger adult sea urchins attain a size refuge from all but largest lobsters, cancrid crabs, sea stars (*Pycnopodia helianthoides*) and wolffish (*Anarhinchus lupus*) (Elner 1980; Duggins 1983; Hagen and Mann 1992; Keats et al. 1986). In Alaska, sea otters (*Enhydras lutris*) are major predators of large *S. droebachiensis* and its congeners, but rarely consume urchins <15–20 mm (Duggins 1980; Estes and Duggins 1995). In the St. Lawrence Estuary, Himmelman and Lavergne (1985) observed anemones (*Tealia felina*) eating large sea urchins that apparently had fallen into the tentacles. Seabirds such as gulls and eider ducks prey on intermediate size and large sea urchins in the low intertidal and shallow subtidal zones (Himmelman and Steele 1971; Guillemette et al. 1992; Bustnes and Lnne 1995).

Humans have become significant predators of *S. droebachiensis*, as evidenced by the history of landings from North American fisheries since the mid-1980s. In developing fisheries in British Columbia (Perry et al. 2002) and Maine (Chen et al. 2003), annual landings peaked in the early 1990s (at ~1000 and 1400 Mt respectively) before stocks underwent a marked decline. The fishery in Nova Scotia, which only began in 1989, reported annual landings of up to 1300 Mt, although it collapsed after repeated mass mortalities of sea urchins in the late 1990s (Miller and Nolan 2000). The use of drags or dredges to harvest *S. droebachiensis* in the Bay of Fundy significantly reduces urchin populations and impacts associated fish and invertebrates (Robinson et al. 2001). There is also evidence of prehistoric fishing of *S. droebachiensis* in this area (see Chapter 1).

6.2. Behavioral Responses to Predators

Laboratory and field experiments have shown that *S. droebachiensis* exhibits a flight response induced by chemical cues emanating from predators (lobsters, rock crabs, cunners, Atlantic wolffish), or crushed conspecifics (which signals the presence of predators) (Mann et al. 1984; Vadas et al. 1986; Scheibling and Hamm 1991; Hagen et al. 2002). These experiments suggest that sea urchins can even differentiate predatory crabs and sea stars on the basis of potential threat (Scheibling 1996). Hagen et al. (2002) suggested that the chemical cues emanating from predators are more likely to be diet-dependent alarm signals, meaning that predators are more likely to be labeled by their choice of prey. They found that *S. droebachiensis* showed a greater behavioral response towards predators on echinivorous diets. Although the adaptive significance of a flight response to highly mobile predators such as crabs and fish is unclear, it does allow escape from predatory sea stars (Duggins 1983). Bernstein et al. (1981) found that *S. droebachiensis* in barrens forage more at night in summer than in winter, and attributed this behavioral shift to avoidance of diurnally active predatory fish in summer (but see also Miller 1985a). Hagen et al. (2002) found that *S. droebachiensis* exhibited both flight (repellant) and fright (arrestant) responses to Atlantic wolffish. They suggested that cessation of movement (fright) may be an adaptive response to visual predators (like wolfish), making detection more difficult.

Aggregation has been interpreted as defensive behavior in *S. droebachiensis* (Garnick 1978; Bernstein et al. 1981; 1983), although this interpretation remains equivocal (Vadas et al. 1986; Scheibling and Hamm 1991). Sea urchins tend to aggregate when enclosed with predatory crabs and lobsters in laboratory tanks or field cages (Bernstein et al. 1981; 1983), but Vadas et al. (1986) argued that this is an experimental artifact arising when the natural flight response of sea urchins is impeded by artificial boundaries. Behavioral responses to predators can vary with sea urchin size, density, and nutritional condition, the availability of food (kelp) and spatial refuges, and the type of predator (Scheibling and Hamm 1991; Hagen and Mann 1994). Consequently, sea urchins may simultaneously exhibit different aggregation patterns and defensive behaviors, even within the same habitat (Scheibling 1996).

Juvenile *S. droebachiensis* in cryptic habitats tend to venture out to forage on exposed rock surfaces as they grow larger (Keats et al. 1985; Himmelman 1986). Field experiments on cobble beds in Nova Scotia demonstrated the importance of appropriately scaled refuges for the survival of different sized juveniles (Scheibling and Hamm 1991; Scheibling and Raymond 1990). Sheltering beneath cobbles reduced predation on recent settlers and small juveniles, but as the sea urchins approached adult size they required larger spatial refuges, such as boulders, to escape predation. Intensive predation on intermediate-sized sea urchins, emerging from the spatial refuges occupied as juveniles but not having attained an adult size refuge, may account for bimodal size distributions (with prominent juvenile and adult modes) commonly observed in populations of *S. droebachiensis* (Scheibling and Stephenson 1984; Scheibling 1986; Ojeda and Dearborn 1991) and other temperate species of sea urchins (reviewed by Scheibling 1996).

Juveniles of Strongylocentrotus exhibit varying degrees of conspecific sheltering behavior, where they aggregate under spine canopies of larger adults. This juvenile–adult association is thought to increase recruitment success by providing protection from predators, and to confer a nutritional advantage by facilitating access to food (Tegner and Dayton 1977, Nishizaki and Ackerman 2004). In laboratory experiments, Nishizaki and Ackerman (2004) showed that the frequency of juvenile sheltering was much lower in *S. droebachiensis* (25%) than in a larger congeneric species, *S. franciscanus* (75%). There was no evidence of nutritional gain by juveniles of either species from this association: growth rates were lower in the presence of adults for *S. franciscanus*, and no different from controls without adults for *S. droebachiensis*.

6.3. Predation as a Mechanism of Population Regulation

The role of predation in regulating populations of *S. droebachiensis* in eastern Canada has been intensely debated since Mann and Breen (1972) first proposed that lobsters are keystone predators in this system, and that overfishing them resulted in sea urchin population outbreaks (reviewed by Chapman and Johnson 1990; Elner and Vadas 1990; Scheibling 1996). Various lines of evidence now refute this hypothesis: predator–prey experiments (Evans and Mann 1977; Elner 1980), analyses of lobster diet (Elner and Campbell 1987), estimates of predation rates from empirical data (Miller 1985a), a simulation model of population dynamics (Mohn and Miller 1987), and fluctuations in sea urchin and lobster abundance (Scheibling 1994). The predation hypothesis was subsequently expanded to include predatory crabs and fish (Breen and Mann 1976a; Wharton and Mann 1981) and complex behavioral interactions among predators (Bernstein et al. 1981, 1983). However, Miller (1985a) found no empirical evidence indicating that sea urchin populations could be regulated by any of these predators (but see also Keats and Miller 1986; Hagen and Mann 1992), and Vadas et al. (1986) challenged the validity of experiments examining behavioral interactions. Pringle et al. (1982) proposed that fish and lobsters likely exerted a greater controlling influence on populations of *S. droebachiensis*, and the structure of nearshore communities in eastern Canada, before overfishing markedly reduced the abundance of these predators. Vadas and Steneck (1995) reached a similar conclusion for the Gulf of Maine, where experimental studies have shown that multiple-predator effects and prey-switching behavior can significantly affect sea urchin abundance and community structure (Siddon and Witman 2004). Thus, if sea urchin population outbreaks in the Northwest Atlantic are related to a relaxation of predation pressure, it is likely because of the removal of a complex of predators rather than just one or a few.

In the Northeast Pacific, however, evidence that sea otters are keystone predators of sea urchins is less equivocal (Estes and Duggins 1995). Otter predation severely restricts the abundance of *S. droebachiensis* and its congeners allowing the persistence of rich macroalgal communities. Where otters are absent, sea urchin densities are high and intensive grazing limits the establishment of kelps and other noncoralline macroalgae (Duggins 1980; Estes and Duggins 1995). The response of urchins to changes in otter abundance is related in part to the magnitude of the perturbation. A 50% reduction in otter density after the Exxon Valdez oil spill in Prince William Sound, Alaska, had no detectable impact on density of *S. droebachiensis*, although the proportion of large sea urchins was greater where otter populations had declined (Dean et al. 2000).

The paucity of information about predation during the larval stage or early postsettlement period of *S. droebachiensis* precludes any firm conclusions about critical life-history stages or bottlenecks in the regulation of population abundance. The cryptic behavior of juvenile sea urchins suggests strong selection pressure for predator avoidance, and increased predation on intermediate-sized individuals during an ontogenetic shift to more exposed foraging areas may be an important bottleneck for population growth (Scheibling and Hamm 1991). For adult *S. droebachiensis* mortality due to predation is probably relatively low because few predators (apart from wolffish and sea otters) can handle large sea urchins. However, overharvesting of the sea urchin fishery in eastern Canada and the Gulf of Maine (Hatcher and Hatcher 1997; McNaught 1999) may have greater consequences for sea urchin populations and the structure of benthic communities than any marine predator, past or present.

7. DISEASE AND PARASITISM

7.1. Microbial Pathogens

Acute infections by a pathogenic amoeba, *Paramoeba invadens* (Jones 1985), cause mass mortalities of *S. droebachiensis* in Nova Scotia (reviewed by Scheibling 1988). The amoeba is waterborne and can be cultured on marine bacteria, indicating a free-living existence (Jones and Scheibling 1985). However, it has not been found in coastal waters and sediments, in areas or years without epizootics (Jones et al. 1985; Jellett et al. 1989). The inability of *P. invadens* to survive in culture at temperatures (2 °C) above the winter minimum in coastal waters of Nova Scotia (0 to −2 °C) suggests it is periodically introduced from warmer regions (Jellett and Scheibling 1988, Scheibling and Hennigar 1997).

Temperature is a key factor regulating the transmission and progression of paramoebiasis (Scheibling and Stephenson 1984; Jellett and Scheibling 1988). Epizootics occur during the peak in ambient sea temperature in late summer/fall in unusually warm years, and mortality is directly related to the magnitude and duration of peak temperatures (Miller and Colodey 1983; Scheibling and Stephenson 1984; Miller 1985b; Scheibling 1986; Scheibling and Hennigar 1997). In the laboratory, the disease progresses exponentially between 12 and 20 °C, and it is arrested at 10–12 °C (Scheibling and Stephenson 1984). Infected individuals recover within 30 days by lowering temperature to at least 8 °C. In nature, disease outbreaks are terminated by declining temperatures in the late fall and surviving sea urchins recover over winter (Scheibling and Stephenson 1984; Scheibling and Hennigar 1997). Sea urchins in deeper waters below the thermocline have a refuge from disease (Brady and Scheibling 2005). Nutritional condition does not affect the rate of mortality from paramoebiasis, and both juveniles and adults are susceptible (Scheibling and Stephenson 1984).

Recurrent outbreaks of disease causing mass mortalities of shallow (<20 m) populations of *S. droebachiensis* were recorded along the Atlantic coast of Nova Scotia between 1980 and 1983 (Miller and Colodey 1983; Scheibling 1984, 1986; Miller 1985b), and in 1993, 1995–1997, 1999, and 2000 (Scheibling and Hennigar 1997; Miller and Nolan 2000; Brady and Scheibling 2005). Anecdotal evidence suggests mass mortalities in previous decades (Miller 1985b). Estimated annual mortalities between 1980 and 1983 ranged from 80 to 260 Kt fresh weight in areas of complete die-off (Miller and Colodey 1983; Miller 1985b; Scheibling 1986). Mortalities in the 1990s were of similar magnitude (Miller and Nolan 2000). Ocean currents and hydrodynamic features influence the spread of disease along the coast of Nova Scotia (Scheibling and Stephenson 1984) and may be involved in the periodic reintroduction of *P. invadens* to coastal areas, if it is an exotic species (Scheibling and Hennigar 1997). These epizootics are highly specific to *S. droebachiensis*

(Scheibling and Stephenson 1984; Jellett et al. 1988) and probably sea urchin density dependent (Scheibling 1984, 1988).

Although extensive mortalities of *S. droebachiensis* have not been observed elsewhere in the Northwest Atlantic, there are reports of disease and localized die-offs of sea urchins in Newfoundland (Hooper 1980), the St. Lawrence Estuary and Gulf of St. Lawrence (Himmelman et al. 1983b; Dumont and Himmelman 2004), and the Gulf of Maine (Martin et al. 1988). These mortalities were associated with colder waters, and not likely caused by *P. invadens*.

7.2. Macroparasitic Infections

An endoparasitic nematode *Echinomermella matsi* (Jones and Hagen 1987) is prevalent in populations of *S. droebachiensis* along the Norwegian coast (Hagen 1987, 1992; Skadsheim et al. 1995; Sivertsen 1996; Stein et al. 1998). Hagen (1992) reported an increase in parasite prevalence from 6% to 65% between 1983 and 1991 at a site near Bodø. In surveys between 1982 and 1992, mean prevalence ranged from 2% to 50% among sites across the known range of the parasite (~700 km) (Skadsheim et al. 1995, Sivertsen 1996). Sivertsen (1996) found prevalence was highest in kelp beds and barrens in the Bodø area (40–88 %), the center of the distribution, where sea urchin density was low. Prevalence was greatest in 2- to 7-year-old sea urchins and decreased in older/larger individuals, likely due to host mortality. The smallest infected sea urchins were 10 mm (~1-year old) and the intensity of infection (worm burden) increased with size/age. *Echinomermella matsi* also infects co-occurring *S. pallidus* at a low rate (2.8%), but not *Echinus esculentus* (Sivertsen 1996).

The occurrence of various life stages of *E. matsi* in the coelom of *S. droebachiensis* suggests that the parasite multiplies within its host (Stein et al. 1996). The nematode affects sea urchins by reducing the gonads and heavy infestations can cause death (Hagen 1992, 1996; Sivertsen 1996; Stein et al. 1998). Using mark-recapture techniques, Stein (1999) showed that *E. matsi* infection has only a small effect on sea urchin growth but a pronounced effect on survival, with an estimated reduction in life expectancy of 33–86%. Mathematical models indicate that effects of *E. matsi* on size-specific survival of sea urchins could be more important in determining population reproductive rates of *S. droebachiensis* than effects on gonadal development (Stein et al. 1998; Stein 1999).

Hagen (1987, 1992) proposed that infection by *E. matsi* causes mass mortalities of *S. droebachiensis* that terminate sea urchin outbreaks in Norway. However, further studies have failed to support this hypothesis, and suggested other factors (including a waterborne agent) may be involved in sea urchin mass mortalities in this region (Christie et al. 1995; Skadsheim et al. 1995; Stein et al. 1995). Hagen (1995) too has rejected the "macroparasite hypothesis" of population regulation in view of renewed sea urchin outbreaks in his study area. Although it is clear that *E. matsi* markedly reduces survival and reproductive capacity of *S. droebachiensis* in heavily infected areas, the role of parasitism in the long-term regulation of sea urchin populations in Norway has yet to be determined (Sivertsen 1996; Stein et al. 1998; Stein 1999).

8. MORTALITY DUE TO ABIOTIC FACTORS

Abiotic agents of mortality most commonly affect sea urchins in intertidal and shallow subtidal habitats (reviewed by Lawrence 1996). Abrasion by sea ice periodically removes benthic assemblages, including *S. droebachiensis*, along the shores of the St. Lawrence Estuary and Newfoundland (Hooper 1980; Keats 1991b; Dumont and Himmelman 2004). Gagnon et al. (2004) reported a 65% decrease in urchin density during winter at an ice-scoured site in the Mingan Islands (northern Gulf of St. Lawrence). Ice-scoured areas are rapidly recolonized by algae but recovery of sea urchin populations may take years. Low-salinity layers formed during spring thaws and periods of heavy rainfall cause periodic mass mortalities of *S. droebachiensis* in fiords and inlets in Newfoundland and Labrador (Hooper (1980). Hooper (1980) reported that spring runoff in 1977 killed all sea urchins to a depth of 12–15 m at Kaipokok Bay, Labrador, eliminating 60–75% of the total population. In the St. Lawrence Estuary, Himmelman et al. (1984) showed that low surface salinities kill small sea urchins during tidal fluctuations at shallow depths, which may account for pronounced differences in population structure and abundance of *S. droebachiensis* along the estuarine gradient. In British Columbia, Cameron and Fankboner (1989) noted a mass mortality of recently recruited *S. droebachiensis* in shallow water, which they attributed to runoff from winter rains.

Large accumulations of drift algae in certain depositional environments can create stagnant and hypoxic bottom conditions causing mass mortalities of *S. droebachiensis* and other invertebrates (Scheibling and Raymond 1990; Dumont and Himmelman 2004). At the other extreme, heavy wave action may dislodge sea urchins, or cause abrasion by unstable substrates, which may account for the large numbers that wash up on beaches in some areas (Sebens 1986). Scheibling et al. (1999) observed that sea urchins sheltered in crevices or formed tight two-dimensional clusters during periods of strong wave surge, in contrast to the three-dimensional grazing aggregations that formed during calmer periods.

9. ECOLOGICAL ROLE

Strongylocentrotus droebachiensis plays a pivotal role in determining community structure in the shallow rocky subtidal zone. Throughout the coastal North Atlantic, there are two distinct community states predicated on its abundance: luxuriant seaweed beds dominated by laminarian kelps and with few sea urchins, and the so-called barrens (or barren grounds) dominated by coralline algae and high densities of sea urchins (reviewed by Chapman and Johnson 1990; Vadas and Elner 1992; see also Sivertsen 1997). Either state can be locally stable at decadal time scales or longer (Scheibling 1986; Keats 1991b; Leinaas and Christie 1996). Along the Atlantic coast of Nova Scotia, kelp beds and barrens alternate in a system driven by sporadic outbreaks of disease and sea urchin mass mortality (Miller 1985b; Scheibling 1986; Scheibling et al. 1999); a similar alternation of states may occur off Norway (Sivertsen 1997). In the northern Gulf of St. Lawrence, where sea urchin mass mortality has not been reported, broad-scale pattern analysis (tens of kilometers and years) indicates high overall stability of the two community states, with fluctuations between kelp beds and barrens occurring at smaller spatial (m) and temporal (seasonal) scales (Gagnon et al. 2005b).

Large-scale transitions from the kelp bed to the barrens state, resulting from increases in sea urchin density and destructive grazing, have been documented on the west (Foreman 1977) and east (Breen and Mann 1976b; Scheibling et al. 1999; Gagnon et al. 2004) coasts of Canada, the Gulf of Maine (Witman 1985; Martin et al. 1988), Iceland (Hjörleifsson et al. 1995), and Norway (Hagen 1983, 1995; Sivertsen 1997). Dense feeding aggregations (or fronts) of large (usually >40 mm) individuals migrate across the substratum at rates of 1–4 m per month, consuming all erect algae. These fronts may form due to increases in sea urchin abundance either within kelp beds (Hagen 1983; Bernstein et al. 1981) or along their outer/deeper margins (Martin et al. 1988; Hjörleifsson et al. 1995; Scheibling et al. 1999; Gagnon et al. 2004). Although the mechanism of aggregation remains controversial (Bernstein et al. 1981, 1983; Vadas et al. 1986), it appears that a threshold biomass of ~2 kg sea urchins per square meter is required to initiate destructive grazing in Nova Scotian kelp beds (Breen and Mann 1976b; Scheibling et al. 1999). Once kelp destruction is underway, the ratio of sea urchin to kelp biomass can be important in regulating grazing rate, and increased sea urchin recruitment in the expanding barrens may accelerate a sea urchin outbreak (Chapman and Johnson 1990; Scheibling et al. 1999).

In some areas, kelp beds persist as isolated patches amid coralline barrens (Bernstein et al. 1981; Chapman 1981; Miller 1985b), or the two community states may be juxtaposed along different depth strata (Martin et al. 1988; Sivertsen 1997). Boundaries between the kelp beds and barrens are dynamic and determined by various biological and physical processes that affect sea urchin abundance and foraging behavior (Bernstein et al. 1983; Scheibling et al. 1999; Gagnon et al. 2004). Kelps typically have refugia in shallow water where sea urchin grazing is inhibited by wave action or physiological stress (Keats 1991b). Thus, the lower limit of kelp beds can vary spatially and seasonally with wave exposure, and may be extended by several or tens of meters where sea urchins are periodically killed by low-salinity layers or disease. In Newfoundland and the Gulf of St. Lawrence, less palatable algae, such as the kelp *Agarum cribosum* (= *Agarum clathratum*), co-occur with moderate densities of sea urchins in deeper waters (Keats et al. 1982; Himmelman 1984; Gagnon et al. 2004, 2005a). In Alaska, *S. droebachiensis* and its congener *S. purpuratus* are highly aggregated, resulting in a mosaic of algal patch types including perennial kelps, annual species that temporarily escape grazing, and encrusting corallines (Duggins 1983).

Apart from dramatically altering the macroalgal assemblage (Foreman 1977) and reducing community primary productivity (Chapman 1981), sea urchin fronts decimate populations of small epifaunal invertebrates (e.g. tube-dwelling amphipods) and can cause reductions in the abundance of other herbivores (e.g. isopods, small gastropods) through losses of algal food resources and habitats (Witman 1985). Kelp destruction also is expected to affect the abundance of predatory fish and invertebrates that typically forage or shelter within kelp beds (Wharton and Mann 1981, but see also Elner and Campbell 1987). Sea urchins persist in barrens by grazing corallines and associated epiphytic microalgae (including the microscopic stages of macroalgae), and by capturing drift algae (Chapman 1981; Johnson and Mann 1982; Himmelman 1984, 1986). Even at moderate densities, sea urchins in barrens preclude the re-establishment of most erect macroalgae and limit the abundance of various other invertebrates (Himmelman et al.

1983a; Witman 1985; Ojeda and Dearborn 1989). However, two species of brown algae, *A. cribosum* and *Desmarestia viridis*, which are unpalatable to sea urchins or repel them through wave-induced motion can form large patches in barrens and serve as refugia for associated understory algae (Scheibling et al. 1999; Gagnon et al. 2003; 2005a).

Destructive grazing of kelp beds by sea urchins also affects detrital food webs in nearshore benthic ecosystems through the production of large amounts of particulate organic matter as feces. Mamelona and Pelletier (2005) estimated a fecal production rate of up to 5.6 g dry weight per square meter per day for a moderately dense (150 urchins per square meter) grazing front of *S. droebachiensis* in the St. Lawrence Estuary. This fecal material is rich in macronutrients and provides a significant source of energy (with an estimated production rate of up to 5000 cal per square meter per day) for the surrounding detritivores.

When *S. droebachiensis* are removed from barrens, either naturally or by humans, seaweeds rapidly colonize the substratum (reviewed by Chapman and Johnson 1990; Vadas and Elner 1992; see also McNaught 1999 for effects of sea urchin harvesting). The rate and pattern of algal succession varies with the pool of available colonists and proximity of spore sources, but typically culminates in dominance by Laminaria within two to three years (Chapman 1981; Johnson and Mann 1988; Lienaas and Christie 1996). Himmelman et al. (1983a) found experimental removal of sea urchins from barrens in the St. Lawrence Estuary also led to increased abundance of limpets and other herbivorous gastropods, mussels, and amphipods as a kelp bed developed (but see also Scheibling and Raymond 1990). Sea urchin removal may also allow the establishment and expansion of non-native species in a community. In Maine, Siddon and Witman (2004) found that crab predation on *S. droebachiensis* indirectly increased the abundance of an introduced ascidian (Diplosoma sp.) in experimental cages. In Nova Scotia, a massive recruitment of the invasive green alga *C. fragile* ssp. *tomentosoides* occurred in barrens after a mass mortality of *S. droebachiensis* (Chapman et al. 2002).

Witman (1985, 1987) described a mutualistic interaction between *S. droebachiensis* and horse mussels (*Modiolus modiolus*) that enable the mussels to coexist with kelp at wave-exposed sites in the Gulf of Maine. Sea urchins increase survival of Modiolus by grazing kelp from the mussel's shell, thereby reducing the probability of dislodgement of mussels by storm-generated waves. Mussel patches increase survival of small sea urchins by providing a structural refuge from both predation and dislodgement of sea urchins by wave action. *Strongylocentrotus droebachiensis* also plays an important role in determining community structure on rock walls in the Gulf of Maine (Sebens 1985, 1986). In the absence of sea urchins, these vertical habitats are colonized by a diverse assemblage of sessile invertebrates (low light levels preclude the establishment of seaweeds). Intense disturbance by sea urchins and other predators results in low-diversity assemblages dominated by corallines, similar to those on horizontal surfaces.

Despite intensive study for over three decades, we still have a poor understanding of broad-scale processes that control population size of *S. droebachiensis* and cause major shifts in the structure of sea urchin-dominated communities. Are population outbreaks mediated by relaxation of predation pressure at some critical life-history stage or by regional and local oceanographic conditions that determine larval supply and recruitment? Are epizootics entirely unpredictable or are certain populations predisposed to mass mortality by dint of history or locality? How will changing ocean climate or anthropogenic effects of pollution and overfishing influence *S. droebachiensis*? In areas of recurrent mass mortalities, such as Nova Scotia and Norway, our predictive ability even at small temporal and spatial scales is limited. For example, destructive grazing of Nova Scotian kelp beds in the early 1990s was arrested by outbreaks of disease, causing a reversal in the transition between community states in the latter half of the decade (Scheibling et al. 1999). Sea urchin-kelp dynamics in the Northwest Atlantic have become even more complex with the introduction of an epiphytic bryozoan *Membranipora membranacea*, which overgrows kelp causing damage and loss of fronds, and a green alga *C. fragile* ssp. *tomentosoides*, which appears to be facilitated by the die-back of kelp during infestations of the bryozoan (Harris and Tyrell 2001; Chapman et al. 2002; Levin et al. 2002). The bryozoan increases the nutritional value of kelp for sea urchins (Nestler and Harris 1994) and the supply of kelp detritus to barrens; it also may accelerate the destruction of kelp beds when sea urchins are abundant. The invasive alga is consumed by *S. droebachiensis*, although urchins exhibit a clear preference for kelp in laboratory and field experiments (Prince and LeBlanc 1992; Scheibling and Anthony 2001; Sumi and Scheibling 2005). Kelp beds have been replaced by dense stands of *C. fragile* ssp. *tomentosoides* over large areas off Nova Scotia and the Gulf of Maine (Chapman et al. 2002; Levin et al. 2002; Theriault et al. in press). The ability of *S. droebachiensis* to regulate the spread of this invasion is limited (Sumi and Scheibling 2005), and the loss of preferred food resources may have deleterious effects on urchin production (Scheibling and Anthony 2001; Lyons and Scheibling, unpubl. data) and the local fishery (Scheibling 2000). Given the complexity of interactions with invading species, and the stochastic nature of processes such as recruitment and mass mortality that cause large-scale fluctuations in abundance, the likelihood of predicting sea urchin population and community dynamics at relevant ecological and economical scales is slight.

ACKNOWLEDGMENT

We thank Monica Bravo for assistance with literature research.

REFERENCES

Adams NL (2001) UV radiation evokes negative phototaxis and covering behavior in the sea urchin *Strongylocentrotus droebachiensis*. Mar Ecol Prog Ser 213: 87–95

Addison JA, Hart MW (2004) Analysis of population genetic structure of the green sea urchin (*Strongylocentrotus droebachiensis*) using microsatellites. Mar Biol 144: 243–251

Arnold DC (1976) Local denudation of the sublittoral fringe by the green sea urchin, *Strongylocentrotus droebachiensis* (O.F. Müller). Can Field Nat 90: 186–187

Balch T, Scheibling RE (2000) Temporal and spatial variability in settlement and recruitment of echinoderms in kelp beds and barrens in Nova Scotia. Mar Ecol Prog Ser 205: 139–154

Balch T, Scheibling RE (2001) Larval supply, settlement and recruitment in echinoderms. Echino Stud 6: 1–83

Balch T, Scheibling RE, Harris LG, Chester CM, Robinson SMC (1998) Variation in settlement of *Strongylocentrotus droebachiensis* in the northwest Atlantic: Effects of spatial scale and sampling method. In: Mooi R, Telford M (eds) Echinoderms: San Francisco. Balkema, Rotterdam, pp 555–560

Bazhin AG (1998) The sea urchin genus Strongylocentrotus in the seas of Russia. In: Mooi R, Telford M (eds) Echinoderms: San Francisco. Balkema, Rotterdam, pp 563–566

Bernstein BB, Schroeter SC, Mann KH (1983) Sea urchin (*Strongylocentrotus droebachiensis*) aggregating behavior investigated by a subtidal multifactorial experiment. Can J Fish Aquat Sci 40: 1975–1986

Bernstein BB, Williams BE, Mann KH (1981) The role of behavioural responses to predators in modifying urchins' (*Strongylocentrotus droebachiensis*) destructive grazing and seasonal foraging patterns. Mar Biol 63: 39–49

Bertram DF, Strathmann RR (1998) Effects of maternal and larval nutrition on growth and form of planktotrophic larvae. Ecology 79: 315–327

Boolootian RA, Turner V (1965) The reproductive cycles of *Arbacia punctulata* and *Strongylocentrotus drobachiensis*. Biol Bull 129: 399–400

Brady S (2003) Growth and reproduction of the green sea urchin *Strongylocentrotus droebachiensis* across a depth gradient. MSc thesis, Dalhousie University, Halifax, Canada

Brady SM, Scheibling RE (2005) Repopulation of the shallow subtidal zone by green sea urchins (*Strongylocentrotus droebachiensis*) following mass mortality in Nova Scotia, Canada. J Mar Biol Assn UK 85: 1511–1517

Brady SM, Scheibling RE (2006) Changes in growth and reproduction of green sea urchins *Strongylocentrotus droebachiensis* during repopulation of the shallow subtidal zone after mass morality. J Exp Mar Biol Ecol 335: 227–291

Breen PA, Mann KH (1976a) Changing lobster abundance and the destruction of kelp beds by sea urchins. Mar Biol 34: 137–142

Breen PA, Mann KH (1976b) Destructive grazing of kelp by sea urchins in eastern Canada. J Fish Res Board Can 33: 1278–1283

Briscoe CS, Sebens KP (1988) Omnivory in *Strongylocentrotus droebachiensis* (Müller) (Echinodermata: Echinoidea): Predation on subtidal mussels. J Exp Mar Biol Ecol 115: 1–24

Burdett-Coutts V, Metaxas A (2004) The effect of the quality of food patches on larval vertical distribution of sea urchins *Lytechinus variegatus* (Lamarck) and *Strongylocentrotus droebachiensis* (Mueller). J Exp Mar Biol Ecol 308: 221–236

Burke RD (1980) Podial sensory receptors and the induction of metamorphosis in echinoids. J Exp Mar Biol Ecol 47: 223–234

Bustnes JO, Lønne OJ (1995) Sea ducks as predators on sea urchins in a northern kelp forest. In: Skjoldal HR, Hopkins C, Erikstad KE, Leinaas HP (eds) Ecology of fjords and coastal waters. Elsevier, Amsterdam, pp 599–608

Cameron JL, Fankboner PV (1989) Reproductive biology of the commercial sea cucumber *Parastichopus californicus* (Stimpson) (Echinodermata: Holothuroidea). II. Observations on the ecology of development, recruitment, and juvenile life stage. J Exp Mar Biol Ecol 127: 43–67

Chapman ARO (1981) Stability of sea urchin dominated barren grounds following destructive grazing of kelp in St. Margaret's Bay, eastern Canada. Mar Biol 62: 307–311

Chapman ARO, Johnson CR (1990) Disturbance and organization of macroalgal assemblages in the northwest Atlantic. Hydrobiologia 192: 77–121

Chapman AS, Scheibling RE, Chapman ARO (2002) Species introductions and changes in marine vegetation of Atlantic Canada. In: Claudi R, Nantel P, Muckle-Jeffs E (eds) Alien Invaders in Canada's Waters, Wetlands and Forests. Natural Resources Canada, Canadian Forest Service Science Branch, Ottawa, pp. 133–148.

Chen Y, Hunter M, Vadas R, Beal B (2003) Developing a growth-transition matrix for the stock assessment of the green sea urchin (*Strongylocentrotus droebachiensis*) off Maine. Fish Bull 101: 737–744

Christie H, Leinaas HP, Skadsheim A (1995) Local patterns in mortality of the green sea urchin, *Strongylocentrotus droebachiensis*, at the Norwegian coast. In: Skjoldal HR, Hopkins C, Erikstad KE, Leinaas HP (eds) Ecology of fjords and coastal waters. Elsevier, Amsterdam, pp 573–584

Cocanour B, Allen KR (1967) The breeding cycles of a sand dollar and a sea urchin. Comp Biochem Physiol 20: 327–331

Cuthbert FM, Hooper RG, McKeever T (1995) Sea urchin feeding and ranching experiments. Canadian Centre Fisheries Innovation, St. John's

Daggett TL, Pearce CM, Tingley M, Robinson SMC, Chopin T (2005) Effect of prepared and macroalgal diets and seed stock source on somatic growth of juvenile green sea urchins (*Strongylocentrotus droebachiensis*). Aquaculture 244: 263–281

Dean TA, Bodkin JL, Jewett SC, Monson DH, Jung D (2000) Changes in sea urchins and kelp following a reduction in sea otter density as a result of the Exxon Valdez oil spill. Mar Ecol Prog Ser 199: 281–291

de Jong-Westman M, March BE, Carefoot TH (1995a) The effect of different nutrient formulations in artificial diets on gonad growth in the sea urchin *Strongylocentrotus droebachiensis*. Can J Zool 73: 1495–1502

de Jong-Westman M, Qian P-Y, March BE, Carefoot TH (1995b) Artificial diets in sea urchin culture: effects of dietary protein level and other additives on egg quality, larval morphometrics, and larval survival in the green sea urchin, *Strongylocentrotus droebachiensis*. Can J Zool 73: 2080–2090

Drouin G, Himmelman JH, Beland P (1985) Impact of tidal salinity fluctuations on echinoderm and mollusc populations. Can J Zool 63: 1377–1387

Duggins DO (1980) Kelp beds and sea otters: an experimental approach. Ecology 61: 447–453

Duggins DO (1981) Sea urchins and kelp: the effects of short term changes in urchin diet. Limnol Oceanogr 26: 391–394

Duggins DO (1983) Starfish predation and the creation of mosaic patterns in a kelp-dominated community. Ecology 64: 1610–1619

Dumont CP, Himmelman JH (2004) Sea urchin mass mortality associated with algal debris from ice scour. In: Heinzeller T, Nebelsick J (eds) Echinoderms: München. Balkema, Leiden, pp 177–182

Dumont C, Himmelman JH, Russell MP (2004) Size-specific movement of green sea urchins *Strongylocentrotus droebachiensis* on urchin barrens in eastern Canada. Mar Ecol Prog Ser 276: 93–101

Ebert TA (1983) Recruitment in echinoderms. Echino Stud 1: 169–203

Ebert TA (1988) Allometry, design and constraint of body components and of shape in sea urchins. J Nat Hist 22: 1407–1425

Elner RW (1980) Predation of the sea urchin (*Strongylocentrotus droebachiensis*) by the American lobster (*Homarus americanus*) and the rock crab (Cancer irroratus). In: Pringle JD, Sharp GJ, Caddy JF (eds) Proceedings of the Workshop on the Relationship Between Sea Urchin Grazing and Commercial Plant/Animal Harvesting. Can Tech Rep Fish Aquat Sci 954: 48–65

Elner RW, Campbell A (1987) Natural diets of lobster *Homarus americanus* from barren ground and macroalgal habitats off southwestern Nova Scotia. Canada. Mar Ecol Prog Ser 37: 131–140

Elner RW, Vadas RL (1990) Inference in ecology: the sea urchin phenomenon in the northwestern Atlantic. Amer Nat 136: 108–125

Estes JA, Duggins DO (1995) Sea otters and kelp forests in Alaska: generality and variation in a community ecological paradigm. Ecol Monogr 65: 75–100

Evans PD, Mann KH (1977) Selection of prey by American lobster (*Homarus americanus*) when offered a choice between sea urchins and crabs. J Fish Res Bd Can 34: 2203–2207

Falk-Petersen I-B, Lönning S (1983) Reproductive cycles of two closely related sea urchin species, urchin *Strongylocentrotus droebachiensis* and *Strongylocentrotus pallidus*. Sarsia 68: 157–164

Field JG (1972) Some observations on the release of dissolved organic carbon by the sea urchin, *Strongylocentrotus droebachiensis*. Limnol Oceanogr 17: 759–761

Fong W, Mann KH (1980) Role of gut flora in the transfer of amino acids through a marine food chain. Can J Fish Aquat Sci 37: 88–96

Foreman RE (1977) Benthic community modification and recovery following intensive grazing by *Strongylocentrotus droebachiensis*. Helgol Wiss Meeresunters 30: 468–484

Gagnon P, Himmelman JH, Johnson LE (2003) Algal colonization in urchin barrens: defense by association during recruitment of the brown alga *Agarum cribrosum*. J Exp Mar Biol Ecol 290: 179–196

Gagnon P, Himmelman JH, Johnson LE (2004) Temporal variation in community interfaces: kelp-bed boundary dynamics adjacent to persistent urchin barrens. Mar Biol 144: 1191–1203

Gagnon P, Johnson LE, Himmelman JH (2005a) Kelp patch dynamics in the face of intense herbivory: stability of *Agarum clathratum* (Phaeophyta) stands and associated flora on urchin barrens. J Phycol 41: 498–505

Gagnon P, Nadon M-O, Jones W, Johnson LE, Himmelman JH, Ripley HT (2005b) Assessing temporal variation in the extent of kelp beds in the northern Gulf of St. Lawrence using analog and digital aerial photography and GIS. Altarum Institute, Ann Arbor, Proceedings of the 8th International Conference on Remote Sensing for Marine and Coastal Environments, Halifax, Nova Scotia, pp 1–8

Garnick E (1978) Behavioural ecology of *Strongylocentrotus droebachiensis* (Müller) (Echinodermata: Echinoidea). Aggregating behaviour and chemotaxis. Oecologia 37: 77–84

Garrido CL, Barber BJ (2001) Effects of temperature and food ration on gonad growth and oogenesis of the green sea urchin, *Strongylocentrotus droebachiensis*. Mar Biol 138: 447–456

Grabowski R, Chen Y (2004) Incorporating uncertainty in to the estimation of the biological reference points F0.1 and Fmax for the Maine green sea urchin (*Strongylocentrotus droebachiensis*) fishery. Fish Res 68: 367–371

Green JM, Martel G, Martin DW (1984) Comparisons of the feeding activity and diets of male and female cunners *Tautogolabrus adspersus* (Pisces: Labridae). Mar Biol 84: 7–11

Guerinot ML, Patriquin DG (1981) The association of N2-fixing bacteria with sea urchins. Mar Biol 62: 197–207

Guillemette M, Ydenberg RC, Himmelman JH (1992) The role of energy intake rate in prey and habitat selection of common eiders *Somateria mollissima* in winter: a risk sensitive interpretation. J Anim Ecol 61: 599–610

Hagen NT (1983) Destructive grazing of kelp beds by sea urchins in Vestfjorden, northern Norway. Sarsia 68: 177–190

Hagen NT (1987) Sea urchin outbreaks and nematode epizootics in Vestfjorden, northern Norway. Sarsia 72: 213–229

Hagen NT (1992) Macroparasitic epizootic disease: a potential mechanism for the termination of sea urchin outbreaks in northern Norway. Mar Biol 114: 469–478

Hagen NT (1995) Recurrent destructive grazing of successionally immature kelp forests by green sea urchins in Vestfjorden, northern Norway. Mar Ecol Prog Ser 123: 95–106

Hagen NT (1996) Parasitic castration of the green sea urchin *Strongylocentrotus droebachiensis* by the nematode endoparasite *Echinomermella matsi*: reduced reproductive potential and reproductive death. Dis Aquat Org 24: 215–226

Hagen NT (1997) Out-of-season maturation of echinoid broodstock in fixed light regimes. Bull Aquacul Assoc Canada 97-1: 61

Hagen NT (1998) Effect of food availability and body size on out-of-season gonad yield in the green sea urchin *Strongylocentrotus droebachiensis*. J Shellfish Res 17: 1533–1539

Hagen NT, Andersen A, Stabell OB (2002) Alarm responses of the green sea urchin *Strongylocentrotus droebachiensis*, induced by chemically labeled durophagous predators and simulated acts of predation. Mar Biol 140: 365–374

Hagen NT, Mann KH (1992) Functional response of the predators American lobster *Homarus americanus* (Milne-Edwards) and Atlantic wolffish *Anarhichas lupas* (L.) to increasing numbers of the green sea urchin *Strongylocentrotus droebachiensis* (Müller). J Exp Mar Biol Ecol 159: 89–112

Hagen NT, Mann KH (1994) Experimental analysis of factors influencing the aggregating behaviour of the green sea urchin *Strongylocentrotus droebachiensis* (Müller). J Exp Mar Biol Ecol 176: 107–126

Harris LG, Chester CM (1996) Effects of location, exposure and physical structure on juvenile recruitment of the sea urchin *Strongylocentrotus droebachiensis* in the Gulf of Maine. J Invert Repro Dev 30: 207–215

Harris LG, Rice B, Nestler EC (1994) Settlement, early survival and growth in a southern Gulf of Maine population of *Strongylocentrotus droebachiensis* (Müller). In: David B, Guille A, Féral J-P, Roux M (eds) Echinoderms through time. Balkema, Rotterdam, pp 701–706

Harris LG, Tyrell MC (2001) Changing community states in the Gulf of Maine: synergism between invaders, overfishing and climate change. Biol Invasions 3: 9–21

Harris LG, Witman JD, Rowley R (1985) A comparison of sea urchin recruitment at sites on the Atlantic and Pacific coasts of North America. In: Keegan BF, O'Connor BDS (eds) Echinodermata. Balkema, Rotterdam, p 389

Hart MW, Scheibling RE (1988) Heat waves, baby booms, and the destruction of kelp beds by sea urchins. Mar Biol 99: 167–176

Hatcher BG, Hatcher AI (1997) Research directions and management options for sea urchin culture in Nova Scotia. Bull Aquacul Assoc Canada 97: 62–65

Himmelman JH (1975) Phytoplankton as a stimulus for spawning in three marine invertebrates. J Exp Mar Biol Ecol 20: 199–214

Himmelman JH (1978) Reproductive cycle of the green sea urchin, *Strongylocentrotus droebachiensis*. Can J Zool 56: 1828–1836

Himmelman JH (1984) Urchin feeding and macroalgal distribution in Newfoundland, Eastern Canada. Nat Can 111: 337–348

Himmelman JH (1986) Population biology of green sea urchins on rocky barrens. Mar Ecol Prog Ser 33: 295–306

Himmelman JH, Cardinal A, Bourget E (1983a) Community development following removal of urchins, *Strongylocentrotus droebachiensis*, from the rocky subtidal zone of the St. Lawrence Estuary, Eastern Canada. Oecologia 59: 27–39

Himmelman JH, Guderley H, Vignault G, Drouin G, Wells PG (1984) Response of the sea urchin, *Strongylocentrotus droebachiensis*, to reduced salinities: importance of size, acclimation, and interpopulation differences. Can J Zool 62: 1015–1021

Himmelman JH, Lavergne Y (1985) Organization of rocky subtidal communities in the St. Lawrence Estuary. Nat Can 112: 143–154

Himmelman JH, Lavergne Y, Axelsen F, Cardinal A, Bourget E (1983b) Sea urchins in the St. Lawrence Estuary: their abundance size-structure, and suitability for commercial exploitation. Can J Fish Aquat Sci 40: 474–486

Himmelman JH, Nedelec H (1990) Urchin foraging and algal survival strategies in intensely grazed communities in eastern Canada. Can J Fish Aquat Sci 47: 1011–1026

Himmelman JH, Steele DH (1971) Foods and predators of the green sea urchin *Strongylocentrotus droebachiensis* in Newfoundland waters. Mar Biol 9: 315–322

Hjörleifsson E, Kaasa Ö, Gunnarsson K (1995) Grazing of kelp by green sea urchin in Eyjafjördur, north Iceland. In: Skjoldal HR, Hopkins C, Erikstad KE, Leinaas HP (eds) Ecology of fjords and coastal waters. Elsevier, Amsterdam, pp 593–597

Hooper R (1980) Observations on algal-grazer interactions in Newfoundland and Labrador. Can Tech Rep Fish Aquat Sci 954: 120–124

Hooper RG, Cuthbery FM, McKeever T (1996) Sea urchin aquaculture – phase II: ration size, seasonal growth rates, mixed kelp diets, fish diet, old urchin growth, and baiting confinement experiments. Canadian Centre Fisheries Innovation, St. John's

Jellett JF, Novitsky JA, Cantley JA, Scheibling RE (1989) Non-occurrence of free-living Paramoeba invadens in water and sediments of Halifax Harbour, Nova Scotia, Canada. Mar Ecol Prog Ser 56: 205–209

Jellett JF, Scheibling RE (1988) Effect of temperature and prey availability on growth of Paramoeba invadens in monoxenic culture. Appl Environ Microbiol 54: 1848–1854

Jellett JF, Scheibling RE, Wardlaw AC (1988) Host specificity of Paramoeba invadens, a sea urchin pathogen. In: Burke RD, Mladenov PV, Lambert P, Parsley RL (eds) Echinoderm biology. Balkema, Rotterdam, pp 755–761

Jensen M (1974) The Strongylocentrotidae (Echinoidea), A morphologic and systematic study. Sarsia 57: 113–148

Johnson AS, Ellers O, Lemire J, Minor M, Leddy H (2002) Sutural loosening and skeletal flexibility during growth: determination of drop-like shapes in sea urchins. Proc Roy Soc Biol Sci Ser B 269: 215–220

Johnson CR, Mann KH (1982) Adaptations of *Strongylocentrotus droebachiensis* for survival on barren grounds in Nova Scotia. In: Lawrence JM (ed.) Echinoderms: Proceedings of the International Conference, Tampa Bay. Balkema, Rotterdam, pp 277–283

Johnson CR, Mann KH (1988) Diversity, patterns of adaptation, and stability of Nova Scotian kelp beds. Ecol Monogr 58: 129–154

Jones GM (1985) Paramoeba invadens n. sp. (Amoebida, Paramoebidae), a pathogenic amoeba from the sea urchin, *Strongylocentrotus droebachiensis*, in eastern Canada. J Protozool 32: 564–569

Jones GM, Hagen NT (1987) *Echinomermella matsi* sp.n., an endoparasitic nematode from the sea urchin *Strongylocentrotus droebachiensis* in northern Norway. Sarsia 72: 203–212

Jones GM, Hebda AJ, Scheibling RE, Miller RJ (1985) Histopathology of the disease causing mass mortality of sea urchins (*Strongylocentrotus droebachiensis*) in Nova Scotia. J Invert Pathol 71: 559–565

Jones GM, Scheibling RE (1985) *Paramoeba* sp. (Amoebida: Paramoebidae) as the possible causative agent of sea urchin mass mortality in Nova Scotia. J Parasitol 71: 559–565

Keats DW (1990) Winter flounder, *Psuedopleuronectes americanus*, Pisces: Pleuronectidae) predation in a green sea urchin dominated sublittoral community in eastern Newfoundland. Mar Ecol Prog Ser 60: 13–22

Keats DW (1991a) American plaice, *Hippoglossoides platessoides* (Fabricius), predation on green sea urchins, *Strongylocentrotus droebachiensis* (O.F. Müller), in eastern Newfoundland. J Fish Biol 38: 67–72

Keats DW (1991b) Refugial Laminaria abundance and reduction in urchin grazing in communities in the north-west Atlantic. J Mar Biol Assn UK 71: 867–876

Keats DW, Hooper RG, Steele DH, South GR (1987) Field observations of summer and autumn spawning by *Strongylocentrotus droebachiensis*, green sea urchins, in eastern Newfoundland. Can Field Natur 101: 463–465

Keats DW, Miller RJ (1986) Comments on "Seaweeds, sea urchins, and lobsters: a reappraisal" by R.J. Miller. Can J Fish Aquat Sci 43: 1675–1676

Keats DW, South GR, Steele DH (1982) The occurrence of *Agarum cribosum* (Mert.) Bory (Phaeophyta, Laminariales) in relation to some of its competitors and predators in Newfoundland. Phycologia 21: 189–191

Keats DW, South GR, Steele DH (1985) Ecology of juvenile green sea urchins (*Strongylocentrotus droebachiensis*) at an urchin dominated subtidal site in eastern Newfoundland. In: Keegan BF, O'Connor BDS (eds) Echinodermata. Balkema, Rotterdam, pp 295–302

Keats DW, Steele DH, South GR (1983) Food relations and short-term aquaculture potential of green sea urchin (*Strongylocentrotus droebachiensis*) in Newfoundland. Mar Sci Res Lab Tech Rep No 243, Memorial Univ, Newfoundland

Keats DW, Steele DH, South GR (1984) Depth-dependent reproductive output of the green sea urchin, *Strongylocentrotus droebachiensis* (O.F. Muller), in relation to the nature and availability of food. J Exp Mar Biol Ecol 80: 77–91

Keats DW, Steele DH, South GR (1986) Atlantic wolffish (*Anarhichas lupus* L.; Pisces: Anarhichidae) predation on green sea urchins (*Strongylocentrotus droebachiensis* (O.F. Müller); Echinodermata: Echinoidea) in eastern Newfoundland. Can J Zool 64: 120–125

Keats DW, Steele DH, South GR (1987) Ocean pout (*Macrozoarces americanus* (Block and Schneider) (Pisces: Zoarcidae)) predation on green sea urchins (*Strongylocentrotus droebachiensis* (O.F. Müller); (Echinodermata: Echinoidea)) in eastern Newfoundland. Can J Zool 65: 1515–1521

Kennedy EJ, Robinson SM, Parsons GJ, Castell J (2001) Studies on feed formulations to maximize somatic growth rates of juvenile green sea urchins (*Strongylocentrotus droebachiensis*). Aquacult Assoc Can Spec Publ 4: 68–71

Kuznetzov VV (1946) Nutrition and growth of plant feeding marine invertebrates of the eastern Murmansk. Bull Acad Sci USSR 4: 431–452

Lambert DM, Harris LG (2000) Larval settlement of the green sea urchin, *Strongylocentrotus droebachiensis*, in the southern Gulf of Maine. Invert Biol 119: 403–409

Lang C, Mann KH (1976) Changes in sea urchin populations after the destruction of kelp beds. Mar Biol 36: 321–326

Lange R (1964) The osmotic adjustment in the echinoid *Strongylocentrotus droebachiensis*. Comp Biochem Physiol 13: 205–216

Larson BR, Vadas RL, Keser M (1980) Feeding and nutritional ecology of the sea urchin Strongylocentrotus drobachiensis in Maine, USA. Mar Biol 59: 49–62

Lauzon-Guay J-S, Scheibling RE, Barbeau, MA (2006) Movement patterns in the green sea urchin, *Strongylocentrotus droebachiensis*. J Mar Biol Assn UK 86: 167–174

Lawrence JM (1975) On the relationships between marine plants and sea urchins. Oceanogr Mar Biol Ann Rev 13: 213–286

Lawrence JM (1996) Mass mortality of echinoderms from abiotic factors. Echino Stud 5: 103–137

Lawrence JM (2000) Conflict between somatic and gonadal growth in sea urchins: a review. L'atelier sur la coordination de la recherche sur l'oursin vert au Canada Atlantique. http://crdpm.cus.ca/OURSIN/rapide.htm

Leinaas HP, Christie H (1996) Effects of removing sea urchins (*Strongylocentrotus droebachiensis*): Stability of the barren state and succession of kelp forest recovery in the east Atlantic. Oecologia 105: 524–536

Lemire M, Himmelman JH (1996) Relation of food preference to fitness for the green sea urchin *Strongylocentrotus droebachiensis*. Mar Biol 127: 73–78

Lesser MP, Barry TM (2003) Survivorship, development, and DNA damage in echinoderm embryos and larvae exposed to ultraviolet radiation (290–400nm). J Exp Mar Biol Ecol 292: 75–91

Levin PS, Coyer JA, Petrik R, Good TP (2002) Community-wide effects of nonindigenous species on temperate rocky reefs. Ecology 83: 3182–3193

Levitan DR (2002a) Density-dependent selection on gamete traits in three congeneric sea urchins. Ecology 83: 464–479

Levitan DR (2002b) The relationship between conspecific fertilization success and reproductive isolation among three congeneric sea urchins. Evolution 56: 1599–1609

Lundquist CJ, Botsford LW (2004) Model projections of the Fishery Implications of the Allee Effect in broadcast spawners. Ecol Appl 14: 929–941

MacKay AA (1976) The sea urchin industry of New Brunswick's Bay of Fundy coast. Final Report Department of Fisheries, Fredricton, New Brunswick

Mamelona J, Pelletier E (2005). Green sea urchin as a significant source of fecal particulate organic matter within nearshore benthic ecosystems. J Exp Mar Biol Ecol 314: 163–174

Mann KH (1977) Destruction of kelp-beds by sea-urchins: a cyclical phenomenon or irreversible degradation? Helgol Wiss Meeresunters 30: 455–467

Mann KH (1985) Invertebrate behaviour and the structure of marine benthic communities. In: Sibley RM, Smith RH (eds) Behavioural ecology. Blackwell Scientific Publ, Oxford, pp 227–246

Mann KH, Breen PA (1972) The relation between lobster abundance, sea urchins and kelp beds. J Fish Res Bd Canada 29: 603–609

Mann KH, Wright JLC, Welsford BE, Hatfield E (1984) Responses of the sea urchin *Strongylocentrotus droebachiensis* (O.F. Müller) to water-borne stimuli from potential predators and potential food algae. J Exp Mar Biol Ecol 79: 233–244

Martin PD, Truchon SP, Harris LG (1988) *Strongylocentrotus droebachiensis* populations and community dynamics at two depth-related zones over an 11-year period. In: Burke RD, Mladenov PV, Lambert P, Parsley RL (eds) Echinoderm biology. Balkema, Rotterdam, pp 475–482

McClintock JB (1980) On estimating energetic values of prey: implications in optimal diet models. Oecologia 70: 161–162

McNaught DC (1999) The indirect effects of macroalgae and micropredation on the post-settlement success of the green sea urchin in Maine. PhD thesis, University of Maine, Orono

Meidel SK (1998) Reproductive ecology of the sea urchin *Strongylocentrotus droebachiensis*. PhD thesis, Dalhousie University, Halifax

Meidel SK, Scheibling RE (1998a) Size and age structure of the sea urchin *Strongylocentrotus droebachiensis* in different habitats. In: Mooi R, Telford M (eds) Echinoderms: San Francisco. Balkema, Rotterdam, pp 737–742

Meidel SK, Scheibling RE (1998b) Annual reproductive cycle of the green sea urchin *Strongylocentrotus droebachiensis*, in differing habitats in Nova Scotia, Can Mar Biol 131: 461–478

Meidel SK, Scheibling RE (1999) Effects of food type and ration on reproductive maturation and growth of the sea urchin *Strongylocentrotus droebachiensis*. Mar Biol 134: 155–166

Meidel SK, Scheibling RE (2001) Variation in egg spawning among subpopulations of sea urchins (*Strongylocentrotus droebachiensis*): a theoretical approach. Mar Ecol Prog Ser 213: 97–110

Meidel SK, Scheibling RE, Metaxas A (1999) Relative importance of parental and larval nutrition on larval development and metamorphosis of the sea urchin, *Strongylocentrotus droebachiensis*. J Exp Mar Biol Ecol 240: 161–178

Meidel SK, Yund PO (2001) Egg longevity and time-integrated fertilization in a temperate sea urchin (*Strongylocentrotus droebachiensis*). Biol Bull 201: 84–94

Miller RJ (1985a) Seaweeds, sea urchins, and lobsters: a reappraisal. Can J Fish Aquat Sci 42: 2061–2072

Miller RJ (1985b) Succession in sea urchin and seaweed abundance in Nova Scotia, Canada. Mar Biol 84: 275–286

Miller RJ, Colodey AG (1983) Widespread mass mortalities of the green sea urchin in Nova Scotia, Canada. Mar Biol 73: 263–267

Miller RJ, Mann KH (1973) Ecological energetics of the seaweed zone in a marine bay on the Atlantic coast of Canada. III. Energy transformations by sea urchins. Mar Biol 18: 99–114

Miller RJ, Nolan SC (2000) Management of the Nova Scotia sea urchin fishery: a nearly successful habitat based management regime. Canadian Stock Assessment Secretariat, Research Document 2000/109, Ottawa, pp 1–41

Minor MA, Scheibling RE (1997) Effects of food ration and feeding regime on growth and reproduction of the sea urchin *Strongylocentrotus droebachiensis*. Mar Biol 129: 159–167

Mohn RK, Miller RJ (1987) A ration-based model of a seaweed-sea urchin community. Ecol Modell 37: 249–267

Moore DS, Miller RJ (1983) Recovery of macroalgae following widespread sea urchin mortality with a description of the nearshore hard-bottom habitat on the Atlantic coast of Nova Scotia. Can Tech Rep Fish Aquat Sci 1230: vii + 94 pp

Mortensen TH (1943) A monograph of the Echinoidea. C.A. Reitzels, Copenhagen

Munk JE (1992) Reproduction and growth of the green sea urchin *Strongylocentrotus droebachiensis* (Müller) near Kodiak, Alaska. J Shellfish Res 11: 245–254

Nestler EC, Harris LG (1994) The importance of omnivory in *Strongylocentrotus droebachiensis* (Muller) in the Gulf of Maine. In: David B, Guille A, Féral J-P, Roux M (eds) Echinoderms through time. Balkema, Rotterdam, pp 813–818

Nishizaki MT, Ackerman JD (2004) Juvenile-adult associations in sea urchins *Strongylocentrotus droebachiensis*: Is nutrition involved? Mar Ecol Prog Ser 268: 93–103

Ojeda FP, Dearborn JH (1989) Community structure of macroinvertebrates inhabiting the rocky subtidal zone in the Gulf of Maine: seasonal and bathymetric distribution. Mar Ecol Prog Ser 57: 147–161

Ojeda FP, Dearborn JH (1991) Feeding ecology of benthic mobile predators: experimental analyses of their influence in rocky subtidal communities of the Gulf of Maine. J Exp Mar Biol Ecol 149: 13–44

Paine RT, Vadas RL (1969) The effects of grazing by sea urchins, Strongylocentrotus spp, on benthic algal populations. Limnol Oceanogr 14: 710–719

Palumbi SR, Wilson AC (1990) Mitochondrial DNA diversity in the sea urchins Strongylocentrotus purpuratus and *S. droebachiensis*. Evolution 44: 403–415

Pearce CM, Daggett TL, Robinson SMC (2002a) Effect of protein source ratio and protein concentration in prepared diets on gonad yield and quality of the green sea urchin *Strongylocentrotus droebachiensis*. Aquaculture 214: 307–332

Pearce CM, Daggett TL, Robinson SMC (2002b) Optimizing prepared feed ration for gonad production of the green sea urchin *Strongylocentrotus droebachiensis*. J World Aquacult Soc 33: 268–277

Pearce CM, Daggett TL, Robinson SMC (2002c) Effect of binder type and concentration on prepared feed stability and gonad yield and quality of the green sea urchin *Strongylocentrotus droebachiensis*. Aquaculture 205: 301–323

Pearce CM, Scheibling RE (1990) Induction of metamorphosis of larvae of the green sea urchin, *Strongylocentrotus droebachiensis*, by coralline red algae. Biol Bull 179: 304–311

Pearce CM, Scheibling RE (1991) Effect of macroalgae, microbial films, and conspecifics on the induction of metamorphosis of the green sea urchin *Strongylocentrotus droebachiensis* (Müller). J Exp Mar Biol Ecol 147: 147–162

Pearce CM, Weavers RW, Williams SW (2004) Effect of three kelp species and a prepared diet on somatic growth of juvenile green sea urchins (*Strongylocentrotus droebachiensis*). Aquacult Assoc Can Spec Publ 8: 73–76

Pearce CM, Williams SW, Yuan F, Castell JD, Robinson SMC (2005) Effect of temperature on somatic growth and survivorship of early post-settled green sea urchins, *Strongylocentrotus droebachiensis* (Müller). Aquacult Res 36: 600–609

Pearse JS (1991) Echinodermata: Echinoidea. In: Giese AC, Pearse JS, Pearse, VB (eds) Reproduction of marine invertebrates: Vol. VI. Academic Press, New York, pp 513–662

Pelletreau KN, Muller-Parker G (2002) Sulfuric acid in the phaeophyte alga Desmarestia munda deters feeding by the sea urchin *Strongylocentrotus droebachiensis*. Mar Biol 141: 1–9

Percy JA (1972) Thermal adaptation in the boreo-arctic echinoid *Strongylocentrotus droebachiensis* (O.F. Müller 1776). I. Seasonal acclimation of respiration. Physiol Zool 45: 277–289

Percy JA (1973) Thermal adaptation in the boreo-arctic echinoid *Strongylocentrotus droebachiensis* (O.F. Müller 1776). II. Seasonal acclimatization and urchin activity. Physiol Zool 46: 129–138

Percy JA (1974a) Thermal adaptation in the Boreo-arctic echinoids, *Strongylocentrotus droebachiensis* (O.F. Muller, 1776). III. Seasonal acclimatization and metabolism of tissues in vitro. Physiol Ecol 47: 59–67

Percy JA (1974b) Thermal adaptation in the boreo-arctic echinoid *Strongylocentrotus droebachiensis* (O.F. Müller 1776). IV. Acclimation in the laboratory. Physiol Zool 47: 163–171

Perry RI, Zhang Z, Harbo R (2002) Development of the green sea urchin (*Strongylocentrotus droebachiensis*) fishery in British Columbia, Canada – back from the brink using a precautionary framework. Fish Res 55: 253–266

Prince JS, LeBlanc WG (1992) Comparative feeding preference of *Strongylocentrotus droebachiensis* (Echinoidea) for the invasive seaweed *Codium fragile* ssp. *tomentosoides* (Chlorophyceae) and four other seaweeds. Mar Biol 113: 159–163

Pringle JD (1986) A review of urchin/macro-algal associations with a new synthesis for nearshore, eastern Canadian waters. Monograf Biol 4: 191–218

Pringle JD, Sharp GJ, Caddy JF (1982) Interactions in kelp bed ecosystems in the northwest Atlantic: Review of a workshop. In: Mercer MC (eds) Multispecies approaches to fisheries management advice. Can Spec Publ Fish Aquat Sci 59, pp 108–115

Propp MV (1977) Ecology of the sea urchin *Strongylocentrotus droebachiensis* of the Barents Sea: metabolism and regulation of abundance. Soviet J Mar Biol 3: 27–37

Raymond BG, Scheibling RE (1987) Recruitment and growth of the sea urchin *Strongylocentrotus droebachiensis* (Müller) following mass mortalities off Nova Scotia, Canada. J Exp Mar Biol Ecol 108: 31–54

Robinson SMC, Bernier S, MacIntyre A (2001) The impact of scallop drags on sea urchin populations and benthos in the Bay of Fundy, Canada. Hydobiologia 465: 103–114

Roller RA, Stickle WB (1985) Effects of salinity on larval tolerance and early development rates of four species of echinoderms. Can J Zool 63: 1531–1538

Rowley RJ (1989) Settlement and recruitment of sea urchins (Strongylocentrotus spp.) in a sea-urchin barren ground and a kelp bed: are populations regulated by settlement or post-settlement processes? Mar Biol 100: 485–494

Russell MP (1998) Resource allocation plasticity in sea urchins: rapid, diet induced, phenotypic changes in the green sea urchin, *Strongylocentrotus droebachiensis* (Muller). J Exp Mar Biol Ecol 220: 1–14

Russell MP, Ebert TA, Petraitis PS (1998) Field estimates of growth and mortality of the green sea urchin, *Strongylocentrotus droebachiensis*. Ophelia 48: 137–153

Sabourin TD, Stickle WB (1981) Effects of salinity on respiration and nitrogen excretion in two species of echinoderms. Mar Biol 65: 91–99

Russell MP, Meredith RW (2000) Natural growth lines in echinoid ossicles are not reliable indicators of age: a test using *Strongylocentrotus droebachiensis*. Invert Biol 119: 410–420

Scheibling RE (1984) Echinoids, epizootics and ecological stability in the rocky subtidal off Nova Scotia, Canada. Helgol Wiss Meeresunters 37: 233–242

Scheibling RE (1986) Increased macroalgal abundance following mass mortalities of sea urchins (*Strongylocentrotus droebachiensis*) along the Atlantic coast of Nova Scotia. Oecologia 68: 186–198

Scheibling RE (1988) Microbial control of sea urchins: Achilles' heel or Pandora's box? In: Burke RD, Mladenov PV, Lambert P, Parsley RL (eds) Echinoderm biology. Balkema, Rotterdam, pp 745–754

Scheibling RE (1994) Interactions among lobsters, sea urchins, and kelp in Nova Scotia, Canada. In: David B, Guille A, Féral J-P, Roux M (eds) Echinoderms through time. Balkema, Rotterdam, pp 865–870

Scheibling RE (1996) The role of predation in regulating sea urchin populations in eastern Canada. Oceanol Acta 19: 421–430

Scheibling RE (2000) Species invasions and community change threaten the sea urchin fishery in Nova Scotia. L'atelier sur la coordination de la recherche sur l'oursin vert au Canada Atlantique. http://crdpm.cus.ca/OURSIN/rapide.htm

Scheibling RE, Anthony SX (2001) Feeding, growth and reproduction of sea urchins (*Strongylocentrotus droebachiensis*) on single and mixed diets of kelp (Laminaria spp.) and the invasive alga *Codium fragile* ssp. *tomentosoides*. Mar Biol 139: 139–146

Scheibling RE, Hamm J (1991) Interactions between sea urchins (*Strongylocentrotus droebachiensis*) and their predators in field and laboratory experiments. Mar Biol 110: 105–116

Scheibling RE, Hennigar AW (1997) Recurrent outbreaks of disease in sea urchins *Strongylocentrotus droebachiensis* in Nova Scotia: evidence for a link with large-scale meteorologic and oceanographic events. Mar Ecol Prog Ser 152: 155–165

Scheibling RE, Hennigar AW, Balch T (1999) Destructive grazing, epiphytism, and disease: the dynamics of sea urchin-kelp interactions in Nova Scotia. Can J Fish Aquat Sci 56: 2300–2314

Scheibling RE, Raymond BG (1990) Community dynamics on a subtidal cobble bed following mass mortalities of sea urchins. Mar Ecol Prog Ser 63: 127–145

Scheibling RE, Stephenson RL (1984) Mass mortality of *Strongylocentrotus droebachiensis* (Echinodermata: Echinoidea) off Nova Scotia, Canada. Mar Biol 78: 153–164

Schroeter SC, Dixon JD, Ebert TA, Rankin JV (1996) Effects of kelp forests *Macrocystis pyrifera* on the larval distribution and settlement of red and purple sea urchins *Strongylocentrotus franciscanus* and *S. purpuratus*. Mar Ecol Prog Ser 133: 125–134

Sebens KP (1985) The ecology of the rocky subtidal zone. Amer Sci 73: 548–557

Sebens KP (1986) Spatial relationships among encrusting marine organisms in the New England subtidal zone. Ecol Monogr 56: 73–96

Siddon CE, Witman JD (2003) Influences of chronic, low-level hydrodynamic forces on subtidal community structure. Mar Ecol Prog Ser 261: 99–110

Siddon CE, Witman JD (2004) Behavioral indirect interactions: multiple predator effects and prey switching in the rocky subtidal. Ecology 85: 2938–2945

Sivertsen K (1996) The incidence, occurrence and distribution of the nematode parasite *Echinomerella matsi* in its echinoid host, *Strongylocentrotus droebachiensis*, in northern Norway. Mar Biol 126: 703–714

Sivertsen K (1997) Geographic and environmental factors affecting the distribution of kelp beds and barren grounds and changes in biota associated with kelp reduction at sites along the Norwegian coast. Can J Fish Aquat Sci 54: 2872–2887

Sivertsen K, Hopkins CCE (1995) Demography of the echinoid *Strongylocentrotus droebachiensis* related to biotope in northern Norway. In: Skjoldal HR, Hopkins C, Erikstad KE, Leinaas HP (eds) Ecology of fjords and coastal waters. Elsevier, Amsterdam, pp 549–571

Skadsheim A, Christie H, Leinaas HP (1995) Population reductions of *Strongylocentrotus droebachiensis* (Echinodermata) in Norway and the distribution of its endoparasite Echinomermella matsi (Nematoda). Mar Ecol Prog Ser 119: 199–209

Starr M, Himmelman JH, Therriault JC (1992) Isolation and properties of a substance from the diatom *Phaeodactylum tricornutum* which induces spawning in the sea urchin *Strongylocentrotus droebachiensis*. Mar Ecol Prog Ser 79: 275–287

Starr M, Himmelman JH, Therriault JC (1994) Direct coupling of marine invertebrate spawning with phytoplankton blooms. Science 247: 1071–1074

Stein A (1999) Effects of the parasitic nematode Echinomermella matsi on the growth and survival of its host, the sea urchin *Strongylocentrotus droebachiensis*. Can J Zool 77: 139–147

Stein A, Halvorsen O, Leinaas HP (1995) No evidence of *Echinomerella matsi* (Nematoda) as a mortality factor in a local mass mortality of *Strongylocentrotus droebachiensis* (Echinoidea). In: Skjoldal HR, Hopkins C, Erikstad KE, Leinaas HP (eds) Ecology of fjords and coastal waters. Elsevier, Amsterdam, pp 585–591

Stein A, Halvorsen O, Leinaas HP (1996) Density-dependent sex ratio in *Echinomerella matsi* (Nematoda), a parasite of the sea urchin *Strongylocentrotus droebachiensis*. Parasitology 112: 105–112

Stein A, Leinaas HP, Halvorsen O, Christie H (1998) Population dynamics of the Echinomermella matsi (Nematoda) – *Strongylocentrotus droebachiensis* (Echinoidea) system: effects on host fecundity. Mar Ecol Prog Ser 163: 193–201

Stephens RE (1972) Studies on the development of the sea urchin *Strongylocentrotus droebachiensis*. I. Ecology and normal development. Biol Bull 142: 132–144

Stickle WB, Ahokas R (1974) The effects of tidal fluctuations of salinity on the perivisceral fluid composition of several echinoderms. Comp Biochem Physiol 47A: 469–476

Stickle WB, Denoux GJ (1976) Effects of in situ tidal salinity fluctuations on osmotic and ionic composition of body fluid in southeastern Alaska rocky intertidal fauna. Mar Biol 37: 125–135

Strathman R (1978) Length of pelagic period in echinoderms with feeding larvae from the northwest Pacific. J Exp Mar Biol Ecol 34: 23–27

Sumi CBT, Scheibling RE (2005) Role of grazing by sea urchins (*Strongylocentrotus droebachiensis*) in regulating the invasive alga *Codium fragile* ssp. *tomentosoides* in Nova Scotia. Mar Ecol Prog Ser 292: 203–212

Swan EF (1958) Growth and variation in sea urchins of York, Maine. J Mar Res 17: 505–522

Swan EF (1961) Some observations on the growth rate of sea urchins in the genus Strongylocentrotus. Biol Bull 120: 420–427

Takagi T, Eaton CA, Ackman RG (1980) Distribution of fatty acids of the common Atlantic sea urchin, *Strongylocentrotus droebachiensis*. Can J Fish Aquat Sci 37: 195–202

Tegner MJ, Dayton PK (1977) Sea urchin recruitment patterns and implications of commercial fishing. Science 196: 324–326.

Tegner MJ, Dayton PK (1981) Population structure, recruitment and mortality of two sea urchins (*Strongylocentrotus franciscanus* and *S. purpuratus*) in a kelp forest. Mar Ecol Prog Ser 5: 255–268

Theriault C, Scheibling RE, Hatcher BG, Jones W (in press) Mapping the distribution of an invasive marine alga (*Codium fragile* spp *tomentosoides*) in optically shallow coastal waters using the Compact Airborne Spectrographic Imager (CASI). Can J Remote Sens

Thompson RJ (1979) Fecundity and reproductive effort in the blue mussel (Mytilus edulis), the sea urchin (*Strongylocentrotus droebachiensis*), and the snow crab (Chionoecetes opilio) from populations in Nova Scotia and Newfoundland. J Fish Res Bd Can 36: 955–964

Thompson RJ (1982) The relationship between food ration and reproductive effort in the green sea urchin *Strongylocentrotus droebachiensis*. Oecologia 56: 50–57

Thompson RJ (1984) Partitioning of energy between growth and reproduction in three populations of the green sea urchin *Strongylocentrotus droebachiensis*. Adv Invert Reprod 3: 425–432

Vadas RL (1977) Preferential feeding: an optimization strategy in sea urchins. Ecol Monogr 47: 337–371

Vadas RL (1990) Comparative foraging behaviour of tropical and boreal sea urchins. In: Hughes RN (ed.) Behavioural mechanisms of food selection. NATO ASI series, Vol. G20, Springer-Verlag, Berlin, pp 531–572

Vadas RL, Elner RW (1992) Plant-animal interactions in the north-west Atlantic. In: John DM, Hawkins SJ, Price JH (eds) Plant–animal interactions in the marine benthos. Clarendon Press, Oxford, pp 33–60

Vadas RL, Elner RW, Garwood PE, Babb IG (1986) Experimental evaluation of aggregation behavior in the sea urchin *Strongylocentrotus droebachiensis*. Mar Biol 90: 433–448

Vadas RL, Smith BD, Beal B, Dowling T (2002) Sympatric growth and size bimodality in the green sea urchin (*Strongylocentrotus droebachiensis*). Ecol Monogr 72: 113–132

Vadas RL, Steneck RS (1995) Overfishing and inferences in kelp-sea urchin interactions. In: Skjoldal HR, Hopkins C, Erikstad KE, Leinaas HP (eds) Ecology of fjords and coastal waters. Elsevier, Amsterdam, pp 509–524

Wahle RA, Gilbert AE (2002) Detecting and quantifying male sea urchin spawning with time integrated fertilization assays. Mar Biol 140: 375–382

Wahle RA, Peckham SH (1999) Density-related reproductive trade-offs in the green sea urchin *Strongylocentrotus droebachiensis* Mar Biol 134: 127–137

Walker CW, Lesser MP (1998) Manipulation of food and photoperiod promotes out-of-season gametogenesis in the green sea urchin, *Strongylocentrotus droebachiensis*: implications for aquaculture. Mar Biol 132: 663–676

Wharton WG, Mann KH (1981) Relationship between destructive grazing by the sea urchin, *Strongylocentrotus droebachiensis*, and the abundance of American lobster, *Homarus americanus*, on the Atlantic coast of Nova Scotia. Can J Fish Aquat Sci 38: 1339–1349

Williams CT, Harris LG (1998) Growth of juvenile green sea urchins on natural and artificial diets. In: Mooi R, Telford M (eds) Echinoderms: San Francisco. Balkema, Rotterdam, pp 887–892

Witman JD (1985) Refuges, biological disturbance, and rocky subtidal community structure in New England. Ecol Monogr 55: 421–445

Witman JD (1987) Subtidal coexistence: storms, grazing, mutualism, and the zonation of kelps and mussels. Ecol Monogr 57: 167–187

Yund PO, Meidel SK (2003) Sea urchin spawning in benthic boundary layers: Are eggs fertilized before advecting away from females? Limnol Oceanogr 48: 795–801

Edible Sea Urchins: Biology and Ecology
Editor: John Miller Lawrence

Chapter 19

The Ecology of *Strongylocentrotus franciscanus* and *Strongylocentrotus purpuratus*

Laura Rogers-Bennett

California Department of Fish and Game, Bodega Marine Laboratory, University of California, Davis, Bodega Bay, CA (USA).

1. INTRODUCTION

Red sea urchins, *Strongylocentrotus franciscanus* (Plate 3C), and purple sea urchins, S. *purpuratus,* are arguably the most-studied and well-known echinoid species in the world. Red and purple sea urchins are dominant members of nearshore rocky reef communities along the North American west coast and are capable of structuring subtidal algal communities and influencing community diversity. While we have long appreciated the role sea urchin herbivory exerts on the surrounding algal community and the creation of sea urchin "barrens," we are only starting to appreciate the positive species associations between sea urchins and the invertebrates and fishes that reside in and around their spine canopy. Red and purple sea urchins are also the basis for important fisheries, with purple sea urchins making up a minor component of the fishery. Purple sea urchins are collected extensively for scientific research, including fertilization biology, embryology, genome analysis and fertilization bioassays that are used to assess the toxicity of marine pollutants, silt, pulp-mill effluent and ultraviolet radiation. More work is needed in assessing how removals from large-scale fishery and scientific collecting affect the role sea urchins play in marine ecosystems. As we understand more we will be able to incorporate this knowledge into wise management practices designed to both provide sustainable fisheries and maintain healthy marine ecosystems.

Increased demand for red sea urchins has led to an expansion of the fisheries from their initial exploitation (Sloan 1985) to fully and overexploited fisheries (Keesing and Hall 1998; Andrew et al. 2002; Botsford et al. 2004). Using traditional measures of overfishing, such as spawning stock biomass below 20% of virgin biomass, overfishing has occurred in parts of the West Coast. West Coast fisheries for red sea urchins have been affected not only by decreases in stocks but also by management policies, El Niño events, shifts in effort, a weakening of the Japanese export market and major competition from fisheries in Canada, Russia, South Korea, and Chile (Andrew et al. 2002). In 2004, the United States sold sea urchin as fresh roe and live fresh product internationally valued at $44.7

and \$6.9 million, respectively. The top three importers of fresh roe product from the United States were Japan, Hong Kong, and Taiwan with Japan importing just under 2 million or 80% of the market.

Red sea urchins are important for fisheries in part due to their large body size. Large red sea urchins measure 198 mm in test diameter from British Columbia (Bureau 1996). They are larger than the purple sea urchin, which reaches 100 mm (Morris et al. 1980), but smaller than the giant heart sea urchin *Spherosoma giganteum* from the deep sea measuring 380 mm. The spines of the species differ in length, reaching 50–55 mm in the red sea urchin and less than 20 mm in the purple sea urchin (Tegner and Levin 1983). Spine length, like many other morphological features, is plastic and sea urchins in areas of high-wave action have shorter spines (Rogers-Bennett et al. 1995). The relative size of Aristotle's lantern is larger for sea urchins residing in food-poor habitats (Ebert 1980). Podial coverage however, appears to be conserved and is nearly proportional to test area in 21 species of echinoids (Strathmann and von Dassow 2001) and does not differ for red sea urchins in wave-exposed sites (Rogers-Bennett, unpubl. data). Purple sea urchins are able to withstand greater wave action than red sea urchins and are found in more exposed intertidal habitats.

Red and purple sea urchins reside on rocky substrates with a broad geographic and depth distribution. Red sea urchins range from the tip of Baja California, Mexico north to Kodiak, Alaska (Ebert et al. 1999). Purple sea urchins have a more reduced range from Isla Cedros in Baja California, Mexico north to Cook Inlet, Alaska (Tegner 2001) but span a greater distribution of depths. Red sea urchins are distributed in subtidal habitats to a depth of 90 m while purple sea urchins are common in intertidal and subtidal habitats to 160 m (Morris et al. 1980).

Red sea urchins (Plate 3C) range in color from a light pink to a dark brick red or black, while purple sea urchins are purple in color. Juveniles may be difficult to distinguish by color as they can both be light purple or even white to greenish. The developing echinoplutei of both species are planktotrophic requiring nutritional input (other than the yolk) for metamorphosis. Larval red sea urchins can be distinguished from purple sea urchins because they have a single dorsal pedicellarium which is absent in larval purple sea urchins, as well as other unique skeletal traits (Strathmann 1979).

Recent advances in the study of red and purple sea urchins have increased our awareness of the multifaceted role they play in nearshore subtidal communities as ecosystem engineers. As we learn more about the spatial distribution of sea urchin stocks we see that their populations are divided into metapopulations or microstocks. This spatial structure has important ramifications for population dynamics and exploitation. One surprising recent discovery is that larval purple sea urchins can clone themselves (Eaves and Palmer 2003). This requires that we re-examine estimates of larval-dispersal distances for this species and possibly others. There is a growing awareness of the role oceanographic processes play in the productivity of sea urchin populations. Ocean conditions affect not only the availability of drift-algal food resources required for gonad development (essential for both sea urchin reproduction and the fishery) but also transport processes influencing larval settlement and successful recruitment. Settlement patterns of sea urchins have been studied in California for more than a decade and these patterns are now being linked to what we know about interannual variation in oceanography. Our goal will be to use what we learn about spatial patterns in sea urchin productivity to help us better manage sea urchin fisheries.

2. SEA URCHIN GRAZING AND KELP FOREST ECOSYSTEMS

Sea urchins have long been acknowledged as major structuring forces within subtidal kelp forest communities (Paine and Vadas 1969; Lawrence 1975; Harrold and Pearse 1987; Tegner and Dayton 2000). In some regions, there appear to be two dominate alternate community states (Simenstad et al. 1978): kelp beds with high species diversity and echinoid dominated rocky grounds with low species diversity. Grazing by sea urchins is infamous for altering algal communities from lush, specious kelp forests to "barrens" characterized by crustose coralline algae and the absence of upright fleshy algae (Leighton et al. 1966). When sea urchins overgraze, their foraging behavior changes; they leave cryptic habitats to form dense feeding aggregations or "fronts" which can in turn denude fleshy kelps by eating through the stipes (Dean et al. 1984). In southern California, while there is a range of community types, kelp forests have higher species diversity than sea urchin barrens with more than 90% of the 275 common species more abundant in kelp communities of which 25% are kelp forest obligates (Graham 2004). In some regions overgrazing appears to be related to sea urchin density (Lawrence 1975), with a distinct threshold of sea urchin density ($2\,kg\,m^{-2}$) above which sea urchins overgraze kelp beds (Breen and Mann 1976). Major recruitment events may increase sea urchin densities to levels where they overgraze kelps, such as the recruitment of 1984 in a central California kelp forest which led to deforestation by sea urchins 2 years later (Watanabe and Harrold 1991). Yet there are also places and times when sea urchins and kelps coexist for decades, as in northern California. Sea urchin overgrazing appears to be less common in South American kelp communities (Dayton 1985).

These two states – kelp beds and sea urchin barrens have led to a popular paradigm of a trophic cascade (*sensu* Carpenter and Kitchell 1988) in which the removal of sea urchin predators leads to sea urchin population explosions and kelp deforestation. This paradigm however, has not been well tested (Sala et al. 1998). There is no debating that human fishing has removed many sea urchin predators (Dayton et al 1998), such as, sea otters (Estes et al. 1998), spiny lobsters (Lafferty 2004) and sea urchin competitors such as southern California's abalone (Dayton et al. 1998; Rogers-Bennett et al. 2002). Deforestation by sea urchins, however, may not be as straight forward as the paradigm suggests. Other processes, such as abiotic factors, are important in dynamics of sea urchin deforestation. For example, shortages of drift algae appears to be an important factor in triggering sea urchins to switch feeding modes from primarily sedentary drift feeders residing in sheltered habitats to active foragers that denude standing algae in southern California kelp forests (Harrold and Reed 1985; Ebling et al. 1985). There may also be negative feedback loops such that, as drift becomes less abundant, sea urchins cause more damage to kelp by grazing pits in the holdfast, a condition termed cavitation (Leighton 1971). This causes structural failure when stressed by waves and leads to further kelp loss (Tegner et al. 1995).

While the mechanisms involved in kelp overgrazing by sea urchins remain an important focus of research and discussion, human activity has profoundly affected this dynamic. Human influences span the gamut from fishing sea urchin predators (Dayton et al. 1998; Jackson et al. 2001), species eradications (Estes et al. 1989; Jackson et al. 2001), climate change coupled with fishing (Harley and Rogers-Bennett 2004), disease intensification (Scheibling et al. 1999; Behrens and Lafferty 2004), species introductions (Levin et al. 2002) and pollution (Pearse et al. 1970). At the same time, it is difficult to distinguish anthropogenic changes from environmental changes. For example, many species removals and declines in marine systems have gone unnoticed as they may be economically unimportant or loosely integrated components of the ecosystem (Dayton et al. 1998). The kelp forests of today are clearly different places than they were 200, 100, or even 50 years ago and the dynamics are probably greatly altered (Dayton et al. 1998). Marine protected areas may be useful places for examining the effects of human fishing on trophic interactions and the dynamics of sea urchins and kelp forests (Rogers-Bennett and Pearse 2001; Behrens and Lafferty 2004).

In the last two decades, kelp deforestation by sea urchins appears to be increasing worldwide (Steneck et al. 2002). Increases in fishing pressures on herbivore predators as well as increases in ocean warming both negatively affect kelp abundances and may act synergistically to increase kelp deforestation events, while, at the same time slowing kelp regeneration. Unfortunately, in regions where kelps are sparse, sea urchin fishing may not be useful in reversing deforestation since sea urchin gonads are of poor quality from barrens and the sea urchins are not fished (K Barsky, pers. comm.). Deforestation appears to strongly affect species diversity (Steneck et al. 2002; Graham 2004). This in turn could impair an ecosystem's ability to rebound from perturbations (Kiessling 2005). We will need a better understanding of the processes involved in deforestation by sea urchins if we are to maintain productive and diverse kelp ecosystems.

3. GROWTH AND SURVIVAL

3.1. Growth

More is known about age and growth in strongylocentrotid sea urchins than perhaps any other genus (Ebert and Russell 1993; Smith et al. 1998; Ebert et al. 1999; Morgan et al. 2000a; Rogers-Bennett et al. 2003). Ebert et al (1999) examined red sea urchin growth in a large-scale tag- recapture program that resulted in 1582 recoveries from Alaska to southern California and found that growth rates were slow. Growth did not peak in red sea urchins until they reached 30–40 mm in test diameter. Growth then tapered off dramatically as the sea urchins became larger. These tagging data demonstrate that growth, though very small, continues even in the very largest animals and suggest that red sea urchins may be a very long lived species.

Researchers have also modeled red and purple sea urchin growth. Growth data for sea urchins often lack information from juveniles since they can be rare and difficult to recapture once tagged. This lack of information from growing juveniles may bias growth-rate estimates in undesirable ways, shortening the estimate of the time it takes for sea urchins to enter the fishery (Yamaguchi 1975; Rogers-Bennett et al. 2003). Since growth continues even as sea urchins grow very large (potentially old), growth models have been used that allow for infinite increase, such as the Tanaka model (Tanaka 1982). Growth model estimates range from 6 to 10 years as the time to reach legal size (89 mm) in the California fishery (Ebert and Russell 1993; Ebert et al. 1999; Morgan et al. 2000a; Rogers-Bennett et al. 2003).

Growth is highly variable between individuals. Full sibling purple sea urchins grown in the laboratory under identical conditions for 1 year ranged in size from 10 to 30 mm (Pearse and Cameron 1991) while red sea urchins ranged in size from 4 to 44 mm (Rogers-Bennett, unpubl. data). Variation in growth of purple sea urchins was also examined at three sites along the geographical extent of their distribution on the west coast of North America. Growth was found to be highly variable with more differences within a site than between sites (Russell 1987). This dramatic individual variation in growth can have important implications for modeling since many models are designed to depict the growth of a single animal rather than the mean of many animals (Sainsbury 1980). Use of these models can result in overestimates of the mean size of a cohort (Sainsbury 1980).

Spatial and temporal variation in growth is also significant for red sea urchins. Examining growth increment data from tagged red sea urchins in combination with growth estimates from size–frequency distributions at two reserve sites suggests growth may vary spatially. Growth model estimates were lower for red sea urchins in the Bodega State Marine Reserve compared to the Caspar Sea Urchin Fishing Closure near Fort Bragg, California which has abundant algal resources (Morgan et al. 2000a). Newly settled sea urchins grew faster in kelp beds than sea urchins barrens (Rowley 1990). Identifying spatial patterns in growth is challenging since patterns can be obscured by large differences in individual growth.

3.2. Survival

Survival estimates for sea urchins have been made using a decaying exponential function with the number of individuals (N_t) of a certain age or cohort as the dependent variable and N_0 as the initial number of individuals

$$N_t = N_0 e^{-Zx}$$

This function has one parameter Z the mortality coefficient and x is time (say in years). In this function, the annual survival rate is e^{-Z} and annual mortality rate is $1 - e^{-Z}$ (Ebert 2001). In this simple model, the probability of survival does not change with age. The mortality coefficient, Z can be estimated using a combination of size–frequency data and parameters from a growth function (Ebert 1999). Other studies have also used this method to estimate red sea urchin survival (Ebert 1987; Ebert and Russell 1993; Smith et al. 1998; Ebert et al. 1999).

Survival estimates for red sea urchins, are very high, ranging from $Z\,yr^{-1}$ of 0.82–0.98 for sites from northern California to Alaska and from 0.67 to 0.91 for southern California sites (Ebert et al. 1999). Likewise, estimates of purple sea urchin survival are also very high ranging from 0.85 to 0.90 for northern sites and from 0.72 to 0.89 for southern sites

(Russell 1987). Both red and purple sea urchins are estimated to have higher survival rates in northern sites compared with southern California. For purple sea urchins the suggestion has been made that the lower survival rates in southern California may be due to higher predation rates and/or higher stress levels associated with warm sea water temperatures (Russell 1987). One caveat for these survival estimates is that assumptions of seasonally stable and stationary size distributions are violated. Irregular recruitment dynamics can change mean size from year to year. Examining the effects of changes in the mean size between one high and one low recruitment year for purple sea urchins did little to change the resulting survivorship curve, which suggests that these results are robust to violations of the assumption (Ebert et al. 1999). The overall conclusion from these data is that sea urchins are very long lived.

3.3. Aging

Growth and survival modeling results lead to a startling conclusion – that 150 mm red sea urchins are 100–200 years old (Ebert et al. 1999). This surprising result of very long life has been validated using an alternative aging method: C^{14} radiocarbon dating. Radiocarbon C^{14} emitted during atmospheric nuclear testing provides a time stamp in the calcium carbonate of animals alive during the 1950s. Ebert and Southon (2003) found that growth estimates from the Tanaka growth curve were validated by quantifying the levels of the C^{14} radiocarbon and calibrating these to the known nuclear testing dates. Their work suggests that the largest red sea urchins found in British Columbia 198 mm in test diameter (Bureau 1996) were more than 200 years old. These estimates far exceed longevity estimates for other sea urchin species (Ebert and Southon 2003). Current work on longevity in red sea urchins suggests that telomerase activity may continue even in the tissues of very large, old individuals (A Bodnar, pers. comm.). These findings, taken together with continuous but slow growth and continuous reproduction in the largest, oldest animals, lead researchers to question the onset of senescence in red sea urchins.

4. REPRODUCTION

Reproduction in red and purple sea urchins follows annual cycles (Giese et al. 1958; Gonor 1973). Food needs to be present for the gonads to develop while most of the year primary oocytes and spermatocytes accumulate, aided by the nutritive phagocytes (see Chapter 2). Prior to spawning, the nutritive phagocytes become reduced as the gametes in both sexes develop and mature. Photoperiod appears to play a large role in the regulation of gametogenesis in purple urchins as well as other species of sea urchins (Pearse et al. 1986). Gametogenesis is triggered when photoperiod is less than 12 h (Pearse et al. 1986). Sea urchins artificially maintained 6 months offset from the natural phase, spawned 6 months behind the natural populations. Furthermore, animals maintained in continuous darkness have reduced gonad growth with enhanced somatic growth (J Pearse, pers. comm.). Reproduction appears to continue throughout the life of red and purple sea urchins. There is no evidence for a decrease in reproductive output or reproductive senescence with increasing age or size as found in other phyla (Rogers-Bennett et al. 2004). In fact, large females produce exponentially more eggs than midsize females (Tegner 1989).

Reproduction in sea urchins in many regions varies spatially primarily in response to food abundance. In northern California, red sea urchins in shallow habitats (5 m) with abundant drift-algae resources have significantly larger gonads (mean = 64 g) more than 4 times greater than red sea urchins at intermediate (mean = 14 g) and deep water habitats (mean = 13 g) (Rogers-Bennett et al. 1995). Gonad index values are positively correlated with drift-algae abundance in the shallow habitats but not in the deep where drift is scarce (Rogers-Bennett et al. 1995). Drift-algae abundance is greatest in the shallow habitats throughout the year and most abundant in the fall (Rogers-Bennett et al. 1995). Red sea urchins in southeast Alaska also have larger gonads in shallow habitats where drift is abundant (Carney 1991). In the inland waters of San Juan Island, Washington, where extreme tidal flows regularly deliver abundant drift-algae resources to deep habitats, red sea urchin gonad indexes are comparable in shallow and deep habitats (K. Britton-Simmons, pers. comm.).

5. FERTILIZATION

Red and purple sea urchins have separate sexes and release eggs and sperm into the water column where fertilization takes place. Sperm are activated once they are released and encounter sea water with a short time (10 min) to swim and fertilize an egg before their mitochondria are exhausted (Christen et al. 1986). The acrosomal protein bindin attaches sperm to eggs during fertilization. This requires males and females to spawn somewhat synchronously and to have high enough concentrations of gametes at a range close enough to achieve physical contact and fertilization success. In these sea urchins, spawning by males precedes females (Levitan 2002) and the presence of sperm stimulates females to spawn. Two natural spawning events observed in populations of red and purple sea urchins in British Columbia, Canada, occurred during heavy phytoplankton blooms (Levitan 2002). During these spawning events the sea urchins did not aggregate. Thirty to 44% of the individuals in the area spawned, and during spawning a higher percent of males (88%) spawned than females.

Variations in adult population densities change patterns in gamete release which influence sperm availability as well as sperm competition. In populations where the density of sea urchins is high, sperm are more likely to compete for eggs, whereas eggs may have plenty of sperm. In populations where sea urchin density is low, eggs may be faced with sperm limitation. Reduction of population densities by fishing could have a dramatic effect on fertilization success, perhaps resulting in Allee effects where fertilization decreases precipitously rather than as a smooth linear decline with decreasing density (Allee 1931).

As sea urchin density increases, fertilization success is enhanced, peaking at 1–3 males per square meter (Levitan 2004). Purple sea urchins reside in more densely packed populations than red sea urchins. This results in a greater reproductive success in purple sea urchins (94%) than red sea urchins (64%) at the densities examined (Levitan 2002). At extremely high male densities, fertilization success is again decreased due to polyspermy which can result in abnormal or arrested development (Ernst 1997). Paternity analyses indicate that 98% of the larvae produced in field-spawning experiments with high male densities were sired by more than one male (Levitan 2004).

Red and purple sea urchins differ in egg size, with purple sea urchins having smaller eggs. This makes the eggs more difficult to target for sperm, requiring an order of magnitude more sperm for fertilization, that may even necessitate higher densities of males near females (Levitan et al. 1992). They also differ with respect to their susceptibility to interspecific fertilization and the production of hybrids. Purple sea urchins are less likely to hybridize than red sea urchins. Larval hybrids of both species have higher mortality than nonhybrids. Hybrid crosses with red sea urchin sperm and green sea urchin, *S. droebachiensis* eggs have survived for 3 years; however hybrid crosses with sperm from purple sea urchins, *S. purpuratus,* did not survive past the larval stage (Levitan 2002).

In a model used to examine the dynamics of fertilization success at low stock levels, simulations show differences in recruitment declines based on the spatial distribution of adults and sperm dispersal distributions (Lundquist and Botsford 2004). With random adult distributions and various sperm distributions curves, the simulation model yielded gradually declining sperm distributions as opposed to precipitous declines associated with a set threshold as would occur with an Allee effect. The decline in successful reproduction (zygote production) was more pronounced in high-flow conditions irrespective of adult density. While the shape of the sperm-dispersal curve is unknown, model simulations using broad dispersal curves produced sharp threshold declines (Allee effect) whereas more narrow dispersal curves produced linear declines toward zero (recruitment failure) (Lundquist and Botsford 2004).

6. LARVAE

6.1. Larval Period

Red and purple sea urchins have a feeding lecithotrophic larval stage which must acquire food in order to develop and metamorphose. The larval period ranges from 27 to 131 days, varying with both temperature and food availability. In Washington State, at temperatures ranging from 7 to 13 °C, larvae metamorphosed into benthic juveniles in 62–131 days (Strathmann 1978). Shorter times to settlement (40 days) were found at warmer temperatures (Cameron and Schroeter 1980) and with increased food rations (Paulay et al. 1985). Red sea urchins have been reared through to metamorphosis in 23 days at temperatures of 15 ± 2.1 °C when fed high concentrations (60 000–100 000 cells ml^{-1}) of the unicellular alga, *Rhodomonas lens* (Rogers-Bennett 1994).

Echinoderm larvae are phenotypically plastic with respect to environmental conditions such as food availability and temperature. Larval echinoids in food-limited environments have longer larval arms and reduced stomach diameters compared with larvae in food-rich environments (Boidron-Metairon 1988; Hart and Scheibling1988). Larvae with longer arms have enhanced food-gathering capabilities and thus, ingest more food (Hart and

Strathmann 1994). Moreover, larvae increase the size of their stomachs and shorten their arm lengths prior to their ability to feed suggesting that morphogenesis occurs, rather than simply the extension of a full stomach (Miner 2005). Cold temperatures slow development but do not change the sequence of trait acquisition (Miller and Emlet 1997).

The timing of the presence of echinoderm larvae in the plankton has been examined in southern California. Larvae in the plankton sampled weekly at the Diablo Nuclear Power Plant from December 1996 to June 1998 showed a ratio of 16:1 purple sea urchin larvae to red sea urchin larvae (J Steinbeck, unpubl. data). The timing of the presence of larvae was roughly synchronous between the two species. Purple sea urchins were most abundant in June 1997 with a smaller peak in abundance earlier in April 1997, while red sea urchin larval abundance peaked in March 1997 with another peak in July 1997. This peak in larval abundance in the spring coincided with the start of the strong warm water El Niño event in 1997 (NOAAa).

6.2. Blastulae and Larval Behavior

Early stage larvae may be particularly susceptible to predation due to their small size and limited movement capability (Rumrill 1990). Swimming behavior may help in evading predators and/or be used to migrate vertically in the water column. Blastulae begin to rotate within the fertilization envelop. After hatching, blastulae continue to rotate and begin to swim up in a helical path in the water column. The angle of inclination of the rotation along the animal-vegetal pole differs between species. Red sea urchins have a steep angle whereas purple sea urchins do not (McDonald 2004). Swimming and sinking studies in the laboratory using hatched swimming blastulae and unhatched sinking blastulae suggest sinking speeds frequently exceed swimming rates and that sinking rates increase with decreasing sea water temperature (McDonald 2004). The smaller purple sea urchin blastulae (105 μm diameter) had greater swimming speeds of $0.4\,mm\,s^{-1}$ than the larger red sea urchin blastulae (170 μm diameter) swimming at $0.2\,mm\ s^{-1}$ (McDonald 2004). These speeds dropped when temperatures increased 3 °C. The decrease in upward swimming velocity with increased temperature could be a useful physiological survival response for embryos entering an unfavorable portion of the water column which exceeds optimal temperatures.

6.3. Larval Cloning

Our understanding of echinoid larval biology and ecology has recently been radically altered with the discovery of asexual reproduction by larval cloning in purple sea urchins and two other echinoids (Eaves and Palmer 2003). Cloning or larval budding, was first described in sea stars, but not until 1988 (Bosch 1988, Bosch et al. 1989). Was this oversight simply because larvae were cultured in batches and individuals were seldom observed or that our paradigm of "normal" development caused us to ignore or remove larvae that looked different? We now know that cloning occurs in all echinoderm classes with the possible exception of crinoids, which have not been investigated. To make this oversight worse, cloning is quite common, occurring in 10–90% of sea star larvae collected from the field (Bosch 1988) and up to 12% in cultured sea cucumber larvae

(Eaves and Palmer 2003). Clones can be generated from different larval body parts. Larvae reared with high concentrations of food that was high in quality, cloned significantly more then poorly fed sea star larvae (Vickery and McClintock 2000). Investigations into the environmental conditions which promote cloning suggest that cultured purple sea urchins clone more in warm-water, high-salinity conditions (A Eaves, pers. comm.).

Larval cloning challenges the notion of fixed, or "set-aside," larval cells since they can differentiate into juvenile cells once cloning is initiated (Eaves and Palmer 2003). The evolutionary consequences of cloning need to be further investigated to determine if it is an ancestral trait, perhaps retained in other deuterostomes such as acorn worms and sea squirts.

Cloning has important implications for larval dispersal, in the context of both gene flow and larval dispersal distances. Larvae that disperse a given distance in the "normal" developmental time of 2–4 weeks can then bud and travel another 2–4 weeks, spreading that genome further than previously estimated. This can occur because clones can subsequently clone multiple times (Balser 1998). Larval dispersal estimates that do not consider cloning need to be re-examined. It has been suggested that there may be an entirely pelagic bauplan (Eaves and Palmer 2003). Indeed, along the California coast, and particularly in the southern California bight, newly settled sea urchins can be found on settlement substrates year round (Ebert et al. 1994). Is this the result of multiple spawnings throughout the year, clones that continue to reproduce asexually, or both? Dispersal distances have important marine conservation applications, such as in the determination of the size and spacing of Marine Protected Areas (Shanks et al. 2003).

7. SETTLEMENT AND RECRUITMENT

Sea urchin (0.5–5.0 mm diameter) settlement has been examined in California on artificial substrates. Ebert et al. (1994) found that settlement is highly seasonal, occurring predominantly from February to July. Settlement is higher and more regular in southern California, within the retention zone of the Bight, compared with northern California, a region with maximum upwelling and strong offshore advection (Ebert et al. 1994). Other studies have found that settlement in northern California and Oregon is favored in years when conditions in June and July are warm with increased salinity and low alongshore windstress (Miller and Emlet 1997; Morgan et al. 2000b; Wing et al. 2003). Settlement of sea urchins on collectors throughout California appears to be highest following warm-water events which may not only reduce offshore transport of larvae but also provide for enhanced larval food and growth (S. Schroeter, pers. comm.). Ebert et al. (1994) also found that purple sea urchins settled in higher numbers and with more geographic coherence than red sea urchins. Furthermore, settlement does not appear to be hindered by the presence of dense kelp forests (Schroeter et al. 1996).

Recruitment dynamics can be inferred from size–frequency distributions so that the absence of juveniles indicates a lack of recruitment success over the past 5–7 years. Examination of tidepools from central California to Oregon revealed few juvenile (5–50 mm) purple sea urchin recruits at sites near headlands and capes (Ebert and Russell 1988). Similarly, in northern California, an examination of the size–frequency distribution revealed more small sea urchins in areas away from headlands that had onshore and poleward movement of water during relaxation events (Morgan et al. 2000b). The mechanism proposed for these patterns is that larvae are advected away from the coast during upwelling and strong offshore jets associated with headlands (Ebert and Russell 1988; Wing et al. 1995). These conclusions are based on the assumption that differences in recruitment, as observed in the size–frequency distribution, arise from differences in settlement, as opposed to post-settlement processes (cf. Connell 1985), over large spatial scales. Post-settlement processes, however, are important for sea urchins at intermediate spatial scales such as on a single reef (Rowley 1989). This highlights the need to distinguish larval settlement (0.5–5.0 mm) from juvenile recruitment (5.0–50 mm).

On smaller spatial scales (1–10 m), juvenile red and purple sea urchins have unique spatial distributions. In some areas, juvenile sea urchins shelter under the spine canopy of adult conspecifics (Low 1975; Tegner and Dayton 1977; Breen et al. 1985; Rogers-Bennett et al. 1995). Adult red sea urchins are more frequently found sheltering juveniles than are adult purple sea urchins, whose spines are shorter. The spatial association of red sea urchins with adults is not due to preferential larval settlement near adults (Cameron and Schroeter 1980). Juvenile red and green sea urchins do actively move toward adult conspecifics contributing to this unique distribution (Rogers-Bennett 1989). Another post-settlement process, differential mortality, is likely responsible for the abundance of juveniles with adults in barrens habitats as compared with kelp forests (Rowley 1989). In laboratory studies, juvenile red and green sea urchins did not prefer feeding adults (Breen et al. 1985) nor did they gain a feeding advantage under the spine canopy of adults. In fact, their growth was significantly reduced compared with juveniles alone (Nishizaki and Ackerman 2004). In preference trials, the number of juveniles moving toward and sheltering under adults increased with increasing water flow and with the presence of potential predators (Nishizaki and Ackerman 2001). Juvenile red sea urchins in northern California were 12 times more likely to shelter under adults in shallow wave-exposed habitats where adults were in rock "scars" compared with deeper-water habitats where adults did not reside in rock "scars" (Rogers-Bennett 1994). Juveniles reacted by sheltering under adults when adults released a secondary chemical cue signaling the presence of predators, while large sea urchins did not react to this cue (Nishizaki and Ackerman 2005).

Recruitment patterns have been observed to differ between southern and northern California. Sea urchin recruitment of juveniles (5–50 mm) was examined in artificial modules made of cinder blocks (surface area 2.6 m^2) at Van Damme State Park in northern California and at three of the Channel Islands in southern California. In 2001–2004, recruitment of juvenile red sea urchins in the modules was 20 times greater in the south, while recruitment of purple sea urchins was nearly 300 times greater in the south (Rogers-Bennett and Kushner, unpubl.). Adult red and purple sea urchins densities on the surrounding natural reef were an order of magnitude lower at the site in northern California than in southern California (Rogers-Bennett and Kushner, unpubl.). Ebert et al. (1999) suggested that the fishery in the north may be driven by unusually successful year classes which persist in the population for many years. These discrepancies in rates of natural recruitment between southern and northern California may have important implications for levels of sustainable fishing between the two regions, suggesting northern California may not be able to sustain as much fishing pressure.

8. POPULATION REGULATION

Regulation of sea urchin populations remains an important research topic in echinoderm studies. Sea urchins populations can boom and bust and we still have more to learn in terms of the mechanisms responsible for these large swings in population density. Certainly the consequences of large populations can be profound and community-wide (see Section 2). A number of biotic factors, such as, competition, predation and disease play key roles in regulating sea urchin populations. Abiotic factors affect sea urchin populations at various stages throughout their life history. For example, fertilization success can be influenced by small-scale hydrodynamics and substrate topography (see Section 5). Larval transport is forced by physical transport processes moving patches of water close to and away from suitable settlement sites (see Section 7). Water flow continues to play a major role in the juvenile stage influencing distribution (Nishizaki and Ackerman 2001) as well as growth and survival. Winter storms can not only dislodge and kill juveniles and adults (Ebling et al. 1985) but also rip out standing kelp beds, thereby negatively affecting drift food supply (Tegner and Dayton 1991).

8.1. Competition

While intraspecific competition occurs in both red and purple sea urchins for limited resources such as food and habitat, there are very few studies examining this process. Within sea urchin barrens there are high densities of sea urchins in close proximity to one another and scarce food resources lead to optimal conditions for competition. There is indirect evidence of intraspecific competition in purple sea urchins: size–frequency distributions show that 90% of the sea urchins in high-density sites (outside reserves) are mid-size and small (< 50 mm) while inside reserves where densities are low, purple sea urchins are much larger (30–70 mm) (D Kushner, KFMP data). There is also evidence for intercohort competition since juvenile red sea urchins (8 mm) in the presence of adults and food have significantly lower growth rates than juveniles without adult conspecifics (Nishizaki and Ackerman 2004). There is no evidence for intraspecific competition between larvae because they are so scarce (Strathmann 1996).

Interspecific competition occurs between red and purple sea urchins in southern California (Ebert 1977; Schroeter 1978). Schroeter (1978) found that red sea urchin are competitive dominants over purple sea urchins because they are able to use their long spines to actively fence with purple sea urchins and exclude them from optimal habitats. Conversely, in Alaska, there is interspecific facilitation between red sea urchins and other congeners. Red sea urchins snag drift kelp with their long spines and provide defense against predatory sun stars *Pycnopodia helianthoides* (Duggins 1981). Little is known about interactions between red and purple sea urchins and the black sea urchin *Centrostephanus coronatus* in southern California.

Abalones are potential competitors with sea urchins (Leighton 1968; Tegner and Levin 1982). Abalones and sea urchins share similar resources: they consume primarily drift algae and live on rocky substrates. Sea urchins are more resistant to starvation than abalones and can even utilize dissolved nutrients living in areas near sewer outfalls (Pearse et al. 1970). In northern California, north of the sea otters' range, there are four species of abalone in the genus *Haliotis*. However, densities are very low for all but the red abalone (*H. rufescens*) (Raimondi et al. 2002; Rogers-Bennett et al. 2002). Red abalone can be very abundant in portions of the subtidal zone in northern California, even at fished sites such as Van Damme State Park (density = 7600 individuals per hectare) (California Department of Fish and Game 2005). At this density adult red abalone and red sea urchins compete for available rocky reef space and for food.(Karpov et al. 2001) During warm-water El Niño events, when drift algae is scarce, hungry red sea urchins move from deeper water (20 m) to shallow water (10 m) to feed (Rogers-Bennett, pers. observ.). When drift food is limiting and competition is intensified, hungry red abalone climb kelp stipes to feed on the blades (Rogers-Bennett, pers. observ.). Red abalone and red sea urchins segregate at the scale of a 60 m^2 transect in the subtidal kelp beds (Karpov et al. 2001). Adult red sea urchin and adult red abalone abundance was negatively correlated on transects at sites in northern California, while purple sea urchin abundance was not correlated with red abalone abundance (Karpov et al. 2001). In sharp contrast, juvenile abalone distribution is facilitated by the presence of the red sea urchin spine canopy. They are more abundant inside reserves with adult red sea urchins than in fished areas with out sea urchins (see Section 11) (Rogers-Bennett and Pearse 2001). In central California where sea otters are present, abalones and sea urchins may compete for deep-crevice habitat which is severely limited (Lowry and Pearse 1973; Hines and Pearse 1982).

In southern California, abalone populations have declined dramatically due to a combination of intense fishing (Rogers-Bennett et al. 2002) followed by disease (Moore et al. 2002). Due to the large-scale removal of abalone biomass, competition between abalones and sea urchins may be greatly reduced. There is some evidence that the dynamics of competition for algal resources may be changing as the red sea urchin fishery in northern California decreases grazing pressure, thereby enhancing kelp beds at one site (Karpov et al. 2001). In southern California, red sea urchin fishing and sea urchin wasting disease may be affecting the competition between red and purple sea urchins (Richards and Kushner 1994).

8.2. Predation

There is a wide range of sea urchin predators. Predation affects both sea urchin population density and the size–frequency distribution. Large sea urchins may reach a refuge in size from predation. Little work has been done to examine the effect of micropredators on sea urchin populations although this appears to be an important source of predation on newly-settled sea urchins (Rowley 1989; McNaught 1999).

Sea otters, *Enhydra lutris,* have the largest predation effect on sea urchins. They were hunted to near extinction, but remnant populations have expanded and today exist in central California, Washington, British Columbia and Alaska (Estes and Duggins 1995). When sea otters move into a habitat, sea urchin and other shellfish populations decline dramatically and individuals are restricted to cryptic microhabitats. The sea otter/sea urchin predation link is well documented (Bertness et al. 2001). Sea otters have been referred to as a "keystone species" (Paine 1966) whose presence is instrumental in shaping the structure of the surrounding nearshore community (Estes and Palmisano 1974; Simenstad et al. 1978). The "keystone" paradigm emerged from studies in Alaska in which habitats with sea otters had lush diverse kelp forests that were visually distinct from barrens habitats which lacked sea otters and were dominated by echinoderms (Estes and Duggins 1995). This is thought of as a three-level trophic cascade with sea otters, invertebrate herbivores (sea urchins), and algae. Another level has been added to this cascade, in which the killer whale, a sea otter predator, indirectly influences sea urchin populations through the trophic cascade (Estes et al. 1998). In central California, the sea otters range now extends from San Francisco south to Santa Barbara (Vogel 2000). Along the pipeline offshore of Santa Barbara sea otters have reduced sea urchin and shellfish populations dramatically (L Rogers-Bennett, pers. obs.).

There is little debate about the widespread effects of sea otters on echinoid populations within their range. However, outside the range of sea otters in California and Baja California, there is a suite of potential community types (e.g. only 10% of more than 200 sites surveyed were sea urchin barrens) (Foster and Schiel 1988). Foster and Schiel's (1988) work suggests (1) there may be a suite of intermediate community states outside the sea otters' range as opposed to two alternative stable states, (2) care must be taken in applying this mutually exclusive states concept to the entire west coast of North America and elsewhere, and (3) a number of other factors have an important role in driving sea urchin-mediated deforestation (see Section 2). More work is needed at sites where sea urchins and kelps routinely coexist.

In southern California, two sea urchin predators are important in regulating sea urchin populations: the spiny lobster, *Panulirus interruptus*, and the sheephead, *Semicossyphus pulcher,* a labrid fish (Tegner and Dayton 1981). Spiny lobsters prefer purple to red sea urchins and juveniles to adults (Tegner and Levin 1983). By comparing sea urchin abundances at sites inside and outside reserves in the northern Channel Islands, where spiny lobsters are protected from fishing, Lafferty (2004) was able to assess the effects of lobsters on sea urchin populations. Inside the reserves, lobster abundance was high and purple sea urchin abundance was low (<100 red and purple sea urchins per square meter) and kelp was abundant (Lafferty 2004). This suggests that lobster predation may be important in regulating sea urchin densities in this region. Sheephead have massive jaws and eat juvenile and adult sea urchins. In a sheephead removal experiment in southern California, in an area devoid of lobsters, sheephead regulated red sea urchin populations and drove sea urchins into cryptic microhabitats (Cowan 1983). At that time it was estimated that sheephead consume more than 8000 sea urchins per hectare per year (Cowan 1983). Making modern estimates of predation rates is difficult since sheephead abundances do not differ significantly inside and outside the reserves in the Channel Islands (Behrens and Lafferty 2004). In the last 3500–4500 years, Indian middens show that the size of sheephead bones decrease with increasing human fishing pressure (Salls 1995). Above the sheephead bone layer in the middens is a lens of purple sea urchin remains (Salls 1995). This finding led to the hypothesis that prehistoric overfishing of sheephead caused a decline in this important predator, which in turn led to extreme sea urchin population increases (Erlandson et al. 1996).

Red and purple sea urchins are also vulnerable to predation by the white sea urchin *Lytechinus anamesus* (=pictus) in southern California. Groups of the small white sea urchin completely consume large red sea urchins. White sea urchins attack both red and purple sea urchins in the field and laboratory, although they prefer to eat kelp (Coyer et al. 1987). Despite this, the abundance of the predatory white sea urchin declined from the mid 1980s in a mixed-species sea urchin barrens at Anacapa Island, in southern California, suggesting other dynamics are operating to regulate sea urchin populations in these barrens (Carroll et al. 2000).

In the north, the sun star (*P. helianthoides*) consumes both red and purple sea urchins <90 mm in test diameter (Duggins 1983; Lafferty and Kushner 2000). Therefore, red sea urchins >90 mm reach a refuge in size from sun star predation while purple sea urchins do not (Duggins 1983).

8.3. Disease

Sea urchin diseases, such as the bald sea urchin disease, affect sea urchin populations around the world (see Chapters 11, 13, and 18). Diseased red and purple sea urchins have been documented in both central and southern California (Richards and Kushner 1994). A red sea urchin mortality event was seen in Santa Cruz in the 1970s, killing an estimated 14 000 sea urchins (Pearse et al. 1977). In the Channel Islands diseased sea urchins exhibit spine loss and dark patches of necrotic tissue (Richards and Kushner 1994). These symptoms are consistent with a *Vibrio* bacterial disease (Gilles and Pearse 1986). However, the causative agent of this disease in the Channel Islands has not been identified. While this disease is fatal, sea urchins do appear to be able to recover as evidenced by the presence of sea urchins with regenerated spines.

The Kelp Forest Monitoring Program has been recording the health and density of sea urchins in the Channel Islands since 1982. In 1992, during a strong El Niño, disease was first recorded in the sea urchin populations (Richards and Kushner 1994) suggesting this disease may have come from elsewhere (Lafferty 2004). Using data from 1992 to 2001, the prevalence of the disease has been shown to have a significant negative effect on population growth in purple sea urchins as measured by r, the logarithm of the relative change in density N, between years t and $t+1$, where (Lafferty 2004)

$$r = \ln\left(\frac{N_{t+1}}{N_t}\right)$$

Disease at these Channel Island sites over the decade did not result in a widespread mass mortality event as has been observed elsewhere for other species (Scheibling 1986; Lessios 1988).

8.4. Physical Factors and Ocean Warming

Wave forces and water motion can have a large direct and indirect affect on sea urchin populations. Red sea urchins may suffer greater mortality due to wave-induced movement of boulders in the subtidal zone (Schroeter 1978). Purple sea urchins appear to be more resistant to disturbance, perhaps due to their smaller size and shorter spines. Purple sea urchins are also more tolerant of the wave forces, heat and desiccation found in intertidal tidepools than red sea urchins. In sheltered subtidal habitats, purple sea urchins can also withstand low oxygen and high-silt concentrations while red sea urchins cannot

(Schroeter 1978). Little work has been done to examine the effects of these physical factors on earlier life history stages although, presumably siltation could affect respiration and survival of newly settled sea urchins.

Wave forces ripping up intact kelp beds can cause food shortages for sea urchins. Sea urchin grazing can further exacerbate this damage leading to total structural failure if kelps are weakened anywhere along the stipe or within the holdfast (Tegner et al. 1995). Even small amounts of herbivore damage when combined with otherwise innocuous current and wave forces can be enough to cause catastrophic failure and total kelp loss (Duggins et al. 2001).

Ocean warming can have multiple indirect affects as warmer water conditions reduce the nitrogen content of the seawater thereby negatively affecting kelp quantity and quality. The appearance of the bald sea urchin disease during the 1992 El Niño and increased incidence of the disease in the warmer easternmost Channel Islands, such as Anacapa and Santa Cruz Islands (Richards and Kushner 1994), suggest a direct link between ocean temperature and this sea urchin disease. In a warm reserve site at Anacapa Island, where no fishing was permitted and lobster abundances were high, sea urchins were at low densities and were not impacted by the disease (Lafferty 2004). This suggests that while sea water temperatures may play a role in the outbreak of the disease, population density, as regulated by predators, also influences the onset of sea urchin epidemics (Lafferty 2004). More work is needed on the interactions between ocean warming and other factors regulating populations, such as, food abundance, disease outbreaks, and predator abundance.

Ocean warming trends have been correlated with sea urchin-predator abundances. The Pacific Decadal Oscillation, defined as the leading principal component of North Pacific monthly sea surface temperature variability (north of 20° N for the 1900–1993 period), indicates ocean conditions have been anomalously warmer from 1977 to 1999 as compared with long-term (100 year) averages (Mantua et al. 1997). Since 1976–1977 the commercial fishery for spiny lobsters has increased threefold, from less than 100 t per year to 375 t in 2004 (Sweetnam 2005). While these data are suggestive, implying warmer ocean conditions favor lobster populations, it should also be noted that in 1977 escape ports were first required in lobster traps to decrease retention of undersize lobsters. The usual caveats associated with landings data also apply here. In particular, there is no information on fishing effort. Fishery-independent data also suggest warm water events enhance the recruitment success of spiny lobsters and sheephead in southern California (Cowan 1985).

Ocean warming also influences the intensity of human fishing for sea urchins, in that there is a negative-feedback loop maintaining sea urchin barrens. Warm ocean conditions can lead to poorly fed sea urchins with decreased sea urchin gonad quality, which would result in decreased human fishing effort. Less sea urchin fishing will maintain high sea urchin abundances, thereby potentially facilitating barrens formation and overgrazing. Patterns of kelp harvest are correlated with landings patterns in the sea urchin fishery in South Korea (Andrew et al. 2002). It has been suggested that the strong El Niño event of 1982–1983 was responsible for decreased red sea urchin landings in southern California (Kato and Schroeter 1985) and Baja California, Mexico (Hammann et al. 1995). If the frequency and intensity of El Niño warming events (defined by NOAA as positive Oceanic

Niño Index greater or equal to +0.5 °C for at least five consecutive months) increases, it will be important to determine if this feedback loop is coincident with an increase in the temporal and spatial extent of sea urchin barrens and kelp deforestation.

9. GENETICS

The systematics and phylogenies of sea urchins in the genus *Strongylocentrotus* remain poorly understood despite abundant ecological, fishery and developmental studies on this group. Selection of the purple sea urchin as a target species for genome analysis (Pennisi 2002) may increase attention to this. Molecular phylogenies with mitochondrial DNA sequences reveal that the genus is divided into two distinct clades, with red sea urchins in one and purple sea urchins in the other, that diverged 13–19 million years ago (Lee 2003). However, recent mitochondrial DNA sequencing data suggest that a major revision of the genus *Strongylocentrotus* may be in order (Biermann 1998). The phylogeny resulting from this work supports the inclusion of three additional species into the group *Strongylocentrotus* which were previously thought to be closely related (Biermann et al. 2003).

Red and purple sea urchins are model organisms to study genetic variation in marine invertebrate populations with potentially broadly dispersing planktonic larvae and a benthic adult stage. The genetic structure of purple sea urchins along the coast of California and Baja California was examined using allozyme and mitochondrial DNA. Neighboring purple sea urchins had as diverse allozyme and DNA structure as sea urchins from geographically distant sites (Edmands et al. 1996). Similarly, neighboring red sea urchins in California had as much or more diverse allozyme structure as sea urchins from distant populations (Moberg and Burton 2000). Northern California populations were not distinguishable from southern California populations based on the six polymorphic loci examined (Moberg and Burton 2000) despite the large geographic distance and potential barriers to dispersal. Surprisingly, juveniles (<30 mm diameter) differed from adults collected from the same location, and genetic variation among the juveniles, in both space and time, was greater than would be predicted with a well-mixed larval pool (Moberg and Burton 2000). This suggests that although there is apparently sufficient gene flow to prevent genetic divergence of populations along the California coast, the larval pool is likely not always homogenous across the geographic range despite the long larval period which Strathmann (1978) indicated is 4–20 weeks.

Debenham et al. (2000) examined DNA sequence data from the binding gene in populations across the species' range from Alaska to Baja California. Their work and previous studies suggest the binding locus is an appropriate marker with sufficient polymorphism to detect genetic structure. The bindin marker revealed sea urchins at six locations had at least four alleles suggesting that, at least for this marker, they are highly polymorphic (Debenham et al. 2000). Multiple microsatellites ($N = 14$) were isolated for paternity studies and were highly polymorphic for red sea urchins from British Columbia (McCartney et al. 2004). Eleven polymorphic, di- and trinucleotide microsatellite loci for three sites in British Columbia show heterozygosities of the loci ranged from 0.39 to

0.85, showing that variability is high (Miller et al. 2004). More work needs to be done to compare allelic frequencies among populations.

Sea urchins have been used as a direct test of the "sweepstakes" hypothesis for marine invertebrates, which states that for free-spawning organisms with high fecundity and high larval mortality, it is possible that only a few adults reproduce successfully each year (Hedgecock 1994). Small effective population sizes relative to stock sizes are observed in many marine invertebrate populations including sea urchins (Hedgecock 1994). In this scenario, where chance plays a large role and few adults reproduce, juvenile sea urchins would have less genetic differentiation than the genetically mixed pool of adults. Using mitochondrial DNA, Flowers et al. (2002) determined that there was no evidence for reduced genetic variation in newly settled (1–14 days post-settlement) purple sea urchins and only slight evidence to suggest cohorts were genetically distinct. This leaves the importance of the sweepstakes hypothesis for sea urchins in question.

10. FISHERIES

Red sea urchins are the primary target of sea urchin fisheries on the west coast of North America while purple sea urchins make up less than 1% of the landings (for reviews see Kalvass and Hendrix 1997; Keesing and Hall 1998; Andrew et al. 2002). Japan is the primary market for sea urchin "roe"; the gonads of both males and females. Sea urchin roe is a specialty food eaten year round but is especially popular during New Year celebrations; a tradition which started as early as the 1600s (Andrew et al. 2002). In Japan, domestic production of wild sea urchins peaked in 1969 leading to a demand for imported sea urchin products both frozen and fresh. Sea urchin imports have increased over the last 30 years with imports of fresh products having risen from 7000 metric tons (t) in 1999 to over 13 000 t in 2004, while imports of roe in brine have also increased. There is no suggestion of a decline in future demand (NOAAb).

10.1. West Coast Fisheries

The fishery in southeast Alaska is in open coastal waters with a small portion of the catch from the inland straights. Commercial exploitation began in the mid 1980s but the fishery did not have landings greater than 500 t until the mid 1990s. In 1994, the Alaskan fishery expanded into the Ketchikan area, away from sea otters. The fishery peaked in the 1997–1998 season at 2235 t but since then has declined nearly threefold to 817 t for the 2004–2005 season (M Pritchard, pers. comm.).

In British Columbia, Canada the fishery remained small in the 1980s and then peaked at 13 000 t in 1992. Since 1994, the fishery has remained stable with approximately 5000–6000 t per year taken. The TAC for the 2004–2005 season was 4884 t of which 4359 t were landed worth an estimated CN$7.8 million (J Rogers, pers. comm.). Catch per unit effort has remained fairly stable for the decade at or around 0.575 t per diver hour (J Rogers, pers. comm.).

In Washington State, sea urchin fishing increased in the 1980s to peak at 3658 t in 1988 and has since declined to less than 10% of this amount (Bradbury 2000). By

2000, red sea urchins made up approximately 60% of the catch with green sea urchins *S. droebachienensis* making up the remainder (Carter and VanBlaricom 2002). Fishery-independent surveys of red sea urchin size indicated that only 5% of the population was lower then the minimum legal size (102 mm). This suggests that the population is an accumulation of large, old individuals with poor recruitment success (Carter and Van Blaricom 2002). Fishery-independent surveys also suggest that sea urchin density has declined since the 1980s although catch per unit effort estimates do not reflect a decline because fishers are able to exploit new subpopulations (Pfister and Bradbury 1996). Density estimates using underwater video and dive surveys were used to establish biomass estimates and set total allowable catches for the state, however these were terminated following budget problems in 1997 (Bradbury 2000).

In Oregon, the red sea urchin fishery followed a similar course as that in Washington State, with a rapid rise in landings in the late 1980s to a peak of 4222 t in 1990, and then a dramatic collapse to less than 5% of the peak by the end of the 1990s. The mean size of red sea urchins has declined as has the proportion of large sea urchins in the catch (Richmond et al. 1997). Purple sea urchins make up a small proportion (<10%) of the fishery (Richmond et al. 1997)

California's red sea urchin fishery has dominated the sea urchin producing regions along the west coast of North America. Despite California's dominance it has not been immune to declines in sea urchin landings. The fishery began in 1971 as an experimental fishery in southern California (Kalvass 2000). The fishery quickly rose to 4540 t by 1980 and then experienced a decline precipitated by a strong El Niño event which affected landings for 3 years. In 1985, the fishery expanded into northern California where landings increased to 13620 t in just 3 years. However, northern California has not seen landings higher than 2724 t for the last decade (Kalvass 2000). The majority of landings in the north from 1988 to 1994 came from a small (65 km) section of the rocky coast (Kalvass and Hendrix 1997). In 2004, an estimated 5.36 t were landed statewide worth $7.1 million, of which 4.75 t were landed in southern California (Sweetnam 2005). Even in the southern portion of the state where landings had been fairly stable, landings are now below the long-term (1975–2004) yearly average (7.53 t) (Sweetnam 2005). Sea urchins are not landed in central California from Point Conception to San Francisco Bay where predation by sea otters precludes a fishery.

In the southern portion of the red sea urchins' range, the fishery is located along the Pacific coast in the northern third of the Mexican Baja peninsula. The red sea urchin fishery began in the early 1970s and landings rose to peak in 1986 at 8493 t. Landings fell drastically the next year to only 1590 t due to a strong El Niño event which affected roe quality and led to decreased effort (Andrew et al. 2002). Landings in 1999–2000 have been just under 2200 t. Purple sea urchins make up a small portion of the total catch in Baja California.

10.2. Fishery Experiments

Experimental fishing studies inside a marine reserve examined how red sea urchins recover from various levels of fishing treatments such as existing regulations (lower size limits only), proposed selective fishing (upper and lower size limits), and fishing reserves (no fishing). The population of highly mobile sea urchins at intermediate depths recovered quickly (1 month) from fishing by the migration of adults from neighboring areas. Sedentary sea urchins from shallow sites recovered slowly or not at all depending on experimental fishing treatment (Rogers-Bennett et al. 1998). Existing regulations led to a decline over 10 years in sea urchin density and poor recruitment of juveniles, while fishing treatments that protected adults led to enhanced recruitment (Rogers-Bennett et al. 1998). Similarly in Washington State, recovery from fishing was via migration of neighboring adults as opposed to juvenile recruitment (Carter and Van Blaricom 2002). Recovery rates observed in small-scale fishing experiments should be considered as maximum rates since commercial fisheries operate at larger scales leaving few neighboring adults available for recolonization (Rogers-Bennett et al. 1998; Carter and VanBlaricom 2002).

10.3. Fishery Enhancement

Fishery enhancement experiments, including stocking juvenile red sea urchins, pioneered in Japan (Omi 1987) have been examined on the west coast of the United States (Tegner 1989). Juvenile red sea urchins cultured in aquaculture facilities were released in California. In one study, 5000 juveniles were tagged with calcein and stocked into two northern and two southern sites. While recovery rates after 1 year were spatially variable (1–22%) with no discernable latitudinal patterns, juvenile size did influence recovery. The largest juveniles (12–18 mm) were recovered more than the two smaller size classes of 3–7 mm and 8–12 mm at all sites (Dixon et al. 1992). Clearly, there are tradeoffs between the cost of producing larger juveniles prior to stocking and higher rates of recovery. In a second study in northern California, red sea urchins reared in the laboratory for 1 year to a mean size of 18 mm were tagged with calcein and stocked into shallow (5 m) and intermediate depth (15 m) habitats inside a sea urchin marine reserve. Twice as many juvenile sea urchins (21%) were recovered from the shallow than the intermediate depth habitat (11%) suggesting that spatial patterns related to depth may be important in selecting enhancement sites (Rogers-Bennett 2001).

10.4. Gonad Enhancement

Another fishery enhancement method is that of gonad or roe enhancement. This involves transplantating of wild juveniles or adults from habitats with poor food quantity or quality to optimal habitats in the ocean at sea ranches or land-based aquaculture facilities (Tegner 1989). Small red sea urchins (33 000) in southern California were transplanted from a barren area to a kelp forest with low densities of large adult red sea urchins (Dixon et al. 1999) Growth and survival were good at the transplant site 1 year later with an estimated 58% (±30%) of the sea urchins surviving. The source site continued to have high recruitment even after the removal of >30 000 juveniles. However recruitment was sporadic and transplant outcomes may not be as good during periods of poor natural recruitment (Dixon et al. 1999). Despite the vagaries of natural recruitment, this method, in addition to the movement of underfed adult sea urchins with poor gonad quality to kelp-rich habitats, may be more promising than the labor-intensive methods required to culture larvae for stocking.

11. FISHERY MANAGEMENT

Stock assessment estimates have been made for a number of red sea urchin stocks using fishery-independent data. Surplus production models are used because they are simplistic with few data requirements, but these models have a number of assumptions that must be examined closely. Densities are determined for a given area from fishery-independent dive surveys along transects or video transects. Densities of sea urchins are generally patchy and include areas with zero densities. Confounding estimates of biomass for red sea urchins are a subset of the population that may be counted in the surveys, but are poorly fed and do not yield marketable gonads. How to treat these poor quality sea urchins in estimates of stock biomass remains an open question.

Another method for estimating red sea urchin biomass is based on fishery-dependent data using a Leslie depletion model. A series of catch-per-unit-effort estimates for each year is plotted and then fit with a regression line. Using this model for stocks in northern California, the prefishing (prior to 1988) biomass was estimated at 76 290 t of which 50 800 t was removed by the fishery from 1988 to 1994 with 13 846 t taken in 1988 alone (Kalvass and Hendrix 1997). For 1988 this is equivalent to an instantaneous fishing mortality rate of $F = 0.2$. This fishing rate was not sustainable and resulted in a boom and bust fishery in the north (Kalvass and Hendrix 1997) and landings have not rebounded in the last decade (Sweetnam 2005). Another method for estimating fishing mortality (F) is based on size distributions of individuals greater than 90 mm (to reduce the influence of new recruits) from fished sites and vital rate parameters from unfished sites. Estimates of F ranged from 1.87 to 0.11 across 11 sites in northern California (Morgan et al. 2000a). Estimates of population parameters can aid in developing fishery management strategies that control fishing intensity and sustain fishery yields. Estimates of lifetime egg production (R_0) based on age or size can help guide fishery managers to set target and limit reference points in an effort to fish sustainably despite inherent uncertainty in the population dynamics (Botsford et al. 2004).

Fisheries management traditionally used size limits to allow certain size classes to spawn before they enter the fishery. All fisheries, expect Alaska, have lower size limits. The processors and market prefer midsize red sea urchins over the largest individuals. Since the largest red sea urchins potentially contribute the most toward reproduction this suggests that a maximum legal size could protect spawners while allowing fishing for high quality product (Rogers-Bennett et al. 1995). This, in combination with the density-dependent mechanisms of adults sheltering juveniles (Tegner and Dayton 1977) and enhanced spawning efficiency (Levitan et al. 1992) supports implementation of a maximum legal size to protect the largest individuals which may have enhanced reproductive output (Birkeland and Dayton 2005). Population growth rate in a size-structured population model was most influenced by changes in the growth and mortality of the largest size classes suggesting again that large sea urchins are important to protect (Ebert 1998). Incorporating maximum legal sizes into a population model of red sea urchins in Washington State resulted in more stable fishery yields and red sea urchin populations (Lai and Bradbury 1998).

Management of sea urchin fisheries on the west coast of North America has focused on single species with few attempts to incorporate multi-species management strategies despite the variety of potential benefits (Pikitch et al. 2004). There is no plan to manage red sea urchins as part of the nearshore communities in which they live. Multi-species resource assessments are only now being initiated in some regions. Despite our knowledge of fishery interactions between sea urchins and abalone (Rogers-Bennett and Pearse 2001) and sea urchins and lobster (Lafferty 2004) we still do not use this information in management. Furthermore, species interactions involving red sea urchins may differ significantly throughout the species' range. Similarly, oceanographic influences on productivity are not measured or incorporated into sea urchin fishery management. In the end, we do not fully understand the effects of large-scale removals of sea urchin by the fishery and what ramifications they may have on the broader nearshore subtidal community (Tegner and Dayton 2000). To assist in understanding the drivers of sea urchin population dynamics we need to look closely at the complex role sea urchins play as members of marine communities throughout the range.

12. CONSERVATION

Conservation of sea urchin populations, at first, may sound unnecessary to some since in the 1960s and 1970s kelp bed management in southern California included the eradication of sea urchins using quick-lime (calcium oxide), hammers, and dredges (Wilson and North 1983). Yet today, we see that sea urchins also have positive interspecific interactions within kelp communities. While sea urchins sometimes viewed as pests in kelp forests when they are starved, more frequently (90%) they are an integral, vital component of nearshore ecosystems (Foster and Schiel 1988). As we strive to understand more about destructive grazing by sea urchins, we see this is a complex phenomenon involving multispecies interactions, recruitment dynamics, and large-scale oceanographic processes. Ocean climate change operating on several temporal scales further compounds the complexities of this herbivore-algal coevolutionary relationship (Steneck et al. 2002).

Sea urchins are also economically important as valuable targets for expanding fisheries worldwide (Andrew et al. 2002). If we consider that sea urchins, like other valuable fisheries, can be overfished then we will be more inclined to examine sustainable fishing strategies. In fact, there are a number of life history features which make red sea urchins particularly susceptible to fishing and these include (1) extreme long life (Ebert and Southon 2003), (2) high densities in close proximity are required for fertilization success (Levitan et al. 1992), (3) successful recruitment of juveniles is temporally and spatially patchy (Pearse and Hines 1987), (4) legal size adults shelter juveniles (Tegner and Dayton 1977), and (5) populations are structured as metapopulations (Rogers-Bennett et al. 1995). These unique life history features highlight the need for vigilance in maintaining sustainable sea urchin fisheries.

12.1. Metapopulation Dynamics

Sea urchins exist in spatially discrete sets of populations. These combined population patches make up a metapopulation (Levins 1969). Patches may be unique with respect to growth, reproduction, and survival rates, while some are sources of reproduction others are sinks (Pulliam 1988). These spatially segregated subpopulations are well known in both the ecology literature and by fishers exploiting different reefs. Fishing "hot spots" can be temporally stable features lasting decades or may blink on and off depending on environmental factors such as oceanographic circulation patterns (Ebling and Hixon 1991). Dispersal between patches, the source of juveniles, and the fishing intensity per patch are all important to our understanding of sea urchin metapopulation dynamics, however they are difficult to quantify. Shallow habitats in northern California (Rogers-Bennett et al. 1995) and southeast Alaska (Carney 1991) have the qualities of source habitats with large fecund animals, an abundance of drift food resources, and high densities of juveniles. These shallow areas are also more susceptible to fishing pressure and respond differently than intermediate and deep habitats (Rogers-Bennett et al. 1998). Spatially explicit metapopulation models with fishing reserves and high exchange rates maximized economic gains and yield when sink areas were fished while source areas were protected in reserves, however, this did not always maximize spawning stock biomass (Tuck and Possingham 2000).

More work is needed on small-scale (reef) variability in life history information relevant to population dynamics. One such study using fishery-independent data has examined spatial variability in productivity at three unfished sites in northern California. Growth was similar at the three sites but estimates of natural mortality varied with some estimates difficult to interpret due to zero and negative values (Morgan et al. 2000a). Fishery-dependent data such as catch-per-unit-effort estimates will be a poor predictor of population trends in metapopulations since fishers can move to exploit new subpopulations (Keesing and Baker 1998). Spatial heterogeneity of fished populations can have important consequences for the dynamics of the fishery and this heterogeneity has been thought about and incorporated into fishery models for some time (Hilborn and Walters 1987).

12.2. Sea Urchins as Ecosystem Engineers

Sea urchin grazing directly and indirectly modulates the availability of algal resources for other species within the community thereby meeting the definition of an ecosystem engineer (Jones et al. 1994; Rogers-Bennett and Pearse 2001). Along with their ability to structure surrounding algal communities through direct herbivory (see Section 2), sea urchins can create cavities within kelp holdfasts and make cryptic microhabitats. In the process they significantly weaken kelps increasing their chances of loss during storms (Tegner et al. 1995). Sea urchins also excrete ammonia which can be important at microscales when nitrogen is limited. In temperate rocky reefs, red and purple sea urchins are effective rock borers scouring out rock pits. Red sea urchins residing in rock pits have restricted movement (Rogers-Bennett 1994) as do purple sea urchins (B Grupe, pers. comm.). In addition, purple sea urchins inside pits are significantly smaller, have larger lantern/body weight indexes, and longer jaw lengths compared to urchins outside of pits, suggesting they are food limited in pits (B Grupe, pers. comm.). Comparatively little work has been done on rock pit formation and bioerosion by purple sea urchins in temperate rocky reefs as compared to sea urchin mediated bioerosion in coral reef systems (Carreiro-Silva and McClanahan 2001; Griffin et al. 2003). More work is needed that examines the role purple sea urchins play in creating large beds of scoured rock pits in the intertidal. Furthermore, more work is needed on the effects of fishing sea urchin ecosystem engineers and the consequences of this on biodiversity and ecosystem functions (Rogers-Bennett and Pearse 2001; Coleman and Williams 2002).

Sea urchins and their extensive spines are themselves structuring components of subtidal communities that modulate the availability of sheltered microhabitat. The spine canopy of red sea urchins and purple sea urchins is structurally complex making an ideal microhabitat for a wide variety of small invertebrates and fishes. Juvenile conspecifics are known to reside under the spine canopy (see Section 7). Likewise, juvenile red and flat abalones are frequently found under the spine canopy of red sea urchins in areas where sea urchins are protected from fishing (Rogers-Bennett and Pearse 2001). Juveniles may be protected from predation and wave action under the spine canopy of sea urchins (Tegner and Dayton 1977), although experiments reveal they do not have enhanced access to food resources (Nishizaki and Ackerman, 2004). While the flow of algal resources as mediated by sea urchins has been examined, less is known about the potential benefits that associated organisms may receive under the spine canopy. Adult sea urchins, by sheltering juvenile sea urchins and abalones under their spines, may be acting as "essential fish habitat" as defined in the Magnuson-Stevens Fishery Conservation and Management Act (Rogers-Bennett and Pearse 2001). This sheltering relationship is not restricted to red sea urchins and has been found in a variety of sea urchin species elsewhere in the world making a strong case for ecosystem management (Mayfield and Branch 2000; Hartney and Grorud 2002).

12.3. Ecosystem Management

Federal agencies in both the United States and Canada have encouraged development of ecosystem-based management (Fluharty 2000; Pikitch et al. 2004), but it has not be implemented for red sea urchins on the West Coast of North America. While there is good biological rationale for such an approach, implementation is confounded by competing fishery interests within ecosystems, determinations of the spatial bounds for ecosystems, and even definitions of goals such as "ecosystem health" or "ecosystem integrity" (Simberloff 1998). Despite these challenges, nearshore kelp beds on rocky substrates are ideal candidates for pilot programs exploring ecosystem management for several reasons (1) they support a diverse assemblage of relatively benthic fished species including invertebrates and fishes, (2) the habitat type and community is well defined by the substrate, (3) they are close to shore facilitating assessment, and (4) wise management will be more important as human populations along the coast expand increasing fishing pressures. Sea urchins within these systems are also prime candidates for ecosystem management as they modulate the flow of resources through the community, interact with other valuable fisheries, and are key species that drive community structure.

Marine protected areas (MPAs) can be used as a tool for ecosystem based management. MPAs have a wide array of uses and designations ranging from no access to multiple uses. No-take MPAs that prohibit fishing have been recommended as one tool for managing sea urchin fisheries with a vast array of goals including natural recruitment areas, buffers in case of population catastrophes, reducing volatility in fishery catch, enhancement of fishery yields, and ecosystem restoration (Botsford et al. 1997; Murray et al. 1999). Algorithms that optimize biodiversity, ecological processes, and socioeconomic factors have been used as tools for the development of plans for MPA networks (Sala et al. 2002; Airame et al. 2004). No-take MPAs might have a wide range of scenarios for sea urchins including population explosions that denude algal forests, robust populations that enhance multiple species, small cryptic populations within the sea otters range or even areas devoid of sea urchins due to disease. This range of outcomes is due in part to trophic cascades (Sala et al. 1998) reflected as conflicts between shellfish fisheries and protected species such as sea otters (Gerber et al. 1999) which are challenges to implementing ecosystem-based management. When restored upper trophic levels have dramatic effects on target fisheries, as is the case for sea otters and sea urchins, it may be necessary to have two types of MPAs – one focused on ecosystem restoration and the other on sustaining shellfish fisheries (Fanshawe et al. 2003). In Japan and South Korea ecosystem management includes both the fishing of target species as well as active habitat enhancement designed to increase productivity of sea urchin fisheries and other fisheries (Andrew et al. 2002).

The productivity of nearshore rocky ecosystems that support sea urchins, however, is very different than it was even 100 years ago as animal populations have declined primarily due to anthropogenic factors (Dayton et al. 1998). Baselines have shifted (Pauly 1995) and expectations of what is abundant or even normal have been greatly reduced. Sea otters are absent from some regions, spiny lobsters, sheephead and other sea urchin predators have dramatically altered abundances. Fishing predators can radically change sea urchin size–frequency distributions, converting them from bimodal to unimodel, and affecting their population dynamics (Tegner and Levin 1983; Behrens and Lafferty 2004). Novel fisheries like the live-fish fishery have emerged since 1993 and targets primarily rockfish 200 t in 10 years (Sweetnam 2005). Emerging nearshore fisheries such as the commercial fishery in southern California for Kellet's Whelk *Kelletia kelettia* have also grown from 8 t in 1997 to 32 t in 6 years (CDFG data). Despite the onset of these new fisheries we still do not have a clear understanding of what effect removing metric tons of a species will have on species interactions and ecosystem functioning. Along with the realization that fishing changes community interactions we have grown more aware of and are starting to quantify the effects of ocean climate on productivity (Tegner and Dayton 1991; Mantua et al. 1997). Ecosystem management will only be feasible with more knowledge of the biological and physical processes involved in population regulation.

ACKNOWLEDGMENTS

This chapter is dedicated to the memory of my friend and colleague Dr Mia J Tegner whose work has played a central role in helping us understand sea urchin population dynamics. I wish to thank numerous scientists in the echinoderm research community for supplying me with references and especially the status of work in progress. I thank the many sea urchin fishery managers along the west coast of North America for providing me with up-to-date sea urchin fishery landings data and S Shoffler for editing References the manuscript. Lastly, I thank Dr John M Lawrence for references, encouragement, and the invitation to write this chapter. Thank you all.

REFERENCES

Airame S, Dugan JE, Lafferty KD, Leslie H, McArdle DA, Warner RR (2004) Applying ecological criteria to marine reserve design: a case study in California Channel Islands. Ecol Appl 13: S170–S184

Allee WC (1931) Animal Aggregations: A Study in General Sociology. Univ. Chicago Press, Chicago, IL

Andrew NL, Agatsuma Y, Ballesteros E, Bazhin AG, Creaser EP, Barnes DKA, Botsford LW, Bradbury A, Campbell A, Dixon JD, Einarsson S, Gerring P, Hebert K, Hunter M, Hur SB, Johnson CR, Junio-Menez MA, Kalvass P, Miller RJ, Moreno CA, Palleiro JS, Rivas D, Robinson SML, Schroeter SC, Steneck RS, Vadas RI, Woodby DA, Xiaoqi Z (2002) Status and management of world sea urchin fisheries. Oceanogr Mar Biol Annu Rev 40: 343–425

Balser E J (1998) Cloning by ophiuroid echinoderm larvae. Biol Bull 194: 187–193

Behrens MD, Lafferty KD (2004) Effects of marine reserves and urchin disease on southern Californian rocky reef communities. Mar Ecol Prog Ser 279: 129–139

Bertness MD, Gaines SD, Hay ME (2001) Marine Community Ecology. Sinauer Assoc. Inc., Sunderland

Biermann CH (1998) The molecular evolution of sperm bindin in six species of sea urchins (Echinoidea: Strongylocentrotidae). Mol Biol Evol 15: 1761–1771

Biermann CH, Kessing BD, Palumbi SR (2003) Phylogeny and development of marine model species: strongylocentrotid sea urchins. Evol Develop 5: 360–371

Birkeland C, Dayton PK (2005) The importance in fishery management of leaving the big ones. Trends Ecol Evol 20: 356–358

Boidron-Metairon IF (1988) Morphological plasticity in laboratory-reared Echinoplutei of *Dendraster excentricus* (Eschscholtz) and *Lytechinus variegatus* (Lamarck) in response to food conditions. J Exp Mar Biol Ecol 119: 31–41

Bosch I (1988) Reproduction by budding in natural populations of bipinaria larvae from the sea star genus Luidia. In: Burke RD, Mladinov PV, Lambert P, Parsley RL (eds) Echinoderm Biology. AA Balkema, Rotterdam. pp 728

Bosch I, Rivkin RB, Alexander SP (1989) Asexual reproduction by oceanic planktotrophic echinoderm larvae. Nature 337: 169–170

Botsford LW, Campbell A, Miller R (2004) Biological reference points in the management of North American sea urchin fisheries. Can J Fish Aquat Sci 61: 1325–1337

Botsford LW, Castilla JC, Peterson CH (1997) The management of fisheries and marine ecosystems. Science 277: 509–515

Botsford LW, Morgan LE, Lockwood DR, Wilen JE (1999) Marine reserves and management of the northern California red sea urchin fishery. CalCOFI Rep 40: 87–93

Bradbury A (2000) Stock assessment and management of red sea urchin (*Strongylocentrotus franciscanus*) in Washington. J Shellfish Res 19: 618–619

Breen PA, Carolsfeld W, Yamanaka KL (1985) Social behavior of juvenile red sea urchins, *Strongylocentrotus franciscanus* (Agassiz). J Exp Mar Biol Ecol 92: 45–61

Breen PA, Mann KH (1976) Changing lobster abundance and the destruction of kelp beds by sea urchins. Mar Biol 34: 137–142

Bureau D (1996) Relationship between feeding, reproductive condition, jaw size and density in the red sea urchin, *Strongylocentrotus franciscanus*. MS Thesis, Simon Fraser Univ., Burnaby, BC, Canada

California Department of Fish and Game (2005) Abalone Recovery and Management Plan

Cameron RA, Schroeter SC (1980) Sea urchin recruitment: effect of substrate selection on juvenile distribution. Mar Ecol Prog Ser 2: 243–247

Carney D (1991) A comparison of densities, size distribution, gonad and total gut indices, and the relative movement of red sea urchins, *Strongylocentrotus franciscanus*, in two depth regimes. MS Thesis, Univ. California, Santa Cruz, CA, USA

Carpenter SR, Kitchell JF (1988) Consumer control of lake productivity. BioScience 38: 764–769

Carreiro-Silva M, McClanahan TR (2001) Echinoid bioerosion and herbivory on Kenyan coral reefs: the role of protection from fishing. J Exp Mar Biol Ecol 262: 133–153

Carroll JC, Engle JM, Coyer JA, Ambrose RF (2000) Long-term changes and species interactions in a sea urchin-dominated community at Anacapa Island, California. Proceeding of the 5th California Islands Symposium. pp 370–378

Carter SK, Van Blaricom GR (2002) Effects of experimental harvest on red sea urchins (*Strongylocentrotus franciscanus*) in northern Washington. Fish Bull 100: 662–673

Christen R, Schackmann RW, Shapiro BM (1986) Ionic regulation of sea-urchin sperm motility metabolism and fertilizing capacity. J Physiol 379: 347–366

Coleman FC, Williams SL (2002) Overexploiting marine ecosystem engineers: potential consequences for biodiversity. Trends Ecol. Evol. 17: 40–44

Connell JH (1985) The consequences of variation in initial settlement vs. post-settlement mortality in rocky intertidal communities. J Exp Mar Biol Ecol 93: 11–45

Cowan RK (1983) The effect of sheephead (*Semicossuphus pulcher*) predation on red sea urchin (*Strongylocentrotus franciscanus*) populations: an experimental analysis. Oecologia 58: 249–255

Cowan RK (1985) Large scale pattern of recruitment by the labrid *Semicossuphus pulcher*: causes and implications. J Mar Res 43: 719–742

Coyer JA, Engle JM, Ambrose RF, Nelson BV (1987) Utilization of purple and red sea urchins (*Strongylocentrotus purpuratus* Stimpson and *S. franciscanus* Agassiz) as food by the white sea urchin (*Lytechinus anamesus* Clark) in the field and laboratory. J Exp Mar Biol Ecol 105: 21–38

Dayton PK (1985) The structure and regulation of some South American kelp forest communities. Ecol Monogr 55: 447–468

Dayton PK, Tegner MJ, Edwards PB, Riser KL (1998) Sliding baselines, ghosts, and reduced expectations in kelp forest communities. Ecol Appl 8: 309–322

Dean TA, Schroeter SC, Dixon JD (1984) Effects of grazing by two species of sea urchins (*Strongylocentrotus franciscanus* and *Lytechinus anamesus*) on recruitment and survival of two species of kelp (*Macrocystis pyrifera* and *Pterygophora californica*). Mar Biol 78: 301–313

Debenham P, Brzezinski M, Foltz K, Gaines S (2000) Genetic structure of populations of the red sea urchin *Strongylocentrotus franciscanus*. J Exp Mar Biol Ecol 253: 49–62

Dixon JD, Schroeter CS, Ebert TA (1992) Experimental outplant of juvenile red sea urchins. Sea urchins, abalone and kelp: their biology, enhancement and management. Sea grant Workshop. Bodega Bay, CA. 19–21 March, p 21

Dixon JD, Schroeter SC, Ebert TA (1999) Increasing the growth, survival, and gonadal quality of red sea urchins by transplanting to areas of high food availability, and use of large scale transplantation of juvenile red sea urchins as a management tool. Tech Rep California Dept Fish and Game, Sacramento

Duggins D, Eckman JE, Siddon CE, Klinger T (2001) Interactive roles of mesograzers and current flow in survival of kelp. Mar Ecol. Prog Ser 223: 143–155

Duggins DO (1983) Starfish predation and the creation of mosaic patterns in a kelp-dominated community. Ecology 64: 1610–1619

Duggins DO (1981) Interspecific facilitation in a guild of benthic marine herbivores. Oecologia 48: 157–163

Eaves AA, Palmer A (2003) Widespread cloning in echinoderm larvae. Nature 425: 146

Ebert TA (1977) An experimental analysis of sea urchin dynamics and community interactions on a rocky jetty. J Exp Mar Biol Ecol 27: 1–22

Ebert TA (1980) Relative growth of sea urchin jaws: an example of plastic resource allocation. Bull Mar Sci 30: 467–474

Ebert TA (1987) Estimating growth and mortality parameters by non-linear regression using average size in catches. ICLARM Conf Proceedings 13: 35–44

Ebert TA (1998) An analysis of the importance of Allee effects in management of the red sea urchin *Strongylocentrotus franciscanus*. In: Mooi R, Telford M (eds), Echinoderms: San Francisco. AA Balkema, Rotterdam, pp 619–627

Ebert TA (1999) Plant and Animal Populations: Methods in Demography. San Diego: Academic Press

Ebert TA (2001) Growth and survival of post-settlement sea urchins. In: Lawrence J (ed.) Edible Sea Urchins. Elsevier, Amsterdam, pp 79–102

Ebert TA, Dixon JD, Schroeter SC, Kalvass PE, Richmond NT, Bradbury WA, Woodby DA (1999) Growth and mortality of red sea urchins across a latitudinal gradient. Mar Ecol Prog Ser 190: 189–209

Ebert TA, Russell MP (1988) Latitudinal variation in size structure of the west coast purple sea urchin: a correlation with headlands. Limnol Oceaogr 33: 286–294

Ebert TA, Russell MP (1993) Growth and mortality of subtidal red sea urchins *Strongylocentrotus franciscanus* at San Nicolas Island, California, USA: problems with models. Mar Biol 117: 79–89

Ebert TA, Schroeter SC, Dixon JD, Kalvass P (1994) Settlement patterns of red and purple sea urchins (*Strongylocentrotus franciscanus* and *S. purpuratus*) in California, USA. Mar Ecol Prog Ser 111: 41–52

Ebert TA, Southon JR (2003) Red sea urchins (*Strongylocentrotus franciscanus*) can live over 100 years: confirmation using A-bomb ^{14}carbon. Fishery Bull 101: 915–922

Ebling AW, Hixon MA (1991) Tropical and temperate reef fishes: comparison of community structure. In: Sale PF (ed.) The Ecology of Fish on Coral Reefs. Academic Press, San Diego. pp 509–563

Ebling AW, Laur DR, Rowley RJ (1985) Severe storm disturbances and reversal of community structures in a southern California kelp forest. Mar Biol 84: 287–294

Edmands S, Moberg PE, Burton RS (1996) Allozyme and mitochondrial DNA evidence of population subdivision in the purple sea urchin *Strongylocentrotus purpuratus*. Mar Biol 126: 443–450

Erlandson JM, Kennett DJ, Ingram, BL, Gutherie DA, Morris DP, Tveskov MA, West GJ, Walker PL (1996) An archaeological and palaeontological chronology for Daisy Cove (CA-SMI-261), San Miguel Island, California. Radiocarbon 38: 355–373

Ernst SG (1997) A century of sea urchin development. Amer Zool 37: 250–259

Estes JA, Duggins DO (1995) Sea otters and kelp forests in Alaska: generality and variation in a community ecology paradigm. Ecol Monogr 65: 75–100

Estes JA, Duggins DO, Rathbun GM (1989) The ecology of extinctions in kelp forest communities. Conserv Biol 3: 252–264

Estes JA, Palmisano JF (1974) Sea otters: their role in structuring nearshore communities. Science 185: 1058–1060

Estes JA, Tinder MT, Williams TM, Doak DF (1998) Killer whale predation on sea otters linking oceanic and nearshore ecosystems. Science 282: 473–476

Fanshawe S, VanBlaricom GR, Shelly AA (2003) Restored top carnivores as detriments to the performance of Marine Protected Areas intended for fishery stability: a case study with abalones and sea otters. Conserv Biol 17: 273–283

Flowers JM, Schroeter SC, Burton RS (2002) The recruitment sweepstakes has many winners: genetic evidence from the sea urchin *Strongylocentrotus purpuratus*. Evolution 56: 1445–1453

Fluharty D (2000) Habitat protection, ecological issues, and implementation of the Sustainable Fisheries Act Ecol Appl 10: 325–337

Foster MS, Schiel DR (1988) Kelp communities and sea otters: keystone species or just another brick in the wall? In: VanBlaricom GR, Estes JA (eds) The Community Ecology of Sea Otters. Springer-Verlag, Berlin. pp 92–115

Gerber LR, Wooster WS, DeMaster DP, Van Blaricom GR (1999) Marine mammals: new objectives in U.S. fishery management. Rev Fish Sci 7: 23–38

Giese AC, Greenfield L, Huang H, Farmanfarmanaian, A, Boolootian RA, Lasker R (1958) Organic productivity in the reproductive cycle of the purple sea urchin. Biol Bull 116: 49–58

Gilles KW, Pearse JS (1986) Desease in sea urchins (*Strongylocentrotus purpuratus*): experimental infection and bacterial virulence. Dis Aquat Org 1: 105–114

Gonor JJ (1973) Reproductive cycles in Oregon populations of the echinoid *Strongylocentrotus purpuratus* (Stimpson) I: annual gonad growth and ovarian gametogenic cycles. J Exp Mar Biol Ecol 12: 45–64

Graham MH (2004) Effects of local deforestation on the diversity and structure of southern California giant kelp forest food webs. Ecosystems 7: 341–357

Griffin SP, García RP, Weil E (2003) Bioerosion in coral reef communities in southwest Puerto Rico by the sea urchin *Echinometra viridis*. Mar Biol 143: 79–84

Hammann MG, Palliero-Nayar JS, Sosa-Nishizaki O (1995) The effects of the 1992 El Niño on the fisheries of Baja California, Mexico. CalCOFI Rep 36: 127–133

Harley CDG, Rogers-Bennett L (2004) The potential synergistic effects of climate change and fishing pressure on exploited invertebrates on rocky intertidal shores. CalCOFI Rep 45: 98–110

Harrold C, Pearse JS (1987) The ecological role of echinoderms in kelp forests. Echinoderm Studies 2: 137–234

Harrold C, Reed DC (1985) Food availability, sea urchin (*Strongylocentrotus franciscanus*) grazing and kelp forest community structure. Ecology 66: 1160–1169

Hart MW, Scheibling RE (1988) Heat waves, baby booms, and the destruction of kelp beds by sea urchins. Mar Biol 99: 167–176

Hart MW, Strathmann RR (1994) Functional consequences of phenotypic plasticity in echinoid larvae. Biol Bull 186: 291–299

Hartney KI, Grorud KI (2002) The effects of sea urchins as biogenic structures on the local abundance of a temperate reef fish. Oecologia 131: 506–513

Hedgecock D (1994) Does variance in reproductive success limit effective population size of marine organisms? In: Beaumont A (ed.) Genetics and Evolution of Aquatic Organisms. Chapman and Hall, London. pp 122–134

Hilborn R, Walters CJ (1987) A general model for simulation of stock and fleet dynamics in spatially heterogeneous fisheries. Can J Fish Aquat Sci 44: 1366–1370

Hines AH, Pearse JS (1982) Abalones, shells and sea otters: dynamics of prey populations in Central California. Ecology 63: 1547–1560

Jackson JBC, Kirby MX, Berger WG, Bjorndal KA, Botsford LW, Bourque BJ, Bradbury RH, Cooke R, Erlandson J, Estes JA, Hughes TP, Kidwell S, Lange CB, Lenihan HS, Pandolfi JM, Peterson CH, Steneck RS, Tegner MJ, Warner RR (2001) Historical overfishing and the recent collapse of coastal ecosystems. Science 293: 629–638

Jones CG, Lawton JH, Shachak M (1994) Positive and negative effects of organisms as physical ecosystem engineers. Ecology 78: 1946–1957

Kalvass PE (2000) Riding the roller coaster: boom and decline in the California red sea urchin fishery. J Shellfish Res 19: 621–622

Kalvass PE, Hendrix JM (1997) The California red sea urchin, *Strongylocentrotus franciscanus,* fishery: catch, effort and management trends. Mar Fish Rev 59: 1–17

Karpov K, Tegner MJ, Rogers-Bennett L, Kalvass P, Taniguchi I (2001) Interactions among red abalones and sea urchins in fished and reserve sites in northern California: implications of competition to management. J Shellfish Res 20: 743–753

Kato S, Schroeter SC (1985) Biology of the red sea urchins, *Strongylocentrotus franciscanus*, and its fishery in California. Mar Fish Rev 47: 1–20

Keesing JK, Baker JL (1998) The benefits of *catch* and *effort* data at a fine spatial scale in the South Australian abalone (*Haliotis laevigata* and *H. rubra*) fishery. Proceedings of the North Pacific Symposium on Invertebrate Stock Assessment and Management. Can Spec Publ Fish Aquat Sci Publ Spec 125: 179–186

Keesing JK, Hall KC (1998) Review of harvests and status of world sea urchin fisheries point to opportunities for aquaculture. J Shellfish Res 17: 1505–1506

Kiessling W (2005) Long-term relationships between ecological stability and biodiversity in Phanerozoic reefs. Nature 433: 410–413

Lafferty KD (2004) Fishing for lobsters indirectly increases epidemics in sea urchins. Ecol Appl 14: 1566–1573

Lafferty KD, Kushner D (2000) Population regulation of the purple sea urchin, Strongylocentrotus purpuratus, at the California Channel Islands. In: Chang W (ed.) 5th California Islands Symp. Minerals Management Service, Santa Barbara, CA.

Lai HL, Bradbury A (1998) A modified catch-at-size analysis model for a red sea urchin (*Strongylocentrotus franciscanus*) population. Proceedings of the North Pacific Symposium on Invertebrate Stock Assessment and Management. Can Spec Publ Fish Aquat Sci 125: 85–96

Lawrence JM (1975) On the relationship between marine plants and sea urchins. Oceanogr Mar Biol Ann Rev 13: 213–286

Lee YH (2003) Molecular phylogenies and divergence times of sea urchin species of Strongylocentrotidae, Echinoida. Mol Biol Evol 20: 1211–1221

Leighton DL (1968) A comparative study of food selection and nutrition in the abalone, *Haliotis rufescens* (Swainson) and the sea urchin, *Strongylocentrotus purpuratus* (Stimpson). PhD Thesis, Univ. California, San Diego, CA, USA, 197pp.

Leighton DL (1971) Grazing activities of benthic invertebrates in Southern California kelp beds. Pac. Sci. 20: 104–113

Leighton DL, Jones LG, North W (1966) Ecological relationships between giant kelp and sea urchins in southern California. In: Young EG, McClachlan JL (eds) Proceedings 5th International Seaweed Symposium, Oxford, UK. Pergamon Press, Oxford. pp 141–153

Lessios HA (1988) Mass mortality of *Diadema antillarum* in the Caribbean: what have we learned? Ann Rev Ecol Syst 19: 371–393

Levin PS, Coyer JA, Petrik R, Good TP (2002) Community-wide effects of nonindigenous species on temperate rocky reefs. Ecology 83: 3182–3193

Levins R (1969) Some demographic and genetic consequences of environmental heterogeneity for biological control. Bull Entomol Soc Amer. 15: 237–240

Levitan DR (2002) Density dependent selection on gamete traits in three congeneric sea urchins. Ecology 83: 464–479

Levitan DR (2004) Density-dependent sexual selection in external fertilizers: variances in male and female fertilization success along the continuum from sperm limitation to sexual conflict in the sea urchin *Strongylocentrotus franciscanus*. Am Nat 164: 298–309

Levitan DR, Sewell MA, F-S Chia F-S (1992) How distribution and abundance influence fertilization success in the sea urchin (*Strongylocentrotus franciscanus*). Ecology 73: 248–254

Low CG (1975) The effect of grouping of *Strongylocentrotus franciscanus*, the giant red sea urchin, on its population biology. PhD Thesis, Univ. British Columbia, Vancouver, BC, Canada

Lowry L, Pearse JS (1973) Abalones and sea urchins in an area inhabited by sea otters. Mar Biol 23: 213–219

Lundquist CJ, Botsford LW (2004) Model projections of the fishery implications of the Allee effect in broadcast spawners. Ecol Appl. 14: 929–941

Mantua NJ, Hare SR, Zhang Y, Wallace JM, Francis RC (1997) A Pacific-interdecadal climate oscillation with impacts on salmon production. Bull Amer Meteorol Soc 78: 1069–1079

Mayfield S, Branch GM (2000) Interrelations among rock lobsters, sea urchins, and juvenile abalone: implications for community management. Can J Fish Aquat Sci 57: 2175–2185

McCartney MA, Brayer K, Levitan DR (2004) Polymorphic microsatellite loci from the red sea urchin, *Strongylocentrotus franciscanus*, with comments on heterozygote deficit. Mol Ecol Notes 4: 226–228

McDonald K (2004) Patterns in early embryonic motility: effects of size and environmental temperature on vertical velocities of sinking and swimming echinoid blastulae. Biol Bull 207: 93–102

McNaught DC (1999) The indirect effects of macroalgae and micropredation on the post-settlement success of the green sea urchin in Maine. PhD Dissertation, Univ. Maine, Orono, ME, USA

Miller BA, Emlet RB (1997) Influence of nearshore hydrodynamics on larval abundance and settlement of sea urchins *Strongylocentrotus franciscanus* and *S. purpuratus* in the Oregon upwelling zone. Mar Ecol Prog Ser 148: 83–94

Miller KM, Kaukinen KH, Laberee K, Supernault KJ (2004) Microsatellite loci from red sea urchins (*Strongylocentrotus franciscanus*). Mol Ecol Notes 4: 722–724

Miner BG (2005) Evolution of feeding structure plasticity in marine invertebrate larvae: a possible trade-off between arm length and stomach size. J Exp Mar Biol Ecol 315: 117–125

Moberg PE, Burton RS (2000) Genetic heterogeneity among adult and recruit red sea urchins, *Strongylocentrotus franciscanus*. Mar Biol 136: 773–784

Moore JD, Finley CA, Robbins TT, Friedman CS (2002) Withering syndrome and restoration of Southern California abalone populations. CalCOFI Rep 43: 112–117

Morgan LE, Botsford LW, Wing SR, Smith BD (2000a) Spatial variability in growth and mortality of the red sea urchin, *Strongylocentrotus franciscanus*, in northern California. Can J Fish Aquat Sci 57: 980–992

Morgan LE, Wing, SR, Botsford LW, Lundquist CJ, Diehl JM (2000b) Spatial variability in red sea urchin (*Strongylocentrotus franciscanus*) recruitment in northern California. Fish Oceanogr. 9: 83–98

Morris RH, Abbott DP, Haderlie EC (1980) Intertidal invertebrates of California. Stanford University Press, Stanford.

Murray SN, Ambrose RF, Bohnsack JA, Botsford LW, Carr MH, Davis GE, Dayton PK, Gotshall D, Gunderson DR, Hixon MA, Lubchenco J, Mangel M, MacCall A, McArdle DA (1999) No-take Reserve Networks: sustaining fishery populations and marine ecosystems. Fisheries 24: 11–25

Nishizaki MT, Ackerman JD (2001) Gimme Shelter: factors influencing juvenile sheltering in *Strongylocentrotus franciscanus*. In: Barker M (ed.) Echinoderms 2000. Swets and Zeitlinger, Lisse. pp 515–520

Nishizaki MT, Ackerman JD (2004) Juvenile-adult associations in sea urchins *Strongylocentrotus franciscanus* and *S. droebachiensis*: is nutrition involved? Mar Ecol Prog Ser 268: 93–103

Nishizaki MT, Ackerman JD (2005) A secondary chemical cue facilitates juvenile-adult postsettlement associations in red sea urchins. Limnol Oceanogr 50: 354–362

NOAAa http://www.cpc.ncep.noaa.gov/products/precip/CWlink/MJO/enso.shtml#forecast

NOAAb http://www.swr.nmfs.noaa.gov/fmd/sunee/imports/jimp.htm

Omi T (1987) Results of sea urchin releases. In: Summary of lectures for 1986 of the Fisheries Culture Research Society: techniques for collecting sea urchin seed and results of releases. (Public Corp. Promot. Aquacult. Hokkaido). Transl. (extended summary) By Madelon Mottet, Japan. Sci. Liaison 595 Tucker Ave. No.39, Friday Harbor, WA. 98250

Paine RT (1966) Food webs complexity and species diversity. Amer Nat 100: 65–75

Paine RT, Vadas RL (1969) The effects of grazing by sea urchin, *Strongylocentrotus* spp. on benthic algal populations. Limnol Oceanogr 14: 710–719

Paulay G Boring LP, Strathmann RR (1985) Food limited growth and development of larvae: experiments with natural sea water. J Exp Mar Biol Ecol 85: 1–10.

Pauly D (1995) Anecdotes and the shifting baseline syndrome of fisheries. Trends Ecol Evol 10: 430.

Pearse JS, Cameron A (1991) Echinodermata: Echinoidea. In: Giese AC, Pearse JS, Pearse VE (eds) Reproduction of Marine Invertebrates. Vol. VI. Echinoderms and Lophophorates. Boxwood Press, Pacific Grove, CA. pp 513–662

Pearse JS, Clark ME, Leighton DL, Mitchell CT, North WJ (1970) Marine waste disposal and sea urchin ecology. In: North WJ (ed.) Kelp Habitat Improvement Project, Annual Report 1969–1970. California Institute of Technology, Pasadena, pp 1–93

Pearse JS, Costa DP, Yellin MB, Agegian CR (1977) Localizad mass mortality of red sea urchins, *Strongylocentrotus franciscanus*, near Santa Cruz, California. Fish Bull US 75: 645–648

Pearse JS, Hines AH (1987) Long-term population dynamics of sea urchins in a central California kelp forest: rare recruitment and rapid decline. Mar Ecol Prog Ser 39: 275–283

Pearse JS, Pearse VB, Davis KP (1986) Photoperiodic regulation of gametogenesis and growth in the sea urchin *Strongylocentrotus purpuratus*. J Exp Zool 237: 107–118

Pennisi E (2002) Chimps and fungi make genome "Top Six". Science 296: 1589–1591

Pfister CA, Bradbury A (1996). Harvesting red sea urchins: recent effects and future predictions. Ecol Appl 6: 298–310

Pikitch EK, Santora C, Babcock EA, Bakun A, Bonfil R, Conover DO, Dayton PK, Doukakis P, Fluharty D, Heneman B, Houde ED, Link J, Livingston PA, Mangel M, McAllister MK, Pope J, Sainsbury KJ. (2004) Ecosystem-based fishery management. Science 305: 346–347

Pulliam HR (1988) Sources, sinks and population regulation. Amer Nat 132: 652–661

Raimondi PT, Wilson CM, Ambrose RF, Engle JM, Minchinton TE (2002) Continued declines of black abalone along the coast of California: are mass mortalities related to El Nino events? Mar Ecol Prog Ser 242: 143–152

Richards D, Kushner D (1994) Kelp Forest Monitoring 1992 Annual Report. National Park Service, Channel Islands National Park, Tech Rep CHIS -94-01.

Richmond NT, Schaefer J, Wood C, McRae J (1997) History and status of the Oregon sea urchin fishery. Oregon Dept. of Fish and Wildlife, Marine Resources Program.

Rogers-Bennett L (1989) The Spatial Association of Juvenile Green Sea Urchins. MS Thesis. Univ. Massachusetts, Boston, MA, USA

Rogers-Bennett L (1994) Spatial patterns in the life history characteristics of red sea urchins, *Strongylocentrotus franciscanus*: implications for recruitment and the California fishery. PhD Dissertation Univ. California, Davis, CA, USA

Rogers-Bennett L (2001) Evaluating stocking as an enhancement strategy for the red sea urchin, Strongylocentrotus franciscanus: depth-specific recoveries. In: Barker M (ed.) Echinoderms 2000. AA Balkema, Rotterdam. pp 527–531

Rogers-Bennett L, Allen BL, Davis GE (2004) Measuring abalone (*Haliotis* spp.) recruitment in California to examine recruitment overfishing and recovery criteria. J Shellfish Res 23: 1201–1207

Rogers-Bennett L, Bennett WA, Fastenau HC, Dewees CM (1995) Spatial variation in red sea urchin reproduction and morphology: implications for harvest refugia. Ecol App 5: 1171–1180

Rogers-Bennett L, Fastenau HC, Dewees CM (1998) Recovery of red sea urchin beds following experimental harvest. In: Mooi R, Telford M (eds) Echinoderms: San Francisco. AA Balkema, Rotterdam, 805–809

Rogers-Bennett L, Haaker PA, Huff TO, Dayton PK (2002) Estimating baseline abundances of abalone in California for restoration. CalCOFI Rep 43: 97–111

Rogers-Bennett L, Pearse JS (2001) Indirect benefits of Marine Protected Areas for juvenile abalone. Conser Biol 15: 642–647

Rogers-Bennett L, Rogers DW, Bennett WA, Ebert TA (2003) Modeling red sea urchin growth using six growth models. Fish Bull 101: 614–626

Rowley RJ (1989) Settlement and recruitment of sea urchins (*Strongylocentrotus* spp.) in a sea-urchin barren ground and a kelp bed: are populations regulated by settlement or post-settlement processes? Mar Biol 100: 485–494

Rowley RJ (1990) Newly settled sea urchins in a kelp bed and sea urchin barren ground: a comparison of growth and mortality. Mar Ecol Prog Ser 62: 229–240

Rumrill SS (1990) Natural mortality of marine larvae. Ophelia 32: 163–198

Russell MP (1987) Life history traits and resource allocation in the purple sea urchin *Strongylocentrotus purpuratus* (Stimpson). J. Exp. Mar. Biol. Ecol. 108: 199–216

Sainsbury KJ (1980) Effect of individual variability on the von Bertalanffy growth equation. Can J Fish Aquat Sci 37: 1337–1351

Sala E, Aburto-Oropeza O, Paredes G, Para I, Barrera JC, Dayton PK (2002) A general model for designing networks of marine reserves. Science 298: 1991–1993

Sala E, Boudouresque CF, Harmelin-Vivien M (1998) Fishing trophic cascades, and the structure of algal assemblages: evaluation of an old but untested paradigm. Oikos 82: 425–439

Salls RA (1995) Ten-thousand years of fishing; the evidence for alternate stable states in nearshore ecosystems as the result of overexpoitation of the California sheephead (*Semiscossyphus pulcher*) by prehistoric fishermen on San Clemente Island, California. In: Memorias IX Simposium International Biol Mar. pp 205–214

Scheibling RE (1986) Increased macroalgal abundance following mass mortalities of sea urchins (*Strongylocentrotus franciscanus*) along the Atlantic coast of Nova Scotia. Oecologia 68: 186–198

Scheibling RE, Hennigar AW, Balch T (1999) Destructive grazing, epiphytism, and disease: the dynamics of sea urchin-kelp interactions in Nova Scotia. Can J Fish Aquat Sci 56: 1–15

Schroeter SC (1978) Experimental studies of competition as a factor affecting the distribution and abundance of purple sea urchins (*Strongylocentrotus purpuratus* (Stimpson). PhD Dissertation Univ. California, Santa Barbara, CA, USA

Schroeter SC, Dixon JD, Ebert TA, Rankin JV (1996) Effects of kelp forests *Macrocystis pyrifera* on the larval distribution and settlement of red and purple sea urchins *Strongylocentrotus franciscanus* and *S. purpuratus*. Mar Ecol Prog Ser 133: 125–134

Shanks AL Grantham BA, Carr MH (2003) Propagule dispersal distance and the size and spacing of marine reserves. Ecol Appl 13: S159–S169

Shpigel M, McBride SC, Marciano S, Lupatsch I (2004) The effect of photoperiod and temperature on the reproduction of European sea urchin *Paracentrotus lividus*. Aquaculture 232: 343–355

Simberloff D (1998) Flagships, umbrellas and keystones: is single-species management passé in the landscape era? Biol Conserv 83: 247–257

Simenstad CA, Estes JA, Kenyon KW (1978) Aleuts, sea otters, and alternative stable-states communities. Science 200: 403–411

Sloan NA (1985) Echinoderm fisheries of the world: a review. In: Keegan BE, O'Connor BDS (eds.) Echinodermata. AA Balkema, Rotterdam. pp 109–124

Smith BD, Botsford LW, Wing SR (1998) Estimation of growth and mortality parameters from size frequency distributions lacking age patterns: the red sea urchin (*Strongylocentrotus franciscanus*) as an example. Can J Fish Aquat Sci 55: 1236–1247

Steneck RS, Graham MH, Bourque BJ, Corbett D, Erlandson JM, Estes JA, Tegner MJ (2002) Kelp forest ecosystems: biodiversity, stability, resilience and future. Env Cons 29: 436–459

Strathmann RR (1979) Echinoid larvae from the northeast Pacific with a key and comment on an unusual type of planktotrophic development. Can J Zool 57: 610–616

Strathmann RR (1978) Length of the pelagic period in echinoderms with feeding larvae from the northeast Pacific. J Exp Mar Biol Ecol 34: 23–27

Strathmann RR (1996) Are planktonic larvae of marine benthic invertebrates too scarce to compete within species? Oceanol Acta 19: 399–407

Strathmann RR, von Dassow M (2001) Podial coverage and test size of regular echinoids. In: Barker M (ed.) Echinoderms 2000. AA Balkema, Rotterdam. pp 543–550

Sweetnam D (2005) Review of some California fisheries for 2004. CalCOFI Rep 46: 10–31

Tanaka M (1982) A new growth curve which expresses infinite increase. Publ Amakusa Mar Biol Lab 6: 167–177

Tegner MJ (1989) The feasibility of enhancing red sea urchin, *Strongylocentrotus franciscanus*, stocks in California: an analysis of the options. Mar Fish Rev 51: 1–22

Tegner MJ (2001) The ecology of *Strongylocentrotus franciscanus* and *Strongylocentrotus purpuratus*. In: Lawrence JM (ed.) Edible Sea Urchins: Biology and Ecology. Elsevier, New York. pp 307–331

Tegner MJ, Dayton PK (1977) Sea urchin recruitment patterns and implications of commercial fishing. Science 196: 324–326

Tegner MJ, Dayton PK (1981) Population structure, recruitment and mortality of two sea urchins (*Strongylocentrotus franciscanus* and *S. purpuratus*) in a kelp forest. Mar Ecol Prog Ser 5: 255–268

Tegner MJ, Dayton PK (1991) Sea urchins, El Niños, and the long term stability of Southern California kelp forest communities. Mar Ecol Prog Ser 77: 49–63

Tegner MJ, Dayton PK (2000) Ecosystem effects of fishing in kelp forest communities. ICES J Mar Sci 57: 579–589

Tegner MJ, Dayton PK, Edwards PB, Riser KL (1995) Sea urchin cavitation of giant kelp (*Macrocystis pyrifera* C. Agardh) holdfasts and its effects on kelp mortality across a large California forest. J Exp Mar Biol Ecol 191: 83–99

Tegner MJ, Levin LA (1982) Do sea urchins and abalones compete in the California kelp forest communities? In: Lawrence JM (ed.) Echinoderms: Proceedings of the International Conference, Tampa Bay. AA Bakema, Rotterdam. pp 265–271

Tegner MJ, Levin LA (1983) Spiny lobsters and sea urchins: analysis of a predator–prey interaction. J Exp Mar Biol Ecol 73: 125–150

Tuck GN, Possingham HP (2000) Marine protected areas for spatially structured exploited stocks. Mar Ecol Prog Ser 192: 89–101

Vickery MS, McClintock JB (2000) Effects of food concentration and availability on the incidence of cloning in planktotrophic larvae of the sea star *Pisaster ochraceus*. Biol Bull 199: 298–304

Vogel G (2000) Migrating otters push law to limit. Science 289: 1271–1273

Watanabe JM, Harrold C (1991) Destructive grazing by sea urchins *Strongylocentrotus* spp. in a central California kelp forest: potential roles of recruitment, depth and predation. Mar Ecol Prog Ser 71: 125–141

Wilson KC, North WJ (1983) A review of kelp bed management in southern California. J World Maricult Soc 14: 347–359

Wing SR, Botsford LW, Morgan LE, Dile JM, Lundquist CL (2003) Inter-annual variability in larval supply to populations of three invertebrate taxa in the northern California Current. Estuar Coastal Shelf Sci 57: 859–872

Wing SR, Largier JL, Botsford LW, Quinn JF (1995) Settlement and transport of benthic invertebrates in an intermittent upwelling region. Limnol Oceangr 40: 316–329

Yamaguchi M (1975) Estimating growth parameters from growth rate data: problems with marine sedentary invertebrates. Oecologia 20: 321–332

Edible Sea Urchins: Biology and Ecology
Editor: John Miller Lawrence

Chapter 20

Ecology of *Strongylocentrotus intermedius*

Yukio Agatsuma

Laboratory of Marine Plant Ecology, Graduate School of Agricultural Science, Tohoku University, Miyagi (Japan).

1. INTRODUCTION

Strongylocentrotus intermedius (A Agassiz) (Plate 2D) is found commonly on the rocky bottom in shallow waters around Hokkaido. It is harvested commercially in Iwate, Aomori, and Hokkaido, Japan. The roe flesh is eaten mainly as sashimi, sushi, and served in a bowl with rice (*Uni Don* in Japanese). Salted raw urchin (*Shio Uni*) is a familiar food in Hokkaido.

2. GEOGRAPHIC DISTRIBUTION

Strongylocentrotus intermedius is found on intertidal and subtidal rocky bottoms in northern regions in the Pacific coastal waters of Choshi, Chiba, the Sea of Japan around Toyama, the Korean peninsula, northeastern China, Sakhalin, and Vladivostok (Shigei 1995).

3. REPRODUCTION

3.1. Size at Maturity

The size at maturity is 30–35 mm test diameter at 2 years age in Funka Bay (Fuji 1960b; Kawamura 1973). Juveniles fed algae *ad libitum* in the laboratory mature at 1 year (Agatsuma and Momma 1988).

3.2. Difference in Reproductive Cycle among Localities

The gametogenic process is classified into five stages: recovery stage, growth stage, prematuration stage, maturation stage, and spent stage (Fuji 1960a). Gonadal indices of sea urchins off the Sea of Japan coast to the western Tsugaru Strait increase in spring, reach a maximum in summer, and then sharply decrease to <10 in September and October, the spawning season (Kawamura 1967, 1973; Tajima et al. 1978; Takahashi 1980; Agatsuma et al. 1989). In contrast, the spawning season of sea urchins off the eastern Pacific coast around Hidaka and the Okhotsk Sea coast continues from June to October and the gonadal indices differ significantly among individuals (Tomita et al. 1984, 1986; Tomita and Tada 1988). Furthermore, spawning off the coast of the eastern Tsugaru Strait, the Pacific Ocean, and Funka Bay occurs in April to May and in August to November (Kawamura et al. 1983; Agatsuma and Momma 1988; Agatsuma et al. 1989, 1994). Thus, spawning of *S. intermedius* in Hokkaido is classified into three types: Sea of Japan type with intensive spawning in autumn, Funka Bay type with spawning in spring and autumn, and Okhotsk Sea and Pacific Ocean type with spawning from spring to autumn. According to Fuji (1960c), who examined seasonal changes in gonadal indices and histological gonadal changes, the spawning season of *S. intermedius* off the eastern Tsugaru Strait coast and in Funka Bay during 1956 and 1959 is September to November, the same as in the Sea of Japan. This may be attributable to larval transportation and juvenile recruitment caused by differences over time in the variability of hydrographic conditions, the Tsushima Warm Current, and the Oyashio Current.

3.3. Fixation of Reproductive Cycle in Each Area

The reproductive cycle of *S. intermedius*, which was introduced from the Sea of Japan to the Pacific coastal waters, does not vary. The spawning season is always autumn (Tomita et al. 1986). *Strongylocentrotus intermedius* artificially produced from adults from the Sea of Japan and transplanted to the Pacific and eastern Tsugaru Strait coastal waters maintain their own fixed reproductive cycle (Agatsuma and Momma 1988; Agatsuma et al. 1994). As a result, the gonadal indices of the transplanted sea urchins are lower than those of native sea urchins during harvesting season (November and March). Conversely, gonads of sea urchins produced from adults in the Pacific coastal waters and transplanted to the Sea of Japan also maintain their own reproductive cycle (Agatsuma et al. 1994). The gonadal size of the transplanted individuals is the same as that of native sea urchins during harvesting season (June to August), although gonads maturing in June and oozing gametes have no commercial value. These results suggest that the artificially produced juveniles must be returned to the same area as their parents (Agatsuma et al. 1995). The results also suggest that the two populations are genetically separated (Tomita et al. 1986; Agatsuma and Momma 1988; Agatsuma et al. 1994; Hoshikawa and Agatsuma 1999).

3.4. Spawning Structure

Kawamura (1973) reported that *S. intermedius* in Oshoro Bay along the southwestern Sea of Japan coast of Hokkaido spawn three or four times in a spawning season. This occurs sequentially in populations from shallow to deep waters.

4. LARVAL ECOLOGY

4.1. Occurrence

Kawamura (1970) studied the morphological features of each developmental stage of *S. intermedius* and *S. nudus*. The larvae of *S. intermedius* occur during September and January off the Sea of Japan and the western Tsugaru Strait (Kawamura 1973; Agatsuma et al. 1989; Hokkaido Central Fisheries Experimental Station et al. 1984), and during May and January off the eastern Tsugaru Strait (Sawada and Miki 1982; Agatsuma et al. 1989). In Oshoro Bay, three or four cohorts of four-armed plutei of 200 μm body length occur from late September to late October, corresponding to three or four spawnings in a spawning season (Kawamura 1973) (see Section 3.4). The four-armed plutei off the coast of Soya, the most northern region in Hokkaido, occur from May to September (Tajima et al. 1978). Metamorphosed larvae occur from late November to January in Oshoro Bay and off the coast of Shakotan Peninsula (Kawamura 1973; Hokkaido Central Fisheries Experimental Station et al. 1984), from August to November off the Okhotsk Sea coast of Abashiri, and from June to July off the coast of Mori in Funka Bay (Kawamura 1993). Thus, geographical differentiation of the periods of larval occurrence is due to spatial variations in spawning seasons. In the area of the Tumen River estuary and in the Far Eastern State Marine Reserve, larvae of *S. intermedius* mainly occur in the 0–15 m water layer from July to September (Dautov 2000).

4.2. Distribution

According to researchers of the Hokkaido Central Fisheries Experimental Station et al. (1984), four-armed larvae are found <2 km off the coast of Shakotan Peninsula in October, eight-armed larvae are found at >3 km in November, and metamorphosed individuals are found at <3 km in December. The successive changes in distribution of each developmental stage are greatly influenced by seasonal current strength and direction.

4.3. Length of Larval Life and Survival

The larval period of *S. intermedius* is two to three months in the field (Kawamura 1973), but about 20 days when larvae are raised on *Chaetoceros gracilis* at water temperatures of 15–18 °C in tanks (Hokkaido Institute of Mariculture 1992). The survival rate of larvae in nature is extremely low. The highest mortality occurs before the six-armed pluteus stage (Kawamura 1973). Larvae from undeveloped gonads usually do not develop normally. Larvae from developed gonads also do not develop to six-armed plutei under poor food conditions (Kawamura 1973). Kashenko (2000) reported that slow development, small size and low survival of *S. intermedius* larvae in seawater from Nakhodka Bay are due to heavy pollution.

5. SETTLEMENT AND METAMORPHOSIS

The test diameter of juveniles immediately after metamorphosis is 320–355 μm (Kawamura 1970, 1973). The percentage of larval settlement is affected by the water temperature at the localities where the parents occur. The highest percentage is found in juveniles from the southern Pacific coast at a water temperature of 18 °C and at 12 °C for those from the Sea of Japan (Tajima et al. 1991). The green alga *Ulvella lens* has been used as a collector for settlement in artificial production because it induces metamorphosis and settlement more strongly than attached diatoms (Kawamura et al. 1983; Hokkaido Institute of Mariculture 1992). Glycoglycerolipids (sulfoquinovosyl monoacylglycerols, sulfoquinovosyl diacylglycerols, monogalactosyl monoacylglycerols, monogalactosyl diacylglycerols, digalactosyl monoacylglycerols, and digalactosyl diacylglycerols) isolated from *U. lens* have been identified as the chemical inducers of larval settlement and metamorphosis of *S. intermedius* (Takahashi et al. 2002). Rugged substrata with attached diatoms induce larval settlement; mud accumulation inhibits settlement (Kawamura et al. 1974). According to Naidenko (1996), glutamine contained in epiphytic calcareous algae, *Melobesia* spp., that colonize older leaves of the sea grass *Zostera marina* induces larval settlement. Agatsuma et al. (2006) reported that the highest metamorphosis percentage, >80%, was found after 1 h exposure to 1/2 dilution of saturated dibromomethane. At this concentration, more than 80% of *S. intermedius* larvae metamorphose 1 h after only 5 min exposure (34–43 ppm). Thus, dibromomethane has an instantaneous effect on high success of metamorphosis of larvae of *S. intermedius* as shown in *S. nudus*.

6. FOOD AND FEEDING AFTER SETTLEMENT

6.1. Food Habit

The principal food of *S. intermedius* in the field is algae (Kawamura 1964, 1973; Kawamura and Taki 1965; Fuji 1967). Food preferences change as the sea urchins grows. Sea urchins of test diameter <5 mm prefer detritus with calcareous algae and diatoms, individuals 5–10 mm in diameter prefer mixed small algae, and those >10 mm prefer macroalgae such as *Laminaria* spp. (Fuji 1967; Kawamura 1973). In culture, this sea urchin also eats terrestrial plants (Hokkaido Central Fisheries Experimental Station et al. 1984) and sessile animals such as *Mytilus edulis*, *Balanus* spp., and *Celleporina* spp. (Kittaka and Imamura 1981). In particular, the terrestrial plant *Reynoutria sachalinensis* is commonly used in culture (Hokkaido Central Fisheries Experimental Station et al. 1984).

6.2. Food Ingestion and Absorption

According to Fuji (1962, 1967), the daily amount of food intake of an adult *S. intermedius* (ca. 50 mm in test diameter) varies greatly among different kinds of algae. The largest amount of food intake was about 3 g wet weight when individuals were fed *L. japonica*, and the lowest was about 0.2 g wet weight with *Ulva pertusa*. These values correspond to 6% and 1% of the body weight of the sea urchins, respectively. The daily amount of food intake by adults declined markedly during July and October, but not in juveniles. This decline in adults may be attributed to some physiological factors correlated with gonadal development. The daily feeding rate decreased as the sea urchins grew larger. However, there was no significant difference among sea urchins of different sizes in their ability to absorb the food. The absorption efficiency also varied widely with the kind of algae, ranging from 83% for *Scytosiphon lomentaria* to 57% for *L. japonica* and *Condrus ocellatus*. Absorption efficiency of individuals fed *L. japonica* increased with rising water temperature; the mean value of absorption efficiency was about 70% in summer and about 55% in winter. The dry weight of food (*L. japonica*) absorbed remained high (ca. 270 mg per day per individual) during February to June and fell to about 50 mg per day per individual in September and October.

6.3. Diurnal Changes in Food Intake

This sea urchin does not feed much when the light is >6000 lux, but it compensates by feeding more actively at night after a bright day. The amount of food intake during 24 h is not affected by the intensity of the light in the daytime (Fuji 1967). In contrast, starved adult *S. intermedius* do not show any diurnal pattern of activity rhythm under 12L–12D (Ito and Hayashi 1999).

6.4. Chemical Stimulus on Feeding

Machiguchi (1987) studied the foraging behavior of *S. intermedius* for *L. longissima* in a selective experiment using a Y-shaped chamber. Sea urchins approached the kelp only when others were grazing on it. This suggests that a chemical substance released from the kelp by sea urchin grazing is an attractant. The glycerolipids digalactosyldiacylglycerol, phosphatidylcholine, 6-sulfoquinovosyldiacylglycerol, and monogalactosyldiacylglycerol, extracted from the brown alga *Eisenia bicyclis*, are potent feeding stimulants for *S. intermedius* (Sakata et al. 1989). Aplysiaterpenoid A and macrocyclic r-pyrone isolated from two red algae, *Plocamium leptophyllum* and *Phacelocarpus labillardieri,* respectively, are potent feeding inhibitors (Sakata et al. 1991a, 1991b). Bromophenols isolated from methanol extracts of the red alga *Odonthalia corymbifera* are feeding deterrent chemicals against *S. intermedius.* These chemicals enable this alga to grow despite dense populations of transplanted hatchery-raised juvenile *S. intermedius* off the Pacific coast in Hokkaido (Kurata et al. 1997). Two phenylpropanoic acid derivatives, tichocarpol A and tichocarpol B, and floridoside isolated from the red alga *Tichocarpus crinitus* exhibit feeding-deterrent activity against *S. intermedius* (Ishii et al. 2004).

7. GROWTH

7.1. Longevity and Growth Rings

The longevity of *S. intermedius* is estimated to be 6–10 years (Agatsuma 1991; Kawamura 1993). Kawamura (1973) estimated the age of *S. intermedius* based on the methods of

Jensen (1969) and Moore (1935) by the number of black rings in charred genital plates. The rings are formed once a year from June to October. Because the width of a one-year ring of a transplanted sea urchin is greater than that of native sea urchins, reflecting their rapid growth until transplantation by supply of food in excess, it is possible to discriminate between the two populations (Agatsuma 1987; Ichinose et al. 2005).

7.2. Energy Transformation to Growth

Fuji (1967) studied energy transformation of food from ingestion to growth. Test growth of adult *S. intermedius* >2 years old is restricted to late winter and early spring and almost ceases during the rest of the year. Juveniles <2 years old grow continuously throughout the year. The rate of juvenile growth is markedly higher than that of the adults. The ability of sea urchins of different sizes to absorb protein nitrogen from food does not differ. Their ability to use protein nitrogen for growth appears to decline in a curvilinear fashion as their diameter increases. When *L. japonica* is supplied as food, use of protein nitrogen by small individuals 18.5 mm in test diameter is about 61%; the corresponding value for large individuals is 37% (54.8 mm diameter). The growth efficiency based on caloric value differs considerably according to the kind of food: in adults (ca. 50 mm diameter) this efficiency ranges from 4.5% (on *Agarum cribrosum*) to 28.6% (on *Alaria crassiforia*). No significant difference was found in their ability to assimilate the food. When *L. japonica* was supplied, the assimilation efficiency averaged about 60%. The growth efficiency decreased with size, ranging from 51% for small individuals (about 20 mm diameter) to about 15% for large individuals (about 55 mm).

The efficiency of growth is expressed as the percentage of the energy consumed that is used for production (gross efficiency) or the percentage of energy absorbed that is converted into production (net efficiency). The percentage of the energy stored over the energy consumed decreases from 19.7% at 1 year to 8.5% at 4 years, and the net efficiency of growth decreases from 28.1% to 12.4%. The amount of food required for maintenance ranges from 13 to 17 mg (dry weight) of algae per day, and the amount of protein nitrogen required for maintenance ranges from 0.4 to 0.7 mg per day for sea urchins of 45–50 mm diameter. The exponential function calculated for individuals with a diameter (D) and total nitrogen excretion per day per individual (N) fed a non-nitrogen diet is $N = 0.000\ 09324D^{2.2570}$. This formula suggests that a sea urchin of about 50 mm test diameter excretes 0.4–0.7 mg of nitrogen per day.

Thus, food intake of *S. intermedius* >2 years old gradually increases during November and February when the gonads begin to recover from spawning. About 70% of the energy stored is converted to test growth during this season. The most active grazing and rapid gonadal growth are found during March and May when about 70% of the energy stored is converted to gonadal growth (gonadal growth season). A decrease in food intake associated with gonadal maturation results in little growth during June and August (gonadal maturation season). During September and October, the spawning season, food intake is minimal and test growth ceases (spawning season).

7.3. Differences in Growth among Localities

The sea urchins grow to 40 mm diameter, the minimum fishing size in Hokkaido, within 2–4 years, but the growth differs among localities and depth (Fuji 1967; Kawamura 1973; Taki 1986; Taki et al. 1992; Abe and Tada 1994). In contrast, 8-year-old sea urchins at Oumu in the Okhotsk Sea living at high densities of 50–100 individuals per square meter have a test diameter of <40 mm (Abe and Tada 1994).

Growth rates of artificially produced juveniles also differ after transplantation to different localities (Agatsuma et al. 1995). Juveniles with a 15 mm diameter transplanted to shallow waters off the Sea of Japan coast in southwestern Hokkaido grow to 40 mm in twenty months after transplantation, occasionally 2 years. In contrast, juveniles with a 20 mm diameter transplanted to shallow waters off the Tsugaru Strait coast and in Funka Bay grow rapidly and reach 40 mm within one year after transplantation and 50 mm within 2 years. Furthermore, growth is rapid off the eastern Tsugaru Strait coast and the Pacific coast of Hidaka in southern Hokkaido, and juveniles of <10 mm test diameter grow to 50 mm within 2 years after transplantation.

The growth rate of sea urchins >3 years old artificially raised from adults from the Sea of Japan coast in Hokkaido and transplanted to the south Pacific coast of Hokkaido is lower than that of native populations of the same year class (Agatsuma and Momma 1988). This growth difference may be caused by genetic differences between populations.

7.4. Food and Growth

Somatic growth of *S. intermedius* is affected by the kind and quantity of marine algae. In particular, it is thought that somatic growth is greatly influenced by the biomass of large brown algae (Laminariaceae), as for *S. nudus* (Agatsuma 1997). Taki (1978) reported that the growth rate of sea urchins fed *L. ochotensis* is higher that that of urchins fed *Rhodoglossum japonicum* or *U. pertusa*. Increase in diameter and body weight of sea urchins fed *Macrocystis integrifolia* is slower than on *L. longissima*. In particular, the feed conversion efficiency of small urchins (<35 mm diameter) fed *M. integrifolia* was lower than that of those fed *L. longissima* (Machiguchi 1992).

Agatsuma (2000) studied seasonal changes in food intake, digestion, and growth of juvenile *S. intermedius* fed the large brown algae *Sargassum confusum* and *L. religiosa*. The food intake and the proportion of food digested were higher on *Sargassum* than on kelp during July and October. They were low during November and March when only thalli without main or lateral branches of the *Sargassum* were consumed and the alga's growth rate was low. The feed conversion efficiency for *Sargassum* was higher than that for kelp during May and June, and lower during August and October.

7.5. Water Temperature and Growth

The growth after settlement of artificially raised juvenile *S. intermedius* of <2 mm test diameter is strongly affected by water temperature. It slows markedly at a water temperature of <5 °C in winter. Juveniles of >2 mm test diameter are not strongly affected by low water temperature (Tajima and Fukuchi 1989). Taki et al. (1992) reported that

S. intermedius off the Pacific coast in eastern Hokkaido grow continually until 8 years old, but that the sea urchins off the coast of Sea of Japan in northern Hokkaido grow continually until 3 years old then cease growth and show a high catabolism coefficient. They concluded that the effect of temperature on assimilation and catabolism of sea urchins at each locality can be attributed to differences in the growth curves between each population.

7.6. Gonadal Growth

The maximum gonadal index before spawning increases with age (Kawamura 1973). However, the gonadal index of very old sea urchins decreases, indicating a low reproductive effort (Kawamura 1993). The gonadal growth of the sea urchins differs among localities and fisheries grounds. Generally, the gonadal indices are low and maturation decreases with depth, being affected by the kinds and biomass of algae (Kawamura 1964, 1965, 1973). The amount of gonad production is greater for individuals fed prepared feed than those fed *L. japonica* (Chang et al. 2005).

8. HABITAT

8.1. Juvenile Habitat

Kawamura (1973, 1993) reported the following suitable conditions for juveniles:

1. low hydrodynamics;
2. rocks of 30 cm diameter with growth of attached diatoms and small algae or accumulation of detritus at their surface;
3. no sand or mud;
4. topographic transportation of abundant metamorphosed larvae;
5. constant circulation of seawater preventing rising water temperature in summer; and
6. absence of predators such as starfish and crabs.

8.2. Habitat Structure

Fuji and Kawamura (1970a) and Kawamura (1973) studied seasonal changes in the distribution of each age of *S. intermedius* at a rocky shore at Ikantai in southern Hokkaido. This shore consisted of an extensive flat rock, a boulder field, an eelgrass area, and a sandy area, and was classified into five major habitats according to the nature of the bottom. The boulder field supported more than 90% of the sea urchin population. The flat rock and eelgrass areas were very poor habitats. Juveniles were densest in the field of small boulders. The 2-year-old urchins were also more abundant in this boulder field than in the other areas, but they dispersed gradually toward the field with large boulders from late spring to late autumn. The field with large boulders was the major habitat for the adult >3 years old. Some adults, during the cold months, extended their distribution over the field covered with small boulders. However, during the rest of the year, they were clumped within the large boulder field.

9. COMMUNITY ECOLOGY

9.1. Bio-economy

Fuji and Kawamura (1970b) and Kawamura (1973) studied the bio-economy of a native population of *S. intermedius* at a rocky shore at Ikantai in southern Hokkaido. The population metabolism expressed as units of dry matter or nitrogen for the population was positive: an initial biomass of 60.7 g m^{-2} yr^{-1} or 1.1 g-N m^{-2} yr^{-1}; population growth of 43.1 g m^{-2} yr^{-1} or 1.3 g-N m^{-2} yr^{-1}; immigration of 46.1 g m^{-2} yr^{-1} or 0.9 g-N m^{-2} yr^{-1}; and recruitment of only 0.14 g m^{-2} yr^{-1} or 0.003 g-N m^{-2}. This sum total was divided in the following way on the debit side of the balance: loss of population due to emigration and natural mortality, 72.6 g m^{-2} yr^{-1} or 1.6 g-N m^{-2}; gametes ejected, 2.9 g m^{-2} yr^{-1} or 0.2 g-N m^{-2}; and final biomass, 51.3 g m^{-2} yr^{-1} or 0.9g-N m^{-2} yr^{-1}. Because the population growth of the sea urchin was supported by 412.4 g m^{-2} yr^{-1} or 8.0 g-N m^{-2} yr^{-1} of *L. angustata* and by 12.2 g m^{-2} yr^{-1} or 0.5 g-N m^{-2} yr^{-1} of *U. pertusa*, the growth coefficient was about 10% in dry matter, and the production efficiency was about 4%. The annual food consumption of the sea urchin population corresponds to about half of the annual growth (production) of the algal community.

9.2. Grazing Effect on Algal Communities

The removal of dense colonies of *S. intermedius* at Esashi off the Okhotsk Sea coast (65 individuals per square meter, 1.7 kg m^{-2}) and at Oumu (99 individuals per square meter, 1.2 kg m^{-2}) in autumn resulted in the growth of large populations of *L. ochotensis* (16 kg m^{-2}, 7.9 kg m^{-2}) from spring to early summer of the next year (Agatsuma 1999). This indicates that growth of the kelp was inhibited by intensive grazing by *S. intermedius*.

10. POPULATION DYNAMICS

10.1. Fluctuation in Larval Occurrence

From the successive changes in the numbers of larvae at each developmental stage off the coast of Shakotan Peninsula, it is thought that 75%–95% of the larvae are lost. A marked annual fluctuation in the numbers of larvae occurs, the maximum being 40 times the minimum (Hokkaido Central Fisheries Experimental Station et al. 1984).

10.2. Juvenile Recruitment

The density of juvenile *S. intermedius* every 12 and 8 months after spawning has been investigated since 1960 at 26 fixed sites on an intertidal rocky bottom in Oshoro Bay. In the 1960s, the maximum density was 550 individuals per square meter. Since 1981, the density has been constant at <9 individuals per square meter (Kawamura 1993). The number of eight-armed and metamorphosed larvae in December and the number of juveniles settling on plates hung in the sea off the Shakotan Peninsula 4–5 months after spawning during 1974 and 1982 showed a statistically significant relationship (Hokkaido Central Fisheries Experimental Station et al. 1984). The number of juveniles settling on the plates and their densities 8 months after spawning in Oshoro Bay also had a statistically significant relationship (Mizushima 1992). This shows that juvenile density is determined by abundance of larvae just before settlement.

Off the Pacific coast of Hidaka, poor recruitment has continued since 1980, although abundant recruitment occurred in late 1960s (Fuji and Kawamura 1970b; Kawamura 1993). Similarly, recruitment has been poor since 1975 off the eastern Pacific coast of Hokkaido (Taki 1986). In contrast, abundant recruitment occurred in 1989 and 1992 off the coast of the Okhotsk Sea (Kawamura 1993; Abe and Tada 1994). Thus, abundant recruitment is now found only in the population of the Okhotsk Sea.

10.3. Fluctuation in Gonadal Growth

Annual fluctuations in the biomass of algae in a fishing ground off the Rebun Island in the Sea of Japan, northern Hokkaido cause fluctuations in gonadal development (Kawamura and Taki 1965). Compared to 1984, significant decrease in gonad index and histopathological gonad changes in 1997 at Alekseev Bight in Amursky Bay are due to anthropogenic pollution (Vaschenko et al. 2001a). Significant histopathological changes in the gonads are also found in the polluted areas in Amur Bay, the Sea of Japan (Vaschenko et al. 2001b).

10.4. Effect of Fisheries on Population Size

A great decrease in adult population size is caused by fishing. At one fishing ground, the number caught during a fishing season reached 76% of the individuals above the fishing size limit (Kawamura 1973). At the fishing ground of Akkeshi off the Pacific coast of eastern Hokkaido, the number caught was equal to the number of recruits to the population over fishing size in 1980 and exceeded it in 1979 (Taki 1986).

10.5. Predation

The survival rate of juveniles transplanted to fishing grounds before summer decreased markedly by predation by rock crabs and sea stars, whereas the survival of juveniles that grew to >15 mm test diameter and transplanted in autumn improved by lack of predation

(Hokkaido Central Fisheries Experimental Station et al. 1984). Saito and Miyamoto (1983) reported that the survival of juvenile *S. intermedius* in an artificial channel in tidal flat rock off the coast of Sea of Japan in southwestern Hokkaido decreased to 20.5%–30.6% during summer.

The principal predator of juvenile *S. intermedius* >15 mm test diameter is the rock crab *Pugettia quadridens*, which is common in shallow waters of the southwestern Sea of Japan off Hokkaido. The male crabs migrate to the juveniles immediately after they are transplanted and prey on them (Kawai and Agatsuma 1995). Experimental evidence indicates the covering behavior of the juvenile *S. intermedius* reduces the predation from the male crabs (Agatsuma 2001). Off the Tsugaru Strait and the Pacific coast of southern Hokkaido, the principal predators are this crab and the sea star *Lysastrosoma anthosticta*. Two crabs, *Paralithodes brevipes* and *Telmessus cheiragonus* prey even on juveniles >20 mm diameter off the Pacific coast of eastern Hokkaido (Agatsuma et al. 1995). Miyamoto et al. (1985) reported that the two rock crabs *Hemigrapsus sanguineus* and *Gaetice depressus* prey on juveniles <13 mm diameter. Their predation decreases at water temperatures of <10 °C. The sea star *Asterina pectinifera* preys on juveniles <10 mm test diameter with an effectiveness that increases greatly with an increase in juvenile density (Miyamoto et al. 1985). Sea urchins in the intertidal zone are also eaten by seagulls and crows (Kawamura 1993).

10.6. Mortality from Sedimentation, High Water Temperature, and Disease

Juveniles produced artificially and transplanted are lost by burial in sand or transportation with inflow of sand (Momma et al. 1992). Occasionally, >50% of the juveniles in shallow waters die at water temperatures of >23 °C in summer (Hokkaido Central Fisheries Experimental Station et al. 1984).

In recent years, spotting disease has caused mass mortality of juvenile *S. intermedius* cultured in tanks at a water temperature of about 20 °C in summer. This disease is caused by a *Flexibacter* sp. that invades the tank in seawater and attaches to the surface of the sea urchins (Tajima et al. 1997, 1998a). In contrast, *Vibrio* sp. caused mass mortality of juveniles in tanks from the end of May to the beginning of June 1995 at low water temperatures of 11–13 °C (Tajima et al. 1998b).

11. CONCLUSIONS

Strongylocentrotus intermedius is adapted to cold waters. As with other strongylocentrotids, it feeds mainly on kelp. Population in the eastern Pacific Ocean and Sea of Okhotsk differ from those in the Sea of Japan. Current recruitment is high in the Sea of Okhotsk and is poor in the Sea of Japan and Pacific Ocean. This may be a result of changes in hydrographic conditions or of fisheries. The reproductive cycles also differ. Ecological and genetic studies should be done to investigate the differences between these populations.

REFERENCES

Abe E, Tada M (1994) The ecology of a sea urchin *Strongylocentrotus intermedius* (A. Agassiz) on the coast of Okhotsk Sea in Hokkaido. Sci Rep Hokkaido Fish Exp Stn 45: 45–56 (in Japanese with English abstract).

Agatsuma Y (1987) On the releasing of the cultured seeds of sea urchin, *Strongylocentrotus intermedius* (A. Agassiz), at Shikabe in the coast of Funka Bay. I. Discrimination between releasing group and aboriginal group by rings formed in genital plate. Hokusuishi Geppo 44: 118–126 (in Japanese).

Agatsuma Y (1991) *Strongylocentrotus intermedius* (A. Agassiz). In: Nagasawa K, Torisawa M (eds) Fishes and marine invertebrates of Hokkaido: Biology and fisheries. Kita-Nihon Kaiyo Center Co. Ltd, Sapporo, pp 324–329 (in Japanese).

Agatsuma Y (1997) Ecological studies on the population dynamics of the sea urchin *Strongylocentrotus nudus*. Sci Rep Hokkaido Fish Exp Stn 51: 1–66 (in Japanese with English abstract).

Agatsuma Y (1999) Marine afforestaion off the Japan Sea coast in Hokkaido. In: Taniguchi K (ed) The ecological mechanism of "Isoyake" and marine afforestation. Kouseisha Kouseikaku, Tokyo, pp 84–97 (in Japanese).

Agatsuma Y (2000) Food consumption and growth of the juvenile sea urchin *Strongylocentrotus intermedius*. Fisheries Sci 66: 467–472.

Agatsuma Y (2001) Effect of the covering behavior of the juvenile sea urchin *Strongylocentrotus intermedius* on predation by the spider crab *Pugettia quadridens*. Fisheries Sci 67: 1181–1183.

Agatsuma Y, Hayashi T, Uchida M (1989) Seasonal larval occurrence and spawning season of two sea urchins, *Strongylocentrotus intermedius* and *S. nudus*, in southern Hokkaido. Sci Rep Hokkaido Fish Exp Stn 33: 9–20 (in Japanese with English abstract).

Agatsuma Y, Kawamata K, Motoya S (1994) Reproductive cycle of cultured seeds of the sea urchin, *Strongylocentrotus intermedius* produced from geographically separated population. Suisanzoshoku 42: 63–70 (in Japanese with English abstract).

Agatsuma Y, Momma H (1988) Release of cultured seeds of the sea urchin, *Strongylocentrotus intermedius* (A. Agassiz), in the Pacific coastal waters of southern Hokkaido. I. Growth and reproductive cycle. Sci Rep Hokkaido Fish Exp Stn 31: 15–25 (in Japanese with English abstract).

Agatsuma Y, Sakai Y, Matsuda T (1995) Manual for transplantation of the sea urchin seeds *Strongylocentrotus intermedius*. Hokkaido Central Fisheries Experimental Station, Otaru (in Japanese).

Agatsuma Y, Seki T, Kurata K, Taniguchi K (2006) Instantaneous effect of dibromomethane on metamorphosis of larvae of the sea urchins *Strongylocentrotus nudus* and *Strongylocentrotus intermedius*. Aquaculture 251: 549–557.

Chang Y-Q, Lawrence JM, Cao X-B, Lawrence AL (2005) Food consumption, absorption, assimilation and growth of the sea urchin *Strongylocentrotus intermedius* fed a prepared feed and the alga *Laminaria japonica*. J World Aqua Soc 36: 68–75.

Dautov SSh (2000) Distribution, species composition and dynamics of echinoderm larvae in the area of the Tumen River estuary and in the Far Eastern State Marine Reserve. Biol Morya (Mar Biol Vladivostok) 26: 16–21.

Fuji A (1960a) Studies on the biology of the sea urchin. I. Superficial and histological gonadal changes in gametogenic process of two sea urchins, *Strongylocentrotus nudus* and *S. intermedius*. Bull Fac Fish Hokkaido Univ 11: 1–14.

Fuji A (1960b) Studies on the biology of the sea urchin. II. Size at first maturity and sexuality of two sea urchins, *Strongylocentrotus nudus* and *S. intermedius*. Bull Fac Fish Hokkaido Univ 11: 43–48.

Fuji A (1960c) Studies on the biology of the sea urchin. III. Reproductive cycle of two sea urchins, *Strongylocentrotus nudus* and *S. intermedius*, in southern Hokkaido. Bull Fac Fish Hokkaido Univ 11: 49–57.

Fuji A (1962) Studies on the biology of the sea urchin. V. Food consumption of *Strongylocentrotus intermedius*. Jap J Ecol 12: 181–186.

Fuji A (1967) Ecological studies on the growth and food consumption of Japanese common littoral sea urchin, *Strongylocentrotus intermedius* (A. Agassiz). Mem Fac Fish Hokkaido Univ 15: 83–160.

Fuji A, Kawamura K (1970a) Studies on the biology of the sea urchin. VI. Habitat structure and regional distribution of *Strongylocentrotus intermedius* on a rocky shore of southern Hokkaido. Bull Jap Soc Sci Fish 36: 755–762.

Fuji A, Kawamura K (1970b) Studies on the biology of the sea urchin. VII. Bio-economics of the population of *Strongylocentrotus intermedius* on a rocky shore of southern Hokkaido. Bull Jap Soc Sci Fish 36: 763–775.

Hokkaido Central Fisheries Experimental Station, Shiribeshihokubu Fisheries Extension Office, Hokkaido Institute of Mariculture (1984) On the natural seeds collection, intermediate culture and release of the sea urchin, *Strongylocentrotus intermedius*. Hokusuishi Geppo 41: 270–315 (in Japanese).

Hokkaido Institute of Mariculture (1992) Manual for artificial seeds production of the sea urchin *Strongylocentrotus intermedius*. Hokkaido Institute of Mariculture, Shikabe, Hokkaido (in Japanese).

Hoshikawa H, Agatsuma Y (1999) Geographical variation in the reproductive cycle of the sea urchin *Strongylocentrotus intermedius* in Hokkaido, Japan: Implications for the selection of juveniles for release. Suisan-ikushu 27: 45–56 (in Japanese with English abstract).

Ichinose H, Tokuda K, Kudo T, Agatsuma Y (2005) Characteristics of annual bands formation as age marker of the hatchery-raised and wild sea urchin *Strongylocentrotus intermedius* in northern Hokkaido, Japan. Aqua Sci 53: 75–82 (in Japanese with English abstract).

Ishii T, Okino T, Suzuki M, Machiguchi Y (2004) Tichocarpols A and B, two novel phenylpropanoids with feeding-deterrent activity from the red alga *Tichocarpus crinitus*. J Nat Prod 67: 1764–1766.

Ito Y, Hayashi I (1999) Behavior of the sea urchin, *Strongylocentrotus nudus* and *S. intermedius*, observed under experimental conditions in the laboratory. Otsuchi Mar Sci 24: 24–29 (in Japanese with English abstract).

Jensen ML (1969) Age determination of echinoids. Sarsia 37: 41–44.

Kashenko SD (2000) Effect of water from Nakhodka Bay (Peter the Great Bay, Sea of Japan) on early development of the sea urchin *Strongylocentrotus intermedius*. Biol Morya (Mar Biol Vladivostok) 26: 320–323.

Kawai T, Agatsuma Y (1995) Predator on released seed of the sea urchin *Strongylocentrotus intermedius* in Shiribeshi, Hokkaido, Japan. Fisheries Sci 62: 317–318.

Kawamura K (1964) Ecological studies on sea urchin *Strongylocentrotus intermedius* on the coast of Funadomari in the north region of Rebun Island. Sci Rep Hokkaido Fish Exp Stn 2: 39–59 (in Japanese with English abstract).

Kawamura K (1965) Ecological studies on sea urchin *Strongylocentrotus intermedius* on the coast of Funadomari in the north region of Rebun Island (II). Sci Rep Hokkaido Fish Exp Stn 3: 19–38 (in Japanese with English abstract).

Kawamura K (1967) Some knowledges on the life cycle of two sea urchins *Strongylocentrotus intermedius* and *S. nudus* off the Japan Sea coast of Yoichi in southwestern Hokkaido. Hokusuishi Geppo 24: 36–45 (in Japanese).

Kawamura K (1970) On the development of the planktonic larvae of Japanese sea urchins *Strongylocentrotus intermedius* and *S. nudus*. Sci Rep Hokkaido Fish Exp Stn 12: 25–32 (in Japanese with English abstract).

Kawamura K (1973) Fishery biological studies on a sea urchin, *Strongylocentrotus intermedius*. Sci Rep Hokkaido Fish Exp Stn 16: 1–54 (in Japanese with English abstract).

Kawamura K (1993) Uni Zouyoushoku to Kakou, Ryutsu. Hokkai Suisan Company, Sapporo (in Japanese).

Kawamura K, Kurotaki S, Takano S, Tachibana G (1974) Some knowledges on the deposited larvae of a sea urchin *Strongylocentrotus intermedius* (A. Agassiz). Hokusuishi Geppo 31: 1–9 (in Japanese).

Kawamura K, Nishihama Y, Yamashita K, Sawazaki M, Kawamata K, Obara A (1983) Experiments on the development of artificial seeds production of the sea urchin *Strongylocentrotus intermedius*. Hokkaido Institute of Mariculture Annual Rep, pp 71–103 (in Japanese).

Kawamura K, Taki J (1965) Ecological studies on sea urchin *Strongylocentrotus intermedius* on the coast of Funadomari in the north region of Rebun Island (III). Sci Rep Hokkaido Fish Exp Stn 4: 22–40 (in Japanese with English abstract).

Kittaka J, Imamura K (1981) Fundamental studies on control of marine fouling organisms by sea urchin. Marine Fouling 3: 53–59 (in Japanese with English abstract).

Kurata K, Taniguchi K, Takashima K, Hayashi I, Suzuki M (1997) Feeding-deterrent bromophenols from *Odonthalia corymbifera*. Phytochemistry 45: 485–487.

Machiguchi Y (1987) Feeding behavior of sea urchin *Strongylocentrotus intermedius* (A. Agassiz) observed in Y-shaped chamber. Bull Hokkaido Reg Fish Res Inst 51: 33–37 (in Japanese with English abstract).

Machiguchi Y (1992) Dietary value of a giant kelp *Macrocystis integrifolia* for Japanese common sea urchin *Strongylocentrotus intermedius* (A. Agassiz). Bull Hokkaido Reg Fish Res Inst 56: 61–70 (in Japanese with English abstract).

Miyamoto T, Ito M, Mizutori Y (1985) Experiments on the qualities for the seeds of the sea urchin *Strongylocentrotus intermedius* collected by the hanging plates in situ. Hokusuishi Geppo 42: 203–221 (in Japanese).

Mizushima T (1992) Sea urchin fisheries and population size. Hokusuishidayori 18: 1–13 (in Japanese).

Momma H, Agatsuma Y, Sawazaki M (1992) Migration with the passage of time and dispersion in the cultured seeds of the sea urchin. Bull Jap Soc Sci Fish 58: 1437–1442 (in Japanese with English abstract).

Moore HB (1935) A comparison of the biology of *Echinus esculentus* in different habitats. Part II. J Mar Biol Assoc 20: 109–128.

Naidenko TKh (1996) Induction of metamorphosis of two species of sea urchin from Sea of Japan. Mar Biol 126: 685–692.

Saito K, Miyamoto T (1983) Ecological studies on sea urchins *Strongylocentrotus intermedius* (A. Agassiz) and *S. nudus* (A. Agassiz) in an artificial channel of tidal rock flat. Sci Rep Hokkaido Fish Exp Stn 25: 21–34 (in Japanese with English abstract).

Sakata K, Iwase Y, Ina K, Fujita D (1991a) Halogenated terpenes isolated from the red alga *Plocamium leptophyllum* as feeding inhibitors for marine herbivores. Nippon Suisan Gakkaishi 57: 743–746.

Sakata K, Iwase Y, Kato K, Ina K, Machiguchi Y (1991b) A simple feeding inhibitor assay for marine herbivorous gastropods and the sea urchin *Strongylocentrotus intermeidus* and its application to unpalatable algal extracts. Nippon Suisan Gakkaishi 57: 261–265.

Sakata K, Kato K, Ina K, Machiguchi Y (1989) Glycerolipids as potent feeding stimulants for the sea urchin, *Strongylocentrotus intermedius*. Agr Biol Chem 53: 1457–1459.

Sawada M, Miki F (1982) Experiment on seeds collection of the sea urchin *Strongylocentrotus nudus* in nature. Aquaculture Cen Aomori Pref Annual Rep 11: 270–273 (in Japanese).

Shigei M (1995) Echinozoa. In: Nishimura S (ed.) Guide to seashore animals of Japan with color pictures and keys Vol. II. Hoikusha, Osaka, pp 538–552 (in Japanese).

Tajima K, Fukuchi M (1989) Studies on the artificial seed production of the sea urchin *Strongylocentrotus intermedius*. I. The growth of early juveniles in winter. Sci Rep Hokkaido Fish Exp Stn 33: 21–29 (in Japanese with English abstract).

Tajima K, Hirano T, Nakano K, Ezura Y (1997) Taxonomical study on the causative bacterium of spotting disease of sea urchin *Strongylocentrotus intermedius*. Fisheries Sci 63: 897–900.

Tajima K, Shimizu M, Miura K, Ohsaki S, Nishihara Y, Ezura Y (1998a) Seasonal fluctuations of *Flexibacter* sp. the causative bacterium of spotting disease of sea urchin *Strongylocentrotus intermedius* in the culturing facilities and coastal area. Fisheries Sci 64: 6–9.

Tajima K, Takeuchi K, Iqbal MM, Nakano K, Shimizu M, Ezura Y (1998b) Studies on a bacterial disease of sea urchin *Strongylocentrotus intermedius* occurring at low water temperature. Fisheries Sci 64: 918–920.

Tajima K, Tomita K, Kudo K, Matsuya M, Yoshida T (1978) A comparison of the gonadal maturation of a sea urchin, *Strongylocentorotus intermedius*, from Soya and Rebun Island in northern Hokkaido. Hokusuishi Geppo 35: 1–9 (in Japanese).

Tajima K, Yamashita K, Fukuchi M (1991) Studies on the artificial seed production of the sea urchin *Strongylocentrotus intermedius*. II. The collection of metamorphosed larvae. Sci Rep Hokkaido Fish Exp Stn 36: 61–70 (in Japanese with English abstract).

Takahashi N (1980) The annual reproductive cycle of the sea-urchin, *Strongylocentrotus intermedius*, at Rishiri island, Hokkaido. Bull Jap Soc Sci Fish 46: 1189.

Takahashi Y, Itoh K, Ishii M, Suzuki M, Itabashi Y (2002) Induction of larval settlement and metamorphosis of the sea urchin *Strongylocentrotus intermedius* by glycoglycerolipids from the green alga *Ulvella lens*. Mar Biol 140: 763–771.

Taki J (1978) Formation of growth lines in test plates of the sea urchin, *Strongylocentrotus intermedius*, reared with different algae. Bull Jap Soc Sci Fish 44: 955–960 (in Japanese with English abstract).

Taki J (1986) Population dynamics of *Strongylocentrotus intermedius* in Akkeshi Bay. Sci Rep Hokkaido Fish Exp Stn 28: 33–43 (in Japanese with English abstract).

Taki J, Tajima K, Taki K (1992) Growth of the sea urchin, *Strongylocentrotus intermedius* in the eastern and northern coasts of Hokkaido. Suisanzoshoku 40: 479–485 (in Japanese with English abstract).

Tomita K, Kishida M, Massaki K, Iizawa N (1984) Seasonal gonad change of the sea urchin *Strongylocentrotus intermedius* in eastern Hokkaido. Hokusuishi Geppo 41: 469–479 (in Japanese).

Tomita K, Mabuchi M, Shirota S, Konda Y (1986) Growth and reproductive cycle of the sea urchin *Strongylocentrotus intermedius* transplanted from the Japan Sea coast of southern Hokkaido to the Pacific coast of eastern Hokkaido. Hokusuishi Geppo 43: 9–19 (in Japanese).

Tomita K, Tada M (1988) Studies on the marine algae and the sea urchin *Strongylocentrotus intermedius*. Hokkaido Abashiri Fish Exp Stn Annual Rep pp 218–223 (in Japanese).

Vaschenko MA, Zhadan PM, Latypova EV (2001a) Long-term changes in the gonad condition of the sea urchin *Strongylocentrotus intermedius* from Alekseev Bight (Amursky Bay) subjected to pollution. Biol Morya (Mar Biol Vladivostok) 27: 60–64.

Vaschenko MA, Zhadan PM, Latypova EV (2001b) Long-term changes in the state of gonads in the sea urchins *Strongylocentrotus intermedius* from Amur Bay, the Sea of Japan. Russian J Ecol 32: 358–364.

Edible Sea Urchins: Biology and Ecology
Editor: John Miller Lawrence

Chapter 21

Ecology of *Strongylocentrotus nudus*

Yukio Agatsuma

Laboratory of Marine Plant Ecology, Graduate School of Agricultural Science, Tohoku University, Miyagi (Japan).

1. INTRODUCTION

Strongylocentrotus nudus (A Agassiz) (Plate 3D) is harvested commercially in the Pacific Ocean from Ibaragi to Hidaka, Hokkaido, and in the Sea of Japan from Toyama to Soya, Hokkaido (Kawamura 1993). The roe is eaten mainly as sashimi, sushi, served in a bowl with rice (*Uni Don* in Japanese), and salted (*Shio Uni* in Japanese) in Tohoku and Hokkaido. A baked casserole of the flesh served on the shells of the Japanese surf clam *Pseudocardium sybillae* or ezo-abalone *Haliotis discus hannai* and called *Kaiyaki Uni* is popular in Fukushima and Iwate prefectures, Tohoku. *Strongylocentrotus nudus* is the most commonly harvested edible sea urchin in Japan and accounts for about 44% of the total commercial harvest (Imai 1995). Juveniles are produced artificially (Okazaki et al. 1975; Otaki et al. 1984; Tenjin and Ishii 1984) and transplanted off the Pacific coast of Tohoku to increase harvest (Okazaki et al. 1976).

2. GEOGRAPHIC DISTRIBUTION

Strongylocentrotus nudus is found on intertidal and subtidal rocky sea bottoms. It is distributed from Dalian in China to Primorskyi Kray in Russia and in Japan. It is found in the Pacific Ocean from Sagami Bay to Cape Erimo, Hokkaido, and in the Sea of Japan from Omi Island, Yamaguchi, to Soya Cape, northern Hokkaido (Kawamura 1993; Shigei 1995).

Interspecific genetic divergence among three sea urchin species from the Sea of Japan shows that *S. nudus* differs significantly from both *S. intermedius* and *S. pallidus* (Manchenko and Yakovlev 2001). An analysis of mitochondrial DNA sequence shows that *S. nudus* and *S. franciscanus* have a large divergence from other strongylocentrotids (Lee 2003).

3. REPRODUCTION

3.1. Reproductive Cycle

Strongylocentrotus nudus matures at 40–45 mm diameter in Funka Bay, Hokkaido (Fuji 1960b). Gonadal development of the adults is classified into five stages: recovery, growth, prematuration, maturation, and spent (Fuji 1960a). In the Sea of Japan and the Tsugaru Straits in southern Hokkaido, the stages change from recovery to growth during January and May, to maturation during June and August, to spawning during September and October when the water temperature falls from 20 to 16 °C. The gonadal indices start to rise in the spring when the gonads are growing, reach a maximum when the gonads are prem aturing or maturing, fall rapidly to the spent stage, and gradually rise again during recovery (Fuji 1960c; Agatsuma et al. 1988; Agatsuma and Sugawara 1988). *Strongylocentrotus nudus* spawns from September to October. No difference is found among localities from Kyoto in southwestern Honshu to Rebun Island in northern Hokkaido (Fuji 1960c; Kawamura 1967; Sugimoto et al. 1982; Odagiri et al. 1984; Agatsuma et al. 1989; Tsuji et al. 1989; Sano et al. 2001).

3.2. Spawning Cues

The spawning season during September and October corresponds to a fall in water temperature from 25 to 20 °C in Wakasa Bay, Kyoto (Tsuji et al. 1989), and from 20 to 15 °C in Tohoku and Hokkaido (Sugimoto et al. 1982; Odagiri et al. 1984; Agatsuma et al. 1988;). Four-armed plutei appeared a few days after a storm off Rishiri Island in the Sea of Japan, northern Hokkaido (Hokkaido 1981). These observations suggest that spawning is synchronized with changes in abiotic factors such as a rapid fall in water temperature caused by storms.

4. LARVAL ECOLOGY

4.1. Occurrence

Larvae of *S. nudus* appear during September and October, reaching a maximum density of hundreds to thousands of individuals per cubic meter (Kawamura 1973; Hokkaido Central Fisheries Experimental Station et al. 1984; Agatsuma et al. 1989). The density of larvae in the Tsugaru Strait, southern Hokkaido, is much lower than that in the Sea of Japan off southern Hokkaido (Agatsuma et al. 1989). It is not known whether larval dispersion or abundance leads to this difference in densities.

In the area of the Tumen River estuary and in the Far Eastern State Marine Reserve, larvae of *S. nudus* mainly occur from 0 to15 m depth from July to September (Dautov 2000).

4.2. Growth and Survival

The larval stage is estimated to last one to two months off the coast of Shakotan Peninsula in the Sea of Japan, southwestern Hokkaido (Hokkaido Central Fisheries Experimental Station et al. 1984). *Chaetoceros gracilis* and *C. calcitrans* are valuable food for larval growth and survival (Tenjin and Ishii 1984). The larval period of plutei-fed *Phaeodactylum tricornutum* is 19–22 days at 24.9 °C, 14–18 days at 22.2 °C, and 18–25 days at 19.4 °C (Tsuchida 1970). The period when larvae are fed *C. gracilis* is shortened to about 15 days at 20–22 °C (Takahashi et al. 1991).

5. METAMORPHOSIS AND SETTLEMENT

5.1. Algal Communities

The diameter of juveniles immediately after metamorphosis is 350–400 μm (Kawamura 1970). Zonation of communities of *Sargassum yezoense*, *Eisenia bicyclis*, and crustose coralline algae from shallow to deep waters is found commonly on subtidal rocky bottoms off the coast of Sanriku to Joban, Tohoku (Taniguchi 1991). Cobblestones and boulders are the major substratum near shore areas where crustose coralline algae predominate, although no foliaceous algae grow on their unstable substrata (Taniguchi 1996). Sano et al. (1998) surveyed the abundance of larval settlement from 2.0 to 5.5 m depth as correlated with the areas of boulders, community of *E. bicyclis*, and crustose coralline community off the Oshika Peninsula, Miyagi. The juveniles occurred in all areas. However, the highest density of the juveniles immediately after settlement was found on crustose coralline algae from 4.5 to 5.5 m. Three times as many juveniles settled on shells of *Patinopecten yessoensis* that were sunk on crustose coralline algae as collectors as on *E. bicyclis*. Metamorphosed juveniles near shore-boulder areas also had a low density, caused by the low density of larvae there.

5.2. Chemical Inducer

The high density of metamorphosed juvenile on crustose coralline algae is promoted by dibromomethane, a secondary metabolic volatile product. This chemical is produced abundantly by crustose coralline algae and strongly induces larval metamorphosis and settlement (Taniguchi et al. 1994). The highest metamorphosis percentage, more than 80%, is found after 1 h exposure to 1/2 dilution of saturated dibromomethane. With this dilution, more than 80% of *S. nudus* larvae metamorphose 1 h after only 10 min exposure, which correspond to the low concentrations of 52–61 ppm dibromomethane (Agatsuma et al. 2006). Thus, dibromomethane has an instantaneous effect on high success of metamorphosis of larvae of *S. nudus*. Dibromomethane is also found in other red algae of the Corallinaceae, in the green alga *Ulvella lens*, and in attached diatoms that have been used as collectors for larval settlement (Taniguchi et al. 1994). Crustose coralline communities secrete abundant dibromomethane into the water all the time, inducing urchin metamorphosis. This increase in density of sea urchins allows the crustose coralline algae to maintain dominance over other algae.

6. FOOD AND FEEDING AFTER SETTLEMENT

6.1. Food

The principal foods of *S. nudus* <10 mm diameter are detritus, attached diatoms, small algae, and crustose coralline algae; individuals >10 mm diameter graze large algae (Sawada, unpubl. data). In areas with low algal biomass, sand, shells, and crustose coralline algae are also found in their stomachs (Kawamura 1964, 1966a; Akimoto and Tenjin 1974; Sawada et al. 1978; Sano et al. 2001). *Strongylocentrotus nudus* also grazes sessile animals *Mytilus edulis, Balanus* spp., and *Celleporina* spp. (Kittaka 1977; Kittaka and Imamura 1981), and fish (Agatsuma and Nishikiori 1991; Agatsuma 1998; Hoshikawa et al. 1998).

6.2. Food Ingestion and Absorption

The amount of food intake differs with type of algae (Machiguchi et al. 1994; Nabata et al. 1999). Food intake as the percent of body weight of individuals about 5 cm diameter is 4.45% on *Laminaria longissima*, 1.55% on *Agarum cribrosum*, 1.26% on *Ulva pertusa*, and 0.1% on *Neodilsea yendoana* (Machiguchi et al. 1994). The feeding rates on *L. japonica, L. angustata, Undaria pinnatifida*, and *Costaria costata* are high. Those on *Polysiphonia morrowii, Dictyopteris divaricata, Desmarestia viridis*, and *S. confusum* are low (Nabata et al. 1999).

From a rearing experiment in Yoichi, Hokkaido, the food intake, amount digested, and feeding rate increases from April to June. Decreases in food intake were caused by low water temperature during January and February and by maturation of the gonads and spawning during July and September. Small *S. nudus* (1 year old) have a greater feeding rate per individual than large ones throughout the year. Digestibility (amount of food digested × 100/amount of food ingested) is about 50% from spring to summer from 10 to 25 °C, but decreases markedly to 10% in winter at about 6 °C (Agatsuma et al. 1993, 1997). In Kyoto, the high feeding rates are found during November to January and June to August when water temperatures fall from 22 to 14 °C and rise from 17 to 23 °C, respectively (Douke and Hamanaka 2001).

The feeding activity of *S. nudus* in the field varies chiefly with the annual reproductive cycle. In Oshoro Bay, feeding activity is high in May when gonads are in the growth stage but low in September when the gonads are in the prematuration stage (Agatsuma et al. 1996).

6.3. Food Selectivity

Because the weight of each algal species in the stomach of *S. nudus* and the biomass of each alga where the sea urchins live are similar, algal feeding selectivity in the field is not considered important (Sawada et al. 1978). Kittaka et al. (1983) and Hayakawa and Kittaka

(1984) simulated the foraging behavior of *S. nudus* on *U. pinnatifida* and *L. japonica* and concluded it is random. However, Machiguchi et al. (1994) carried out multiple-prey experiments with *L. longissima, A. cribrosum, U. pertusa*, and *N. yendoana* in aquaria and found that *S. nudus* selectively grazed *L. longissima*. Furthermore, behavioral responses of *S. nudus* in multiple-prey tanks indicated that stimuli from *L. longissima* strongly attract the sea urchins, but that *A. cribrosum, N. yendoana*, and *Phyllospadix iwatensis* did not. The authors concluded that the intensity of food selectivity in the field is affected by the density of food algae, distribution of algae, algal species composition, and density of sea urchins.

6.4. Chemical Defense of Algae

The amount of food intake is greatly affected by secondary metabolites. Some species of small perennial brown algae (Dictyotaceae) and red algae (Rhodomelaceae) grow throughout the year on the coralline flats where *S. nudus* is dense. Some diterpenes with feeding-deterrent activity for *S. nudus* were isolated from *Dilophus okamurai, D. divaricata*, and *Laurencia saitoi* (Kurata et al. 1988, 1989, 1990, 1998; Shiraishi et al. 1991a, 1991b). Bromophenols with feeding-deterrent activity were isolated from ethanol extracts of *Odonthalia corymbifera* (Kurata et al. 1997). These algae grow on coralline flats among abundant herbivores by producing chemical defenses to secure exclusive possession of sites. By excluding herbivores, these algae ensure a succession to large perennial algae and a marine forest (Taniguchi 1996). Because *Ecklonia stolonifera, E. kurome, E. cava*, and *E. bicyclis* avoid being grazed by producing water-soluble phlorotannins, sea urchins do not graze on growing thalli directly but on their drift fragments (Taniguchi et al. 1991, 1992a, 1992b).

6.5. Abiotic Factors on Feeding

Machiguchi (1993) observed a decrease in food intake by *S. nudus* in winter in eastern Hokkaido. Food intake decreased markedly at water temperatures $<5\,°C$ (Machiguchi et al. 1994).

Grazing activity is also affected by wave action. Yano et al. (1994) observed the feeding behavior of *S. nudus* in the field for 160 days from December to February (winter) and from May to July (spring to summer), correlated with water temperature and water velocity. Grazing activity is affected not by water temperature but by water velocity at water temperatures $>7\,°C$ in winter. The sea urchins ceased moving at a water velocity >40 cm s^{-1}, and moved little but did not feed at water velocities of 30–40 cm s^{-1}. They fed actively only at water velocities $<20\,cm\,s^{-1}$. At water temperatures $>10\,°C$ and current velocities $<20\,cm\,s^{-1}$, the sea urchins moved more actively when presented with food than without. Kawamata (1998) observed the relationship between feeding activity and water velocity in an oscillating-flow tank. The feeding rate of sea urchins 80 mm in diameter was markedly reduced at a peak velocity of 30 cm s^{-1} and virtually ceased at >40 cm s^{-1}. They could barely move at a water velocity $>70\,cm\,s^{-1}$. *Strongylocentrotus nudus* appears not to feed in winter because of the low water temperature (ca. 5 °C) and high wave action in the field (Agatsuma et al. 1996).

6.6. Foraging

Strongylocentrotus nudus >20 mm in diameter (1 year old) showed remarkable seasonal migrations to use available algal foods. *Strongylocentrotus nudus* with growth-stage gonads were found in the crustose coralline-dominated subtidal zone from April to July, where abundant detrital drift algae derived from the intertidal zone where *L. religiosa* dominated was the principal food source. The sea urchins migrated in search of algal food to the intertidal zone from July to October when their gonads were in the growth, prematuration, or maturation stages, and returned to the subtidal zone from November to March when their gonads were in the recovery stage. This return migration is due to the loss of algal food from the intertidal zone and by high wave action. Yearling sea urchins inhabited the subtidal zone throughout the year because of their preference for attached diatoms or small algae (Agatsuma and Kawai 1997).

7. GROWTH

7.1. Somatic Growth

Strongylocentrotus nudus lives for 14–15 years (Agatsuma 1991). Their ages have been calculated from the number of rings formed in the genital plates, as for *S. intermedius* (Kawamura 1973; Jensen 1969; Kawamura 1966b; Sawada 1977).

The seasonal growth rates of 1- to 5-year-old *S. nudus* fed *L. religiosa* were examined in a rearing experiment in Yoichi, southwestern Hokkaido (Agatsuma et al. 1993). The growth rate of small *S. nudus* was high. In particular, that of the 0- to 1-year-old juveniles, which have undeveloped gonads, was very high throughout the year. The food is converted mainly to test growth. The growth rate of the adults decreased in winter, corresponding to a decrease in food intake and digestion, and in summer, corresponding to maturation and spawning. It increased in spring, corresponding to an increase in food intake, and in autumn, corresponding to postspawning. The food is converted mainly to gonadal growth in spring and test growth in autumn. This seasonal growth pattern of *S. nudus* is also found in field populations in Hokkaido (Agatsuma 1997). The high growth rate is found from November to January from a rearing experiment in Kyoto (Douke and Hamanaka 2001). Agatsuma et al. (2002) fed juvenile *S. nudus* boiled and fresh stipes, blades, and sporophylls of the brown alga *U. pinnatifida*. Growth was greatest with fresh blades.

The growth of *S. nudus*, correlated with biomass of food algae, was investigated in geographically separate fishing grounds and along the same coast (Agatsuma 1997). The growth rate was greatly affected by the kind and abundance of algae. Differences were notable between the Sea of Japan and the Pacific Ocean and between fishing grounds along the same coast. The growth rate of *S. nudus* inhabiting areas dominated by the large perennial brown algae *L. japonica*, *L. angustata*, *Kjellmaniella crassifolia*, and *Alaria crassifolia* is high, and sea urchins reach 5 cm diameter after 2–4 years. The growth rate is very low on coralline flats, taking 7–8 years to reach the same size. The growth difference is also found between large annual alga *L. religiosa*-bed and crustose coralline-bed in southwestern Hokkaido (Dotsu et al. 2004).

The feeding efficiency with *E. stolonifera* for growth of *S. nudus* is as high as that with *L. japonica* (Sato and Notoya 1988). One-year-old *S. nudus* showed high growth rates on *L. angustata, U. pinnatifida*, and *C. costata* during July and August. Other algae gave low growth rates (Nabata et al. 1999). Thus, brown laminarians promote somatic growth of *S. nudus*.

7.2. Gonadal Growth and Color

Gonadal growth, which is also affected by the kinds and abundance of algae, differed among locations within a fishing ground and between geographically separate fishing grounds (Agatsuma 1997). Agatsuma et al. (2005) studied the factors causing brown colored gonads in *S. nudus* in *E. bicyclis* beds and crustose coralline beds in Miyagi prefecture, Japan. Gonad indices of the sea urchins in *Eisenia* beds are significantly higher than those in crustose coralline habitats. In *Eisenia* beds, brown colored gonads are found in sea urchins with a test diameter >7 cm, which correspond to an age of >7 years, and is associated with a decrease in gonad indices. In crustose coralline beds, brown colored gonads are similarly observed in sea urchins >7 years old and in those <7 years old with low gonad indices of <5%. Agatsuma et al. (2005) suggested that the brown colorization is correlated with aging and/or gonad size as determined by food availability. In this study, the main causative pigment of brown colorization is likely to be contained in water- and fat-insoluble residues.

Gonadal growth of *S. nudus* fed *L. angustata* is markedly higher than those fed *S. confusum* and *P. morrowii* (Nabata et al. 1999). Thus, large brown laminarians strongly promote gonadal growth of *S. nudus* (Nabata et al. 1999; Agatsuma 1997). Fish flesh also strongly promotes gonadal growth (Agatsuma and Nishikiori 1991; Agatsuma 1998; Hoshikawa et al. 1998).

Agatsuma et al. (2002) fed juvenile *S. nudus* boiled and fresh stipes, blades, and sporophylls of the brown alga *U. pinnatifida*. Sporophylls promoted the best gonadal growth although the intake of boiled stipes with low carbohydrate and crude protein content was higher than that of fresh stipe and sporophyhlls.

Undeveloped gonads of adult *S. nudus* sampled from trophically poor coralline flats and fed on *L. religiosa* in excess from April to August grew rapidly during the first 20 days of the experiment. The gonad index increased from 6.5 at the start of the experiment to 18 (the minimum size for commercial landing) in 2 months (Agatsuma 1999a).

8. HABITAT

8.1. Juvenile Habitat

Distribution and migration of *S. nudus* is affected by substrata, wave action, algal vegetation, and algal drifts (Kawamura 1964, 1966a; Akimoto and Tenjin 1974; Sawada et al. 1978; Nabata et al. 1992). In particular, high densities of *S. nudus* occur on coralline flats (Taniguchi 1991; Nabata et al. 1992; Agatsuma 1997). Akimoto and Tenjin (1974) found juveniles under gravel and boulders dominated by crustose coralline algae in intertidal areas off Nagasaki, Fukushima. Off the coast of Tsugaru Strait, Fukushima juveniles are found on immobile rocks covered with crustose coralline algae from 3 to 9 m depth. The range of their habitats gradually expands vertically after 1 year of age. The density of the juveniles and adults at these depths is high throughout the year (Agatsuma et al. 1986).

Sano et al. (1998) investigated larval settlement in an area of boulders, *E. bicyclis* communities, and crustose coralline communities off the Pacific coast of Oshika Peninsula. One-year-old juveniles survive at a high rate in the boulders and the crustose coralline communities. The survival of juveniles in the communities of *E. bicyclis* was very low. High mortality of juveniles of *Evechinus chloroticus* (Andrew and Choat 1985) and *S. purpuratus* (Rowley 1990) also occurs in kelp forests.

8.2. Movement

Ito and Hayashi (1999) monitored diel movement of *S. nudus* with 47.5 mm diameter without food under 12 L–12 D. Individuals repeatedly moved for short periods of time at 1–5 cm min^{-1} and were more active in light. Moving reached 2–3 m for 2 h per day. Individual interference is more intense for *S. nudus* than *S. intermedius*.

Large sea urchins >4 cm in diameter (>2 years old) inhabit communities of *E. bicyclis* off the coast of Oshika Peninsula. It is thought that individuals 1–2 years old migrate from crustose coralline communities to the kelp forest, particularly rapidly growing 1-year-old individuals (Sano et al. 2001). Individuals >20 mm in diameter (1 year old) migrate in search of algal food to the intertidal zone from July to October and return to the subtidal zone from November to March owing to the loss of algal food from the intertidal zone and the high wave action in Oshoro Bay (Agatsuma and Kawai 1997).

According to Tsuji et al. (1989) and Douke et al. (2003), *S. nudus* in Wakasa Bay, Kyoto, close to the southern limit of the geographic distribution, lives at a depth of >4 m. The sea urchins migrate to deeper waters and occur under rocks in summer, but live in shallow water on the surface of rocks where food algae are abundant from autumn to spring. It is thought that the migration to deeper waters in summer is to avoid high water temperatures.

Dotsu et al. (1997) investigated seasonal changes in the density of *S. nudus* on the vertical surface of two breakwaters, one exposed to high wave action in winter and one not exposed to wave action throughout the year. The density on the former was extremely low in winter, but that on the latter was high throughout the year. They concluded that the strength of wave action determines the density of the sea urchins.

9. COMMUNITY ECOLOGY

Coralline flats (Ayling 1981) where crustose coralline red algae dominate and no large algae grow in subtidal regions off the Sea of Japan coast off southwestern Hokkaido have expanded from the 1960s (Fujita 1987). The maintenance of coralline flats leads to reduced production by fisheries (Taniguchi 1996). The cause has been considered to be intensive grazing by *S. nudus* at densities of >7 individuals per square meter (Nabata et al. 1992; Fujita 1989; Agatsuma et al. 1997). Nabata et al. (1992) reported that communities of *L. religiosa* raised on artificial stone beds on coralline flats in autumn disappeared under intensive grazing by *S. nudus* until the next spring. Off the coast of Kyoto, growth of Sargassaceae brown algae are reduced by intensive grazing of *S. nudus* (Douke et al. 2004).

Field enclosure experiments with *S. nudus* at several densities showed that sea urchins at a biomass of $>200\,g\,m^{-2}$ have a great impact on algal vegetation (Kikuchi and Uki 1981). In Aomori prefecture, the removal of dense populations of *S. nudus* and the culturing of *L. japonica* resulted in formation of a kelp forest. After the disappearance of communities of *Pachymeniopsis elliptica* by intensive grazing by transplanted *S. nudus*, recapture of sea urchins with enhanced gonads resulted in the formation of a kelp forest (Sawada et al. 1981).

Agatsuma et al. (1997) studied the succession of marine algae after the removal of dense populations of *S. nudus* from coralline flats in Suttsu Bay, southwestern Hokkaido. At Yaoi, facing the open sea, attached diatoms, small annual algae such as *U. pertusa* and *P. morrowii*, large annual algae such as *D. viridis* and *U. pinnatifida*, and small perennial algae such as *D. divaricata* invaded the coralline flats in that order. They were followed by the large perennial alga *S. confusum* as the climax, a species endemic to the Sea of Japan. In contrast, at Rokujo in the inner part of the bay, dominance of the small annual alga *U. pertusa*, which colonized in the early phase of algal succession, continued owing to the influence of inflow of freshwater and sand from a river near the site. The community of *S. confusum*, the climax phase of algal succession, persists over long periods owing probably to the warm temperatures caused by the Tsushima Warm Current.

Off the coast of Rishiri Island in the Sea of Japan, northern Hokkaido, removal of sea urchins resulted in a climax community of *L. ochcotensis* (Agatsuma 1999b). Dotsu et al. (1999) reported that the growth of a community of *L. religiosa* on mounds dotted on the seabed in rough topography is due to a decline in the density of sea urchins caused by high wave action during autumn and winter when zoospores of the kelp are released.

A field experiment demonstrated the upper surface of the scallop *Patinopecten yessoensis* serves as mobile refugia for epibenthic macroalgae from the grazing of *S. nudus* and *S. intermedius* (Ozolinsh and Kupriyanova 2000).

Over 35% of the egg masses spawned from the sea raven *Hemitripterus villosus* in Peter the Great Bay, the Sea of Japan are eaten by echinoderms, primarily *Patiria pectinifera* and *S. nudus*. Their predation can greatly affect the abundance of this species (Markevich 2000).

10. POPULATION DYNAMICS

10.1. Recruitment of Juveniles

The density of larvae off the coast of Shakotan Peninsula fluctuates annually by a factor of 10 (Kawamura 1993). Researchers monitored the recruitment of *S. nudus* off the coast of Okushiri Island and Shimamaki in the Sea of Japan, southwestern Hokkaido, for 16 years (Agatsuma et al. 1998). They recorded high recruitment levels (14.1 individuals per square meter) in 1984 and continuous annual recruitment after 1990. The relationship between juvenile densities and average water temperature in September of the previous year was statistically significant. They concluded high water temperature in September contributes to an improvement in the juvenile survival rate by shortening the larval period.

10.2. Annual Fluctuations

The recruitment and growth of juvenile *S. nudus* and algal biomass were monitored off the coast of Fukushima, Tsugaru Strait, Hokkaido (Agatsuma 1994). The growth of abundant year classes of 1982 and 1983 greatly increased in 1984 when an abundant *L. japonica* community formed from shallow to deep water on coralline flats owing to low water temperatures caused by the inflow of coastal Oyashio waters in 1984 and 1985 (Ohtani 1987). As a result, the commercial landing of *S. nudus* in Fukushima increased abruptly from 1985 to 1987. Abundant recruitment of juveniles occurred all around the Sea of Japan coast of Hokkaido. Abundant *L. religiosa* and *L. ochcotensis* grew from the southern to the northern regions of the Sea of Japan in 1984 at low water temperatures in winter owing to a decrease in the flow of the Tsushima Warm Current. This resulted in an abrupt increase in commercial landings of *S. nudus* in Hokkaido (Agatsuma 1997). Sano et al. (2001) reported that growth of *S. nudus* was also promoted by the expansion of *E. bicyclis* to deep waters owing to the strong influence of the Oyashio Current off the Pacific coast of Tohoku.

Annual hydrographic conditions also affect the foraging activity of *S. nudus*. The length of time *S. nudus* forages actively during spring and summer changes owing to annual fluctuations in gonadal development attributable to the biomass of the algal community on which they feed (Agatsuma et al. 2000).

Thus, feeding, gonadal development, and somatic growth of *S. nudus* off the coast of the Sea of Japan to the Tsugaru Straits in southwestern to southern Hokkaido are attributable to both the kinds and abundance of algae, these being related to the Tsushima Warm Current and its variability. The annual recruitment level of juveniles is also affected by hydrographic conditions. Abundant juvenile recruitment due to high water temperature in September and successive formation of kelp forest due to low water temperature in winter results in an increase in the population size of the sea urchins and promotion of somatic and gonadal growth. These lead to an increase in commercial landings.

Abundant biomass of *L. religiosa* and high densities of *S. nudus* were observed before the 1950s when coralline flats were not found in shallow waters. An analysis of long-term water temperatures showed that abundant recruitment of the juvenile *S. nudus* occurred frequently and subsequently the kelp forest grew. Thus, it seems that sea urchins and kelps coexisted before the 1950s (Agatsuma 1999b).

10.3. Decrease in Population Size

The population size of *S. nudus* is reduced by biotic and abiotic factors. Mass mortality of *S. nudus* occurred in Wakasa Bay in the summer of 1994 owing to high water temperatures of 26–28 °C (Tsuji et al. 1994).

Laboratory experiments showed that the predators of *S. nudus* are the sea stars *Solaster paxillatus, Asterina pectinifera*, and *Lysastrosoma anthostictha* and the rock crabs *Pugettia quadridens* and *Telmessus acutidens* (Amagaya 1987). *Plazaster borealis* is also a predator (Takahashi et al. 2000). Avoidance reaction of *S. nudus* to *P. borealis* is induced by (3β, 5α, 6α, 23)-cholest-9(11)-ene-3,6,23-triol 3,6-disulfate disodium salt released from the sea star (Takahashi et al. 2000). Predation by *P. quadridens* on the sea urchins is reduced at temperatures of <10 °C, especially <5 °C (Shiraishi 1997). Musashi et al. (1997) reported that *S. nudus* dies after parasitization by the snail *Pelseneeria castanea*. Up to 43% of *S. nudus* were parasitized off the Pacific coast of Iwate. The snails' geographic distribution has been expanding north, and about 5% of the sea urchin population is parasitized off the Sea of Japan coast in southwestern Hokkaido (Kawai and Agatsuma 1994). The influence of predation and parasitism on sea urchin population size has not been measured.

11. CONCLUSIONS

Strongylocentrotus nudus is a dominant sea urchin species in the northwest Pacific. The habitats of recruits and adults differ. They recruit onto coralline flats and migrate to kelp forests when 1 year old or greater. Somatic and gonadal growth are supported by brown macroalgae. *Strongylocentrotus nudus* forms extremely high population densities that can result in overgrazing. Its high growth rate, great longevity, high reproductive efforts, reproduction after a period of maximum feeding and growth responses to food availability, are characteristics of a competitive species. The population biology of *S. nudus* is not well known and studies on the effect of biotic and abiotic factors must be done.

REFERENCES

Agatsuma Y (1991) *Strongylocentrotus nudus* (A. Agassiz). In: Nagasawa K, Torisawa M (eds) Fishes and marine invertebrates of Hokkaido: Biology and fisheries. Kita-Nihon Kaiyo Center Co. Ltd, Sapporo, pp 330–333 (in Japanese)

Agatsuma Y (1994) Factors contributing to changes in the population size of the sea urchin, *Strongylocentrotus nudus* off the coast of Tsugaru Strait, southern Hokkaido. Suisanzoshoku 42: 207–213 (in Japanese with English abstract)

Agatsuma Y (1997) Ecological studies on the population dynamics of the sea urchin *Strongylocentrotus nudus*. Sci Rep Hokkaido Fish Exp Stn 51: 1–66 (in Japanese with English abstract)

Agatsuma Y (1998) Aquaculture of the sea urchin (*Strongylocentrotus nudus*) transplanted from coralline flats in Hokkaido, Japan. J Shellfish Res 17: 1541–1547

Agatsuma Y (1999a) Gonadal growth of the sea urchin, *Strongylocentrotus nudus*, from trophically poor coralline flats and fed excess kelp, *Laminaria religiosa*. Suisanzoshoku 47: 325–330

Agatsuma Y (1999b) Marine afforestation off the Japan Sea coast in Hokkaido. In: Taniguchi K (ed.) The ecological mechanism of "Isoyake" and marine afforestation. Kouseisha Kouseikaku, Tokyo, pp 84–97 (in Japanese)

Agatsuma Y, Hayashi T, Uchida M (1989) Seasonal larval occurrence and spawning season of two sea urchins, *Strongylocentrotus intermedius* and *S. nudus*, in southern Hokkaido. Sci Rep Hokkaido Fish Exp Stn 33: 9–20 (in Japanese with English abstract)

Agatsuma Y, Kawai T (1997) Seasonal migration of the sea urchin *Strongylocentrotus nudus* in Oshoro Bay of southwestern Hokkaido, Japan. Nippon Suisan Gakkaishi 63: 557–562 (in Japanese with English abstract)

Agatsuma Y, Matsuyama K, Nakata A (1996) Seasonal changes in feeding activity of the sea urchin *Strongylocentrotus nudus* in Oshoro Bay, southwestern Hokkaido. Nippon Suisan Gakkaishi 62: 592–597 (in Japanese with English abstract)

Agatsuma Y, Matsuyama K, Nakata A, Kawai T, Nishikawa N (1997) Marine algal succession on coralline flats after removal of sea urchins in Suttsu Bay on the Japan Sea coast of Hokkaido, Japan. Nippon Suisan Gakkaishi 63: 672–680 (in Japanese with English abstract)

Agatsuma Y, Motoya S, Sasaki Y, Kato T (1986) Ecology of the sea urchin *Strongylocentrotus nudus* (A. Agassiz) at Urawa, Fukushima in southern Hokkaido. Hokusuishi Geppo 43: 126–136 (in Japanese)

Agatsuma Y, Motoya S, Sugawara Y (1988) Reproductive cycle and food ingestion of the sea urchin, *Strongylocentrotus nudus* (A. Agassiz), in southern Hokkaido. I. Seasonal changes of gonad. Sci Rep Hokkaido Fish Exp Stn 30: 33–41 (in Japanese with English abstract)

Agatsuma Y, Nakao S, Motoya S, Tajima K, Miyamoto T (1998) Relationship between year-to-year fluctuations in recruitment of juvenile sea urchins *Strongylocentrotus nudus* and seawater temperature in southwestern Hokkaido. Fisheries Sci 64: 1–5

Agatsuma Y, Nakata A, Matsuyama K (1993) Feeding and assimilation of the sea urchin, *Strongylocentrotus nudus* for *Laminaria religiosa*. Sci Rep Hokkaido Fish Exp Stn 40: 21–29 (in Japanese with English abstract)

Agatsuma Y, Nakata A, Matsuyama K (2000) Seasonal foraging activity of the sea urchin *Strongylocentrotus nudus* on coralline flats in Oshoro Bay in southwestern Hokkaido, Japan. Fisheries Sci 66: 198–203

Agatsuma Y, Nishikiori T (1991) Gonad development of the northern sea urchin *Strongylocentrotus nudus* experimentally fed fishes. I. Gonad development. Sci Rep Hokkaido Fish Exp Stn 37: 59–66 (in Japanese with English abstract)

Agatsuma Y, Sato M, Taniguchi K (2005) Factors causing brown-colored gonads of the sea urchin *Strongylocentrotus nudus* in northern Honshu, Japan. Aquaculture 249: 449–458

Agatsuma Y, Seki T, Kurata K, Taniguchi K (2006) Instantaneous effect of dibromomethane on metamorphosis of larvae of the sea urchins *Strongylocentrotus nudus* and *Strongylocentrotus intermedius*. Aquaculture 251: 549–557

Agatsuma Y, Sugawara Y (1988) Reproductive cycle and food ingestion of the sea urchin, *Strongylocentrotus nudus* (A. Agassiz), in southern Hokkaido. II. Seasonal changes of the gut content and test weight. Sci Rep Hokkaido Fish Exp Stn 30: 43–49 (in Japanese with English abstract).

Agatsuma Y, Yamada Y, Taniguchi K (2002) Dietary effect of the boiled stipe of brown alga *Undaria pinnatifida* on the growth and gonadal enhancement of the sea urchin *Strongylocentrotus nudus*. Fisheries Sci 68: 1274–1281

Akimoto Y, Tenjin A (1974) Ecological studies on the sea urchin *Strongylocentrotus nudus* (A. Agassiz) in the prohibition on the coast of Nagasaki in the south region of Fukushima. Bull Fukushima Pref Fish Exp Stn 2: 19–29 (in Japanese with English abstract)

Amagaya A (1987) Predation of the sea urchin *Strongylocentrotus nudus* by sea stars. Rep Tohoku Mariculture Res Conference, pp 37–41 (in Japanese)

Andrew NL, Choat JH (1985) Habitat related differences in the survivorship and growth of juvenile sea urchins. Mar Ecol Prog Series 27: 155–161

Ayling AM (1981) The role of biological disturbance in temperate subtidal encrusting communities. Ecology 62: 830–847

Dautov SSh (2000) Distribution, species composition and dynamics of echinoderm larvae in the area of the Tumen River estuary and in the Far Eastern State Marine Reserve. Biol Morya (Mar Biol Vladivostok) 26: 16–21

Dotsu K, Nomura H, Ohta M, Iwakura Y (1999) Factors causing formation of *Laminaria religiosa* bed on coralline flats along the southwest coast of Hokkaido. Nippon Suisan Gakkaishi 65: 216–222 (in Japanese with English abstract)

Dotsu K, Nomura H, Ohta M, Kashiwagi M (2004) A comparison of the growth of the sea urchin *Strongylocentrotus nudus* in two habitats, a kelp bed and its adjacent coralline flat. Suisanzoshoku 52: 215–219

Dotsu K, Ohta M, Saito J, Yamashita K (1997) Relationships between wave forces and distribution patterns of the sea urchin *Strongylocentrotus nudus* on the coastal artificial structures. Suisanzoshoku 45: 445–450 (in Japanese with English abstract)

Douke A, Hamanaka Y (2001) Growth of a sea urchin *Strongylocentrotus nudus* reared in laboratory. Bull Kyoto Inst Ocean Fish Sci 23: 25–29 (in Japanese with English abstract)

Douke A, Yoshiya M, Itani M (2003) Seasonal vertical movement of the sea urchin *Strongylocentrotus nudus* in the coastal waters off Kyoto Prefecture. Bull Kyoto Inst Ocean Fish Sci 25: 19–25 (in Japanese with English abstract)

Douke A, Yoshiya M, Itani M (2004) Prevalence and decay of Sargassaceae seaweed grazed by herbivorous species. Bull Kyoto Inst Ocean Fish Sci 26: 15–20 (in Japanese with English abstract)

Fuji A (1960a) Studies on the biology of the sea urchin. I. Superficial and histological gonadal changes in gametogenic process of two sea urchins *Strongylocentrotus nudus* and *S. intermedius*. Bull Fac Fish Hokkaido Univ 11: 1–14

Fuji A (1960b) Studies on the biology of the sea urchin. II. Size at first maturity and sexuality of two sea urchins, *Strongylocentrotus nudus* and *S. intermedius*. Bull Fac Fish Hokkaido Univ 11: 43–48

Fuji A (1960c) Studies on the biology of the sea urchin. III. Reproductive cycle of two sea urchins, *Strongylocentrotus nudus* and *S. intermedius*, in southern Hokkaido. Bull Fac Fish Hokkaido Univ 11: 49–57

Fujita D (1987) The report of interview to fishermen on "Isoyake" in Taisei-cho, Hokkaido. Suisanzoshoku 35: 135–138 (in Japanese)

Fujita D (1989) Marine algal distribution in the "Isoyake" area at Taisei, Hokkaido. Nankiseibutsu 31: 109–114 (in Japanese with English abstract)

Hayakawa Y, Kittaka J (1984) Simulation of feeding behavior of sea urchin *Strongylocentrotus nudus*. Bull Jap Soc Fish 50: 233–240

Hokkaido (1981) Report of large-scale development of propagation ground. North Rishiri Island, *Strongylocentrotus intermedius*. Hokkaido, 1–88 (in Japanese)

Hokkaido Central Fisheries Experimental Station, Shiribeshihokubu Fisheries Extension Office, Hokkaido Institute of Mariculture (1984) On the natural seeds collection, intermediate culture and release of the sea urchin, *Strongylocentrotus intermedius*. Hokusuishi Geppo 41: 270–315 (in Japanese)

Hoshikawa H, Takahashi K, Sugimoto T, Tsuji K, Nobuta S (1998) The effects of fish meal feeding on the gonad quality of cultivated sea urchins, *Strongylocentrotus nudus* (A. Agassiz). Sci Rep Hokkaido Fish Exp Stn 52: 17–24 (in Japanese with English abstract)

Imai T (1995) Studies on the sea urchin propagation in central Japan. Bull Kanagawa Fish Exp Stn 6: 1–90 (in Japanese with English abstract)

Ito Y, Hayashi I (1999) Behavior of the sea urchin, *Strongylocentrotus nudus* and *S. intermedius*, observed under experimental conditions in the laboratory. Otsuchi Mar Sci 24: 24–29 (in Japanese with English abstract)

Jensen ML (1969) Age determination of echinoids. Sarsia 37: 41–44

Kawai T, Agatsuma Y (1994) Distribution and occurrence of the parasitic snail *Pelseneeria castanea* off the coast of Shiribeshi in western Hokkaido. Saibaigiken 23: 7–9 (in Japanese)

Kawamata S (1998) Effect of wave-induced oscillatory flow on grazing by a subtidal sea urchin *Strongylocentrotus nudus* (A. Agassiz). J Exp Mar Biol Ecol 224: 31–38

Kawamura K (1964) The sea urchin *Strongylocentrotus nudus* off the coast of Touei, Urakawa in Hokkaido. Hokusuishi Geppo 21: 510–524

Kawamura K (1966a) Ecological studies and some discussions about the methods of conservation of the sea urchin *Strongylocentrotus nudus* on the coast of Urakawa, Hokkaido. Sci Rep Hokkaido Fish Exp Stn 5: 7–30 (in Japanese with English abstract)

Kawamura K (1966b) On the age determining character and growth of a sea urchin *Strongylocentrotus nudus*. Sci Rep Hokkaido Fish Exp Stn 6: 56–61 (in Japanese with English abstract)

Kawamura K (1967) Some information on ecology of two sea urchins *Strongylocentrotus intermedius* and *S. nudus* off the Japan Sea coast of Yoichi in southwestern Hokkaido. Hokusuishi Geppo 24: 36–45 (in Japanese)

Kawamura K (1970) On the development of the planktonic larvae of Japanese sea urchcins *Strongylocentrotus intermedius* and *S. nudus*. Sci Rep Hokkaido Fish Exp Stn 12: 25–32 (in Japanese with English abstract)

Kawamura K (1973) Fishery biological studies on a sea urchin, *Strongylocentrotus intermedius*. Sci Rep Hokkaido Fish Exp Stn16: 1–54 (in Japanese with English abstract)

Kawamura K (1993) Uni Zouyoushoku to Kakou, Ryutsu. Hokkai Suisan Co, Sapporo (in Japanese)

Kikuchi S, Uki N (1981) Productivity of benthic grazers, abalone and sea urchin in *Laminaria* beds. In: Japanese Society of Fisheries, Science (ed.) Seaweed beds. Fisheries Science Series 23. Kouseisha-Kouseikaku, Tokyo, pp 9–23 (in Japanese)

Kittaka J (1977) Control of attached organisms on cultured scallop by predators. Kaiyo Kagaku 9: 241–246 (in Japanese).

Kittaka J, Imamura K (1981) Fundamental studies on control of marine fouling organisms by sea urchin. Marine Fouling 3: 53–59 (in Japanese with English abstract)

Kittaka J, Nishimura K, Yamada K, Hayakawa Y (1983) Experimental analysis on feeding behavior of sea urchin. Marine Fouling 4: 5–9 (in Japanese with English abstract)

Kurata K, Shiraishi K, Takato T, Taniguchi K, Suzuki M (1988) A new feeding-deterrent diterpenoid from the brown alga *Dilophus okamurai* Dawson. Chemistry Letter 10: 1629–1632

Kurata K, Taniguchi K, Shiraishi K, Suzuki M (1989) Structures of secospatane-type diterpenes with feeding-deterrent activity from the brown alga *Dilophus okamurai*. Tetrahedron Lett 30: 1567–1570

Kurata K, Taniguchi K, Shiraishi K, Suzuki M (1990) Feeding-deterrent diterpenes from the brown alga *Dilophus okamurai*. Phytochemistry 29: 3453–3455

Kurata K, Taniguchi K, Takashima K, Hayashi I, Suzuki M (1997) Feeding-deterrent bromophenols from *Odonthalia corymbifera*. Phytochemistry 45: 485–487

Kurata K, Taniguchi T, Agatsuma Y, Suzuki M (1998) Diterpenoid feeding-deterrents from *Laurencia saitoi*. Phytochemistry 47: 363–369

Lee Y-H (2003) Molecular phylogenies and divergence times of sea urchin species of Strongylocentrotidae, Echinoida. Mol Biol Evol 20: 1211–1221

Machiguchi Y (1993) Growth and feeding behavior of the sea urchin *Strongylocentrotus nudus* (A. Agassiz). Bull Hokkaido Natl Fish Res Inst 57: 81–86 (in Japanese with English abstract)

Machiguchi Y, Mizutori S, Sanbonsuga Y (1994) Food preference of sea urchin *Strongylocentrotus nudus* in laboratory. Bull Hokkaido Natl Fish Res Inst 58: 35–43 (in Japanese with English abstract)

Manchenko GP, Yakovlev SN (2001) Genetic divergence between three sea urchin species of the genus *Strongylocentrotus* from the Sea of Japan. Biochem System Ecol 29: 31–44

Markevich AI (2000) Spawning of the sea raven *Hemitripterus villosus* in Peter the Great Bay, Sea of Japan. Biol Morya (Mar Biol Vladivostok) 26: 272–274

Musashi T, Saido T, Uchida T (1997) The effects of the parasitic snail *Pelseneeria castanea* on the sea urchin *Strongylocentrotus nudus* along the Iwate coast of northern Japan. Bull Iwate Pref Fish Tech Center 1: 27–35 (in Japanese with English abstract)

Nabata S, Abe E, Kakiuchi M (1992) On the Isoyake condition in Taisei-cho, southwestern Hokkaido. Sci Rep Hokkaido Fish Exp Stn 38: 1–14 (in Japanese with English abstract)

Nabata S, Hoshikawa H, Sakai Y, Funaoka T, Ohori, Imamura T (1999) Food value of several algae for growth of the sea urchin, *Strongylocentrotus nudus*. Sci Rep Hokkaido Fish Exp Stn 54: 33–40 (in Japanese with English abstract)

Odagiri A, Asuke M, Sato K (1984) Gonadal maturation of the sea urchin *Strongylocentrotus nudus*, inhabiting in the deep water off shore of Okoppe, Aomori Prefecture. Sci Rep Aquaculture Cen Aomori Pref 3: 1–7 (in Japanese with English abstract)

Ohtani K (1987) Westward inflow of the coastal Oyashio Water into Tsugaru Strait. Bull Fac Fish Hokkaido Univ 38: 209–220 (in Japanese with English abstract)

Okazaki K, Akimoto Y, Isogami K (1976) On the pursuit research of the released artificial sea urchin *Strongylocentrotus nudus*. Bull Fukushima Pref Fish Exp Stn 4: 61–68 (in Japanese)

Okazaki K, Tenjin A, Akimoto Y (1975) Studies on the artificial maturation of the sea urchin *Strongylocentrotus nudus*. Bull Fukushima Pref Fish Exp Stn 3: 51–55 (in Japanese)

Otaki K, Shimozono S, Tenjin A (1984) Studies on the artificial production of the sea urchin *Strongylocentrotus nudus* (A. Agassiz). I. Practical mass culturing technique of plutei and metamorphic larvae. Bull Fukushima Pref Fish Farm Exp Stn 1: 1–18 (in Japanese)

Ozolinsh AV, Kupriyanova EK (2000) Hitch-hiking on scallops: Grazing avoidance by macrophytes. J Mar Biol Assn UK 80: 743–744

Rowley RJ (1990) Newly settled sea urchins in a kelp bed and urchin barren ground: A comparison of growth and mortality. Mar Ecol Prog Ser 62: 229–240

Sano M, Omori M, Taniguchi K, Seki T (2001) Age distribution of the sea urchin *Strogylocentrotus nudus* (A. Agassiz) in relation to algal zonation in a rocky coastal area on Oshika Peninsula, northern Japan. Fisheries Sci 67: 628–639.

Sano M, Omori M, Taniguchi K, Seki T, Sasaki R (1998) Distribution of the sea urchin *Strongylocentrotus nudus* in relation to marine algal zonation in the rocky coastal area of the Oshika Peninsula, northern Japan. Benthos Research 53: 79–87

Sato K, Notoya M (1988) Food value of *Ecklonia stolonifera* for growth of topshell, sea urchin and abalone. Nippon Suisan Gakkaishi 54: 1451 (in Japanese with English abstract)

Sawada M (1977) Ecological study on the sea urchin *Stronglylocentrotus nudus*. I. Age and growth. Aomori Pref Aqua Res Center Reports 7: 1–7 (in Japanese)

Sawada M, Miki F, Asuke M (1978) Ecological study on the sea urchin *Stronglylocentrotus nudus*. II. Migration and food habit. Aomori Pref Aqua Res Center Reports 11: 1–9 (in Japanese)

Sawada M, Miki F, Asuke M (1981) Aquatic afforestation with *Laminaria*. In: Japanese Society of Fisheries, Science (ed.) Seaweed beds. Fisheries Science Series 23. Kouseisha-Kouseikaku, Tokyo, pp 130–141 (in Japanese)

Shigei M (1995) Echinozoa. In: Nishimura S (ed.) Guide to seashore animals of Japan with color pictures and keys. Vol. II. Hoikusha, Osaka, pp 538–552 (in Japanese)

Shiraishi K (1997) Effect of water temperature on the predation of the sea urchin, *Strongylocentrotus nudus*. Suisanzoshoku 45: 321–325 (in Japanese with English abstract)

Shiraishi K, Taniguchi K, Kurata K, Suzuki M (1991a) Feeding deterrent effect of the methanol extract from the brown alga *Dilophus okamurai* against the sea urchin *Strongylocentrotus nudus*. Nippon Suisan Gakkaishi 57: 1591–1595 (in Japanese with English abstract)

Shiraishi K, Taniguchi K, Kurata K, Suzuki M (1991b) Effects of the methanol extracts from the brown alga *Dictyopteris divaricata* on feeding by the sea urchin *Strongylocentrotus nudus* and the abalone *Haliotis discus hannai*. Nippon Suisan Gakkaishi 57: 1945–1948 (in Japanese with English abstract)

Sugimoto T, Tajima K, Tomita K (1982) Reproductive cycle of the sea urchin, *Strongylocentrotus nudus*, on the northern coast of Hokkaido. Sci Rep Hokkaido Fish Exp Stn 24: 91–99 (in Japanese with English abstract)

Takahashi N, Ojika M, Ejima D (2000) Isolation and identification of a trihydroxysteroid disulfate from the starfish *Plazaster borealis* which induces avoidance reaction-inducing substance in the sea urchin *Strongylocentrotus nudus*. Fisheries Sci 66: 412–413

Takahashi H, Yamaguchi H, Ohta K (1991) Artificial production of juvenile sea urchins. Iwate Pref Aqua Cen Annual Rep pp 5–31 (in Japanese)

Taniguchi K (1991) Marine afforestation of *Eisenia bicyclis* (Laminaria; Phaeophyta). NOAA Tech Rpt NMFS 102: 47–57

Taniguchi K (1996) Fundamental and practice of marine afforestation. Jpn J Phycol (Sorui) 44: 103–108 (in Japanese with English abstract)

Taniguchi K, Akimoto Y, Kurata K, Suzuki M (1992a) Chemical defense mechanism of the brown alga *Eisenia bicyclis* against marine herbivores. Nippon Suisan Gakkaishi 58: 571–575 (in Japanese with English abstract)

Taniguchi K, Kurata K, Maruzoi T, Suzuki M (1994) Dibromomethane, a chemical inducer on settlement and metamorphosis of the sea urchin larvae. Fisheries Sci 60: 795–796

Taniguchi K, Kurata K, Suzuki M (1991) Feeding-deterrent effect of phlorotannins from the brown alga *Ecklonia stolonifera* against the abalone *Haliotis discus hannai*. Nippon Suisan Gakkaishi 57: 2065–2071 (in Japanese with English abstract)

Taniguchi K, Kurata K, Suzuki M (1992b) Feeding-deterrent activity of some Laminariaceous brown algae against the Ezo-abalone. Nippon Suisan Gakkaishi 58: 577–581 (in Japanese with English abstract)

Tenjin A, Ishii T (1984) Studies on the living food for planktonic larvae of sea urchin and bivalve. Bull Fukushima Pref Fish Farm Exp Stn 1: 29–34 (in Japanese)

Tsuchida K (1970) Experiment on settlement of sea urchins. Iwate Pref Fish Exp Stn Annual Rep pp 60–61 (in Japanese)

Tsuji S, Munekiyo M, Itani M, Douke A (1994) Mass biolysis of a sea urchin in the western part of Wakasa Bay. Bull Kyoto Inst Ocean Fish Sci 17: 51–54 (in Japanese)

Tsuji S, Yoshiya M, Tanaka M, Kuwahara A, Uchino K (1989) Seasonal changes in distributions and ripeness of gonad of a sea urchin *Strongylocentrotus nudus* in the western part of Wakasa Bay. Bull Kyoto Inst Ocean Fish Sci 12: 15–21 (in Japanese)

Yano K, Akeda S, Satoh J, Matsuyama K, Agatsuma Y, Nakata A (1994) Influence of current and water temperature on feeding behavior of sea urchin. In: Techno-ocean 94 International Symposium Proceedings. 1: 195–198 (in Japanese with English abstract)

Edible Sea Urchins: Biology and Ecology
Editor: John Miller Lawrence

Chapter 22

Ecology of *Hemicentrotus pulcherrimus*, *Pseudocentrotus depressus*, and *Anthocidaris crassispina*

Yukio Agatsuma

Laboratory of Marine Plant Ecology, Graduate School of Agricultural Science, Tohoku University, Miyagi (Japan).

1. INTRODUCTION

Hemicentrotus pulcherrimus (A Agassiz) (Plate 2B), *Pseudocentrotus depressus* (A Agassiz) (Plate 2C), and *Anthocidaris crassispina* (A Agassiz) are important fisheries resources in shallow waters in southern Japan. As *H. pulcherrimus* have a bitter taste, they are not eaten raw but are preserved in small bottles mixed in brine or alcohol. In particular, the *Echizen Uni* brand, processed from the 1600s in Fukui, and *Shimonoseki Uni*, from the 1800s in Yamaguchi, are highly regarded.

2. GEOGRAPHIC DISTRIBUTION

Hemicentrotus pulcherrimus is found in intertidal and subtidal zones from Kyushu to Ishikari Bay in Hokkaido, Japan, and in Korea and China (Shigei 1995). In recent years, its distribution has spread to northern Hokkaido and Rebun Island in the Sea of Japan closely related to an increased influx of the Tsushima Warm Current (Agatsuma, unpubl. data). It is commercially harvested mainly from northern Kyushu to Fukui in the Sea of Japan.

Pseudocentrotus depressus is found in intertidal and subtidal zones from Kyushu to Choshi, Chiba on the Pacific Ocean, in the southern Sea of Japan around Tsugaru Strait; and on Cheju Island, Korea (Shigei 1995). It is commercially harvested around Kyoto and Chiba.

Anthocidaris crassispina is found in intertidal and subtidal zones in the Pacific coastal regions around Ibaragi, in the southern Japan Sea around Akita, and in Taiwan and southeastern China (Shigei 1995). It is harvested around Kyoto in the Sea of Japan and around Wakayama in the Pacific Ocean.

Biermann et al. (2003) reported the phylogenetic relationships of ten strongylocentrotid sea urchin species using mitochondrial DNA sequences. Their results support the inclusion of *H. pulcherrimus* and *P. depressus* into the genus *Strongylocentrotus*. In particular, *H. pulcherrimus* and *P. depressus* are closely related to *S. intermedius* and *S. nudus*, respectively.

3. REPRODUCTION

3.1. Reproductive Cycle

Hemicentrotus pulcherrimus matures in eastern Fukui at 26 mm diameter (Kawana 1938). In this region, spawning occurs from December to April when the water temperature falls from 13 to 10 °C and rises again to 13 °C. Spawning is vigorous from January to March when the water temperature is about 10 °C (Kawana 1938). In Genkai, Saga, in northern Kyushu, spawning occurs from December to May when the water temperature falls to <15 °C and rises again to 20 °C. Spawning is vigorous from January to March (Shimazaki et al. 1987). The gonadal index (gonadal weight $\times 100$ body weight^{-1}) off the southwestern Sea of Japan coast in Shimane increases after August and reaches a maximum in December to January. During this period, the gonads recover from spawning and mature. Spawning occurs from late January to late February when water temperature falls from 13 to 10 °C (Isemura 1991). Off the Pacific coast of Shirahama in Wakayama, the spawning season is January to March (Kobayashi and Konaka 1971).

In southwestern and southern coastal waters of the Sea of Japan in Hokkaido, gonadal maturation of *H. pulcherrimus* begins in November and is maintained for long periods. The spawning season is March to June when the water temperature rises from 6 to 13 °C, later than in Honshu. The delay in spawning is attributed to low water temperature in winter (Agatsuma 1992; Agatsuma and Nakata 2004). The ripe ovaries have a bitter taste (Murata et al. 1998). The bitterness is due to a novel sulfur-containing amino acid (pulcherrimine) with the structure 4-(2′-carboxy-2′-hydroxy-ethylthio)-2-piperidinecarboxylic acid (Murata and Sata 2000; Murata et al. 2001). Individuals with a low pulcherrimine content in their ovaries increase from February to August and decrease from August to November (Murata et al. 2002).

Anthocidaris crassispina spawns in July to August off the coast of Kyoto and Nagasaki in the southwestern Sea of Japan and along the Pacific coast of Chiba in central Japan (Masuda and Dan 1977; Tsuji et al. 1989; Yamasaki and Kiyomoto 1993). In Wakasa Bay, Kyoto, the gonadal index increases after December and reaches a maximum in June to July, then abruptly decreases in July or August when water temperature rises from 23 to 29 °C (Tsuji et al. 1989). From histological observation of this sea urchin at Hirato Island, Nagasaki, Yamasaki and Kiyomoto (1993) classified gonadal development into the five stages of Fuji (1960) (1) recovering (September to January), (2) growing (February to March), (3) prematuring (April), (4) maturing (May to June), and (5) spent (July to August). In Hong Kong, the sea urchins spawn from June to October. The gonads start to grow immediately afterwards, from November or December (Chiu 1988a).

The spawning season of *P. depressus* at Nansei, Mie is October to December (Unuma et al. 1996).

3.2. Abiotic Factors and Maturation

Water temperature has an important role in gonadal maturation of *H. pulcherrimus*, *A. crassispina*, and *P. depressus*, but light does not (Yamamoto et al. 1988; Ito et al. 1989; Sakairi et al. 1989). According to Ito et al. (1989), *H. pulcherrimus* that had experienced a period of rising water temperature to 26 °C could mature and spawn about 45 days after the temperature was again lowered to 15 °C. In mass artificial production of juveniles, the warmer season is more desirable for larval survival and juvenile growth. Accordingly, lowering the water temperature to 15 °C in late July leads to successful mass production from October in the warmer season.

Gonadal maturation of *P. depressus* is promoted at a constant water temperature of 19–20 °C, but not at 13 °C (Yamamoto et al. 1988; Noguchi et al. 1995). Juvenile *P. depressus* have been mass-produced during late October and early November, corresponding to the spawning season of native populations. Noguchi et al. (1995) reported that gonadal maturation of the adult sea urchins is promoted by a rise in water temperature to 25 °C from February to June, and then a constant water temperature of 20 °C after July. This causes mass production to begin in middle or late August and avoids mass mortality of the juveniles by bacterial infection in winter (Masaki et al. 1988).

Gonadal maturation of *A. crassispina* is promoted at constant water temperatures of 20 °C and 25 °C, but not at 15 °C (Yamamoto et al. 1988; Sakairi et al. 1989).

3.3. Spawning Cue

Spawning of *H. pulcherrimus* synchronizes with the new and full moon (Kobayashi 1992). Kobayashi (1969) measured changes in the gonad index (gonadal weight $\times$100 test weight^{-1}) of *A. crassispina* during spawning season in July to August at Shirahama, Wakayama, and concluded that the sea urchins spawn at both new and full moon (semilunar periodicity). However, the sea urchins at Minamiizu, Shizuoka, spawn only at full moon during spawning season in June to August (lunar periodicity) (Horii 1997).

4. LARVAL ECOLOGY

The ecology of the larvae of *H. pulcherrimus*, *P. depressus*, and *A. crassispina* have not been studied. However, larval growth, survival, and time to metamorphosis have been studied in the laboratory.

Ito et al. (1986) examined the survival rates of larvae of *H. pulcherrimus* fed *Chaetoceros gracilis* from 24 h after fertilization at water temperatures from 16 to 24 °C. The time to eight-armed plutei decreased as the water temperature rose from 17 to 20 °C. At 20 °C, 90% developed to that stage in 14 days. At 17 °C, only 67% developed to that stage. Survival at 16 °C and 21–24 °C was extremely low, and larvae were abnormally shaped.

Yamanobe (1962) reported that larvae of *P. depressus* fed *C. simplex* at a water temperature of 20 °C metamorphosed 40 days after hatching, and that 42% survived to the six-armed plutei stage. At 19 °C, 84% metamorphosed (Kakuda and Nakamura 1975). All larvae fed on *C. gracilis* alone or with *C. cerratosporum* at 19 °C metamorphosed

19–27 days after hatching (Kakuda 1978a). Ito et al. (1987) concluded that the optimum water temperature for culturing larvae is 18–20 °C, leading to a high survival rate of >80%. *Dunalielle salina* for four-armed plutei and *Phaeoductylum tricornutum* in addition to *D. salina* for six- and eight-armed plutei are used as valuable foods for successful metamorphosis as *C. gracilis* (Kume 2002).

Larvae of *A. crassispina* fed *C. gracilis* at a constant water temperature of 28 °C metamorphose 13–17 days after hatching (Kakuda and Nakamura 1974). *C. gracilis* and *C. calcitrans* are valuable foods for larval growth and survival (Noda and Ito 1987).

5. SETTLEMENT AND METAMORPHOSIS

5.1. Induction with Algae

According to Ito et al. (1991), the brown alga *Hizikia fusiformis* induced settlement and metamorphosis of larvae of *H. pulcherrimus,* more than attached diatoms or the algae *Sargassum thunbergii, Undaria pinnatifida, Eisenia bicyclis*, and *Ulva pertusa*. The rate of larval metamorphosis with *E. bicyclis* was lowest, and most larvae died. Many more larvae of *H. pulcherrimus* and *P. depressus* settled with crustose coralline algae than with attached diatoms (Kakuda 1978b). Ito et al. (1991) examined the induction of metamorphosis of fully developed eight-armed plutei of *H. pulcherrimus* and *P. depressus* with attached diatoms, *H. fusiformis*, attached diatoms with *H. fusiformis*, attached diatoms with an extract of *H. fusiformis*, and extract of *H. fusiformis* alone. Both species of plutei were strongly induced by attached diatoms with *H. fusiformis* and attached diatoms with an extract of *H. fusiformis*.

The percentage of metamorphosis of larvae of *P. depressus* with attached diatoms increases with density (Tani and Ito 1979). Metamorphosis of larvae of *H. pulcherrimus* and *P. depressus* with dense attached diatoms is greatly increased by the introduction of *H. fusiformis* (Ito et al. 1991). The induction of metamorphosis is attributable to direct attachment to the alga or attached diatoms (Ito et al. 1991). Rahim et al. (2004a) reported that *P. depressus* and *A. crassispina* metamorphose in the laboratory both on natural microbial films and diatom-based film. Diatom-based films could promote larval metamorphosis of *P. depressus*, but are less important for *A. crassispina*. For laboratory-cultured diatom-based film, both species of sea urchins show a similar response, in which reduction in diatom and bacteria density result in a decrease in the original inducing activity. They suggest a synergistic effect between diatom and bacteria in inducing larval metamorphosis. They also reported that the inductive activity of diatoms alone in larvae of both species is significantly less than that of the diatom-based films and that no larvae metamorphose on bacterial films alone (Rahim et al. 2004b). The metamorphosis-inducing cues in diatom-based films are unstable. None of five isolated species of periphytic diatom induced larval metamorphosis alone.

Metamorphosis of fully developed eight-armed larvae of *P. depressus* and *A. crassispina* is induced at a high rate by their attachment to attached diatoms, *H. fusiformis*, *S. thunbergii*, crustose coralline algae, and foliose coralline algae (Tani and Ito 1979; Ito 1984; Ito et al. 1991). Li et al. (2004a) reported water conditioned with fifteen different species of macroalgae in combination with the diatom *Navicula ramosissima* induces metamorphosis of *A. crassispina* larvae, but does not induce metamorphosis of *P. depressus* larvae. In particular, high inductive activity for larval metamorphosis of *A. crassispina* is found in coralline red algae and brown algae conditioned waters. The activity of *Corallina pilulifera* conditioned water is relatively low in February or March and in September, corresponding to the nongrowing season. *U. pertusa*-conditioned water does not induce metamorphosis except in spring or early summer.

5.2. Chemical Inducer

Potassium chloride induces larval metamorphosis of *P. depressus*. All larvae settle and metamorphose within 24 h after treatment in 100 mM KCl for 5 min at water temperatures of 14–26 °C (Kawahara et al. 1995). The larvae of *P. depressus* are also induced to metamorphose by 10^{-5}–10^{-4} L-glutamine (Yazaki and Harashima 1994).

Kitamura et al. (1992) reported that benzene, ether, and methanol in lipophilic extracts from *C. pilulifera* induce settlement and metamorphosis of eight-armed plutei of *P. depressus* and *A. crassispina*. Eicosapentaenoic acid (20:5) in free fatty acids was isolated as a chemical inducer (Kitamura et al. 1993). Carbohydrates do not play a major role in inducing their metamorphosis (Rahim et al. 2004b). Li et al. (2004b) reported that the active substances in water conditioned by the *C. pilulifera* in combination with *N. ramosissima* as inducers of metamorphosis in larvae of *A. crassispina* are relatively heat stable, nonvolatile and polar with estimated molecular weights of about 100 daltons or less. They suggested that more than two metamorphosis-inducing substances may be present in *C. pilulifera*-conditioned water.

6. FOOD AND FEEDING AFTER SETTLEMENT

6.1. Food

The principal food of *H. pulcherrimus* in the wild is marine algae (Kawana 1938). Benthic animals are also found in their stomachs (Matsui 1966) but little is eaten (Oshima et al. 1957).

The principal stomach contents of *P. depressus* in shallow waters at Miura, Kanagawa are the drift brown algae *E. bicyclis* and *Ecklonia cava* in summer, these two algae and *Hypnea* spp. in autumn, and juvenile *E. bicyclis* and *E. cava* in winter and spring (Imai and Arai 1986).

Imai and Kodama (1986) reported that marine algae account for >90% of the total dry weight of stomach contents of *A. crassispina* in shallow waters of Miura, Kanagawa. Drift *E. bicyclis*, *E. cava*, and coralline algae are abundant in the contents. According to Chiu (1984), *A. crassispina* along the coast of Hong Kong *Ulva conglobata* commonly feeds on 9 algal species. In winter (December to April), urchins at all localities graze heavily on *Colpomenia sinuosa*, *S. hemiphyllum*, and *Hypnea* spp. in autumn, and juvenile *E. bicyclis* and *E. cava* in winter and spring (Imai and Arai 1986). In summer (July to October), they graze on *Sargassum* spp. and *Corallina* spp. in localities where macrophytes are perennial. In areas with a summer macrophytic decline, they depend upon coralline (*Corallina* spp.), filamentous (*Cladophora delicatula* and *Gelidium divaricatum*), and encrusting (*Lithophyllum*) algae.

Yatsuya and Nakahara (2004b) reported that the main components in the gut contents of *A. crassispina* in *Sargassum* beds and adjacent *Corallina* beds are *Sargassum* spp. in Kodomari, in the western part of Wakasa Bay, Kyoto. The percentages in *Corallina* beds are less than those in *Sargassum* beds. Stable isotope analysis showed that gut contents in the two habitats are significantly different but gonad is not. This suggests that *Sargassum* spp. as the main food for *A. crassispina* is obtained as drift from *Sargassum* beds.

6.2. Feeding and Food Selectivity

One-to-two-year-old *H. pulcherrimus* have high feeding rates (food intake $\times$100 body weight^{-1}) on brown algae (*S. thunbergii, S. confusum, E. bicyclis*, and *U. pinnatifida*) and low feeding rates on green algae (*U. pertusa*) (Kakuda et al. 1970). They feed heavily on *E. cava* and *S. serratifolium* (Oshima et al. 1957).

Futashima et al. (1990) fed *P. depressus* simultaneously on *E. bicyclis, U. pinnatifida, U. pertusa, S. ringgoldianum, Codium fragile, S. thunbergii*, and *G. amansii*. The sea urchins preferred *E. bicyclis*, followed by *U. pinnatifida* and *Ulva pertusa*. Large urchins much preferred *E. bicyclis* (Futashima et al. 1990; Kakuda et al. 1995).

Oshima et al. (1957) fed *A. crassispina* on *E. cava*, coralline algae, *S. serratifolium, S. thunbergii, G. amansii, U. pertusa*, and *Zostera marina*. The sea urchins preferred *S. serratifolium, S. thunbergii*, and *E. cava*. They also eat a small amount of animal food (Oshima et al. 1957). Food preference experiments performed on six common algal species in the laboratory showed that, in terms of percentage intake, *C. sinuosa* was most preferred, followed by *Petalonia fascia, S. hemiphyllum, Pandia australis, C. pilulifera*, and *U. conglobata*. The feeding rate was highest for *C. sinuosa* (Chiu 1984).

Kaneko et al. (1981) reported that the stomach of *A. crassispina* became empty 3 days after removal of food, but that the stomach of *H. pulcherrimus* still retained food after 6 days. *H. pulcherrimus* survived starvation for 49 days, but stomach and gonad weight decreased markedly.

7. GROWTH

7.1. Longevity and Growth Rings

Longevity of *H. pulcherrimus* is estimated to be 7 or 8 years (Matsui 1966) based on the number of black rings in charred genital plates according to the method of Jensen (1969) and Kawamura (1973). The rings are formed once a year from September to November in Oshoro Bay, Hokkaido (Agatsuma and Nakata 2004).

Black annual rings of *P. depressus* appear in charred genital plates during June and October, corresponding to cessation of growth (Kakuda 1989).

Longevity of *A. crassispina* is estimated to be about 9 years (Chiu 1990) based on the number of growth rings in the nonmadreporic genital plates according to Dix (1972).

7.2. Somatic Growth

Hirano (1968) reported that the growth rate of 2-year-old *H. pulcherrimus* is high from October to January (gonad maturation) and March to July (recovery). Conversely, the growth rate is low from July to October, when water temperature is high (gonadal growth), and January to March (spawning). In contrast, Agatsuma and Nakata (2004) reported that the sea urchin grew rapidly for one and a half months from late August to late September or early October when water temperature reached a maximum of >20 °C, then began to fall (gonadal growth prior to maturation) in Oshoro Bay where brown algae *Sargassum* spp. grew throughout the year. Growth ceased from October to November (initiation of gonadal maturation). They estimated diameters to be 10.9 mm at 1 year, 23.9 mm at 2 years, 31.1 mm at 3 years, and 37.2 mm at 4 years based on age indicated by growth rings (see Section 7.1). Diameters of *H. pulcherrimus* in shallow waters at Taki, Shimane are estimated to be 15.0 mm at 1.5 years, 28.0 mm at 2.5 years, and 35.8 mm at 3.5 years (Isemura 1991). Fuji (1963) analyzed the growth of *H. pulcherrimus* in shallow waters at Misaki, Kanagawa by Petersen's size frequency analysis, consecutive observation of the Petersen curves, and tagging. He estimated diameters to be 10 mm at 0.5 years, 21 mm at 1.5 years, 28 mm at 2.5 years, and 34 mm at 3.5 years. Agatsuma et al. (2005) investigated growth of *H. pulcherrimus* on fucoid beds and algal turfs in Akita Prefecture in northern Japan. The most rapid growth occurred in a large perennial fucoid bed. Slower growth occurred in a small perennial *Chondrus ocellatus*-dominated bed. Slowest growth occurred in the small perennial *Dictyopteris divaricata*- and *Laurencia* spp.-dominated beds. These algae possess chemicals which act as feeding deterrents against the sea urchins. Diameters are estimated to be 5–8 mm at 1 year, ca. 20 mm at 2 years, and ca. 30 mm at 3 years in fucoid beds.

The growth and survival of juvenile *H. pulcherrimus* of 0.36 mm diameter immediately after metamorphosis were higher when fed attached diatoms than when fed *H. fusiformis* and *U. pertusa* (Ito et al. 1991). However, the growth of juveniles with diameters of 4, 6, and 15 mm fed *H. fusiformis*, *U. pertusa*, *S. patens*, and *S. serratifolium* were highest on *H. fusiformis*, followed by *U. pertusa* (Noguchi and Kawahara 1993). Juvenile *H. pulcherrimus* immediately after metamorphosis and settlement in April were fed attached diatoms until they reached 8–10 mm diameter and then were fed *U. pertusa*, *E. bicyclis*, *U. pinnatifida*, and *S. thunbergii*. They grew to 9.8 mm after 6 months and 21.5 mm after 1 year (Kakuda et al. 1970).

Ito et al. (1987) studied the growth and survival of juvenile *P. depressus* of 1 and 1–2 mm diameter fed 9 species of diatoms, and individuals of 2 to 5 mm diameter fed 6 species of diatoms and 3 species of algae (*U. pertusa, H. fusiforme*, and *E. bicyclis*). The diatom *N. ramosissima* was the most valuable food for the juvenile. The growth and survival of juveniles fed algae were low.

Pseudocentrotus depressus fed *E. bicyclis* has a high growth rate (Futashima et al. 1990; Kakuda et al. 1995). The rings in charred genital plates were used to estimate that the diameters of *P. depressus* at Kiwado, Yamaguchi were 16.6 mm at 1 year, 36.2 mm at 2 years, 50.2 mm at 3 years, and 60.1 mm at 4 years (Kakuda 1989).

Growth of *A. crassispina* in *Sargassum* beds (*Myagropsis myagroides*, *S. patents* and *S. piluliferum*) is higher that that in *Corallina* beds (*C. pilulifera* and *Amphiroa* spp.) in Kyoto (Yatsuya and Nakahara 2004a).

Chiu (1990) reported that high growth rates (14.3–15.7 mm yr^{-1}) occur in *A. crassispina* 2–4 years of age. The diameters are 14.3–19.5 mm at 1 year, 26.9–33.3 mm at 2 years, 37.0–44.2 mm at 3 years, and 45.2–52.5 mm at 5 years in Hong Kong. Growth of *A. crassispina* in Cheju Island, Korea mainly occurs from December to March, with no obvious growth during April to June when annual rings are formed (Hong and Chung 1998).

7.3. Gonadal Growth

Off the Sea of Japan coast of Akita, the gonad index of *H. pulcherrimus* in a fucoid bed where standing crops exceeded 3 kg m^{-2} is significantly higher than that in a *Laurencia* bed (Agatsuma et al. 2005). Gonadal growth of *H. pulcherrimus* fed a prepared diet with high protein is rapid (Nagai and Kaneko 1975). *Hemicentrotus pulcherrimus* fed green algae have fast gonadal growth and reddish yellow gonads, while those fed brown algae have slow gonadal growth and brown gonads (Kawana 1938).

Hur (1988) reported that spinach, radish leaves, and lettuce promote better gonadal growth in *H. pulcherrimus* than *U. pinnatifida*, *Grateloupia prolongata*, *S. thunbergii*, *Chondrus* sp., and *S. sagamianum*. He suggested that vegetables are a suitable diet for intensive sea urchin aquaculture.

Anthocidaris crassispina in *Sargassum* beds have higher gonad indices than those in *Corallina* beds (Yatsuya and Nakahara 2004a).

8. HABITAT

Hemicentrotus pulcherrimus is stenohaline, with an optimum salinity range from 30 to 35‰, and prefers low light levels (<50 lux), especially under starving conditions (You et al. 2003). Juvenile *H. pulcherrimus* 0.8–4.5 mm in diameter inhabit the roots of *Amphiroa dilatata* and *Phyllospadix iwatensis* in summer in the shallow waters of the eastern regions of Fukui (Kawana 1938). Kawana (1938) reported that adult *H. pulcherrimus* commonly live under rocks (rarely in crevices), but are found on the surface of rocks or algae from January to April, their spawning season. Higher densities of *H. pulcherrimus* are found on rocks with gravel or sand and mud under them (Kawana 1938; Isemura 1991). Many brittle stars (*Ophioplocus japonicus*) live there (Kawana 1938). The density of *H. pulcherrimus* is low under rocks with a base of <0.1 m^2 (Isemura 1991).

Juvenile and adult *H. pulcherrimus* are dense off the Japan Sea coast of Toyokita, Yamaguchi at a depth of <2.1 m where large brown algae (*S. piluliferum*, *S. patents*, and *S. serratifolium*) grow (Inoue et al. 1969). Off the coast of Yamaguchi, *H. pulcherrimus* is found at a depth of <3 or 4 m, mainly in areas not exposed to strong wave action. However, *A. crassispina* lives on the rocky bottom at depths of >10 m facing the open sea (Nakamura and Yoshinaga 1962). Imai (1980a) reported that *H. pulcherrimus* and *A. crassispina* off the Pacific coast of Jogashima, Kanagawa are dense at a depth of

0.5–1.0 m where *C. sinuosa*, *E. bicyclis*, and *S. ringgoldianum* grow. *Hemicentrotus pulcherrimus* lives there under rocks 0.3–0.5 m in diameter. In contrast, *A. crassispina* inhabit crevices or notches of rocks. Imai (1980b) investigated their distribution all around Miura Peninsula. They were dense in small inlets at depths of 0–1 m where *H. fusiformis* grows and sparse at 1–2 m where *G. amansii* grows. They live there mainly under rocks 50–60 cm in diameter. In contrast, *A. crassispina* lives in shallow waters facing the open sea at depths of >1 m where *E. bicyclis* and *E. cava* grow. They live there in crevices or ledges of rocks both vertical and horizontal.

According to Imai and Arai (1994), juvenile *P. depressus* live under rocks 20–30 cm diameter scattered on the sandy bottom at depths of 8–12 m where crustose coralline algae and algae of the family Squamariaceae grow. In contrast, adults live at depths of 3–8 m where *E. cava*, *Peyssonelia caulifera*, and *Callophyllis japonica* grow. The microhabitats of adults were classified as (1) gravel bottom and rocks 30–60 cm in diameter, (2) rock piles, and (3) crevices in rocks. Gravel bottom and rocks was the most densely inhabited microhabitat.

Anthocidaris crassispina inhabit pits and crevices on rocks or shelters on cliffs off the coast of Miura (Imai and Kodama 1994). According to Tsuji et al. (1989), *A. crassispina* in Wakasa Bay live predominantly on the flat rocky bottom at depths of <4 m where foliose coralline species, algae of the family Sargassaceae, *Dictyota dichotoma*, *C. sinuosa*, and *Laurencia* spp. grow. In Kodomari in the western part of Wakasa Bay, Kyoto, the density of *A. crassispina* in *Sargassum* beds (0.3 individuals per square meter) is lower than that in *Corallina* beds (11.1 individuals per square meter) (Yatsuya and Nakahara 2004a). Yusa and Yamamoto (1994) studied the movement of *A. crassispina* in the intertidal zone at Shirahama, Wakayama. This sea urchin lives in or about small pits on the rock surface. Sea urchins living inside the pits never leave but those living outside the pits move in summer and winter, especially during the night. By transplanting individuals between the inside and outside of pits, Yusa and Yamamoto concluded that this sea urchin prefers sheltered microhabitats. A morphological and physiological investigation showed that sea urchins inside the pits had similar gut weight (including contents), shorter lateral spines, and heavier gonads than sea urchins of similar diameter outside the pits. Yusa and Yamamoto considered that living in pits saves energy through immobility, reduces the breakage of spines, and results in a higher allocation to reproduction. Yamanishi and Tanaka (1971) reported that the strong clinging power of *A. crassispina* enables them to inhabit tide pools exposed to strong wave action.

Chiu (1988b) investigated the effect of substrate type, habitat topography, hydrography, macroalgae, and human activity on the occurrence of *A. crassispina* in shallow waters of Hong Kong. The population there lives on rocky or coral substrate where *Sargassum* spp. or *Lythophyllum* spp. grow and salinity is >20%. In particular, Chiu (1988b) showed that suitable habitats for this species are eradicated by foreshore reclamation for urban development. Freeman (2003) reported that *A. crassispina* range from a high abundance of ~16 individuals per square meter on steeply inclined rocky outcrops exposed to strong onshore wave surges to a complete absence on gravel and sandy substrata in Cape d'Aguilar Marine Reserve, Hong Kong. On the steep rocky slopes *A. crassispina* exhibits a size-dependent gradient where size increases in a down-shore direction as water depth increases. An experiment translocation of large and small size classes showed that each size class re-established its original location within 3–5 days. Freeman (2003) also observed that *A. crassispina* is predominately nocturnal with almost 100% of the population moving between dusk and dawn. Locomotory activity patterns are strongly correlated with changes in seawater depth and changes in the direction of water flow during tidal cycles. A camera monitor of the sea urchin's behavior under constant laboratory conditions ascertained that the pattern of locomotory activity is endogenously controlled and synchronized with changes in the tidal cycles.

9. COMMUNITY ECOLOGY

Growth and gonad production of *H. pulcherrimus* in northern Japan are affected by algal community stage and differ among species of small perennial algae in the community stage with or without chemical defense (Agatsuma et al. 2005).

Chiu (1988c) conducted a field enclosure experiment with several densities of *A. crassispina* to determine the role of grazing by the urchin in structuring the epibenthic macroalgal community. He showed that the natural density of the sea urchin (0.5–22.3 individuals per square meter) has no effect on the macroalgal community structure.

At subtidal regions in Wakasa Bay where *A. crassispina* lives at depths of <4 m and *S. nudus* lives at >4 m (Tsuji et al. 1989), mass mortality of *S. nudus* occurred in the summer of 1994 owing to high water temperature of >28 °C (Tsuji et al. 1994). Juvenile *A. crassispina* recruited there the next year. *A. crassispina* also enlarged its habitat into the regions previously inhabited by *S. nudus*. This shows that *S. nudus* had been restricting the distribution of *A. crassispina*.

Yotsui and Maesako (1993) reported that coralline flats where *A. crassispina* and herbivorous marine snails are dense are widespread in subtidal regions at the eastern end of Tsushima Island, Nagasaki. The removal of these individuals and the introduction of mature *E. bicyclis* promoted kelp growth. Yotsui et al. (1994) showed that presence of a marine kelp forest in the subtidal zone of the western coast of this island is due to reduced grazing of *A. crassispina* caused by abundant harvesting of the sea urchins.

Juveniles of wild and hatchery-raised *Haliotis discus discus* are found under the spine canopy of *A. crassispina* (Kiyomoto and Yamasaki 1999; Kojima 2005). Because *H. pulcherrimus*, the abalones *H. discus discus*, and *H. diversicolor aquatilis* occupy the same habitat, they may compete for space and food (Imai and Arai 1994). However their relationship has not been studied.

10. POPULATION DYNAMICS

No studies have been done on the population dynamics of *H. pulcherrimus*, *P. depressus*, or *A. crassispina*. Kawana (1954) reported that the number of years when commercial harvest of *H. pulcherrimus* in eastern Fukui was high was equal to the number of years when recruitment of juvenile herring (*Clupea pallasii*) off the Sea of Japan in Hokkaido was high, and that the latter occurred 2 years after the former. This may be caused by variability in the Tsushima Warm Current.

11. CONCLUSIONS

Hemicentrotus pulcherrimus, *P. depressus*, and *A. crassispina* co-occur in the field in temperate waters of southern Japan. Among them only *A. crassispina* seems to form extreme population densities that result in overgrazing. *H. pulcherrimus* has low mortality, short time to maturity, and great decrease in growth in response to low food availability. These are characteristics of a competitive species. Insufficient information is available to characterize *P. depressus*. In the future, the annual life cycle at each developmental stage of these sea urchins should be described.

REFERENCES

Agatsuma Y (1992) Annual reproductive cycle of the sea urchin, *Hemicentrotus pulcherrimus* in southern Hokkaido. Suisanzoshoku 40: 475–478 (in Japanese with English abstract)

Agatsuma Y, Nakabayashi N, Miura N, Taniguchi K (2005) Growth and gonad production of the sea urchin *Hemicentrotus pulcherrimus* in the fucoid bed and algal turf in northern Japan. PSZN Mar Ecol 26: 1–10

Agatsuma Y, Nakata A (2004) Age determination, reproduction and growth of the sea urchin *Hemicentrotus pulcherrimus* in Oshoro Bay, Hokkaido, Japan. J Mar Biol Ass UK 84: 401–405

Biermann CH, Kessing BD, Palumbi SR (2003) Phylogeny and development of marine model species: Strongylocentrotid sea urchins. Evol Dev 5: 360–371

Chiu ST (1984) Feeding biology of the short-spined sea urchin *Anthocidaris crassispina* (A. Agassiz) in Hong Kong. In: Keegan BF, O'Connor BDS (eds) Echinodermata. Balkema, Rotterdam, pp 223–232

Chiu ST (1988a) Reproductive biology of *Anthocidaris crassispina* in Hong Kong. In: Burke RD, Mladenov PV, Lambert P, Parsley RL (eds) Echinoderm biology. Balkema, Rotterdam, pp 193–204

Chiu ST (1988b) The distribution and habitat of *Anthocidaris crassispina* (Echinodermata: Echinoidea) in Hong Kong. Asian Mar Biol 5: 115–122

Chiu ST (1988c) *Anthocidaris crassispina* (Echinodermata: Echinoidea) grazing epibenthic macroalgae in Hong Kong. Asian Mar Biol 5: 123–132

Chiu ST (1990) Age and growth of *Anthocidaris crassispina* (Echinodermata: Echinoidea) in Hong Kong. Bull Mar Sci 47: 94–103

Dix TG (1972) Biology of *Evechinus chloroticus* (Echinodermata: Echinometridae) from different localities. 4. Age, growth and size. N Z J Mar Freshwater Res 6: 48–68

Freeman SM (2003) Size-dependent distribution, abundance and diurnal rhythmicity patterns in the short-spined sea urchin *Anthocidaris crassispina*. Estuar Coast Shelf Sci 58: 703–713

Fuji A (1960) Studies on the biology of the sea urchin. I. Superficial and histological gonadal changes in gametogenic process of two sea urchins, *Strongylocentrotus nudus* and *S. intermedius*. Bull Fac Fish Hokkaido Univ 11: 1–14

Fuji R (1963) On the growth of the sea urchin, *Hemicentrotus pulcherrimus* (A. Agassiz). Bull Jap Soc Sci Fish 29: 118–126

Futashima K, Ito T, Ezaki O (1990) Study on development of technologies in marine cultivation of several beneficial animals. Competition of algal animals. Bull Fukuoka Fish Exp Stn 16: 31–36 (in Japanese)

Hirano O (1968) On the culture of sea urchin, *Hemicentrotus pulcherrimus*. I. The morphological features of the sea urchin collected at each location and its growth variation. J Shimonoseki Univ Fish 16: 109–116 (in Japanese with English abstract)

Hong S-W, Chung S-C (1998) Age and growth of the purple sea urchin, *Anthocidaris crassispina* in Cheju Island. Bull Korean Fish Soc 31: 302–308

Horii T (1997) The annual reproductive cycle and lunar spawning rhythms of the purple sea urchin *Anthocidaris crassispina*. Nippon Suisan Gakkaishi 63: 17–22 (in Japanese with English abstract)

Hur SB (1988) Feeding habits and growth of the sea urchin *Strongylocentrotus pulcherrimus* (A. Agassiz) reared in the laboratory. J Aquacul 1: 121–133

Imai T (1980a) On the sea urchins off Miura city. I. The study of distribution, environment, growth and gonad in Jogashima. Bull Kanagawa Pref Fish Exp Stn 1: 35–79 (in Japanese)

Imai T (1980b) On the sea urchins off Miura city. II. The study of distribution, environment, growth and gonad to Kamimiyata from Hatsuse. Bull Kanagawa Pref Fish Exp Stn 2: 27–36 (in Japanese)

Imai T, Arai S (1986) Feeding behavior and grazing rate of the sea urchin *Pseudocentrotus depressus* (A. Agassiz). Suisanzoshoku 34: 157–166 (in Japanese with English abstract)

Imai T, Arai S (1994) Local peculiarities as living for the red sea urchin *Pseudocentrotus depressus* of Bishamon waters off southern Miura Peninsula, Kanagawa Prefecture, Japan. Suisanzoshoku 42: 307–313 (in Japanese with English abstract)

Imai T, Kodama K (1986) Feeding behavior of the sea urchin *Anthocidaris crassispina* (A. Agassiz). Suisanzoshoku 34: 147–155 (in Japanese with English abstract)

Imai T, Kodama K (1994) Population density of the purple sea urchin, *Anthocidaris crassispina* in relation to its habitats environment. Suisanzoshoku 42: 321–327 (in Japanese with English abstract)

Inoue Y, Nakamura T, Kakuda N, Terao Y, Shigemune S, Nishimura T (1969) Studies of ecology of sea urchins and environment. Bull Yamaguchi Pref Gaikai Fish Exp Stn 10: 1–46 (in Japanese)

Isemura H (1991) Ecology of the sea urchin *Hemicentrotus pulcherrimus* off the central coast of Shimane Prefecture. Saibaigiken 19: 67–74 (in Japanese)

Ito S, Kobayakawa J, Tani Y (1986) Optimum water temperature for the echinopluteus larvae *Hemicentrotus pulcherrimus*. Saibaigiken 15: 9–12 (in Japanese)

Ito S, Kobayakawa J, Tani Y (1987) Optimum water temperature for the echinopluteus larvae *Pseudocentrotus depressus*. Bull Saga Pref Fish Exp Stn 1: 12–14 (in Japanese)

Ito S, Kobayakawa J, Tani Y, Nakamura N (1991) The combined effect of *Hizikia fusiformis* with attaching diatoms for larval settlement and metamorphosis of *Hemicentrotus pulcherrimus* and *Pseudocentrotus depressus*. Saibaigiken 19: 61–66 (in Japanese)

Ito S, Shibayama M, Kobayakawa J, Tani Y (1989) Promotion of maturation and spawning of sea urchin *Hemicentrotus pulcherrimus* by regulating water temperature. Nippon Suisan Gakkaishi 55: 757–763 (in Japanese with English abstract)

Ito Y (1984) The effect of benthic diatom on the acceleration of the metamorphosis of sea urchin larvae. Mar Foul 5: 15–18 (in Japanese)

Ito Y, Ito S, Kanamaru H, Masaki K (1987) Dietary effect of attaching diatom *Navicula ramosissima* on mass production of young sea urchin *Pseudocentrotus depressus*. Nippon Suisan Gakkaishi 53: 1735–1740 (in Japanese with English abstract)

Jensen ML (1969) Age determination of echinoids. Sarsia 37: 41–44

Kakuda N (1978a) Studies on the artificial production on the sea urchin. Practical mass culturing technique of planktonic larvae of two species of sea urchin *Hemicentrotus pulcherrimus* (A. Agassiz) and *Pseudocentrotus depressus* (A. Agassiz). Aquaculture (Jpn J Aqua) 25: 121–127 (in Japanese)

Kakuda N (1978b) Studies of the artificial production on the sea urchin. Mass culturing of young of *Hemicentrotus pulcherrimus* (A. Agassiz) and *Pseudocentrotus depressus* (A. Agassiz). Aquaculture (Jpn J Aqua) 25: 128–133 (in Japanese)

Kakuda N (1989) Age determining character and growth of a sea urchin *Pseudocentrotus depressus*. Nippon Suisan Gakkaishi 55: 1899–1905 (in Japanese with English abstract)

Kakuda N, Nakamura T (1974) Studies on the artificial production of sea urchins. Aquaculture (Jpn J Aqua) 22: 49–54 (in Japanese)

Kakuda N, Nakamura T (1975) Studies on the artificial production of sea urchins. Aquaculture (Jpn J Aqua) 22: 56–60 (in Japanese)

Kakuda N, Suizu H, Yurano N (1995) Food value of three species of brown algae for growth of the red sea urchin, *Pseudocentrotus depressus*. Bull Yamaguchi Pref Gaikai Fish Exp Stn 25: 30–34 (in Japanese)

Kakuda N, Terano Y, Nakamura T, Inoue Y (1970) Growth and food intake of artificially produced juvenile sea urchin *Hemicentrotus pulcherrimus*. Aquaculture (Jpn J Aqua) 17: 155–165 (in Japanese)

Kaneko I, Ikeda Y, Ozaki H (1981) Biometrical relations between body weight and organ weights in freshly sampled and starved sea urchin. Bull Jap Soc Fish 47: 593–597 (in Japanese with English abstract)

Kawahara I, Hirose S, Ito S, Miyazaki Y, Kitamura H (1995) Effect of KCl on the larval settlement and metamorphosis of the sea urchin, *Pseudocentrotus depressus*. Suisanzoshoku 43: 237–241 (in Japanese with English abstract)

Kawamura K (1973) Fishery biological studies on a sea urchin, *Strongylocentrotus intermedius*. Sci Rep Hokkaido Fish Exp Stn 16: 1–54 (in Japanese with English abstract)

Kawana T (1938) Propagation of *Hemicentrotus pulcherrimus*. Suisan Kenkyushi 33: 104–116 (in Japanese)

Kawana T (1954) Relationship between commercial landing of *Hemicentrotus pulcherrimus* and recruitment level of juvenile herring. Aquaculture (Jpn J Aqua) 4: 55–56 (in Japanese)

Kitamura H, Kitahara S, Hirayama K (1992) Lipophilic inducers extracted from *Corallina pilulifera* for larval settlement and metamorphosis of two sea urchins *Pseudocentrotus depressus* and *Anthocidaris crassispina*. Nippon Suisan Gakkaishi 58: 75–78

Kitamura H, Kitahara S, Koh HB (1993) The induction of larval settlement and metamorphosis of two sea urchins, *Pseudocentrotus depressus* and *Anthocidaris crassispina*, by free fatty acids extracted from the coralline red alga *Corallina pilulifera*. Mar Biol 115: 387–392

Kiyomoto S, Yamasaki M (1999) Size dependent changes in habitat, distribution and food habit of juvenile disc abalone *Haliotis discus discus* on the coast of Nagasaki Prefecture, southwest Japan. Bull Tohoku Natl Fish Res Inst 62: 71–81

Kobayashi N (1969) Spawning periodicity of sea urchins at Seto. III. *Tripneustes gratilla*, *Echinometra mathaei*, *Anthocidaris crassispina* and *Echinostrephus aciculatus*. Sci Eng Rev Doshisha Univ 9: 254–269 (in Japanese with English abstract)

Kobayashi N (1992) Spawning periodicity of sea urchins at Seto. *Hemicentrotus pulcherrimus*. Publ Seto Mar Biol Lab 35: 335–345

Kobayashi N, Konaka K (1971) Studies on periodicity in oogenesis of sea urchin. Relation between the oogenesis and the appearance and disappearance of nutritive phagocytes detected by some histochemical methods. Sci Eng Rev Doshisha Univ 12: 131–149 (in Japanese with English abstract)

Kojima H (2005) Ecological study of the population of *Haliotis discus discus* (Gastropoda: Haliotidae) for fishery management. Bull Tokushima Pref Fish Res Ins 3: 1–120 (in Japanese with English abstract)

Kume H (2002) Examination of substitute feed for *Chaetoceros* sp. at the larval stage of sea urchin, *Pseudocentrotus depressus*. Suisanzoshoku 50: 91–96 (in Japanese with English abstract)

Li J-Y, Rahim SAKA, Satuito CG, Kitamura H (2004a) Combination of macroalgae-conditioned water and periphytic diatom *Navicula ramosissima* as an inducer of larval metamorphosis in the sea urchins *Anthocidaris crassispina* and *Pseudocentrotus depressus*. Sessile Organisms 21: 1–6

Li J-Y, Rahim SAKA, Satuito CG, Kitamura H (2004b) Characterization of the active substances in water conditioned by the coralline red alga *Corallina pilulifera* as inducers of metamorphosis in larvae of the sea urchin *Anthocidaris crassispina*. Sessile Organisms 21: 41–46 (in Japanese with English abstract)

Masaki K, Noguchi K, Kanamaru H (1988) Mass mortality of artificially produced juvenile sea urchin *Pseudocentrotus depressus*. Seikai-block Sorui-Kairui Kenkyukaihou 5: 45–59 (in Japanese)

Masuda R, Dan JC (1977) Studies on the annual reproductive cycle of the sea urchin and the acid phosphatase activity of relict ova. Biol Bull 153: 577–590

Matsui I (1966) The propagation of the sea urchins. Suisanzoshokusosho 12 Nihonsuisanshigenhogokyokai, Tokyo (in Japanese)

Murata Y, Sata NU (2000) Isolation and structure of Pulcherrimine, a novel bitter-tasting amino acid from the sea urchin (*Hemicentrotus pulcherrimus*) ovaries. J Agri Food Chem 48: 5557–5560

Murata Y, Sata NU, Yokoyama M, Kuwahara R, Kaneniwa M, Oohara I (2001) Determination of a novel bitter amino acid, pulcherrimine, in the gonad of the green sea urchin *Hemicentrotus pulcherrimus*. Fisheries Sci 67: 341–345

Murata Y, Yamamoto T, Kaneniwa M, Kuwahara R, Yokoyama M (1998) Occurrence of bitter gonad in *Hemicentrotus pulcherrimus*. Nippon Suisan Gakkaishi 64: 477–478 (in Japanese with English abstract)

Murata Y, Yokoyama M, Unuma T, Sata NU, Kuwahara R, Kaneniwa M (2002) Seasonal changes of bitterness and pulcherrimine content in gonads of green sea urchin *Hemicentrotus pulcherrimus* at Iwaki in Fukushima Prefecture. Fisheries Sci 68: 184–189

Nagai Y, Kaneko K (1975) Culture experiments on the sea urchin *Strongylocentrotus pulcherrimus* fed an artificial diet. Mar Biol 29: 105–108

Nakamura T, Yoshinaga H (1962) Sea urchins off the coast of Gaikai in Yamaguchi Prefecture. Aquaculture (Jpn J Aqua) 9: 189–200 (in Japanese)

Noda S, Ito Y (1987) Foods of the echinopluteus larvae of *Anthocidaris crassispina*. Bull Saga Pref Sea Farm Center 1: 45–48 (in Japanese)

Noguchi K, Kawahara I (1993) Valuable diets of the juvenile sea urchin *Hemicentrotus pulcherrimus*. Bull Saga Pref Sea Farm Center 2: 77–79 (in Japanese)
Noguchi K, Kawahara I, Goto M, Masaki K (1995) Promotion of gonadal maturation by regulating water temperature in the sea urchin *Pseudocentrotus depressus*. Bull Saga Pref Sea Farm Center 4: 101–107 (in Japanese)
Oshima Y, Ishiwatari N, Tanaka J (1957) Food habits of two sea urchins *Anthocidaris crassispina* and *Hemicentrotus pulcherrimus*. Aquaculture (Jpn J Aqua) 5: 26–30 (in Japanese)
Rahim SAKA, Li J-Y, Kitamura H (2004a) Larval metamorphosis of the sea urchins, *Pseudocentrotus depressus* and *Anthocidaris crassispina* in response to microbial films. Mar Biol 144: 71–78
Rahim SAKA, Li J-Y, Satuito CG, Kitamura H (2004b) The role of diatom-based film as an inducer of metamorphosis in larvae of two species of sea urchin, *Pseudocentrotus depressus* and *Anthocidaris crassispina*. Sessile Organisms 21: 7–12 (in Japanese with English abstract)
Sakairi K, Yamamoto M, Ohtsu K, Yoshida M (1989) Environmental control of gonadal maturation in laboratory-reared sea urchins, *Anthocidaris crassispina* and *Hemicentrotus pulcherrimus*. Zool Sci 6: 721–730
Shigei M (1995) Echinozoa. In: Nishimura S (ed) Guide to seashore animals of Japan with color pictures and keys. II. Hoikusha, Osaka, pp 538–552 (in Japanese)
Shimazaki D, Igata K, Goto M (1987) Spawning season of the sea urchin *Hemicentrotus pulcherrimus* off the coast of Genkai, Saga Prefecture. Bull Saga Pref Fish Exp Stn 1: 22–24 (in Japanese)
Tani Y, Ito Y (1979) The effect of benthic diatoms on the settlement and metamorphosis of *Pseudocentrotus depressus* larvae. Suisanzoshoku 27: 148–150 (in Japanese)
Tsuji S, Munekiyo M, Itani M, Douke A (1994) Mass biolysis of a sea urchin in the western part of Wakasa Bay. Bull Kyoto Inst Ocean Fish Sci 17: 51–54 (in Japanese)
Tsuji S, Yoshiya M, Tanaka M, Kuwahara A, Uchino K (1989) Seasonal changes in distribution and ripeness of gonad of a sea urchin *Strongylocentrotus nudus* in the western part of Wakasa Bay. Bull Kyoto Inst Ocean Fish Sci 12: 15–21 (in Japanese)
Unuma T, Konishi K, Furuita H, Yamamoto T, Akiyama T (1996) Seasonal changes in gonads of cultured and wild red sea urchin, *Pseudocentrotus depressus*. Suisanzoshoku 44: 169–175
Yamamoto M, Ishine M, Yoshida M (1988) Gonadal maturation independent of photic conditions in laboratory-reared sea urchins, *Pseudocentrotus depressus* and *Hemicentrotus pulcherrimus*. Zool Sci 5: 979–988
Yamanishi R, Tanaka A (1971) Contributions to the biology of littoral sea urchins. I. Measurements of clinging power and observations on stability of sea urchin colonies. Publ Seto Mar Biol Lab 19: 2–15
Yamanobe A (1962) On the culture of the echinopluteus larvae *Pseudocentrotus depressus*. Aquaculture (Jpn J Aqua) 10: 213–219 (in Japanese)
Yamasaki M, Kiyomoto S (1993) Reproductive cycle of the sea urchin *Anthocidaris crassispina* from Hirado Island, Nagasaki Prefecture. Bull Seikai Natl Fish Res Inst 71: 33–40 (in Japanese with English abstract)
Yatsuya K, Nakahara H (2004a) Density, growth and reproduction of the sea urchin *Anthocidaris crassispina* (A. Agassiz) in two different adjacent habitats, the *Sargassum* area and *Corallina* area. Fisheries Sci 70: 233–240
Yatsuya K, Nakahara H (2004b) Diet and stable isotope rations of gut contents and gonad of the sea urchin *Anthocidaris crassispina* (A. Agassiz) in two different adjacent habitats, the *Sargassum* area and *Corallina* area. Fisheries Sci 70: 285–292
Yazaki I, Harashima H (1994) Induction of metamorphosis in the sea urchin, *Pseudocentrotus depressus*, using L-glutamine. Zool Sci 11: 253–260
Yotsui T, Maesako N (1993) Restoration experiments of *Eisenia bicyclis* beds on barren grounds at Tsushima Islands. Suisanzoshoku 41: 67–70 (in Japanese with English abstract)
Yotsui T, Maesako N, Niiyama H (1994) Isoyake, barrens of macrophytes, on coastal region of Tsushima Islands. Bull Nagasaki Pref Inst Fish 20: 73–77 (in Japanese with English abstract)
You K, Zeng X, Liu H, Zhang X, Liu Q (2003) Selectivity and tolerance of sea urchin (*Hemicentrotus pulcherrimus*) to environmental change. Chin J Appl Ecol 14: 409–412
Yusa Y, Yamamoto T (1994) Inside or outside the pits: Variable mobility in conspecific sea urchin, *Anthocidaris crassispina* (A. Agassiz). Publ Seto Mar Biol Lab 36: 255–266

Edible Sea Urchins: Biology and Ecology
Editor: John Miller Lawrence

Chapter 23

Ecology of *Lytechinus*

Stephen A Watts[a], James B McClintock[a], and John M Lawrence[b]

[a]*Department of Biology, University of Alabama at Birmingham, Birmingham, AL (USA).*
[b]*Department of Biology, University of South Florida, Tampa, FL (USA).*

1. THE GENUS *LYTECHINUS*

Zigler and Lessios (2004) divided the shallow-water species of *Lytechinus* A Agassiz into two clades: an Atlantic clade composed of *Lytechinus williamsi* Chesher and *Lytechinus variegatus* (Lamarck) (Plate 5D) and a Pacific clade containing *Lytechinus pictus* (Verrill), *Lytechinus anamesus* HL Clark, *Lytechinus semituberculatus* (Valenciennes), and *Lytechinus panamensis* Mortensen.

The synonymy of *L. pictus* and *L. anamesus* has been known for some time (Durham et al. 1980; Larrain, in Maluf 1988). Zigler and Lessios (2004) confirmed from mitochondrial DNA and bindin evidence *L. anamesus* and *L. pictus* are a single species. Pearse and Mooi (in press) designate the species as *L. pictus.*

Lytechinus semituberculatus occurs along the northwest coast of South America and at the Galápagos. *Lytechinus panamensis* is endemic to the Bay of Panama. Zigler and Lessios (2004) found *L. semituberculatus* and *L. panamensis* to be indistinguishable genetically despite distinct morphologies and concluded the status of these two nominal species is uncertain.

Lytechinus williamsi has been reported from the Colombian Atlantic coast, Panamanian Caribbean coast, Belize, and the Florida Keys (Chesher 1968; Hendler et al. 1995; Zigler and Lessios 2004). *Lytechinus variegatus* has been divided into four nominal subspecies (Mortensen 1943): (1) *Lytechinus v. pallidus* HL Clark, found in the east Atlantic at the Cape Verde Islands; (2) *Lytechinus v. variegatus* (Lamarck), found throughout the West Indies to Brazil; (3) *Lytechinus v. atlanticus* A Agassiz, endemic to Bermuda; and (4) *Lytechinus v. carolinus* A Agassiz, found along the Atlantic coast from northern Carolina to the Florida Keys (but not in the Keys) and the Gulf of Mexico. Serafy (1973) raised *L. pallidus* to full species status and concluded *L. v. variegatus*, *L. v. atlanticus*, and *L. v. carolinus* are valid subspecies. Pawson and Miller (1982) gave evidence for a genetic basis for the phenotypic differences between *L. v. atlanticus* and *L. v. variegatus*. However, isozyme variation suggests little basis for separating these two subspecies (Rosenberg and Wain 1982). Zigler and Lessios (2004) found no molecular evidence of differentiation between *L. v. variegatus* from the Caribbean and *L. v. atlanticus* from Bermuda. Zigler and Lessios (2004) concluded their mitochondrial DNA data, but not their bindin data, indicate *L. v. carolinus* and *L. v. variegatus* are distinct.

Two deep-water species occur: *L. callipeplus* HL Clark in the Caribbean and the Gulf of Guinea and *L. euerces* HL Clark in the Caribbean. Their molecular evidence led Zigler and Lessios (2004) to suggest removing *L. euerces* from the genus *Lytechinus* and making it a sister species to *Sphaerechinus granularis*.

Lytechinus variegatus has the most extensive range of the species, occurring from North Carolina to southern Brazil. *Lytechinus variegatus* is the largest species, reaching 92 mm diameter at Ragged Key, Florida (Moore et al. 1963), 87 mm diameter in the West Indies (Clark 1933), and 75 mm in southern Brazil (Juqueira et al. 1997). The largest diameters recorded for the other species are <35 mm. The largest *L. variegatus* still does not begin to reach the size of the closely related, sympatric toxopneustid *Tripneustes ventricosus* (150 mm in diameter, Mortensen 1943). The biological and ecological significance of the difference in sizes is not known.

2. HABITATS

2.1. *Lytechinus variegatus*

Lytechinus variegatus occurs in a great variety of habitats in shallow water. *Lytechinus variegatus* is found in beds of the seagrasses *Thalassia testudinum, Syringodium filiforme, Cymodocea manatorum, Halodule wrightii*, and *Halimeda* sp., on hard bottoms covered with algae, rock, shell hash or sand (Field 1892; Clark 1933; Sharp and Gray 1962; Moore et al. 1963; Kier and Grant 1965; Meyer and Birkeland 1974; Greenway 1977; Lowe and Lawrence 1976; Serafy 1979; Aseltine 1982; Engstrom 1982; Vadas et al. 1982; Klinger 1984; Thayer et al. 1984; Lessios 1985; Beddingfield and McClintock 2000; Noriega et al. 2002; Cobb and Lawrence 2005). In some regions *L. variegatus* can be found on rock, shell hash, sand, or on hard bottoms covered with algae (Beddingfield and McClintock 2000). *Lytechinus variegatus* is found on mussels (*Modiolus americanus*) within seagrass meadows (*T. testudinum* and *H. wrightii*). Moore et al. (1963) indicated that *L. variegatus* is intolerant of suspended silt and abandons areas of turbidity. Aseltine (1982) reported *T. ventricosus* was found in a barrier reef lagoon of the Bahamas on limestone shelves and within exposed limestone beds with high water flow and large grain sediments, and *L. variegatus* was found in seagrass beds with fine grain sediments. She postulated that sediments and hydraulic conditions characterized the use of habitat, and that *L. variegatus* moves with the flow of water. In fact, in high water flows *L. variegatus* reorients its spines, possibly to affect water flow. Moore and McPherson (1965) reported similarly that *L. variegatus* prefers to live in regions of high tidal flows.

Although both *L. variegatus* and *T. ventricosus* occur in Discovery Bay, Jamaica (Keller 1983), only *L. variegatus* is found in the lush meadows of *T. testudinum* in Jamaica Bay (Greenway 1995). Lewis (1958) pointed out it is of interest that *L. variegatus* is not found at Barbados along with *T. ventricosus*. This may be because *T. ventricosus* is most abundant on areas of broken rock and rock flats. Similarly, *T. ventricosus* is found on the reef flats at St. Croix and *L. variegatus* is not (Ogden and Lobel 1978).

Lytechinus variegatus occurs on sand between limestone outcroppings at 20 m depth off the central Florida Gulf Coast (Hill and Lawrence 2003; Cobb and Lawrence 2005). A variety of attached and drift algae are present.

2.2. *Lytechinus semituberculatus* and *Lytechinus pictus*

Glynn and Wellington (1983) reported *L. semituberculatus* on flat or gently sloping rock surfaces. Lawrence and Sonnenholzner (2004) found *L. semituberculatus* in a variety of habitats in central Galápagos: (1) sheltered Academy Bay with medium hydrodynamics and large rocks on sand; (2) exposed North Seymour Island with high current and hydrodynamics and a steep slope with large rocks, pockets of sand, and some coral; (3) Itabaca Channel with high current and large rocks on sand; and (4) sheltered Muelle with high current and small flat rocks with brown and green algae. They were not found on the steep slope at Caamaño Islet with low-medium hydrodynamics and large rocks with pockets of sand or at exposed Punta Estrada with little current and medium hydrodynamics and large rocks on sand.

Lytechinus anamesus and *L. pictus* are the same species (see Section 1). Durham et al. (1980) summarized the habitats of *L. anamesus* as "sandy bottoms at depths of 2–300 m, often in large herds, and on seagrass in bays." This is similar to the shallow-water habitats of *L. variegatus*. Coyer et al. (1987, 1993) reported *L. anamesus* on extensive sand plains and barren grounds at Anacapa Island, California. *Lytechinus anamesus* also occurs outside or near margins of kelp beds at (Coyer et al. 1993; Dean et al. 1984; Leighton 1971; Schroeter et al. 1983) but have been noted inside a kelp bed (Coyer et al. 1993; Schroeter et al. 1983). JS Pearse (pers. comm.) observed *L. pictus* to a depth of ca. 100 m in Monterey Bay, California, and is the northernmost range reported along the California coast.

Zigler and Lessios (2004) noted no *Lytechinus* have been reported in shallow waters between the Sea of Cortez and the Gulf of Panama, a distance of 3500 km. The basis for this absence is not known.

3. ABUNDANCE

3.1. *Lytechinus variegatus*

The densities of *L. variegatus* have been measured most often in shallow water, primarily in seagrass beds. Densities are variable in populations, normally ranging from 0 to 40 individuals per square meter (e.g. Moore et al. 1963; Glynn 1968; Meyer and Birkeland 1974; Meyer et al. 1974; Rivera 1978; Engstrom 1982; Keller 1983; Vadas et al. 1982; Oliver 1987; Greenway 1995; Junqueira et al. 1997; Beddingfield and McClintock 2000), as would be expected in a vagile species with variable recruitment and a short lifespan. Moore et al. (1963) found a correlation of density with body size, with densities ranging from 43 to 250 individuals per square-meter. Individuals 4.8 cm in diameter had a mean of 43 individuals per square meter-while those 6.5 cm in diameter had a mean of 1.9 individuals per square meter. This may reflect mortality rather than interaction. Densities ranged from 3 to 636 individuals per square meter off Dixie County in west-central Florida (Camp et al. 1973), the latter value representing a sea urchin front. Another front of *L. variegatus* (>300 individuals per square meter) was documented in seagrass meadows of *S. filiforme* in the Florida Keys in August 1997 (Peterson et al. 2002). Biweekly changes of up to 23 individuals in specific 1 m^2 quadrats on a *Thalassia*-reef flat at Galeta, Panama, was greater than the yearly change (Meyer and Birkeland 1974; Meyer et al. 1974), and undoubtedly resulted from movement. Rivera and Vicente (1975) reported densities of 13.4 individuals per square meter in *T. testudinum* beds found in the inner Jobos Bay, Puerto Rico. In Discovery Bay, Jamaica, Keller (1976) reported densities of 0.4 individuals per square meter in a region dominated by *T. testudinum* encrusted with coralline red algae, and densities of 3.8 individuals per square meter in protected sites of *T. testudinum* covered with blue-green algae. Junqueira (1998) found stable populations (2.5 individuals per square meter) on subtidal hardbottoms, but seasonally varying populations from 0 to 0.6 individuals per square meter in an intertidal sand flat with patches of the seagrass *H. wrightii* off the coast of Rio de Janeiro. Beddingfield and McClintock (2000) found annual changes in the density of *L. variegatus* at St. Joseph Bay in the northeast Gulf of Mexico result from seasonal recruitment and mortality. Densities there were habitat specific, reaching 140 individuals per square meter in beds of *T. testudinum*, 37 in beds of *S. filiforme*, and <1 on sand. Enormous densities reaching more than 500 individuals per square meter (Camp et al. 1973; Maciá and Lirman 1999; Rose et al. 1999; Peterson et al. 2002) that result from unusual episodic recruitment are not permanent. The high density of 364 individuals per square meter in August 1997 at the mouth of Florida Bay decreased to 20–50 individuals per square meter in December 1998 (Rose et al. 1999). Noriega et al. (2002) compared seagrass habitats in Mochima Bay, Venezuela and found a significant correlation between density of *L. variegatus* and various parameters of *T. testudinum* including percent cover, leaf density, leaf length, shoot density, and total biomass. However, only total seagrass biomass had a significant effect on sea urchin abundance and distribution.

In most shallow-water habitats densities of *L. variegatus* can vary, both seasonally and annually, and densities are generally less than 15 individuals per square meter. Moreover, in some areas populations may display a highly patchy spatial distribution (e.g. Sánchez-Jérez et al. 2001). The absence of *L. variegatus* in an appropriate habitat may be the result of episodic biotic or abiotic factors, particularly in nearshore populations such as in a bay or lagoon.

A single study in deeper water reported density of *L. variegatus* off the central Florida Gulf Coast shelf at 13 and 20 m depths were <1 individuals per square meter (Hill and Lawrence 2003).

3.2. *Lytechinus semituberculatus* and *Lytechinus pictus*

Glynn and Wellington (1983) found densities of *L. semitubuculatus* up to 70 individuals per square meter in the western sector of the Galápagos. These are greater than those found by Lawrence and Sonnenholzner (2004) in central Galápagos.

L. anamesus (=*pictus*) lives in aggregations (Dean et al. 1984). Dean et al. found they occurred in broad bands at least 20 m wide on the edge of a kelp bed. Densities of

L. anamesus reached ca. 80 individuals per square meter near and ca. 40 individuals per square meter inside a kelp bed at San Onofre, California in 1978–1979 (Schroeter et al. 1983). The same pattern was observed at Anacapa Island, California (Coyer et al. 1993), with lower (calculated) densities of 23 individuals per square meter near a kelp bed and 13 individuals per square meter inside a kelp bed. The (calculated) density was highest (36 individuals per square meter) on barren grounds with coralline algae outside the kelp bed. Schroeter et al. (1983) found an exception to this pattern at a site at San Onofre where *L. anamesus* was more abundant inside (ca. 40 individuals per square meter) than outside (ca. 10 individuals per square meter) the kelp bed. Although Dean et al. (1984) found temporal changes in density of *L. anamesus* in the San Onofre kelp bed, they were always present. In contrast, Olsenberg et al. (1994) reported migration of *L. anamesus* at ca. 27 m depth off the southern California coast, present in winter and spring but absent in summer and fall, affects seasonal abundance.

4. FACTORS INFLUENCING DISTRIBUTION AND ABUNDANCE

4.1. Abiotic Factors

Temperature may be the most important factor influencing the distribution and abundance of *L. variegatus*, both latitudinally and locally. *Lytechinus variegatus* survives temperatures from at least 11 to 35 °C in St. Joseph Bay, Florida (Beddingfield 1997). Moore et al. (1963) recorded a mortality event near Miami Beach, Florida, and attributed the event to a decrease in average water temperatures of 22 °C to a low of 15.8 °C. Similarly, Beddingfield and McClintock (1994) reported that a mass mortality event in St. Joseph Bay, Florida, was most likely the result of low water temperatures during winter months. These studies indicate that the thermal history of the individual will influence the lower lethal limits. Mayer (1914) indicated that *L. variegatus* were killed at 37–38 °C, and suggested that they live near the high end of their thermal tolerance. Meyer and Birkeland (1973) indicated a mass mortality event in Galeta Point, Panama, was the result of high temperatures (39.5 °C) in association with exposure. Rivera (1978) reported high temperature tolerances of 34–35 °C. This upper thermal tolerance limit was reduced to less than 34 °C when individuals were simultaneously exposed to reduced salinities (Lawrence 1975). Similarly, Glynn (1968) reported that exsiccation of *L. variegatus* at low tide reduced survivorship at high temperatures, another indication of the interactions of physical stresses. Oliver (1987) reported that temperature and day length had the largest influence on population densities, and argued that abiotic factors were more important in regulating populations than biotic factors. Development occurs from 17 to 33 °C (Eisen and Inoue 1979).

Sublethal temperature extremes are important as they affect functioning. Kleitman (1941) found that *L. variegatus* did not right themselves at 10 °C, and Lawrence (1975) found that individuals from an environmental temperature of 30 to 33 °C were near their upper lethal limit as their righting time at 34 °C was twice as long as at 28 °C. A decrease in feeding and growth at high temperatures has been noted (Moore et al. 1963; Moore and McPherson 1965). Development rate increases with temperature from 17 to 27 °C

(Eisen and Inoue 1979). *Lytechinus anamesus* (=*pictus*) has an optimal temperature for feeding at 16–18.5 °C (Leighton 1971).

L. variegatus does not survive well at low salinities as it does not regulate the osmotic pressure of its coelomic fluid. Changes in osmolality and ions generally follow the environment (Bishop et al. 1994). Roller and Stickle (1993) observed that the LC 50 occurs between 18 and 20 ppt salinity. Below this threshold all individuals die within 4 days. Lawrence (1975) indicated that reduced salinities decreased the functional well-being of *L. variegatus*. The effect was greater when individuals were simultaneously exposed to high temperatures. Transient exposure to 2 ppt salinity for 30 min resulted in 100% mortality, although individuals survived exposure to 18 ppt for the same period (Irlandi et al. 1997). Mortality resulting from heavy rains occurred in Jamaica (Goodbody 1961). Seasonal rains that dilute nearshore seagrass flats resulted in mortalities in first-year class *L. variegatus* from May to November (Allain 1975, 1977). Boettger et al. 2002) recorded mass mortality events in April and September 1998 in St. Andrews Bay, Florida, following tropical storms that decreased nearshore salinities to at least 19 ppt. Greenway (1977) attributed a mass mortality event to a depth of 2 m to reduced salinities following a hurricane in 1973 and tropical storms in 1974 in Kingston Harbour, Jamaica. In some cases, populations recovered within several months due to immigration (e.g. Greenway 1977), others report populations had not recovered after 18 months (Boettger, pers. comm.). Even transient salinity reductions due to freshwater canal discharge will influence the abundance and feeding characteristics of nearshore *L. variegatus* (Irlandi et al. 1997). If seasonal or episodic rains occur during periods of natural spawning, recruitment may also be significantly decreased.

Low salinities also affect development. Hintz (1993) found that fertilization success is minimal at 15 ppt salinity. Fertilization success was greatly reduced if gametes acclimated to a higher salinity were combined at even slightly reduced salinities, further demonstrating the sensitivity of fertilization to decreased salinity. Roller and Stickle (1993) found that survival of plutei to metamorphosis decreased at salinities <35 ppt. These data show that reductions in salinity can influence the distribution and abundance of both larval and adult *L. variegatus*, but these reductions would most likely have an effect in shallow (1–2 m) seagrass beds that are prone to above-normal rainfall runoff.

Sunlight or, more specifically, ultraviolet light has been reported to induce behavioral changes and possibly influence abundance of *L. variegatus*, particularly in nearshore environments. Tennent (1942) found short-term exposure of eggs of *L. variegatus* to sunlight had no effect on early embryonic development. Sharp and Gray (1962) indicated that *L. variegatus* responds negatively to direct sunlight and shortwave UV in the laboratory. Behaviorally, the sea urchins would "cover" themselves during the day and dropped the cover at night. They further reported that the response to unfiltered light was "violent" and that the sea urchins buried themselves in shells and other cover. Lees and Carter (1972) indicated that UV light causes mortality in *L. anamesus* (=*pictus*); however, they used levels that were higher than those found in nature. Little empirical evidence has demonstrated direct effects of UV on *L. variegatus*.

The covering reaction of *L. variegatus* has long been known (Field 1892) and often attributed to protection from light. Millott (1956) described covering by *L. variegatus* as a function of the tube feet that moves debris from the bottom to the aboral part of the individual. Lawrence (1976) described the reversal of the process in *L. variegatus*, movement of debris by the tube feet from the aboral to the oral part by inverted individuals. Covering in *L. variegatus* also has been proposed to function as camouflage for protection from predators (Boone 1928; Kier and Grant 1965), or it may basically be a reflex behavior (Lawrence 1976). Experimental studies of the effect of light on covering in *L. variegatus* (Millott 1956; Glynn 1968; Sharp and Gray 1972) often had a small sample size and number of replications. However, the evidence of Sharp and Gray (1972) indicate that *L. variegatus* is negatively phototactic to ultraviolet light and covers in response to it. Millott (1975) pointed out the difficulty of separating the response of the tube feet to light for covering from stimuli (including light) that promote locomotion. Observations that *L. variegatus* cover during night in the field (Allain 1975) or in low light intensity in the laboratory (Boone 1928) do not necessarily mean covering does not provide protection from light. Although Millott (1956) discounted a stabilizing effect of covering, Lees and Carter (1972) found increased covering and stability of *L. anamesus* with increased hydrodynamics. No data document protection from predation by covering.

L. variegatus can survive in very shallow water, particularly during low tides. Wave action can have a pronounced effect on abundance in these habitats. Although *L. variegatus* are commonly found on sand flats, they cannot attach and individuals can be easily swept from surfaces during high wave action (Sharp and Gray 1962). Rivera (1977) indicated that wind-caused waves could sweep over the shallow flats and transport all loose individuals to the shore where they die. *Lytechinus anamesus* cover themselves with debris during periods of tidal surge (Lees and Carter 1972), but this has not been reported in *L. variegatus*. Whether the amount of covering is greater than that which occurs normally in the field or whether the debris is sufficient to increase the mass of the individual is not known. Certainly, the level of impact of wave action is hard to determine, but may be a causal factor in that sea urchins may be moved by wave action to shallower regions where the individuals could be ultimately killed by high temperatures or exposure.

Whether initiated by wave action washing *L. variegatus* onshore or by water receding from tidal flats during low tide, emersion (desiccation) can cause mortality in *L. variegatus*. Glynn (1968) reported that local fishermen on Caribbean reef flats in Panama consider midday, low tidal exposure kills a common phenomenon. Mortality has reached 64% of the reef populations on Caracoles Reef. Even covered, *L. variegatus* lose 35% of their initial weight in 4 h when exsiccated, and cannot recover after 2 h exposure. This problem would be exacerbated at high temperatures (Meyer and Birkeland 1973). Hendler (1977) reported that exposure, occurring twice yearly along the reef flats and Punta Galeta, Panama, caused mass mortalities of *L. variegatus* and other species of sea urchins. He noted that fast growing species like *L. variegatus* were very sensitive to emersion, a result that would be predicted for a ruderal species (Lawrence 1991). Cubit et al. (1986) and Junqueira et al. (1997) also found that exposure negatively impacted *L. variegatus* populations on shallow reef flats. It is likely that individuals weakened by exposure to other physical factors could be exposed by wave action or tidal fluctuations and ultimately die from desiccation.

The effect of low oxygen concentration on nearshore populations of *L. variegatus* is not known. Hypoxia can influence the distribution of a number of echinoderm species (Lawrence 1996), but there are few reports on sea urchins (Stachowitsch 1983). *Lytechinus variegatus* would probably not survive transient periods of hypoxia or anoxia. Red tide blooms resulting in oxygen concentrations of 0–2 $mg\,l^{-1}$ have been reported in the Gulf of Mexico where *L. variegatus* occurs (Fish and Wildlife Research Institute 2005).

With the exception of exposure to phosphates, limited information exists on the effects of pollutants on *L. variegatus*. Chesher (1975) reported that copper released from a desalination plant caused a mass mortality of *L. variegatus* in Key West, Florida. Individuals exposed to domestic and industrial wastes had irregular growth and a large number of deformities including lack of pentamery and effects on the ambulacra (Allain 1975). Boettger and McClintock (1999) and Boettger et al. (2001) found that organic and inorganic phosphates can negatively impact behavior, feeding, absorption efficiency, and nutrient allocation in *L. variegatus*. Inorganic and organic phosphate exposure also had a negative effect on the gonad index and spawning activity, while also altering the biochemical composition of gonads, reducing sizes of oocytes, and decreasing sperm output (Boettger and McClintock 2001a, 2002). Moreover, both phosphates have a significant negative effect on fertilization success and cause abnormal development in embryonic and early larval development in *L. variegatus* (Boettger and McClintock 2001b). *Lytechinus semituberculatus* was adversely affected by the Jessica oil spill in Galápagos (Edgar et al. 2002). Since many pollutants cause sublethal stresses in larvae and adults that ultimately affect fitness, investigations of anthropogenic effects are necessary.

4.2. Biotic Factors

Rivera (1978) reported that predation by *Cassis tuberosa* accounted for heavy mortality of *L. variegatus* in Jobos Bay, Puerto Rico. Predation by wading birds such as the Louisiana heron *Hydranassa tricolor*, the little blue heron *Florida caerulea*, and the common egret *Casmerodium albus* placed additional pressure on *L. variegatus* in shallow water. Schroeter et al. (1984) concluded the distribution of *L. anamesus* (=*pictus*) in southern California kelp forests can be controlled by predation by *Asterina* (=*Patiria*) *miniata*.

Rivera (1978) further indicated that populations could be influenced by the ingestion of certain toxic algal species such as *Microcoleus* sp. or by a black fungus (characterized by a black "fungal net"; no mention of species), as consumption caused death. Serafy (1979) attributed mortality of *L. variegatus* off the central Florida Gulf Coast in 1996 to a red tide. Mass mortality of *L. variegatus* in the same location in the summer of 2005 was associated with a red tide (Dent, Lawrence, unpubl. obs.). Red tide lysate affects sperm mobility, fertilization, and embryonic development of *L. variegatus* (Moon and Morrill 1976). While *L. variegatus* exposed to gram negative pathogenic bacteria isolated from diseased individuals of the heart urchin *Meoma ventricosa* show disease traits similar to those observed in infected *M. ventricosa* in the field (Curaçao, Antilles), no signs of this ultimately fatal infection were noted in field populations of *L. variegatus* (Nagelkerken et al. 1999). In spite of recorded observations of biotic factors, Rivera (1977) and Oliver (1987) both concluded that abiotic factors are more important than biotic factors in regulating distribution and abundance of *L. variegatus* in local populations. Rivera (1977) found that biological mechanisms of mortality could account for 16% of all mortalities, physical mechanisms 58%, and nondetermined mechanisms 26%. Rivera suggested that physical mechanisms are important sources of mortality, but may vary in their level of influence over time, and biological mechanisms were small but a relatively constant source of mortality. We hypothesize that biotic factors, including food availability and predation, may have a greater impact on distribution and abundance of offshore *L. variegatus* than abiotic factors. Schroeter et al. (1983) found *L. anamesus* (=*pictus*) more abundant inside than outside a kelp bed in southern California, in contrast to the usual situation. They concluded *P.* (=*Asterina*) *miniata* controls distribution of *L. anamesus.* As the distribution of *P. miniata*, but not that of *L. anamesus*, is controlled by suitable substrates, they suggested biotic control by *P. miniata* accounts for the distribution of *L. anamesus.* Tewfik et al. (2005) reported abundance of *L. variegatus* was greatly increased as a result of high anthropogenic nutrient enrichment even in the absence of seagrass detritus. They suggested nutrient enrichment enhanced phytoplankton, epiphytes, and detritus associated with these to provide alternative food to the sea urchin.

Future studies should also address the role of sublethal stress on distribution, abundance, and production of *Lytechinus*. Changes in many factors, including those discussed previously, could produce responses that do not kill the individual, but affect growth, gonad production, gamete production, endocrine responses, immune responses, or other parameters associated with the physiological well-being of the organism and, as a consequence, influence the fitness of the population. From a practical standpoint, understanding sublethal stresses is very important for those species of sea urchins that are fished or cultured.

5. FOOD AND FEEDING

Omnivory is the most common strategy used by *L. variegatus* (Beddingfield and McClintock 2000). A variety of plant and animal food are consumed (Boone 1928; Mortensen 1943; Moore et al. 1963; Moore and McPherson 1965; Kier and Grant 1965; Camp et al. 1973; Prim 1973; Lowe 1975; Greenway 1977; Bach 1979; Klinger 1984; McClintock et al. 1982; Klinger and Lawrence 1984; Valentine and Heck 1991; Montague et al. 1991; Valentine and Heck 1993; Greenway 1995; Beddingfield and McClintock 1998; Peterson et al. 2002; Cobb and Lawrence 2005). In nearshore environments, *L. variegatus* feeds most commonly on the turtle grass *T. testudinum* but will feed on a variety of seagrasses including *S. filiforme*, *C. manatorum,* and *Haladule wrightii.* Montaque et al. (1991) found that sea urchins ingested decayed blades of *T. testudinum* at a significantly higher rate than green blades. *L. variegatus* consumes preferentially detrital blades of *T. testudinum* that are epiphytized in preference to nonepiphytized or epiphytized fresh blades (Greenway 1995). In fact, *L. variegatus* prefers the epibionts found on the seagrasses to the seagrass itself (Beddingfield and McClintock 1998). Where *L. variegatus* occurs in association with beds of the mussel *M. americanus* (Valentine and Heck 1993), it preys on the mussel, consuming the shell and flesh, as well as, the epibionts on the shell (Sklenar 1994). Detritus and other organic material in the sand may also be an important component of the diet (Lowe 1975; Lowe and Lawrence 1976; Rivera 1978; Vadas et al. 1982; McClintock et al. 1982; Beddingfield 1997). Beddingfield and McClintock (1998) agreed with Vadas et al. (1982) that the diet of *L. variegatus* is determined by both the availability and palatability of available food. Beddingfield (1997) found aggregative behavior in *L. variegatus* to be associated with food patches regardless of season in Saint Joseph Bay, Florida. In combined laboratory and field experiments, *L. variegatus* formed aggregations in response to food (the seagrass *Syringodium)*, and also to a combination of food and exposure to a potential predator (the spiny lobster *Panulirus argus*) (Vadas and Elner 2003). *Lytechinus variegatus* at depths 13 m and greater off the central Florida Gulf Coast consumed a diversity of algae (34 species) (Cobb and Lawrence 2005). Most, except for *Liththamnion* spp., were consumed infrequently, indicating opportunistic feeding and no strong preference.

Valentine and Heck (2001) found that nitrogen-enriched *T. testudinum* (higher nutritive quality) was not consumed in greater amounts than a non-nitrogen-enriched seagrass diet by *L. variegatus*. In contrast, sea urchins consumed a non-nitrogen-enriched seagrass diet in higher amounts than a nitrogen-enriched seagrass diet, indicating they compensate for low quality food by ingesting greater amounts of these lower quality foods. Ferguson (1982a, 1982b) indicated that dissolved organic matter could be an important source of energy for *L. variegatus*. Lowe (1975), Klinger (1984) and Beddingfield and McClintock (1998) suggested that a diverse diet is beneficial for growth by supplying a variety of nutrients.

Lowe (1975) and Beddingfield and McClintock (1998) found no relationship between food preferences and the organic content of foods. Klinger (1984) pointed out that organic content is not a good indication of accessible nutrients. This conclusion is supported by the variation of the absorption efficiency of *L. variegatus* fed different foods (Beddingfield and McClintock 1998). McGlathery (1995) found that *L. variegatus* did not show increased grazing of nutrient-enriched seagrass blades in fertilized seagrass meadows. Klinger (1984) and Beddingfield and McClintock (1998) concluded that *L. variegatus* has little distance chemoreceptive abilities and requires contact and ingestion of potential foods for detection. Klinger (1984) reported that the amount of time the food was fed upon after contact depended directly on the organic content.

The rates of consumption of various natural foods have been estimated in a variety of studies (see Beddingfield and McClintock 1998). These rates vary with the quality of the food, shape, size, texture, palatability, density, and manipulability (Lowe 1975; Klinger 1984; Montague et al. 1991). Maximal rates are generally <1.5 g wet weight per individual per day for large individuals (for natural diets). Direct comparisons among studies are difficult, as some studies have used wet weight and others, dry weight. In addition, studies rarely report feeding rates in terms of both amount ingested per individual and amount ingested per unit weight. Since size affects feeding rate (Moore and McPherson 1965), consumption rates in terms of body weight are necessary to establish the allometric relation. Additional insight can be obtained by calculating consumption rates in terms of organic and energy content.

In the field, temperature and hydrodynamics influence consumption rate. Moore and McPherson (1965) and Aseltine (1982) showed that *L. variegatus* generally feeds in areas of high tidal flow and that consumption in the laboratory is directly related to flow rate (Moore and McPherson 1965). This suggests that hydrodynamics, up to a point, influence production. Consumption is temperature-dependent and is affected by the thermal history of the individual (Moore and McPherson 1965). Winter-acclimatized individuals fed at temperatures as low as 14 °C but not at 30 °C; summer-acclimatized individuals fed minimally at 14 °C and maximally at 30 °C. Summer individuals could feed at 32 °C but not at 34 °C. Laboratory studies have corroborated temperature acclimation of the consumption rate for *L. variegatus* (Klinger et al. 1986, Hofer 2002). Feeding by *L. variegatus* is markedly depressed during the winter in the northeast Gulf of Mexico (Beddingfield and McClintock 1999).

Leighton (1971) observed grazing on kelp by *L. anamesus* (=*pictus*). His laboratory studies indicate a preference for *Macrocystis, Laminaria,* and *Gigartina* over *Egregia, Pterygophora*, and *Cystoseira*. Dean et al. (1984) noted *L. anamesus* seldom graze adult kelp like *Strongylocentrotus franciscanus*, but graze primarily small life stages.

6. GROWTH AND SURVIVAL

As a ruderal species (Lawrence and Bazhin 1998) *L. variegatus* has a fast growth rate. Pawson and Miller (1982) found that postmetamorphic juveniles increased from a diameter of 0.42 mm to 8.4 mm in 36 weeks in the laboratory and reached 23 mm in 84 weeks. Wallace (pers. comm.) found that postmetamorphic juveniles, feeding on algal assemblages in aquaria, grew to 10 mm diameter in approximately 20 weeks, and then doubled to approximately 20 mm diameter by 24 weeks. Powell et al. (2005) determined that competent larvae will settle, metamorphose to juveniles, and feed on a diatom biofilm. They isolated a diatom (*Amphora* sp.) that supports early growth in newlymetamorphosed juveniles (400−500 μm). Feeding is initiated within 2 days of metamorphosis. Initial survivorship is dependent on the quality of the broodstock (egg quality). However, this diatom will support high survivorship and growth only during the first 4 weeks post-metamorphosis. The diatom will not support growth and survivorship after this time (ca. 1–1.5 mm diameter).

Once urchins feeding on diatoms reach 1–1.5 mm in diameter they experience high mortality unless an alternative feed is available. Juvenile sea urchins were successfully reared past this critical size on a mixed-taxa biofilm containing green filamentous algae, red algae, diatoms, and their associated mucilage (Powell et al. 2005). This diet easily supports growth of juveniles to 10 mm diameter. Specific growth rates range from 8% to 10% body wet weight gain per day at this stage. The growth rate of small juveniles (<5 mm diameter) in the field is not known.

Moore et al. (1963) reported that individuals in the field grew from 12 to 55 mm within 1 year at Miami, Florida. Growth occurred primarily between January and June. They postulated that high summer temperatures inhibited growth rates. They suggested reduced growth rates in larger individuals resulted from reduced absorption of nutrients from the gut (which does not change proportionally to the overall size), not a limitation in food availability. They also suggested that growth might decrease as individuals began to produce gonads at approximately 40 mm diameter. They calculated individuals 62 mm in diameter eating seagrass have a production efficiency of 6.7%.

Camp et al. (1973) estimated that 40 mm diameter individuals were 1 year old. Rivera (1978) estimated that first year individuals grew at a rate of 4.1 mm per 30 days in the field in Jobos Bay, Puerto Rico, but only 1.4 mm per 30 days when fed *T. testudinum* in the laboratory. Similarly, Beddingfield and McClintock (2000) found that *L. variegatus* from the northern Gulf of Mexico attain a size of 35–40 mm in the first year, with a few individuals reaching 50 mm. Engstrom (1982) reported lower rates of growth along the coast of Puerto Rico, in which individuals attained a diameter of 24–29 mm in the first year and only 34–43 mm in the second year.

Based on growth rate estimates in the field, Moore et al. (1963) suggested that most individuals are 1–2 years of age, with larger individuals being 3–4 years. Allain (1975) suggested the lifespan of *L. variegatus* was 3 years. Beddingfield and McClintock (2000) correlated sea urchin diameters with growth bands in the demipyramids of the lanterns. They indicated that most do not live beyond 3–4 years. Most of these estimates of lifespan are derived from nearshore populations that are most likely impacted by episodic environmental stresses. The maximal age of larger individuals found in deep-water habitats is not known. In contrast to occurrence of growth bands in the demipyramids (Beddingfield and McClintock 2000), growth bands in the test of *L. variegatus* are not well defined and may be related to food availability (Hill et al. 2004).

7. REPRODUCTION

7.1. *Lytechinus variegatus*

Lytechinus variegatus matures at a diameter of 40 mm at the age of 1 year (Moore et al. 1963). Maximal relative gonad size (volume) is at 50–55 mm diameter. *Lytechinus variegatus* is gonochoric and hermaphroditism is rare. Moore et al. (1963) found an increase in the abundance of hermaphrodites following a cold winter in Miami, Florida, suggesting temperature-influenced sex differentiation.

The reproductive cycle of *L. variegatus* varies widely, depending on the location and, most likely, environmental conditions. *Lytechinus variegatus* does not always show pronounced annual fluctuations in the gonad index, although a definitive gonad cycle is apparent in some populations. Typically, gonads of *L. variegatus* increase in volume in spring and sometimes again to a lesser extent in late summer or fall (Moore et al. 1963; Moore and Lopez 1972; Ernst and Blake 1981; Cameron 1986; Lessios 1991; Beddingfield and McClintock 2000; McCarthy and Young 2002). However, timing of the peak spawning season within a population can vary as much as 4 months (Moore and Lopez 1972) and the magnitude of the increase in gonad volume can vary dramatically from year to year (Beddingfield and McClintock 2000). Some populations show no marked or synchronized reproductive cycle (Lessios 1985; Cameron 1986; Junqueira 1998), and even those with distinct reproductive seasons may exhibit frequent sporadic small-scale spawning events as evidenced by substantial variations in oocyte size frequencies in censused ripe females (McCarthy and Young 2002). In the Florida panhandle, gametogenesis in spring was synchronous, but as the reproductive season progressed, asynchrony of reproductive stage was observed. Image analysis of gonad sections indicated that average production of gametes was synchronous for females and males, suggesting asynchrony of stage is relatively transient, perhaps due to differences in nutrition (Cunningham and Watts 2003). These variations in apparent gonad development cycles are often mirrored by spawning patterns. Marked variations can occur even among populations that are in close proximity (Beddingfield and McClintock 2000) and are thought to result from food availability or abiotic factors. There is little evidence of aggregative behavior in adult *L. variegatus* being associated with reproductive behavior. McCarthy and Young (2002) found that on spatial scales of 1–2 m aggregation was somewhat more pronounced just prior to the reproductive season. However, higher amounts of food in the guts of individuals at this time suggest feeding activity could also explain this aggregative behavior (McCarthy and Young 2002). In a subsequent study (McCarthy and Young 2004) found water-borne gametes had no effect on aggregation in adult *L. variegatus* regardless of reproductive status or the presence/absence of sperm in the water column.

Gonad indices of individuals in the field are less than those fed formulated diets in the laboratory (Hammer et al. 2004; Hammer et al. in press). This suggests that either the quantity or quality of food may be limiting or that biotic or abiotic stress reduces gonad production in the field. Pearse and Cameron (1991) suggested that temperature and photoperiod greatly influence gonad development in sea urchins, however, Moore and Lopez (1972) found little correlation of gonad development with seasonal or annual changes in temperature. They reported correlations with annual rainfall and suggested that rainfall patterns indirectly influence nutrient availability and, thus, gonad development. Photoperiod has not been examined as a proximate cue.

In general, spawning occurs when temperatures are rising or near their annual peak. Beddingfield and McClintock (2000) described definitive spring and summer peaks of spawning in several populations in the northern Gulf of Mexico, corresponding generally to peaks in gonad indices. Similarly, Moore et al. (1963) reported that individuals at Bermuda spawn primarily in the spring, but show a lunar spawning rhythm. Those individuals found in south Florida spawned primarily in the summer, but were assumed to spawn year round (based on the presence of ripe individuals in the samples). They attributed this regional difference to differences in temperature. Lessios (1991) found *L. variegatus* spawned year round on a semilunar cycle, generally spawning every full and new moon. The pattern of gonad development and spawning reported for *L. variegatus* indicates a greater seasonal synchrony in individuals at high latitudes than in those in more tropical regions, and that spawning may be influenced by the lunar cycle. This suggests that temperature and/or photoperiod have a major role in regulating the timing and magnitude of reproduction, but no studies to support this hypothesis have been done.

In deeper water off the central Florida Gulf Coast shelf, *L. variegatus* has low gut and gonad indices and showed no growth (Hill and Lawrence 2003). This suggests food limitation and low production. Despite decreased production, *L. variegatus* extends its distribution and reproduction potential by existing in the habitat.

7.2. *Lytechinus pictus*

Although *L. pictus* has been used frequently as a model system for developmental study, information on reproduction seems limited to the report that spawning of *L. pictus* occurs from June through September (Pearse and Cameron 1991).

8. LARVAL ECOLOGY AND RECRUITMENT

Although the pattern of larval development in *L. variegatus* has been well described (Tennent 1911; Mortensen 1921; Mazur and Miller 1971; Michel 1984; McEdward and Herrera 1999; George et al. 2000), larval development in the field has not been well studied. The larvae probably stay in the water column for several weeks to several months, depending on environmental conditions and settlement cues, until they become competent. Laboratory studies indicate that echinoplutei larvae of *L. variegatus* actively aggregate in the water column in response to food patches (mixed phytoplankton) and that this behavior is likely chemosensory based (Burdett-Coutts and Metaxas 2004). These chemosensory larval behaviors suggest that similar to many marine invertebrate planktotrophic larvae, chemical information may be important in facilitating patterns of larval dispersal and settlement in *L. variegatus*. However, there have been no studies of the mechanisms mediating larval settlement in this species. Cameron (1986) almost always found larvae from June to October (plankton samples were not taken in winter and spring) in Puerto Rico, which was correlated longitudinally with spawning patterns. In contrast, Leonard et al. (1995, unpubl. data) found no larvae in the water column during 1994–1995 at St. Joseph Bay in the northern Gulf of Mexico, despite large populations of *L. variegatus* and significant juvenile recruitment in the surrounding seagrass beds during the same period (Beddingfield 1997). Based on genetic similarities among distinct populations, Pawson and Miller (1982) concluded that larvae of *L. variegatus* recruit locally and do not disperse widely.

Larval settlement and juvenile recruitment can vary both temporally and spatially. In Discovery Bay, Jamaica, Keller (1976) found little evidence of substantial recruitment. Following a mass mortality of *L. variegatus* in Galeta Bay, Panama, the population became dominated by new recruits (Hendler 1977). Rivera (1978) found small but constant recruitment at three of the four sites at Jobos Bay, Puerto Rico, with peak recruitment occurring from October to December. Cameron (1986) defined recruits as those individuals < 15 mm diameter, and reported that settlement was asynchronous and that recruitment could occur at any season. At Biscayne Bay, Florida, Oliver (1987) found that recruitment was evident year round, with major periods in the fall and early spring, and variable between two collection sites. Junqueira et al. (1997) found recruitment of *L. variegatus* at Araruama Lagoon, Brazil was continuous throughout the year, but reached a peak from August to October. They related this peak to patterns of upwelling in the region. Following annual catastrophic mortality events, summer recruitment events were necessary to replenish populations, and population recovery depended on the magnitude of recruitment. At St. Joseph Bay, northeastern Gulf of Mexico, recruitment had both interannual and site-specific variability, being highest in seagrass habitats in fall and spring (Beddingfield and McClintock 2000). Interestingly, most of the recruits <10 mm diameter were found under the spine canopies of adults, suggesting behavior that can limit predation of new recruits, and chemical communication between juveniles and adults.

9. POPULATION ECOLOGY

9.1. Predation

Rivera (1978) found that wave action occasionally washed *L. variegatus* onto shore where they were eaten by rats (*Rattus norvegicus*). Moore et al. (1963) reported that shore birds are a major predator on individuals in very shallow water, usually during low tides. Herring gulls pick up a sea urchin with their beaks and drop it onto wet sand (sometimes 4 or 5 times) to break it, and then eat the gonads and gut. Small wading birds like the semipalmated plover clean the remains. Rivera (1978) observed predation by wading birds such as the Louisiana heron *H. tricolor*, the little blue heron *F. caerulea*, and the common egret *C. albus*. Hendler (1977) observed predation by birds only during low water exposures and suggested that they have a limited effect on the population.

The primary predators of *L. variegatus* are fish and several macroinvertebrates (Table 1). Beddingfield and McClintock (1998) suggested that *L. variegatus* use seagrass beds as a refuge to hide from predators. In spite of a number of predatory species that

Table 1 Predators of *Lytechinus*.

Predator	Location	References
Lytechinus variegatus		
Fish, reef grunt, *Haomulon albrum*	Caribbean Sea	Randall (1967)
Porgy, *Calamus bajonado*		
Wrasses, *Halichoeres bivittatus*, *H. poeyi*, and *H. radiatus*		
Trunk fish, *Lactophrys trigonus*		
Reef fish	Discovery Bay, Jamaica	Keller (1976)
Fish, *Archosargus rhomboidalis*, *Haemulon macrostomum*, and *Anisotremus virginicus*	Jobos Bay, Puerto Rico	Rivera (1977, 1978)
Gastropod, *Cassis turberosa*	Puerto Rico	Rivera (1977, 1978), Engstrom (1982)
Spider crab, *Libinia* sp.	Northern Gulf of Mexico	Beddingfield (1997)
Blue crab, *Callinectes sapidus*		
Lytechinus anamesus (=*pictus*)		
Starfish, *Patiria* (=*Asterina*) *miniata*	Southern California	Schroeter et al. (1983)

consume *L. variegatus*, Rivera (1978) indicated that predation or other biological mechanisms of death only accounted for 16% of the observed mortalities of *L. variegatus* in nearshore populations at Puerto Rico. The role of predation in influencing the abundance and distribution of *L. variegatus* in offshore habitats is not known.

In field studies, *L. variegatus* exposed to damaged conspecifics (simulated predation) demonstrated a response comprised of two distinct phases; initially a short (2 min) rapid alarm response (rapid flight) followed by a slower sustained flight response (Vadas and Elner 2003). Moreover, sea urchins aggregated on food patches and exposed to a predator (the lobster *P. argus*) reduced their aggregated behavior, an indication of risk-aversion behavior.

The starfish *A. (=Patiria) miniata* elicits an escape response by *L. anamesus* (=*pictus*) that is directly related to size of the sea urchin (Schroeter et al. 1983). Small *L. anamesus* are preyed on more heavily than large ones. Undoubtedly, other predators of *L. pictus* exist.

9.2. Competition

Lytechinus variegatus is dispersed generally at random within seagrass beds or other habitats. Although aggregation of *L. variegatus* has been reported (Petersen et al. 1965), most studies indicate aggregations occur infrequently (Moore et al. 1963; Oliver 1987; Beddingfield and McClintock 2000; McCarthy and Young 2002) except for mass recruitments (e.g. Camp et al. 1973; Peterson et al. 2002) their benefits/consequences are not well known, but seem likely to be associated with exploiting food patches. A significant degree of intraspecific competition would be predicted in sea urchin fronts where densities are much higher than normal. Keller (1976) found density-dependent mortality among individuals held in cages. These densities greatly exceeded those in the field and the mortalities may have been the result of a cage effect. Greenway (1977) indicated that feeding rates of individuals decreased at high densities, an occurrence that may limit overgrazing of the seagrass beds in most populations.

Ogden and Lobel (1978), Aseltine (1982), and Greenway (1995) reported that *T. ventricosus* and *L. variegatus*, although found in the same general location, are separated by microhabitat. Keller (1976) found little or no competition between *L. variegatus* and *T. ventricosus. Lytechinus variegatus* and *Arbacia punctulata* co-occur along the central Florida Gulf Coast but usually occupy distinct microhabitats, *L. variegatus* on sand and *A. punctulata* on rock (Hill and Lawrence 2003; Cobb and Lawrence 2005). Cobb and Lawrence found *A. punctulata* consumed sessile invertebrates except when macroflora are unusually high while *L. variegatus* consumed macroflora except when it was very limited. As diets do not overlap, there would be little interspecific competition for food.

Lytechinus anamesus co-occured with *Centrostephanus coronatus, S. franciscanus,* and *Stronglocentrotus purpuratus* at Anacapa Island, although they had different patterns of abundance (Coyer et al. 1987, 1993). Competitive interaction has not been investigated.

10. COMMUNITY ECOLOGY

Most studies have examined the impact of *L. variegatus* on seagrass production. Moore et al. (1963) suggested that feeding equaled production in seagrass (*T. testudinum*) beds in southern Florida. Keller (1983) found that field densities of *L. variegatus* were below the carrying capacity of *T. testudinum* meadows at Discovery Bay, Jamaica. Similarly, Maciá (2000) found that *L. variegatus* did not significantly affect biomass of *T. testudinum* at Biscayne Bay, Florida. Valentine and Heck (1991) found that field densities of 10–40 individuals per square meter could impact turtlegrass from fall to spring, but not in the summer, at St. Joseph Bay in the northeast Gulf of Mexico. Grazing during summer stimulates seagrass production that seems to compensate for leaf loss (Valentine et al. 1997, Valentine et al. 2000). A similar seasonal effect of *L. variegatus* on seagrass occurs in the Florida Keys at field densities of 20 individuals per square meter in shallow (<2 m) water but not at field densities of 0–8 individuals per square meter in deeper (6–7 m) water (Valentine et al. 2000). Ibarra-Obando et al. (2004) found isolated and inconsistent effects of simulated sea urchin (*L. variegatus*) herbivory (clipping blades in half by hand) on *T. testudinum* below-ground biomass, shoot density, average numbers of leaves per shoot, and leaf length and width in Perdido Bay, Florida.

Peterson et al. (2002) followed the resultant community-level effects of a massive grazing front of *L. variegatus* that effectively eradicated a large-scale monospecific community of the seagrass *S. filiforme* in the Florida Keys. They found that the loss of seagrass biomass initiated community-wide cascading effects, including altered resource utilizations and species diversity patterns. The disappearance of the seagrass canopy and subsequent below-ground loss of seagrass biomass, destabilized sediments and increased water column turbidity causing a decline in light availability and sediment nitrogen levels and a depleted seed bank. The resultant community consisted of remnant seagrass survivors and weedy macroalgal colonizers. Peterson et al. (2002) suggest that this higher species diversity combined with decreased light levels may prevent the return of this community to its seagrass-dominated state.

Lytechinus variegatus may also influence the structure of animal communities associated with seagrass systems. Valentine and Heck (1993) found dense populations of *L. variegatus* in seagrass meadows in Saint Joseph Bay, Florida, dominated by the mussel *M. americanus*. Sklenar (1994) found that *L. variegatus* negatively impacted mussel populations by preying upon them.

Consumption by *L. vareigatus* of epibionts (both epiphytes and epizoa) of mussels (Sklenar 1994) and seagrass blades (Beddingfield and McClintock 1998) likely has ramifications on the dynamics of epibiotic communities. A significant trophodynamic role exists for *L. variegatus* in the turnover of decaying material. Although *L. variegatus* consumes fresh blades of *T. testudinum*, it prefers detrital blades (Montague et al. 1991; Greenway 1995). This would provide a pathway for conversion of abundant detrital-based energy. Feces and ammonium production by *L. variegatus* would provide nutrients to the environment. These trophic pathways have not been documented. Greenway (1995) pointed out that *L. variegatus* has an important role as their feces are a food source for detritovores. Interestingly, Goodbody (1970) suggested that *L. variegatus* provides an important link between seagrass production and pelagic communities. Spawning releases copious amounts of gametes into the water column where they provide food for fish and nekton.

Overgrazing events are the result of massive sea urchin fronts. One front occurred in the northeast Gulf of Mexico in late summer, 1971, when densities of *L. variegatus* reached 636 individuals per square meter (Camp et al. 1973) (Plate 5C). The front was 0.5–9 m across, several kilometers wide. Individuals were piled 2–8 deep in some aggregations. Approximately 20% of a large seagrass community comprised of *T. testudinum* mixed with *S. filiforme* and *Diplanthera wrightii* was consumed. Seagrass regrowth occurred behind the aggregations. The front was short lived, with the aggregations dispersed and many dead individuals in the area in October (Camp, pers. comm.), apparently the result of salinity reduction (Lawrence 1975). Another large front consumed several hectares in Card Sound, Florida (Bach 1979). A massive overgrazing event was documented by Rose et al. (1999) and Maciá and Lirman (1999) in outer Florida Bay. Unusually dense populations of *L. variegatus* overgrazed at least 0.8 km^2 of seagrass (primarily *S. filiforme)* habitat. The initial density of 364 individuals per square meter declined over 1.5 years to 20–50 individuals per square meter (Rose et al. 1999). More than 95% of the short shoot meristems of the seagrass were eaten. This would make recovery of the seagrass unlikely. Indeed, Maciá and Lirman (1999) saw no recovery of seagrass after 9 months at a site where the rhizomes had been eaten. Associated animal communities were also impacted including the depletion of both epifaunal and infaunal mollusc assemblages (Rose et al. 1999). Moreover, resuspension of fine-grained surface sediments resulted in significant changes in community structure and in the physical properties of the sediments. Ultimately, the changes in the community were great enough to cause loss of essential fishery habitat and structural refugia for juvenile fish and invertebrates, reduction in both primary and secondary production, degradation of water quality, and perhaps indirect effects. The influence of these changes could exceed the local boundary of the grazed habitat (Rose et al. 1999; Maciá and Lirman 1999). Maciá and Lirman (1999) noted that grazing front formation in *L. variegatus* can last up to 2.5 years, a considerable period of time when considering these community level responses.

Top-down control of community structure in southern California kelp beds may result from predation by the starfish *A. miniata* on *L. anamesus* (=*pictus*) (Schroeter et al. 1983). As the distribution of *A. miniata* is related to substrate characteristics, Schroeter et al. concluded predation could explain why *L. anamesus* is more abundant outside than inside kelp beds. Feeding by *L. anamesus* on early developmental stages of kelps may prevent seaward expansion of the bed at San Onofre (Dean et al. 1984). Dean et al. (1988) experimentally demonstrated *L. anamesus* inhibited recruitment of laminarian algae by killing gametophyte of microscopic sporophyte lifestages. They found less effect on *Cystoseira osmundacea*, an expected result since this is not a preferred alga for *L. anamesus*. Dean et al. suggested this differential grazing may have a great effect on community composition. Experimental evidence led Sala and Graham (2002) to conclude *L. anamesus* could be important in removing microscopic sporophytes of *Macrocystis pyrifera* and could affect kelp recruitment.

Lytechinus anamesus was observed nibbling, grazing, or mobbing *S. franciscanus* and *S. purpuratus*, with the frequency of the behaviors directly correlated with the density of *L. anamesus* (Coyer et al. 1987). Up to 9% of *L. anamesus* were observed eating strongylocentrotids. Where densities of *L. anamesus* were highest, up to 6% of *S. purpuratus* and 25% of *S. franciscanus* were being attacked. This effect on populations of the strongylocentrotids could have a community effect.

11. CONCLUSIONS

Lytechinus species are dominant in many habitats. The ecological role is primarily known for *L. variegatus*. As a ruderal genus, *L. variegatus* exhibits rapid growth, early reproductive maturity, and short longevity, characteristics of a ruderal species. Whether these characteristics occur in the other species is not known. The basis for the different asymptotic sizes is not known. The basis for the extensive distribution of *L. variegatus* in contrast to the other species is not known. Information is needed on all life-history stages of individuals of the species from a variety of habitats. There is minimal information on the postmetamorphic, early juvenile stage, including knowledge about nutrition, environmental influences, natural predators and community interactions. Additional information is needed to understand the effects of abiotic factors as well as anthropogenic stresses.

ACKNOWLEDGMENTS

We would like to thank the Mississippi–Alabama Sea Grant Consortium for support. We thank D Pawson, G Hendler, JS Pearse, and H Lessios for comments on the taxonomy of *Lytechinus*.

REFERENCES

Allain JY (1975) Contribution à la biologie de *Lytechinus variegatus* (Lamarck) de la baie de Carthagene (Colombie) (Echinodermata: Echinoidea). Union Océanographes France 7: 57–62

Allain JY (1977) Mortalidad natural de *Lytechinus variegatus* (Lamarck), (Echinodermata: Echinoidea) en la Bahia de Cartagena, Colombia. Museo del Mar Boletin No 7: 51–60

Aseltine DA (1982) *Tripneustes ventricosus* and *Lytechinus variegatus* (Echinoidea: Toxopneustidae): Habitat differences and the role of water turbulence. MSc thesis, Ohio State University, Columbus

Bach SD (1979) Standing crop, growth, and production of calcarious Siphonales (Chlorophyta) in a south Florida lagoon. Bull Mar Sci 29: 191–201

Beddingfield SD (1997) The nutrition, growth, reproduction and population dynamics of *Lytechinus variegatus* (Lamarck) from contrasting habitats in St. Joseph Bay, Florida. PhD dissertation, University of Alabama at Birmingham, Birmingham

Beddingfield SD, McClintock JB (1994) Environmentally-induced catastrophic mortality of the sea urchin *Lytechinus variegatus* in shallow seagrass habitats of St. Josephs Bay, Florida. Bull Mar Sci 55: 235–240

Beddingfield SD, McClintock JB (1998) Differential survivorship, reproduction, growth, and nutrient allocation in the regular echinoid *Lytechinus variegatus* (Lamarck) fed natural diets. J Exp Mar Biol Ecol 226: 195–215

Beddingfield SD, McClintock JB (2000) Demographic characteristics of *Lytechinus variegatus* (Echinodermata: Echinoidea) from three habitats in a north Florida bay, Gulf of Mexico. PSZN Mar Ecol 21: 17–40

Bishop CD, Lee KJ, Watts SA (1994) A comparison of osmolality and specific ion concentrations in the fluid compartments of the regular sea urchin *Lytechinus variegatus* Lamarck (Echinodermata: Echinoidea) in varying salinities. Comp Biochem Physiol 108A: 497–502

Boettger SA, McClintock JB (1999) Behavioral responses of *Lytechinus variegatus* during chronic exposure to inorganic and organic phosphates. In: Candia Carnevali MD, Bonasoro F (eds) Echinoderm Research 1998. Balkema, Rotterdam, p 55

Boettger SA, McClintock JB (2001a) The effects of chronic inorganic and organic phosphate exposure on development and biochemical composition of gonads in the sea urchin *Lytechinus varigatus* (Echinodermata: Echinoidea). In: Féral JP, David B (eds) Echinoderm Research. Swets & Zeitlinger, Lisse, pp 265–268

Boettger SA, McClintock JB (2001b) The effects of organic and inorganic phosphates on fertilization and early development in the sea urchin *Lytechinus variegatus* (Echinodermata: Echinoidea). Comp Biochem Physiol 129C: 307–315

Boettger SA, McClintock JB (2002) Effects of inorganic and organic phosphate exposure on aspects of reproduction in the common sea urchin *Lytechinus variegatus* (Echinodermata: Echinoidea). Exp Zool 7: 660–671

Boettger SA, McClintock JB, Klinger TS (2001) Effects of inorganic and organic phosphates on feeding, feeding absorption, nutrient allocation, growth, and righting responses of the sea urchin *Lytechinus variegatus*. Mar Biol 138: 741–751

Boettger SA, Thompson LE, Watts SA, McClintock JB, Lawrence JM (2002) Episodic rainfall influences the distribution and abundance of the regular sea urchin *Lytechinus variegatus* in Saint Andrew Bay, Northern Gulf of Mexico. Gulf Mexico Sci 20: 67–74

Boone L (1928) Echinodermata from tropical east American seas. Bull Bingham Oceanogr Lab 1: 1–22

Burdett-Coutts V, Metaxas A (2004) The effect of the quality of food patches on larval vertical distribution of the sea urchins *Lytechinus variegatus* (Lamarck) and *Strongylocentrotus droebachiensis* (Mueller). J Exp Mar Biol Ecol 308: 221–236

Cameron RA (1986) Reproduction, larval occurrence and recruitment in Caribbean sea urchins. Bull Mar Sci 39: 332–346

Camp DK, Cobb SP, van Breedveld JF (1973) Overgrazing of seagrasses by a regular urchin, *Lytechinus variegatus*. BioScience 23: 37–38

Chesher RH (1968) *Lytechinus williamsi*, a new sea urchin from Panama. Breviora 305: 1–13

Chesher RH (1975) Biological impact of a large-scale desalination plant at Key West, Florida. In: Ferguson Wood EJ, Johannes RE (eds) Tropical Marine Pollution. Elsevier, Amsterdam, pp 99–153

Clark HL (1933) Scientific survey of Porto Rico and the Virgin Islands. XVI(1). A handbook of the littoral echinoderms of Porto Rico and the other West Indians islands. New York Academy of Sciences, New York

Cobb J, Lawrence JM (2005) Diets and coexistence of the sea urchins *Lytechinus variegatus* and *Arbacia punctulata* (Echinodermata) along the central Florida gulf coast. Mar Ecol Prog Ser 295: 171–182

Coyer JA, Ambrose RF, Engle JM, Carroll JC (1993) Interactions between corals and algae on a temperate zone rocky reef: Mediation by sea urchins. J Exp Mar Biol Ecol 167: 21–37

Coyer JA, Engle JM, Ambrose RF, Nelson BV (1987) Utilization of purple and red sea urchins (*Strongylocentrotus purpurtus* Stimpson and *S. franciscanus* Agassiz) as food by the white sea urchin (*Lytechinus anamesus* Clark) in the field and laboratory. J Exp Mar Biol Ecol 105: 21–38

Cubit JD, Windsor DM, Thompson RC, Burgett JM (1986) Water-level fluctuations, emersion regimes, and variations of echinoid populations on a Caribbean reef flat. Estuar Coast Shelf Sci 22: 719–737

Cunningham A, Watts SW (2003) The reproductive cycle of the sea urchin *Lytechinus variegatus* from the Florida panhandle. In: Heinzeller T, Nebelsick JH (eds) Echinoderms: München. Proceedings of the 11th International Echinoderm Conference, Munich, Germany, pp 575

Dean TA, Jacobsen RF, Thies K, Lagos SL (1988) Differential effects of grazing by white sea urchins on recruitment of browth algae. Mar Ecol Prog Ser 48: 99–102

Dean TA, Schroeter SC, Dixon JD (1984) Effects of grazing by two species of sea urchins (*Strongylocentrotus franciscanus* and *Lytechinus anamesus*) on recruitment and survival of two species of kelp (*Macrocystis pyrifera* and *Pterygophora californica*). Mar Biol 78: 301–313

Durham JW, Wagner CD, Abbott DP (1980) Echinoidea. In: Morris RH, Abbott DP, Haderlie EC (eds) Intertidal Invertebrates of California. Stanford University Press, Stanford, pp 160–176

Edgar G, Kerrison L, Shepherd S, Toral V (2002) Impacts of the Jessica oil spill on intertidal and shallow subtidal plants and animals. In: Longhead LW, Edgar GJ (eds) Biological Impacts of the Jessica Oil Spill on the Galápagos Environment. Charles Darwin Foundation for the Galápagos Islands, Galápagos, pp 58–65

Eisen A, Inoue S (1979) Temperature-dependent timing of mitosis and cleavage in *Lytechinus variegatus*. Biol Bull 157: 367–368

Engstrom NA (1982) Immigration as a factor in maintaining populations of the sea urchin *Lytechinus variegatus* (Echinodermata: Echinoidea) in seagrass beds on the southwest coast of Puerto Rico. Stud Neotrop Fauna Env 17: 51–60

Ernst RG, Blake NJ (1981) Reproductive patterns within subpopulations of *Lytechinus variegatus* (Lamarck) (Echinodermata: Echinoidea). J Exp Mar Biol Ecol 55: 25–37

Ferguson JC (1982a) A comparative study of the net metabolic benefits derived from the uptake and release of free amino acids by marine invertebrates. Biol Bull 162: 1–17

Ferguson JC (1982b) Support of metabolism of superficial structures through direct uptake of dissolved primary amines in echinoderms. In: Lawrence JM (ed.) Echinoderms: Proceedings of the International Conference, Tampa Bay. Balkema, Rotterdam, pp 345–351

Field GW (1892) The echinoderms of Kingston Harbor. Johns Hopkins Univ Circ 11: 83

Fish and Wildlife Research Institute (2005) http://research.myfwc.com

George SB, Lawrence JM, Lawrence AL, Ford J (2000) Fertilization and development of eggs of the sea urchin *Lytechinus variegatus* maintained on an extruded feed. J World Aqua Soc 31: 232–238

Glynn PW (1968) Mass mortalities of echinoids and other reef flat organisms coincident with midday, low water exposures in Puerto Rico. Mar Biol 1: 226–243

Glynn PW, Wellington GM (1983) Corals and Coral Reefs of the Galápagos Islands. University of California Press, Berkeley

Goodbody I (1961) Mass mortality of a marine fauna following tropical rains. Ecology 42: 150–155

Goodbody IM (1970) The biology of Kingston Harbour. J Sci Council Jamaica 1: 10–34

Greenway M (1977) The production and utilization of *Thalassia testudinum* König in Kingston Harbour, Jamaica. PhD dissertation, University of the West Indies, Kingston

Greenway M (1995) Trophic relationships of macarofauna within a Jamaican seagrass meadow and the role of the echinoid *Lytechinus variegatus* (Lamarck). Bull Mar Sci 56: 719–736

Hammer BW, Hammer HS, Watts SA, Desmond RA, Lawrence JM, Lawrence AL (2004) The effects of dietary protein concentration on feeding and growth of small *Lytechinus variegatus* (Echinodermata: Echinoidea). Mar Biol 145: 1143–1157

Hammer HS, Watts SA, Desmond R, Lawrence AL, Lawrence JM (2006). The effect of dietary protein concentration on the consumption, survival, growth and production of the sea urchin *Lytechinus variegatus.* Aquaculture 254: 483–495

Hendler G (1977) The differential effects of seasonal stress and predation on the stability of reef-flat echinoid populations. In: Taylor DL (ed.) Proceedings of the Third International Corral Reef Symposium, Rosenstiel School of Marine and Atmospheric Science, Miami, pp 217–223

Hendler G, Miller JE, Pawson DL, Kier PM (1995) Sea Stars, Sea Urchins, and Allies. Smithsonian Institution Press, Washington

Hill SK, Aragona JB, Lawrence JM (2004) Growth bands in test plates of the sea urchins *Arbacia punctulata* and *Lytechinus variegtus* (Echinodermata) on the central Flordia gulf coast shelf. Gulf Mexico Sci 22: 96–100

Hill SK, Lawrence JM (2003) Habitats and characteristics of the sea urchins *Lytechinus variegatus* and *Arbacia punctulata* (Echinodermata) on the Florida gulf-coast shelf. PSZN Mar Ecol 24: 15–30

Hintz JL (1993) The effect of salinity on early development of several species of echinoids and asteroids. MSc thesis, University of South Florida, Tampa

Hofer SC (2002) The effect of temperature on feeding and growth characteristics of the sea urchin *Lytechinus variegatus* (Echinodermata: Echinoidea). MSc thesis, University of Alabama at Birmingham, Birmingham

Ibarra-Obando SE, Heck KL, Spitzer PM (2004) Effects of simultaneous changes in light, nutrients, and herbivory levels, on the structure and function of a subtropical turtlegrass meadow. J Exp Mar Biol Ecol 301: 193–224

Irlandi E, Maciá S, Serafy J (1997) Salinity reductions from freshwater canal discharge: Effects on mortality and feeding of an urchin (*Lytechinus variegatus*) and a gastropod (*Lithopoma tectum*). Bull Mar Sci 61: 869–879

Junqueira AdeOR (1998) Biologia populacional de *Lytechinus variegatus* (Lamarck, 1816) em habitats contrastantes do litoral do Rio de Janeiro, Brasil. PhD dissertation, Universidade of Sao Paulo, Brazil

Junqueira AdeOR, Ventura CRR, de Carvalho ALPS, Schmidt AJ (1997) Population recovery of the sea urchin *Lytechinus variegatus* in a seagrass falt (Araruama Lagoon, Brazil): The role of recruitment in a disturbed environment. Invert Reprod Develop 31: 143–150

Keller BD (1976) Sea urchin abundance patterns in seagrass meadows: The effects of predation and competitive interactions. PhD dissertation, Johns Hopkins University, Baltimore

Keller BD (1983) Coexistence of sea urchins in seagrass meadows: An experimental analysis of competition and predation. Ecology 64: 1581–1598

Kier PM, Grant RE (1965) Echinoid distribution and habits, Key Largo Coral Reef Preserve, Florida. Smithsonian Misc Coll 149(6): 1–68

Kleitman N (1941) The effect of temperature on the righting of echinoderms. Biol Bull 80: 292–298

Klinger TS (1984) Feeding of a marine generalist grazer: *Lytechinus variegatus* (Lamarck) (Echinodermata: Echinoidea). PhD dissertation, University of South Florida, Tampa

Klinger TS, Hsieh HL, Pangallo RA, Chen CP, Lawrence JM (1986) The effect of temperature on feeding, digestion and absorption of *Lytechinus variegatus* (Lamarck) (Echinodermata: Echinoidea). Physiol Zool 59: 332–336

Klinger TS, Lawrence JM (1984) Phagostimulation of *Lytechinus variegatus* (Lamarck) (Echinodermata: Echinoidea). Mar Behav Physiol 11: 49–67

Lawrence J (1991) Analysis of characteristics of echinoderms associated with stress and disturbance. In: Yanagisawa T, Yasumasu I, Oguro C, Suzuki N, Motokawa T (eds) Biology of Echinodermata. Balkema, Rotterdam, pp 11–26

Lawrence JM (1975) The effect of temperature-salinity combinations on the functional well-being of adult *Lytechinus variegatus* (Lamarck) (Echinodermata: Echinoidea). J Exp Mar Biol Ecol 18: 271–275

Lawrence JM (1976) Covering response in sea urchins. Nature 262: 490–491

Lawrence JM (1996) Mass mortality of echinoderms from abiotic factors. Echinoderm studies 5: 103–137

Lawrence JM, Bazhin A (1998) Life-history strategies and the potential of sea urchins for aquaculture. J Shellfish Res 17: 1515–1522

Lawrence JM, Sonnenholzner J (2004) Distribution and abundance of asteroids, echinoids, and holothuroids in Galápagos. In: Heinzeller T, Nebelsick JH (eds). Echinoderms: München. Taylor & Francis, London, pp 239–244

Lees DC, Carter GA (1972) The covering response to surge, sunlight, and ultraviolet light in *Lytechinus anamesus* (Echinoidea). Ecology 53: 1127–1133

Leighton DL (1971) Grazing activities of benthic invertebrates in southern California kelp beds. Nova Hedwigia 32: 421–453

Leonard NB, McClintock JB, Marion KR (1995) Early recruitment studies of the regular urchin *Lytechinus variegatus* in different benthic habitats. J Ala Acad Sci 66: 5

Lessios HA (1985) Annual reproductive periodicity in eight echinoid species on the Caribbean coast of Panama. In: Keegan BF, O'Connor BDS (eds) Echinodermata. Balkema, Rotterdam, pp 303–311

Lessios MA (1991) Presence and absence of monthly reproductive rhythms among eight Carribean echinoids off the coast of Panama. J Exp Mar Biol Ecol 153: 27–47

Lewis JB (1958) The biology of the tropical sea urchin *Tripneustes esculentus* Leske in Barbados, British West Indies. Can J Zool 36: 607–621

Lowe EF (1975) Absorptive efficiencies, feeding rates, and food preferences of *Lytechinus variegatus* (Echinodermata: Echinoidea) for selected marine plant. MSc thesis, University of South Florida, Tampa

Lowe EF, Lawrence JM (1976) Absorption efficiencies of *Lytechinus variegatus* (Lamarck) (Echinodermata: Echinoidea) for selected marine plants. J Exp Mar Biol Ecol 21: 223–234

Maciá S (2000) The effects of sea urchin grazing and drift algal blooms on a subtropical seagrass bed. J Exp Mar Biol Ecol 246: 53–67

Maciá S, Lirman D (1999) Destruction of Florida Bay seagrasses by a grazing front of sea urchins. Bull Mar Sci 65: 593–601

Maluf LY (1988) Composition and distribution of the central eastern Pacific echinoderms. Technical Report Number 2. Natural History Museum of Los Angeles County, pp 1–242

Mayer AG (1914) The effects of temperature on tropical marine animals. Carnegie Institution of Washington, Publication 539. Papers from the Tortugas Laboratory 6: 1–24

Mazur JE, Miller JW (1971) A description of the complete metamorphosis of the sea urchin *Lytechinus variegatus* cultured in synthetic seawater. Ohio J Sci 71: 30–36

McCarthy DA, Young CM (2002) Gametogenesis and reproductive behavior in the echinoid *Lytechinus variegatus*. Mar Ecol Prog Ser 233: 157–168

McCarthy DA, Young CM (2004) Effects of water-borne gametes on the aggregation behavior of *Lytechinus variegatus*. Mar Ecol Prog Ser 238: 191–198

McClintock JB, Klinger TS, Lawrence JM (1982) Feeding preferences of echinoids for plant and animal food models. Bull Mar Sci 32: 365–369

McEdward LR, Herrera JC (1999) Body form and skeletal morphometrics during larval development of the sea urchin *Lytechinus variegatus* (Lamarck). J Exp Mar Biol Ecol 232: 151–176

McGlathery KJ (1995) Nutrient and grazing influences on a subtropical seagrass community. Mar Ecol Prog Ser 122: 239–252

Meyer DL, Birkeland C (1974) Marine Studies – Galeta Point. In: Rubinoff RW (ed.) Smithsonian Institution Environmental Sciences Program, Smithsonian Institution, Washington, D.C., pp 129–226

Meyer DL, Birkeland C, Hendler G (1974) Marine Studies – Galeta Point. In: Windsor DM (ed.) Smithsonian Institution Environmental Sciences Program, Smithsonian Institution, Washington, D.C., pp 223–351

Michel HB (1984) Culture of *Lytechinus variegatus* (Lamarck) (Echinodermata: Echinoidea) from egg to young adult. Bull Mar Sci 34: 312–314

Millott N (1956) The covering reaction of sea-urchins. I. A preliminary account of covering in the tropical echinoid *Lytechinus variegatus* (Lamarck), and its relation to light. J Exp Biol 33: 508–523

Millott N (1975) The photosensitivity of echinoids. Adv Mar Biol 13: 1–52

Montague JR, Aguinaga JA, Ambrisco KA, Vassil DL, Collazo W (1991) Laboratory measurement of ingestion rate for the sea urchin *Lytechinus variegatus* (Lamarck). Fla Scientist 54: 129–134

Moon RT, Morrill JB (1976) The effects of *Gymnodinium breve* lysate on the larval development of the sea urchin *Lytechinus variegatus*. J Environ Sci Health A11: 673–683

Moore HB, Jutare T, Bauer JC, Jones JA (1963) The biology of *Lytechinus variegatus*. Bull Mar Sci 13: 23–53

Moore HB, Lopez NN (1972) Factors controlling variation in the seasonal spawning pattern of *Lytechinus variegatus*. Mar Biol 14: 275–280

Moore HB, McPherson BF (1965) A contribution to the study of the productivity of the urchins *Tripneustes esculentus* and *Lytechinus variegatus*. Bull Mar Sci 15: 855–871

Mortensen T (1921) Studies of the development and larval forms of echinoderms. GEC Gad, Copenhagen

Mortensen T (1943) A Monography of the Echinoidea 3(2). CA Reitzel, Copenhagen

Nagelkerken I, Smith GW, Snelder E, Kaarel M, James S (1999) Sea urchin *Meoma ventricosa* dieoff in Curaçao (Netherlands Antilles) associated with a pathogenic bacterium. Dis Aquat Org 38: 71–74

Noriega N, Cróquer A, Pauls SM (2002) Población de *Lytechinus variegatus* (Echinoidea: Toxopneustidae) y característica estructurales de las praderas de *Thalassia testudinum* en la Bahía de Mochima, Venezuela. Rev Biol Trop 50: 49–56

Ogden JC, Lobel PS (1978) The role of herbivorous fishes and urchins in coral reef communities. Env Biol Fish 3: 49–63

Oliver GD (1987) Population dynamics of *Lytechinus variegatus*. MSc thesis, University of Miami, Corral Gables

Olsenberg CW, Schmitt J, Holdbrook SJ, Abu-Saba KE, Flegal AR (1994) Detection of environmental impacts: Natural variability, effect size, and power analysis. Ecol Appl 4: 16–30

Pawson DL, Miller JE (1982) Studies of genetically controlled phenotypic characters in laboratory-reared *Lytechinus variegatus* (Lamarck) (Echinodermata: Echinoidea) from Bermuda and Florida. In: Lawrence JM (ed.) Echinoderms: Proceedings of the International Conference, Tampa Bay. Balkema, Rotterdam, pp 165–171

Pearse JS, Cameron RA (1991) Echinodermata: Echinoidea. In: Giese AC, Pearse JS, Pearse VB. Reproduction of marine invertebrates. Volume VI. Echinoderms and Lophophorates. The Boxwood Press, Pacific Grove, pp 513–662

Pearse JS, Mooi R (in press) Echinoidea. Light & Smith Manual: Intertidal Invertebrates from Central to Oregan (4th edition). In: Carlton JT (ed.) University of California Press, Berkeley and Los Angeles

Petersen JA, Sawaya P, Liu P-Y (1965) Alguns aspectos da ecologia de Echinodermata. Ann Acad Brasil Cienc 37 (suppl): 167–170

Peterson BJ, Rose CD, Rutten LM, Fourqurean JW (2002) Disturbance and recovery following catastrophic grazing: Studies of a successional chronosequence in a seagrass bed. Oikos 97: 361–370

Powell ML, Morris AL, D'Abramo LR, Lawrence AL, Watts SA (2005) Advances in juvenile culture of sea urchins. Aquaculture America 340

Prim PP (1973) Utilization of marine plants and their constituents by enteric bacteria of echinoids (Echinodermata). MSc thesis, University of South Florida, Tampa

Randall JE (1967) Food habits of reef fishes of the West Indies. Stud Trop Oceanogr 5: 665–847

Rivera JA (1977) Echinoid mortality at Jobos Bay, Puerto Rico. Proc Assoc Isl Mar Lab Carib 13: 9

Rivera JA (1978) Aspects of the biology of *Lytechinus variegatus* (Lamarck, 1816) at Jobos Bay, Puerto Rico (Echinodea: Toxopneustidae). MSc thesis, University of Puerto Rico, Mayaguez

Rivera JA, Vicente VP (1975) Aspects of the ecology and ethology of *Lytechinus variegatus* in Jobos Bay, Puerto Rico. Proc Assoc Isl Mar Lab Carib 11: 13

Roller RA, Stickle WB (1993) Effects of temperature and salinity acclimation of adults on larval survival, physiology, and early development of *Lytechinus variegatus* (Echinodermata: Echinoidea). Mar Biol 116: 583–591

Rose CD, Sharp WC, Kenworthy WJ, Hunt JH, Lyons WG, Prager EG, Valentine JF, Hall MO, Whitfield PE, Fourqurean JW (1999) Overgrazing of a large seagrass bed by the sea urchin *Lytechinus variegatus* in Outer Florida Bay. Mar Ecol Prog Ser 190: 211–222

Rosenberg VA, Wain RP (1982) Isozyme variation and genetic differentiation in the decorator sea urchin *Lytechinus variegatus* (Lamarck, 1816) In: Lawrence JM (ed.) Echinoderms: Proceedings of the International Conference, Tampa Bay. Balkema, Rotterdam, pp 193–197

Sala E, Graham MH (2002) Community-wide distribution of predator-prey interaction strength in kelp forests. Proc Nat Acad Sci USA 99: 3678–3683

Sánchez-Jérez P, Cesar A, Cortez FS, Pereira CDS, Silva SLR (2001) Spatial distribution of the most abundant sea urchin populations on the southeastern coast of São Paulo (Brazil). Cienc Mar 27: 139–153

Schroeter SC, Dixon J, Kasatendiek J (1983) Effects of the starfish *Patiria miniata* on the distribution of the sea urchin *Lytechinus anamesus* in a southern California kelp forest. Oecologia 56: 141–147

Serafy DK (1973) Variation in the polytypic sea urchin *Lytechinus variegatus* (Lamarck, 1816) in the western Atlantic (Echinodermata: Echinoidea). Bull Mar Sci 23: 525–534

Serafy DK (1979) Echinoids (Echinodermata: Echinoidea). Memoirs of the Hourglass Cruises. Vol. V, Part III. Florida Department of Natural Resources, St. Petersburg, FL

Sharp DT, Gray IE (1962) Studies on factors affecting local distribution of two sea urchins, *Arbacia punctulata* and *Lytechinus variegatus*. Ecology 43: 309–313

Sklenar SA (1994) Interactions between sea urchin grazers (*Lytechinus variegatus* and *Arbacia punctulata*) and mussels (Modiolus americanus): A mutualistic relationship? MSc thesis, University of South Alabama, Mobile

Stachowitsch M (1984) Mass mortality in the Gulf of Triest: The course of community destruction. PSZNI Mar Ecol 5: 243–264

Tennent DH (1911) Echinoderm hybridization. Pap Mar Biol Lab Tortugas 12: 119–151

Tennent DH (1942) The photodynamic action of dyes on the eggs of the sea urchin, Lytechinus variegatus. Carnegie Institute of Washington Publication 539. Papers from the Tortugas Laboratory 35: 1–153

Tewfik A, Rasmussen JB, McCann KS (2005) Anthropogenic enrichment alters a marine benthic food web. Ecology 86: 2726–2736

Thayer CW, Kensorthy WJ, Fonseca MS (1984) The ecology of eelgrass meadows of the Atlantic coast: A community profile. US Fish Wildl Serv FWS/OBS–84/02, Offices of Biological Sciences, Washington, D.C.

Vadas RL, Elner RW (2003) Responses to predation cues and food in two sympatric, tropical sea urchins. PSZN Mar Ecol 24: 101–121

Vadas RL, Fenchel T, Ogden JC (1982) Ecological studies on the sea urchin *Lytechinus variegatus*, and the algal-seagrass communities of the Miskito Cays, Nicaragua. Aqua Bot 14: 109–125

Valentine JF, Heck Jr KL (1991) The role of sea urchin grazing in regulating subtropical seagrass meadows: Evidence from field manipulations in the northern Gulf of Mexico. J Exp Mar Biol Ecol 154: 215–230

Valentine JF, Heck Jr KL (1993) Mussels in seagrass meadows: Their influence on macroinvertebrate abundance and secondary production in the northern Gulf of Mexico. Mar Ecol Prog Ser 96: 63–74

Valentine JF, Heck Jr KL, Busby J, Webb D (1997) Experimental evidence that herbivory increases shoot density and productivity in a subtropical turtle grass (*Thalassia testudinum*) meadow. Oecologia 112: 193–200

Valentine JF, Heck Jr KL, Kirsch KD, Webb D (2000) Role of sea urchin *Lytechinus variegatus* grazing in regulating subtropical turtlegrass *Thalassia testudinum* meadows in the Florida Keys (USA). Mar Ecol Prog Ser 200: 213–228

Valentine JF, Heck KL (2001) The role of leaf nitrogen content in determining turtlegrass (*Thalassia testudinum*) grazing by a generalized herbivore in the northeastern Gulf of Mexico. J Exp Mar Biol Ecol 258: 65–86

Zigler KS, Lessios HA (2004) Speciation on the coasts of the new world: Phylogeography and the evolution of bindin in the sea urchin genus *Lytechinus*. Evolution 58: 1225–1241

Edible Sea Urchins: Biology and Ecology
Editor: John Miller Lawrence

Chapter 24

Ecology of *Tripneustes*

John M Lawrence[a] and Yukio Agatsuma[b]

[a]*Department of Biology, University of South Florida, Tampa, FL (USA).*
[b]*Laboratory of Marine Plant Ecology, Graduate School of Agricultural Sciences, Tohoku University, Miyagi (Japan).*

1. THE GENUS *TRIPNEUSTES*

Traditionally, the three extant species of the genus *Tripneustes* L Agassiz 1841 have been *T. gratilla* (Linnaeus 1758) (Plate 5A), *T. ventricosus* (Lamarck 1816), and *T. depressus* A Agassiz 1863. Clark (1912) concluded the characteristics supposedly distinguishing the species are so slight that they probably were a single species. Mortensen (1943) also pointed out the high similarity in general morphology in the genus. The sperm protein bindin functions in sperm–egg attachment and is usually considered to be species specific. Bindin of the Caribbean *T. ventricosus* and the eastern Pacific *T. depressus*, separated by the Isthmus of Panama for 3 myr, are small but distinct while that of *T depressus* and the Indo-Pacific *T. gratilla* have no fixed differences (Zigler and Lessios 2003). Similarly, analysis of mitochondrial DNA for cytochrome oxidase I showed the eastern Pacific and Indo-Pacific *Tripneustes* did not differ but the two did differ from Caribbean and Brazilian *Tripneustes* (Lessios et al. 2003). *Tripneustes ventricosus* is probably the first sea urchin whose fisheries were managed, with a closed season during the peak of the breeding season implemented at Barbados since 1879 (Scheibling and Mladenov 1987). The attempt failed and the fishery had virtually collapsed by the 1980s, attributed to over fishing and possibly pollution (Scheibling and Mladenov 1987). *Tripneustes gratilla* is cultured in Okinawa (Shimabukuro 1991) (Plate 5B).

2. DISTRIBUTION

Tripneustes is a circumtropical genus that extends into the subtropics. Mortensen (1943) reported *T. gratilla* throughout the Indo-West Pacific from east Africa (Red Sea to Natal), the South Sea Islands (from the Norfolk and Kermadec Islands to the Marquesas and Hawaii), and from Australia (to Port Jackson on the east coast and Sharks Bay on the west), to southern Japan (with the Bonin Islands). Clark and Rowe (1971) excepted only the Persian Gulf and the west Indian and Pakistani coasts from its distribution in the Indo-Pacific Ocean, and Clark (1946) noted its absence in the Torres Strait, the Arafura Sea, and the northern coast of Australia. Similarly, it is found in the Gulf of Aqaba (Dafni

1992) but not in the Gulf of Suez (James and Pearse 1969). It extends to Transkei on the coast of South Africa (Marshall et al. 1991). *Tripneustes depressus* was listed for the Gulf of California, the west coast of Mexico, and the Galapagos and Clarion Islands (Caso 1974b; Maluf 1988; Mortensen 1943).

Tripneustes ventricosus is found in shallow water extending from Bermuda and southern Florida, throughout the West Indies south to Brazil, at Trinidad and Ascension Islands and on the west coast of Africa from the Gulf of Guinea to Walfish Bay (Mortensen 1943). In Brasil, *T. ventricosus* extends to Rio de Janeiro (Tommasi 1972). It is found at Panama (Lessios 1985) and at Quintana Roo (Caso 1974a). It is not found in the Gulf of Mexico (Serafy 1979).

This tropical species is limited in its poleward distribution by temperature. The population of *T. gratilla* was greatly affected by temperature of $<10\,^{\circ}C$ in 1962–1963 (Tokioka 1963) and had not recovered by 1965 (Tokioka 1966). It almost completely disappeared from the shallow waters at Seto, Japan after water temperatures of $<20\,^{\circ}C$ from mid-November to mid-April 1965. *Tripneustes gratilla* is one of the most common sea urchins of the intertidal and littoral zones of the coral reefs of the Bonin Islands where the water temperature ranges from 23.6 to 26.8 °C (Shigei 1970). It is found rarely and the tests are small at its northern limits on the Miura and Izu Peninsula (Shigei 1973). *Tripneustes gratilla* experiences similar temperature extremes at Mauritius in the Indian Ocean (21.6–27.2 °C) (Baissac et al. 1962) and at Madagascar (26–32 °C) (Maharavo 1993); and *T. ventricosus*, in southern Florida (20–31 °C) (McPherson 1965). Mass mortality of *T. ventricosus* in very shallow seagrass beds occurs in south Florida when winter seawater temperature is unusually low (Moore et al. 1963).

Small *T. gratilla* from the Gulf of Aqaba are unable to move when transferred from 27 to 22 °C (Lawrence 1973). This indicates inactivity during the winter as the seasonal temperature ranges between 21 and 27 °C (Dafni 1992). Dafni stated that *T. gratilla* could tolerate $<15\,^{\circ}C$. However, Moore and McPherson (1965) found complete acclimation of *T. ventricosus* for food consumption. The highest latitude of *T. gratilla* is in the southeastern Pacific at Easter Island (Fell 1974), which has a temperature range of 17.5–24 °C (DeSalvo et al. 1988). Populations can be near their upper limit in shallow waters during the summer, as Colón-Jones (1993) reported that *T. ventricosus* becomes sluggish at 33 °C and dies within 4 h at 40 °C. Of the 24 species of echinoderms found on a reef at Bahia, northeastern Brazil, *T. ventricosus* was one of the few species not affected by 1997–1998 El Niño (Attrill et al. 2004). However, the population was subtidal where temperatures reached only 27 °C. *Tripneustes ventricosus* has a relatively narrow temperature range for embryonic development, with 100% mortality of the plutei at 20–30 °C in contrast to about 10% at 25 °C (Cameron et al. 1985).

3. HABITATS

Tripneustes are most common in very shallow waters although *T. gratilla* can be found at 75 m and *T. ventricosus*, at 30 m (Mortensen 1943). They occur on a variety of habitats, including seagrass and algal beds, sand with rubble, rock, and coral reef flats (Table 1).

Table 1 *Tripneustes*: habitats.

Tripneustes gratilla	
Japan. Intertidal and subtidal of coral reefs.	Shigei (1970, 1973)
Okinawa. Sand bottoms below mean low water of sheltered reefs	Nakasone (1974)
Okinawa. Boulders, sandy gravel, reef rocks. Algal and seagrass areas of reef flats.	Shimabukuro (1991)
Papua New Guinea. *Thalassia hemprichii* beds.	Nojima & Mukai (1985), Koike et al. (1987), Mukai et al. (1987)
Philippines. Shallow sand flat with dead coral heads, seagrass and algae; coral	Guieb (1981)
Philippines. Community with patches of seagrass; mud–sand with seagrass and algae.	Klumpp et al. (1993)
Northern Great Barrier Reef. Reef flat, sand with coral; intertidal reef flat, sand, rubble, seagrass, coral outcrops.	Byrne et al. (2004)
Solomon Islands. Limestone or rubble of reef flat.	Morton (1973)
Mauritius. Lagoon.	Baissac et al. (1962)
Seychelles. Sheltered sandy bottom on reef flats, seaward edge of *T. hemprichii* beds, algal ridge of reef flat.	Lewis and Taylor (1966), Taylor (1968), Taylor and Lewis (1970)
Réunion Island. Depressions (0.5–1 m) in coral reef.	Faure & Montaggioni (1970)
Réunion Island. Back reefs.	Lison de Loma et al. (1999)
Réunion Island. Back reef, oligotrophic, corals.	Lison de Loma et al. (2002)
Back reef, eutrophic, macroalgae.	
Lord Howe Island. Reef flats.	Clark (1938)
Gulf of Aqaba. Open reef flat, boulder surfaces.	Schumacher (1974)
Gulf of Aqaba. *Halodule stipulacea* beds, sand.	Jafari and Mahasneh (1984)
Gulf of Aqaba. Rocks, coral knolls, *H. stipulacea* beds, sand.	Dafni and Togol (1986)
Gulf of Aqaba. Reef flat.	Dotan (1990)
Gulf of Aqaba. Rock with algal turf *(gazon)*.	Dafni (1992)
Red Sea. Reef flats.	James and Pearse (1969)
Red Sea. Shallow-water seagrass beds, reef flats.	Nebelsick (1992)
Kenya. Fringing reefs.	McClanahan (1998) Alcoverro and Mariani (2002)

(*Continued*)

Table 1 Continued

Kenya. Seagrass beds dominated by *Thalassodendron ciliatum*	
Zanzibar. *Cymodocea ciliata* and *T. hemprichii* beds.	Herring (1972)
Moçambique. *T. hemprichii* and algal beds.	MacNae and Kalk (1963)
Madagascar. Sandy pools of reef flat.	Petit (1930)
Madagascar. Shallow seagrass beds of reef flat. Detrital ridges.	Régis and Tomassin (1982), Maharavo (1993), Maharavo et al. (1994)
Easter Island. Sand at 5 m depth. Easter Island.	Codoceo R (1974)
Hawaii. Low relief without living coral.	Ebert (1971)
Hawaii. Leeward reef with heavy surge.	Ogden et al. (1989)
Tripneustes depressus (=*gratilla*)	
Ecuador. Rubble and open rocky flat with coral rubble	Sonnenholzner and Lawrence (2002)
Tripneustes ventricosus	
Venezuela. *T. testudinum* beds.	Caycedo (1979)
Venezuela. Primarily *T. testudinum beds*, also rocks and coral.	Gallo (1988)
Netherlands Antilles. Reef, rock pools, coral rock, sandy bottom.	Engel (1939)
Grenadines. *T. testudinum* beds.	Scheibling (1982)
Barbados. Most abundant on loose broken rock and on rock flats with abundant algae. Absent from sandy bottoms and living coral reef	Lewis (1958)
Barbados. *T. testudinum* beds, coral rubble.	Lilly (1975)
Barbados. Coral rubble; also scattered coral, algae, or *T. testudinum* beds.	Scheibling and Mladenov (1987)
Barbados. Coral rubble without or with abundant fleshy algae (*Dictyota* sp.) to 4–5 m depth, rubble and sand bottom with rubble at 5–7 m.	Scheibling and Mladenov (1988)
Jamaica. *T. testudinum* beds.	Keller (1976), (1983)
Jamaica. Fore reef, primarily 3–5 m depth.	Moses and Bonem 2001
	Ogden (1976)
Virgin Islands. *T. testudinum* beds.	
Virgin Islands. *T. testudinum* beds, patch reefs.	Tertschnig (1989)
Puerto Rico. *T. testudinum* beds behind reef front, 1–2 m depth; dead coral and rocks exposed to open sea with *Padina*.	Stevenson and Ufret (1966)
Puerto Rico. Inner platform reefs with *T. testudinum* and *Halimeda opuntia*.	Cameron (1986)
Puerto Rico. *T. testudinum* and zoanthid beds with intermittent patches of sand.	Colón–Jones (1993)

Table 1 Continued

Florida. *T testudinum* beds.	Voss and Voss (1955) Kier and Grant (1965) Moore et al. (1963), Moore and McPherson (1965)
Florida. *T. testudinum* beds, rocks, rubble, coral reefs, rock ledges.	McPherson (1965)
Bahamas. Primarily on limestone shelves, also *T. testudinum* beds.	Aseltine (1982)
Panama. Reef flat.	Hendler (1977)
Honduras. *T. testudinum* beds.	Lessios (1998)
Mexico. *T. testudinum* beds, coral reefs.	Caso (1974a)
Bermuda. *T. testudinum* beds.	Tertschnig (1985)

The relation between density and depth distribution of *T. gratilla* at Hawaii is variable (Ebert 1971). He stated that it probably does best in protected areas although it survives a wide range of hydrodynamics and concluded the pattern of distribution has little pattern and results from chance. Ebert (1982) subsequently ranked its hydrodynamics exposure as ranging from protected to relatively high. Ogden et al. (1989) noted that *T. gratilla* occupied exposed sites on a leeward reef at Hawaii with a heavy surge. Dotan (1990) similarly found the distribution of *T. gratilla* in the Gulf of Aqaba was irregular, unrelated to exposure, and with great variation in populations over time. Ogden and Lobel (1978) found *T. ventricosus* could occupy algal beds with high hydrodynamics in the Caribbean. Seagrass beds and limestone shelves were the habitat of *T. ventricosus* in the Bahamas, particularly with good water flow (Aseltine 1982).

Tertschnig (1989) found the density of *T. ventricosus* decreased with increasing distance from the reefs (50 m) and that the distribution pattern in the lagoon was not correlated with vegetational parameters of the seagrass. They occurred on seagrass beds between patch reefs and mangrove islands at Puerto Rico (Colón-Jones 1993). Scheibling and Mladenov (1988) reported a variety of habitats for *T. ventricosus* at Barbados. With one exception, they were at a depth of 4–5 m within about 50 m of an offshore barrier reef with strong currents and wave surge.

The density of *Tripneustes* within habitats with suitable hydrodynamics shows great variability (Table 2). In the Seychelles, they are most common at the seaward edges of seagrass beds (Taylor and Lewis 1970). Mukai et al. (1987) found *T. gratilla* aggregated in seagrass beds <1 m deep at Papua-New Guinea, with highest densities where seagrass coverage was >80%. Alcoverro and Mariani (2002) found aggregations of *T. gratilla* with densities averaging ca. 1.5 individuals per square meter in Mombassa lagoon (Kenya).

Lilly (1975) found *T. ventricosus* most common at the seaward edge of grass flats at Barbados and Moore et al. (1963) reported that *T. ventricosus* at Miami was mainly confined to the outer edge of seagrass beds. In contrast, Aseltine (1982) found *T. ventricosus* at the Bahamas within seagrass beds.

Table 2 *Tripneustes*: density, individuals per square meter.

T. gratilla		
Papua New Guinea	0.098, seagrass beds	Mukai et al. (1985)
	0.1–0.33, seagrass beds	Nojima and Mukai (1985)
Philippines	0.1–1, seagrass beds	Klumpp et al. (1993)
	0–2, seagrass beds	Junio–Meñez et al. (1998)
	1.6, reef flat, sand with seagrass	
Philippines	1.5, reef flat, silty sand with seagrass and algae	Uy et al. (2000)
Réunion Island	6.8, back reef, coral	Lison de Loma et al. (2002)
	5.0, back reef, macroalgae	
Easter Island	4	Codoceo (1974)
Gulf of Aqaba	8	Mastaller (1979)
	0–0.1, fore reef, reef slope	Dotan (1990)
	0–4.9, reef flat	
Kenya	3.47–6.11, reef lagoons	Muthiga (2005)
Madagascar	3–20, rubble ridge; 5–10, algal beds; 10–20 seagrass beds	Régis and Thomassin (1982)
	0.8–1.25, seagrass beds	Maharavo (1993)
Hawaii	0–<1, coral and rubble	Ebert (1971)
	3.7, reef flat	Ogden et al. (1989)
T. depressus (=*gratilla*)		
Baja California Sur	Maximum: 28.6; average 0.039	Holguin Quiñones et al. (2000)
Ecuador	Maximum: 2.18, decreases with depth	Sonnenholzner and Lawrence (2002)
T. ventricosus		
Bermuda	1.2	Tertschnig (1985)
Bahamas	0.68	Aseltine (1982)
Jamaica	0.6–0.8	Keller (1983)
Jamaica	<0.25, fore reef 1993–1996	Aronson and Precht (2000)
Jamaica	<0.17–2.00, forereef 1996–1998	Haley & Solandt (2001)
	0.45–2.35, backreef 1998–1999	
	0.5–0.75, fore reef 1998–1999	
Jamaica	Ca. 0.5–0.7, forereef.	Moses and Bonem (2001)
Virgin Islands	0.18, seagrass bed	Tertschnig (1989)
Barbados	<1, coral rubble 1–5, sand bottom with coral rubble	Scheibling and Mladenov (1988)
Puerto Rico	2.5, seagrass bed	Colón–Jones (1993)

4. BEHAVIOR

Petit (1930) usually found *T. gratilla* in groups of three or four, often touching and even overlapping. Nojima and Mukai (1985) also found a tendency for pairing. In contrast, Codoceo (1974) reported that *T. gratilla* were usually isolated and Shimabukuro (1991) rarely found *T. gratilla* touching one another, even at high densities. Maharavo (1993) found *T. gratilla* at Madagascar have a random distribution. Tertschnig (1985) proposed that the dispersed distribution of *T. ventricosus* on seagrass beds resulted from their feeding behavior. As they depleted the epiphytized seagrass leaves, they increased activity and moved to adjacent areas rather than eat the proximal green, nonepiphytized portions of the leaves. Maharavo et al. (1993) similarly reported that *T. gratilla* on the reef flats at Nosy Bé, Madagascar feed on seagrass leaves in a clump until it is depleted and then move to another ungrazed clump.

Ogden et al. (1989) reported *T. gratilla* on an exposed site with large rocks and extensive cover of crustose coralline algae and microalgae of a leeward reef moved little as a mini-halo the size of the individuals was cleared of algae (Ogden et al. 1989). Although they considered this a "homesite," a cleared area was recolonized quickly, suggesting vagility.

Nojima and Mukai (1985) found *T. gratilla* moved randomly in seagrass beds about 1.3 m day^{-1}. Tertschnig (1989) reported movement of *T. ventricosus* averaged 21 cm h^{-1} in the day and 53 cm h^{-1} in the night. Movement on a patch reef was less, 5 cm h^{-1} in the day and 26 cm h^{-1} in the night. Solandt and Campbell (1998) found that the occasional *T. ventricosus* on the forereef at Jamaica were always cryptic and did not move for long periods of time.

During March and April, *T. ventricosus* at Barbados move from upper surfaces of rocks and on grassy flats to form groups of several to a dozen under rocks and ledges (Lewis 1958). Closely spaced or touching clusters have been reported at Barbados in June (Scheibling and Mladenov 1988). Aggregations consisted of up to 143 individuals with an average distance between clumps of slightly more than 3 m on coral rubble. Scheibling (pers. comm.) observed a massive aggregation several meters in diameter on a featureless sand bottom within the barrier reef at Watering Bay on the windward side of Carriacou. Scheibling and Mladenov suggested these might be reproductive aggregations as they occurred at the appropriate season. Aggregation of *T. gratilla* of 10–20 individuals per square meter at Madagascar occurs during the spawning periods (Régis and Thomassin 1982).

The covering or masking behavior of *Tripneustes* may have a number of functions that are not mutually exclusive. Petit (1930) interpreted the behavior in *T. gratilla* as photodefensive. Lewis (1958) also interpreted the covering response of *T. ventricosus* as a light-avoiding reaction as individuals from under rocks or in shaded areas seldom covered. Moore et al. (1963) reported more covering of *T. ventricosus* in the summer, which they considered a response to more intense sunlight. Kier and Grant (1965) concluded that the covering behavior of *T. ventricosus* is much less than that of *Lytechinus variegatus,* noting that *T. ventricosus* are normally nearly uncovered while nearby *L. variegatus* may be sparsely to almost completely covered. Pigmented *T. ventricosus* covered significantly less when sunlight was decreased and nonpigmented ones covered themselves significantly more than pigmented ones (Kehas et al. 2005). Kehas et al. suggested covering is a defense against radiation and the nonpigmented individuals have a greater susceptibility to radiation.

Tripneustes may catch drift seagrass and algae for consumption. Nojima and Mukai (1985) reported that inactive *T. gratilla* held material close to their test while active individuals, feeding or moving, did not cover. In contrast, Tertschnig (1989) reported that *T. ventricosus* fed even when completely covered, but that the amount of cover was much less for moving individuals than for stationary ones. On a leeward reef at Hawaii, 89% of the inshore *T. gratilla* carried drift material, while only 14% further offshore did (Ogden et al. 1989). Most of the individuals at sites without macroalgae began consuming the drift material within minutes after capture.

Lewis (1958) reported a shadow reflex in *T. ventricosus* that consisted of an immediate erection of the spines followed by a circular waving motion and movement of the individual. A light reflex involved not only the spines and movement, but violent waving of the pedicellariae. These may be appropriate responses to shifting light patterns produced by a diurnal fish predator. Tertschnig (1989) attributed the nocturnal movement of *T. ventricosus* on patch reefs at the Virgin Islands to diurnal fish predators known to prey on sea urchins in the West Indies (Randall 1967; Keller I 1976). Tertschnig also noted that *T. ventricosus* become inactive when wrasse feed on their tube feet. Nocturnal feeding by *T. gratilla* at Réunion Island (Lison de Loma et al. 1999) and diurnal grazing by *T. gratilla* in the Gulf of Aqaba (Schumacher 1974) may indicate a diel predator behavior.

Tripneustes ventricosus shows a flight response initiated prior to contact by *Oreaster reticulatus* (Scheibling 1982) and an alarm response in response to body extracts (Parker and Shulman 1986, Vadas and Elner 2003). *Tripneustes ventricosus* exposed to the coelomic fluid of other individuals autotomize their pedicellariae into the water (Vásquez unpubl.; Lawrence, pers. obs.), which could be a defensive mechanism against predators.

5. FOOD

As *Tripneustes* are found in a variety of habitats, it is not surprising their food varies (Table 3). Field observations in the Philippines by Klumpp et al. (1993) showed that *T. gratilla* foraged close to the substratum, ingesting primarily live leaves of *Thalassia hemprichii*, and to a lesser extent on *Halimeda, Syringodium isoetifolium*, and coral rubble where available. They fed equally on young and on old, more epiphytized leaves. Only small amounts of dead seagrass leaves were eaten, and the algae *Amphiroa fragilissima* and *Sargassum crassfolium* were avoided. *Tripneustes gratilla* in seagrass at Papua New Guinea primarily eats living leaves of *T. hemprichii* (Mukai and Nojima 1985). Very small *T. gratilla* eat sessile diatoms and large individuals eat macroalgae *(Sargassum* spp., seagrasses, and microflora) (Shimabukuro 1991). Surprisingly, although only seagrass species were eaten by *T. gratilla* off Toliara, Madagascar, *T. hemprichii* was clearly not favored (Vaïtilingon et al. 2003). Algae were the primary diet of *T. gratilla* at La Réunion although considerable detritus also was eaten (Lison de Loma et al. 2002). Lison de Loma et al. reported strong selectivity, with the brown algal *Turbinaria ornata* most selected and calcareous algae strongly avoided. Interestingly, selectivity varied with site suggested interaction with other factors that affect feeding.

Food of *T. gratilla* on a leeward reef at Hawaii varied with site (Ogden et al. 1989), being coralline algae with encrusting microalgae, drift algae, or macroalgae. Maharavo

Table 3 *Tripneustes*: food.

Tripneustes gratilla	
Philippines. *Hydroclathrus clathratus*, eelgrass.	Tuason and Gomez (1979)
Philippines. *Hydroclathrus clathratus*, also seagrasses and other algae.	Guieb (1981)
Philippines. Seagrass, filamentous green algae, macroalgae, epiphytes, animals.	Gomez et al. (1983)
Philippines. Usually *Thalassia hemprichii*, live leaves more than macroalgae or dead leaves.	Klumpp et al. (1993)
Philippines. *Sargassum*	Juinio–Meñez et al. (1998)
Papua New Guinea. *Thalassia hemprichii.*	Mukai and Nojima (1985)
Papua New Guinea. Detrital leaves of *Thalassia hemprichii.*	Nojima et al. (1987)
Gulf of Aqaba. *Halodule stipulacea.*	Jafari and Mahasneh (1984)
Mauritius. Mostly *Cymodocea ciliata.* Also algae, coral, mud, sand, calcareous algae, foraminiferans, polyzoa, molluscs.	Herring (1972)
Zanzibar. Mostly *C. ciliata.* Also coral, mud and silt, also foraminiferans.	Herring (1972)
Madagascar. Mainly leaves (including epiphytes) of *Thalassodendron* and *Syringodium isoetifolium*; also *T hemprichii, Halodule uninervis* and algae.	Maharavo (1993), Maharavo et al. (1994)
Madagascar. Only seagrasss.	Vaïtilingon et al. (2003)
Kenya. Seagrass beds dominated by *Thalassodendron ciliatum.*	Alcoverro and Mariani (2002)
Hawaii. Crustose coralline algae, macroalgae, sand with macroflora.	Ogden et al. (1989)
Tripneustes ventricosus	
Barbados. Usually algae, although some sand.	Lewis (1958)
Barbados. *Thalassia testudinum.*	Lilly (1975)
Barbados. Epilithic coralline algae, fleshy algae.	Scheibling and Mladenov (1988)
Virgin Islands. Primarily *T. testudinum. Padina, Syringodium, Dictyota* and other algae.	Kitting et al. (1974)
Virgin Islands. Usually old, epiphytized portions of leaves of *T. testudinum;* also *Syringodiun filiforme,* some *Halimeda* sp.	Tertschnig (1989)
Virgin Islands. *T. testudinum, Syringodium,* macroalgae.	Abbot and Ogden, in Ogden and Lobel (1978)
Puerto Rico. *T. testudinum, Padina.*	Stevenson and Ufret (1966)
Florida. *T. testudinum.*	Moore and McPherson (1965)

(1993) suggested *T. gratilla* on the reef flats at Madagascar also ingests seston caught by the spines.

Lewis (1958) found the guts of *T. ventricosus* at Barbados almost always contained only algae, although some were packed with sand. In the laboratory, broad leafy forms such as *Padina* or *Dictyota* and *Halimeda* and *Enteromorpha* were rejected. Lilly (1975) reported

T. ventricosus overgrazed *Thalassia testudinum* to the short shoot in several locations at Barbados. Leaf regeneration by *T. testudinum* was negligible for several months until the density of *T. ventricosus* became less. Tertschnig (1989) reported *T. ventricosus* at the Virgin Islands primarily ate *T. testudinum* with the brown, epiphytized blade much preferred over the green portions. The seagrass *Syringodium filiforme* was eaten, but *Halimeda* only rarely. Keller (1983) observed *T. ventricosus* at Jamaica fed almost exclusively on leaves of *T. testudinum*. However, Scheibling and Mladenov (1988) found the food of *T. ventricosus* varied with habitat at Barbados, consisting of epilithic coralline algae for individuals on coral rubble and fleshy algae at another location. At one site, the gut contained algae and sand. Maharavo et al. (1994) concluded the food of *T. gratilla* was a function of availability. The conclusion of Ogden and Lobel (1978) and Keller (1983) that *T. ventricosus* is more specialized than the co-occurring *L. variegatus* seems unjustified.

Schumacher (1974) stated *T. gratilla* grazed during the entire day but provided no data. Nojima and Mukai (1985) found the diel activity of *T. gratilla* was not determined by light as it fed mainly during daytime, being inactive from midnight to early morning. Klumpp et al. (1993) also reported continuous feeding. In contrast, *T. gratilla* feeds nocturnally at Réunion Island (Lison de Loma et al. 1999), which suggests predation. A diel rhythm in feeding by *T. gratilla* was found at Madagascar, less pronounced during the period of active feeding (winter) (Vaïtilingon et al. 2003). This indicates several factors can affect feeding and their interaction may obscure patterns.

6. GROWTH

Growth of *Tripneustes* is rapid and similar rates for first-year growth (diameter) have been reported for both species: 60–80 mm (Lewis 1958), 75 mm (maximum 100 mm) (McPherson 1965), and 70 mm (Scheibling and Mladenov 1988) for *T. ventricosus*; and 50 mm (Maharavo 1993) to 60–70 mm (Shimabukuro 1991) for *T. gratilla*. Most of this growth occurs early, with *T. ventricosus* reaching about 50 mm in 5 months (McPherson 1965) and *T. gratilla* reaching 60 mm in 5 months in the Philippines (Bacolod and Dy 1986) and in the Gulf of Aqaba (Dafni 1992). *Tripneustes gratilla* fed algal turf in the laboratory at Eilat grew from an average 0.246 g in October to 25.4 g in October (see Chapter 1).

Lewis (1958) reported a slowing of growth in *T. ventricosus* in early summer and McPherson (1965) found no growth of large individuals during the summer. Moore et al. (1963) recorded a negative correlation between temperature and growth. This could be the result of a negative energy budget if energy utilization resulting from an increase in respiration with temperature exceeded the energy intake resulting from an increase in feeding as reported by Moore and McPherson (1965). Certainly this would occur at >30 °C as the feeding rate decreases precipitously at this temperature. The decreased growth rate of the large individuals in summer also could have resulted from the usual decrease in growth rate with an increase in body size. Maharavo (1993) similarly attributed the decrease in growth of *T. gratilla* during the summer at Madagascar to stress.

Growth of *T. ventricosus* was greater on brown algal beds than on seagrass beds at Barbados (Lilly 1975). Overgrazing by *T. ventricosus* there reduced their growth rate. The growth of *T. ventricosus* is greatly affected by characteristics of seagrass beds in the Virgin Islands (Tertschnig 1986).

Shimabukuro (1991) reported an increase in growth rate of very small *T. gratilla* was associated with a change in food from sessile diatoms to macroalgae. These individuals reached 10 mm diameter by May and June and 60–70 cm by November. They are food limited in the field, as individuals fed *Sargassum* spp. *ad libitum* in culture grew more rapidly and attained sizes to 139 mm diameter in 22 months.

The largest individuals reported from the field include 124 mm (Tertschnig 1986), 127 mm (McPherson 1965), 150 mm (Mortensen 1943), and 160 mm (Caso 1974a) for *T. ventricosus;* 108 mm (Liao and Clark 1995), 145 mm (Clark 1912), 155 mm (Baker 1968), and 160 mm (Codoceo 1974, Caso 1974b) for *T. gratilla.* Such large individuals are evidence of lack of major disturbance for 4–5 years and/or an absence of fishing.

7. REPRODUCTION

The sex ratio of *T. ventricosus* at three locations in the Virgin Islands was 1:1 (Gilliken et al. 1974) and that of *T. gratilla* was 1:1 in the Gulf of Aqaba and northern Red Sea (Fouda and Hellal 1990) and in Kenyan reef lagoons (Muthiga 2005).

Gonads develop in *T. ventricosus* at a test diameter of 20–45 mm (Lewis 1958; Moore, in McPherson 1965; McPherson 1965) and in *T. gratilla* at a diameter of about 50 mm (Dafni and Tobol 1986; Maharavo 1993; Juinio-Meñez et al.1998) at age less than 1 year. The gonad index of *T. gratilla* increased with size to about 70 mm and did not decrease with further size increase to 100 mm (Maharavo 1993).

The maximum gonad size reported for *T. gratilla* in the field is about 10–15% of the wet weight body (Tuason and Gomez 1979; Fouda and Hellal 1990; Maharavo 1993) and about 20% of the test volume (Kobayashi 1969). Maharavo reported an increase in gonad production by *T. gratilla* with size until about 70 mm, with no indication of a decrease in gonad index to a size of 100 mm. Muthiga (2005) reported no significant relationship between gonad index and test diameter of *T. gratilla* but a significant relationship between gonad weight and test diameter. The maximum gonad size for *T. ventricosus* is 14% of the test volume (McPherson 1965).

Gonad production by male and female sea urchins seems to differ little (Lawrence 1987), but this has not been investigated for *Tripneustes*. Gonad production in the field is greatly affected by food. Gonad production by *T. gratilla* on sand bottoms in the Gulf of Aqaba was much less than in seagrass beds (Jafari and Mahasneh 1984). In the Solitary Islands (New South Wales, Australia), The maximum gonad size in *T. gratilla* (ca. 15% wet body weight) peaked in late summer-early autumn with little difference between sexes (O'Connor et al. 1976, 1978).

The season of spawning by *T. gratilla* varies. It has been reported in spring and fall at the Great Barrier Reef (Stephenson 1934), winter in the northern Red Sea (Pearse 1974) and Gulf of Aqaba (Kidron et al. 1972), summer off Japan (Kobayashi 1969; Onoda 1936), summer to autumn off Kenya (Muthiga 2005), autumn in the Philippines and at Taiwan

(Chen and Chang 1981; Tuason and Gomez 1979). Mature ova and sperm throughout the year in *T. gratilla* in the Solitary Islands with high production of ova in autumn and winter. However, they could induce fertilization in the autumn only (O'Connor et al. 1976, 1978).

Most gonad production in *T. ventricosus* at Barbados occurs in the spring and summer (Lewis 1958). Gonad production in *T. ventricosus* at southern Florida may occur in the spring and also in the fall (Moore et al. 1963; McPherson 1965). Most gonad production in *T. gratilla* occurs in the spring with spawning in the fall in the Philippines and at Taiwan (Tuason and Gomez 1979; Chen and Chang 1981, respectively). Shimabukuro (1991) reported a gonadal growth phase in *T. gratilla* associated with an accumulation of "granules" from March to June at Okinawa that began to diminish with gamete development in the summer, with mature gonads filled with eggs and sperm by September. Spawning occurs in the summer off Japan (Kobayashi 1969; Onoda 1936). Mature gonads occur in the winter in the northern Red Sea (Pearse 1974) and the Gulf of Aqaba (Kidron et al. 1972). At Madagascar, gametogenic activity is seen in the populations of *T. gratilla* throughout the year (Maharavo 1993). Active sperm and mature ova occur in *T. ventricosus* in the winter but are most abundant in late spring and through the summer at Barbados (Lewis 1958). McPherson (1965) found some *T. ventricosus* had large gonads and active sperm or mature eggs every month, and that most individuals had some active sperm or mature eggs every month. Cameron (1986) also found the gonads of some *T. ventricosus* at Puerto Rico oozed gametes every month, but the percentage was highest from January through the spring. Mature gametes are found throughout the year in Panamian *T. ventricosus* (Lessios 1985) and spawning shows no lunar periodicity (Lessios 1991). Muthiga (2005) found a significant relationship between gonad index and lunar day, with spawning occurring between lunar day 7 and 21. Pearse (1974) concluded reproduction of *T. gratilla* had no relation to temperature. Chen and Chang (1981) concluded the geographical variation in spawning is related to temperature.

8. RECRUITMENT

Lewis (1958) found small *T. ventricosus* (10–30 mm diameter) at Barbados in late summer and early fall under rocks and in crevices. McPherson (1965) found small individuals in July in south Florida. Aseltine (1982) found small individuals in August in the Bahamas, but they were few. Keller (1983) reported recruits at Jamaica during July. Tertschnig (1989) also first found small individuals (10–30 mm diameter) in August in the Virgin Islands while Rivera (1979) found them from October to December at Puerto Rico. In contrast, Cameron (1986) found recruits in November and Colón-Jones (1993) found them from March to May in Puerto Rico. Small *T. ventricosus* at Barbados are found on sand flats instead of the coral rubble reefs where large individuals are found (Scheibling and Mladenov 1987). Small *T. ventricosus* have been found in shallow wave-exposed areas in the Bahamas (Moore et at. 1963), Virgin Islands (Scheibling and Mladenov 1987), and seagrass beds in Bermuda (McPherson 1965; Tertschnig 1985). Colonization by *T. ventricosus* of newly exposed surfaces was greater on the backreef (1 m depth) than the forereef

(5 m depth) at Jamaica (Solandt and Campbell 1998). Nevertheless, the data of Haley and Solandt (2001) suggest recruitment of *T. ventricosus* on both the forereef and backreef.

Abundant, small *T. gratilla* occurred in March 1965 at Seto, Japan (Tokioka 1966). Lawrence (unpubl.) found numerous 8–22 mm diameter *T. gratilla* in rock crevices <1 m depth at Eilat, Gulf of Aqaba in November 1969. *Tripneustes gratilla* <10 mm diameter were found at the same locality from May to June 1980, 1981, August 1982, January 1983, and February 1984 (Dafni 1992; Dafni and Tobol 1986). Recruitment of *T. gratilla* in the Philippines had a major peak in November with a minor one in March (Bacolod and Dy 1986) that matched the spawning periods indicated by Tuason and Gomez (1978). Klumpp et al. (1993) found a major recruitment in the Philippines to be in April–May. Mukai et al. (1987) reported only one recruitment of *T. gratilla* in seagrass beds at Papua New Guinea. For Okinawa, Shimabukuro (1991) found reported recruitment of 0.3–0.9 mm *T. gratilla* in February, reaching 2–3 mm by May. He found small individuals in a narrow zone 0–0.5 m depth. Individuals <2 cm diameter were beneath boulders, gravel, and algal turf. They migrated in November to the deeper areas where the larger, older individuals occurred.

9. MORTALITY

Bacolod and Dy (1986) found an annual mortality of 99% of *T. gratilla* in the central Philippines. Ebert (1982) concluded most of the populations of *T. gratilla* he studied were annuals. Shimabukuro (1991) reported *T. gratilla* seldom lived more than 2 years at Okinawa. High mortality may have resulted in poor recovery of tagged *T. gratilla* at Hawaii, Kenya, Seychelles, and Israel after 1 year by Ebert (1982) although migration cannot be discounted. Colón-Jones (1993) also reported a 90% decrease in a population of *T. ventricosus* in 1 year at Puerto Rico.

Dafni (1992) and Dafni and Tobol (1986) reported mass mortality of exposed populations of *T. gratilla* at Eilat in two successive winters. Lilly (1975) noted *T. ventricosus* were swept from coral rubble reefs and many were washed ashore, especially after heavy seas. Rivera (1977) also reported *T. ventricosus* from shallow seagrass were swept onshore from wind-generated waves at Puerto Rico. Glynn (1968) and Hendler (1977) reported death of *T. ventricosus* on reef flats in the Caribbean from subaerial exposure. Lilly (1975) noted *T. ventricosus* were swept from coral rubble reefs and many were washed ashore, especially after heavy seas. A mass mortality of small *T. gratilla* occurred from rain runoff in shallow water on a reef flat at Okinawa (Shimabukuro 1991). Haley and Solandt (2001) reported a population of *T. ventricosus* in Discovery Bay, Jamaica was ephemeral with a 10-fold increase in one season followed by a decline to usual densities within 2 years.

Rivera (1979) found *T. ventricosus* at Puerto Rico covered with a black network (believed to be a fungus) that produced heavy mortality. *Tripneustes ventricosus* were not affected by the pathogen that caused mass mortality of *Diadema antillarum* (Lessios et al. 1984). Mass mortalities of *T. ventricosus* of unknown origin occurred at Puerto Rico in autumn 1990 (Colón-Jones 1993) and winter 1995 (Williams et al. 1996). Maharavo (1993) found no evidence of disease in *T. gratilla* at Madagascar.

10. COMMUNITY

10.1. Effects of Feeding

As *Tripneustes* has a high rate of feeding (Gomez et al. 1983; Guieb 1981; Keller 1983; Klumpp et al. 1993; Lilly 1975; Maharavo 1993; Moore and McPherson 1965; Mukai and Nojima 1985; Tertschnig 1989), one might expect major consequences on the community. *Tripneustes gratilla* is considered a pest in seaweed farms in the Philippines (Bacolod and Dy 1986; Gomez et al. 1983).

Although *T. ventricosus* can overgraze *T. testudinum* in its immediate vicinity at Barbados (Lilly 1975), the area recovers within several months after the sea urchins move away. Seagrasses similarly affected by *T. gratilla* at Madagascar also recover, and the beds as a whole are not overgrazed (Maharavo 1993). Tertschnig (1989) calculated the consumption of *T. testudinum* by *T. ventricosus* was only 3.6% of the daily production in the Virgin Islands. Zieman et al. (1979) calculated that a parrot fish, *D. antillarum* and *T. ventricosus* consumed 5–10% of the daily blade production in the same locality. The grazing impact by *T. gratilla* on the seagrass bed at Papua New Guinea was estimated to be 0.04% day^{-1} of the standing crop (Mukai and Nojima 1985). Klumpp et al. (1993) calculated the estimated annual grazing rate by *T. gratilla* and *Salmacis sphaeroides* at a site in the Philippines was 24% of the annual seagrass production. They noted variation in sea urchin population structure and density could result in the grazing varying from <5% to >100% seasonally.

Alcoverro and Mariani (2002) reported aggregations of *T. gratilla* in seagrass beds in a Kenyan lagoon dominated by *Thalassodendron ciliatum*. The aggregations were linear in structure (mean density, 10.4 individuals per square meter). Mean estimated consumption ranged from 1.8 to 5 shoots $m^{-2}d^{-1}$. Behind their direction of movement was an area of defoliated seagrass rhizomes (85% dead shoots). Calculated frequency of return of an aggregation to an area ranged from 34 to 99 months, with a recovery time of the seagrass of 44 months. This remarkable observation is similar to that reported for another toxopneustid, *L. variegatus*, off the gulf coast of Florida (Camp et al. 1973).

An indirect effect of feeding is detachment of leaves. Nojima et al. (1987) estimated *T. gratilla* at Papua New Guinea detached about one-third of the amount of leaves of *T. hemprichii* consumed. Almost all of this was deposited in place to form detritus.

A second indirect effect of feeding is production of feces that re-enters the community in fragmented or soluble forms available to other consumers. The organic and energy content of feces is less than that of the food (Lilly 1975; Lawrence 1976; Moore and McPherson 1965; Klumpp et a1. 1993). The C : P ratio of the feces is very high, while the C : N ratio is very low (Koike et al. 1987; Vink and Atkinson. 1985). Dy and Yap (2000) and Dy et al. (2002) found high rates of ammonium excretion by *T. gratilla* and emphasized regeneration of inorganic nutrients in the community.

The only report of top-down control of community as a result of feeding is that coral recruitment from most horizontal or low-sloped substrata is limited in the Gulf of Aqaba by grazing of *T. gratilla* as well as *D. setosum* (Schumacher 1974). Benayahu and Loya (1977) subsequently found no *T. gratilla* in the region and attributed primary control to *D. setosum*. Haley and Solandt (2001) suggested ephemeral appearance and decline of

T. ventricosus on the coral covered with macroalgae can act as a successional stage for recruitment and establishment of *D. antillarum*.

10.2. Competition

Shimabukuro (1991) suggested both intra- and interspecific competition involving *T. gratilla* for food or habitat. He reported *Echinometra mathaei* invaded a fishing ground of *T. gratilla* and eventually surplanted them. *Echinometra mathaei* is not known for its migratory habits. Shimabukuro suggested the echinoids *Toxopneustes pileolus* and *Pseudoboletia maculata* had little competition with *T. gratilla* as their densities were low.

Lytechinus variegatus does not occur at Barbados where *T. ventricosus* is abundant (Lewis 1958). Aseltine (1982) also found habitat separation for *T. ventricosus* and *L. variegatus* in the Bahamas. Keller (1983) found no evidence of strong interaction between the two species in the seagrass beds in Discovery Bay, Jamaica and attributed their co-occurrence to diet differences. Colón-Jones (1993) also found the occurrence of *T. ventricosus* and *L. variegatus* in seagrass beds at Puerto Rico to be positively correlated.

More information is available on the interaction between *T. ventricosus* and *D. antillarum*. Ogden (1976) reported density of *T. ventricosus* increased on patch reefs at St. Croix, USVI after *D. antillarum* had been removed. Moses and Bonem (2001) concluded increase in density of *T. ventricosus* on the forereef along the north coast of Jamaica in 1996–1998 may have resulted from the mass mortality of *D. antillarum* that opened up this habitat to the former species.

Aronson and Precht (2000) found an increase in density on the forereef at Discovery Bay, Jamaica from nearly none in 1996 to ca. 70 individuals per 100 square meter for *T. ventricosus* and ca. 50 individuals per 100 square meters for *D. antillarum* 1998. While the density of *T. ventricosus* was the same in 1999, that of *D. antillarum* increased to ca. 130 individuals per 100 square meters. From 1998 to 1999 the macroalgal cover decreased from ca. 60% to 15% and crustose/microturf/bare substratum increased from ca. 18% to 60%. Aronson and Precht emphasized the key role of herbivory in structuring shallow reef communities in the Caribbean. These increases in density of the two species could have resulted from migration or recruitment. Whether the continued increase in density of *D. antillarum* but not that of *T. ventricosus* was association with competition or the change in habitat from macroalgae to crustose/microturf/bare substratum is speculative.

Haley and Solandt (2001) also reported great yearly changes in density of *T. ventricosus* and *D. antillarum* in the western part of Discovery Bay. A ten-fold increase in density to 2.00 individuals per square meter of *T. ventricosus* on the primarily coralline substrate of the forereef (5 m depth) occurred in 1998, followed by a decline to normal levels of <0.17 individuals per square meter in 2000. *T. ventricosus* showed a similar change in density on the backreef (1–2 m depth). The density increased from 0.45 individuals per square meter in 1998 to 2.35 individuals per square meter in 1999 followed by a decline to 1.31 individuals per square meter in 1999. These two observations suggest successful recruitment and a cohort lifespan of 2 years. Data on changes in size-frequency distribution over time rather than maximum test size would have been useful. Solandt and

Haley noted migration or mortality could have contributed to the observed changes in density. They also noted potential interaction between the two species. On the backreef, as the two species occupied different microhabitats with *T. ventricosus* present in seagrass and *D. antillarum* present on patch reefs and coral rubble, interaction would be small. On the forereef, the density of *D. antillarum* on the forereef decreased in 1998 and then recovered by 2000, a reciprocal pattern to that of *T. ventricosus*. This possibility was not considered as the 1998 increase in density of *D. antillarum* is not significant. Instead, Haley and Solandt proposed *T. ventricosus* acts as a successional stage for *D. antillarum* by browsing down the macroalgae to a state more appropriate for feeding by the latter.

McClanahan (1998) found evidence for partitioning of spatial refuge resource by *E. mathaci, D. setosum*, and *T. gratilla* at low to intermediate levels of predation by fish on fringing reefs at Kenya.

10.3. Predation

The helmet conch *Cassis tuberosa* preys on *T. ventricosus* (Moore 1956; Hughes and Hughes 1971; Engstrom 1976; Keller 1976; Tertschnig 1989) as does another gastropod *Charonia variegata* (Percharde 1972). Many fish prey upon sea urchins in the Caribbean Sea (Randall 1967; Keller 1976). The absence of small *T. ventricosus* on coral rubble flats could result from predation by the queen triggerfish *Batistes vetula* (Scheibling and Mladenov 1987). Wrasse prey on small *T. gratilla* in the Gulf of Aqaba. The starfish *O. reticulus* preys heavily on *T. ventricosus* in seagrass beds in the Grenadines (Scheibling 1982). Wading birds prey on *T. ventricosus* at Puerto Rico (Rivera 1979). The effects of predation on *Tripneustes* populations are not known.

11. CONCLUSIONS

Tripneustes appears to be a generalist in habitat and food requirements in tropical and subtropical, shallow-water habitats with high hydrodynamics subject to disturbance. *Tripneustes* has early maturation, rapid growth, high gonadal production, sporadic recruitment, and short longevity. All are characteristics of a ruderal species.

REFERENCES

Alcoverro T, Mariani S (2002). Effects of sea urchin grazing on seagrass (*Thalassodendron ciliatum*) beds of a Kenyan lagoon. Mar Ecol Prog Ser 226: 255–363

Aronson RB, Precht WF (2000). Herbivory and algal dynamics on the coral reef at Discovery Bay, Jamaica. Limnol Oceanogr 45: 251–255

Aseltine D (1982). *Tripneustes ventricosus* and *Lytechinus variegatus* (Echinoidea: Toxopneustidae): habitat differences and the role of water turbulence. MS thesis, Ohio State University. Columbus

Attrill MJ, Kelmo F, Jones MB (2004). Impact of the 1997–98 El Niño event on the coral reef-associated echinoderm assemblage from northern Bahia, northeastern Brazil. Climate Res 28: 151–158.

Bacolod PT, Dy DT (1986). Growth, recruitment pattern and mortality rate of the sea urchin, *Tripneustes gratilla* Linnaeus, in a seaweed farm at Danahon Reef, Central Philippines. Philip Sci 23: 1–14

Baissac JdeB, Lubet PE, Michel CM (1962). Les biocoenoses benthiques littorales de l'Ile Maurice. Rec Tray St Mar Endoume Bull 25: 253–291

Baker AN (1968). The echinoid fauna of northeastern New Zealand. Trans Roy Soc N Z Zool 8: 239–245

Benayahu Y, Loya Y (1977). Seasonal occurrence of benthic-algae communities and grazing regulation by sea urchins at the coral reefs of Eilat, Red Sea. In: Taylor DL (ed.) Proc Third Internat Coral Reef Symp, Rosenstiel School of Marine and Atmospheric Science, Miami, pp 383–389

Byrne M, Cisternas P, Hoggett A, O'Hara T, Uthicke S (2004). Diversity of echinoderms at Raine Island, Great Barrier Reef. In: Heinzeller T, Nebelsick J (eds) Echinoderms: München. Taylor & Francis, London, pp 159–163

Cameron RA (1986). Reproduction, larval occurrence and recruitment in Caribbean sea urchins. Bull Mar Sci 39: 332–346

Cameron RA, Boidron-Metairon I, Monterrosa O (1985). Does the embryonic response to temperature and salinity by four species of Caribbean sea urchins parallel the reproductive synchrony. Proc Fifth Internat Coral Reef Symp, Tahiti 5: 273–278

Camp DL. Cobb SP, van Breedveld JF (1973). Overgrazing of seagrasses by a regular urchin, *Lytechinus variegatus*. BioScience 23: 37–38.

Caso ME (1974a). Contribución al estudio de los equinoideos de México el género, *Tripneustes* Agassiz. Morfología y ecología de *Tripneustes ventricosus* (Lamarck). An Centro Cienc del Mar y Limnol Univ Nat Autón México 1: 1–24

Caso ME (1974b). Contribución al estudio de los equinoideos de México, Morfología de *Tripneustes depressus* Agassiz y estudio comparativo entre *T ventricosus* y *T. depressus*. An Centro Cienc del Mar y Limnol. Univ Nat Autón México 1: 25–40

Caycedo IE (1979). Observaciones de los equinodermos en las Islas del Rosario. An Inst Invest Mar Punta Betin 11: 39–47

Chen C-P, Chang K-H (1981). Reproductive periodicity of the sea urchin, *Tripneustes gratilla* (L.) in Taiwan compared with other regions. Internat J Invert Reprod 3: 309–319

Clark AM, Rowe FWE (1971). Monograph of shallow-water Indo-West Pacific echinoderms. Trustees of the British Museum (Natural History), London

Clark HL (1912). Hawaiian and other Pacific Echini. Mem Mus Comp Zool Harvard College 34: 205–283

Clark HL (1938). Echinoderms from Australia. Mem Mus Comp Zool Harvard College 55: 1–596

Clark HL (1946). The echinoderm fauna of Australia. The Carnegie Institution of Washington Publication 566: 1–567

Codoceo RM (1974). Equinodermos de Ia Isla de Pascua. Bol Mus Nac Hist Nat, Santiago 33: 53–63

Colón-Jones DE (1993). Size (age) factors controlling the distribution and population size of the white-spined sea urchin, *Tripneustes ventricosus* (Lamarck, 1816). MS thesis, University of Puerto Rico, Mayaguez

Dafni J (1992). Growth rate of the sea urchin *Tripneustes gratilla elatensis*. Isr J Zool 38: 25–33

Dafni J, Tobol R (1986). Population structure patterns of a common Red Sea echinoid (*Tripneustes gratilla eilatensis*). Isr J Zool 34: 191–204

DeSalvo LH, Randall SE, Cea A (1988). Ecological reconnaissance of the Easter Island sublittoral marine environment. Nat Geogr Res 4: 451–473

Dotan A (1990). Distribution of regular sea urchins on coral reefs near the south-eastern tip of the Sinai Peninsula, Red Sea. Isr J Zool 37: 15–29

Dy DT, Uy FA, Corrales CM (2002). Feeding, respiration, and excretion by the tropical sea urchin *Tripneustes gratilla* (Echinodermata: Echinoidea). from the Philippines. J Mar Biol Assn UK 82: 299–302

Dy DT, Yap HT (2000). Ammonium and phosphate excretion in three common echinoderms from Philippine coral reefs. J Exp Mar Biol Ecol 251: 227–238

Ebert TA (1971). A preliminary quantitative survey of the echinoid fauna of Kealakekua and Honaunau Bays, Hawaii. Pac Sci 25: 112–131

Ebert TA (1982). Longevity, life history, and relative body wall size in sea urchins. Ecol Monogr 52: 353–394

Engel H (1939). Echinoderms from Aruba, Curaçao and northern Venezuela. Capita Zoologica. 8: 1–12

Engstrom NA (1976). Predation by the helmet shell *Cassis tuberosa* upon sea urchins in *Thalassia* beds. Proc Assoc Island Mar Labs Carib 11: 14

Faure G, Montaggioni L (1970). Le récif corallien de Saint-Pierre de La Réunion (Ocean Indien): géomorphologie et répartition des peuplements. Rec Trav Sta Mar Endoume Hors Ser No 10: 271–284

Fell FJ (1974). The echinoids of Easter Island (Rapa Nui). Pac Sci 28: 147–158

Fouda MM, Hellal AM (1990). Reproductive biology of *Tripneustes gratilla* (L.) from Gulf of Aqaba and northern Red Sea. In: De Ridder C, M Lahaye M, Jangoux M (eds) Echinoderm research. Balkema, Rotterdam, pp 77–82

Gallo NJ (1988). Contribucion al conociemiento de los equinodennos del Parque Nacional Natural Tayrona. I. Echinoidea. Trianea 1: 99–110

Gillikin S, Jaffe S, Mengel L, Taylor R (1974). Sex ratio, sexual dimorphism and spawning activity in *T. ventricosus.* Unpubl student report, West Indies Laboratory, St. Croix

Glynn P (1968). Mass mortalities of echinoids and other reef flat organisms coincident with midday, low water exposures in Puerto Rico. Mar Biol 1: 226–243

Gomez ED, Guieb RA, Aro E (1983). Studies on the predators of commercially important seaweeds. Fish Res J Philipp 8: 1–17

Guieb RA (1981). Studies on the diet and food preference of *Diadema setosum* Leske 1778 and *Tripneustes gratilla* Linnaeus 1758 (Echinodermata: Echinoidea) in Calatagan, Batangas. MS thesis, University of the Philippines, Diliman

Haley MP, Solandt J-L (2001). Population fluctuations of the sea urchins *Diadema antillarum* and *Tripneustes ventricosus* at Discovery Bay, Jamaica: a case of biological succession. Carib J Sci 37: 239–245

Hendler G (1977). The differential effects of seasonal stress and predation on the stability of reef-flat echinoid populations. Proc Third Internat Coral Reef Symp, Miami, pp 217–223

Herring PJ (1972). Observations on the distribution and feeding habits of some littoral echinoids from Zanzibar. J Nat Hist 6: 169–175

Holguin Quiñones O, Wright López H, Solís Marín F (2000). Asteroidea, Echinoidea y Holothuroidea en fondos someros de la Bahía, Baja California Sur, México. Rev Biol Trop 48: 749–758

Hughes RN, Hughes HPI (1971). A study of the gastropod *Cassis tuberosa* (L.) preying upon sea urchins. J Exp Mar Biol Ecol 7: 305–3 15

Jafari RD, Mahasneh DM (1984). The effect of seagrass grazing on the sexual maturity of the sea urchin *Tripneustes gratilla* in the Gulf of Aqaba (Jordan). Am'man al'J'ami'ah al'Urdun'iyah. 14: 127–136

James DB, Pearse JS (1969). Echinoderms from the Gulf of Suez and the northern Red Sea. J Mar Bio Assoc India 11: 78–125

Juinio-Meñez MA, Macawaris NND, Bangi HGP (1998). Community-based sea urchin (*Tripneustes gratilla*) grow-out culture as a resource management tool. Can Spec Publ Fish Aquat Sci 125: 393–399

Kehas AJ, Theoharides KA, Gilbert JJ (2005). Effect of sunlight intensity and albinism on the covering response of the Caribbean sea urchin *Tripneustes ventricosus.* Mar Biol 146: 1111–1117

Keller BD (1976). Sea urchin abundance patterns in seagrass beds: the effects of predation and competitive interactions. PhD thesis, John Hopkins University, Baltimore

Keller BD (1983). Coexistence of sea urchins in seagrass beds: an experimental analysis of competition and predation. Ecology 64: 1581–1598

Kidron J, Fishelson L, Moav B (1972). Cytology of an unusual case of hermaphroditic gonads in the tropical sea urchin *Tripneustes gratilla* from Eilat (Red Sea). Mar Biol 14: 260–263

Kier PM, Grant RE (1965). Echinoid distribution and habits, Key Largo Coral Reef Preserve, Florida. Smithsonian Misc Coll 149(6): 1–64

Kitting C, Mroz A, Tufford D, Pyun 0 (1974). Food and feeding activity of the sea urchin, *Tripneustes ventricosus.* Unpubl student report, West Indies Laboratory, St. Croix

Klumpp DW, Salita-Espinosa JT, Fortes MD (1993). Feeding ecology and trophic role of sea urchins in a tropical seagrass community. Aquat Bot 45: 205–229

Kobayashi N (1969). Spawning periodicity of sea urchins at Seto. III. *Tripneustes gratilla, Echinometra mathaei, Anthocidaris crassipina* and *Echinostrephus aciculatus.* Sci Eng Rev Doshisha Univ 9: 254–269

Koike I, Mukai H, Nojima S (1987). The role of the sea urchin, *Tripneustes gratilla* (Linnaeus) in decomposition and nutrient cycling in a tropical seagrass bed. Ecol Res 19–29

Lawrence JM (1973). Temperature tolerances of tropical shallow-water echinoids (Echinodermata) at Elat (Red Sea). Isr J Zool 22: 143–150

Lawrence JM (1976). Absorption efficiencies of four species of tropical echinoids fed *Thalassia testudinum*. Thal Jugoslavica 12: 201–205

Lawrence JM (1987). A functional biology of echinoderms. The Johns Hopkins University Press, Baltimore

Lessios HA (1985). Annual reproductive periodicity in eight echinoid species on the Caribbean coast of Panama. In: Keegan BF, O'Connor BDS (eds) Echinodermata. Balkema, Rotterdam, pp 303–311

Lessios HA (1991). Presence and absence of monthly reproductive rhythms among eight Caribbean echinoids off the coast of Panama. J Exp Mar Biol Ecol 153: 27–47

Lessios HA (1998). Shallow water echinoids of Cayos Cochinos, Honduros. Rev Biol Trop 46 Suppl (4): 95–101

Lessios HA, Cubit SD, Robertson DR, Shulman MJ, Parker MR, Garrity SD, Levings SC (1984). Mass mortality of *Diadema antillarum* on the Caribbean coast of Panama. Coral Reefs. 3: 173–182

Lessios HA, Kane J, Robertson DR (2003). Phylogeography of the pantropical sea urchin *Tripneustes*: contrasting patterns of population structure between oceans. Evolution 57: 2926–2036

Lewis JB (1958). The biology of the tropical sea urchin *Tripneustes esculentus* Leske in Barbados, British West Indies. Can J Res 36: 607–621

Lewis MS, Taylor JD (1966). Marine sediments and bottom communities of the Seychelles. Phil Trans Roy Soc Lond B 549: 179–190

Liao Y, Clark AM (1995). The echinoderms of southern China. Science Press, Beijing

Lilly GR (1975). The influence of diet on growth and bioenergetics of the tropical sea urchin, *Tripneustes ventricosus*. PhD thesis, University of British Columbia, Vancouver

Lison de Loma T, Conand C, Harmelin-Vivien M, Ballesteros E (2002). Food selectivity of *Tripneustes gratilla* (L.) (Echinodermata: Echinoidea) in oligotrophic and nutrient-enriched coral reefs at La Réunion (Indian Ocean). Bull Mar Sci 70: 927–938

Lison de Loma T, Harmelin-Vivien ML, Conand C (1999). Diel feeding rhythm of the sea urchin *Tripneustes gratilla* (L.) on a coral reef at La Reunion, Indian Ocean. In: Candia Carnavali MD, Bonasoro F (eds) Echinoderm research 1998. Balkema, Rotterdam, pp. 87–92

MacNae W, Kalk M (1963). The faunal and flora of sand flats at Inhaca Island, Moçambique. J An Ecol 31: 93–128

Maharavo J (1993). Etude de l'oursin comestible *Tripneustes gratilla* (L. 1758) dans la région de Nosy-Bé (côte nord-ouest de Madagascar): Densité, morphometrie, nutrition, croissance, processus réproducteurs, impact de l'exploitation sur les populations. Thesis. Doc en Sci, Université de Aix-Marseille III, Marseille.

Maharavo J, Régis M-B, Thomassin BA (1994). Food preference of *Tripneustes gratilla* (L.) (Echinoidea) on fringing reef flats off the NW coast of Madagascar (SW Indian Ocean). In: David B, Guille A, Féral J-P, Roux M (eds) Echinoderms through time. Balkema, Rotterdam, pp 769–774

Maluf LY (1988). Composition and distribution of the central eastern Pacific echinoderms. Tech. Reports, Number 2. Natural History Museum of Los Angeles County

Marshall DJ, Hodgson AN, Pretorius RA (1991). New southern geographical records of intertidal sea urchins (Echinodermata: Echinoidea), with notes on abundance. S Afr Tydskr Dierk 26: 204–205

Mastaller M (1979). Beitrage zur Faunistik und Őkologie der Mollusken und Echinodermen in der Korallenriffen bei Aqaba, Rotes Meer. PhD thesis, Universitat Ruhr, Bochum

McClanahan TR (1998). Predation and the distribution and abundance of tropical sea urchin populations. J Exp Mar Biol Ecol 221: 231–255

McPherson BF (1965). Contributions to the biology of the sea urchin *Tripneustes ventricosus*. Bull Mar Sci Gulf Carib 15: 228–244

Moore DR (1956). Observations of predation on echinoderms by three species of Cassididae. Nautilus 69: 73–76

Moore HB, Jutare T, Jones JA, McPherson BF, Roper CFE (1963). A contribution to the biology *of Tripneustes esculentus*. Bull Mar Sci Gulf Carib 13: 267–281

Moore HB, McPherson BF (1965). A contribution to the study of the productivity of the urchins *Tripneustes esculentus* and *Lytechinus variegatus*. Bull Mar Sci 15: 855–871

Mortensen T (1943). A monograph of the Echinoidea. 11.2. Camarodonta. I. Copenhagen. CA Reitzel, Copenhagen

Morton JE (1973). The intertidal ecology of the British Solomon Islands. I. The zonation patterns of the weather coasts. Phil Trans Roy Soc Lond 255B: 491–542

Moses CS, Bonem RM (2001). Recent population dynamics of *Diadema antillarum* and *Tripneustes ventricosus* along the north coast of Jamaica, W.I. Bull Mar Sci 68 327–336

Mukai H, Nishihira M, Nojima S (1987). Distribution and biomass of predominant benthic animals. In: Hattori A (ed.) Studies on dynamics of the biological community in tropical seagrasss ecosystems in Papua New Guinea: the second report. Ocean Research Institute, Tokyo, pp 62–75.

Mukai H, Nojima S (1985). A preliminary study on grazing and defecation rates of a seagrass grazer, *Tripneustes gratilla* (L.) (Echinodermata: Echinoidea), in Papua New Guinean seagrass beds. Spec Publ Mukaishima Mar Biol Sta 1985: 173–183

Muthiga NA (2005). Testing for the effects of seasonal and lunar periodicity on the reproduction of the edible sea urchin *Tripneustes gratilla* (L) in Kenyan coral reef lagoons. Hydrobiologia 549: 57–64

Nakasone Y, Yamazato K, Nishihira M, Kamura S, Aramoto Y (1974). Preliminary report on the ecological distribution of benthic animals on the coral reefs of Sesoko Island, Okinawa. Ecol Stud Nat Cons Ryukyu Isl 1: 2 13–236

Nebelsick JH (1992). The northern Bay of Safaga (Red Sea, Egypt); an actuopaläontological approach. III. Distribution of echinoids. Beitr Palaont Osterr 17: 5–79

Nojima S, Mukai H (1985). A preliminary report on the distribution pattern, daily activity and moving pattern of a seagrass grazer, *Tripneustes gratilla* (L.) (Echinodermata: Echinoidea), in Papua New Guinean seagrass beds. Spec Publ Mukaishima Mar Biol Sta 1985. pp. 173–183

Nojima S, Nishihira M, Mukai H (1987). Outflow of seagrass leaves and effects of sea urchin grazing. In: Hattori A (ed.) Studies on dynamics of the biological community in tropical seagrass ecosystems in Papua New Guinea: the second report. Ocean Research Institute, University of Tokyo, Tokyo, pp 109–115

O'Connor C, Riley G, Bloom D (1976). Reproductive periodicities of the echinoids of the Solitary Islands in light of some ecological variables. Thal Jugoslavica 12: 245–267

O'Connor C, Riley G, Lefebvre 5, Bloom D (1978). Environmental influences on histological changes in the reproductive cycle of four New South Wales sea urchins. Aquaculture. 15: 1–17

Ogden JC (1976). Some aspects of herbivore-plant relationships on Caribbean reefs and seagrass beds. Aquat Bot 2: 103–116

Ogden JC, Lobel PS (1978). The role of herbivorous fishes and urchins in coral reef communities. Env Biol Fish 3: 49–63

Ogden N, Ogden SC, Abbot IA (1989). Distribution, abundance, and food of sea urchins on a leeward Hawaiian reef. Bull Mar Sci 45: 539–549

Onoda K (1936). Notes on the development of some Japanese echinoids with special reference to the structure of the larval body. Jap J Zool 6: 637–654

Parker DA, Shulman MJ (1986). Avoiding predation: alarm responses of Caribbean sea urchins to simulated predation on conspecific and heterospecific sea urchins. Mar Biol 93: 201–208

Pearse JS (1974). Reproductive patterns of tropical reef animals: three species of sea urchins. Proc 2nd Int Coral Reef Symp, pp 235–240

Percharde PL (1972). Observations on the gastropod, *Charonia variegata*, in Trinidad and Tobago. Nautilus 85: 84–92

Petit G (1930). Localisation et comportement de quelques especes d'échinodermes sur les récifs madréporiques de Tuléar (Madagascar). C R Soc Biogeog, Paris 56: 56–59

Randall JE (1967). Food habits of reef fishes of the West Indies. Stud Trop Oceanogr 5: 665–847

Régis M-B, Thomassin BA (1982). Écologie des échinoïdes réguliers dans les récifs coralliens de la région de Tuléar (s.w. de Madagascar): adaptation de la microstructure des piquants. Ann Inst Océanogr 58: 117–158

Rivera JA (1977). Echinoid mortality at Jobos Bay, Puerto Rico. Proc Assoc Is Mar Labs Carib 13: 9

Rivera JA(1979). Aspects of the biology of *Lytechinus variegatus* (Lamarck, 1816) at Jobos Bay, Puerto Rico (Echinoidea: Toxopneustidae). MS thesis, University of Puerto Rico, Mayaguez

Scheibling RE (1982). Feeding habits of *Oreaster reticulatus* (Echinodermata: Asteroidea). Bull Mar Sci 32: 504–510

Scheibling RE, Mladenov PV (1987). The decline of the sea urchin, *Tripneustes ventricosus*, fishery of Barbados: a survey of fishermen and consumers. Mar Fish Rev 49: 62–69

Scheibling RE, Mladenov PV (1988). Distribution, abundance and size structure of *Tripneustes ventricosus* on traditional fishing grounds following the collapse of the sea urchin fishery in Barbados. In: Burke RD, Mladenov PV, Lambert P, Parsley RL (eds) Echinoderm biology. Rotterdam, Balkema, pp 449–446

Schumacher H (1974). On the conditions accompanying the first settlement of corals on artificial reefs with special reference to the influence of grazing sea urchins (Eilat, Red Sea). Proc 2 Internat Coral Reef Symp 1: 257–267
Serafy DK (1979). Echinoids (Echinodermata: Echinoidea). Mem Hourglass Cruises V(111). Florida Dept Nat Res, St. Petersburg
Shigei M (1970). Echinoids of the Bonin Islands. J Fac Sci Univ Tokyo, Sec IV. 12: 1–22
Shigei M (1973). A check list of echinoids found in Sagami Bay with brief notes on each species. J Fac Sci Univ Tokyo, Sec IV 13: 1–33
Shimabukuro S (1991). *Tripneustes gratilla* (sea urchin). In: Shokit S, Kakazu K, Tomori A, Toma T (eds), Yamaguchi M (English ed) Aquaculture in tropical areas. Midori Shobo Co, Ltd, Tokyo, pp. 3 13–328
Solandt J-L, Campbell AC (1998). Habitat selection in Jamaican echinoids. In: Mooi R, Telford M (eds) Echinoderms: San Francisco. Balkema, Rotterdam, pp 821–827
Sonnenholzner JI, Lawrence J (2002). A brief survey of the echinoderms communities of the central and southern marine-coastal wetlands of the continental coast off Ecuador. Bol Ecotropica: Ecosistemas Trop No 36: 27–35
Stephenson A (1934). The breeding of reef animals. Part II. Invertebrates other than corals. Great Barrier Reef Expeditions 1928–29, Sci Rep 3: 247–272
Stevenson RA, Ufret SL (1966). Iron, manganese, and nickel in skeletons and food of the sea urchins *Tripneustes esculentus* and *Echinometra lucunter*. Linmol Oceanogr 11: 11–17
Taylor JD (1968). Coral reef and associated invertebrate communities (mainly molluscan) around Mahé, Seychelles. Proc Roy Soc Lond 254B: 129–206
Taylor JD, Lewis MS (1970). The flora,fauna and sediments of the marine grass beds of Mahé, Seychelles. J Nat Hist 4: 199–220
Tertschnig W (1985). Sea urchins in seagrass communities: resource management as functional perspective of adaptive strategies. In: Keegan BF, O'Connor BDS (eds) Echinodermata. Balkema, Rotterdam, pp 361–367
Tertschnig W (1986). Populationsdynamik und Okologie von Seeigeln in Mediterranen und Tropisch-Atlantischen Seegrasbestanden. PhD thesis, Universität Wien, Vienna
Tertschnig WP (1989). Diel activity patterns and foraging dynamics of the sea urchin *Tripneustes ventricosus* in a tropical seagrass community and a reef environment (Virgin Islands). PSZN: Mar Ecol 10: 3–21
Tokioka T (1963). Supposed effects of the cold weather of the winter 1962–63 upon the intertidal fauna in the vicinity of Seto. Pubi Seto Mar Biol Lab 9: 415–424
Tokioka T (1966). Recovery of the *Echinometra* populations in the intertidal zone in the vicinity of Seto, with a preliminary note on the mass mortality of some sea urchins in the summer season. Publ Seto Mar Biol Lab 14: 7–16
Tommasi LR (1972). Equinodermes de regiâo entre o Amapâ (Brasil) e a Flórida (E.U.A.). II. Echinozoa. Bolm Inst Oceanogr 21, 15–67
Tuason AY, Gomez ED. (1979). The reproductive biology of *Tripneustes gratilla* (Linnaeus) (Echinoidea: Echinodermata) with some notes on *Diadema setosum* Leske. Proc Int Symp. Mar Biogeogr Evol Southern Hemisphere. 2: 707–716
Uy FA, Pacifico KP, Dy DT (2000). The distribution of *Tripneustes gratilla* (Linnaeus) (Echinodermata: Echinoidea) in a small embayment of eastern Mactan Island, Cebu, central Philippines. Philipp Sci 37: 42–50
Vadas RL Sr, Elner RW (2003). Responses to predation cues and food in two species of sympatric, tropical sea urchins. PSZN: Mar Ecol 24: 101–121
Vaïtilingon D, Rasolofonirina R, Jangoux M (2003). Feeding preferences, seasonal gut repletion indices, and diel feeding patterns of the sea urchin *Tripneustes gratilla* (Echinodermata: Echinoidea) on a coastal habitat off Tolilara (Madagascar) Mar Biol 143: 451–458
Vink S, Atkinson MJ (1985). High dissolved C:P excretion ratios for large benthic marine invertebrates. Mar Ecol Prog Ser 21: 191–195
Voss GL, Voss NA (1955). An ecological survey of Soldier Key, Biscayne Bay, Florida. Bull Mar Sci Gulf Carib 5: 203–229
Williams EG Jr, Bunkley-Williams L, Bruckner RJ, Bruckner AW, Ortiz-Corps AR, Bowden-Kerby WA, Colón-Jones DE (1996). Recurring mass mortalities of the white-spined sea urchin, *Tripneustes ventricosus*, (Echinodermata: Echinoidea) in Puerto Rico. Carib J Sci 32: 111–112

Zieman JC, Thayer GW, Robblee MB, Zieman RT (1979). Production and export of seagrasses from a tropical bay. In: Livingston RI (ed.) Ecological processes in coastal and marine systems. Plenum Press, NewYork, pp 21–34

Zigler KS, Lessios HA (2003). Evolution of bindin in the pantropical sea urchin *Tripneustes*: comparisons to bindin of other genera. Mol Biol Evol 20: 220–231

Edible Sea Urchins: Biology and Ecology
Editor: John Miller Lawrence

Chapter 25

Sea Urchin Roe Cuisine

John M Lawrence

Department of Biology, University of South Florida, Tampa, FL (USA).

Sea urchin roe is eaten most simply raw on bread, perhaps with lemon added. I have eaten it in Marseille, raw on the halfshell along with the coelomic fluid, gut, and the algal gut contents. It can be embellished. According to Soyer (1853), the ancient Greeks mixed the roe with vinegar, sweet cooked wine, parsley, and mint. He said the Romans had a recipe for "sluggish" appetites that involved stuffing sea urchins with a cooked mixture of oil, garum, sweet wine, and pepper and then cooking them on a slow fire.

Today many think of eating sea urchin roe in the Japanese manner as *sushi*. But the Japanese have a variety of ways of preparing sea urchins (Yokota 2002). *Uni no Kanten* is prepared by suspending roe in *kanten* (agar-agar) flavored with saké, soy sauce, and dashi. Yukio Agatsuma has provided recipes of traditional Japanese methods for preparing sea urchins for eating. *Echizen Uni* is from Fukui Prefecture and uses *Hemicentrotus pulcherrimus*. The roe are removed from the test, washed in seawater, drained, sprinkled with salt (20% of the total roe), and placed in wooden casks. *Shimonoseki Uni* from Yamaguchi Prefecture also uses *H. pulcherrimus*. The roe are removed, washed, drained, and sprinkled with salt (8–12% of the total roe) as before. However, alcohol (12–15%) is added to the roe and the mixture stirred before it is bottled. The mixture is allowed to ripen for about 1–2 weeks at room temperature. It will keep for a year at room temperature. *Strongylocentrotus nudus* is used in *Kaiyaki Uni* from Iwate Prefecture. The roe are removed, washed with seawater, and baked in an iron pot on a charcoal fire for about 20 min. The cooked roe are cooled and then placed on the shell of the abalone *Haliotis discus hannai* and wrapped in cellophane.

Tatsuya Unuma has described several additional Japanese recipes. *Ichigo-ni* is a clear soup containing sea urchin roe and sliced abalone. *Ichigi* means "strawberry" and *ni* means "boil." The name was given because the roe in the soup looks like a wild berry. *Uni-don* is a popular dish in which uncooked roe is slightly flavored with soy sauce and horseradish and then put on cooked rice in a bowl. *Uni-meshi* is similar except the roe are mixed with rice, soy sauce, saki, and seasonings and cooked. Roe is also used to make a special paste by mixing roe with egg yolk and seasoning. The paste is spread on grilled fish or shellfish. Okinawan people usually eat roe raw or with soy sauce although sometimes the roe are grilled in the test (Y Hiratsuka, pers. comm.).

Traditional preparations of *Paracentrotus lividus* are found in Provence. An omelet can be prepared by cooking the beaten roe quickly in very hot oil (Jouveau 1976).

An *oursinado* is eaten with bread or potatoes, as a soup base, or as a *crème d'oursin* served with fish (Gruénais-Vanverts 1982). It is prepared by mashing roe in a mortar and pestle and adding olive oil little by little to make a mayonnaise.

The *chardons, Tripneustes ventricosus,* are used in the Antilles for a *blaff* and a *tarte* (Négre 1978). For a *blaff d'oursins,* roe are marinaded in a mixture of salt, pepper, garlic, lemon juice, and crushed allspice. This is then added to minced, browned onions with tomatoes, garlic, chives, logwood (*bois d'Inde*), cloves, and a bouquet garni and simmered. For a *tarte,* a short pastry is prepared as usual for a crust. The filling is a mixture of browned onions, laurel, chives, logwood, garlic, lemon, salt and pepper, tomatoes, and roe. This is simmered briefly, poured into the crust and baked.

Helen Bangi has described the preparation of *T. gratilla* in the Philippines. They may be steamed, boiled, or grilled in an open fire before removing the roe, to which lemon juice (*calamansi*) is added. In the far northwestern province of Ilocos of Luzon, the gut and coelomic fluid are also eaten. A common dish in the Mindanao provinces involves removing the Aristotle's lantern, gut, and coelomic fluid and leaving the roe. Uncooked rice is then placed inside the test. After steaming or boiling to cook the rice, the test is removed, leaving the sea urchin shaped rice with the five strips of roe beautifully designed around it. Salt and *calamansi* are added.

Cookbooks in New Zealand typically have recipes for the roe of *kina, Evechinus chiororicus.* Goode and Wilson (1979) suggested eating the roe raw with lemon or on toast after lightly boiling them in salt water. They also described a roe sauce for fish prepared by mixing roe with a small amount of olive oil and Hollandaise Sauce. According to Burton (1982), early European settlers in New Zealand made a pie by alternating roe and breadcrumbs in a greased dish. The pie was covered with breadcrumbs, chopped bacon, and baked. The Maori prepare *kina poha* by filling tests with roe and roasting them. The most exotic Maori recipe given by Burton and by Fuller (1978), *kina kotero* ("cured kina"), involves placing the roe in a bag in a stream or large container of freshwater for several days before eating the roe as a relish or on slices of bread and butter. Burton said this is an "acquired" taste. Hudson et al. (2001) described two methods Maori used for preservation by fermentation of kina in spring. In one, kina are stored under freshwater for up to 3 weeks before eating. In the other, they are buried underground for 3–4 months and then removed and eaten or packed in leaves for reburying.

Interestingly, Ransom (1948) reported Fox Island Aleuts ate *S. droebachiensis* raw in winter but believed soaking sea urchins in freshwater for several minutes was required during summer. These Aleuts did not cook the sea urchins in any way.

A recipe for the Chilean *Loxechinus albus* was given to me by Laura Prat of Concepción. She cooks roe with cream, salt, and nutmeg. This is mixed with pancakes cut into thin ribbons like spaghetti and served hot with very cold white Chilean wine.

ACKNOWLEDGMENTS

I thank Y Agatsuma, H Bangi, L Prat, and T Unuma for providing descriptions of national cuisine and D Corbett for providing literature information.

REFERENCES

Burton D (1982) Two hundred years of New Zealand food and cookery. AR & AW Reed Ltd, Wellington

Fuller D (1978) Maori food and cookery. AR & AW Reed Ltd, Wellington

Goode J, Wilson CW (1979) The original Australian and New Zealand fish cookbook (revised edition). AR & AW Reed Pty Ltd, Sydney

Gruénais-Vanverts MA (1982) Cuisine de Provence Luberon. Editions Denoël, Paris, France

Hudson JA, Hasell S, Whyte R, Monson S (2001) Preliminary microbiological investigation of the preparation of two traditional Maori foods (Kina and Tiroi). J Appl Microbiol 91: 814–821

Jouveau R (1976) La cuisine provençale de tradition populaire. Imprimerie Bene, Nimes

Négre A (1978) La cusine Antillaise. Les Editions du Pacifique, Papeeteë

Ransom JE (1948) The Aleut natural food economy. Amer Anthropologist 48: 607–623

Soyer A (1853) The Pantropheon: or, a history of food and its preparation in ancient times. Simkin and Marshall, London

Yokota Y (2002) Fishery and consumption of the sea urchin in Japan. In: Yokota Y, Matranga V, Smolenicka Z (eds). The sea urchin: from basic biology to aquaculture. AA Balkema Publishers, Lisse. pp 129–138

Index

D

E

F

G

H

I

J

K

L

M

N

O

P

R

S